Bacterial Regulatory Networks

Edited by

Alain A.M. Filloux

Centre for Molecular Microbiology and Infection
Division of Cell and Molecular Biology
Imperial College London
London
UK

Caister Academic Press

Caister Academic Press
Norfolk, UK

www.caister.com

British Library Cataloguing-in-Publication Data
A catalogue record for this book is available from the British Library

ISBN: 978-1-908230-03-4

Cover design adapted from Figure 6.4 and 12.7a

Printed and bound in Great Britain

Contents

Contributors

Simon C. Andrews
School of Biological Sciences
University of Reading
Reading
UK

s.c.andrews@reading.ac.uk

Judith P. Armitage
Oxford Centre for Integrative Systems Biology and Department of Biochemistry
University of Oxford
Oxford
UK

judith.armitage@bioch.ox.ac.uk

Karlijn C. Bastiaansen
Department of Molecular Microbiology
VU University
Amsterdam
the Netherlands

k.c.j.t.bastiaansen@vu.nl

Wilbert Bitter
Department of Molecular Microbiology and Department of Medical Microbiology
VU University and VU University Medical Center
Amsterdam
the Netherlands

w.bitter@vumc.nl

Martin Buck
Department of Life Sciences
Faculty of Natural Sciences
Imperial College London
London
UK

m.buck@imperial.ac.uk

Patricia C. Burrows
Department of Life Sciences
Faculty of Natural Sciences
Imperial College London
London
UK

patricia.c.burrows00@imperial.ac.uk

Delphine Caly
BIOMERIT Research Centre
Department of Microbiology
University College Cork
Ireland

d.caly@ucc.ie

Pierre Cornelis
Department of Bioengineering Sciences
Research Group Microbiology
VIB Structural Biology
Vrije Universiteit Brussel
Brussels
Belgium

pcornel@vub.ac.be

Giulia Devescovi
International Centre for Genetic Engineering and Biotechnology
Trieste
Italy

devescov@icgeb.com

Charles J. Dorman
Department of Microbiology
School of Genetics and Microbiology
Trinity College Dublin
Dublin
Ireland

cjdorman@tcd.ie

J. Maxwell Dow
BIOMERIT Research Centre
Department of Microbiology
University College Cork
Cork
Ireland

m.dow@ucc.ie

Thomas Geissmann
Université de Lyon
INSERM, U851
Hospices Civils de Lyon
Centre National de Référence des Staphylocoques
Lyon
France

tom.geissmann@univ-lyon1.fr

Juan F. González
International Centre for Genetic Engineering and Biotechnology
Trieste
Italy

gonzales@icgeb.org

Benjamin A. Hall
Centre for Computational Science
Department of Chemistry
University College London
London
UK

benjamin.a.hall@ucl.ac.uk

Elisabeth Härtig
Institute for Microbiology
Technical University of Brunswick
Brunswick
Germany

e.haertig@tu-bs.de

Regine Hengge
Institute of Biology – Microbiology
Freie Universität Berlin
Berlin
Germany

Rhenggea@zedat.fu-berlin.de

Martijn Herber
Department of Genetics
Groningen Biomolecular Sciences and Biotechnology Institute
University of Groningen
Groningen
the Netherlands

martijn.herber@rug.nl

Dieter Jahn
Institute for Microbiology
Technical University of Brunswick
Brunswick
Germany

d.jahn@tu-bs.de

Elizabeth E. Jefferys
Oxford Centre for Integrative Systems Biology
University of Oxford
Oxford
UK

elizabeth.jefferys@new.ox.ac.uk

Eberhard Klauck
Institute of Biology – Microbiology
Freie Universität Berlin
Berlin
Germany

ebklauck@zedat.fu-berlin.de

Oscar P. Kuipers
Department of Genetics
Groningen Biomolecular Sciences and Biotechnology Institute
University of Groningen
Groningen;
Kluyver Centre for Genomics of Industrial Fermentation
Delft/Groningen
the Netherlands

o.p.kuipers@rug.nl

Efthimia Lioliou
Architecture et réactivité de l'ARN, UPR 9002 CNRS
Université de Strasbourg
Strasbourg
France

e.lioliou@ibmc-cnrs.unistra.fr

María A. Llamas
Department of Environmental Protection
Estación Experimental del Zaidín – CSIC
Granada
Spain

marian.llamas@eez.csic.es

Yvonne McCarthy
BIOMERIT Research Centre
Department of Microbiology
University College Cork
Cork
Ireland

y.mccarthy@ucc.ie

Mario Alberto Martinez-Nuñez
Department of Molecular Microbiology
Instituto de Biotecnologia
Universidad Nacional Autónoma de México
Morelos
Mexico

malberto@uxdea4.iimas.unam.mx

Karen O'Donovan
BIOMERIT Research Centre
Department of Microbiology
University College Cork
Cork
Ireland

kod@student.ucc.ie

Zhensheng Pan
Section of Microbiology
Faculty of Medicine and Centre for Molecular Microbiology & Infection
Imperial College London
London
UK

z.pan@imperial.ac.uk

Ernesto Perez-Rueda
Departamento de Ingeniería Celular y Biocatálisis
Instituto de Biotecnologia
Universidad Nacional Autónoma de México
Morelos
Mexico

erueda@ibt.unam.mx

Nancy Rivera-Gomez
Departamento de Ingeniería Celular y Biocatálisis
Instituto de Biotecnologia
Universidad Nacional Autónoma de México
Morelos
Mexico

nrivera@ibt.unam.mx

Mark A.J. Roberts
Department of Biochemistry
University of Oxford
Oxford
UK

mark.roberts@lincoln.ox.ac.uk

Pascale Romby
Architecture et réactivité de l'ARN, UPR 9002 CNRS
Université de Strasbourg
Strasbourg
France

p.romby@ibmc-cnrs.unistra.fr

Cédric Romilly
Architecture et réactivité de l'ARN, UPR 9002 CNRS
Université de Strasbourg
Strasbourg
France

c.romilly@ibmc-cnrs.unistra.fr

Robert P. Ryan
BIOMERIT Research Centre
Department of Microbiology
University College Cork
Cork
Ireland

r.ryan@ucc.ie

Max Schobert
Institute for Microbiology
Technical University of Brunswick
Brunswick
Germany

m.shobert@tu-bs.de

Kathryn A. Scott
Oxford Centre for Integrative Systems Biology and Department of Biochemistry
University of Oxford
Oxford
UK

kathryn.scott@bioch.ox.ac.uk

Jeroen Siebring
Department of Genetics
Groningen Biomolecular Sciences and Biotechnology Institute
University of Groningen
Groningen
the Netherlands

j.siebring@rug.nl

Robin A. Sorg
Department of Genetics
Groningen Biomolecular Sciences and Biotechnology Institute
University of Groningen
Groningen
the Netherlands

r.a.sorg@rug.nl

Zulma R. Suárez-Moreno
International Centre for Genetic Engineering and Biotechnology
Trieste
Italy

zrsuarezm@gmail.com

Silvia Tenorio-Salgado
Departamento de Ingeniería Celular y Biocatálisis
Instituto de Biotecnologia
Universidad Nacional Autónoma de México
Morelos
Mexico

s.tenorio.salgado@gmail.com

Petra Tielen
Institute for Microbiology
Technical University of Brunswick
Brunswick
Germany

p.tielen@tu-bs.de

François Vandenesch
Université de Lyon;
INSERM, U851;
Hospices Civils de Lyon;
Centre national de référence des staphylocoques
Lyon
France

francois.vandenesch@univ-lyon1.fr

Vittorio Venturi
International Centre for Genetic Engineering and Biotechnology
Trieste
Italy

venturi@icgeb.org

David E. Whitworth
Institute of Biological, Environmental and Rural Sciences
Aberystwyth University
Aberystwyth
UK

dew@aber.ac.uk

Simone C. Wiesler
Department of Life Sciences
Faculty of Natural Sciences
Imperial College London
London
UK

s.wiesler@imperial.ac.uk

Sivaramesh Wigneshweraraj
Section of Microbiology
Faculty of Medicine and Centre for Molecular Microbiology & Infection
Imperial College London
London
UK

s.r.wig@imperial.ac.uk

Current books of interest

Bacterial Gene Regulation and Transcriptional Networks	2013
Bioremediation of Mercury: Current Research and Industrial Applications	2012
Neurospora: Genomics and Molecular Biology	2012
Rhabdoviruses	2012
Horizontal Gene Transfer in Microorganisms	2012
Microbial Ecological Theory: Current Perspectives	2012
Two-Component Systems in Bacteria	2012
Foodborne and Waterborne Bacterial Pathogens	2012
Yersinia: Systems Biology and Control	2012
Stress Response in Microbiology	2012
Bacterial Regulatory Networks	2012
Systems Microbiology: Current Topics and Applications	2012
Quantitative Real-time PCR in Applied Microbiology	2012
Bacterial Spores: Current Research and Applications	2012
Small DNA Tumour Viruses	2012
Extremophiles: Microbiology and Biotechnology	2012
Bacillus: Cellular and Molecular Biology (Second edition)	2012
Microbial Biofilms: Current Research and Applications	2012
Bacterial Glycomics: Current Research, Technology and Applications	2012
Non-coding RNAs and Epigenetic Regulation of Gene Expression	2012
Brucella: Molecular Microbiology and Genomics	2012
Molecular Virology and Control of Flaviviruses	2012
Bacterial Pathogenesis: Molecular and Cellular Mechanisms	2012
Bunyaviridae: Molecular and Cellular Biology	2011
Emerging Trends in Antibacterial Discovery: Answering the Call to Arms	2011
Epigenetics: A Reference Manual	2011
Metagenomics: Current Innovations and Future Trends	2011
Nitrogen Cycling in Bacteria: Molecular Analysis	2011
The Biology of Paramyxoviruses	2011
Helicobacter pylori	2011
Microbial Bioremediation of Non-metals: Current Research	2011
Lactic Acid Bacteria and Bifidobacteria: Current Progress in Advanced Research	2011
Viruses and Interferon: Current Research	2011
Essentials of Veterinary Parasitology	2011
Hepatitis C: Antiviral Drug Discovery and Development	2011
Streptomyces: Molecular Biology and Biotechnology	2011
Alphaherpesviruses: Molecular Virology	2011
Recent Advances in Plant Virology	2011

Full details at www.caister.com

Preface

Higher eukaryotic organisms, including humans, use vision, hearing, smell, touch or taste to promptly analyse the surrounding environment and react appropriately to avoid danger or take benefit of a favourable situation. In humans, perception of a danger or stress may result in adrenaline being released in the blood. Adrenaline contributes to increase speed and strength or to decrease pain sensitivity, thus making your body and brain ready to take quick decisions such as running away or fighting. Similarly, when you smell or see something that is likely to be good food, you will start to salivate and prepare your digestive organs to process the food that you will inevitably be tempted to swallow.

One may think that prokaryotic organisms do not have such sensory sophistication, but a closer look at their molecular equipment indicates that they have this capacity. In bacteria, the decision-making process is all about a quick change in gene expression. Such change starts with the synthesis of new determinants/molecules/proteins that will be more appropriate to cope with the newly encountered environment. This is combined with the arrest in synthesis of other molecules that became useless in the new context, since carrying on with their production will be nothing else than a waste of precious time and energy. For example, in an iron-limiting environment, and iron is an essential element of many key biological processes, bacteria will switch on the expression of genes that will allow synthesis and release of siderophores (Chapter 10). These molecules are scavengers that will actively hunt and trap iron wherever available in order to bring it back into the cells. This internalization procedure will also involve receptors and transporters, the expression of which is also induced upon perception of low iron levels in the environment. The production of all the components involved in active iron capture is not required in other conditions than iron limitation and therefore a tight control in expression of dozens of genes prevents unnecessary waste of energy.

Iron limitation is only one of many other examples that are described in this book. The multitude and complexity of control events happening in one bacterial cell over its lifetime are tied into a hugely sophisticated 'bacterial regulatory network', which within seconds drastically change the transcription profile in the cell. The control events may involve master regulatory elements that sit on top of a hierarchical cascade, and which control global responses by affecting expression of hundreds of genes. These are for example alternative sigma factors that come into play to influence selectivity of the RNA polymerase in response to general stress situation (Chapters 1–3). Another example is the global regulator H-NS, a DNA-binding protein that keeps hundreds of genes silent (Chapter 7). These factors may in turn control expression of other regulators whose role will be more specific and that will induce change in expression of very few or even only one gene. A notorious regulatory system involved in responding to very specific changes in the environment is the two-component system (Chapter 8). This system involves a sensor protein whose role is to identify precisely what the environmental conditions are and in many cases it will be responsible for monitoring the presence/absence of only one specific stimulus. It is also important to highlight that in many cases the bacterium would like to respond to environmental changes not only at a single cell level but would like to organize a coordinate change in the whole bacterial population. One such phenomenon has been known for years as quorum sensing (Chapter 4). This system is pinpointing to the fact that bacteria might not behave in a selfish manner but thanks to appropriate regulatory mechanisms and to the production of diffusible molecules have established their own communication/language skills. Quorum sensing has given a new vision of the microbiological world, and the Science behind this social behaviour was fancily quoted by Peter Greenberg (University of Washington) as 'sociomicrobiology'.

It is remarkable that in many cases some of the changes induce by bacterial regulatory network are hardly visible to the human eyes. However, in several cases these changes in bacterial behaviour are striking. One example is the series of regulatory events, in response to harsh nutritional stress, which leads to sporulation of bacteria such as *Bacillus subtilis*. Such drastically different state of the bacteria may cause the emergence of two distinct subpopulations, and has been linked to so-called bistable regulatory networks (Chapter 12). In other cases, bacteria may switch from a single cell and motility behaviour to a sessile lifestyle within a compact but organized and active bacterial population called biofilm. Sometimes, the morphotype of a bacterial colony primed to form a biofilm is clearly identifiable on agar plates. Such a colony may display a remarkable wrinkly aspect often associated with the production of an exopolysaccharide whose aim is to glue bacterial cells together. The switch between motile and sessile lifestyles is also the result of sophisticated regulatory networks, and in some cases these regulatory cascades result in the production of a simple secondary messenger, such as the c-di-GMP (Chapter 5), whose mechanism of action and impact on cell behaviour is yet far to be clearly understood. Finally, chemotactic networks sensing gradient of attractive nutrients or repelling toxic compounds guide mobility and directional movement of a single bacterium (Chapter 9). Observing bacteria swimming and tumbling is something that was earlier catch by the human eye looking through rudimentary microscopes and whose mechanism and control are now understood with exquisite molecular details.

The regulatory networks are not necessarily linear and could be seen as trees or microchips with integrated circuits involving multiple routes and ramifications. Signalling into one branch will spread into many other routes that in a way or another have a connection with the initial input. Moreover, the signalling does not necessarily go exclusively into one direction but involves feedback loops control for example once the initial stimulus or signal is gone. These multiconnections and ramifications very often encourage scientists to present highly sophisticated diagram in which stimuli, regulators and effectors are connected by a multitude of arrows that challenge our analytical capacity. It is also important to realize that regulatory controls may not only be exerted at transcriptional levels and the importance of small regulatory RNA in post-transcriptional control of gene expression is nowadays a very fashionable field of investigation (Chapter 6).

The book that has been assembled here is far to be comprehensive but is meant to cover a wide array of examples, which hopefully will reflect this complexity while providing a useful tool to gain better grasp in understanding bacterial regulatory networks, how they have evolved (Chapter 13) and how they could further be studied and understood. One should be aware that there is probably a gap in between how these regulatory networks work and are inter-connected under circumstances that have been created in our laboratories, and how they really behave when the bacteria are thriving in their natural environment or host. This is certainly a challenge for further research to understand adaptive behaviour of bacteria in their natural niches, which may be of fairly higher complexity than our test tubes, not only in terms of nutrient or oxygen availabilities (Chapter 11) but also in terms of promiscuity with a multitude of other organisms.

I would like to thank the authors that have contributed to the writing of this book. They are all leaders in their field and their commitment to this task has been greatly appreciated. I also would like to thank all the colleagues, whose names are listed here below, that have helped to review the chapters and provided insightful suggestions to the authors. Finally, I wish the owner of this book an enjoyable time reading it and hope he/she will recommend it to many colleagues in the field be it a microbiologist or any curious neophyte eager to broaden his/her vision of a microscopic but rich and sophisticated world.

Alain Filloux (pictured)

Acknowledgements

Thanks to Udo Bläsi, Emmanuelle Bouveret, Jeffrey Cole, Remus T. Dame, Elena Fabiano, Jörgen Johansson, Michael Hecker, John D. Helmann, Vincent Méjean, Helga Mikkelsen, Gabriel Moreno-Hagelsieb, Steven L. Porter, Gary Sawers, Eric Stewart, David Studholme, Tim Tolker-Nielsen, Gottfried Unden and Paul Williams.

σ^S-Controlling Networks in *Escherichia coli*

1

Eberhard Klauck and Regine Hengge

Abstract

The σ^S (RpoS) sigma subunit is the master regulator of the general stress response in *Escherichia coli*, which controls the expression of more than 500 genes during entry into stationary phase or upon exposure to many different stress conditions. σ^S is present at very low levels only in rapidly growing cells, but multiple stress signals are integrated in a way that results in strong σ^S accumulation and efficient σ^S-containing RNAP holoenzyme (Eσ^S) formation. The first part of this review summarizes the molecular control mechanisms of switching from the 'low-σ^S' to 'high-σ^S' state, which operate at the levels of *rpoS* transcription, *rpoS* mRNA turnover and translation, σ^S proteolysis and Eσ^S formation, and outlines multiple stress signal integration into these highly interconnected regulatory processes. We then show that, despite its complexity, the σ^S control network essentially is an intricate combination of a few typical network motifs. These are several feed-forward loops that control σ^S expression, a central and homeostatic negative feedback loop that integrates post-transcriptional σ^S control mechanisms, mutual inhibition of sigma factors competing for RNAP core enzyme governing σ^S activity control, and a series of smaller positive feedback loops that seem to stabilize the 'high-σ^S' state.

Introduction

The σ^S or RpoS subunit of RNA polymerase is the master regulator of the general stress response in *Escherichia coli* and other γ-proteobacteria (Hengge, 2009b, 2011). σ^S is one of seven sigma subunits of RNA polymerase in *E. coli*. Sigma factors are essential for specific promoter recognition and also play an important role in open complex formation during transcription initiation (Helmann, 2011). The regulation of the amount of the RNA polymerase containing a specific sigma factor in response to various signals is an effective way to change the global gene expression pattern (Sharma and Chatterji, 2010; Helmann, 2011).

σ^S controls the expression of approximately 500 genes which corresponds to about 10% of the *E. coli* genes (Weber *et al.*, 2005; Dong and Schellhorn, 2009). In response to different stress conditions, the amount of σ^S increases rapidly in fast-growing cells and σ^S can replace the essential housekeeping σ^{70} factor on the RNA polymerase, which leads to the expression of genes that help the cell to cope with the actual stress situation. σ^S inducing stress conditions include sudden starvation for various nutrients, heat shock, high osmolarity or low pH (Lange and Hengge-Aronis, 1991b, 1994a; Bearson *et al.*, 1996; Hengge-Aronis, 1996a; Muffler *et al.*, 1996c, 1997a; Zgurskaya *et al.*, 1997; Ruiz *et al.*, 2001; Mandel and Silhavy, 2005). Besides responding to an actual stress condition in exponentially growing cells, σ^S is the master regulator of the general stress response, which is induced when cells enter into stationary phase, as a consequence of high cell density and/or depletion of an essential nutrient (Lange and Hengge-Aronis, 1994a). This response governs the preventative expression of different stress protection mechanisms resulting in cross protection (Jenkins *et al.*, 1988, 1990; Lange and Hengge-Aronis, 1991b; McCann *et al.*, 1991; Hengge, 2011) and the transition from a 'fast growth' metabolism to a 'maintenance' metabolism. Depending on the specific stress condition and the growth phase, the molecular mechanisms leading to an increased σ^S level in the cell are different. Moreover, the subsets of genes that are expressed under σ^S control can be modulated by the specific situation of the cell (Hengge-Aronis, 2002b; Hengge, 2011). This is managed by additional regulators, which for their own expression and/or activity may require σ^S and/or specific additional signals.

The regulation of σ^S itself, which is the focus of this review, is extraordinarily complex. In order to understand the regulatory cycles involved, it is necessary to first give an overview of the different levels of σ^S regulation and the factors involved therein. In the second part of this review we try to emphasize the network aspect of this regulation. This turned out to be not so easy because most work concerning bacterial regulatory networks emphasizes transcriptional networks only. In the regulation of the cellular σ^S content, however, transcriptional regulation is only one level of control and the contribution of posttranscriptional regulatory mechanisms is clearly essential. In fact, these levels of control are also interconnected by various positive and negative feedback loops. The physiological significance and dynamic behaviour of these feedback loops can only be envisaged when transcriptional and post-transcriptional modes of regulation are integrated. Currently, we are clearly in a position where we have a rather precise picture of the architecture of the σ^S control network, including its components, network motifs and the relevant interconnections (Hengge, 2011). However, a precise description of the dynamic behaviour of the σ^S control network is still a task for the future.

The physiological context of σ^S induction and action

Owing to strong competition for nutrients of different species and the high metabolic and reproductive efficiency of bacteria, the natural environment of bacteria is generally limited or even exhausted for various nutrients. Thus, life under stress conditions is the normal state rather than an exception (Koch, 1971). The reaction of *E. coli* to decreasing nutrients is rather complex and includes different strategies. Major principles guiding these strategies are the competition of various sigma subunits for a limiting amount of RNAP core enzyme, which serves to globally reorganize 'transcriptional space' (Gruber and Gross, 2003) and the global signalling power of the nucleotide second messengers 5′,3′-cyclic adenosine monophosphate (cAMP) and 5′,3′-bis-guanosine penta/tetraphosphate [(p)ppGpp] (Magnusson *et al.*, 2005; Potrykus and Cashel, 2008; Pesavento and Hengge, 2009). Both regulatory principles directly feed into the general stress response via different pathways, which will be described in detail below.

The first response of *E. coli* to decreasing amounts of an essential nutrient in rich medium is the expression of certain sets of gene products that increase the uptake efficiency of the limiting nutrient or allow the utilization of alternative substrates that contain the missing nutrient (for instance alternative sugars as a carbon source; Ferenci, 2001). In a complex medium such as LB, this corresponds to the beginning of the post-exponential phase (Sezonov *et al.*, 2007). The alarmones cAMP and (p)ppGpp are crucial for the complex responses elicited. Even in minimal medium, in which the growth rate is not yet affected when the intracellular concentration of remaining glucose, i.e. a carbon source preferred by *E. coli*, decreases below 300 μM, the level of the alarmone cAMP increases and the glucose-specific 'hunger response' is turned on (Notley-McRobb *et al.*, 1997; Ferenci, 2001). Binding of cAMP to its dimeric receptor protein CRP (cAMP receptor protein, also called catabolite gene activator protein or CAP) leads to the transcriptional activation of genes for the utilization of alternative carbon sources (for further information, see Stülke and Hillen, 1999; Harman, 2001; Brückner and Titgemeyer, 2002). If such carbon sources are available, this strategy might be successful and the cells keep growing. In addition, cAMP-CRP stimulates the expression of genes involved in flagella synthesis, motility and chemotaxis (Helmann, 1991; Soutourina *et al.*, 1999), which gives the bacteria the opportunity to find an environment with higher nutrient supply. The other alarmone that begins to accumulate during the post-exponential phase is (p)ppGpp, which overall redirects RNAP away from the expression of ribosomal genes towards amino acid biosynthetic and other genes, whose gene products are required to balance growth with limiting resources (Balsalobre, 2011; Bouveret, 2011; Traxler *et al.*, 2011). Overall, in the post-exponential phase of the growth cycle cells no longer maximize growth, but still optimize growth in response to limiting resources. Moreover, the post-exponential phase is also a 'foraging' phase (Zhao *et al.*, 2007).

However, if these specific responses to nutrient limitation are not sufficient to maintain growth, and the medium is successively depleted, the growth rate is gradually reduced further and finally the cells enter into stationary phase. Under these conditions, the 'optimizing growth' strategy is replaced by a 'stress protection and survival' strategy, which is built upon the general stress response (for reviews see Hengge-Aronis, 2002a; Klauck *et al.*, 2007; Hengge, 2011). In fact, the master regulator of this response, σ^S, gradually accumulates already during the post-exponential or foraging phase of the growth cycle, but does not yet compete well with the other sigma factors for RNAP core enzyme. During entry into stationary phase and with the help of several auxiliary factors, $E\sigma^S$ becomes a major form of RNAP holoenzyme which now drives σ^S-dependent gene expression. Besides limitation for various nutrients, i.e. conditions associated with entry into stationary phase, numerous other stress conditions can affect the cellular level of σ^S, some actually as strongly as starvation.

Regulation of the cellular σ^S content and activity

The cellular σ^S content and activity are regulated in response to a multitude of different environmental stimuli like starvation for different nutrients, high or low temperature, high cell density, reduced growth rate, acidic pH or high osmolarity (reviewed in Hengge-Aronis, 2002b; Klauck *et al.*, 2007; Hengge, 2011). But how does the cell manage to integrate so many signals on one single parameter, the amount of active σ^S? Part of the answer is, that different signals affect different levels of regulation (Fig. 1.1). In addition to this distribution of signals on the levels of transcription, transcript turnover, translation, protein stability and activity, each of these regulatory levels can be affected by many signals. Though many proteins and their molecular mechanisms involved in the regulation of *rpoS*/σ^S are known, the question of how the integration of the environmental signals is achieved at the molecular level is still not fully understood (at least in quantitative terms). The current picture of this complex integration has recently been summarized (Hengge, 2011).

The different levels of regulation, which are described in detail below, do not always contribute equally to the increase of the cellular σ^S content upon starvation or other stress situations. While increased *rpoS* transcription is substantial during entry into stationary phase in complex medium, increased translation of *rpoS* mRNA and/or a stop of σ^S degradation play the major role for σ^S accumulation under many acute and potentially life-threatening stress conditions. Different levels of regulation can also compensate for each other. For instance, when cells that grow in minimal medium run out of glucose as a sole carbon source, there is a simultaneous reduction of σ^S expression *and* turnover, that saves energy but has little consequence for the overall cellular σ^S level. Last but not least, also regulation of σ^S activity or the amount of $E\sigma^S$ in the cell is important. Although we initially present all these levels of control separately for the sake of clarity, it will also become apparent that they are interconnected in multiple ways.

Control of rpoS transcription

A basal level of *rpoS* transcription maintains a certain level of *rpoS* mRNA in the cell, even under conditions of fast growth and very low σ^S levels, when *rpoS* mRNA is not translated efficiently and σ^S protein is degraded rapidly. Transcriptional regulation does not seem to play an important role in the induction of σ^S under specific stress conditions but plays a role during a gradual decrease in growth rate (summarized in Hengge-Aronis, 2002b; Klauck *et al.*, 2007; Hengge, 2011).

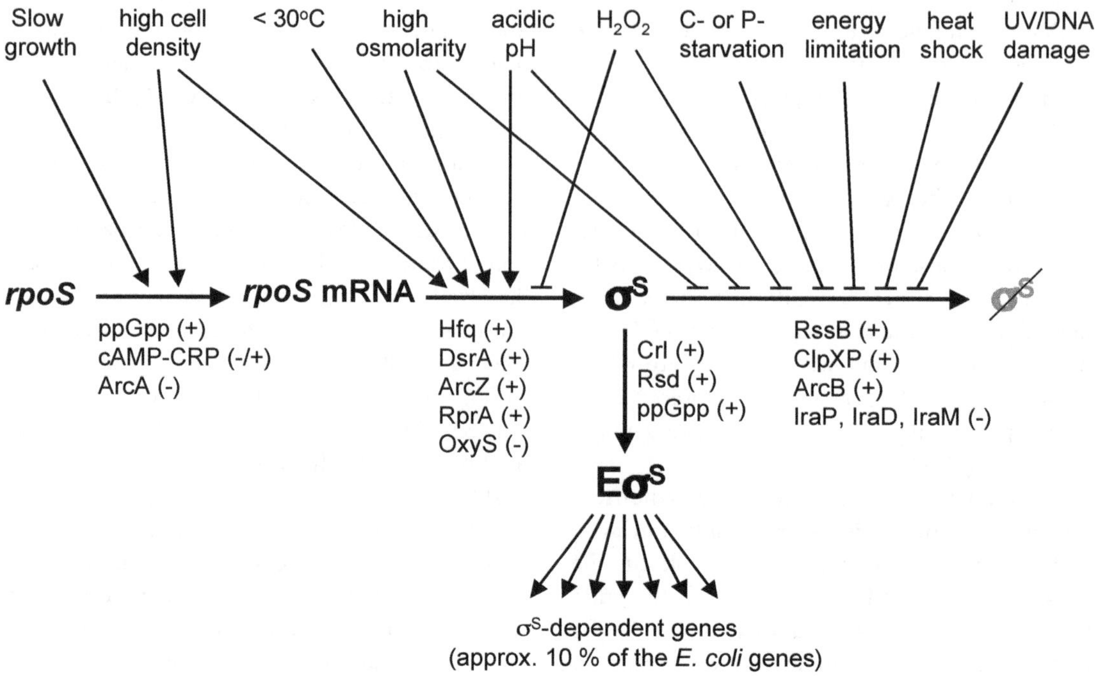

Figure 1.1 Diverse stress conditions affect different levels of σ^S control. These levels are *rpoS* transcription, *rpoS* mRNA translation, σ^S proteolysis and $E\sigma^S$ holoenzyme formation, with the regulatory components indicated that affect these mechanisms. This figure has previously been published in Hengge (2011) and is used here with permission.

The *rpoS* promoter region

The *rpoS* gene is located in an operon with the *nlpD* gene (Lange and Hengge-Aronis, 1994b; Uehara *et al.*, 2010). A polycistronic *nlpD-rpoS* mRNA is expressed from two promoters (*nlpDp1*, *nlpDp2*) which contributes to the basal levels of *rpoS* expression. The third and main promoter (*rpoSp*) is located within the *nlpD* gene and is activated during entry into stationary phase (Lange *et al.*, 1995). Transcription from all *rpoS* promoters is mediated by $E\sigma^{70}$ (Takayanagi *et al.*, 1994; Lange and Hengge-Aronis, 1994b).

Different reports describe an increase in the *rpoS* transcription rate in response to a reduction of growth (Lange and Hengge-Aronis, 1994a; Notley and Ferenci, 1996; Teich *et al.*, 1999; Ihssen and Egli, 2004; Liu *et al.*, 2005). Transcription from the *rpoSp* promoter is gradually induced during the postexponential phase in rich medium (Lange and Hengge-Aronis, 1991b, 1994a), when growth starts to decrease. In minimal medium where doubling time is slower, *rpoS* transcript and σ^S levels are relatively high already in exponential phase and transcription is not further activated during entry into stationary phase (Lange and Hengge-Aronis, 1991b, 1994a; Mandel and Silhavy, 2005).

cAMP-CRP

The cAMP–CRP complex plays a complex dual role in *rpoS* transcription. During exponential growth, cAMP-CRP interferes with *rpoS* transcription probably due to an indirect mechanism (Lange and Hengge-Aronis, 1991a, 1994a; F. Mika and R. Hengge, unpublished data). By contrast, in late post-exponential phase, cAMP-CRP directly binds to a class I activator site at position −62.5 upstream of the *rpoSp* transcriptional start site and contributes to transcriptional activation (Mika and Hengge, 2005).

The histone-like protein Fis

A strong increase in *rpoS* transcription during the logarithmic growth phase was described in a *fis*-mutant strain of *Salmonella typhimurium*, whereas no effect was detectable in stationary phase (Hirsch and Elliott, 2005a). The same is true for *Escherichia coli* (A. Typas and R. Hengge, unpublished results). Fis is an abundant histone-like nucleoid-associated protein in early exponential phase cells and is nearly undetectable in stationary phase (for reviews on Fis see Finkel and Johnson, 1992; Browning *et al.*, 2010). It strongly up-regulates ribosomal gene expression and represses genes that are nonessential during fast growth. The binding site of Fis was identified at −50 with respect to the *rpoS* promoter, at least for *Salmonella typhimurium* (Hirsch and Elliott, 2005a). Thus, Fis seems to contribute to down-regulate *rpoS* transcription in the logarithmic growth phase.

Two-component systems

The phosphorylated ArcA response regulator was shown to bind directly around −63 upstream and +23 downstream of the transcriptional start site of *rpoSp*. The −63 binding site overlaps with the binding site for cAMP-CRP (see above), which activates *rpoS* transcription and might, at least in part, relieve the inhibitory effect of ArcA in stationary phase (Mika and Hengge, 2005; F. Mika and R. Hengge, unpublished). ArcA is phosphorylated via the histidine sensor kinase ArcB, which in addition also phosphorylates the response regulator and proteolytic σ^S-recognition factor RssB (Mika and Hengge, 2005). Thus, the ArcB/ArcA/RssB system down-regulates σ^S at the levels of transcription and proteolysis and does so in response to oxygen availability and the energy status of the cell (for details, see below).

The BarA/UvrA two-component system, which was shown to play a role in the regulation of carbon metabolism, motility, biofilm formation and virulence (Pernestig *et al.*, 2003; Teplitski *et al.*, 2003; Herren *et al.*, 2006; Tomenius *et al.*, 2006), seems to positively influence *rpoS* transcription, though no details about its mode of action are known (Mukhopadhyay *et al.*, 2000; Sugiura *et al.*, 2003; Hengge, 2008).

The alarmone (p)ppGpp

(p)ppGpp is a global signalling molecule that increases in response to nutritional stress and its cellular concentration is inversely correlated with growth rate (Venturi, 2003; Magnusson *et al.*, 2005; Potrykus and Cashel, 2008; Bouveret, 2011). Cellular levels of (p)ppGpp and σ^S closely correlate at different growth rates in glucose-limited cultures under different cultivation strategies. Especially under conditions of fast changes in nutrient availability in a chemostat this correlation is quite obvious (Teich *et al.*, 1999). (p)ppGpp-free *relA spoT* mutants exhibit only very low residual σ^S levels (Gentry *et al.*, 1993; Lange *et al.*, 1995; Webb *et al.*, 1999). It has been reported that (p)ppGpp acts positively on *rpoS* transcription (Gentry *et al.*, 1993; Lange *et al.*, 1995; Hirsch and Elliott, 2002), and it most likely affects both transcription initiation and transcriptional elongation (Lange *et al.*, 1995). How (p)ppGpp acts on *rpoS* transcription is not clear, since this molecule exerts a lot of direct and indirect effects in the cell (Magnusson *et al.*, 2005). For example, it positively affects polyphosphate accumulation, which (probably directly) stimulates *rpoS* transcription (Gentry *et al.*, 1993; Shiba *et al.*, 1997). In addition, (p)ppGpp plays direct and indirect roles in *rpoS* translation and σ^S stability (see below).

Control of *rpoS* mRNA turnover and translation

rpoS mRNA turnover

The half-life of the *rpoS* mRNA lies within the range of several minutes, which does not differ from the average mRNA stability in *E. coli*. RNase E and RNase III are the most important RNases for *rpoS* mRNA decay. While *rpoS* transcript stability can be influenced by overproducing the small regulatory RNAs that also stimulate translation (McCullen *et al.*, 2010), changes in *rpoS* mRNA half-life in wild-type strains do not seem to contribute significantly to the accumulation of σ^S during entry into stationary phase (Basineni *et al.*, 2009).

rpoS mRNA secondary structure

By contrast, activation of *rpoS* mRNA translation is essential for the rapid increase of the cellular σ^S content upon shifts to high osmolarity and low pH (Fig. 1.1). In addition, translational induction of *rpoS* occurs in the post-exponential growth phase and precedes the stabilization of the σ^S protein (Lange and Hengge-Aronis, 1994a). A strong increase in the rate of *rpoS* translation might also be involved in σ^S stabilization due to the titration of the proteolytic system and the σ^S/RssB homeostatic positive feedback regulation (see below).

The *rpoS* mRNA transcribed from the promoter *rpoSp* exhibits an unusually long 5′-untranslated region (UTR) of 565 bp (Lange *et al.*, 1995). Under non-stress conditions, this region folds into a secondary structure, in which the translation initiation region (TIR) is not accessible for ribosomes (Brown and Elliott, 1997; Lease *et al.*, 1998; Majdalani *et al.*, 1998; Cunning and Elliott, 1999; Hirsch and Elliott, 2005b). Under translation-inducing conditions (see Fig. 1.1), a change of secondary structure releases the TIR from basepairing. In contrast to the *rpoH* mRNA, which changes its structure without the aid of proteins upon increased temperature (Narberhaus, 2006), induction of *rpoS* mRNA translation requires additional proteins and small regulatory RNAs (Hengge, 2011).

The role of Hfq and small regulatory RNAs

The RNA-binding protein Hfq is essential for efficient *rpoS* mRNA translation (Brown and Elliott, 1996; Muffler *et al.*, 1996b). Hfq is a pleiotropic regulator (Tsui *et al.*, 1994; Muffler *et al.*, 1997b) and functions as a platform that facilitates the interactions between mRNAs and small regulatory RNAs to regulate the translation of the mRNAs (Repoila *et al.*, 2003; Zhang *et al.*, 2003; Valentin-Hansen *et al.*, 2004; Brennan and Link, 2007). Hfq is a hexameric protein with a central pore, around which the RNA is bound (Schumacher, 2002; Sauter, 2003). Hfq favours the interaction with at least three small regulatory RNAs, DsrA, ArcZ and RprA (Sledjeski *et al.*, 1996; Majdalani *et al.*, 2002; Mandin and Gottesman, 2010). These small RNAs activate *rpoS* translation by base-pairing with the 'anti-TIR' region of *rpoS* mRNA. This resolves the intramolecular base-pairing of the mRNA and makes the TIR accessible for the ribosome (Sledjeski *et al.*, 1996; Lease and Belfort, 2000; Majdalani *et al.*, 2001, 2002; Arluison *et al.*, 2007; Updegrove *et al.*, 2008). Additional proteins like the ribosomal protein S1 seem to be involved in the interaction between *rpoS* mRNA and DsrA (Koleva *et al.*, 2006). In addition, DsrA and RprA protect *rpoS* mRNA from degradation by RNase E (McCullen *et al.*, 2010). While DsrA, ArcZ and RprA activate *rpoS* translation by the same principal mechanism, they are induced under different conditions: DsrA is essential for *rpoS* induction at low temperature (Sledjeski *et al.*, 1996; Majdalani *et al.*, 1998), ArcZ is under negative control of the ArcB/ArcA two-component system (Mandin and Gottesman, 2010), and the physiological role of RprA remain to be fully identified, though a minor role in hyperosmotic induction of *rpoS* translation and its accumulation in stationary phase have been reported (Majdalani *et al.*, 2002; Repoila *et al.*, 2003; Madhugiri *et al.*, 2010).

An additional small RNA, OxyS, induced under H_2O_2 stress, inhibits *rpoS* translation by an unknown mechanism (Zhang *et al.*, 1998). Since σ^S is also stabilized under these conditions (Merrikh *et al.*, 2009), there is hardly any change in the net σ^S level (Zhang *et al.*, 1998).

Additional factors controlling *rpoS* translation

The histone-like protein, HU, which plays an important role in the nucleoid architecture (reviewed in Dorman and Deighan, 2003), was shown to specifically bind to a RNA fragment containing the TIR region of *rpoS* (Balandina *et al.*, 2001). So, HU might directly stimulate *rpoS* translation. Another histone-like protein, H-NS (for a review, see Dorman, 2004), shows the opposite effect on *rpoS* translation and additionally stimulates σ^S proteolysis (Barth *et al.*, 1995; Yamashino *et al.*, 1995). It preferentially binds to curved DNA, which is commonly found at promoters. This might be the reason for its function as a global repressor of transcription of at least 100 genes (Hommais *et al.*, 2001; Dorman, 2004). H-NS was shown to bind to *rpoS* mRNA and DsrA with low affinity and seems to influence their stability by enhancing their cleavage by ribonucleases (Brescia *et al.*, 2004).

Other components were found to somehow influence *rpoS* translation indirectly. These include the LysR-like transcription factors LeuO, which affects *rpoS* translation by repressing the *dsrA* gene at low temperature (Klauck *et al.*, 1997; Repoila and Gottesman, 2001), and LrhA, which inhibits *rpoS* translation by an unclarified mechanism (Peterson *et al.*, 2006) and seems itself repressed by the Rcs-phosphorelay system, which also activates *rprA* transcription (Majdalani *et al.*, 2002; Peterson *et al.*, 2006). Again by unknown mechanisms, the SsrA-tagging system (tmRNA) can positively affect *rpoS* translation, while Lon protease seems to interferes with *rpoS* translation at low temperature (Ranquet and Gottesman, 2007). Also mutants defective for the DnaK chaperone (Rockabrand *et al.*, 1995; Muffler *et al.*, 1997a; Rockabrand *et al.*, 1998), the PTS component EIIA(Glc) (Ueguchi *et al.*, 2001), the RNA polymerase-binding protein DksA (Webb *et al.*, 1999) showed effects on the accumulation of σ^S or the expression of translational *rpoS::lacZ* fusions, but nothing is known about how these proteins influence *rpoS* translation.

Regulation of σ^S stability

σ^S has long been known as a paradigm of a regulator subject to controlled proteolysis (Lange and Hengge-Aronis, 1994a; Jenal and Hengge-Aronis, 2003). Today, all essential components of the σ^S degradation machinery are known. Moreover, we understand the basic regulatory mechanisms of this machinery and even multiple signal input fairly well (for a recent summary, see Hengge, 2009b; Fig. 1.2).

In fast-growing cells, newly synthesized σ^S is rapidly degraded with a half-life of a few minutes (Lange and Hengge-Aronis, 1994a; Hengge-Aronis, 2002b), depending on the actual growth conditions. This keeps the σ^S level low as even in the relative absence of stress there is a basal rate of transcription and translation (Lange and Hengge-Aronis, 1994a). There are many different stress conditions that lead to a sudden stabilization and therefore a sudden increase in the σ^S level, including carbon or phosphorus starvation, the diauxic lag phase, shift to high osmolarity or to an acidic pH (Bearson *et al.*, 1996; Hengge-Aronis, 1996a; Muffler *et al.*, 1996c; Fischer *et al.*, 1998; Mandel and Silhavy, 2005; Peterson *et al.*, 2005; Merrikh

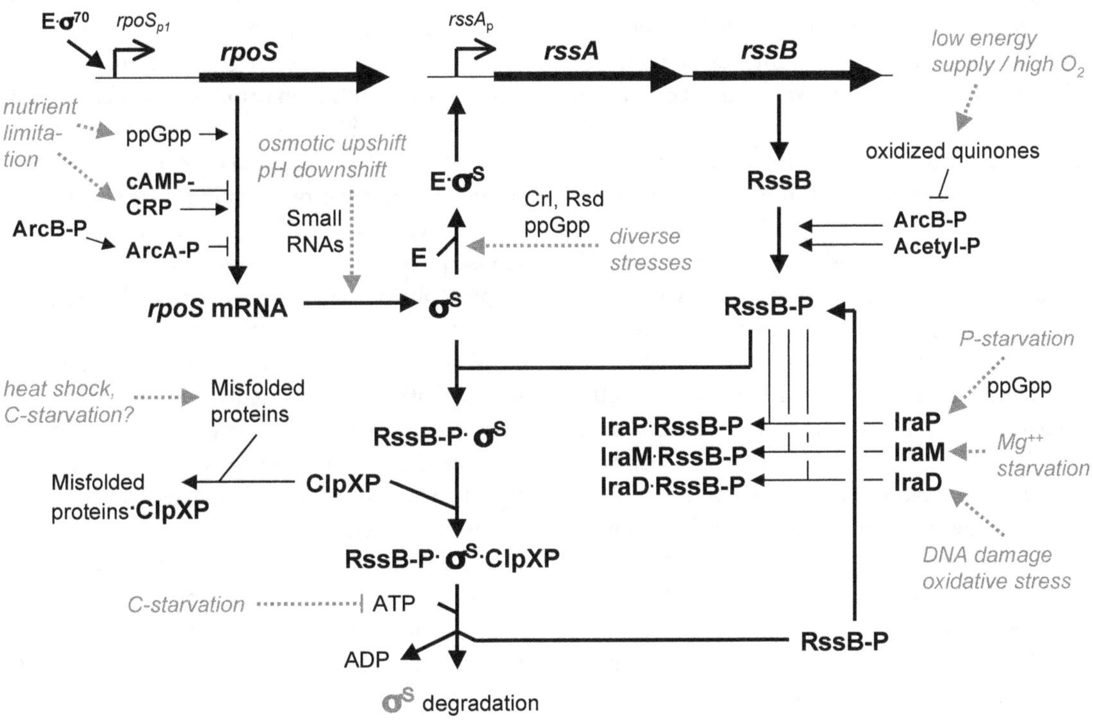

Figure 1.2 Molecular mechanisms and signal input in the control of σ^S expression and proteolysis. An overview integrating *rpoS* transcription and translation with the homeostatic σ^S/RssB FBL and the RssB/ClpXP-mediated catalytic cycle of σ^S proteolysis is shown, with differential stress input into all these interlinked processes (for details, see text). This figure has previously been published in Hengge (2011) and is used here with permission.

et al., 2009). By contrast, the rise of σ^S in complex medium during entry into stationary phase is a more gradual one (Lange and Hengge-Aronis, 1994a; Zgurskaya *et al.*, 1997).

Role of the targeting factor RssB in ClpXP-mediated σ^S proteolysis

The ATP dependent protease ClpXP (for a review, see Baker and Sauer, 2006) is responsible for the degradation of σ^S (Schweder *et al.*, 1996). But, unlike other ClpXP substrates, σ^S needs a specific protein that serves as a recognition factor for σ^S degradation, the response regulator RssB (also called SpreE) (Muffler *et al.*, 1996a; Pratt and Silhavy, 1996). With these essential proteins (and ATP for ClpX and acetyl phosphate as a phosphodonor for RssB, see below) it was possible to establish an *in vitro* σ^S degradation system (Zhou *et al.*, 2001; Bougdour *et al.*, 2006; Typas *et al.*, 2007a), which reproduced σ^S half-life of several minutes as observed *in vivo*, depending on the amount and activity of the components. Phosphorylated RssB binds with high affinity to domain 3 of σ^S (previously called region 2.5) with K173 being essential for this interaction (Becker *et al.*, 1999). The first α-helix of domain 3, which contains K173, was shown to be sufficient for the interaction with RssB (Stüdemann *et al.*, 2003). On the other hand, both domains of the RssB protein are necessary to fulfil its function as a targeting factor for σ^S proteolysis (Klauck *et al.*, 2001). Phosphorylation of RssB seems to increase its affinity for σ^S, which is required to effectively degrade σ^S *in vivo* (Bouché *et al.*, 1998; Moreno *et al.*, 2000; Klauck *et al.*, 2001; Peterson *et al.*, 2004), since the cellular RssB level is about 20-fold lower than σ^S (Klauck *et al.*, 2001; Pruteanu and Hengge-Aronis, 2002). *In vitro*, σ^S degradation can also be achieved with high concentrations of unphosphorylated RssB (E. Klauck and R. Hengge, unpublished results). Moreover, the non-phosphorylatable $RssB^{D58P}$ and $RssB^{D58A}$ mutant proteins can support some σ^S degradation, suggesting an affinity for σ^S that is intermediate between that of phosphorylated and non-phosphorylated wild-type RssB (for a full discussion, see Hengge, 2009b).

The binding of RssB to σ^S induces the conformational change required to produce accessibility of a N-terminal ClpX recognition region in σ^S. An essential part of this region lies between amino acids 7 and 35, because a σ^S mutant protein with a deletion in this region normally binds to RssB *in vitro* but is not degraded by ClpXP *in vivo* (Stüdemann *et al.*, 2003). In the presence of a non-hydrolysable ATP analogue and acetyl phosphate a stable quaternary σ^S–RssB–ClpXP complex was detectable (Zhou *et al.*, 2001). In the absence of ClpXP (in *clpP* mutants) or when ClpXP might be bound mainly to other substrates *in vivo*, RssB inhibits σ^S activity by sequestering it and thereby acts like an anti-sigma factor as this prevents σ^S from binding to RNAP core enzyme. Artificial overproduction of RssB in stoichiometric amounts with σ^S also showed its anti-sigma factor potential (Zhou and Gottesman, 1998; Becker *et al.*, 2000; Moreno *et al.*, 2000; Klauck *et al.*, 2001).

Slow stabilization of σ^S during entry into stationary phase

Although only a few proteins are directly involved in σ^S proteolysis, the discovery of additional regulatory features revealed a complex and effective control of proteolysis in response to different environmental conditions (see also Hengge, 2009b). Interestingly, the main regulation or σ^S proteolysis occurs at the level of specific substrate recognition, which means it affects σ^S–RssB interaction. Consequently, stress-induced down-regulation of σ^S proteolysis does not necessarily affect ClpXP-mediated proteolysis of other proteins.

Many stress signals have to be integrated in the regulation of σ^S proteolysis (Fig. 1.1). It was easy to imagine that the phosphorylation status of RssB might change in response to environmental signals, since only phosphorylated RssB is able to bind σ^S efficiently *in vivo* and *in vitro* (Becker *et al.*, 1999; Zhou *et al.*, 2001; Peterson *et al.*, 2004). However, RssB phosphorylation does not seem a major target for regulation for three reasons: (i) phosphorylated RssB is, compared with other response regulators, extremely stable and does not show any intrinsic phosphatase activity (Klauck *et al.*, 2001); (ii) during the proteolytic cycle, RssB is apparently not dephosphorylated (Hengge, 2009b); and (iii) mutant searches have never come up with a phosphatase for RssB.

Therefore, the phosphorylation status of RssB is unlikely to play a role in the fast 'emergency' stabilization of σ^S. However, phosphorylation of (newly synthesized) RssB seems involved in fine tuning σ^S degradation during transition into stationary phase in complex medium. Under these conditions, the cellular concentrations of both *in vivo* phosphodonors for RssB decrease. These are acetyl phosphate (McCleary and Stock, 1994; Bouché *et al.*, 1998; Klein *et al.*, 2007) and phosphorylated ArcB, whose autophosphorylation is inhibited by oxidized quinones which most likely arise in response to energy limitation in the presence of oxygen (Fig. 1.2) (Malpica *et al.*, 2004, 2006; Mika and Hengge, 2005). Since the cellular concentration of phosphorylated RssB is rate-limiting for σ^S proteolysis (Pruteanu and Hengge-Aronis, 2002), also a gradual reduction in acetyl phosphate and phosphorylated ArcB can contribute to the gradual stabilization of σ^S during entry into stationary phase.

Role of changes in the σ^S–RssB ratio for the rapid control of σ^S proteolysis in response to acute stresses

Sudden acute and potential life-threatening stress conditions result in an instantaneous stop of σ^S proteolysis. The basis of these immediate reactions are rapid changes in the cellular ratio of σ^S to RssB that are available for interaction. In fast growing cells, the cellular level of RssB is about twenty times lower than that of σ^S (Becker *et al.*, 2000). This ratio is maintained by a homeostatic negative feedback regulation (Fig. 1.2), because RssB expression is σ^S-dependent (Ruiz *et al.*, 2001; Pruteanu and Hengge-Aronis, 2002). This feedback regulation and the fact that RssB is the rate limiting factor of σ^S proteolysis (Pruteanu and Hengge-Aronis, 2002) is the key feature of the entire σ^S control system since it has important consequences: (i) it maintains constant σ^S degradation rates even during changes of σ^S expression (up to certain threshold where the σ^S-dependent *rssAB* promoter is maximally activated); and (ii) it sets the stage for the rapid emergency stabilization by mechanisms that suddenly change the RssB:σ^S ratio, which are based upon competition, titration and sequestration of the molecular players, i.e. σ^S, RssB, ClpXP and RNAP core enzyme. These mechanisms provide the basis for the integration of the different stress signals in the rapid control of σ^S proteolysis.

Titration of RssB by increased σ^S synthesis

For instance, a strong increase of *rpoS* mRNA translation in response to a hyperosmotic upshift or a acid shock simply 'overwhelms' the proteolysis system, since it out-titrates the amount of RssB available to target σ^S to the ClpXP protease (Muffler *et al.*, 1996c; Heuveling *et al.*, 2008). Although the proteolytic machinery remains active for degrading σ^S, an *in vivo* measurement of *total* σ^S half-life shows nearly stable σ^S, because a large fraction of σ^S cannot be degraded. Thus, titration of RssB and therefore overall stabilization of σ^S can be a secondary consequence of stress conditions that strongly activate *rpoS* mRNA translation. Moreover, the σ^S-RssB feedback cycle should provide for adaptive behaviour, such that in the somewhat longer run, σ^S then becomes unstable again (Muffler *et al.*, 1996c; Heuveling *et al.*, 2008).

Sequestration or 'trapping' of RssB

The σ^S:RssB ratio can not only be increased by accumulating σ^S but also by decreasing the level of RssB available for interaction with σ^S. This is the function of three anti-RssB proteins, IraP, IraM and IraD. These three small proteins sequester and thereby inhibit activity of RssB by direct interaction, but are induced under different conditions. IraP specifically inhibits RssB under phosphate starvation conditions (Bougdour *et al.*, 2006). IraM is magnesium starvation-induced via the PhoP/Q two-component system in *E. coli* (Bougdour *et al.*, 2008), while IraD is induced by hydrogen peroxide or DNA damage, independently of the well known SOS response and the OxyR system (Merrikh *et al.*, 2009). While the Ira proteins are not similar with respect to primary sequence, they are thus functionally related. The Ira proteins can also be seen as 'connector' proteins which affect two-component system output in an unorthodox way (Mitrophanov and Groisman, 2008).

A functionally similar state is the trapping of RssB in proteolytically inactive complexes with

σ^S. This might be the case during heat shock or carbon starvation, when ClpXP is sequestered either by misfolded, heat-denatured proteins, or by aberrant proteins formed because of translational errors at the ribosome as a consequence of starvation for carbon sources (Fredriksson *et al.*, 2007).

Trapping of RssB in proteolytically inactive ternary σ^S/RssB/ClpXP complexes could be a consequence of decreased ATP levels during starvation of essential nutrients (Martin *et al.*, 2008). Indeed, strains carrying different σ^S-regulated *lacZ* gene fusions show lower β-galactosidase activities (relative to σ^S levels) in stationary phase cells than otherwise isogenic strains with mutations in *clpP* or *rssB*. This means that, in the presence of RssB and ClpXP, probably all RssB as well as a minority fraction of σ^S is trapped and therefore inactive in stationary phase. The major fraction of σ^S would then not be degraded and would be free to interact with RNAP core enzyme (Becker *et al.*, 2000; Zhou *et al.*, 2001).

A role for competition between RssB and core RNAP for σ^S in the control of σ^S proteolysis

Binding of σ^S to either RssB or RNA polymerase core enzyme (E) is mutually exclusive. This means that σ^S as a component of the RNAP holoenzyme is stable (Zhou *et al.*, 2001). Consequently, any mechanisms that favour the formation of $E\sigma^S$ reduce the σ^S degradation rate. This was directly shown for Crl, a protein that binds σ^S (Bougdour *et al.*, 2004) and stimulates its interaction with core RNA polymerase, thereby also reducing σ^S proteolysis (Typas *et al.*, 2007a). The same might be true for Rsd, an anti σ^{70} factor, that reduces the amount of free σ^{70} in stationary phase and thus might indirectly promote the formation of σ^S-containing RNA polymerase (Jishage and Ishihama, 1998). Also (*p*)ppGpp seems to promote $E\sigma^S$ holoenzyme formation and may thereby also contribute to σ^S stabilization (Jishage *et al.*, 2002a). That processes influencing this competition can indirectly modulate σ^S proteolysis is illustrated by the recent finding that σ^S degradation is reduced in a mutant defective in *rpoN*, which codes for σ^{54}, an alternative sigma factor responsible for the utilization of nitrogen sources (Dong *et al.*, 2011).

Additional factors controlling σ^S degradation

A number of other components have been reported to somehow influence the rate of σ^S proteolysis under some conditions. In general, these factors seem to feed into the σ^S recognition pathway. Thus, the CsgD regulatory protein apparently can activate the expression of IraP, which sequesters RssB (Gualdi *et al.*, 2007). By a still unknown mechanism, H-NS promotes σ^S proteolysis since σ^S is not degraded in *hns* mutants (Barth *et al.*, 1995; Yamashino *et al.*, 1995; Zhou and Gottesman, 2006). DnaK seems to have a positive influence on σ^S stabilization during entry into stationary phase (Muffler *et al.*, 1997a). Finally, a RssB/ClpXP independent degradation of σ^S promoted by the response regulator RstA (which also exists in *E. coli*) was described in *Salmonella* (Cabeza *et al.*, 2007), but neither the protease involved in this process nor some mechanistic details were reported.

Regulation of σ^S activity

The activity of sigma factors is dependent on their binding to RNAP core enzyme (E). The formation of the σ^S containing RNAP holoenzyme ($E\sigma^S$) is a target for regulation (Sharma and Chatterji, 2010; Helmann, 2011). In contrast to many other sigma factors, e.g. σ^E or σ^{28} (Helmann, 1999), no anti-sigma factor is known for σ^S that directly inhibits its activity. Although RssB has the potential to act as one, this function seems to be irrelevant *in vivo* owing to the low physiological RssB:σ^S ratio (Becker *et al.*, 2000).

At the centre of sigma factor activity control is the competition of various sigma factors for core RNAP, which is present in a limiting amount that is determined by the overall biosynthetic potential of the cell, i.e. the available resources (Gruber and Gross, 2003). At first sight, σ^S seems ill-suited for this competition. Although nearly 10% of the genes in the *E. coli* genome are under σ^S control (Weber *et al.*, 2005) and σ^S is the predominantly active sigma factor in stationary phase, its level reaches one-third of the σ^{70} level only (Jishage and Ishihama, 1995) and its affinity to RNA polymerase is the lowest of all sigma factors (Maeda *et al.*, 2000; Colland *et al.*, 2002). Therefore, mechanisms must exist that favour the formation of RNAP holoenzyme containing alternative sigma factors and in particular, that of Eσ^S.

(p)ppGpp, the anti-σ^{70} factor Rsd and 6S RNA

The alarmone (p)ppGpp, together with DksA protein, affects RNAP holoenzyme formation in favour of alternative sigma factors and enhances transcriptional activity from genes driven by alternative holoenzymes (Kvint *et al.*, 2000; Jishage *et al.*, 2002b; Laurie *et al.*, 2002). These activities add to the (p)ppGpp-mediated stringent response, which consists in the reduction of transcription at ribosomal and other growth-rate-regulated genes, thereby releases RNA polymerase to activate many other σ^{70}-dependent genes, including *rpoS* (reviewed in Magnusson *et al.*, 2005; Srivatsan and Wang, 2008; Sharma and Chatterji, 2010).

Another reason for increased formation of RNAP holoenzyme with alternative sigma factors is the inactivation of σ^{70}. The anti-σ^{70}-protein Rsd, that accumulates in stationary phase, was shown to bind σ^{70} specifically *in vitro* with a 1:1 stoichiometry (Jishage and Ishihama, 1998, 1999; Patikoglou *et al.*, 2007). *In vivo* experiments showed that Rsd is able to sequester σ^{70}, leading to up-regulation of σ^S-dependent genes (Jishage and Ishihama, 1999; Mitchell *et al.*, 2007). However, the function of Rsd might go beyond simply sequestering σ^{70} because it was shown at least *in vitro* that Rsd also binds to core RNAP and that it can displace σ^{70} from the holoenzyme (Ilag *et al.*, 2004).

Yet another stationary phase induced factor that binds to σ^{70}-containing RNA polymerase holoenzyme and inhibits its activity is the 6S RNA (Wassarman and Storz, 2000). 6S RNA structure is actually a mimic of an open complex (Wassarman, 2007), which binds best to Eσ^{70} (Gildehaus *et al.*, 2007).

Crl and FliZ

While (p)ppGpp, Rsd and 6S RNA all reduce formation of vegetative RNAP, they do not seem to discriminate between the alternative sigma factors. By contrast, two factors have been identified to date, that affect the activity specifically of σ^S. One is Crl protein, which binds to σ^S (Bougdour *et al.*, 2004; England *et al.*, 2008) and actively and specifically promotes the formation of σ^S-containing RNAP holoenzyme during entry into stationary phase, which is especially important as long as σ^S levels are still rather low (Robbe-Saule *et al.*, 2007; Typas *et al.*, 2007a).

By contrast, FliZ protein negatively affects σ^S-dependent gene expression, with its pattern of target genes mirroring that of Crl (Pesavento *et al.*, 2008). The *fliAZY* operon is expressed under FlhDC control during the post exponential phase, when cells become motile. During that period, despite already elevated σ^S-levels, FliZ interferes with σ^S-dependent gene expression. FliZ does not interact with σ^S, but it directly binds to extended −10 promoter regions of certain σ^S-dependent genes and thereby prevents Eσ^S from activating these genes (Pesavento and Hengge, 2012). Overall, FliZ gives transient priority to flagella gene expression during the post-exponential phase, before it is itself down-regulated during entry into stationary phase.

Regulatory network motifs in the control of cellular σ^S levels and activity

In this section we try to describe the architecture of the σ^S regulatory network by pointing out the nature and combination of the network motifs involved. This is a rather daunting task for several reasons. First, the cellular amount of σ^S is regulated by an large number of different factors and involves all possible levels of regulation on the way from the gene to an active protein. Yet, research on regulatory networks usually emphasizes the function of transcriptional networks only (see for instance the ground-breaking compilation by Alon, 2007a). Nevertheless, we will show that network motif analysis, despite having been developed for transcriptional networks, is fruitful also for analysing the σ^S control network with its massive dependence on post-transcriptional control. Another complication in applying network analysis to the description of σ^S control is the fact that we are used to classical transcription factors as the major players in networks, i.e. regulatory factors that in hierarchical relationships either turn on and off genes. However, important regulatory mechanisms in σ^S control are based on the non-hierarchical principles of titration, competition and sequestration, which also operate in a graded or analogue manner rather than on a digital 'on–off' basis (Hengge, 2009b, 2011). Also the regulatory input of histone-like regulatory proteins such as Fis is often analogue rather than digital (Blot *et al.*, 2006; Marr *et al.*, 2008). Finally, it is well established that complex networks are built from a limited number of small network motifs and possible dynamic properties of several isolated motifs have been described (Bren and Alon, 2011). However, owing to its unprecedented signal integration power, the σ^S control network contains so many intertwined and nested network motifs that predicting the precise contributions of specific motifs or the reaction of the entire network to many different stress conditions is quite a challenge. Yet, there is hardly any other complex regulatory network in bacteria for which as many qualitative and even quantitative data are available. This makes the σ^S control network obviously a paradigm for future research on the functioning of complex cellular regulatory networks in bacteria.

Network motifs

Transcriptional regulatory networks have been analysed extensively (for summaries of the current views, see Bren and Alon, 2011; Chalancon and Babu, 2011). These networks are hierarchical and modular in structure (Babu *et al.*, 2004). In fact, these principles were already pointed out for the architecture of the complex σ^S regulon (Hengge-Aronis, 1996b) long before they became common place in the network scientific community. Moreover, networks are composed of a limited number of network motifs (Milo *et al.*, 2002; Bren and Alon, 2011). These are (i) positive or negative autoregulation, which can be direct or indirect and set up feedback loops (FBL); (ii) feed-forward loops (FFL), in which a regulator X controls an output Z both directly and indirectly (via an intermediate regulator Y; Fig. 1.3); (iii) single input modules (SIM), in which a single regulator controls many targets; (iv) dense overlapping regulons (DOR), in which many regulators control many targets in an overlapping manner; and (v) regulatory cascades, in which several regulators control the following regulators (besides other target genes) in a sequential arrangement (Milo *et al.*, 2002).

These motifs confer a wide variety of dynamic behaviours (summarized by Bren and Alon, 2011). Negative autoregulation can reduce noise in gene expression, speed up response time, provide adaptation of a response and thereby increase the dynamic range over which a system can respond. Positive autoregulation can result in bursts of gene expression and can lead to bistable gene expression and population heterogeneity. Coherent FFLs, in which the direct and indirect 'branches' of regulation act in the same direction (either positive or negative), can set up AND gates, which render the response dependent on two signals occurring sequentially. This can result in noise reduction, a pronounced difference in the kinetics of switching the response on or off, or 'priming' in developmental pathways. Incoherent FFL, in which direct regulation by X is positive, but the indirect 'branch' acts negatively, can generate pulses of gene expression. Alternatively, this motif can also 'prime' the cell for a response, which then can be triggered by a second signal that inhibits or eliminates the inhibitory secondary regulator Y. SIM motifs are the basis of classical global regulation in bacteria, but owing to increasing knowledge, are more and more replaced by DORs, since most bacterial promoters seem to respond to multiple regulatory input. Regulatory cascades, usually in combination with various types of FBLs and FFLs, are the hallmark of bacterial developmental pathways (see, for instance, Eichenberger *et al.*, 2004). Another intriguing motif is mutual inhibition which produces switching behaviour and can generate bistability. This motif tends to be overlooked because it can involve the 'cross-talk' of entire motifs (for instance cascades) rather than of single regulators (in the latter case it would be double negative and therefore positive autoregulation). Also, mutual inhibition of motifs or entire pathways occurs less frequently. However, it tends to occur at the top hierarchical levels of networks and therefore can have drastic impact on the behaviour of an entire network (an example is reported in Pesavento *et al.*, 2008).

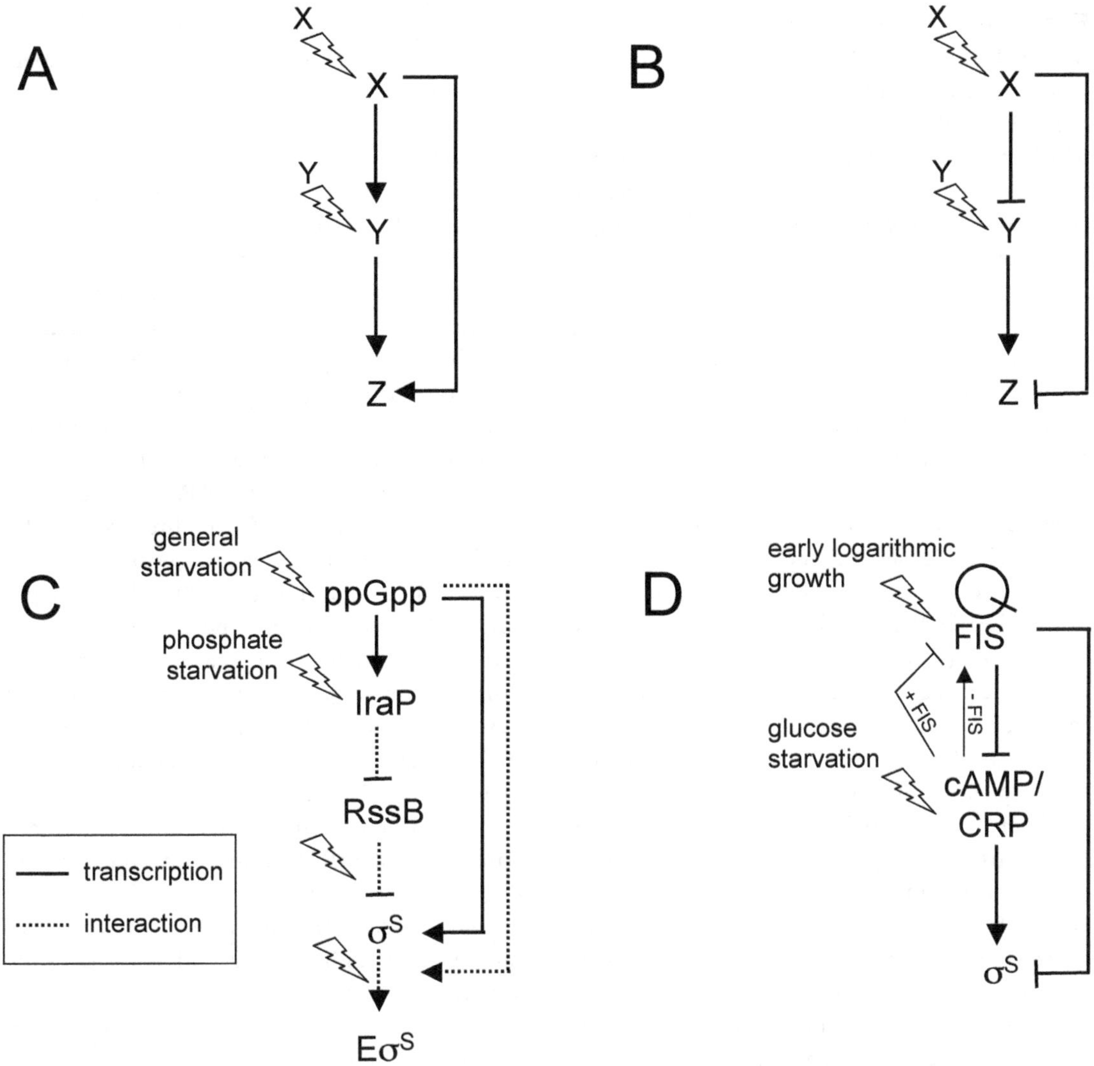

Figure 1.3 Feed-forward loops in the control of σ^S expression. Theoretical coherent FFLs of type 1 (both 'branches' activating the output Z; A) and type 2 (both 'branches' inhibiting the output Z; B) are shown. Practical examples in the σ^S control network for both types of FFLs are the role of the master input molecule {*p*)ppGpp (C) and the master regulator Fis (D). For mechanistic details, see text.

All these motifs are found in large numbers in the entire σ^S network, both in the upstream part that controls σ^S levels and activity as well as in the downstream part, meaning the σ^S regulon. The σ^S regulon actually seems to have been the first system, in which the pattern of complex overlapping control was pointed out (Hengge-Aronis, 1996b) that is now referred to as a DOR. Complex subnetworks within the huge σ^S regulon, which consist of intertwined cascades and multiple FBLs and FFLs, control for instance σ^S-dependent curli and biofilm formation or acid stress resistance (recently summarized in Hengge, 2011). The control of σ^S itself is mainly dependent on three types of network motifs. These are several key FFLs (using (*p*)ppGpp, FIS, and the Arc system), a central and homeostatic negative FBL (the σ^S-RssB feedback cycle), and mutual inhibition (of sigma factors competing for RNAP core enzyme). In addition, numerous smaller positive FBLs seem to contribute to stabilizing the 'high-σ^S-activity' state.

Feed-forward loops in the control of σ^S expression

One of the most frequent motifs in the *E. coli* transcriptional network is the feed-forward loop, which can be coherent (both 'branches' of control are of the same sign) or incoherent (one 'branch' acts positively, the other negatively) (Mangan and Alon, 2003). In σ^S regulation at least three complex FFLs are obvious. These FFLs show the coherent configuration (either of type 1 or type 2; see Fig. 1.3A and B) and they are integrated with additional patterns of regulation. Moreover, they show intricate combinations of transcriptional and post-transcriptional regulation.

The general stress alarmone (p)ppGpp

The nucleotide second messenger (p)ppGpp, which directly binds to RNAP core enzyme, positively affects σ^S levels as well as σ^S activity (see above). (p)ppGpp is situated at the top of a complex coherent FFL governing σ^S regulation, in which it acts by several mechanisms (Fig. 1.3C): (i) during the post-exponential phase in complex medium or in response to various kinds of nutrient limitation, ppGpp stimulates *rpoS* transcription, probably at the levels of both transcription initiation and elongation (Gentry *et al.*, 1993; Lange *et al.*, 1995; Hirsch and Elliott, 2002); (ii) under conditions of phosphate starvation, it also induces the expression of IraP, which by sequestering RssB stabilizes σ^S (Boughour and Gottesman, 2007); and (iii) it affects sigma factor competition in favour of the formation of Eσ^S (and other RNAP alternative holoenzymes), which not only activates (Jishage *et al.*, 2002a) but should also stabilize σ^S (see above and Fig. 1.3C). Mechanistically all these influences add up independently to elevate σ^S levels and activity and thus correspond to an SUM function (Alon, 2007b). This complex (p)ppGpp response is also a graded response reacting differentially to different (p)ppGpp concentrations (Traxler *et al.*, 2011). During the growth cycle, this probably translates into a temporal succession of processes, with *rpoS* transcription initiation being stimulated already by the moderate (p)ppGpp levels required to reduce ribosomal gene expression, which redirects Eσ^{70} to other promoters (including *rpoS*p), and higher levels of (p)ppGpp driving the formation of alternative holoenzymes such Eσ^S.

Fis/Crp

In rapidly growing cells, σ^S levels are low. This is due to basal transcription of *rpoS* only, a mRNA secondary structure that allows only inefficient translation and fast degradation of the residual σ^S protein produced. In addition, the abundant regulatory protein Fis represses *rpoS* transcription (Hirsch and Elliott, 2005a); (A. Typas and R. Hengge, unpublished results). Moreover, FIS also represses transcription of *crp* (Gonzalez-Gil, 1998). CRP, when bound by its allosteric activator cAMP, directly activates *rpoS* transcription during the postexponential phase (Mika and Hengge, 2005). In principle, this pattern sets up a type II coherent feed forward loop (Fig. 1.3D), which ensures that during rapid growth, σ^S levels are low. As σ^S is nevertheless produced at a certain low rate, this FFL-mediated repression is not at odds with the need to rapidly increase σ^S levels in response to sudden acute stress, since these conditions stimulate *rpoS* mRNA translation and inhibit σ^S degradation (see above).

Nevertheless, this FFL system is obviously more complex. Fis not only influences *crp* transcription (Gonzalez-Gil, 1998), but CRP also acts as transcriptional activator or repressor for the *fis* gene, depending on the availability of Fis (Nasser, 2001). In addition, Fis itself is negatively autoregulated (Ball *et al.*, 1992; Ninnemann, 1992; Osuna *et al.*, 1995). It appears that this Fis-CRP FBL is negative and homeostatic under conditions, where Fis levels are low (i.e. during the post-exponential phase; here CRP has a positive effect on the expression of Fis), but mutually negative and therefore switch-promoting under conditions where Fis is present at high levels (this constellation may occur when rapidly growing cells are suddenly starved for a carbon source, which results in a sharp increase in cAMP levels and therefore CRP activity). It is hard to predict how this system might react *in vivo*, especially under conditions when different parameters change in parallel.

The ArcB/ArcA/ArcZ system

The histidine sensor kinase ArcB acts as a phosphodonor for two response regulators, ArcA and RssB (Georgellis *et al.*, 1997; Mika and Hengge, 2005), and thereby down-regulates σ^S levels via two 'branches' of a FFL, since ArcA represses *rpoS* transcription and RssB promotes σ^S proteolysis (Fig. 1.4). As a consequence, both *arcB* as well as *arcA* mutants have strongly elevated σ^S level in log phase (Mika and Hengge, 2005).

Together with the FNR system, the ArcB/ArcA system has long been known to play a key role in the regulation of the energy metabolism in response to oxygen availability (for a recent review, see Kiley and Donohue, 2011). In contrast to FNR, which is a direct oxygen sensor, the Arc system directly responds to the redox state of the quinone pool of the respiratory chain (Malpica *et al.*, 2004). Oxidized quinones inhibit the autophosphorylase activity of ArcB and thereby ArcA phosphorylation which allows expression of many genes involved in aerobic respiratory metabolism (Malpica *et al.*, 2006). However, the redox state of quinones is not only determined by the availability of oxygen as a terminal electron acceptor, but is rather a consequence of the *balance* between electron influx and efflux in the respiratory chain, with the former depending on energy supply (Fig. 1.4). Thus, high energy supply can be expected to maintain the activity of the ArcB/ArcA system even in the presence of moderate oxygen levels (as in typical laboratory batch cultures). Overall, the activity of the ArcB/ArcA system thus integrates the two parameters oxygen availability and energy supply, consistent with its role in the regulation of σ^S (Mika and Hengge, 2005).

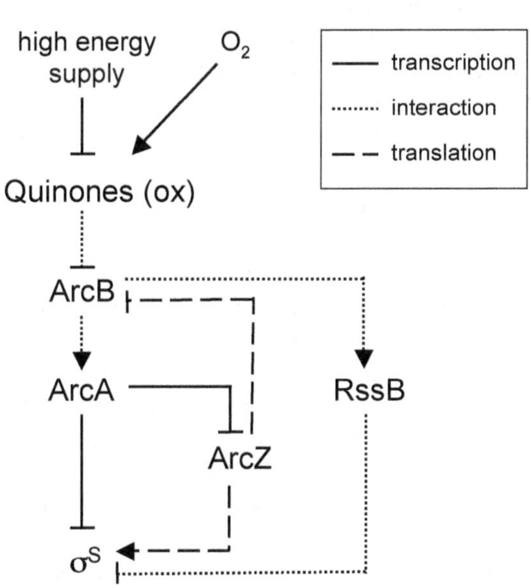

Figure 1.4 Role of the ArcB/ArcA two-component system and the small ArcZ RNA in σ^S control. In response to energy supply and oxygen availability, which determine the redox state of the quinones, the sensor kinase ArcB represses *rpoS* transcription (via ArcA) and promotes σ^S proteolysis (via RssB) in a FFL. ArcB/ArcA also down-regulate the small regulatory RNA ArcZ, which counteracts ArcB/ArcA effects by feedback-repressing ArcB and promoting *rpoS* mRNA translation (see text for details).

ArcZ, a recently described small regulatory RNA (Mandin and Gottesman, 2010), further adds to this control circuit. In contrast to ArcB/ArcA, ArcZ is a positively acting factor in σ^S control as it can stimulate *rpoS* translation. ArcZ expression is repressed by ArcB/ArcA, and in a feedback cycle, ArcZ represses the transcription of *arcB* (Mandin and Gottesman, 2010). This circuit probably operates a positive FBL, which contributes to switching between the 'low-σ^S' state (high ArcB/ArcA activity, low ArcZ) and the 'high-σ^S' state (low ArcB/ArcA activity, high ArcZ) (Fig. 1.4). The effect of ArcA on σ^S expression actually also sets up a small type 2 coherent FFL (see Figs. 1.3B and 1.4), with ArcA not only directly repressing *rpoS*, but also indirectly down-regulating σ^S by repressing *arcZ* expression (Fig. 1.3).

The σ^S-RssB negative feedback loop as the central circuit in the post-transcriptional control of σ^S

Titration or sequestration of RssB as a means for rapid stabilization of σ^S requires (i) that RssB expression is fine-tuned such that its level is rate-limiting for σ^S turnover (Pruteanu and Hengge-Aronis, 2002) and (ii) that under non-stress conditions a distinct σ^S:RssB ratio (which is about 20:1) (Becker *et al.*, 2000) and therefore a distinct overall rate of proteolysis has to be maintained. However, the rate of synthesis of σ^S in growing cells can fluctuate owing to stochastic variations (noise) of the rates of *rpoS* transcription or translation or in response to weak or transient signals. Robust maintenance of σ^S degradation and its regulatability in response to acute stress conditions therefore requires that RssB levels be coupled to σ^S levels. This is achieved by σ^S-dependent regulation of RssB expression (Ruiz *et al.*, 2001; Pruteanu and Hengge-Aronis, 2002), or in other words, by σ^S activating the expression of its own 'destruction factor' (see Fig. 1.2). This sets up a homeostatic negative FBL, which extends the dynamic range (of σ^S synthesis) in which the system maintains rapid σ^S proteolysis (despite increasing σ^S levels) and therefore its ability to react to acute stress conditions by rapidly stabilizing and therefore further accumulating σ^S (Backfisch *et al.*, 2005). In addition, this homeostatic FBL provides adaptive behaviour. This can be seen as *transient* rapid stabilization of σ^S in response to certain stress conditions such as shift to mild acidic or hyperosmotic conditions, which allows a resumption of growth. The basis of this adaptive behaviour is the difference in the kinetics of RssB titration or sequestration (rapid) and of σ^S-stimulated RssB expression (more slowly).

Obviously, the homeostatic σ^S-RssB FBL can operate only up to the point, where the *rssAp* promoter (which controls *rssAB* transcription) becomes saturated. Further increasing σ^S levels will then titrate the proteolytic system, rates of σ^S degradation will be successively reduced and σ^S accumulates, as has been experimentally demonstrated and mathematically modelled with a system in which σ^S expression was under *tac* promoter control (Backfisch *et al.*, 2005). In wild-type cells, this may be the case during entry into stationary phase in complex medium, where high levels of active σ^S drive up also RssB levels. Owing to other mechanisms that stabilize σ^S under these conditions (probably low ATP levels and reduced ArcB/RssB phosphorylation), this accumulating RssB is not immediately promoting σ^S degradation, but is available to do so instantaneously when new nutrients become available (Pruteanu and Hengge-Aronis, 2002).

In conclusion, the σ^S-RssB regulatory FBL is *the* central circuit in σ^S regulation. It is the prerequisite for rapid and efficient regulation of σ^S proteolysis by titration, competition and sequestration mechanisms that integrate numerous stress signals and lead to the sudden increase of σ^S in acutely stressed cells (Fig. 1.2) and it allows to rapidly remove the σ^S accumulated in starving cells upon outgrowth (Hengge, 2009b, 2011).

Mutual inhibition of sigma factors by competition for RNAP core enzyme

As described above, sigma factors have to compete for a limiting amount of free RNAP core enzyme in the cell (Gruber and Gross, 2003). The cellular levels of sigma factors as well as their availability and affinity for interaction with RNAP core enzyme are intricately regulated. From a network perspective, each sigma factor interferes with activity of the other ones, i.e. this is a pattern of complex mutual inhibition. The major competitor of σ^S is the vegetative σ^{70} (Fig. 1.5A), but the full picture is more complicated. Thus, in the post-exponential phase, when σ^S begins to accumulate, and during entry into stationary phase, when active σ^S-containing holoenyzme ($E\sigma^S$) is formed, additional competitors are σ^{28} (FliA) as well as σ^{54} and, in particular σ^E, which also shows stationary phase induction (Costanzo and Ades, 2006; Barembruch and Hengge, 2007; Dong *et al.*, 2011).

Mutual inhibition by sigma factor competition for RNAP core enzyme is a key regulatory motif that primarily controls σ^S activity. Yet, since σ^S within the holoenzyme complex is protected from RssB, any accessory components that affect the efficiency of σ^S binding to RNAP core in the 'competition game' also have an effect on the rate of σ^S degradation and thereby can affect the cellular σ^S level.

Mutual inhibition corresponds to a double-negative, i.e. effectively positive FBL which can produce switch-like behaviour. In the case of $E\sigma^S$ formation, this pattern is further modulated by at least two feedback circuits (all integrated in Fig. 1.5D). One of these is a negative FBL that involves regulation of σ^S expression: owing to sigma factor competition, increasing formation of $E\sigma^S$ down-regulates levels of the vegetative $E\sigma^{70}$, which is required to transcribe all the promoters known to contribute to *rpoS* expression

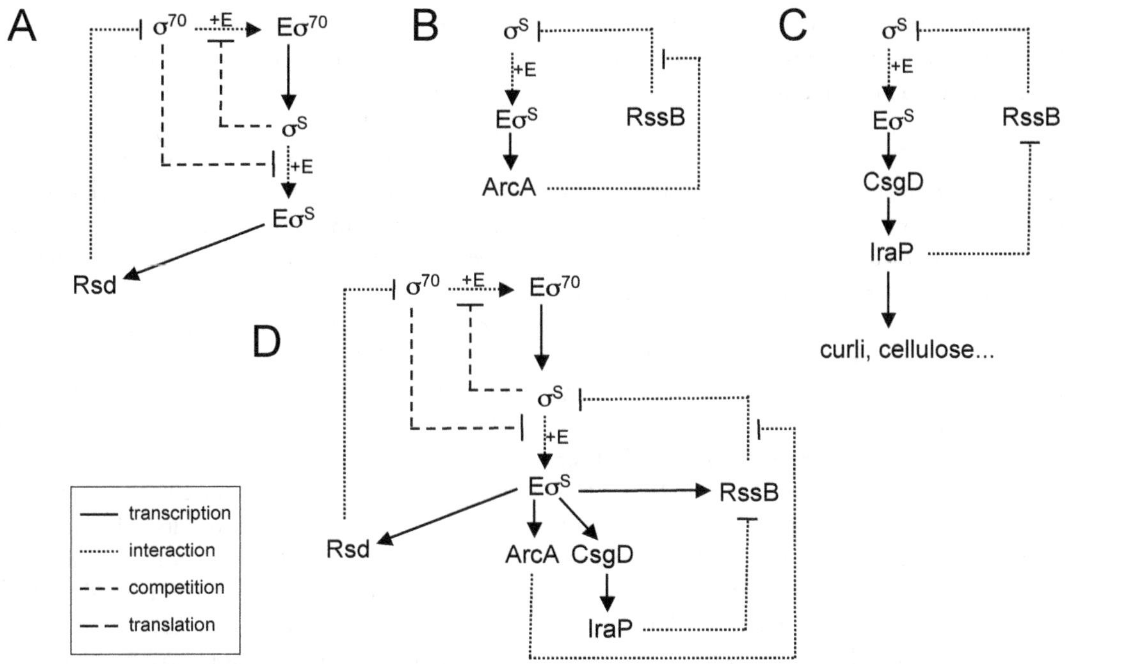

Figure 1.5 Feedback loops in the control of σ^S proteolysis and activity. Mutual inhibition of $E\sigma^S$ and $E\sigma^{70}$ holoenzyme formation is interlinked with $E\sigma^{70}$ driving the expression of *rpoS* and the positive FBL set up by σ^S-dependent expression of the anti-σ^{70} factor Rsd (A). The transcription factors ArcA (B) and CsgD (C), the expression of which is partly or completely σ^S-dependent, respectively, control additional factors that feed back into the control of σ^S proteolysis. Finally, all these FBLs are integrated into a single system (D).

(see above) (Takayanagi *et al.*, 1994; Lange and Hengge-Aronis, 1994b; Lange *et al.*, 1995). This negative σ^S-σ^{70} FBL should have homeostatic function, thus making sure that σ^S levels do not accumulate too much. For instance, extremely high σ^S levels could be a disadvantage when cells have to rapidly shift back from a 'maintenance' metabolism to a 'growth' metabolism (at the expense of stress resistances), and therefore to a gene expression scenario that is mainly dependent on the vegetative sigma factor. The second feedback circuit that modulates σ^{70}/σ^S competition for RNAP core enzyme, is the positive FBL set up by σ^S-mediated expression of the anti-σ^S factor Rsd (see next paragraph).

Positive feedback loops in the post-transcriptional control of σ^S

Several feedback loops support and fine-tune stabilization, accumulation and activity of σ^S under stress conditions. These FBLs involve components whose expression is directly or indirectly affected by σ^S. These components can influence *rpoS* mRNA turnover, sigma factor competition, RssB phosphorylation or the ratio of σ^S:RssB available for interaction. Overall, these systems are highly integrated and can contribute to stabilize the 'high-σ^S-activity' state of global gene expression (Fig. 1.5).

The σ^S-regulated RNaseIII inhibitor YmdB

RNaseIII contributes significantly to the turnover of *rpoS* mRNA in growing cells (Basineni *et al.*, 2009). However, during entry into stationary phase or upon shift to low temperature, σ^S activates the expression of YmdB, which directly binds and inhibits the activity of RNaseIII (Kim *et al.*, 2009). This may set up a positive FBL by which σ^S could stabilize its own transcript. Moreover, RNaseIII activity is also reduced in cells exposed to increased osmolarity (Sim *et al.*, 2010). It is conceivable that this is equally due to σ^S-mediated expression of YmdB and it is consistent with a

2-fold stabilization of *rpoS* mRNA that can be observed upon osmotic upshift (T. Kolmsee and R. Hengge, unpublished results).

The σ^S-Rsd FBL

The Rsd protein sequesters free σ^{70} and thereby prevents it from binding to RNA polymerase and even might actively displace it from RNA polymerase (Jishage and Ishihama, 1998; Ilag *et al.*, 2004; Westblade *et al.*, 2004). Thus, Rsd reduces the concentration of free cellular σ^{70} available for RNAP interaction, which indirectly promotes $E\sigma^S$ formation and therefore σ^S activity (Fig. 1.5A). Since expression of *rsd* is further activated by σ^S (Jishage and Ishihama, 1999), the short-term effect of this positive FBL is to boost σ^S-dependent gene expression. In the long run, however, the effect should be attenuated by the reduced $E\sigma^{70}$ levels which support a low rate of *rpoS* transcription only (see above).

The σ^S-ArcA-RssB FBL

Besides acetyl phosphate (Bouché *et al.*, 1998) also the sensor kinase ArcB acts as an *in vivo* phosphodonor for RssB. *In vitro*, it was demonstrated that the response regulators ArcA and RssB compete for ArcB phosphorylation, with phosphotransfer between ArcB and RssB being about ten times slower than between ArcA and ArcB (Mika and Hengge, 2005). As a consequence, ArcA highly efficiently competes with RssB for phosphorylation by ArcB and therefore increasing cellular amounts of ArcA inhibit RssB activity and thus σ^S proteolysis. The *in vivo* relevance of this competitive inhibition can be seen in stationary phase cells, where the expression of ArcA is enhanced by σ^S, which sets up a positive FBL that further stabilizes σ^S (Mika and Hengge, 2005) (Fig. 1.5B). It is interesting to note that this effect is a *positive* contribution of ArcA that is relevant during stationary phase, whereas in growing cells ArcA acts as a transcriptional repressor of σ^S expression (Mika and Hengge, 2005) (see above).

The σ^S-CsgD-IraP-RssB FBL

CsgD is a key regulator for biofilm-related components such as adhesive curli fimbriae and the matrix component cellulose. In enteric bacteria, these factors are essential for the transition from the planktonic to sessile lifestyle (Hengge, 2009c). In most *E. coli* strains, *csgD* expression is restricted to temperatures below 30°C (Olsen *et al.*, 1989; Arnqvist *et al.*, 1992) and has long been known to be under σ^S control (Hammar *et al.*, 1995). Only recently the details of this most complex subnetwork within the σ^S regulon known to date have been worked out (summarized in Hengge, 2009a). One of the CsgD-activated target genes seems to be *iraP* (Gualdi *et al.*, 2007), the product of which is a direct antagonist of RssB and thereby can stabilize σ^S (Bougdour *et al.*, 2006) (Fig. 1.5C). Indeed, *csgD* mutants seem to stabilize σ^S somewhat less during entry into stationary phase at 28°C (Gualdi *et al.*, 2007) (J. Vogt and R. Hengge, unpublished results). Interestingly, CsgD-dependent *iraP* expression was shown in phosphate-rich medium (Gualdi *et al.*, 2007), while ppGpp-mediated induction of IraP was observed upon starvation for phosphate and other nutrients (Bougdour *et al.*, 2006), suggesting that IraP can accumulate under rather different conditions and then lead to σ^S stabilization.

The σ^S activity inhibitor FliZ

FliZ negatively affects the expression of many σ^S-dependent genes (Pesavento *et al.*, 2008) (and see above). FliZ is expressed in late exponential phase and, as a class II motility gene, is under the control of FlhDC and $E\sigma^{70}$ (Girgis *et al.*, 2007). At a distinct point during this phase, when more and more $E\sigma^S$ accumulates at the expense of $E\sigma^{70}$, expression of flagella genes and also of *fliZ* comes to an end (Pesavento *et al.*, 2008). So, σ^S negatively affects the expression of its own functional antagonist, FliZ, probably by reducing the amount of $E\sigma^{70}$ in the cell via sigma factor competition.

Conclusion and perspectives

Twenty years after the first description of σ^S as a central regulator of stationary phase gene expression in *E. coli* (Lange and Hengge-Aronis, 1991b), it is now possible to integrate the extremely complex regulation of σ^S into a wider systems picture.

From a molecular biology perspective, research on σ^S has provided key paradigms, for instance for regulated proteolysis and proteolysis as a principle for cellular regulation in bacteria (Jenal and Hengge-Aronis, 2003; Hengge, 2009b), for regulation by small regulatory RNAs (Gottesman, 2004; Waters and Storz, 2009) or for sigma factor specificity and evolution of promoter recognition (Typas *et al.*, 2007b). From a physiological perspective, it has become clear that σ^S is *the* major player in deciding between very different bacterial lifestyles, which either optimize growth in response to remaining nutrients or ensure survival and multiple stress resistance when conditions no longer support growth (for a comprehensive recent review, see Hengge, 2011). Also here, regulation and function of σ^S have provided the conceptual blueprint for studying these fundamental processes in other bacterial species.

From a network perspective, studying the regulation of σ^S has provided key concepts such as the hierarchical organization of bacterial regulatory networks or the motif of densely overlapping regulons (Hengge-Aronis, 1996b; Alon, 2007a). As summarized in this review, we are now in a position to essentially describe the intricate σ^S control network in qualitative terms despite its rather overwhelming complexity. The quantitative prediction of the dynamics of this highly integrated network under different environmental conditions remains a challenge for the future.

Acknowledgements

Research in the Hengge laboratory has been generously funded by the Deutsche Forschungsgemeinschaft, an ERC Advanced Investigator Grant and the Fonds der Chemischen Industrie.

References

Alon, U. (2007a). An introduction to systems biology: design principles of biological circuits (Chapman and Hall/CRC Press, London).

Alon, U. (2007b). Network motifs: theory and experimental approaches. Nat. Rev. *8*, 450–461.

Arluison, V., Hohng, S., Roy, R., Pellegrini, O., Régnier, P., and Ha, T. (2007). Spectroscopic observation of RNA chaperone activities of Hfq in post-transcriptional regulation by a small non-coding RNA. Nucleic Acids Res. *35*, 999–1006.

Arnqvist, A., Olsen, A., Pfeifer, J., Russel, D., and Normark, S. (1992). The Crl protein activates cryptic genes for curli formation and fibronectin binding in *Escherichia coli*. Mol. Microbiol. *6*, 2443–2452.

Babu, M.M., Luscombe, N.M., Aravind, L., Gerstein, M., and Teichmann, S.A. (2004). Structure and evolution of transcriptional regulatory networks. Curr. Opin. Struct. Biol. *14*, 283–291.

Backfisch, T., Pruteanu, M., Hengge, R., and Gilles, E.-D. (2005). Mathematical modeling of RpoS regulation in *E. coli*. Found. Syst. Biol. Eng. *160*, 239–242.

Baker, T.A., and Sauer, R.T. (2006). ATP-dependent proteases of bacteria: recognition logic and operating principles. Trends Biochem. Sci. *31*, 647–653.

Balandina, A., Claret, L., Hengge-Aronis, R., and Rouvière-Yaniv, J. (2001). The *Escherichia coli* histone-like protein HU regulates *rpoS* translation. Mol. Microbiol. *39*, 1069–1079.

Ball, C.A., Osuna, R., Ferguson, K.C., and Johnson, R.C. (1992). Dramatic changes in Fis levels upon nutrient upshift in *Escherichia coli*. J. Bacteriol. *174*, 8043–8056.

Balsalobre, C. (2011). Concentration matters! ppGpp, from a whispering to a strident alarmone. Mol. Microbiol. *79*, 827–829.

Barembruch, C., and Hengge, R. (2007). Cellular levels and activity of the flagellar sigma factor FliA of *Escherichia coli* are controlled by FlgM-modulated proteolysis. Mol. Microbiol. *65*, 76–89.

Barth, M., Marschall, C., Muffler, A., Fischer, D., and Hengge-Aronis, R. (1995). A role for the histone-like protein H-NS in growth phase-dependent and osmotic regulation of σ^S and many σ^S-dependent genes in *Escherichia coli*. J. Bacteriol. *177*, 3455–3464.

Basineni, S.R., Madhugiri, R., Kolmsee, T., Hengge, R., and Klug, G. (2009). The influence of Hfq and ribonucleases on the stability of the small non-coding RNA OxyS and its target *rpoS* in *E. coli* is growth phase dependent. RNA Biol. *6*, 584–594.

Bearson, S.M.D., Benjamin, W.H. Jr., Swords, W.E., and Foster, J.W. (1996). Acid shock induction of RpoS is mediated by the mouse virulence gene *mvia* of *Salmonella typhimurium*. J. Bacteriol. *178*, 2572–2579.

Becker, G., Klauck, E., and Hengge-Aronis, R. (1999). Regulation of RpoS proteolysis in *Escherichia coli*: The response regulator RssB is a recognition factor that interacts with the turnover element in RpoS. Proc. Natl. Acad. Sci. U.S.A. *96*, 6439–6444.

Becker, G., Klauck, E., and Hengge-Aronis, R. (2000). The response regulator RssB, a recognition factor for σ^S proteolysis in *Escherichia coli*, can act like an anti-σ^S factor. Mol. Microbiol. *35*, 657–666.

Blot, N., Mavathur, R., Geertz, M., Travers, A., and Muskhelishvili, G. (2006). Homeostatic regulation of supercoiling sensitivity coordinates transcription of the bacterial genome. EMBO Rep. *7*, 710–715.

Bouché, S., Klauck, E., Fischer, D., Lucassen, M., Jung, K., and Hengge-Aronis, R. (1998). Regulation of RssB-dependent proteolysis in *Escherichia coli*: a role for acetyl phosphate in a response regulator-controlled process. Mol. Microbiol. *27*, 787–795.

Boughour, A., and Gottesman, S. (2007). ppGpp regulation of RpoS degradation via anti-adaptor protein IraP. Proc. Natl. Acad. Sci. U.S.A. *104*, 12896–12901.

Bougdour, A., Lelong, C., and Geiselmann, J. (2004). Crl, a low temperature-induced protein in *Escherichia coli* that binds directly to the stationary phase sigma subunit of RNA polymerase. J. Biol. Chem. *279*, 19540–19550.

Bougdour, A., Wickner, S., and Gottesman, S. (2006). Modulating RssB activity: IraP, a novel regulator of σ^S stability in *Escherichia coli*. Genes Dev. *20*, 884–897.

Bougdour, A., Cunning, C., Baptiste, P.J., Elliott, T., and Gottesman, S. (2008). Multiple pathways for regulation of σ^S (RpoS) stability in *Escherichia coli* via the action of multiple anti-adaptors. Mol. Microbiol. *68*, 298–313.

Bouveret, E., and A. Battesti (2011). The stringent response. In Bacterial Stress Responses, 2nd edn, Storz, G., and Hengge, R., eds (ASM Press ,Washington DC), pp. 231–250.

Bren, A., and Alon, U. (2011). Architecture and dynamics of transcriptional networks. In Bacterial Stress Responses, 2nd edn, G. Storz, and R. Hengge, eds (Washington DC, ASM Press), pp. 17–30.

Brennan, R.G., and Link, T.M. (2007). Hfq structure, function and ligand binding. Curr. Opin. Microbiol. *10*, 125–133.

Brescia, C., Kaw, M., and Sledjeski, D. (2004). The DNA binding protein H-NS binds to and alters the stability of RNA *in vitro* and *in vivo*. J. Mol. Biol. *339*, 505–514.

Brown, L., and Elliott, T. (1996). Efficient translation of the RpoS sigma factor in *Salmonella typhimurium* requires host factor I, an RNA-binding protein encoded by the *hfq* gene. J. Bacteriol. *178*, 3763–3770.

Brown, L., and Elliott, T. (1997). Mutations that increase expression of the *rpoS* gene and decrease its dependence on *hfq* function in *Salmonella typhimurium*. J. Bacteriol. *179*, 656–662.

Browning, D.F., Grainger, D.C., and Busby, S. (2010). Effects of nucleoid-associated proteins on bacterial chromosome structure and gene expression. Curr. Opin. Microbiol. *13*, 773–780.

Brückner, R., and Titgemeyer, F. (2002). Carbon catabolite repression in bacteria: choice of the carbon source and autoregulatory limitation of sugar utilization. FEMS Microbiol. Lett. *209*, 141–148.

Cabeza, M.L., Aguirre, A., Soncini, F.C., and Vescovi, E.G. (2007). Induction of RpoS degradation by the two-component system regulator RstA in *Salmonella*. J. Bacteriol. *189*, 7335–7342.

Chalancon, G., and Babu, M.M. (2011). Structure and evolution of transcriptional regulatory networks. In Bacterial Stress Responses 2nd edn, Storz, G., and Hengge, R., eds (ASM Press, Washington DC).

Colland, F., Fujita, N., Ishihama, A., and Kolb, A. (2002). The interaction between σ^S, the stationary phase σ factor, and the core enzyme of *Escherichia coli* RNA polymerase. Genes Cells 7, 233–247.

Costanzo, A., and Ades, S. (2006). Growth phase-dependent regulation of the extracytoplasmic stress factor, sigmaE by guanosine 3′,5′-bispyrophosphate (ppGpp). J. Bacteriol. *188*, 4627–4634.

Cunning, C., and Elliott, T. (1999). RpoS synthesis is growth rate regulated in *Salmonella typhimurium* but its turnover is not dependent on acetyl phosphate synthesis or PTS function. J. Bacteriol. *181*, 4853–4862.

Dong, T., and Schellhorn, H.E. (2009). Control of RpoS in global gene expression of *Escherichia coli* in minimal medium. Mol. Genet. Genomics *281*, 19–31.

Dong, T., Yu, R., and Schellhorn, H. (2011). Antagonistic regulation of motility and transcriptome expression of RpoN and RpoS in *Escherichia coli*. Mol. Microbiol. *79*, 375–386.

Dorman, C.J. (2004). H-NS: a universal regulator for a dynamic genome. Nat. Rev. Microbiol. *2*, 391–400.

Dorman, C., and Deighan, P. (2003). Regulation of gene expression by histone-like proteins in bacteria. Curr. Opin. Gen. Dev. *13*, 179–184.

Eichenberger, P., Fujita, M., Jensen, S.T., Conlon, E.M., Rudner, D.Z., Wang, S.T., Ferguson, C., Haga, K., Sato, T., Liu, J.S., *et al.* (2004). The program of gene transcription for a single differentiating cell type during sporulation in *Bacillus subtilis*. PLoS Biol. 2, e328.

England, P., Westblade, L.F., Karimova, G., Robbe-Saule, V., Norel, F., and Kolb, A. (2008). Binding of the unorthodox transcription activator, Crl, to the components of the transcription machinery. J. Biol. Biochem. *283*, 33455–33464.

Ferenci, T. (2001). Hungry bacteria – definition and properties of a nutritional state. Environ. Microbiol. *3*, 605–611.

Finkel, S.E., and Johnson, R.C. (1992). The Fis protein: it′s not just for DNA inversion anymore. Mol. Microbiol. *6*, 3257–3265.

Fischer, D., Teich, A., Neubauer, P., and Hengge-Aronis, R. (1998). The general stress sigma factor σ^S of *Escherichia coli* is induced during diauxic shift from glucose to lactose. J. Bacteriol. *180*, 6203–6206.

Fredriksson, A., Ballesteros, M., Peterson, C.N., Persson, Ö., Silhavy, T.J., and Nyström, T. (2007). Decline of ribosomal fidelity contributes to the accumulation and stabilization of the master stress regulator σ^S upon carbon starvation. Genes Dev. *21*, 862–874.

Gentry, D.R., Hernandez, V.J., Nguyen, L.H., Jensen, D.B., and Cashel, M. (1993). Synthesis of the stationary-phase sigma factor σ^S is positively regulated by ppGpp. J. Bacteriol. *175*, 7982–7989.

Georgellis, D., Lynch, A.S., and Lin, E.C.C. (1997). *In vitro* phosphorylation study of the Arc two-component signal transduction system of *Escherichia coli*. J. Bacteriol. *179*, 5429–5435.

Gildehaus, N., Neusser, T., Wurm, R., and Wagner, R. (2007). Studies on the function of the riboregulator 6S RNA from *E. coli*: RNA polymerase binding, inhibition of *in vitro* transcription and synthesis of RNA-directed *de novo* transcripts. Nucleic Acids Res. *35*, 1885–1896.

Girgis, H.S., Liu, Y., Ryu, W.S., and Tavazoie, S. (2007). A comprehensive genetic characterization of bacterial motility. PLoS Genet. *3*, 1644–1660.

Gonzalez-Gil, G., Kahmann, R. Muskhelishvili G. (1998). Regulation of *crp* transcription by oscillation between distinct nucleoprotein complexes. EMBO J. *17*, 2877–2885.

Gottesman, S. (2004). The small RNA regulators of *Escherichia coli*: roles and mechanisms. Annu. Rev. Microbiol. *58*, 303–328.

Gruber, T., and Gross, C.A. (2003). Multiple sigma subunits and the partitioning of bacterial transcription space. Annu. Rev. Microbiol. *57*, 441–466.

Gualdi, L., Tagliabue, L., and Landini, P. (2007). A biofilm formation-gene expression relay system in *Escherichia coli*: modulation of σ^S-dependent gene expression by the CsgD regulatory protein via σ^S protein stabilization. J. Bacteriol. *189*, 8034–43.

Hammar, M., Arnquist, A., Bian, Z., Olsén, A., and Normark, S. (1995). Expression of two *csg* operons is required for production of fibronectin- and Congo red-binding curli polymers in *Escherichia coli* K-12. Mol. Microbiol. *18*, 661–670.

Harman, J.G. (2001). Allosteric regulation of the cAMP receptor protein. Biochim. Biophys. Acta *1547*, 1–17.

Helmann, J. (2011). Regulation by Alternative σ Factors. In Bacterial Stress Responses, Storz, G., and Hengge, R., eds (ASM Press, Washington DC), pp. 31–43.

Helmann, J.D. (1991). Alternative sigma factors and the regulation of flagellar gene expression. Mol. Microbiol. *5*, 2875–2882.

Helmann, J.D. (1999). Anti-sigma factors. Curr. Opin. Microbiol. *2*, 135–141.

Hengge, R. (2008). The two-component network and the general stress sigma factor RpoS (σ^S) in *Escherichia coli*. In Bacterial Signal Transduction: Network and Drug Targets, Utsumi, R., ed. (Landes Bioscience, Austin, TX), pp. 40–53.

Hengge, R. (2009a). Principles of cyclic-di-GMP signalling in bacteria. Nat Rev Microbiol. *7*, 263–73.

Hengge, R. (2009b). Proteolysis of σ^S (RpoS) and the general stress response in *Escherichia coli*. Res. Microbiol. *160*, 667–676.

Hengge, R. (2009c). Role of cyclic-di-GMP in the regulatory networks of *Escherichia coli*. In The Second Messenger Cyclic-Di-GMP, Wolfe, A.J., and Visick, K.L., eds (ASM Press, Washington DC), pp. 230–252.

Hengge, R. (2011). The general stress response in Gram-negative bacteria. In Bacterial Stress Responses Storz, G. and R. Hengge, eds (Washington DC, ASM Press), pp. 251–289.

Hengge-Aronis, R. (1996a). Back to log phase: σ^S as a global regulator in the osmotic control of gene expression in *Escherichia coli*. Mol. Microbiol. *21*, 887–893.

Hengge-Aronis, R. (1996b). Regulation of gene expression during entry into stationary phase. In *Escherichia coli* and *Salmonella typhimurium*: Cellular and Molecular Biology, Neidhardt, F.C., Curtiss III, R., Ingraham, J.L., Lin, E.C.C., Low, K.B., Magasanik, B., Reznikoff, W.S., Riley, M., Schaechter, M., and Umbarger, H.E., eds (ASM Press, Washington DC), pp. 1497–1512.

Hengge-Aronis, R. (2002a). Recent insights into the general stress response regulatory network in *Escherichia coli*. J. Mol. Microbiol. Biotechnol. *4*, 341–346.

Hengge-Aronis, R. (2002b). Signal transduction and regulatory mechanisms involved in control of the σ^S subunit of RNA polymerase in *Escherichia coli*. Microbiol. Mol. Biol. Rev. *66*, 373–395.

Herren, C., Mitra, A., Palaniyandi, S., Coleman, A., Elankumaran, S., and Mukhopadhyay, S. (2006). The BarA-UvrY two-component system regulates virulence in avian pathogenic *Escherichia coli* O78:K80:H9. Infect. Immun. *74*, 4900–4909.

Heuveling, J., Possling, A., and Hengge, R. (2008). A role for Lon protease in the control of the acid resistance genes of *Escherichia coli*. Mol. Microbiol. *69*, 534–547.

Hirsch, M., and Elliott, T. (2002). Role of ppGpp in *rpoS* stationary-phase regulation in *Escherichia coli*. J. Bacteriol. *184*, 5077–5087.

Hirsch, M., and Elliott, T. (2005a). Fis regulates transcriptional induction of RpoS in *Salmonella enterica*. J. Bacteriol. *187*, 1568–1580.

Hirsch, M., and Elliott, T. (2005b). Stationary-phase regulation of RpoS translation in *Escherichia coli*. J. Bacteriol. *187*, 7204–7213.

Hommais, F., Krin, E., Laurent-Winter, C., Soutourina, O., Malpertuy, A., Le Caer, J.-P., Danchin, A., and Bertin, P. (2001). Large-scale monitoring of pleiotropic regulation of gene expression by the prokaryotic nucleoid-associated protein H-NS. Mol. Microbiol. *40*, 20–36.

Ihssen, J., and Egli, T. (2004). Specific growth rate and not cell density controls the general stress response in *Escherichia coli*. Microbiology *150*, 1637–1648.

Ilag, L.L., Westblade, L.F., Deshayes, C., Kolb, A., Busby, S.J., and Robinson, C.V. (2004). Mass spectrometry of *Escherichia coli* RNA polymerase: interactions of the core enzyme with sigma70 and Rsd protein. Structure *12*, 269–275.

Jenal, U., and Hengge-Aronis, R. (2003). Regulation by proteolysis in bacterial cells. Curr. Opin. Microbiol. *6*, 163–172.

Jenkins, D.E., Chaisson, S.A., and Matin, A. (1990). Starvation-induced cross-protection against osmotic challenge in *Escherichia coli*. J. Bacteriol. *172*, 2779–2781.

Jenkins, D.E., Schultz, J.E., and Matin, A. (1988). Starvation–induced cross-protection against heat or H_2O_2 challenge in *Escherichia coli*. J. Bacteriol. *170*, 3910–3914.

Jishage, M., and Ishihama, A. (1995). Regulation of RNA polymerase sigma subunit synthesis in *Escherichia coli*: intracellular levels of σ^{70} and σ^{38}. J. Bacteriol. *177*, 6832–6835.

Jishage, M., and Ishihama, A. (1998). A stationary phase protein in *Escherichia coli* with binding activity to the major σ subunit of RNA polymerase. Proc. Natl. Acad. Sci. U.S.A. *95*, 4953–4958.

Jishage, M., and Ishihama, A. (1999). Transcriptional organization and *in vivo* role of the *Escherichia coli rsd* gene, encoding the regulator of RNA polymerase sigma D. J. Bacteriol. *181*, 3768–3776.

Jishage, M., Kvint, K., Shingler, V., and Nystrom, T. (2002a). Regulation of sigma factor competition by the alarmone ppGpp. Genes Dev. *16*, 1260–1270.

Jishage, M., Kvint, K., Shingler, V., and Nyström, T. (2002b). Regulation of sigma factor competition by the alarmone ppGpp. Genes Dev. *16*, 1260–1270.

Kiley, P.J., and Donohue, T.J. (2011). Global responses of bacteria to oxygen deprivation. In Bacterial Stress Responses Storz, G. and Hengge, R. eds (ASM Press, Washington DC), pp. 175–204.

Kim, K.S., Manasherob, R., and Cohen, S.N. (2009). YmdB: a stress-responsive ribonuclease-binding regulator of *E. coli* RNase III activity. Genes Dev. *22*, 3497–3508.

Klauck, E., Böhringer, J., and Hengge-Aronis, R. (1997). The LysR-like regulator LeuO in *Escherichia coli* is involved in the translational regulation of *rpoS* by affecting the expression of the small regulatory DsrA-RNA. Mol. Microbiol. *25*, 559–569.

Klauck, E., Lingnau, M., and Hengge-Aronis, R. (2001). Role of the response regulator RssB in σ^S recognition and initiation of σ^S proteolysis in *Escherichia coli*. Mol. Microbiol. *40*, 1381–1390.

Klauck, E., Typas, A., and Hengge, R. (2007). The sigmaS subunit of RNA polymerase as a signal integrator and network master regulator in the general stress response in *Escherichia coli*. Sci. Prog. *90*, 103–127.

Klein, A.H., Shulla, A., Reimann, S.A., Keating, D.H., and Wolfe, A.J. (2007). The intracellular concentration of acetyl phosphate in *Escherichia coli* is sufficient for direct phosphorylation of two component response regulators. J. Bacteriol. *189*, 5574–5581.

Koch, A.L. (1971). The adaptive responses of *Escherichia coli* to a feast and famine existence. Adv. Microb. Physiol. *6*, 147–217.

Koleva, R.I., Austin, C.A., Kowaleski, J.M., Neems, D.S., Wang, L., Vary, C.P.H., and Schlax, P.J. (2006). Interactions of ribosomal protein S1 with DsrA and *rpoS* mRNA. Biochem. Biophys. Res. Commun. *348*, 662–668.

Kvint, K., Farewell, A., and Nystrom, T. (2000). RpoS-dependent promoters require guanosine tetraphosphate for induction even in the presence of high levels of σ^S. J. Biol. Chem. *275*, 14795–14798.

Lange, R., and Hengge-Aronis, R. (1991a). Growth phase-regulated expression of *bolA* and morphology of stationary phase *Escherichia coli* cells is controlled by the novel sigma factor σ^S (*rpoS*). J. Bacteriol. *173*, 4474–4481.

Lange, R., and Hengge-Aronis, R. (1991b). Identification of a central regulator of stationary-phase gene expression in *Escherichia coli*. Mol. Microbiol. *5*, 49–59.

Lange, R., and Hengge-Aronis, R. (1994a). The cellular concentration of the σ^S subunit of RNA-polymerase in *Escherichia coli* is controlled at the levels of transcription, translation and protein stability. Genes Dev. *8*, 1600–1612.

Lange, R., and Hengge-Aronis, R. (1994b). The *nlpD* gene is located in an operon with *rpoS* on the *Escherichia coli* chromosome and encodes a novel lipoprotein with a potential function in cell wall formation. Mol. Microbiol. *13*, 733–743.

Lange, R., Fischer, D., and Hengge-Aronis, R. (1995). Identification of transcriptional start sites and the role of ppGpp in the expression of *rpoS*, the structural gene for the σ^S subunit of RNA-polymerase in *Escherichia coli*. J. Bacteriol. *177*, 4676–4680.

Laurie, A.D., Bernardo, L.M., Sze, C.C., Skarfstad, E., Szalewska-Palasz, A., Nyström, T., and Shingler, V. (2002). The role of the alarmone (*p*)ppGpp in sigma N competition for core RNA polymerase. J. Biol. Chem. *278*, 1494–1503

Lease, R.A., and Belfort, M. (2000). Riboregulation by DsrA RNA: trans-actions for global economy. Mol. Microbiol. *38*, 667–672.

Lease, R.A., Cusick, M.E., and Belfort, M. (1998). Riboregulation in *Escherichia coli*: DsrA RNA acts by RNA:RNA interaction at multiple loci. Proc. Natl. Acad. Sci. U.S.A. *95*, 12456–12461.

Liu, M., Durfee, T., Cabrera, J.E., Zhao, K., Jin, D.J., and Blattner, F.R. (2005). Global transcriptional programs reveal a carbon source foraging strategy by *Escherichia coli*. J. Biol. Chem. *280*, 15921–15927.

McCann, M.P., Kidwell, J.P., and Matin, A. (1991). The putative σ factor KatF has a central role in development of starvation-mediated general resistance in *Escherichia coli*. J. Bacteriol. *173*, 4188–4194.

McCleary, W.R., and Stock, J.B. (1994). Acetyl phosphate and the activation of two-component response regulators. J. Biol. Chem. *269*, 31567–31572.

McCullen, C.A., Benhammou, J.N., Majdalani, N., and Gottesman, S. (2010). Mechanism of positive regulation by DsrA and RprA small noncoding RNAs: pairing increases translation and protects rpoS mRNA from degradation. J. Bacteriol. *192*, 5559–5571.

Madhugiri, R., Basineni, S.R., and Klug, G. (2010). Turn-over of the small non-coding RNA RprA in *E. coli* is influenced by osmolarity. Mol. Genet. Genomics *284*, 307–318.

Maeda, H., Fujita, N., and Ishihama, A. (2000). Competition among seven *Escherichia coli* sigma subunits: relative binding affinities to the core RNA polymerase. Nucleic Acids Res. *28*, 3497–3503.

Magnusson, L.U., Farewell, A., and Nystrom, T. (2005). ppGpp: a global regulator in *Escherichia coli*. Trends Microbiol. *13*, 236–242.

Majdalani, N., Cunning, C., Sledjeski, D., Elliott, T., and Gottesman, S. (1998). DsrA RNA regulates translation of RpoS message by an anti-antisense mechanism, independent of its action as an antisilencer of transcription. Proc. Natl. Acad. Sci. U.S.A. *95*, 12462–12467.

Majdalani, N., Chen, S., Murrow, J., St. John, K., and Gottesman, S. (2001). Regulation of RpoS by a novel small RNA: the characterization of RprA. Mol. Microbiol. *39*, 1382–1394.

Majdalani, N., Hernandez, D., and Gottesman, S. (2002). Regulation and mode of action of the second small RNA activator of RpoS translation, RprA. Mol. Microbiol. *46*, 813–826.

Malpica, R., Franco, B., Rodriguez, C., Kwon, O., and Georgellis, D. (2004). Identification of quinone-sensitive redox switch in the ArcB sensor kinase. Proc. Natl. Acad. Sci. U.S.A. *101*, 13318–13323.

Malpica, R., Sandoval, G.R., Rodriguez, C., Franco, B., and Georgellis, D. (2006). Signaling by the arc two-component system provides a link between the redox state of the quinone pool and gene expression. Antioxid. Redox Signal. *8*, 781–795.

Mandel, M.J., and Silhavy, T.J. (2005). Starvation for different nutrients in *Escherichia coli* results in differential modulation of RpoS levels and stability. J. Bacteriol. *187*, 434–442.

Mandin, P., and Gottesman, S. (2010). Integrating anaerobic/aerobic sensing and the general stress response through the ArcZ small RNA. EMBO J. *29*, 3094–3104.

Mangan, S., and Alon, U. (2003). Structure and function of the feed-forward loop network motif. Proc. Natl. Acad. Sci. U.S.A. *100*, 11980–11985.

Marr, C., Geertz, M., Hütt, M.T., and Muskhelishvili, G. (2008). Dissecting the logical types of network control in gene expression profiles. BMC Syst. Biol. *2*, 18.

Martin, A., Baker, T.A., and Sauer, R.T. (2008). Protein unfolding by a AAA+ protease is dependent on ATP-hydrolysis rates and substrate energy landscapes. Nat. Struct. Mol. Biol. *15*, 139–145.

Merrikh, H., Ferrazzoli, A.E., and Lovett, S.T. (2009). Growth phase and (*p*)ppGpp control of IraD, a regulator of RpoS stability, in *Escherichia coli*. J. Bacteriol. *191*, 7436–7446.

Mika, F., and Hengge, R. (2005). A two-component phosphotransfer network involving ArcB, ArcA and RssB coordinates synthesis and proteolysis of σ^S in *E. coli*. Genes Dev. *19*, 2770–2781.

Milo, R., Shen-Orr, S., Itzkovitz, S., Kashtan, N., Chklovskii, D., and Alon, U. (2002). Network motifs: simple building blocks of complex networks. Science *298*, 824–827.

Mitchell, J.E., Oshima, T., Piper, S.E., Webster, C.L., Westblade, L.F., Karimova, G., Ladant, D., Kolb, A., Hobman, J.L., Busby, S.J., *et al.* (2007). The *Escherichia coli* regulator of sigma 70 protein, Rsd, can up-regulate some stress-dependent promoters by sequestering sigma 70. J. Bacteriol. *189*, 3489–3495.

Mitrophanov, A.Y., and Groisman, E.A. (2008). Signal integration in bacterial two-component regulatory systems. Genes Dev. *22*, 2601–2611.

Moreno, M., Audia, J.P., Bearson, S.M.D., Webb, C., and Foster, J.W. (2000). Regulation of σ^S degradation in *Salmonella enterica* var *typhimurium*: *in vivo* interactions between σ^S, the response regulator MviA (RssB) and ClpX. J. Mol. Microbiol. Biotechnol. *2*, 245–254.

Muffler, A., Fischer, D., Altuvia, S., Storz, G., and Hengge-Aronis, R. (1996a). The response regulator RssB controls stability of the σ^S subunit of RNA polymerase in *Escherichia coli*. EMBO J. *15*, 1333–1339.

Muffler, A., Fischer, D., and Hengge-Aronis, R. (1996b). The RNA-binding protein HF-I, known as a host factor for phage Qβ RNA replication, is essential for the translational regulation of *rpoS* in *Escherichia coli*. Genes Dev. *10*, 1143–1151.

Muffler, A., Traulsen, D.D., Lange, R., and Hengge-Aronis, R. (1996c). Posttranscriptional osmotic regulation of the σ^S subunit of RNA polymerase in *Escherichia coli*. J. Bacteriol. *178*, 1607–1613.

Muffler, A., Barth, M., Marschall, C., and Hengge-Aronis, R. (1997a). Heat shock regulation of σ^S turnover: a role for DnaK and relationship between stress responses mediated by σ^S and σ^{32} in *Escherichia coli*. J. Bacteriol. *179*, 445–452.

Muffler, A., Traulsen, D.D., Fischer, D., Lange, R., and Hengge-Aronis, R. (1997b). The RNA-binding protein HF-I plays a global regulatory role which is largely, but not exclusively, due to its role in expression of the σ^S subunit of RNA polymerase in *Escherichia coli*. J. Bacteriol. *179*, 297–300.

Mukhopadhyay, S., Audia, J.P., Roy, R.N., and Schellhorn, H.E. (2000). Transcriptional induction of the conserved alternative sigma factor RpoS in *Escherichia coli* is dependent on BarA, a probable two-component regulator. Mol. Microbiol. *37*, 371–381.

Narberhaus, F. (2006). RNA thermometers. FEMS Microbiol. Rev. *30*, 3–16.

Nasser, W., Schneider, R., Travers, A., and Muskhelishvili, G. (2001). CRP modulates fis transcription by alternate formation of activating and repressing nucleoprotein complexes. J. Biol. Chem. *276*, 17878–17886.

Ninnemann, O., Koch, C., and Kahmann, R. (1992). The *E. coli* fis promoter is subject to stringent control and autoregulation. EMBO J. *10*, 1593–1603.

Notley, L., and Ferenci, T. (1996). Induction of RpoS-dependent functions in glucose-limited continuous culture: what level of nutrient limitation induces the stationary phase of *Escherichia coli*. J. Bacteriol. *178*, 1465–1468.

Notley-McRobb, L., Death, A., and Ferenci, T. (1997). The relationship between external glucose concentration and cAMP levels inside *Escherichia coli*: implications for models of phosphotransferase-mediated regulation of adenylate cyclase. Micobiology *143*, 1909–1918.

Olsen, A., Jonsson, A., and Normark, S. (1989). Fibronectin binding mediated by a novel class of surface organelles on *Escherichia coli*. Nature *338*, 652–655.

Osuna, R., Lienau, D., Hughes, K.T., and Johnson, R.C. (1995). Sequence, regulation, and functions of *fis* in *Salmonella typhimurium*. J. Bacteriol. *177*, 2021–2032.

Patikoglou, G.A., Westblade, L.F., Campbell, E.A., Lamour, V., Lane, W.J., and Darst, S.A. (2007). Crystal Structure of the *Escherichia coli* Regulator of sigma(70), Rsd, in Complex with sigma(70) Domain 4. J. Mol. Biol. *372*, 649–659.

Pernestig, A.-K., Georgellis, D., Romeo, T., Suzuki, K., Tomenius, H., Normark, S., and Melefors, Ö. (2003). The *Escherichia coli* BarA-UvrY two-component system is needed for efficient switching between glycolytic and gluconeogenic carbon sources. J. Bacteriol. *185*, 843–853.

Pesavento, C., Becker, G., Sommerfeldt, N., Possling, A., Tschowri, N., Mehlis, A., and Hengge, R. (2008). Inverse regulatory coordination of motility and curli-mediated adhesion in *Escherichia coli*. Genes Dev. *22*, 2434–2446.

Pesavento, C., and Hengge, R. (2009). Bacterial nucleotide-based second messengers. Curr. Opin. Microbiol. *12*, 170–176.

Pesavento, C., and Hengge, R. (2012). The global repressor FliZ antagonizes gene expression by σS-containing RNA polymerase due to overlapping DNA binding specificity. Nucleic Acids Res. doi: 10.1093/nar/gks055.

Peterson, C.N., Ruiz, N., and Silhavy, T.J. (2004). RpoS proteolysis is regulated by a mechanism that does not require the SprE (RssB) response regulator phosphorylation site. J. Bacteriol. *186*, 7403–7410.

Peterson, C.N., Mandel, M.J., and Silhavy, T.J. (2005). *Escherichia coli* starvation diets: essential nutrients weigh in distinctly. J. Bacteriol. *187*, 7549–7553.

Peterson, C.N., Carabetta, V.J., Chowdhury, T., and Silhavy, T.J. (2006). LrhA regulates *rpoS* translation in response to the Rcs phosphorelay system in *Escherichia coli*. J. Bacteriol. *188*, 3175–3181.

Potrykus, K., and Cashel, M. (2008). (*p*)ppGpp: still Magical? Annu. Rev. Microbiol. *62*, 35–51.

Pratt, L.A., and Silhavy, T.J. (1996). The response regulator, SprE, controls the stability of RpoS. Proc. Natl. Acad. Sci. U.S.A. *93*, 2488–2492.

Pruteanu, M., and Hengge-Aronis, R. (2002). The cellular level of the recognition factor RssB is rate-limiting for σ^S proteolysis: Implications for RssB regulation and signal transduction in σ^S turnover in *Escherichia coli*. Mol. Microbiol. *45*, 1701–1714.

Ranquet, C., and Gottesman, S. (2007). Translational regulation of the *Escherichia coli* stress factor RpoS: a role for SsrA and Lon. J. Bacteriol. *189*, 4872–4879.

Repoila, F., and Gottesman, S. (2001). Signal transduction cascade for regulation of *rpoS*: temperature regulation of *dsrA*. J. Bacteriol. *183*, 4012–4023.

Repoila, F., Majdalani, N., and Gottesman, S. (2003). Small non-coding RNAs, co-ordinators of adaptation processes in *Escherichia coli*: the RpoS paradigm. Mol. Microbiol. *48*, 855–861.

Robbe-Saule, V., Lopes, M.D., Kolb, A., and Norel, F. (2007). Physiological effects of Crl in *Salmonella* are modulated by σ^S level and promoter specificity. J. Bacteriol. *189*, 2976–2987.

Rockabrand, D., Arthur, T., Korinek, G., Livers, K., and Blum, P. (1995). An essential role for the *Escherichia coli* DnaK protein in starvation-induced thermotolerance, H_2O_2 resistance, and reductive division. J. Bacteriol. *177*, 3695–3703.

Rockabrand, D., Livers, K., Austin, T., Kaiser, R., Jensen, D., Burgess, R., and Blum, P. (1998). Roles of DnaK and RpoS in starvation-induced thermotolerance of *Escherichia coli*. J. Bacteriol. *180*, 846–854.

Ruiz, N., Peterson, C.N., and Silhavy, T.J. (2001). RpoS-dependent transcriptional control of *sprE*: regulatory feedback loop. J. Bacteriol. *183*, 5974–5981.

Sauter, C. (2003). Sm-like protein in eubacteria: the crystal structure of the Hfq protein from *Escherichia coli*. Nucl Acids Res. *31*, 4091–4098.

Schumacher, M.A. (2002). Structures of the pleiotropic translational regulator Hfq and an Hfq–RNA complex: a bacterial Sm-like protein. EMBO J. *21*, 3546–3556.

Schweder, T., Lee, K.-H., Lomovskaya, O., and Matin, A. (1996). Regulation of *Escherichia coli* starvation sigma factor (σ^S) by ClpXP protease. J. Bacteriol. *178*, 470–476.

Sezonov, G., Joseleau-Petit, D., and D'Ari, R. (2007). *Escherichia coli* physiology in Luria-Bertani broth. J. Bacteriol. *189*, 8746–8749.

Sharma, U.K., and Chatterji, D. (2010). Transcriptional switching in *Escherichia coli* during stress and starvation by modulation of σ^{70} activity. FEMS Microbiol Rev. *34*, 646–657.

Shiba, T., Tsutsumi, K., Yano, H., Ihara, Y., Kameda, A., Tanaka, K., Takahashi, H., Munekata, M., Rao, N.N., and Kornberg, A. (1997). Inorganic polyphosphate and the induction of *rpoS* expression. Proc. Natl. Acad. Sci. U.S.A. *94*, 11210–11215.

Sim, S.H., Yeom, J.H., Shin, C., Song, W.S., Shin, E., Kim, H.M., Cha, C.J., Han, S.H., Ha, N.C., Kim, S.W., *et al.* (2010). *Escherichia coli* ribonuclease III activity is down-regulated by osmotic stress: consequences for the degradation of bdm mRNA in biofilm formation. Mol. Microbiol. *75*, 413–425.

Sledjeski, D.D., Gupta, A., and Gottesman, S. (1996). The small RNA, DsrA, is essential for the low temperature expression of RpoS during exponential growth in *E. coli*. EMBO J. *15*, 3993–4000.

Soutourina, O., Kolb, A., Krin, E., Laurent-Winter, C., Rimsky, S., Danchin, A., and Bertin, P. (1999). Multiple control of flagellum biosynthesis in *Escherichia coli*: role of H-NS protein and the cyclic AMP-catabolite activator protein complex in transcription of the flhDC master operon. J. Bacteriol. *181*, 7500–7508.

Srivatsan, A., and Wang, J.D. (2008). Control of bacterial transcription, translation and replication by (*p*)ppGpp. Curr. Opin. Microbiol. *11*, 100–105.

Stüdemann, A., Noirclerc-Savoye, M., Klauck, E., Becker, G., Schneider, D., and Hengge, R. (2003). Sequential recognition of two distinct sites in σ^S by the proteolytic targeting factor RssB and ClpX. EMBO J. *22*, 4111–4120.

Stülke, J., and Hillen, W. (1999). Carbon catabolite repression in bacteria. Curr. Opin. Microbiol. *2*, 195–201.

Sugiura, M., Aiba, H., and Mizuno, T. (2003). Identification and classification of two-component systems that affect *rpoS* expression in *Escherichia coli*. Biosci. Biotechnol. Biochem. *67*, 1612–1615.

Takayanagi, Y., Tanaka, K., and Takahashi, H. (1994). Structure of the 5′ upstream region and the regulation of the *rpoS* gene of *Escherichia coli*. Mol. Gen. Genet. *243*, 525–531.

Teich, A., Meyer, S., Lin, H.Y., Andersson, L., Enfors, S.O., and Neubauer, P. (1999). Growth rate related concentration changes of the starvation response regulators σ^S and ppGpp in glucose-limited fed-batch and continuous cultures of *Escherichia coli*. Biotechnol. Progr. *15*, 123–129.

Teplitski, M., Goodier, R.I., and Ahmer, B.M. (2003). Pathways leading from BarA/SirA to motility and virulence gene expression in *Salmonella*. J. Bacteriol. *185*, 7257–7265.

Tomenius, H., Pernestig, A.K., Jonas, K., Georgellis, D., Mollby, R., Normark, S., and Melefors, O. (2006). The *Escherichia coli* BarA-UvrY two-component system is a virulence determinant in the urinary tract. BMC Microbiol. *6*, 27.

Traxler, M.F., Zacharia, V.M., Marquardt, S., Summers, S.M., Nguyen, H.T., Stark, S.E., and Conway, T. (2011).

Discretely calibrated regulatory loops controlled by ppGpp partition gene induction across the 'feast to famine' gradient in *Escherichia coli*. Mol. Microbiol. *79*, 830–845.

Tsui, H.-C.T., Leung, H.-C.L., and Winkler, M.E. (1994). Characterization of broadly pleiotropic phenotypes caused by an *hfq* insertion mutation in *Escherichia coli* K-12. Mol. Microbiol. *13*, 35–49.

Typas, A., Barembruch, C., Possling, A., and Hengge, R. (2007a). Stationary phase reorganisation of the transcription machinery by Crl protein, a fine-tuner of σ^S activity and levels. EMBO J. *26*, 1569–78.

Typas, A., Becker, G., and Hengge, R. (2007b). The molecular basis of selective promoter activation by the sigmaS subunit of RNA polymerase. Mol. Microbiol. *63*, 1296–1306.

Ueguchi, C., Misonou, N., and Mizuno, T. (2001). Negative control of *rpoS* expression by phosphoenolpyruvate: carbohydrate phosphotransferase system in *Escherichia coli*. J. Bacteriol. *183*, 520–527.

Uehara, T., Parzych, K.R., Dinh, T., and Bernhardt, T.G. (2010). Daughter cell separation is controlled by cytokinetic ring-activated cell wall hydrolysis. EMBO J. *29*, 1412–1422.

Updegrove, T., Wilf, N., Sun, X., and Wartell, R.M. (2008). Effect of Hfq on RprA-rpoS mRNA pairing: Hfq-RNA binding and the influence of the 5′ rpoS mRNA leader region. Biochemistry *47*, 11184–11195.

Valentin-Hansen, P., Eriksen, M., and Udesen, C. (2004). The bacterial Sm-like protein Hfq: a key player in RNA transactions. Mol. Microbiol. *51*, 1525–1533.

Venturi, V. (2003). Control of *rpoS* transcription in *Escherichia coli* and *Pseudomonas*: why so different? Mol. Microbiol. *49*, 1–9.

Wassarman, K.M. (2007). 6S RNA: a regulator of transcription. Mol. Microbiol. *65*, 1425–1431.

Wassarman, K.M., and Storz, G. (2000). 6S RNA regulates *E. coli* RNA polymerase activity. Cell *101*, 613–623.

Waters, L.S., and Storz, G. (2009). Regulatory RNAs in bacteria. Cell *136*, 615–628.

Webb, C., Moreno, M., Wilmes-Riesenberg, M., Curtiss III, R., and Foster, J.W. (1999). Effects of DksA and ClpP protease on sigma S production and virulence in *Salmonella typhimurium*. Mol. Microbiol. *34*, 112–123.

Weber, H., Polen, T., Heuveling, J., Wendisch, V., and Hengge, R. (2005). Genome-wide analysis of the general stress response network in *Escherichia coli*: σ^S-dependent genes, promoters and sigma factor selectivity. J. Bacteriol. *187*, 1591–1603.

Westblade, L.F., Ilag, L.L., Powell, A.K., Kolb, A., Robinson, C.V., and Busby, S.J. (2004). Studies of the *Escherichia coli* Rsd–sigma70 complex. J. Mol. Biol. *335*, 685–692.

Yamashino, T., Ueguchi, C., and Mizuno, T. (1995). Quantitative control of the stationary phase-specific sigma factor, σ^S, in *Escherichia coli*: involvement of the nucleoid protein H-NS. EMBO J. *14*, 594–602.

Zgurskaya, H.I., Keyhan, M., and Matin, A. (1997). The σ^S level in starving *Escherichia coli* cells increases solely as a result of its increased stability, despite decreased synthesis. Mol. Microbiol. *24*, 643–651.

Zhang, A., Altuvia, S., Tiwari, A., Argaman, L., Hengge-Aronis, R., and Storz, G. (1998). The OxyS regulatory RNA represses *rpoS* translation and binds the Hfq (HF-I) protein. EMBO J. *17*, 6061–6068.

Zhang, A., Wassarman, K.M., Rosenow, C., Tjaden, B.C., Storz, G., and Gottesman, S. (2003). Global analysis of small RNA and mRNA targets of Hfq. Mol. Microbiol. *50*, 1111–1124.

Zhao, K., Liu, M., and Burgess, M.M. (2007). Adaptation in bacterial flagellar and motility systems: from regulon members to "foraging"-like behavior in. Nucleic Acids Res. *35*, 4441–4452.

Zhou, A.N., and Gottesman, S. (1998). Regulation of proteolysis of the stationary-phase sigma factor RpoS. J. Bacteriol. *180*, 1154–1158.

Zhou, Y., and Gottesman, S. (2006). Modes of regulation of RpoS by H-NS. J. Bacteriol. *188*, 7022–7025.

Zhou, Y., Gottesman, S., Hoskins, J.R., Maurizi, M.R., and Wickner, S. (2001). The RssB response regulator directly targets σ^S for degradation by ClpXP. Genes Dev. *15*, 627–637.

Bacterial Virulence Gene Expression Contributed by the Alternative σ Factor, σ^{54}

2

Patricia C. Burrows, Simone C. Wiesler, Zhensheng Pan, Martin Buck and Sivaramesh Wigneshweraraj

Abstract

Amongst the many accessory factors that bind RNA polymerase (RNAp) and serve to control its activities, sigma (σ) factors ubiquitously feature in programming of gene expression in response to abiotic and biotic cues. Here we review the role of the major variant σ factor, σ^{54}, in the expression of gene sets used for establishing the virulence of a wide range of pathogenic bacteria. The tight coupling of σ^{54}-dependent transcription to signalling pathways underpins the regulation of such systems, and allows a wide dynamic range of gene expression.

Background

All biological processes rely on the expression of genes, the products of which act together to mediate cellular function. The tractability of bacteria has allowed rapid progress in elucidating the strategies that underpin the regulation of gene expression at genetic, molecular and global levels. Many bacteria respond to physiological needs by using complex networks of signal transduction pathways to alter patterns of gene expression. The coordinated and regulated expression of genes is important for the ability of bacteria to successfully and rapidly mount an adaptive response in order to develop, differentiate, and survive in changing environmental conditions. This is particularly relevant, for example, in the context of a bacterial pathogen, which can be exposed to a series of hostile host induced changes to its environment that can make its chances of survival far from optimal. To survive, the bacteria must initiate the appropriate adaptive and protective responses, which are primarily and essentially reflected in transcriptional changes of specific sets of genes. Transcription is the first step during gene expression at which RNA is synthesized from a DNA template. Since transcription represents a major regulatory checkpoint of gene expression, the control of transcription often involves the interplay of signalling pathways with the molecular machinery of transcription, the DNA-dependent RNA polymerase (RNAp), the central enzyme that catalyses transcription.

Regulation of bacterial transcription occurs primarily at the initiation step (although regulation also takes place in the elongation and termination steps) and can be affected by multiple factors, which include transcription regulatory proteins, small effector molecules (for example, guanosine pentaphosphate ppGpp) and nucleic acid elements (for example, upstream activating sequences, UP-elements, promoter composition, small non-coding RNAs) that either directly or indirectly modulate the activity of the RNAp (Browning and Busby, 2004; Haugen *et al.*, 2008). All RNA synthesis in bacteria is accomplished by a single multisubunit RNAp, which consists of highly conserved (in terms of sequence and structure) 'catalytic core' (subunit composition $\alpha_2\beta\beta'\omega$; abbreviated as E) and an interchangeable sigma (σ) factor subunit. The σ factor is responsible for the distribution of RNAp molecules between the promoters of different transcription units. Even though the primary function of a σ factor is to confer promoter specificity upon the RNAp, σ factors also participate in the transcription initiation process (for example, in promoter DNA strand-separation for RNA synthesis) and also serve as interaction sites for transcription regulatory proteins (Borukhov and Severinov, 2002; Browning and Busby, 2004). All bacterial genomes encode a primary σ factor devoted to the transcription of housekeeping genes, and most also encode several alternative σ factors that direct the RNAp to specific promoters, thus driving the expression of genes whose products are involved in different adaptive processes, including virulence, stress response and developmental gene expression (Gruber and Gross, 2003). The number of σ factors encoded by a bacterial genome varies from 1 (in *Mycoplasma genitalium*) to 63 (in *Streptomyces coelicolor*).

In general, the repertoire of alternative σ factors used to coordinate and regulate the transcriptional responses to changing cellular demands varies widely between different bacterial species and often reflects the lifestyle of the bacteria. Since the regulon of a single σ factor often consists of several transcription units, σ factors represent an effective and powerful regulatory 'handle' for the execution of complex genetic programmes in response to specific stress conditions, growth transitions, morphological and lifestyle changes (Gruber and Gross, 2003). In addition, since the expression of alternative σ factors is often under direct or indirect control of other (alternative) σ factors, the regulation of gene expression can also be affected by complex networks of regulatory interactions among σ factors (see later). *Escherichia coli*, for example, can choose from a pool of six alternative σ factors to fine-tune its transcriptional output (Ishihama, 2000). The total cellular concentration of σ factors within an *E. coli* cell exceeds that of the RNAp concentration and results in competition between σ factors for limiting levels of RNAp (Ishihama, 2000). Therefore, competition of σ factors for RNAp also probably represents an important mechanism for altering the pattern of gene expression. Anti-σ factors represent a further means to control gene expression via the σ factor: Anti-σ factors are σ factor binding proteins that regulate the availability and activity of a cognate σ factor (Helmann, 1999). Therefore, the σ factor composition of the RNAp represents an important and efficient means to reprogram the pattern of gene expression at the transcriptome level in response to the different environmental conditions encountered by bacteria.

An alternative σ factor in a class of its own

Seven different species of σ factors, σ^{70}, σ^{54}, σ^{38}, σ^{32}, σ^{28}, σ^{24} and σ^{18}, have been identified in *E. coli*, each participating in the transcription of a specific set of genes (Ishihama, 2000). The essential genes expressed during exponential-phase growth are transcribed by the RNAp containing the 'housekeeping' σ^{70} factor, which is the primary σ factor encoded by *E. coli*. Five of the six alternative *E. coli* σ factors (σ^{38}, σ^{32}, σ^{28}, σ^{24} and σ^{18}) are required for the transcription of genes under different growth and stress conditions (Ishihama, 2000). Therefore, although the concentration of the catalytic core form of RNAp stays constant at a level that is characteristic of the rate of cell growth, the concentration of each alternative σ factor is subject to variation depending on growth conditions. For instance, the intracellular concentration is highest for the σ^{70} factor in both exponential and stationary phases of growth and under various stress conditions, whereas the alternative σ factor, σ^{38}, also termed σ^{S}, which is primarily associated with the transcription of genes required in the stationary phase of growth, only becomes available in the stationary phase (Jishage *et al.*, 1996; Jishage and Ishihama, 1997). Most alternative bacterial σ factors are related in sequence and structure to *E. coli* σ^{70} and therefore are grouped in a class represented by the *E. coli* σ^{70} factor (Paget and Helmann, 2003). However, the genomes of several phylogenetically diverse bacteria encode a second distinct variant-type of σ factor, called σ^{54} (and also termed σ^{N}), encoded by the *rpoN* gene and reviewed in Buck *et al.* (2000); Studholme and Buck (2000a,b); Wigneshweraraj *et al.* (2005, 2008). In this chapter, we review σ^{54}-dependent gene expression in bacteria with a particular emphasis on the expression of virulence and virulence-associated genes. σ^{54} is the sole representative of its class and shares no detectable sequence similarity with σ factors belonging to the σ^{70} class. The σ^{54} class also differs at the level of promoter-specificity and functional and regulatory properties. For the sake of simplicity, hereafter, σ^{70} or Eσ^{70} is used to refer to σ factors or Eσ belonging to the σ^{70} class.

Insights from biochemical studies: regulation of Eσ^{54} dependent transcription

Even though σ^{54} and σ^{70} utilize a common cellular RNAp and interact with overlapping surfaces on the RNAp catalytic β and β′ subunits (Datwyler and Meares, 2000; Wigneshweraraj *et al.*, 2000), the holoenzymes formed are regulated differently (Buck *et al.*, 2000; Wigneshweraraj *et al.*, 2005, 2008). At σ^{70}- and σ^{54}-dependent promoters, transcription starts with the σ-directed binding of the RNAp, which results in the formation of the initial RNAp–promoter complex, the so-called closed promoter complex (RPc). At most promoters used by Eσ^{70}, the specificity of this process is determined by two conserved hexanucleotide sequences centred at positions −35 and −10 with respect to the transcription start site at +1 (+1 site) (Fig. 2.1A). The −35 and −10 elements represent the major 'docking' sites for the σ factor at promoters. In contrast, σ^{54} directs the RNAp to promoters characterized by two highly conserved dinucleotide sequences, GG and GC, located at positions −24 and −12 with respect to the +1 site. In the last survey, Barrios *et al.* (1999) aligned 186 σ^{54}-dependent promoters from 47 bacterial species and revealed that there are eight and five conserved nucleotides around the −24 and −12 elements, respectively (Fig. 2.1B) (Barrios *et al.*, 1999). The length of the non-conserved sequence separating the GG and GC elements is essential for promoter function, since there appears to be a stringent requirement that the GG and GC are on the same face of the DNA helix for the effective binding and usage of the promoter by Eσ^{54} (Buck, 1986). It seems that the interaction between σ^{54} and the −24 element has a role in 'anchoring' the RNAp to the promoter, whereas the interaction at the −12 element has a regulatory function (Guo *et al.*, 1999, 2000; Wigneshweraraj *et al.*, 2001; Burrows *et al.*, 2004). In stark contrast to the RPc formed at σ^{54}-dependent promoters, the RPc formed at most σ^{70}-dependent promoters is short-lived and often isomerises spontaneously, via several intermediate states, to the transcriptionally proficient state, known as the open promoter complex (RPo). In the RPo, the double-stranded promoter DNA between the −10 element and the +1 site are locally melted (i.e. in single-stranded form) and the template DNA strand loaded into the catalytic cleft of the RNAp. For σ^{70}-dependent promoters, regulation often occurs at the RPc and/or RPo formation steps and is typically mediated by transcription regulatory proteins that facilitate (transcription activators) or antagonise (transcription repressors) RPc and/or RPo formation. Transcription activators improve the performance of a σ^{70}-dependent promoter by improving its affinity for RNAp and often bind to specific sites that are proximal or overlap the −35 element in order to help 'recruit' RNAp to the promoter through interactions between the transcription activator and the subunits of RNAp (Browning and Busby, 2004). As discussed below, RPo formation at σ^{54}-dependent promoters is strictly dependent on a specialized class of transcriptional activator proteins which are ATPases (Rappas *et al.*, 2007; Ghosh *et al.*, 2010).

Transcription activation at Eσ^{54}-dependent promoters

The activity of a bacterial promoter is often adjusted by a wide variety of *cis*- and *trans*-acting transcriptional regulatory proteins that sense the environment and the physiological state of the cell. Most σ^{70} promoters can be regarded as transcriptionally proficient since they retain at least a minimal level of activity in the absence of accessory factors. Most transcriptional repressors function by blocking access of Eσ^{70} to their target promoter. In contrast, transcription activators that operate at σ^{70} promoters often function to increase promoter activity above that of basal activity. Thus, σ^{70}-dependent promoters can be subject to positive regulation at RPc and/or RPo formation (Browning and Busby, 2004). Transcription activators often improve the performance of a σ^{70}-dependent promoter by improving its affinity for RNAp to promote RPc formation and/or facilitating structural transitions in the RNAp and promoter DNA to allow efficient RPo formation. Many transcription activators bind to specific sites

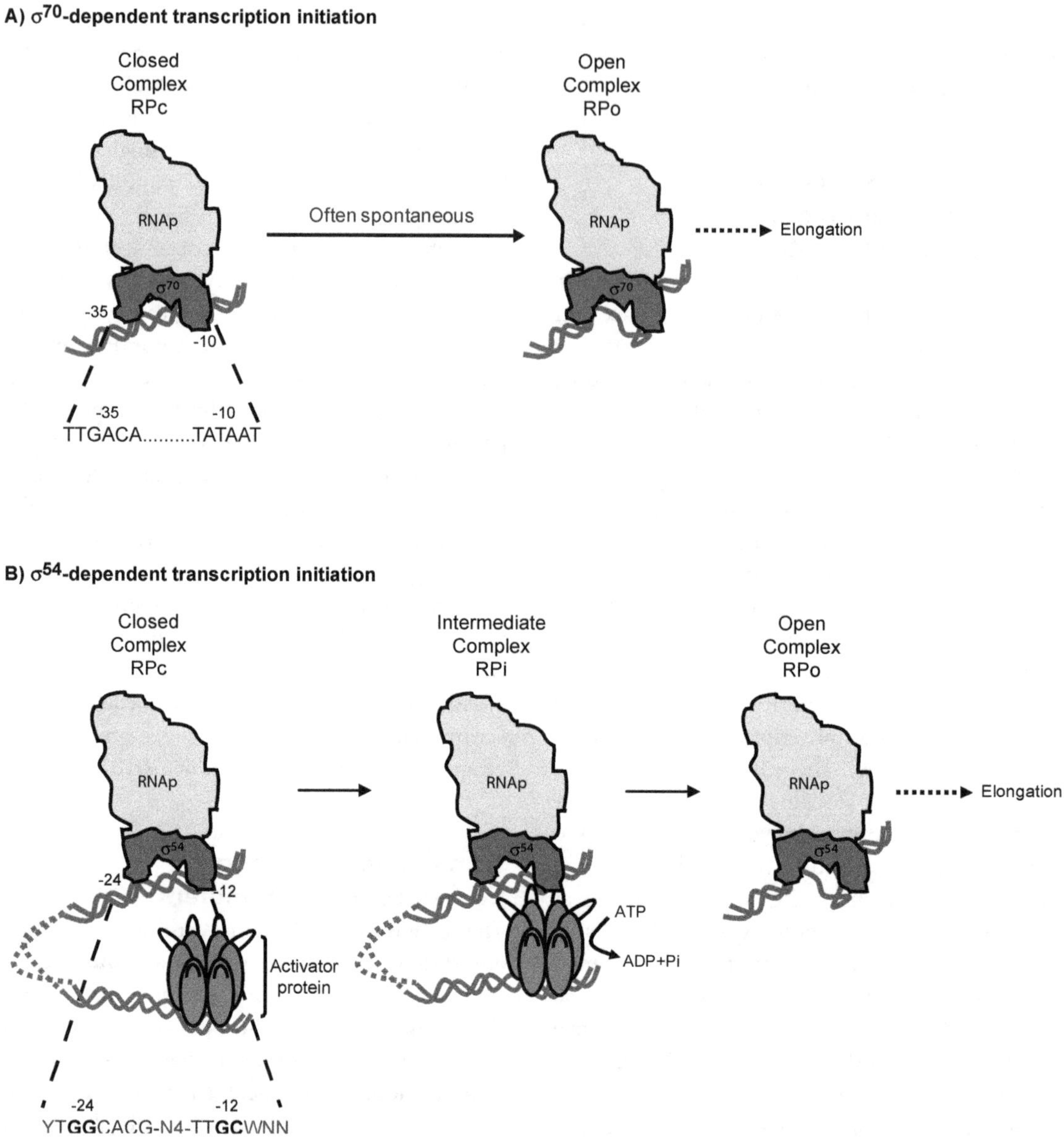

Figure 2.1 Schematic representation of the events leading to open complex formation (RPo) by (A) Eσ^{70} and (B) Eσ^{54}. The typical promoter recognition sequences are given. In the case of the σ^{54} promoter, W represents either A or T; Y represents pyrimidines and N represents non-significant sites (as indicated in Barrios *et al.*, 1999). A colour version of this figure is available in the plate section at the back of the book.

that are proximal or overlap the −35 element and as such often interact with at least one of the RNAp subunits. In *E. coli*, the majority of transcriptional activators are sequence-specific DNA binding proteins, only a small fraction of which (~10%) directly associate with Eσ^{70} and modulate its activity independent of DNA binding (Ishihama, 2010). A critical function of transcriptional activators that bind DNA is how they interact with Eσ^{70}. This can occur either 'on-DNA', in the case of activators that function by classical recruitment or 'off-DNA and on-DNA', in the case of activators that function by pre-recruitment, whereby the transcriptional activator interacts with Eσ^{70}

'off DNA' and subsequently recruits the Eσ^{70} to the promoter (Shah and Wolf, 2004; Ishihama, 2010). The complexity of transcription activation at many Eσ^{70} promoters is compounded by the action of more than one transcription activator or in cases where a promoter is regulated by a repressor and activator – allowing integration of multiple signals at one promoter (Browning and Busby, 2004).

In contrast to Eσ^{70}, Eσ^{54} adopts a transcriptionally silent configuration upon forming the RPc. σ^{54} imposes kinetic and thermodynamic constraint on RPo formation owing to a repressive set of interactions between σ^{54} and the −12 promoter element, where local DNA melting for RPo formation nucleates (Fig. 2.1B) (Guo *et al.*, 1999, 2000; Cannon *et al.*, 2000). Thus, activation of transcription from the Eσ^{54}-RPc is an energy consuming process and strictly requires a specialized class of transcription activator proteins, which are ATPases that belong to the AAA+ (ATPases associated with various cellular activities) protein family (hereafter referred to as σ^{54} activators) (Rappas *et al.*, 2007; Ghosh *et al.*, 2010). The vast majority of σ^{54} activators are sequence specific DNA binding proteins (Studholme and Dixon, 2003). Activation of RPo formation at σ^{54}-dependent promoters occurs in response to regulatory cues from upstream signalling events, which often 'activate' the σ^{54} activator, thereby allow it to bind as higher-order oligomers (usually hexamers) to DNA sites, called 'enhancers', that are often located ~100–150 nucleotides upstream, but rarely downstream from the σ^{54} promoter (Buck *et al.*, 2000, 2006; Wigneshweraraj *et al.*, 2008; Ghosh *et al.*, 2010; Joly *et al.*, 2010; Shingler, 2010). The interaction between the enhancer-bound σ^{54} activator ATPase and the Eσ^{54}-RPc occurs via a DNA looping event and is often facilitated by DNA-bending proteins, such as the integration host factor (IHF). In a reaction consuming ATP, the σ^{54} activator functions to remodel the Eσ^{54}-RPc to remove the repressive set of interaction between σ^{54} and the −12 element and so to facilitate DNA strand separation and RPo formation (Guo *et al.*, 1999, 2000; Cannon *et al.*, 2000; Wigneshweraraj *et al.*, 2001). In clear contrast to the role of transcription activators at Eσ^{70}-dependent promoters, which merely function to up-regulate basal promoter activity, the mechanism of transcription activation at Eσ^{54}-dependent promoters allows gene expression to be effectively 'switched on' only in response to a particular physiological or environmental signal and thereby ensuring effective management of cellular resources – ultimately these signals unlock the ATPase activity of the activator, often via promoting self-assembly (Shingler, 2010). The widespread distribution of σ^{54} in diverse bacterial species and the usage of σ^{54}-dependent promoters for the transcription of genes, the products of which are required for a wide variety of cellular functions, further underscores the regulatory advantages that accompany the dependence on Eσ^{54} for the transcription of genes.

Distribution and biology of Eσ^{54}-dependent gene expression in bacteria

While historically associated with a role in the transcription of genes required for nitrogen assimilation, nitrogen metabolism and nitrogen fixation in a wide-range of enteric bacteria, σ^{54} represents at most 8% of the total stock of σ factor molecules in *E. coli* and is the only alternative σ factor to be expressed at comparable levels to σ^{70} during exponential growth in rich media (Jishage *et al.*, 1996; Jishage and Ishihama, 1997). In *E. coli*, σ^{54} is not essential for growth in rich media, but σ^{54}-dependent gene products are required for survival under various stress conditions. Transcriptome expression profiling analysis and genome-wide location studies in *E. coli* indicate that ~49 transcription units (comprising in excess of 100 genes) are likely to depend on Eσ^{54} for their expression (Fig. 2.2). Of these transcription units, 30 are classified as having 'strong' evidence for σ^{54} dependency; the remainder have been identified by transcriptome data (classified as 'weak' evidence) and have yet to be functionally verified (Reitzer and Schneider, 2001; Zhao *et al.*, 2010). However, it seems that ~13 (σ^{38}-dependent) genes in *E. coli* are subject to negative regulation by σ^{54} through σ^{38}, the stationary phase σ factor – via an as yet not fully understood mechanism in which σ^{54} appears to negatively affect σ^{38} levels and activity (Dong *et al.*, 2011). The complexity

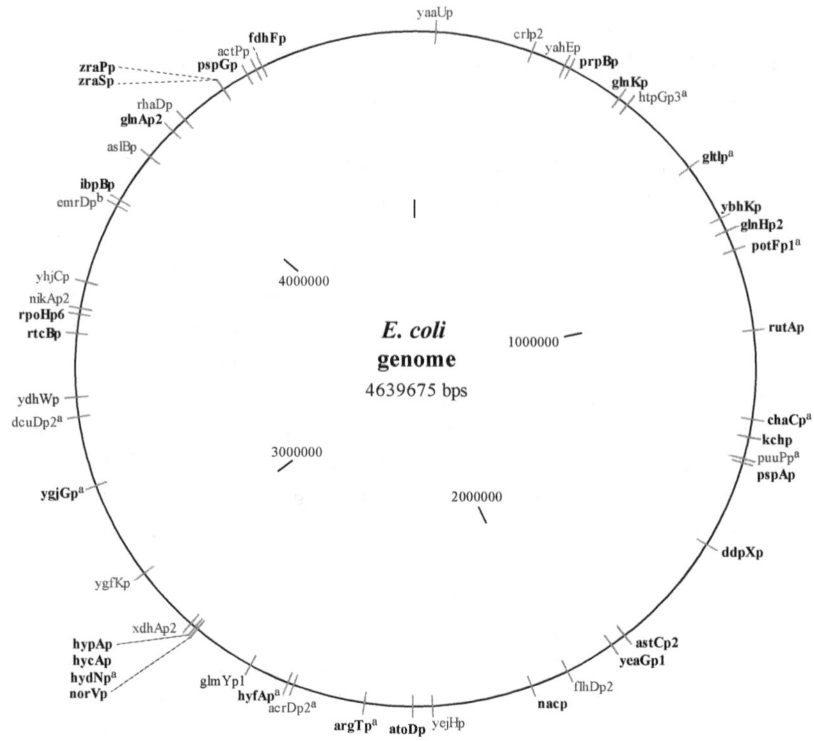

Figure 2.2 Position of the transcription units regulated by $E\sigma^{54}$ in the *E. coli* genome. The bold lettering refers to promoters that have been identified by (Reitzer and Schneider, 2001) and are present as part of the sigmulon on RegulonDB (http://regulondb.ccg.unam.mx/); grey lettering refers to promoters identified in transcriptomic studies (Zhao *et al.*, 2010); black lettering refers to the glmYp1 promoter identified by (Reichenbach *et al.*, 2009); [a] indicates promoters with less than 80% identity with the consensus sequence. A colour version of this figure is available in the plate section at the back of the book.

of the regulation and expression of the σ^{54} regulon in *E. coli* appears to be compounded by the fact that about 60% of the σ^{54} regulon is, either directly or indirectly, under reciprocal control by σ^{38}, i.e. many genes positively regulated by σ^{54} are subject to negative regulation by σ^{38} and *vice versa*. It seems that the functional interplay between σ^{54} and σ^{38} represents a paradigmatic example of how gene expression patterns can be affected at the transcriptional level by regulatory interactions between σ factors (Dong *et al.*, 2011). Further, examples exist (such as the small RNA GlmY) of dual control by perfectly overlapping σ^{70} and σ^{54} promoters that adjust expression in response to different environmental signals (Reichenbach *et al.*, 2009; Gopel *et al.*, 2011). Since many genes of the σ^{54} regulon are involved in counteracting nutritional stresses, the expression of the σ^{54} regulon is also affected by levels of ppGpp, the alarmone which is involved in the stringent response in bacteria (Nystrom, 2004; Shingler, 2010). In *E. coli*, for example, increased ppGpp levels lead to increased transcription from the *glnA* promoter, which drives expression of the enzyme glutamine synthetase (used for catalysing the production of glutamine from glutamate and ammonia under conditions of nitrogen deprivation) (Bernardo *et al.*, 2006, 2009). However, unlike its effect on the general stress sigma factor, σ^{24} (or σ^{E}), the effect of ppGpp on σ^{54}-dependent (and σ^{38}-dependent) transcription is indirect and is achieved by altering the competition for core RNAp between the alternative σ factors and the general housekeeping σ^{70} factor (Costanzo and Ades, 2006; Costanzo *et al.*, 2008).

The increased availability of complete bacterial genome sequences has substantiated the view that σ^{54} is required for the expression of genes whose products have a range of different functions (Wosten, 1998; Studholme and Buck, 2000b; Kazmierczak *et al.*, 2005). Inspection of the 672 complete bacterial genome sequences[1] demonstrated that σ^{54} was present (as an annotated gene that contains the characteristic RpoN box motif; Burrows *et al.*, 2003; Merrick, 1993) in 336 organisms (Fig. 2.3 and Table 2.1) – all of which contained at least one σ^{54} activator (hence we infer that these organisms contain an 'active' form of σ^{54} rather than *rpoN* pseudogenes). Of this number, 207 organisms were classified as pathogens (Fig. 2.3 and Table 2.1), suggesting that σ^{54} is more prevalent in pathogenic bacteria (vs. non-pathogens), which we infer may be a consequence of the role it performs in regulating virulence gene expression. Given the distribution of σ^{54} in bacterial genomes, it is no surprise that σ^{54} is found in a diverse collection of organisms, ranging from proteobacteria (such

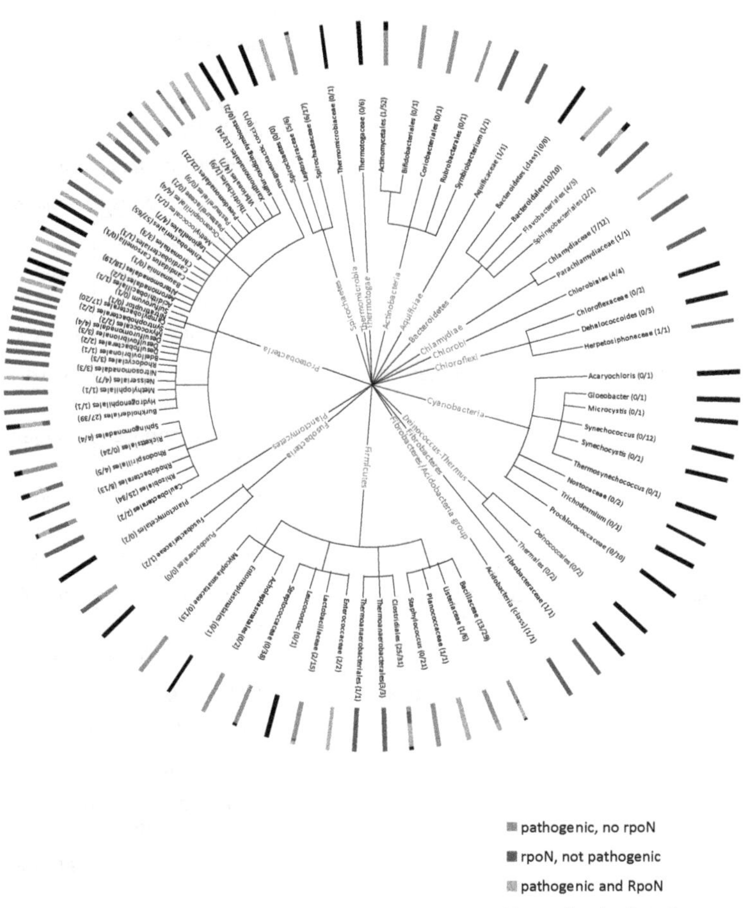

Figure 2.3 A phylogenetic illustration of the class, order and family ranks of the 672 sequenced bacterial genomes (www.jcvi.org). The number of species per family in which the *rpoN* gene was found in relation to the total number of sequenced genomes is given in parentheses. The coloured bars indicate for each family the proportions of species that contain *rpoN* and are pathogens (green) or not (red), and which do not contain *rpoN* and are pathogens (blue) or not (black). A colour version of this figure is available in the plate section at the back of the book.

1 Available in the NCBI database at the time of writing (January 2011).

Table 2.1 All bacterial genomes sequenced to date which contain the *rpoN* gene (as an annotated gene that contains the characteristic RpoN box motif; Merrick, 1993; Burrows *et al.*, 2003) are listed and their pathogenicity (yes or no) indicated (see Fig. 2.3). All of which contained at least one σ^{54} activator (hence we infer that these organisms contain an 'active' form of σ^{54} rather than *rpoN* pseudogenes)

Class	Order	Family	Species name	Pathogen
Actinobacteria	Actinobacteridae	Actinomycetales	*Propionibacterium* sp. F0233	Y
	Symbiobacterium	Symbiobacterium	*Symbiobacterium thermophilum* IAM14863	N
Aquificiae	Aquificales	Aquificaceae	*Aquifex aeolicus* VF5	N
Bacteroidetes	Bacteroidia	Bacteroidales	*Bacteroides fragilis* NCTC9343	Y
			Bacteroides fragilis YCH46	Y
			Bacteroides thetaiotaomicron VPI-5482	Y
			Parabacteroides distasonis ATCC 8503	N
			Porphyromonas endodontalis ATCC 35406	Y
			Porphyromonas gingivalis W83	Y
			Porphyromonas uenonis 60–3	Y
			Prevotella intermedia 17	Y
			Prevotella melaninogenica ATCC 25845	Y
			Prevotella ruminicola 23	Y
	Flavobacteria	Flavobacteriales	*Capnocytophaga gingivalis* ATCC 33624	Y
			Capnocytophaga sputigena Capno	Y
			Flavobacterium johnsoniae UW101	N
			Gramella forsetii KT0803	N
	Sphingobacteria	Sphingobacteriales	*Cytophaga hutchinsonii* ATCC 33406	N
			Salinibacter ruber DSM 13855	N
Chlamydiae	Chlamydiales	Chlamydiaceae	*Chlamydia pneumoniae* CWL029	Y
			Chlamydia trachomatis 434/Bu	Y
			Chlamydia trachomatis A/HAR-13	Y
			Chlamydia trachomatis L2b/UCH-1/proctitis	Y
			Chlamydia trachomatis serovar D	Y
			Chlamydophila abortus S26 3	Y
			Chlamydophila felis Fe/C-56	Y
		Parachlamydiaceae	*Candidatus Protochlamydia amoebophila* UWE25	N
Chlorobi	Chlorobia	Chlorobiales	*Chlorobium chlorochromatii* CaD3	N
			Chlorobium tepidum TLS	N
			Pelodictyon luteolum DSM 273	N
			Prosthecochloris vibrioformis DSM 265	N

Table 2.1 Continued

Class	Order	Family	Species name	Pathogen
Chloroflexi	Herpetosiphonales	Herpetosiphonaceae	*Herpetosiphon aurantiacus* ATCC 23779	N
Cyanobacteria	Chroococcales	Synechocystis	*Synechocystis* sp. PCC6803	N
Fibrobacteres	Fibrobacterales	Fibrobacteraceae	*Fibrobacter succinogenes* S85	N
Fibrobacteres/ Acidobacteria group	Acidobacteria	Acidobacteria (class)	*Acidobacteria bacterium* Ellin345	N
Firmicutes	Bacillales	Bacillaceae	*Bacillus amyloliquefaciens* FZB42	Y
			Bacillus anthracis Sterne	Y
			Bacillus cereus Zk	Y
			Bacillus cereus subsp. cytotoxis NVH 391–98	Y
			Bacillus clausii KSM-K16	Y
			Bacillus licheniformis ATCC 14580	Y
			Bacillus licheniformis ATCC14580 (DSM 13)	Y
			Bacillus pumilus SAFR-032	Y
			Bacillus subtilis 168	Y
			Bacillus thuringiensis Al Hakam	Y
			Bacillus weihenstephanensis KBAB4	Y
			Geobacillus kaustophilus HTA426	N
			Geobacillus thermodenitrificans NG80–2	N
		Listeriaceae	*Listeria welshimeri* serovar 6b str. SLCC5334	Y
		Planococcaceae	*Lysinibacillus sphaericus* C3–41	Y
	Clostridia	Clostridiales	*Alkaliphilus metalliredigens* QYMF	N
			Alkaliphilus oremlandii OhILAs	N
			Carboxydothermus hydrogenoformans Z-2901	N
			Clostridium acetobutylicum ATCC824	Y
			Clostridium beijerinckii NCIMB 8052	Y
			Clostridium botulinum A ATCC 19397	Y
			Clostridium botulinum A Hall	Y
			Clostridium botulinum A2 Kyoto	Y
			Clostridium botulinum A3 strain Loch Maree	Y
			Clostridium botulinum B Eklund 17B	Y
			Clostridium botulinum B1 strain Okra	Y

Table 2.1 Continued

Class	Order	Family	Species name	Pathogen
			Clostridium botulinum Ba4 strain 657	Y
			Clostridium botulinum F Langeland	Y
			Clostridium botulinum Hall A Sanger	Y
			Clostridium difficile 630	Y
			Clostridium kluyveri DSM 555	Y
			Clostridium novyi NT	Y
			Clostridium perfringens ATCC13124	Y
			Clostridium perfringens SM101	Y
			Clostridium tetani E88	Y
			Desulfitobacterium hafniense Y51	N
			Desulfotomaculum reducens MI-1	N
			Heliobacterium modesticaldum Ice1	N
			Pelotomaculum thermopropionicum SI	N
			Syntrophomonas wolfei subsp. *wolfei* str. Goettingen	N
		Thermoanaerobacterales	*Thermoanaerobacter pseudethanolicus* ATCC 33223	N
			Thermoanaerobacter sp. X514	N
			Thermoanaerobacter tengcongensis MB4(T)	N
		Thermoanaerobacteriales	*Moorella thermoacetica* ATCC 39073	N
	Lactobacillales	Enterococcaceae	*Enterococcus faecalis* TUSoD Ef11	Y
			Enterococcus faecalis V583	Y
		Lactobacillaceae	*Lactobacillus plantarum* WCFS1	Y
			Pediococcus pentosaceus ATCC 25745	N
Fusobacteria	Fusobacteriales	Fusobacteriaceae	*Leptotrichia goodfellowii* F0264	N
Proteobacteria	Alphaproteobacteria	Caulobacterales	*Caulobacter crescentus* CB15	N
			Caulobacter sp. K31	N
		Rhizobiales	*Bradyrhizobium* sp. ORS278	N
			Brucella abortus 9–941	Y
			Brucella canis ATCC 23365	Y
			Brucella melitensis 16M	Y
			Brucella melitensis biovar Abortus	Y
			Brucella suis 1330	Y
			Brucella suis ATCC 23445	Y
			Mesorhizobium loti MAFF303099	N

Table 2.1 Continued

Class	Order	Family	Species name	Pathogen
			Mesorhizobium sp. BNC1	N
			Methylobacterium extorquens PA1	N
			Methylobacterium sp. 4–46	N
			Nitrobacter hamburgensis X14	N
			Nitrobacter winogradskyi Nb-255	N
			Ochrobactrum anthropi ATCC 49188	Y
			Oligotropha carboxidovorans OM5	N
			Rhizobium etli CFN 42	Y
			Rhizobium leguminosarum bv. viciae 3841	Y
			Rhodopseudomonas palustris BisA53	N
			Rhodopseudomonas palustris BisB18	N
			Rhodopseudomonas palustris BisB5	N
			Rhodopseudomonas palustris CGA009	N
			Rhodopseudomonas palustris HaA2	N
			Sinorhizobium medicae WSM419	N
			Sinorhizobium meliloti 1021	N
			Xanthobacter autotrophicus Py2	N
		Rhodobacterales	*Citreicella* sp. SE45	N
			Dinoroseobacter shibae DFL 12	N
			Hyphomonas neptunium ATCC 15444	N
			Maricaulis maris MCS10	N
			Rhodobacter sphaeroides 2 4 1	N
			Rhodobacter sphaeroides ATCC 17025	N
			Rhodobacter sphaeroides ATCC 17029	N
			Ruegeria pomeroyi DSS-3	N
		Rhodospirillales	*Gluconobacter oxydans* 621H	Y
			Granulibacter bethesdensis CGDNIH1	Y
			Magnetospirillum magneticum AMB-1	N
			Rhodospirillum rubrum ATCC 11170	N
		Sphingomonadales	*Erythrobacter litoralis* HTCC2594	N
			Novosphingobium aromaticivorans DSM 12444	N

Table 2.1 Continued

Class	Order	Family	Species name	Pathogen
			Sphingopyxis alaskensis RB2256	N
			Zymomonas mobilis ZM4	N
	Betaproteobacteria	Burkholderiales	*Acidovorax avenae* subsp. citrulli AAC00–1–0	Y
			Acidovorax sp. JS42	Y
			Bordetella avium 197N	Y
			Bordetella bronchiseptica RB50	Y
			Bordetella parapertussis 12822	Y
			Bordetella pertussis Tohama I	Y
			Bordetella petrii DSM 12804	Y
			Burkholderia ambifaria AMMD	Y
			Burkholderia ambifaria MC40–6	Y
			Burkholderia cenocepacia AU 1054	Y
			Burkholderia cenocepacia HI2424	Y
			Burkholderia mallei ATCC 23344	Y
			Burkholderia mallei NCTC 10229	Y
			Burkholderia mallei NCTC 10247	Y
			Burkholderia mallei SAVP1	Y
			Burkholderia multivorans ATCC 17616 JGI	Y
			Burkholderia multivorans ATCC 17616 Tohoku	Y
			Burkholderia pseudomallei 1106a	Y
			Burkholderia pseudomallei 1106b	Y
			Burkholderia pseudomallei 1710b	Y
			Burkholderia pseudomallei 668	Y
			Burkholderia pseudomallei K96243	Y
			Burkholderia sp. 383	Y
			Burkholderia vietnamiensis G4	Y
			Burkholderia xenovorans LB400	Y
			Herminiimonas arsenicoxydans str.	N
			Janthinobacterium sp. Marseille	Y
			Leptothrix cholodnii SP-6	N
			Methylibium petroleiphilum PM1	N
			Polaromonas naphthalenivorans CJ2	N
			Polaromonas sp. JS666	N
			Ralstonia eutropha H16	Y
			Ralstonia eutropha JMP134	Y
			Ralstonia metallidurans CH34	Y
			Ralstonia solanacearum GMI1000	Y

Table 2.1 Continued

Class	Order	Family	Species name	Pathogen
			Rhodoferax ferrireducens DSM 15236	N
			Verminephrobacter eiseniae EF01–2	N
		Hydrogenophilales	*Thiobacillus denitrificans* ATCC 25259	N
		Methylophilales	*Methylobacillus flagellatus* KT	N
		Neisseriales	*Chromobacterium violaceum* ATCC12472	N
			Neisseria flavescens SK114	Y
			Neisseria meningitidis 053442	Y
			Neisseria meningitidis FAM18	Y
		Nitrosomonadales	*Nitrosomonas europaea* ATCC 19718	N
			Nitrosomonas eutropha C91	N
			Nitrosospira multiformis ATCC 25196	N
		Rhodocyclales	*Azoarcus* sp. BH72	N
			Azoarcus sp. EbN1	N
			Dechloromonas aromatica RCB	N
	Deltaproteobacteria	Bdellovibrionales	*Bdellovibrio bacteriovorus* HD100	N
		Desulfobacterales	*Desulfococcus oleovorans* Hxd3	N
			Desulfotalea psychrophila LSv54	N
		Desulfovibrionales	*Desulfovibrio desulfuricans* G20	N
			Desulfovibrio vulgaris Hildenborough	N
			Lawsonia intracellularis PHE/MN1–00	N
		Desulfuromonadales	*Geobacter metallireducens* GS-15	N
			Geobacter sulfurreducens PCA	N
			Geobacter uraniireducens Rf4	N
			Pelobacter carbinolicus DSM 2380	N
		Myxococcales	*Anaeromyxobacter dehalogenans* 2CP-C	N
			Myxococcus xanthus DK 1622	N
		Syntrophobacterales	*Syntrophobacter fumaroxidans* MPOB	N
			Syntrophus aciditrophicus SB	N
	Epsilonproteobacteria	Campylobacterales	*Campylobacter concisus* 13826	Y
			Campylobacter curvus 525.92	Y
			Campylobacter fetus subsp. fetus 82–40	Y
			Campylobacter jejuni NCTC 11168	Y
			Campylobacter jejuni RM1221	Y

Table 2.1 Continued

Class	Order	Family	Species name	Pathogen
			Campylobacter jejuni subsp. doylei 269.97	Y
			Campylobacter jejuni subsp. jejuni 81–176	Y
			Campylobacter jejuni subsp. jejuni 81116	Y
			Campylobacter rectus RM3267	Y
			Campylobacter showae RM3277	Y
			Helicobacter acinonychis str. Sheeba	Y
			Helicobacter hepaticus ATCC 51449	Y
			Helicobacter pylori 26695	Y
			Helicobacter pylori HPAG1	Y
			Helicobacter pylori J99	Y
			Thiomicrospira denitrificans ATCC 33889	N
			Wolinella succinogenes DSMZ 1740	N
	Gammaproteobacteria	Acidithiobacillales	*Acidithiobacillus ferrooxidans* ATCC 23270	N
		Aeromonadales	*Aeromonas salmonicida* subsp. salmonicida A449	Y
		Alteromonadales	*Colwellia psychrerythraea* 34H	N
			Idiomarina loihiensis L2TR	N
			Marinobacter aquaeolei VT8	N
			Pseudoalteromonas atlantica T6c	N
			Psychromonas ingrahamii 37	N
			Saccharophagus degradans 2–40	N
			Shewanella baltica OS185	N
			Shewanella baltica OS195	N
			Shewanella denitrificans OS217	N
			Shewanella halifaxensis HAW-EB4	N
			Shewanella loihica PV-4	N
			Shewanella oneidensis MR-1	N
			Shewanella pealeana ATCC 700345	N
			Shewanella sediminis HAW-EB3	N
			Shewanella sp. ANA-3	N
			Shewanella sp. MR-4	N
			Shewanella sp. MR-7	N
			Teredinibacter turnerae T7901	N
		Cardiobacteriales	*Dichelobacter nodosus* VCS1703A	Y
		Chromatiales	*Alkalilimnicola ehrlichei* MLHE-1	N

Table 2.1 Continued

Class	Order	Family	Species name	Pathogen
			Halorhodospira halophila SL1	N
			Nitrosococcus oceani ATCC 19707	N
		Enterobacteriales	*Citrobacter koseri* ATCC BAA-895	Y
			Cronobacter sakazakii ATCC BAA-894	Y
			Enterobacter sp. 638	Y
			Erwinia carotovora atroseptica SCRI1043	Y
			Escherichia coli 536	Y
			Escherichia coli 55989	Y
			Escherichia coli APEC O1	Y
			Escherichia coli ATCC 8739	Y
			Escherichia coli B str. REL606	Y
			Escherichia coli BL21(DE3)	Y
			Escherichia coli BW2952	Y
			Escherichia coli CFT073	Y
			Escherichia coli E24377A	Y
			Escherichia coli ED1a	Y
			Escherichia coli IAI1	Y
			Escherichia coli IAI39	Y
			Escherichia coli K12-MG1655	Y
			Escherichia coli O127:H6 str. E2348/69	Y
			Escherichia coli O157:H7 EDL933	Y
			Escherichia coli O157:H7 VT2-Sakai	Y
			Escherichia coli O157:H7 str. EC4115	Y
			Escherichia coli S88	Y
			Escherichia coli SE11	Y
			Escherichia coli SMS-3–5	Y
			Escherichia coli UMN026	Y
			Escherichia coli UTI89	Y
			Escherichia coli W3110	Y
			Escherichia coli str. K-12 substr. DH10B	Y
			Escherichia fergusonii ATCC 35469	Y
			Klebsiella pneumoniae 342	Y
			Klebsiella pneumoniae subsp. *pneumoniae* MGH 78578	Y
			Photorhabdus luminescens TTO1	Y
			Salmonella enterica Choleraesuis	Y

Table 2.1 Continued

Class	Order	Family	Species name	Pathogen
			Salmonella enterica Paratyphi ATCC9150	Y
			Salmonella enterica serovar Typhi CT18	Y
			Salmonella enterica serovar Typhi Ty2	Y
			Salmonella enterica subsp. *arizonae* serovar 62:z4,z23:–	Y
			Salmonella enterica subsp. *enterica* serovar Heidelberg str. SL476	Y
			Salmonella enterica subsp. *enterica* serovar Paratyphi B str. SPB7	Y
			Salmonella typhimurium LT2 SGSC1412	Y
			Serratia proteamaculans 568	Y
			Shigella boydii Sb227	Y
			Shigella dysenteriae Sd197	Y
			Shigella flexneri 2a 2457T	Y
			Shigella flexneri 2a str. 301	Y
			Shigella flexneri 5 str. 8401	Y
			Shigella sonnei Ss046	Y
			Sodalis glossinidius str. morsitans	N
			Yersinia enterocolitica subsp. *enterocolitica* 8081	Y
			Yersinia pestis Angola	Y
			Yersinia pestis Antiqua JGI	Y
			Yersinia pestis KIM	Y
			Yersinia pestis Nepal516	Y
			Yersinia pestis Pestoides F	Y
			Yersinia pestis biovar *Medievalis* 91001	Y
			Yersinia pseudotuberculosis IP 31758	Y
			Yersinia pseudotuberculosis IP32953	Y
		Legionellales	*Legionella pneumophila* Lens	Y
			Legionella pneumophila Paris	Y
			Legionella pneumophila Philadelphia 1	Y
			Legionella pneumophila str. Corby	Y
		Methylococcales	*Methylococcus capsulatus* Bath	N
		Oceanospirillales	*Alcanivorax borkumensis* SK2	N
			Chromohalobacter salexigens DSM 3043	N

Table 2.1 Continued

Class	Order	Family	Species name	Pathogen
			Hahella chejuensis KCTC 2396	Y
			Marinomonas sp. MWYL1	N
		Pseudomonadales	*Acinetobacter radioresistens* SK82	Y
			Acinetobacter sp. ADP1	Y
			Cellvibrio japonicus Ueda107	N
			Enhydrobacter aerosaccus SK60	N
			Pseudomonas aeruginosa PAO1	Y
			Pseudomonas aeruginosa UCBPP-PA14	Y
			Pseudomonas entomophila L48	Y
			Pseudomonas fluorescens Pf-5	Y
			Pseudomonas fluorescens PfO-1	Y
			Pseudomonas fluorescens SBW25	Y
			Pseudomonas mendocina ymp	Y
			Pseudomonas putida F1	Y
			Pseudomonas putida GB-1	Y
			Pseudomonas putida KT2440	Y
			Pseudomonas stutzeri A1501	Y
			Pseudomonas syringae DC3000	Y
			Pseudomonas syringae pv B728a	Y
			Pseudomonas syringae pv phaseolicola	Y
			Psychrobacter arcticum 273–4	N
			Psychrobacter cryohalolentis K5	N
			Psychrobacter sp. PRwf-1	N
		Thiotrichales	*Thiomicrospira crunogena* XCL-2	N
		Vibrionales	*Photobacterium profundum* SS9	N
			Vibrio cholerae El Tor N16961	Y
			Vibrio cholerae O395	Y
			Vibrio fischeri ES114	Y
		Xanthomonadales	*Xanthomonas axonopodis* pv. *citri* 306	Y
			Xanthomonas campestris 8004	Y
			Xanthomonas campestris pv. *armoraciae* 756C	Y
			Xanthomonas campestris pv. *campestris*	Y
			Xanthomonas campestris pv. *campestris* ATCC33913	Y
			Xanthomonas campestris pv. *vesicatoria* str. 85–10	Y
			Xanthomonas oryzae KACC10331	Y

Table 2.1 Continued

Class	Order	Family	Species name	Pathogen
			Xanthomonas oryzae pv. *oryzae* MAFF 311018	Y
			Xanthomonas oryzae pv. *oryzae* PXO99A	Y
			Xanthomonas oryzae pv. *oryzicola* BLS256	Y
			Xanthomonas vasicola pv. *musacearum*	Y
			Xylella fastidiosa 9a5c	Y
			Xylella fastidiosa M12	Y
Spirochaetes	Spirochaetales	Leptospiraceae	*Leptospira biflexa* serovar Patoc strain Patoc 1 (Paris)	Y
			Leptospira borgpetersenii serovar Hardjo-bovis JB197	Y
			Leptospira borgpetersenii serovar Hardjo-bovis L550	Y
			Leptospira interrogans Copenhageni Fiocruz L1–130	Y
			Leptospira interrogans serovar lai str. 56601	Y
		Spirochaetaceae	*Borrelia afzelii* PKo	Y
			Borrelia burgdorferi B31	Y
			Borrelia burgdorferi ZS7	Y
			Borrelia garinii PBi	Y
			Treponema denticola ATCC 35405	Y
			Treponema pallidum Nichols	Y

as *Brucella melitensis, Burkholderia cenocepacia, Desulfovibrio desulfuricans, Campylobacter jejuni, E. coli* and *Pseudomonas aeruginosa*) (Blattner *et al.*, 1997; Parkhill *et al.*, 2000; Stover *et al.*, 2000; DelVecchio *et al.*, 2002; Holden *et al.*, 2009; Brown *et al.*, 2011), to chlamydias (such as *C. trachomatis*) (Stephens *et al.*, 1998; Kalman *et al.*, 1999) and the spirochaetes (such as *Borrelia burgdorferi*) (Fraser *et al.*, 1997, 1998), as well as extreme thermophiles (*Aquifex aeolicus*) (Deckert *et al.*, 1998; Studholme *et al.*, 2000), the metal reducing bacterium *Geobacter sulfurreducens* (Leang *et al.*, 2009) and the low-GC Gram-positive bacteria (such as *Bacillus subtilis* and *Clostridium difficile*) (Kunst *et al.*, 1997; Shimizu *et al.*, 2002; Sebaihia *et al.*, 2006). Notably, σ^{54} also appears to be present within the phylogenetically distinct planctomycetes such as *Rhodopirellula baltica* (Glockner *et al.*, 2003) and *Planctomyces limnophilus* (Leary *et al.*, 1998). However, σ^{54} is notably absent in *Mycobacterium tuberculosis* (Cole *et al.*, 1998; Brosch *et al.*, 2000), *Streptococcus* (Ferretti *et al.*, 2001, 2004; Tettelin *et al.*, 2001) and *Staphylococcus* species (Takeuchi *et al.*, 2005; Chua *et al.*, 2010; Howden *et al.*, 2010; Tse *et al.*, 2010), as well as the specialized pathogen *Rickettsia prowazekii* (Andersson *et al.*, 1998) and *Mycoplasma pneumoniae* (Dandekar *et al.*, 2000). There is also no evidence of σ^{54} in the photosynthetic *Synechocystis* species (Kaneko *et al.*, 1996) and the thermophile *Thermotoga maritima* (Nelson *et al.*, 1999; Kyrpides *et al.*, 2000). Intriguingly, the genomes of several pathogenic neisseriae (*Neisseria meningitidis* and *N. gonorrhoeae*) harbour regions that partially resemble the *rpoN* gene sequence (so called *rpoN*-like sequences or RLS) (Laskos *et al.*, 1998). However, *N. meningitidis* and *N. gonorrhoeae* do not encode a functional σ^{54} protein. It appears that the RLS may have arisen from an ancestral *rpoN* homologue that underwent a deletion removing the sequence encoding the DNA binding helix–turn–helix (HTH) motif, and changing the subsequent reading frame (Laskos *et al.*, 1998).

Inspection of the genes regulated by σ^{54} does not reveal a unifying theme in terms of cellular function; rather σ^{54}-dependent transcription appears to respond to many different physiological or environmental cues, which are not obviously related. This repertoire of functions includes utilization of arginine and ornithine (*B. subtilis*) (Gardan *et al.*, 1997), ammonia assimilation (*G. sulfurreducens*) (Leang *et al.*, 2009), RNA modification (*E. coli*) (Genschik *et al.*, 1998) and repair (*E. coli*) (Tanaka and Shuman, 2011), chemotaxis (*Rhodospirillum centenum*), flagellation (*Campylobacter coli, E. coli, P. aeruginosa* and *Vibrio cholerae*), electron transport (*Alcaligenes dentrificans*), acid resistance (*E. coli*) (Riordan *et al.*, 2010), regulation of luminescence (*V. harveyi*) (Lilley and Bassler, 2000), quorum sensing (*P. aeruginosa*) (Heurlier *et al.*, 2003; Thompson *et al.*, 2003), biofilm formation (*V. fischeri*) (Visick, 2009; Yildiz and Visick, 2009), antifungal compound production (*P. fluorescens*) (Pechy-Tarr *et al.*, 2005) and expression of alternative σ factors (*P. syringae*) (Hendrickson *et al.*, 2000a,b; Hutcheson *et al.*, 2001; Jovanovic *et al.*, 2011). In *Caulobacter crescentus* the expression of σ^{54} (and the subsequent genes regulated by σ^{54}) is cell cycle regulated (Gober and Marques, 1995; Marques *et al.*, 1997). The *C. crescentus* cell cycle has a precisely controlled differentiation programme in which two unequal daughter cells, a motile swarmer cell and a sessile stalked cell, are generated (Gober and Marques, 1995). In *C. crescentus* σ^{54} is required for expression of late flagellar genes, which are temporally and spatially controlled such that they are only expressed in the swarmer pole of the pre-divisional cell (Gober and Marques, 1995; Marques *et al.*, 1997). Generally it appears that σ^{54} is not essential for survival and growth under favourable environmental conditions, except in the case of *Myxococcus xanthus*, where an *rpoN* null mutation is lethal (Keseler and Kaiser, 1997) and *G. sulfurreducens*, where *rpoN* null mutants could not be obtained owing to the essential nature of σ^{54} and its role on regulating a variety of cellular processes (Leang *et al.*, 2009). *M. xanthus* is seen as a model organism for self-association, but is also considered valuable owing to the large numbers of secondary metabolites it produces during predation, many of which are known to have medicinal properties, such as the antibiotic myxalamid (Gerth *et al.*, 1983). When starved of nutrients, *M. xanthus* undergoes a multicellular developmental programme that results in formation of fruiting bodies and spores (Keseler and Kaiser, 1997; Jakobsen *et al.*, 2004; Kaiser, 2004). Fruiting body formation requires *M. xanthus* to use both its A- and S-gliding motors, rather than rotary motors to drive spinning flagellar (Kaiser, 2003, 2004). S-motility is achieved by *M. xanthus* retracting its type IV pili at the leading pole (Kaiser, 1979; Nudleman and Kaiser, 2004), and A-motility is achieved via slime secretion at its trailing pole (Wolgemuth *et al.*, 2002). Nutrition state is assessed in part by the A-signal (but also potentially involving the stringent response), which enables *M. xanthus* to choose between (i) fruiting body development with differentiation of spores or (ii) slowed growth (at a rate compatible with the level of nutrients remaining) (Kaiser, 2004). Response to the A-signal requires σ^{54} and a number of different activator proteins, at least one, SasR, has been identified as having a role in fruiting body development (Kaiser *et al.*, 1985, 2004; Kroos *et al.*, 1986; Kuspa *et al.*, 1986; Keseler and Kaiser, 1995, 1997). Other σ^{54} regulated genes are required for sporulation, which too appear to be A-signal dependent (Caberoy *et al.*, 2003; Jakobsen *et al.*, 2004; Kaiser, 2004). Whilst extracellular A-signal initiates fruiting body development, subsequent morphological development requires the C-signal – which includes sporulation (and the change from rod-shaped to spheroid-shaped cells) (Kaiser, 2003, 2004; Kaiser and Welch, 2004). Further demonstrating the critical role of σ^{54} in *M. xanthus* development, the level of C-signal protein is itself regulated by σ^{54} (and the σ^{54} activator ActB) in a positive feedback loop (Kaiser, 2004). It has been postulated that the large number of developmental genes regulated by σ^{54} in *M. xanthus* could enable *M. xanthus* to respond to changing nutrient conditions in highly restricted cell cycle-dependent manner.

The role of σ^{54} in virulence gene expression and regulation

Although no obvious single theme unifies σ^{54}-dependent gene expression, all genes regulated by σ^{54} strictly require signal-responsive regulatory systems for their expression. This feature enables gene expression to be effectively 'switched on' only in response to a particular physiological or environmental cue and ensures effective management of cellular resources. The unique features arising from the failure of the Eσ^{54}-RPc to spontaneously isomerize to the RPo and regulatory advantages provided by σ^{54}-dependent gene expression may therefore be of particular significance when considering the role of σ^{54} in regulating expression of virulence and virulence-associated genes in bacterial pathogens (Wosten, 1998; Studholme and Buck, 2000a; Reitzer and Schneider, 2001; Kazmierczak *et al.*, 2005). Virulence and virulence-associated genes encode proteins that contribute to at least one aspect of bacterial pathobiology. Specifically, virulence genes are defined as essential for the bacterium to effectively establish an infection in the host organism. Virulence-associated genes contribute to bacterial survival in a particular environment or within the host organism; therefore activation of virulence-associated genes may enhance transmission (to new individuals) or survival (within a host organism). The many roles σ^{54} can play in plant and animal pathogens, and its contribution to virulence may provide clues about how to effectively manage pathogens. We present findings from several exemplar systems, where σ^{54} is required for the expression of virulence and/or virulence-associated genes in bacteria, but have not sought to be exhaustive.

Regulation of the general stress response

The ability to simply survive under a wide range of environmental conditions (including those found in different host systems) contributes to the transmission potential of a microbial pathogen. Survival under these (often rapidly) changing environmental conditions requires the timely and appropriate alterations in gene expression and protein activity. At the transcription level, alterations in expression patterns are mediated by different alternative σ factors, one of the key factors being σ^{38}, which is required to maintain cell viability during stationary phase (Lange and Hengge-Aronis, 1991). Regulation of σ^{38} expression and activity is complex (Hengge-Aronis, 1996, 2002a,b,c), and in *Borrelia burgdorferi*, the causative agent of Lyme disease, involves σ^{54} (Hubner *et al.*, 2001; Fisher *et al.*, 2005; Smith *et al.*, 2007; Ouyang *et al.*, 2008). *B. burgdorferi* has an intricate enzootic life cycle that involves ticks (*Ixodes scapularis*) and mammals (*Peromyscus leucopus*) (Lane *et al.*, 1991). Adaptation to these diverse host environments requires co-ordinated expression of numerous genes, many of which encode surface lipoproteins, such as the two major virulence genes *ospA* and *ospC* (Hubner *et al.*, 2001; Pal *et al.*, 2004; Yang *et al.*, 2004). Expression of these genes are controlled by σ^{54}, which directly regulates σ^{38} expression (Smith *et al.*, 2007) with the aid of the σ^{54} transcriptional regulator Rrp2 (in the so-called RpoN-RpoS-Rrp2 pathway) (Burtnick *et al.*, 2007; Boardman *et al.*, 2008; Ouyang *et al.*, 2008; Blevins *et al.*, 2009; Xu *et al.*, 2010) and another regulator BosR (Ouyang *et al.*, 2009, 2011). Microarray analyses suggest that more than 200 genes are controlled by σ^{54}-dependent promoters (independent of σ^{38}), the gene products of which are involved in the transmission process (Fisher *et al.*, 2005). However, the finding that BosR, alongside Rrp2, is essential for σ^{54}-dependent transcription of σ^{38} in this system (i.e. that an additional activator is required), may have broader significance in the understanding of σ^{54}-dependent gene expression in pathogenic bacteria (Ouyang *et al.*, 2011).

Regulation of cell envelope stress

The cell envelope of Gram-negative bacteria allows the passage of molecules in and out of the cell and includes structures such as flagella and secretion systems. Given the importance of this barrier, extracytoplasmic stress response systems have evolved to counteract any problems with the cell envelope. One such stress response is the phage shock protein (psp) system (Darwin,

2005; Joly *et al.*, 2010). Highly conserved in many Gram-negative pathogens, its expression (of the *pspABCDE* operon and the *pspG* gene) is dependent on σ^{54} and the transcriptional activator PspF (Weiner *et al.*, 1991; Jovanovic *et al.*, 1996). The Psp response functions to help maintain the energy state of the cell under stress (i.e. conserve the proton motive force) (Kleerebezem *et al.*, 1996; Model *et al.*, 1997 Jovanovic *et al.*, 2006; Darwin, 2007), where it is proposed that PspA serves to prevent proton leakage and maintain the integrity of the cell membrane (Kleerebezem *et al.*, 1996; Kobayashi *et al.*, 2007; Standar *et al.*, 2008). Notably, mutating components of this system (i.e. *pspC*) results in attenuated virulence and impaired growth in *Yersina enterocolitica* (Darwin and Miller, 1999, 2001). In *Salmonella enterica* sv. Typhimurium, PspA is required for virulence and divalent metal transport (Karlinsey *et al.*, 2010) and *psp* gene expression is highly up-regulated during macrophage infection (Eriksson *et al.*, 2003) and swarming (Wang *et al.*, 2004). Notably, *psp* expression is also up-regulated during biofilm formation in *E. coli* (Beloin *et al.*, 2004) and within *E. coli* persister cells (that form biofilm and develop multidrug tolerance) (Keren *et al.*, 2004). Since resistance to antibiotics and the stresses associated with persister cells is facilitated by energy production-dependent mechanisms (Ma *et al.*, 2009) and that some multidrug resistance pumps use the proton motive force to drive drug efflux, it could be inferred that the Psp response (and hence σ^{54}) has a role in microbial drug response.

Regulation of lifestyle

Many intracellular pathogens alternate between different lifestyles depending on the specific environmental conditions. *C. trachomatis* is an important human pathogen that exhibits a unique stage-specific developmental cycle involving the interconversion between two metabolically distinct forms: the infectious elementary body and the metabolically active reticulate body (Moulder, 1991). *C. trachomatis* contains three σ factors (Stephens *et al.*, 1998; Kalman *et al.*, 1999), σ^{66} (a homologue to the house-keeping σ^{70} factor), σ^{28} (also termed FliA) and σ^{54} that are expressed in a temporal developmental stage-specific fashion (Stephens *et al.*, 1998; Mathews *et al.*, 1999; Belland *et al.*, 2003; Nicholson *et al.*, 2003). σ^{54} is expressed late in the developmental cycle (Mathews *et al.*, 1999) and is proposed to regulate, potentially requiring the σ^{54} transcription activator CtcC, conversion of the reticulate body to the elementary body in response to energy levels and redox state (Mathews and Timms, 2000; Koo and Stephens, 2003). Sequence analysis has identified several putative σ^{54}-dependent promoters; however, the functions of the encoded gene products remain to be determined. It is also possible that chlamydiae have compensated for the absence of stress-responsive sigma factors (σ^{38} and σ^{24}), by adapting σ^{54} to respond to, for example, membrane stress.

Regulation of motility

Flagella and flagellar motility are important virulence and colonization factors in a wide range of pathogenic bacteria. The flagellum is a large protein complex encoded by the flagellar regulon genes and their expression are regulated by the alternative σ factor, σ^{28} (also known as the flagella σ factor, FliA) (Helmann, 1991). Multiple mechanisms exist within different bacterial species for regulating the activity of FliA (and hence flagellar gene expression). For example, in *S. enterica* serovar Typhimurium, FliA is regulated by the anti-σ factor FlgM (Gillen and Hughes, 1993). In some enteric bacteria FliA regulation involves σ^{54} and in *E. coli*, σ^{38} acts antagonistically to σ^{54} (Dong *et al.*, 2011). Examples of organisms in which σ^{54} has been shown to contribute to flagellar synthesis and thus virulence are discussed briefly below.

Campylobacter jejuni (a human pathogen) is the most common cause of food-borne illness in the developed world. In *C. jejuni*, σ^{54} is essential for flagellar secretory apparatus gene expression (Nachamkin *et al.*, 1993; Jagannathan *et al.*, 2001; Hendrixson and DiRita, 2003; Wosten *et al.*, 2004; Hendrixson, 2006; Kamal *et al.*, 2007) – and hence secretion of virulence (Cia) proteins (Konkel *et al.*, 2004; Young *et al.*, 2007; Dasti *et al.*, 2010). Microarray analysis has indicated that rod and hook genes associated with the 'middle' stage of the expression cascade (after completion of the earlier MS ring and export apparatus) are controlled by σ^{54} (Kamal *et al.*, 2007). Therefore, in the absence of *rpoN*, Cia protein secretion and host-cell invasion is abolished, adherence reduced and colonization activities impaired, presumably due to being non-flagellate (Fernando *et al.*, 2007; Kamal *et al.*, 2007). *Helicobacter pylori* is the principal cause of chronic inflammation of the mucous membranes of the stomach and peptide ulcers. Whilst it is known that several flagellar genes require σ^{54} and FlgR for their transcription (Suerbaum *et al.*, 1993; Spohn and Scarlato, 1999; Niehus *et al.*, 2004; Pereira *et al.*, 2006; Danielli *et al.*, 2010), it has yet to be conclusively established whether σ^{54} directly regulates σ^{28} (FliA) – despite the presence of a σ^{54} promoter consensus sequence upstream of the *fliA* gene (Niehus *et al.*, 2004; Douillard *et al.*, 2008). Notably, FlgR in *H. pylori* lacks a C-terminal DNA binding domain (Brahmachary *et al.*, 2004) and as such is the only σ^{54} transcriptional activator in this organism. The absence of a DNA binding domain removes the requirement for extensive regulatory regions upstream of the promoter site [hence the upstream activator sequence designates which σ^{54} activator regulates a specific σ^{54}-dependent promoter(s) and hence gene product(s)]. These sites are usually associated with a specific σ^{54} transcriptional activator that serves to regulate the activity of the σ^{54}-dependent promoter (and therefore the associated genes/gene products) in response to changing environmental conditions. Given that CtcC (another σ^{54} transcriptional activator) in *Chlamydia trachomatis* also lacks a C-terminal DNA binding domain (Koo and Stephens, 2003), it would appear that this unusual structural arrangement is not unique and should be considered of regulatory importance.

Legionella pneumophila, the causative agent of legionnaires' disease, is a ubiquitous motile bacterium inhabiting man-made and natural water systems. *L. pneumophila* alternates between two distinct life cycles: an intracellular replicative phase and an infectious transmissive phase (Molofsky and Swanson, 2004). Studies have demonstrated that the flagellum, whilst positively affecting infection, is not needed for intracellular replication (Dietrich *et al.*, 2001); yet is required for 'full' fitness (Heuner *et al.*, 2002). Colonization by the opportunistic pathogen, *P. aeruginosa*, requires formation of the type IV pili (associated with so-called twitching motility), which, like the flagellar regulon, is regulated by σ^{54}, the transcriptional activator FleR and the master regulator FleQ (Totten *et al.*, 1990; Ramphal *et al.*, 1991; Goldberg and Dahnke, 1992; Arora *et al.*, 1997, 1998; Potvin *et al.*, 2008). *rpoN* mutants are defective in cell adhesion and motility, both of which are attributed to disruption of the pili and flagella (Ramphal *et al.*, 1991; Arora *et al.*, 1997, 1998; Jyot *et al.*, 2002; Potvin *et al.*, 2008). Unsurprisingly, other bacterial species (such as *Moraxella lacunata* and *Moraxella bovis* that are human and bovine pathogens, respectively, which primarily infect the conjunctiva and the cornea, and *Kingella kingae*, an important cause of septic arthritis and endocarditis in children) have also been shown to require σ^{54} to form type IV pili, which are used for infection (Heinrich and Glasgow, 1997; Kehl-Fie *et al.*, 2009).

The flagellar genes of *V. cholerae*, which causes the human diarrhoeal disease cholera, are transcribed in a four-tiered transcription hierarchy, with class II and III genes regulated by σ^{54} and the transcriptional activator FlrA (Klose and Mekalanos, 1998; Prouty *et al.*, 2001; Syed *et al.*, 2009). In *V. cholerae*, σ^{54} also regulates expression of the membrane proteins FlgO and FlgP that function in motility, attachment and colonization (Martinez *et al.*, 2009), and expression of virulence factors (Pukatzki *et al.*, 2006). In *V.* (*Listonella*) *anguillarium*, the cause of vibriosis in fish, σ^{54} also regulates flagellum synthesis and, by inference, virulence (Milton *et al.*, 1996; O'Toole *et al.*, 1996, 1997). However, in *V.* (*Listonella*) *anguillarium*, σ^{54} does not appear to regulate any factors required for virulence subsequent to penetration of the fish epithelium, but is important for infection (O'Toole *et al.*, 1997). Given the observation that deletion of *rpoN* results in aflagellate cells with reduced colonization (Klose and Mekalanos, 1998; Klose, 2001; Martinez *et al.*, 2009) in a number of species, it could be inferred that flagellar regulation, rather than motility *per se* affects virulence.

Regulation of secretion (and secreted virulence factors)

Secretion is used by both prokaryotes and eukaryotes to move proteins, DNA and chemical molecules across cell walls and membranes. Gram-negative bacteria have six types of secretion systems, the most extensively studied of which is the type III secretion (T3SS) system. The T3SS (which exhibits sequence similarities with the basal bodies and assembly machinery of flagella; Blocker *et al.*, 2003) is central to the pathogenesis of numerous important clinical pathogens of humans and animals (for example *S. enterica*, *Shigella flexneri*, *Burkholderia mallei*, *Y. pestis*, *E. coli* and *P. aeruginosa*) and to pathogens of crops and ornamental plants (for example *P. syringae*, *Erwinia* spp., *Pantoea* spp. and *Dickeya* spp.). The type III secretion system comprises more than 20 proteins and forms a needle-like structure, which spans the inner membrane, periplasm and outer membrane and is used to secrete virulence factors that are required for host cell infection (Mole *et al.*, 2007). In Group I type plant pathogens (*P. syringae*, *Erwinia* spp., *Pantoea* spp., and *Dickeya* spp.) regulated expression of the majority of T3SS structural and regulatory proteins as well as effectors and other virulence factors, is dependent on the alternative σ factor, HrpL (Xiao *et al.*, 1994; Wei and Beer, 1995). In all these groups expression of *hrpL* has shown to be σ^{54}-dependent, which in *Erwinia* spp. requires the σ^{54} activator HrpS, and in *P. syringae* requires the strictly co-dependent HrpR and HrpS system (Frederick *et al.*, 1993; Wei and Beer, 1995; Hendrickson *et al.*, 2000a; Hutcheson *et al.*, 2001; Chatterjee *et al.*, 2002; Jovanovic *et al.*, 2011). Notably, homologues of HrpRS and their regulatory control proteins have also been identified in *P. mendocina* – a *P. aeruginosa*-related human pathogen (Anzai *et al.*, 2000) – implying that σ^{54} may also regulate T3SS (and hence infection) in these organisms (Jovanovic *et al.*, 2011). It is also clear that HrpL regulates or coordinates expression of other T3SS-independent virulence and growth-promoting factors, which in *P. syringae* include a phytotoxin biosynthesis homologue (indoleacetate-lysine ligase) and a regulator of coronatine biosynthesis (CorR) (Vencato *et al.*, 2006). Notably, the role, if any, of σ^{54} in the regulation of other T3SS gene expression in other organisms has yet to be determined, it certainly does not appear to be required for pathogenicity in *X. campestris* pv. *Vesicatoria* (Horns and Bonas, 1996). However, in *E. coli*, microarray analysis of an *rpoN* mutant, suggests that σ^{54} regulates locus of enterocyte effacement (LEE – a T3SS) expression, and contributes to acid resistance (Riordan *et al.*, 2010).

Recently, a new mechanism for protein transport across membranes, the type VI secretion system (T6SS), has been described for Gram-negative pathogens such as *P. aeruginosa* and *V. cholerae* (Pukatzki *et al.*, 2006; Filloux *et al.*, 2008). Bioinformatics and mutagenesis approaches have identified that T6SS gene expression is regulated by σ^{54} and associated transcriptional activators, such as VasH in *V. cholerae* (Pukatzki *et al.*, 2006; Miyata *et al.*, 2010; Bernard *et al.*, 2011); deletion of *rpoN* in *V. cholerae* resulted in attenuated infection and an inability to kill the bacterial competitor *Dictyostelium discoideum* (Pukatzki *et al.*, 2006). The T6SS genes have also been shown to be regulated by quorum sensing (Ishikawa *et al.*, 2009; Khajanchi *et al.*, 2009; Lesic *et al.*, 2009), a mechanism also employed in T3SS regulation (Mole *et al.*, 2007), implying a complex interplay between different regulatory pathways involving σ^{54} and its cognate activator proteins is central to regulated gene expression and virulence.

Contribution to the intrinsic resistome of bacteria

The intrinsic resistome is defined as the ensemble of genes that contribute to the intrinsic resistance bacteria have to an environment and whose presence in strains of a bacterial species is not the result of recent antibiotic exposure or the result of horizontal gene transfer (Fajardo *et al.*, 2008). In *E. coli* the absence of σ^{54} results in increased susceptibility to the antibiotics gentamicin, tobramycin, neomycin, cephradine, cefoxitin and colistin (Liu *et al.*, 2010), and increased fitness in the presence of the RNAp targeting antibiotic rifampicin (Nichols *et al.*, 2011) – implying that σ^{54} can alter the sensitivity of RNAp to antibiotics. In *Listeria monocytogenes* (a food-borne, facultative pathogen) σ^{54} has been shown to alter the sensitivity of *L. monocytogenes* to antibacterial peptides, specifically the subclass IIa bacteriocin mesentericin Y105 (Robichon *et al.*, 1997; Dalet *et al.*, 2001; Hechard *et al.*, 2001), potentially via changes in the cell membrane (Vadyvaloo *et al.*, 2004; Okada *et al.*, 2006). A similar role for σ^{54}

in susceptibility of *Enterococcus faecalis* to mesentricin Y105 has also been observed (Dalet *et al.*, 2001; Hechard *et al.*, 2001). In *P. aeruginosa*, σ^{54} has also been shown to contribute to tolerance to quinolones and carbapenems antimicrobials, potentially connected to increased pyoverdine synthesis and *vqsR* gene expression (Viducic *et al.*, 2007). Clearly, the intrinsic resistome and the ability to evolve higher levels of resistance to antibiotics via non-classical antibiotic resistance genes involves a complex network of elements that could include σ^{54}.

Final perspectives

Although the mechanistic basis of gene activation by σ^{54} transcriptional activators is fairly well advanced, rather less is known about the specific upstream signals that trigger the activity of these activator proteins. Although exceptions exist, for example in the enteric *ntr* system, the *xyl* and related catabolic systems in pseudomonads, those signals that reflect the recognition of a host environment in pathogen–host interactions are not generally known. Defining environmental conditions for triggering the activity of σ^{54} transcriptional activators now needs to be linked to identifying the specific signalling molecules. Whether such signalling molecules in themselves reflect an integration of more than one environmental cue, or whether the promoter architecture is used to integrate several cues is also not well established outside of the ntr-crp (nitrogen–carbon) relationship studied at a few enteric promoters controlled by NtrC (Tian *et al.*, 2001; Huo *et al.*, 2006; Mao *et al.*, 2007).

Ultimately the gene sets described above will fall into simple or more complex regulons. Linking these implied networks is a challenge, as is arriving at the description of the originating signals that trigger expression. Systems approaches to genome-wide expression, the metabolome and proteome will help unravel the pathways and their relationships. Potentially, antimicrobials may emerge based on interfering with important adaptive responses mediated by σ^{54}, or through exploiting its ability to silence RNAp as a stable closed promoter complex. The contribution of *rpoN* to the cellular localization of key proteins may underpin successful pathogens (Christen *et al.*, 2010).

Acknowledgements

The authors would like to thank Dr Christoph Engl for comments and critical reading of this review. The work in SW and MB laboratories is supported by grants from Biotechnology and Biological Sciences Research Council (BBSRC), Wellcome Trust and Leverhulme Trust. SW is a recipient of a BBSRC David Phillips Fellowship (BB/E023703).

References

Andersson, S.G., Zomorodipour, A., Andersson, J.O., Sicheritz-Ponten, T., Alsmark, U.C., Podowski, R.M., Naslund, A.K., Eriksson, A.S., Winkler, H.H., and Kurland, C.G. (1998). The genome sequence of *Rickettsia prowazekii* and the origin of mitochondria. Nature *396*, 133–140.

Anzai, Y., Kim, H., Park, J.Y., Wakabayashi, H., and Oyaizu, H. (2000). Phylogenetic affiliation of the pseudomonads based on 16S rRNA sequence. Int. J. Syst. Evol. Microbiol. *50*, 1563–1589.

Arora, S.K., Ritchings, B.W., Almira, E.C., Lory, S., and Ramphal, R. (1997). A transcriptional activator, FleQ, regulates mucin adhesion and flagellar gene expression in *Pseudomonas aeruginosa* in a cascade manner. J. Bacteriol. *179*, 5574–5581.

Arora, S.K., Ritchings, B.W., Almira, E.C., Lory, S., and Ramphal, R. (1998). The *Pseudomonas aeruginosa* flagellar cap protein, FliD, is responsible for mucin adhesion. Infect. Immun. *66*, 1000–1007.

Barrios, H., Valderrama, B., and Morett, E. (1999). Compilation and analysis of sigma(54)-dependent promoter sequences. Nucleic Acids Res. *27*, 4305–4313.

Belland, R.J., Zhong, G., Crane, D.D., Hogan, D., Sturdevant, D., Sharma, J., Beatty, W.L., and Caldwell, H.D. (2003). Genomic transcriptional profiling of the developmental cycle of *Chlamydia trachomatis*. Proc. Natl. Acad. Sci. U.S.A. *100*, 8478–8483.

Beloin, C., Valle, J., Latour-Lambert, P., Faure, P., Kzreminski, M., Balestrino, D., Haagensen, J.A., Molin, S., Prensier, G., Arbeille, B., *et al.* (2004). Global impact of mature biofilm lifestyle on *Escherichia coli* K-12 gene expression. Mol. Microbiol. *51*, 659–674.

Bernard, C.S., Brunet, Y.R., Gavioli, M., Lloubes, R., and Cascales, E. (2011). Regulation of Type VI Secretion gene clusters by sigma54 and cognate enhancer binding proteins. J. Bacteriol. *193*, 2158–2167.

Bernardo, L.M., Johansson, L.U., Solera, D., Skarfstad, E., and Shingler, V. (2006). The guanosine tetraphosphate (ppGpp) alarmone, DksA and promoter affinity for RNA polymerase in regulation of sigma-dependent transcription. Mol. Microbiol. *60*, 749–764.

Bernardo, L.M., Johansson, L.U., Skarfstad, E., and Shingler, V. (2009). sigma54-promoter discrimination and regulation by ppGpp and DksA. J. Biol. Chem. *284*, 828–838.

Blattner, F.R., Plunkett, G., 3rd, Bloch, C.A., Perna, N.T., Burland, V., Riley, M., Collado-Vides, J., Glasner, J.D., Rode, C.K., Mayhew, G.F., *et al.* (1997). The complete genome sequence of *Escherichia coli* K-12. Science *277*, 1453–1462.

Blevins, J.S., Xu, H., He, M., Norgard, M.V., Reitzer, L., and Yang, X.F. (2009). Rrp2, a sigma54-dependent transcriptional activator of *Borrelia burgdorferi*, activates rpoS in an enhancer-independent manner. J. Bacteriol. *191*, 2902–2905.

Blocker, A., Komoriya, K., and Aizawa, S. (2003). Type III secretion systems and bacterial flagella: insights into their function from structural similarities. Proc. Natl. Acad. Sci. U.S.A. *100*, 3027–3030.

Boardman, B.K., He, M., Ouyang, Z., Xu, H., Pang, X., and Yang, X.F. (2008). Essential role of the response regulator Rrp2 in the infectious cycle of *Borrelia burgdorferi*. Infect. Immun. *76*, 3844–3853.

Borukhov, S., and Severinov, K. (2002). Role of the RNA polymerase sigma subunit in transcription initiation. Res. Microbiol. *153*, 557–562.

Brahmachary, P., Dashti, M.G., Olson, J.W., and Hoover, T.R. (2004). *Helicobacter pylori* FlgR is an enhancer-independent activator of sigma54-RNA polymerase holoenzyme. J. Bacteriol. *186*, 4535–4542.

Brosch, R., Gordon, S.V., Pym, A., Eiglmeier, K., Garnier, T., and Cole, S.T. (2000). Comparative genomics of the mycobacteria. Int. J. Med. Microbiol. *290*, 143–152.

Brown, S.D., Gilmour, C.C., Kucken, A.M., Wall, J.D., Elias, D.A., Brandt, C.C., Podar, M., Chertkov, O., Held, B., Bruce, D.C., *et al.* (2011). Genome sequence of the mercury-methylating strain *Desulfovibrio desulfuricans* ND132. J. Bacteriol. *193*, 2078–2079.

Browning, D.F., and Busby, S.J. (2004). The regulation of bacterial transcription initiation. Nat. Rev. Microbiol. *2*, 57–65.

Buck, M. (1986). Deletion analysis of the Klebsiella pneumoniae nitrogenase promoter: importance of spacing between conserved sequences around positions −12 and −24 for activation by the nifA and ntrC (glnG) products. J. Bacteriol. *166*, 545–551.

Buck, M., Gallegos, M.T., Studholme, D.J., Guo, Y., and Gralla, J.D. (2000). The bacterial enhancer-dependent sigma(54) (sigma(N)) transcription factor. J. Bacteriol. *182*, 4129–4136.

Buck, M., Bose, D., Burrows, P., Cannon, W., Joly, N., Pape, T., Rappas, M., Schumacher, J., Wigneshweraraj, S., and Zhang, X. (2006). A second paradigm for gene activation in bacteria. Biochem. Soc. Trans. *34*, 1067–1071.

Burrows, P.C., Severinov, K., Ishihama, A., Buck, M., and Wigneshweraraj, S.R. (2003). Mapping sigma 54-RNA polymerase interactions at the −24 consensus promoter element. J. Biol. Chem. *278*, 29728–29743.

Burrows, P.C., Severinov, K., Buck, M., and Wigneshweraraj, S.R. (2004). Reorganisation of an RNA polymerase-promoter DNA complex for DNA melting. Embo J. *23*, 4253–4263.

Burtnick, M.N., Downey, J.S., Brett, P.J., Boylan, J.A., Frye, J.G., Hoover, T.R., and Gherardini, F.C. (2007). Insights into the complex regulation of rpoS in *Borrelia burgdorferi*. Mol. Microbiol. *65*, 277–293.

Caberoy, N.B., Welch, R.D., Jakobsen, J.S., Slater, S.C., and Garza, A.G. (2003). Global mutational analysis of NtrC-like activators in *Myxococcus xanthus*: identifying activator mutants defective for motility and fruiting body development. J. Bacteriol. *185*, 6083–6094.

Cannon, W.V., Gallegos, M.T., and Buck, M. (2000). Isomerization of a binary sigma-promoter DNA complex by transcription activators. Nat. Struct. Biol. *7*, 594–601.

Chatterjee, A., Cui, Y., and Chatterjee, A.K. (2002). *Regulation of Erwinia carotovora* hrpL(Ecc) (sigma-L(Ecc)), which encodes an extracytoplasmic function subfamily of sigma factor required for expression of the HRP regulon. Mol. Plant Microbe Interact. *15*, 971–980.

Christen, B., Fero, M.J., Hillson, N.J., Bowman, G., Hong, S.H., Shapiro, L., and McAdams, H.H. (2010). High-throughput identification of protein localization dependency networks. Proc. Natl. Acad. Sci. U.S.A. *107*, 4681–4686.

Chua, K., Seemann, T., Harrison, P.F., Davies, J.K., Coutts, S.J., Chen, H., Haring, V., Moore, R., Howden, B.P., and Stinear, T.P. (2010). Complete genome sequence of *Staphylococcus* aureus strain JKD6159, a unique Australian clone of ST93-IV community methicillin-resistant *Staphylococcus* aureus. J. Bacteriol. *192*, 5556–5557.

Cole, S.T., Brosch, R., Parkhill, J., Garnier, T., Churcher, C., Harris, D., Gordon, S.V., Eiglmeier, K., Gas, S., Barry, C.E., 3rd, *et al.* (1998). Deciphering the biology of Mycobacterium tuberculosis from the complete genome sequence. Nature *393*, 537–544.

Costanzo, A., and Ades, S.E. (2006). Growth phase-dependent regulation of the extracytoplasmic stress factor, sigmaE, by guanosine 3′,5′-bispyrophosphate (ppGpp). J. Bacteriol. *188*, 4627–4634.

Costanzo, A., Nicoloff, H., Barchinger, S.E., Banta, A.B., Gourse, R.L., and Ades, S.E. (2008). ppGpp and DksA likely regulate the activity of the extracytoplasmic stress factor sigmaE in *Escherichia coli* by both direct and indirect mechanisms. Mol. Microbiol. *67*, 619–632.

Dalet, K., Cenatiempo, Y., Cossart, P., and Hechard, Y. (2001). A sigma(54)-dependent PTS permease of the mannose family is responsible for sensitivity of *Listeria monocytogenes* to mesentericin Y105. Microbiology *147*, 3263–3269.

Dandekar, T., Huynen, M., Regula, J.T., Ueberle, B., Zimmermann, C.U., Andrade, M.A., Doerks, T., Sanchez-Pulido, L., Snel, B., Suyama, M., *et al.* (2000). Re-annotating the *Mycoplasma pneumoniae* genome sequence: adding value, function and reading frames. Nucleic Acids Res. *28*, 3278–3288.

Danielli, A., Amore, G., and Scarlato, V. (2010). Built shallow to maintain homeostasis and persistent infection: insight into the transcriptional regulatory network of the gastric human pathogen *Helicobacter pylori*. PLoS Pathog. *6*, e1000938.

Darwin, A.J. (2005). The phage-shock-protein response. Mol. Microbiol. *57*, 621–628.

Darwin, A.J. (2007). Regulation of the phage-shock-protein stress response in *Yersinia enterocolitica*. Adv. Exp. Med. Biol. *603*, 167–177.

Darwin, A.J., and Miller, V.L. (1999). Identification of *Yersinia enterocolitica* genes affecting survival in an animal host using signature-tagged transposon mutagenesis. Mol. Microbiol. *32*, 51–62.

Darwin, A.J., and Miller, V.L. (2001). The psp locus of *Yersinia enterocolitica* is required for virulence and for growth *in vitro* when the Ysc type III secretion system is produced. Mol. Microbiol. *39*, 429–444.

Dasti, J.I., Tareen, A.M., Lugert, R., Zautner, A.E., and Gross, U. (2010). *Campylobacter jejuni*: a brief overview on pathogenicity-associated factors and disease-mediating mechanisms. Int. J. Med. Microbiol. *300*, 205–211.

Datwyler, S.A., and Meares, C.F. (2000). Protein–protein interactions mapped by artificial proteases: where sigma factors bind to RNA polymerase. Trends Biochem. Sci. *25*, 408–414.

Deckert, G., Warren, P.V., Gaasterland, T., Young, W.G., Lenox, A.L., Graham, D.E., Overbeek, R., Snead, M.A., Keller, M., Aujay, M., *et al.* (1998). The complete genome of the hyperthermophilic bacterium *Aquifex aeolicus*. Nature *392*, 353–358.

DelVecchio, V.G., Kapatral, V., Redkar, R.J., Patra, G., Mujer, C., Los, T., Ivanova, N., Anderson, I., Bhattacharyya, A., Lykidis, A., *et al.* (2002). The genome sequence of the facultative intracellular pathogen *Brucella melitensis*. Proc. Natl. Acad. Sci. U.S.A. *99*, 443–448.

Dietrich, C., Heuner, K., Brand, B.C., Hacker, J., and Steinert, M. (2001). Flagellum of *Legionella pneumophila* positively affects the early phase of infection of eukaryotic host cells. Infect. Immun. *69*, 2116–2122.

Dong, T., Yu, R., and Schellhorn, H. (2011). Antagonistic regulation of motility and transcriptome expression by RpoN and RpoS in *Escherichia coli*. Mol. Microbiol. *79*, 375–386.

Douillard, F.P., Ryan, K.A., Caly, D.L., Hinds, J., Witney, A.A., Husain, S.E., and O'Toole, P.W. (2008). Posttranscriptional regulation of flagellin synthesis in *Helicobacter pylori* by the RpoN chaperone HP0958. J. Bacteriol. *190*, 7975–7984.

Eriksson, S., Lucchini, S., Thompson, A., Rhen, M., and Hinton, J.C. (2003). Unravelling the biology of macrophage infection by gene expression profiling of intracellular *Salmonella* enterica. Mol. Microbiol. *47*, 103–118.

Fajardo, A., Martinez-Martin, N., Mercadillo, M., Galan, J.C., Ghysels, B., Matthijs, S., Cornelis, P., Wiehlmann, L., Tummler, B., Baquero, F., and Martinez, J.L. (2008). The neglected intrinsic resistome of bacterial pathogens. PLoS One 3, e1619.

Fernando, U., Biswas, D., Allan, B., Willson, P., and Potter, A.A. (2007). Influence of *Campylobacter jejuni* fliA, rpoN and flgK genes on colonization of the chicken gut. Int. J. Food Microbiol. *118*, 194–200.

Ferretti, J.J., Ajdic, D., and McShan, W.M. (2004). Comparative genomics of streptococcal species. Indian J. Med. Res. *119*, 1–6.

Ferretti, J.J., McShan, W.M., Ajdic, D., Savic, D.J., Savic, G., Lyon, K., Primeaux, C., Sezate, S., Suvorov, A.N., Kenton, S., *et al.* (2001). Complete genome sequence of an M1 strain of *Streptococcus* pyogenes. Proc. Natl. Acad. Sci. U.S.A. *98*, 4658–4663.

Filloux, A., Hachani, A., and Bleves, S. (2008). The bacterial type VI secretion machine: yet another player for protein transport across membranes. Microbiology *154*, 1570–1583.

Fisher, M.A., Grimm, D., Henion, A.K., Elias, A.F., Stewart, P.E., Rosa, P.A., and Gherardini, F.C. (2005). *Borrelia burgdorferi* sigma54 is required for mammalian infection and vector transmission but not for tick colonization. Proc. Natl. Acad. Sci. U.S.A. *102*, 5162–5167.

Fraser, C.M., Casjens, S., Huang, W.M., Sutton, G.G., Clayton, R., Lathigra, R., White, O., Ketchum, K.A., Dodson, R., Hickey, E.K., *et al.* (1997). Genomic sequence of a Lyme disease spirochaete, *Borrelia burgdorferi*. Nature *390*, 580–586.

Fraser, C.M., Norris, S.J., Weinstock, G.M., White, O., Sutton, G.G., Dodson, R., Gwinn, M., Hickey, E.K., Clayton, R., Ketchum, K.A., *et al.* (1998). Complete genome sequence of *Treponema pallidum*, the syphilis spirochete. Science *281*, 375–388.

Frederick, R.D., Majerczak, D.R., and Coplin, D.L. (1993). *Erwinia stewartii* WtsA, a positive regulator of pathogenicity gene expression, is similar to *Pseudomonas syringae* pv. phaseolicola HrpS. Mol. Microbiol. *9*, 477–485.

Gardan, R., Rapoport, G., and Debarbouille, M. (1997). Role of the transcriptional activator RocR in the arginine-degradation pathway of *Bacillus subtilis*. Mol. Microbiol. *24*, 825–837.

Genschik, P., Drabikowski, K., and Filipowicz, W. (1998). Characterization of the *Escherichia coli* RNA 3'-terminal phosphate cyclase and its sigma54-regulated operon. J. Biol. Chem. *273*, 25516–25526.

Gerth, K., Jansen, R., Reifenstahl, G., Hofle, G., Irschik, H., Kunze, B., Reichenbach, H., and Thierbach, G. (1983). The myxalamids, new antibiotics from *Myxococcus xanthus* (Myxobacterales). I. Production, physico-chemical and biological properties, and mechanism of action. J. Antibiot. (Tokyo) *36*, 1150–1156.

Ghosh, T., Bose, D., and Zhang, X. (2010). Mechanisms for activating bacterial RNA polymerase. FEMS Microbiol. Rev. *34*, 611–627.

Gillen, K.L., and Hughes, K.T. (1993). Transcription from two promoters and autoregulation contribute to the control of expression of the *Salmonella* typhimurium flagellar regulatory gene flgM. J. Bacteriol. *175*, 7006–7015.

Glockner, F.O., Kube, M., Bauer, M., Teeling, H., Lombardot, T., Ludwig, W., Gade, D., Beck, A., Borzym, K., Heitmann, K., *et al.* (2003). Complete genome sequence of the marine planctomycete

Pirellula sp. strain 1. Proc. Natl. Acad. Sci. U.S.A. *100*, 8298–8303.

Gober, J.W., and Marques, M.V. (1995). Regulation of cellular differentiation in *Caulobacter crescentus*. Microbiol. Rev. *59*, 31–47.

Goldberg, J.B., and Dahnke, T. (1992). *Pseudomonas aeruginosa* AlgB, which modulates the expression of alginate, is a member of the NtrC subclass of prokaryotic regulators. Mol. Microbiol. *6*, 59–66.

Gopel, Y., Luttmann, D., Heroven, A.K., Reichenbach, B., Dersch, P., and Gorke, B. (2011). Common and divergent features in transcriptional control of the homologous small RNAs GlmY and GlmZ in Enterobacteriaceae. Nucleic Acids Res. *39*, 1294–1309.

Gruber, T.M., and Gross, C.A. (2003). Multiple sigma subunits and the partitioning of bacterial transcription space. Annu. Rev. Microbiol. *57*, 441–466.

Guo, Y., Wang, L., and Gralla, J.D. (1999). A fork junction DNA-protein switch that controls promoter melting by the bacterial enhancer-dependent sigma factor. Embo J. *18*, 3736–3745.

Guo, Y., Lew, C.M., and Gralla, J.D. (2000). Promoter opening by sigma(54) and sigma(70) RNA polymerases: sigma factor-directed alterations in the mechanism and tightness of control. Genes Dev. *14*, 2242–2255.

Haugen, S.P., Ross, W., and Gourse, R.L. (2008). Advances in bacterial promoter recognition and its control by factors that do not bind DNA. Nat. Rev. Microbiol. *6*, 507–519.

Hechard, Y., Pelletier, C., Cenatiempo, Y., and Frere, J. (2001). Analysis of sigma(54)-dependent genes in Enterococcus faecalis: a mannose PTS permease (EII(Man)) is involved in sensitivity to a bacteriocin, mesentericin Y105. Microbiology *147*, 1575–1580.

Heinrich, D.W., and Glasgow, A.C. (1997). Transcriptional regulation of type 4 pilin genes and the site-specific recombinase gene, piv, in *Moraxella lacunata* and *Moraxella bovis*. J. Bacteriol. *179*, 7298–7305.

Helmann, J.D. (1991). Alternative sigma factors and the regulation of flagellar gene expression. Mol. Microbiol. *5*, 2875–2882.

Helmann, J.D. (1999). Anti-sigma factors. Curr. Opin. Microbiol. 2, 135–141.

Hendrickson, E.L., Guevera, P., and Ausubel, F.M. (2000a). The alternative sigma factor RpoN is required for hrp activity in *Pseudomonas syringae* pv. *maculicola* and acts at the level of hrpL transcription. J. Bacteriol. *182*, 3508–3516.

Hendrickson, E.L., Guevera, P., Penaloza-Vazquez, A., Shao, J., Bender, C., and Ausubel, F.M. (2000b). Virulence of the phytopathogen *Pseudomonas syringae* pv. *maculicola* is rpoN dependent. J. Bacteriol. *182*, 3498–3507.

Hendrixson, D.R. (2006). A phase-variable mechanism controlling the *Campylobacter jejuni* FlgR response regulator influences commensalism. Mol. Microbiol. *61*, 1646–1659.

Hendrixson, D.R., and DiRita, V.J. (2003). Transcription of sigma54-dependent but not sigma28-dependent flagellar genes in *Campylobacter jejuni* is associated with formation of the flagellar secretory apparatus. Mol. Microbiol. *50*, 687–702.

Hengge-Aronis, R. (1996). Back to log phase: sigma S as a global regulator in the osmotic control of gene expression in *Escherichia coli*. Mol. Microbiol. *21*, 887–893.

Hengge-Aronis, R. (2002a). Recent insights into the general stress response regulatory network in *Escherichia coli*. J. Mol. Microbiol. Biotechnol. *4*, 341–346.

Hengge-Aronis, R. (2002b). Signal transduction and regulatory mechanisms involved in control of the sigma(S) (RpoS) subunit of RNA polymerase. Microbiol. Mol. Biol. Rev. *66*, 373–395, table of contents.

Hengge-Aronis, R. (2002c). Stationary phase gene regulation: what makes an *Escherichia coli* promoter sigmaS-selective? Curr. Opin. Microbiol. *5*, 591–595.

Heuner, K., Dietrich, C., Skriwan, C., Steinert, M., and Hacker, J. (2002). Influence of the alternative sigma(28) factor on virulence and flagellum expression of *Legionella pneumophila*. Infect. Immun. *70*, 1604–1608.

Heurlier, K., Denervaud, V., Pessi, G., Reimmann, C., and Haas, D. (2003). Negative control of quorum sensing by RpoN (sigma54) in *Pseudomonas aeruginosa* PAO1. J. Bacteriol. *185*, 2227–2235.

Holden, M.T., Seth-Smith, H.M., Crossman, L.C., Sebaihia, M., Bentley, S.D., Cerdeno-Tarraga, A.M., Thomson, N.R., Bason, N., Quail, M.A., Sharp, S., *et al.* (2009). The genome of *Burkholderia cenocepacia* J2315, an epidemic pathogen of cystic fibrosis patients. J. Bacteriol. *191*, 261–277.

Horns, T., and Bonas, U. (1996). The rpoN gene of *Xanthomonas campestris* pv. *vesicatoria* is not required for pathogenicity. Mol. Plant Microbe Interact. *9*, 856–859.

Howden, B.P., Seemann, T., Harrison, P.F., McEvoy, C.R., Stanton, J.A., Rand, C.J., Mason, C.W., Jensen, S.O., Firth, N., Davies, J.K., *et al.* (2010). Complete genome sequence of *Staphylococcus* aureus strain JKD6008, an ST239 clone of methicillin-resistant *Staphylococcus aureus* with intermediate-level vancomycin resistance. J. Bacteriol. *192*, 5848–5849.

Hubner, A., Yang, X., Nolen, D.M., Popova, T.G., Cabello, F.C., and Norgard, M.V. (2001). Expression of *Borrelia burgdorferi* OspC and DbpA is controlled by a RpoN-RpoS regulatory pathway. Proc. Natl. Acad. Sci. U.S.A. *98*, 12724–12729.

Huo, Y.X., Tian, Z.X., Rappas, M., Wen, J., Chen, Y.C., You, C.H., Zhang, X., Buck, M., Wang, Y.P., and Kolb, A. (2006). Protein-induced DNA bending clarifies the architectural organization of the sigma54-dependent glnAp2 promoter. Mol. Microbiol. *59*, 168–180.

Hutcheson, S.W., Bretz, J., Sussan, T., Jin, S., and Pak, K. (2001). Enhancer-binding proteins HrpR and HrpS interact to regulate hrp-encoded type III protein secretion in *Pseudomonas syringae* strains. J. Bacteriol. *183*, 5589–5598.

Ishihama, A. (2000). Functional modulation of *Escherichia coli* RNA polymerase. Annu. Rev. Microbiol. *54*, 499–518.

Ishihama, A. (2010). Prokaryotic genome regulation: multifactor promoters, multitarget regulators and hierarchic networks. FEMS Microbiol. Rev. *34*, 628–645.

Ishikawa, T., Rompikuntal, P.K., Lindmark, B., Milton, D.L., and Wai, S.N. (2009). Quorum sensing regulation of the two hcp alleles in *Vibrio cholerae* O1 strains. PLoS One *4*, e6734.

Jagannathan, A., Constantinidou, C., and Penn, C.W. (2001). Roles of rpoN, fliA, and flgR in expression of flagella in *Campylobacter jejuni*. J. Bacteriol. *183*, 2937–2942.

Jakobsen, J.S., Jelsbak, L., Jelsbak, L., Welch, R.D., Cummings, C., Goldman, B., Stark, E., Slater, S., and Kaiser, D. (2004). Sigma54 enhancer binding proteins and *Myxococcus xanthus* fruiting body development. J. Bacteriol. *186*, 4361–4368.

Jishage, M., and Ishihama, A. (1997). Variation in RNA polymerase sigma subunit composition within different stocks of *Escherichia coli* W3110. J. Bacteriol. *179*, 959–963.

Jishage, M., Iwata, A., Ueda, S., and Ishihama, A. (1996). Regulation of RNA polymerase sigma subunit synthesis in *Escherichia coli*: intracellular levels of four species of sigma subunit under various growth conditions. J. Bacteriol. *178*, 5447–5451.

Joly, N., Engl, C., Jovanovic, G., Huvet, M., Toni, T., Sheng, X., Stumpf, M.P., and Buck, M. (2010). Managing membrane stress: the phage shock protein (Psp) response, from molecular mechanisms to physiology. FEMS Microbiol. Rev. *34*, 797–827.

Jovanovic, G., Weiner, L., and Model, P. (1996). Identification, nucleotide sequence, and characterization of PspF, the transcriptional activator of the *Escherichia coli* stress-induced psp operon. J. Bacteriol. *178*, 1936–1945.

Jovanovic, G., Lloyd, L.J., Stumpf, M.P., Mayhew, A.J., and Buck, M. (2006). Induction and function of the phage shock protein extracytoplasmic stress response in *Escherichia coli*. J. Biol. Chem. *281*, 21147–21161.

Jovanovic, M., James, E.H., Burrows, P.C., Rego, F.G., Buck, M., and Schumacher, J. (2011). Regulation of the co-evolved HrpR and HrpS AAA+ proteins required for *Pseudomonas syringae* pathogenicity. Nat. Commun. 2, 177.

Jyot, J., Dasgupta, N., and Ramphal, R. (2002). FleQ, the major flagellar gene regulator in *Pseudomonas aeruginosa*, binds to enhancer sites located either upstream or atypically downstream of the RpoN binding site. J. Bacteriol. *184*, 5251–5260.

Kaiser, D. (1979). Social gliding is correlated with the presence of pili in *Myxococcus xanthus*. Proc. Natl. Acad. Sci. U.S.A. *76*, 5952–5956.

Kaiser, D. (2003). Coupling cell movement to multicellular development in myxobacteria. Nat. Rev. Microbiol. *1*, 45–54.

Kaiser, D. (2004). Signaling in myxobacteria. Annu. Rev. Microbiol. *58*, 75–98.

Kaiser, D., and Welch, R. (2004). Dynamics of fruiting body morphogenesis. J. Bacteriol. *186*, 919–927.

Kaiser, D., Kroos, L., and Kuspa, A. (1985). Cell interactions govern the temporal pattern of Myxococcus development. Cold Spring Harb. Symp. Quant. Biol. *50*, 823–830.

Kalman, S., Mitchell, W., Marathe, R., Lammel, C., Fan, J., Hyman, R.W., Olinger, L., Grimwood, J., Davis, R.W., and Stephens, R.S. (1999). Comparative genomes of *Chlamydia pneumoniae* and *C. trachomatis*. Nat. Genet. *21*, 385–389.

Kamal, N., Dorrell, N., Jagannathan, A., Turner, S.M., Constantinidou, C., Studholme, D.J., Marsden, G., Hinds, J., Laing, K.G., Wren, B.W., *et al.* (2007). Deletion of a previously uncharacterized flagellar-hook-length control gene fliK modulates the sigma54-dependent regulon in *Campylobacter jejuni*. Microbiology *153*, 3099–3111.

Kaneko, T., Sato, S., Kotani, H., Tanaka, A., Asamizu, E., Nakamura, Y., Miyajima, N., Hirosawa, M., Sugiura, M., Sasamoto, S., *et al.* (1996). Sequence analysis of the genome of the unicellular cyanobacterium *Synechocystis* sp. strain PCC6803. II. Sequence determination of the entire genome and assignment of potential protein-coding regions. DNA Res 3, 109–136.

Karlinsey, J.E., Maguire, M.E., Becker, L.A., Crouch, M.L., and Fang, F.C. (2010). The phage shock protein PspA facilitates divalent metal transport and is required for virulence of *Salmonella enterica* sv. Typhimurium. Mol. Microbiol. *78*, 669–685.

Kazmierczak, M.J., Wiedmann, M., and Boor, K.J. (2005). Alternative sigma factors and their roles in bacterial virulence. Microbiol. Mol. Biol. Rev. *69*, 527–543.

Kehl-Fie, T.E., Porsch, E.A., Miller, S.E., and St Geme, J.W., 3rd (2009). Expression of *Kingella kingae* type IV pili is regulated by sigma54, PilS, and PilR. J. Bacteriol. *191*, 4976–4986.

Keren, I., Shah, D., Spoering, A., Kaldalu, N., and Lewis, K. (2004). Specialized persister cells and the mechanism of multidrug tolerance in *Escherichia coli*. J. Bacteriol. *186*, 8172–8180.

Keseler, I.M., and Kaiser, D. (1995). An early A-signal-dependent gene in *Myxococcus xanthus* has a sigma 54-like promoter. J. Bacteriol. *177*, 4638–4644.

Keseler, I.M., and Kaiser, D. (1997). sigma54, a vital protein for *Myxococcus xanthus*. Proc. Natl. Acad. Sci. U.S.A. *94*, 1979–1984.

Khajanchi, B.K., Sha, J., Kozlova, E.V., Erova, T.E., Suarez, G., Sierra, J.C., Popov, V.L., Horneman, A.J., and Chopra, A.K. (2009). N-acylhomoserine lactones involved in quorum sensing control the type VI secretion system, biofilm formation, protease production, and *in vivo* virulence in a clinical isolate of *Aeromonas hydrophila*. Microbiology *155*, 3518–3531.

Kleerebezem, M., Crielaard, W., and Tommassen, J. (1996). Involvement of stress protein PspA (phage shock protein A) of *Escherichia coli* in maintenance of the protonmotive force under stress conditions. Embo J. *15*, 162–171.

Klose, K.E. (2001). Regulation of virulence in *Vibrio cholerae*. Int. J. Med. Microbiol. *291*, 81–88.

Klose, K.E., and Mekalanos, J.J. (1998). Distinct roles of an alternative sigma factor during both free-swimming and colonizing phases of the *Vibrio cholerae* pathogenic cycle. Mol. Microbiol. *28*, 501–520.

Kobayashi, R., Suzuki, T., and Yoshida, M. (2007). *Escherichia coli* phage-shock protein A (PspA) binds to membrane phospholipids and repairs proton leakage of the damaged membranes. Mol. Microbiol. *66*, 100–109.

Konkel, M.E., Klena, J.D., Rivera-Amill, V., Monteville, M.R., Biswas, D., Raphael, B., and Mickelson, J. (2004). Secretion of virulence proteins from *Campylobacter jejuni* is dependent on a functional flagellar export apparatus. J. Bacteriol. *186*, 3296–3303.

Koo, I.C., and Stephens, R.S. (2003). A developmentally regulated two-component signal transduction system in Chlamydia. J. Biol. Chem. *278*, 17314–17319.

Kroos, L., Kuspa, A., and Kaiser, D. (1986). A global analysis of developmentally regulated genes in *Myxococcus xanthus*. Dev. Biol. *117*, 252–266.

Kunst, F., Ogasawara, N., Moszer, I., Albertini, A.M., Alloni, G., Azevedo, V., Bertero, M.G., Bessieres, P., Bolotin, A., Borchert, S., *et al.* (1997). The complete genome sequence of the Gram-positive bacterium *Bacillus subtilis*. Nature *390*, 249–256.

Kuspa, A., Kroos, L., and Kaiser, D. (1986). Intercellular signaling is required for developmental gene expression in *Myxococcus xanthus*. Dev. Biol. *117*, 267–276.

Kyrpides, N.C., Ouzounis, C.A., Iliopoulos, I., Vonstein, V., and Overbeek, R. (2000). Analysis of the *Thermotoga maritima* genome combining a variety of sequence similarity and genome context tools. Nucleic Acids Res. *28*, 4573–4576.

Lane, R.S., Piesman, J., and Burgdorfer, W. (1991). Lyme borreliosis: relation of its causative agent to its vectors and hosts in North America and Europe. Annu. Rev. Entomol. *36*, 587–609.

Lange, R., and Hengge-Aronis, R. (1991). Identification of a central regulator of stationary-phase gene expression in *Escherichia coli*. Mol. Microbiol. *5*, 49–59.

Laskos, L., Dillard, J.P., Seifert, H.S., Fyfe, J.A., and Davies, J.K. (1998). The pathogenic neisseriae contain an inactive rpoN gene and do not utilize the pilE sigma54 promoter. Gene *208*, 95–102.

Leang, C., Krushkal, J., Ueki, T., Puljic, M., Sun, J., Juarez, K., Nunez, C., Reguera, G., DiDonato, R., Postier, B., *et al.* (2009). Genome-wide analysis of the RpoN regulon in *Geobacter sulfurreducens*. BMC Genomics *10*, 331.

Leary, B.A., Ward-Rainey, N., and Hoover, T.R. (1998). Cloning and characterization of *Planctomyces limnophilus* rpoN: complementation of a *Salmonella* typhimurium rpoN mutant strain. Gene *221*, 151–157.

Lesic, B., Starkey, M., He, J., Hazan, R., and Rahme, L.G. (2009). Quorum sensing differentially regulates *Pseudomonas aeruginosa* type VI secretion locus I and homologous loci II and III, which are required for pathogenesis. Microbiology *155*, 2845–2855.

Lilley, B.N., and Bassler, B.L. (2000). Regulation of quorum sensing in *Vibrio harveyi* by LuxO and sigma-54. Mol. Microbiol. *36*, 940–954.

Liu, A., Tran, L., Becket, E., Lee, K., Chinn, L., Park, E., Tran, K., and Miller, J.H. (2010). Antibiotic sensitivity profiles determined with an *Escherichia coli* gene knockout collection: generating an antibiotic bar code. Antimicrob. Agents Chemother. *54*, 1393–1403.

Ma, C., Sim, S., Shi, W., Du, L., Xing, D., and Zhang, Y. (2009). Energy production genes sucB and ubiF are involved in persister survival and tolerance to multiple antibiotics and stresses in *Escherichia coli*. FEMS Microbiol. Lett. *303*, 33–40.

Mao, X.J., Huo, Y.X., Buck, M., Kolb, A., and Wang, Y.P. (2007). Interplay between CRP-cAMP and PII-Ntr systems forms novel regulatory network between carbon metabolism and nitrogen assimilation in *Escherichia coli*. Nucleic Acids Res. *35*, 1432–1440.

Marques, M.V., Gomes, S.L., and Gober, J.W. (1997). A gene coding for a putative sigma 54 activator is developmentally regulated in *Caulobacter crescentus*. J. Bacteriol. *179*, 5502–5510.

Martinez, R.M., Dharmasena, M.N., Kirn, T.J., and Taylor, R.K. (2009). Characterization of two outer membrane proteins, FlgO and FlgP, that influence *Vibrio cholerae* motility. J. Bacteriol. *191*, 5669–5679.

Mathews, S.A., and Timms, P. (2000). Identification and mapping of sigma-54 promoters in *Chlamydia trachomatis*. J. Bacteriol. *182*, 6239–6242.

Mathews, S.A., Volp, K.M., and Timms, P. (1999). Development of a quantitative gene expression assay for *Chlamydia trachomatis* identified temporal expression of sigma factors. FEBS Lett. *458*, 354–358.

Merrick, M.J. (1993). In a class of its own–the RNA polymerase sigma factor sigma 54 (sigma N). Mol. Microbiol. *10*, 903–909.

Milton, D.L., O'Toole, R., Horstedt, P., and Wolf-Watz, H. (1996). Flagellin A is essential for the virulence of *Vibrio anguillarum*. J. Bacteriol. *178*, 1310–1319.

Miyata, S.T., Kitaoka, M., Wieteska, L., Frech, C., Chen, N., and Pukatzki, S. (2010). The *Vibrio cholerae* type VI secretion system: evaluating its role in the human disease cholera. Front. Microbiol. *1*, 117.

Model, P., Jovanovic, G., and Dworkin, J. (1997). The *Escherichia coli* phage-shock-protein (psp) operon. Mol. Microbiol. *24*, 255–261.

Mole, B.M., Baltrus, D.A., Dangl, J.L., and Grant, S.R. (2007). Global virulence regulation networks in phytopathogenic bacteria. Trends Microbiol. *15*, 363–371.

Molofsky, A.B., and Swanson, M.S. (2004). Differentiate to thrive: lessons from the *Legionella pneumophila* life cycle. Mol. Microbiol. *53*, 29–40.

Moulder, J.W. (1991). Interaction of chlamydiae and host cells *in vitro*. Microbiol. Rev. *55*, 143–190.

Nachamkin, I., Yang, X.H., and Stern, N.J. (1993). Role of *Campylobacter jejuni* flagella as colonization factors for three-day-old chicks: analysis with flagellar mutants. Appl. Environ. Microbiol. *59*, 1269–1273.

Nelson, K.E., Clayton, R.A., Gill, S.R., Gwinn, M.L., Dodson, R.J., Haft, D.H., Hickey, E.K., Peterson, J.D., Nelson, W.C., Ketchum, K.A., *et al.* (1999). Evidence for lateral gene transfer between Archaea and bacteria from genome sequence of *Thermotoga maritima*. Nature *399*, 323–329.

Nichols, R.J., Sen, S., Choo, Y.J., Beltrao, P., Zietek, M., Chaba, R., Lee, S., Kazmierczak, K.M., Lee, K.J., Wong, A., *et al.* (2011). Phenotypic landscape of a bacterial cell. Cell *144*, 143–156.

Nicholson, T.L., Olinger, L., Chong, K., Schoolnik, G., and Stephens, R.S. (2003). Global stage-specific gene regulation during the developmental cycle of *Chlamydia trachomatis*. J. Bacteriol. *185*, 3179–3189.

Niehus, E., Gressmann, H., Ye, F., Schlapbach, R., Dehio, M., Dehio, C., Stack, A., Meyer, T.F., Suerbaum, S., and Josenhans, C. (2004). Genome-wide analysis of transcriptional hierarchy and feedback regulation in the flagellar system of *Helicobacter pylori*. Mol. Microbiol. *52*, 947–961.

Nudleman, E., and Kaiser, D. (2004). Pulling together with type IV pili. J. Mol. Microbiol. Biotechnol. *7*, 52–62.

Nystrom, T. (2004). Growth versus maintenance: a trade-off dictated by RNA polymerase availability and sigma factor competition? Mol. Microbiol. *54*, 855–862.

O'Toole, R., Milton, D.L., and Wolf-Watz, H. (1996). Chemotactic motility is required for invasion of the host by the fish pathogen *Vibrio anguillarum*. Mol. Microbiol. *19*, 625–637.

O'Toole, R., Milton, D.L., Horstedt, P., and Wolf-Watz, H. (1997). RpoN of the fish pathogen *Vibrio* (*Listonella*) *anguillarum* is essential for flagellum production and virulence by the water-borne but not intraperitoneal route of inoculation. Microbiology *143*, 3849–3859.

Okada, Y., Okada, N., Makino, S., Asakura, H., Yamamoto, S., and Igimi, S. (2006). The sigma factor RpoN (sigma54) is involved in osmotolerance in *Listeria monocytogenes*. FEMS Microbiol. Lett. *263*, 54–60.

Ouyang, Z., Blevins, J.S., and Norgard, M.V. (2008). Transcriptional interplay among the regulators Rrp2, RpoN and RpoS in *Borrelia burgdorferi*. Microbiology *154*, 2641–2658.

Ouyang, Z., Kumar, M., Kariu, T., Haq, S., Goldberg, M., Pal, U., and Norgard, M.V. (2009). BosR (BB0647) governs virulence expression in *Borrelia burgdorferi*. Mol. Microbiol. 74, 1331–1343.

Ouyang, Z., Deka, R.K., and Norgard, M.V. (2011). BosR (BB0647) controls the RpoN-RpoS regulatory pathway and virulence expression in *Borrelia burgdorferi* by a novel DNA-binding mechanism. PLoS Pathog. *7*, e1001272.

Paget, M.S., and Helmann, J.D. (2003). The sigma70 family of sigma factors. Genome Biol. *4*, 203.

Pal, U., Yang, X., Chen, M., Bockenstedt, L.K., Anderson, J.F., Flavell, R.A., Norgard, M.V., and Fikrig, E. (2004). OspC facilitates *Borrelia burgdorferi* invasion of *Ixodes scapularis* salivary glands. J. Clin. Invest. *113*, 220–230.

Parkhill, J., Wren, B.W., Mungall, K., Ketley, J.M., Churcher, C., Basham, D., Chillingworth, T., Davies, R.M., Feltwell, T., Holroyd, S., *et al.* (2000). The genome sequence of the food-borne pathogen *Campylobacter jejuni* reveals hypervariable sequences. Nature *403*, 665–668.

Pechy-Tarr, M., Bottiglieri, M., Mathys, S., Lejbolle, K.B., Schnider-Keel, U., Maurhofer, M., and Keel, C. (2005). RpoN (sigma54) controls production of antifungal compounds and biocontrol activity in *Pseudomonas fluorescens* CHA0. Mol. Plant Microbe Interact. *18*, 260–272.

Pereira, L.E., Brahmachary, P., and Hoover, T.R. (2006). Characterization of *Helicobacter pylori* sigma54 promoter-binding activity. FEMS Microbiol. Lett. *259*, 20–26.

Potvin, E., Sanschagrin, F., and Levesque, R.C. (2008). Sigma factors in *Pseudomonas aeruginosa*. FEMS Microbiol. Rev. *32*, 38–55.

Prouty, M.G., Correa, N.E., and Klose, K.E. (2001). The novel sigma54- and sigma28-dependent flagellar gene transcription hierarchy of *Vibrio cholerae*. Mol. Microbiol. *39*, 1595–1609.

Pukatzki, S., Ma, A.T., Sturtevant, D., Krastins, B., Sarracino, D., Nelson, W.C., Heidelberg, J.F., and Mekalanos, J.J. (2006). Identification of a conserved bacterial protein secretion system in *Vibrio cholerae* using the *Dictyostelium* host model system. Proc. Natl. Acad. Sci. U.S.A. *103*, 1528–1533.

Ramphal, R., Koo, L., Ishimoto, K.S., Totten, P.A., Lara, J.C., and Lory, S. (1991). Adhesion of *Pseudomonas aeruginosa* pilin-deficient mutants to mucin. Infect. Immun. *59*, 1307–1311.

Rappas, M., Bose, D., and Zhang, X. (2007). Bacterial enhancer-binding proteins: unlocking sigma54-dependent gene transcription. Curr. Opin. Struct. Biol. *17*, 110–116.

Reichenbach, B., Gopel, Y., and Gorke, B. (2009). Dual control by perfectly overlapping sigma 54- and sigma 70- promoters adjusts small RNA GlmY expression to different environmental signals. Mol. Microbiol. *74*, 1054–1070.

Reitzer, L., and Schneider, B.L. (2001). Metabolic context and possible physiological themes of sigma(54)-dependent genes in *Escherichia coli*. Microbiol. Mol. Biol. Rev. *65*, 422–444, table of contents.

Riordan, J.T., Tietjen, J.A., Walsh, C.W., Gustafson, J.E., and Whittam, T.S. (2010). Inactivation of alternative sigma factor 54 (RpoN) leads to increased acid resistance, and alters locus of enterocyte effacement (LEE) expression in *Escherichia coli* O157: H7. Microbiology *156*, 719–730.

Robichon, D., Gouin, E., Debarbouille, M., Cossart, P., Cenatiempo, Y., and Hechard, Y. (1997). The rpoN (sigma54) gene from *Listeria monocytogenes* is involved in resistance to mesentericin Y105, an antibacterial peptide from *Leuconostoc mesenteroides*. J. Bacteriol. *179*, 7591–7594.

Sebaihia, M., Wren, B.W., Mullany, P., Fairweather, N.F., Minton, N., Stabler, R., Thomson, N.R., Roberts, A.P., Cerdeno-Tarraga, A.M., Wang, H., *et al.* (2006). The multidrug-resistant human pathogen *Clostridium difficile* has a highly mobile, mosaic genome. Nat. Genet. *38*, 779–786.

Shah, I.M., and Wolf, R.E., Jr. (2004). Novel protein–protein interaction between *Escherichia coli* SoxS and the DNA binding determinant of the RNA polymerase alpha subunit: SoxS functions as a co-sigma factor and redeploys RNA polymerase from UP-element-containing promoters to SoxS-dependent promoters during oxidative stress. J. Mol. Biol. *343*, 513–532.

Shimizu, T., Ohtani, K., Hirakawa, H., Ohshima, K., Yamashita, A., Shiba, T., Ogasawara, N., Hattori, M., Kuhara, S., and Hayashi, H. (2002). Complete genome sequence of *Clostridium perfringens*, an anaerobic flesh-eater. Proc. Natl. Acad. Sci. U.S.A. *99*, 996–1001.

Shingler, V. (2010). Signal sensory systems that impact σ^{54}-dependent transcription. FEMS Microbiol. Rev. 35, 425–440.

Smith, A.H., Blevins, J.S., Bachlani, G.N., Yang, X.F., and Norgard, M.V. (2007). Evidence that RpoS (sigmaS) in *Borrelia burgdorferi* is controlled directly by RpoN (sigma54/sigmaN). J. Bacteriol. *189*, 2139–2144.

Spohn, G., and Scarlato, V. (1999). Motility of *Helicobacter pylori* is coordinately regulated by the transcriptional activator FlgR, an NtrC homolog. J. Bacteriol. *181*, 593–599.

Standar, K., Mehner, D., Osadnik, H., Berthelmann, F., Hause, G., Lunsdorf, H., and Bruser, T. (2008). PspA can form large scaffolds in *Escherichia coli*. FEBS Lett. *582*, 3585–3589.

Stephens, R.S., Kalman, S., Lammel, C., Fan, J., Marathe, R., Aravind, L., Mitchell, W., Olinger, L., Tatusov, R.L., Zhao, Q., *et al.* (1998). Genome sequence of an obligate intracellular pathogen of humans: *Chlamydia trachomatis*. Science *282*, 754–759.

Stover, C.K., Pham, X.Q., Erwin, A.L., Mizoguchi, S.D., Warrener, P., Hickey, M.J., Brinkman, F.S., Hufnagle, W.O., Kowalik, D.J., Lagrou, M., *et al.* (2000). Complete genome sequence of *Pseudomonas aeruginosa* PAO1, an opportunistic pathogen. Nature *406*, 959–964.

Studholme, D.J., and Buck, M. (2000a). The biology of enhancer-dependent transcriptional regulation in bacteria: insights from genome sequences. FEMS Microbiol. Lett. *186*, 1–9.

Studholme, D.J., and Buck, M. (2000b). Novel roles of sigmaN in small genomes. Microbiology *146*, 4–5.

Studholme, D.J., and Dixon, R. (2003). Domain architectures of sigma54-dependent transcriptional activators. J. Bacteriol. *185*, 1757–1767.

Studholme, D.J., Wigneshwereraraj, S.R., Gallegos, M.T., and Buck, M. (2000). Functionality of purified sigma(N) (sigma(54)) and a NifA-like protein from the hyperthermophile *Aquifex aeolicus*. J. Bacteriol. *182*, 1616–1623.

Suerbaum, S., Josenhans, C., and Labigne, A. (1993). Cloning and genetic characterization of the *Helicobacter pylori* and *Helicobacter mustelae* flaB flagellin genes and construction of *H. pylori* flaA- and flaB-negative mutants by electroporation-mediated allelic exchange. J. Bacteriol. *175*, 3278–3288.

Syed, K.A., Beyhan, S., Correa, N., Queen, J., Liu, J., Peng, F., Satchell, K.J., Yildiz, F., and Klose, K.E. (2009). The *Vibrio cholerae* flagellar regulatory hierarchy controls expression of virulence factors. J. Bacteriol. *191*, 6555–6570.

Takeuchi, F., Watanabe, S., Baba, T., Yuzawa, H., Ito, T., Morimoto, Y., Kuroda, M., Cui, L., Takahashi, M., Ankai, A., *et al.* (2005). Whole-genome sequencing of *Staphylococcus haemolyticus* uncovers the extreme plasticity of its genome and the evolution of human-colonizing staphylococcal species. J. Bacteriol. *187*, 7292–7308.

Tanaka, N., and Shuman, S. (2011). RtcB is the RNA ligase component of an *Escherichia coli* RNA repair operon. J. Biol. Chem. *286*, 7727–7731

Tettelin, H., Nelson, K.E., Paulsen, I.T., Eisen, J.A., Read, T.D., Peterson, S., Heidelberg, J., DeBoy, R.T., Haft, D.H., Dodson, R.J., *et al.* (2001). Complete genome sequence of a virulent isolate of *Streptococcus pneumoniae*. Science *293*, 498–506.

Thompson, L.S., Webb, J.S., Rice, S.A., and Kjelleberg, S. (2003). The alternative sigma factor RpoN regulates the quorum sensing gene rhlI in *Pseudomonas aeruginosa*. FEMS Microbiol. Lett. *220*, 187–195.

Tian, Z.X., Li, Q.S., Buck, M., Kolb, A., and Wang, Y.P. (2001). The CRP–cAMP complex and downregulation of the glnAp2 promoter provides a novel regulatory linkage between carbon metabolism and nitrogen assimilation in *Escherichia coli*. Mol. Microbiol. *41*, 911–924.

Totten, P.A., Lara, J.C., and Lory, S. (1990). The rpoN gene product of *Pseudomonas aeruginosa* is required for expression of diverse genes, including the flagellin gene. J. Bacteriol. *172*, 389–396.

Tse, H., Tsoi, H.W., Leung, S.P., Lau, S.K., Woo, P.C., and Yuen, K.Y. (2010). Complete genome sequence of *Staphylococcus lugdunensis* strain HKU09-01. J. Bacteriol. *192*, 1471–1472.

Vadyvaloo, V., Arous, S., Gravesen, A., Hechard, Y., Chauhan-Haubrock, R., Hastings, J.W., and Rautenbach, M. (2004). Cell-surface alterations in class IIa bacteriocin-resistant *Listeria monocytogenes* strains. Microbiology *150*, 3025–3033.

Vencato, M., Tian, F., Alfano, J.R., Buell, C.R., Cartinhour, S., DeClerck, G.A., Guttman, D.S., Stavrinides, J., Joardar, V., Lindeberg, M., *et al.* (2006). Bioinformatics-enabled identification of the HrpL regulon and type III secretion system effector proteins of *Pseudomonas syringae* pv. *phaseolicola* 1448A. Mol. Plant Microbe Interact. *19*, 1193–1206.

Viducic, D., Ono, T., Murakami, K., Katakami, M., Susilowati, H., and Miyake, Y. (2007). rpoN gene of *Pseudomonas aeruginosa* alters its susceptibility to quinolones and carbapenems. Antimicrob. Agents. Chemother. *51*, 1455–1462.

Visick, K.L. (2009). An intricate network of regulators controls biofilm formation and colonization by *Vibrio fischeri*. Mol. Microbiol. *74*, 782–789.

Wang, Q., Frye, J.G., McClelland, M., and Harshey, R.M. (2004). Gene expression patterns during swarming in *Salmonella* typhimurium: genes specific to surface growth and putative new motility and pathogenicity genes. Mol. Microbiol. *52*, 169–187.

Wei, Z.M., and Beer, S.V. (1995). hrpL activates *Erwinia amylovora* hrp gene transcription and is a member of the ECF subfamily of sigma factors. J. Bacteriol. *177*, 6201–6210.

Weiner, L., Brissette, J.L., and Model, P. (1991). Stress-induced expression of the *Escherichia coli* phage shock protein operon is dependent on sigma 54 and modulated by positive and negative feedback mechanisms. Genes Dev. *5*, 1912–1923.

Wigneshweraraj, S., Bose, D., Burrows, P.C., Joly, N., Schumacher, J., Rappas, M., Pape, T., Zhang, X., Stockley, P., Severinov, K., and Buck, M. (2008). Modus operandi of the bacterial RNA polymerase containing the sigma54 promoter-specificity factor. Mol. Microbiol. *68*, 538–546.

Wigneshweraraj, S.R., Fujita, N., Ishihama, A., and Buck, M. (2000). Conservation of sigma-core RNA polymerase proximity relationships between the enhancer-independent and enhancer-dependent sigma classes. EMBO J. *19*, 3038–3048.

Wigneshweraraj, S.R., Chaney, M.K., Ishihama, A., and Buck, M. (2001). Regulatory sequences in sigma 54 localise near the start of DNA melting. J. Mol. Biol. *306*, 681–701.

Wigneshweraraj, S.R., Burrows, P.C., Bordes, P., Schumacher, J., Rappas, M., Finn, R.D., Cannon, W.V., Zhang, X., and Buck, M. (2005). The second paradigm for activation of transcription. Prog. Nucleic Acid Res. Mol. Biol. *79*, 339–369.

Wolgemuth, C., Hoiczyk, E., Kaiser, D., and Oster, G. (2002). How myxobacteria glide. Curr. Biol. *12*, 369–377.

Wosten, M.M. (1998). Eubacterial sigma-factors. FEMS Microbiol. Rev. *22*, 127–150.

Wosten, M.M., Wagenaar, J.A., and van Putten, J.P. (2004). The FlgS/FlgR two-component signal transduction system regulates the fla regulon in *Campylobacter jejuni*. J. Biol. Chem. *279*, 16214–16222.

Xiao, Y., Heu, S., Yi, J., Lu, Y., and Hutcheson, S.W. (1994). Identification of a putative alternate sigma factor and characterization of a multicomponent regulatory cascade controlling the expression of *Pseudomonas syringae* pv. syringae Pss61 hrp and hrmA genes. J. Bacteriol. *176*, 1025–1036.

Xu, H., Caimano, M.J., Lin, T., He, M., Radolf, J.D., Norris, S.J., Gheradini, F., Wolfe, A.J., and Yang, X.F. (2010). Role of acetyl-phosphate in activation of the Rrp2-RpoN-RpoS pathway in *Borrelia burgdorferi*. PLoS Pathog. *6*, e1001104

Yang, X.F., Pal, U., Alani, S.M., Fikrig, E., and Norgard, M.V. (2004). Essential role for OspA/B in the life cycle of the Lyme disease spirochete. J. Exp. Med. *199*, 641–648.

Yildiz, F.H., and Visick, K.L. (2009). Vibrio biofilms: so much the same yet so different. Trends Microbiol. *17*, 109–118.

Young, K.T., Davis, L.M., and Dirita, V.J. (2007). *Campylobacter jejuni*: molecular biology and pathogenesis. Nat. Rev. Microbiol. *5*, 665–679.

Zhao, K., Liu, M., and Burgess, R.R. (2010). Promoter and regulon analysis of nitrogen assimilation factor, sigma54, reveal alternative strategy for *E. coli* MG1655 flagellar biosynthesis. Nucleic Acids Res. *38*, 1273–1283.

Extracytoplasmic Function Sigma Factors: From Stress Management to Iron Uptake

3

Karlijn C. Bastiaansen, Wilbert Bitter and María A. Llamas

Abstract

Gene expression in bacteria is mainly controlled at the level of transcription initiation. To achieve this process a number of different mechanisms have evolved, one of which is the utilization of alternative sigma factors. Sigma factors are small proteins that associate with the RNA polymerase core enzyme (RNAPc) and direct it to specific promoter sequences, where they initiate gene transcription. Bacteria are able to regulate transcription initiation by synthesizing and activating different sigma factors that recognize different promoter consensus sequences. The largest group of alternative sigma factors consists of the so-called extracytoplasmic function (ECF) sigma factors that regulate gene expression in response to cell envelope stresses or environmental stimuli. The activity of ECF sigma factors is controlled by anti-sigma factors and a complex cascade of regulated (proteolytic) modifications. In Gram-negative bacteria, ECF sigma factors are also controlled by cell-surface signalling (CSS), a regulatory system that includes an outer membrane receptor in the signal transduction pathway. In this chapter we will discuss the general composition and function of ECF sigma factors and their role in cell envelope stress responses and CSS.

Introduction

Regulation of gene expression is an essential mechanism that allows bacteria to rapidly adapt to alterations in their environment. In this way, bacteria preserve energy and profit by expressing particular genes only when they are required. Gene expression in bacteria is regulated primarily at the level of transcription initiation by adjusting promoter recognition of the RNA polymerase (RNAP) holoenzyme. The RNAP holoenzyme of bacteria comprises a five-subunit core enzyme (RNAPc; subunit composition $\alpha_2\beta\beta'\omega$) and a dissociable sigma (σ) subunit (Murakami and Darst, 2003). The sigma factor contains most promoter recognition determinants and confers promoter specificity to the RNA polymerase. The sigma subunit is also required for transcription initiation, which is a stepwise process. First, the RNAP holoenzyme forms a closed complex around the promoter in which DNA is double-stranded. After isomerisation (melting) of the DNA around the transcription start point, an open complex will be formed and transcription commences (Browning and Busby, 2004). After transcription initiation the sigma factor dissociates from the RNAP and is available to interact with a next core enzyme complex, while the remaining RNAPc complex drives RNA transcription until a transcription termination site is encountered (Mooney *et al.*, 2005). All bacteria contain a primary sigma factor that directs transcription of general housekeeping genes. Moreover, most bacteria also encode several alternative sigma factors, which can redirect RNAP to initiate transcription from specific sets of alternative promoters (Helmann and Chamberlin, 1988). By alternating the use of standard and alternative sigma factors bacteria can adequately regulate general cell functions as well as responses to specific signals. Therefore, the versatility and adaptability of bacteria is partially reflected by the number of alternative sigma factors they produce (Paget and Helmann, 2003).

Bacterial sigma factors have been classified into two main and unrelated families, namely the σ^{70} and the σ^{54} families. σ^{54}-type sigma factors are structurally different from σ^{70} family members and utilize a different mechanism of open complex formation (Merrick, 1993). Differences between these families are also observed in promoter recognition. σ^{54} promoters are highly conserved short sequences located at positions −24 and −12 upstream the transcription initiation site, while σ^{70} promoters are typically located at −35 and −10 upstream (Merrick, 1993). Whereas σ^{70}-related sigma factors are present in all bacterial genomes, there are no known representatives of the σ^{54} family in any GC-rich Gram-positive bacteria or cyanobacteria. The σ^{70} family can be divided into four different groups based on phylogenetic relationship and modular structure (Lonetto *et al.*, 1992; Helmann, 2002). This review will focus on Group 4 σ^{70} factors, which includes the so-called extracytoplasmic function (ECF) sigma factors that play a key role in the bacterial response to the environment.

The σ70 family of sigma factors

The σ^{70} family of sigma factors consists of a large number of primary and alternative sigma factors that are all related to the primary σ^{70} sigma factor of *Escherichia coli*, which is the model protein of this family. The *E. coli* σ^{70} is encoded by *rpoD* (RNA polymerase subunit) and its superscript reflects its molecular weight (70 kDa). Originally Lonetto *et al.* (1992) proposed to divide this family into three distinct groups or subfamilies based on sequence identity and functional characteristics: the homologues of the *E. coli* primary sigma factor σ^{70} (Group 1), a group of highly related but non-essential sigma factors (Group 2), and the more distantly related alternative sigma factors (Group 3). Later this classification was refined and an additional group was added, namely the extracytoplasmic function (ECF) sigma factors (Group 4) (Lonetto *et al.*, 1994; Helmann, 2002).

The primary *E. coli* σ^{70} factor and its homologues in other bacterial species are allocated to Group 1. These sigma factors are highly conserved and recognize similar target promoter sequences: TTGACA near the −35 and TATAAT near the −10 elements (Helmann, 2002). Primary sigma factors are usually between 40 and 70 kDa in size and contain four conserved regions (Fig. 3.1) (Lonetto *et al.*, 1992). These regions can be divided into subregions, each of which plays a different role in the interaction of the sigma factor with the RNAPc and with the target DNA, i.e. the promoter region. These subregions correspond to three distinct structural domains (domain 2 to 4,

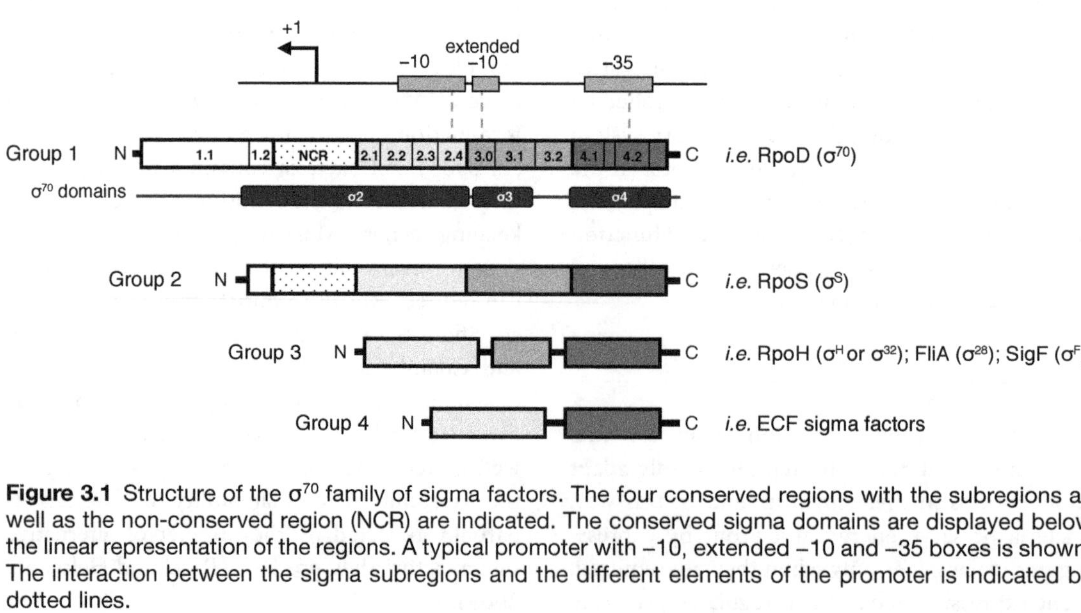

Figure 3.1 Structure of the σ^{70} family of sigma factors. The four conserved regions with the subregions as well as the non-conserved region (NCR) are indicated. The conserved sigma domains are displayed below the linear representation of the regions. A typical promoter with −10, extended −10 and −35 boxes is shown. The interaction between the sigma subregions and the different elements of the promoter is indicated by dotted lines.

Fig. 3.1) that are flexibly joined by linkers, which facilitate conformational changes of the sigma subunit (Gruber and Gross, 2003). Conserved regions 1.2 up to 2.4 constitute domain 2 (σ_2), regions 3.0 and 3.1 comprise domain 3 (σ_3) and regions 4.1 to 4.2 form domain 4 (σ_4). Region 1.1 is considered to form an N-terminal extension and is only present in primary σ^{70} factors (Helmann and Chamberlin, 1988; Lonetto *et al.*, 1992; Murakami *et al.*, 2002b). When associated with the RNAPc region 1.1 either inhibits or stimulates open complex formation depending on the promoter (Vuthoori *et al.*, 2001). Structural data suggest that this region is located in the active site channel of the RNAP holoenzyme and that it acts as a DNA mimic because of its negative charge. Upon promoter binding, region 1.1 is displaced by double-stranded promoter DNA and moves away from the inside of this cleft (Mekler *et al.*, 2002; Murakami *et al.*, 2002b). In addition, region 1.1 might stabilize the interactions between RNA polymerase and the primary σ^{70} factor (Hinton *et al.*, 2006). It has also been proposed that this region is involved in preventing the interaction of free sigma subunits with promoter DNA by inducing compact folding of the sigma factor, which is then unable to bind DNA (Dombroski *et al.*, 1993; Schwartz *et al.*, 2008). Structural domain 2 (σ_2) is formed by region 1.2, a stretch of non-conserved residues (NCR), and regions 2.1 to 2.4. Region 1.2, which is highly conserved amongst all σ^{70} group members, plays an important role in transcription initiation. This region makes sequence-specific contact with the non-template strand downstream of the −10 promoter element and supports the function of domain 2 by stabilizing its conformation, thereby facilitating processes such as promoter binding and DNA melting (Haugen *et al.*, 2008). The four subregions of region 2 are, amongst others, involved in promoter recognition and interaction with the RNAPc. Region 2.4 is mainly implicated in binding of the −10 box of the promoter (Fig. 3.1) (Lonetto *et al.*, 1992). Region 2.3 is essential for promoter melting and for interaction with the −10 promoter element (Young *et al.*, 2004). Region 2.2, the most conserved region within the σ^{70} family of sigma factors, is a very strong RNAP binding determinant and establishes the first interactions between sigma and RNAPc (Sharp *et al.*, 1999; Young *et al.*, 2001). Structural domain 3 (σ_3) is comprised of the regions 3.0 and 3.1 and forms three alpha helices. The helix–turn–helix motif in region 3.1 is mainly involved in the extensive interaction with the RNAPc (Murakami *et al.*, 2002a,b). The conserved region 3.0 interacts with the extended −10 promoter element (Fig. 3.1), which in general substitutes for poorly matched −35 and −10 promoter elements (Sharp *et al.*, 1999). Finally, domain 4 (σ_4) is composed of subregions 4.1 and 4.2. Region 4.2 interacts with the −35 element of the promoter (Fig. 3.1) (Lonetto *et al.*, 1992; Campbell *et al.*, 2002). In addition, region 4 is also involved in RNAPc binding (Sharp *et al.*, 1999). However, for some promoter sequences this domain seems dispensable for transcription initiation (Campbell *et al.*, 2002).

Sigma factors belonging to the other three groups (Groups 2 to 4) always contain domains 2 and 4, which therefore form the minimal requirement for a functional sigma factor. Group 2 sigma factors are closely related to σ^{70} and direct RNAP to similar promoter sequences, although they are not essential for growth and lack most of region 1 (Fig. 3.1) (Lonetto *et al.*, 1992). The best-studied Group 2 sigma factor is RpoS (σ^S) of *E. coli*, which controls the expression of stationary phase and stress survival genes (Hengge, 2009; Navarro Llorens *et al.*, 2010). Group 3 sigma factors have been designated as secondary sigma factors and generally coordinate activation of stress responses or developmental processes (Lonetto *et al.*, 1992; Helmann, 2002). Although related to σ^{70}, these proteins diverge more in sequence and are smaller than primary sigma factors (25–35 kDa) (Fig. 3.1). This group can be divided in different clusters of evolutionary related proteins that often have related functions, including a heat-shock cluster represented by RpoH (σ^H or σ^{32}) of *E. coli*, a flagellar biosynthesis cluster represented by FliA (σ^{28}) of *E. coli* and *Salmonella*, and a sporulation cluster represented by SigF (σ^F) of *Bacillus subtilis*. Finally, Group 4 is composed of ECF sigma factors, which are the smallest sigma factors (Fig. 3.1). They represent the largest and most diverse subfamily, and have been recently considered the third fundamental mechanism of bacterial signal transduction (Staroń *et al.*, 2009).

Extracytoplasmic function sigma factors

The existence of the diverse group of extracytoplasmic function (ECF) sigma factors became evident in a multiple sequence alignment of *Streptomyces coelicolor* σ^E (Lonetto *et al.*, 1994). This analysis showed that, although *S. coelicolor* σ^E shared similarities with primary sigma factors, there were significant sequence differences and certain regions were even missing (Fig. 3.1). Furthermore, σ^E showed a structural composition similar to a range of other sigma factors including *E. coli* FecI, *Pseudomonas aeruginosa* AlgU, *B. subtilis* SigX and *Myxococcus xanthus* CarQ (Lonetto *et al.*, 1994). Prior to this study, many of these proteins were known to positively regulate gene expression, but were considered to act as classical transcriptional activators by enhancing the function of the RNA polymerase holoenzyme. The work of Lonetto *et al.* (1994) altered this assumption and assigned a function as sigma factors to these regulatory proteins. This new subfamily of sigma factors was named ECF sigma factors, referring to their role in the regulation of extracytoplasmic functions, such as secretion, iron transport or stress responses (Lonetto *et al.*, 1994). Since the work of Lonetto and coworkers, an increasing number of ECF sigma factors have been identified and characterized in many different bacteria and it has become clear that the regulation of extracytoplasmic functions is indeed the main role of these sigma factors. In general, ECF sigma factors respond to signals originating from within or outside the cell envelope and activate genes that are involved in cell envelope functions (Missiakas and Raina, 1998; Helmann, 2002). In addition, ECF sigma factors are also involved in the regulation of virulence in pathogenic bacteria. Although this function can be indirect, an increasing number of ECF sigma factors have been identified that play a direct role in this process (Kazmierczak *et al.*, 2005).

ECF sigma factors are significantly smaller than primary sigma factors and consist only of structural domains 2 and 4 (Fig. 3.1) (Lonetto *et al.*, 1994). Region 4 is in structure and in sequence the most conserved region within this group (Lonetto *et al.*, 1994). In contrast, subregion 2.4, which is implicated in the recognition of the −10 promoter element, shows most variation, which probably reflects differences in promoter binding specificity (Lonetto *et al.*, 1994). The lack of almost the complete region 3 implies that extended −10 promoters do not play a role in ECF sigma factor promoter recognition. This reduced region 3 forms a short unstructured region that functions to link domain 2 and 4 (Campbell *et al.*, 2003). Apart from their structure, there are several additional features that characterize ECF sigma factors. They generally regulate quite few operons as opposed to the large number of genes controlled by primary sigma factors. Moreover, most ECF sigma factors, with the exception of the FecI-like group (see below), autoregulate their own expression (Helmann, 2002). In addition, ECF sigma factors usually share a high degree of conservation in their −35 promoter element, implying that promoter specificity of these sigma factors is predominantly determined by the less conserved −10 element sequence (Lane and Darst, 2006). And finally, ECF sigma factors are often co-transcribed with their cognate inhibitor, the so called anti-sigma factor or sigma factor regulator.

Comparative genome analyses have established that ECF sigma factors represent the third most common mechanism for bacterial signal transduction (Staroń *et al.*, 2009). However, the distribution of ECF sigma factors among bacterial genomes varies greatly. A minority of species do not contain any ECF sigma factors, while other species encode over 80 (Butcher *et al.*, 2008; Staroń *et al.*, 2009). A high number of ECF sigma factors, which is often associated with high numbers of one- and two-component systems, generally reflects the diversity of the bacterial living environment as well as the bacterial genome size. Species that exist in a stable environment and have a small genome typically do not contain many ECF sigma factors. In contrast, species with a large genome living in complex habitats encode up to dozens. For example, the versatile opportunistic pathogen *P. aeruginosa* thrives in diverse habitats, ranging from soil to the human airways, and encodes 19 ECF sigma factors (Visca *et al.*, 2002; Llamas *et al.*, 2008). The Gram-positive *S. coelicolor*, that predominantly lives in soil and is known to produce several antibiotics, encodes over 60 σ^{70} family sigma factors, of which more than 50 are classified as ECF sigma factors. The most impressive number of ECF sigma factors up to date is found in the Gram-negative *Sorangium cellulosum*. The genome of this myxobacterium is approximately 13 million base-pairs in size and encodes a striking 83 ECF sigma factors (Staroń *et al.*, 2009). Recently, ECF sigma factors have been classified in 43 major phylogenetically distinct groups (Staroń *et al.*, 2009). Of those, the two well-established types of RpoE-like and FecI-like sigma factors are the predominant groups. The function of these ECF sigma factors, together with a third group containing the designated cytoplasmic sensing ECF sigma factors, will be discussed in detail (Table 3.1).

As mentioned above, most ECF sigma factors are co-transcribed with an anti-sigma factor (or sigma factor regulator) that controls the activity of the sigma factor post-transcriptionally. Similar to the sensor kinase/response regulator pair of two-component systems, the functional unit of ECF-dependent signalling is formed by the ECF sigma factor and its cognate anti-sigma factor. Anti-sigma factors typically have a small N-terminal cytoplasmic domain, which is often linked to a C-terminal periplasmic domain via one or sometimes multiple transmembrane domains. Despite a low degree of sequence similarity, structural homology exists between the N-terminal domains of most Group 4 anti-sigma factors (Campbell *et al.*, 2007). This domain, which contains about 85–90 amino acids is sufficient to bind the ECF sigma factor and has therefore been designated ASD (from anti-sigma domain) (Campbell *et al.*, 2007). Binding of the ECF sigma factor to its anti-sigma factor and to the RNA polymerase is mutually exclusive. In absence of the inducing

Table 3.1 ECF sigma factors described in this chapter

Category	ECF sigma factor	Function
Cell envelope stress management (RpoE-like) sigma factors	*E. coli* RpoE (σ^E)	Maintenance of cell envelope integrity
	P. aeruginosa AlgT (σ^E or σ^{22})	Production of the exopolysaccharide alginate
	B. subtilis SigW (σ^W), SigX (σ^X) and SigM (σ^M)	Response to inhibitors of cell envelope integrity
	M. xanthus CarQ (σ^{CarQ})	Synthesis of carotenoids
Cytoplasmic stress sensing ECF sigma factors	*Mycobacterium tuberculosis* SigE (σ^E)	Response to cytoplasmic and cell envelope stresses
	S. coelicolor SigR (σ^R)	Oxidative stress response
	EcfG-like (σ^{EcfG}) sigma factors (i.e. *Methylobacterium extorquens*)	General stress response in α-proteobacteria
Cell-surface signalling (FecI-like) sigma factors	*E. coli* FecI	Transport of iron citrate
	P. aeruginosa FpvI	Transport of the *P. aeruginosa* endogenous siderophore pyoverdine
	Pseudomonas putida PupI	Uptake of the heterologous siderophores pseudobactin BN7 and BN8
	P. aeruginosa FoxI, FiuI and FemI	Uptake of the heterologous siderophores ferrioxamine B, ferrichrome and mycobactin, respectively
	Bordetella pertussis RhuI	Uptake of haem
	Serratia marcescens HasI	Transport of haem-haemophore
	P. aeruginosa PvdS	Production of pyoverdine, exotoxin A and endoprotease PrpL
	P. aeruginosa VreI	Synthesis of a type II and a two-partner (TPS) secretion system
	Ralstonia solanacearum PrhI	Synthesis of a type III secretion system

signal the anti-sigma factor sequesters the ECF sigma factor, only releasing it in response to the stimulus.

Different mechanisms activating ECF sigma factors have been identified (Fig. 3.2) (Staroń and Mascher, 2010a). One of them is by regulated intramembrane proteolysis (or RIP), a widely used signalling pathway by which a transmembrane anti-sigma factor is subjected to proteolytic degradation resulting in the release and activation of the ECF sigma factor (Fig. 3.2A) (Brown *et al.*, 2000; Heinrich and Wiegert, 2009). Another mechanism involves the function of a two-domain anti-anti-sigma factor. In response to a stimulus this protein is phosphorylated by a histidine kinase and binds to the anti-sigma factor. This results in the release and activation of the ECF sigma factor (Fig. 3.2B). In Gram-negative bacteria, cell-surface signalling (CSS) (Fig. 3.2C) is a common and important mechanism for the control of ECF

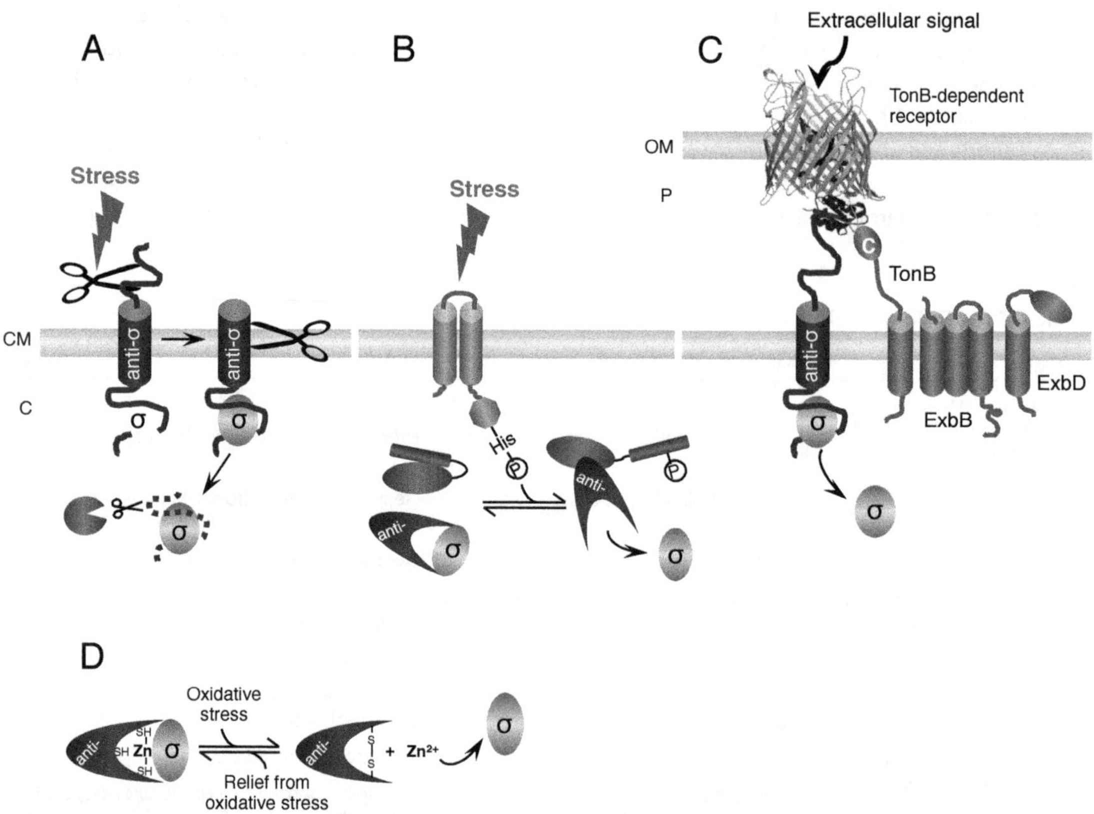

Figure 3.2 Mechanisms activating ECF sigma factors. (A) Regulated intramembrane proteolysis (RIP). The proteolytic cascade, first in the periplasm and in the cytoplasmic membrane, and later in the cytoplasm, responsible for the anti-sigma factor degradation in response to a stimulus (stress) is shown. (B) Anti-anti-sigma factor. The activation of a two-domain anti-anti-sigma factor by phosphorylation via a histidine kinase in response to stress is shown. Upon phosphorylation, this protein interacts with the anti-sigma factor, resulting in the release of the ECF sigma factor. (C) Cell-surface signalling (CSS). The interactions of the TonB-dependent outer membrane receptor with both the periplasmic domain of a transmembrane anti-sigma factor and the TonB protein in response to the environmental signal are shown. In addition, the interaction between the N-terminal region of the anti-sigma factor and the cytosolic ECF sigma factor is shown. Upon activation the sigma factor is released into the cytoplasm either bound or not to the cytoplasmic N-terminal domain of the anti-sigma factor. (D) Oxidation of cysteine residues. The reversible formation of intramolecular disulphide bonds in a zinc-dependent anti-sigma factor in response to oxidative stress and the release of both the Zn^{2+} ion and the ECF sigma factor are shown. OM, outer membrane; P, periplasm; CM, cytoplasmic membrane; C, cytoplasm.

sigma factor activity. In response to an environmental signal the sigma factor is activated via a signalling pathway that involves not only the ECF sigma factor and anti-sigma factor proteins, but also a TonB-dependent outer membrane receptor and the TonB-ExbB-ExbD system in the cytoplasmic membrane. The activity of the FecI-like group of ECF sigma factors is often controlled by CSS, as discussed below. Finally, ECF sigma factors can also be regulated by oxidative stress. Oxidation of cysteine residues within the ASD domain of specific zinc-dependent anti-sigma factors leads to the formation of intramolecular disulphide bonds, which results in the dissociation of the Zn^{2+} ion and release of the ECF sigma factor (Fig. 3.2D). Examples for all these mechanisms are given within this chapter. ECF classification also identified anti-sigma factors that carried conserved domains of still unknown function, suggesting the existence of alternative mechanisms for sigma factor activation (Staroń *et al.*, 2009). Moreover, not all ECF sigma factors are linked to potential anti-sigma factors, which suggests the existence of alternative pathways in the control of sigma factor activity. In some cases the ECF sigma factor might be regulated only at the transcriptional level, as has been described for *S. coelicolor sigE* (σ^E). This sigma factor is regulated by the CseBC two-component system in response to cell envelope stress (Hong *et al.*, 2002). Other ECF sigma factors not associated with anti-sigma factors have long C-terminal extensions that might play a role in the signalling pathway. A final group of ECF sigma factors is associated with completely unrelated proteins, such as putative serine/threonine kinases or other enzymes (Staroń *et al.*, 2009). Functional links between these protein pairs remain to be discovered.

An important question raised by the presence of such a large sigma factor repertoire is how the RNAPc is able to distinguish between different sigma factors to ensure proper gene regulation, especially when alternative sigma factors are activated. The concentration of primary σ^{70} factor within the cell is known to remain stable, even under circumstances that increase the concentration of one or more alternative sigma factors. Different sigma factors have to compete for the limited amount of RNAPc present in the cell (Ishihama, 2000; Maeda *et al.*, 2000a). The primary σ^{70} has the highest binding affinity for RNAPc and is present in the highest amount, which is sufficient to fully saturate the RNAPc (Maeda *et al.*, 2000a; Tiburzi *et al.*, 2008). Then how do alternative sigma factors compete with σ^{70} for the RNAP holoenzyme? A model has been proposed in which the σ^{70} containing RNAP is mainly bound in the cytoplasm to strong promoters, leaving weak promoters, such as ECF-dependent promoters that do not contain strong −10 and −35 consensus sequences, free (Grigorova *et al.*, 2006). When an alternative sigma factor reaches a threshold concentration, which usually needs to exceed the total amount of RNAP in the cell (Grigorova *et al.*, 2006; Tiburzi *et al.*, 2008), it will be able to compete for the RNAPc and direct transcription from weak promoters. In addition, specific regulators can be present or induced that facilitate sigma factor substitution of the RNAP (recently reviewed in Österberg *et al.* 2011). One strategy to achieve this is to decrease the ability of σ^{70} to bind to the RNAPc or by inhibiting RNAP-σ^{70} activity. A well-studied example of this includes ppGpp nucleotides and its co-factor DksA (Costanzo and Ades, 2006; Costanzo *et al.*, 2008) or Rsd, a stationary phase protein that sequesters σ^{70} (Jishage and Ishihama, 1998). An alternative approach is to increase the binding affinity of the alternative sigma factor for RNAPc, as has been shown for the regulatory protein Crl of *S. enterica*. This protein binds the alternative sigma factor σ^S and increases its association rate to the RNAP holoenzyme (England *et al.*, 2008; Monteil *et al.*, 2010).

Role of ECF sigma factors in stress management

All organisms have stress response systems that allow them to sense and respond to damaging conditions by altering gene expression. Bacteria use different regulatory pathways to respond to stresses, and alternative sigma factors play a key role in this process. In response to cytoplasmic stress, both the stationary phase sigma factor RpoS (σ^S) (Group 2) and the heat-shock sigma factor RpoH (σ^H) (Group 3) are considered to be master regulators in several bacteria. Within

Group 4, the ECF sigma factors belonging to the cytoplasmic sensing group also control cytoplasmic stress responses. However, an additional level of complexity is produced when the stress signal is generated outside the cytoplasm and the information must be communicated across the cytoplasmic membrane. This is known as cell envelope or periplasmic stress. Envelope stress responses play an important role in many processes, including protein folding, cell wall biosynthesis, and pathogenesis. ECF sigma factors, together with two-component systems, are known to be essential in this process (Raivio and Silhavy, 2001). The best-characterized ECF sigma factors controlling cell envelope stress are the ones belonging to the RpoE-like group.

Cell envelope stress management ECF sigma factors – RpoE-like sigma factors

RpoE-like sigma factors form a widely distributed group that is found in most bacterial phyla (Staroń *et al.*, 2009). The best-studied RpoE-like sigma factors are RpoE (σ^E) of *E. coli*, AlgT (σ^{AlgT}, also called AlgU) of *Pseudomonas aeruginosa*, SigW (σ^W), SigX (σ^X) and SigM (σ^M) of *Bacillus subtilis*, and CarQ (σ^{CarQ}) of *Myxococcus xanthus*. The corresponding anti-sigma factor of these sigma factors is generally a cytoplasmic membrane protein with one transmembrane region. Upon receiving a stimulus, the anti-sigma factor is usually degraded via the mechanism of regulated intramembrane proteolysis (Fig. 3.2A), which leads to the activation of the ECF sigma factor.

E. coli RpoE (σ^E)

In *E. coli* the ECF sigma factor σ^E plays a central role in maintaining cell envelope integrity (Raivio and Silhavy, 2001; Hayden and Ades, 2008). σ^E is important not only under damaging conditions that generate stress, but also during basic cellular physiology. In fact, *rpoE* is essential for bacterial viability (De Las Peñas *et al.*, 1997). σ^E is activated by stresses that disrupt protein folding in the cell envelope, such as heat shock, overproduction of outer membrane proteins, and mutations that inactivate periplasmic chaperones. The molecular mechanism controlling σ^E activation has been studied in great detail (reviewed in Ades, 2008). In unstressed cells, σ^E is sequestered at the cytoplasmic side of the inner membrane by the anti-sigma factor RseA (Fig. 3.3A). RseA is a transmembrane protein that binds σ^E with high affinity, thereby preventing the interaction of σ^E with the RNA polymerase. The co-crystal structure of the cytoplasmic domain of RseA bound to σ^E shows that the domains of σ^E that contact RseA are also involved in the interaction of σ^E with RNA polymerase. Therefore, binding of RseA and RNA polymerase to σ^E is mutually exclusive (Campbell *et al.*, 2003). The periplasmic protein RseB binds to the periplasmic domain of RseA and enhances the inhibition of σ^E (Fig. 3.3A). Release of σ^E from RseA is regulated by intramembrane proteolysis (Fig. 3.2A). This proteolytic cascade involves two cytoplasmic membrane anchored proteases, DegS and RseP (also known as YaeL) (Fig. 3.3) (Alba *et al.*, 2002; Kanehara *et al.*, 2002). The DegS and RseP proteases act sequentially, cleaving first the periplasmic domain of RseA and subsequently its transmembrane region (Fig. 3.3B). The signal inducing DegS activity is the presence of unfolded porins in the periplasm. The C-terminal peptides of the porins OmpC or OmpF can bind the PDZ domain of DegS, activating its protease function (Fig. 3.3B). After DegS- and RseP-mediated proteolysis, the cytoplasmic domain of RseA bound to σ^E is released into the cytoplasm, where RseA is fully degraded by cytoplasmic proteases such as the ClpXP complex (Fig. 3.3B) (Flynn *et al.*, 2003, 2004). Subsequently, free σ^E can bind the RNAPc and direct the transcription of the σ^E regulon genes.

In *E. coli* K12, the σ^E regulon is composed of 96 genes organized in 50 different transcriptional units (Rhodius *et al.*, 2006). The prediction of the σ^E regulon in nine different *E. coli*-related genomes has identified a 'core' and an 'extended' σ^E regulon (Rhodius *et al.*, 2006). Although the 'core' regulon includes genes that affect many different aspects of the cell, a significant fraction encodes chaperones required for the delivery and assembly of outer membrane proteins and LPS, and periplasmic proteases such as DegP to degrade misfolded proteins. In addition, at least two small RNAs are regulated that target porin-encoding mRNAs (Ruiz *et al.*, 2006). σ^E positively autoregulates its own expression, thereby ensuring a rapid

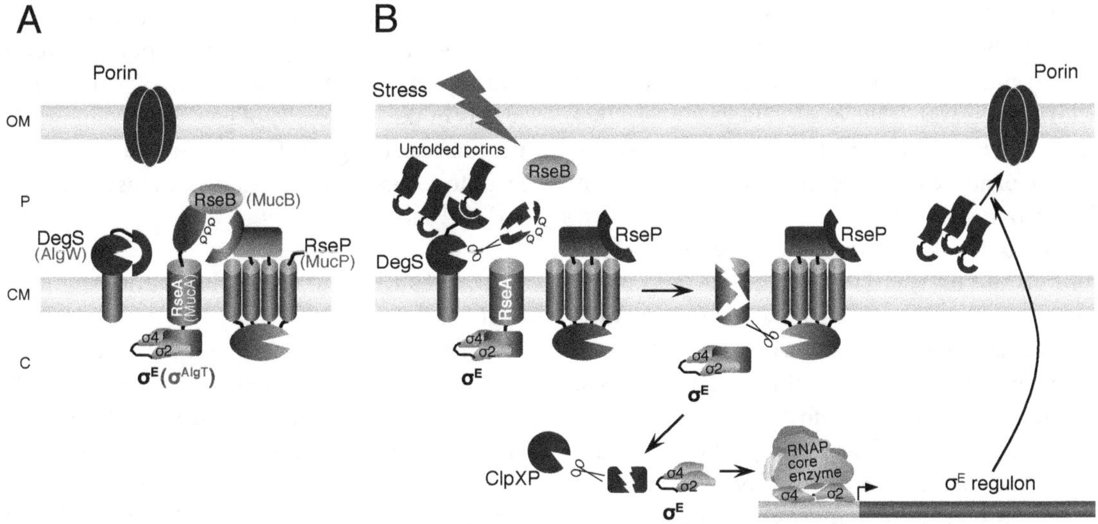

Figure 3.3 Activation of σ^E by cell envelope stress. (A) In the uninduced state, DegS and RseP are inactive by inhibitory interactions involving their corresponding PDZ domains, RseB and a glutamine-rich region of RseA. In this situation, RseB is bound to the periplasmic domain of RseA and σ^E to its cytoplasmic domain. The *P. aeruginosa* σ^E, RseA, RseB, DegS and RseP homologues (σ^{AlgT}, MucA, MucB, AlgW and MucP, respectively) are indicated between brackets. (B) Upon cell envelope stress, the C-terminal peptides of unfolded proteins bind to DegS, activating DegS. RseB is removed from RseA (via a still unknown mechanism) and activated DegS can cleave the periplasmic domain of RseA, which removes the interactions inhibiting the activity of RseP. RseP degrades then the transmembrane region of RseA releasing the cytoplasmic domain of RseA bound to σ^E into the cytoplasm. Finally the ClpX protease degrades the cytoplasmic domain of RseA allowing σ^E to interact with the RNA polymerase and direct transcription of target genes. Many of the genes of the σ^E regulon encode proteins that restore the proper folding of outer membrane proteins. OM, outer membrane; P, periplasm; CM, cytoplasmic membrane; C, cytoplasm.

amplification of the signal required to maintain cell envelope homeostasis. Simultaneously, it also induces the expression of the *rseA* anti-sigma factor, ensuring a negative feedback loop (Rhodius *et al.*, 2006). The majority of the predicted 'extended' regulon members are involved in functions known to be important for adaptation of the organism within its eukaryotic host (Rhodius *et al.*, 2006), and suggests a role for σ^E in facilitating bacterial survival *in vivo*.

P. aeruginosa AlgT (σ^{AlgT} or σ^{22})

The *P. aeruginosa algT mucABCD* locus is homologous to the *E. coli rpoE rseABC* cluster. This locus regulates the production of a capsular polysaccharide, alginate, which is responsible for the mucoid colony morphology found in most *P. aeruginosa* clinical isolates from chronic lung infections (i.e. from cystic fibrosis patients). Isolation of mucoid *P. aeruginosa* is normally an indicator of clinical deterioration and decreased life expectancy (Govan and Deretic, 1996). The accumulation of mucus in the airways produces limitations in pulmonary functions and alginate enables *P. aeruginosa* to grow as compact microcolonies or biofilms, thereby increasing its resistance to host defences and antibiotic treatment. The key to mucoid conversion is the expression of the alginate biosynthetic operon, whose transcription is regulated by the ECF sigma factor σ^{AlgT} (Hershberger *et al.*, 1995). The regulation of σ^{AlgT} activity follows a pathway similar to that of *E. coli* σ^E (Qiu *et al.*, 2007, 2008; Wood and Ohman, 2009) (Fig. 3.3). σ^{AlgT} activity is inhibited by the anti-sigma factor MucA, a RseA homologue (Fig. 3.3A). This inhibition is enhanced by the periplasmic protein MucB. Most mucoid *P. aeruginosa* isolates contain missense or nonsense mutations in *mucA* (Anthony *et al.*, 2002), which lead to the release of active σ^{AlgT} and concomitant overproduction of alginate. Like its *E. coli* homologue, MucA is also subjected to proteolysis, which is initially mediated by the intramembrane proteases AlgW (a DegS homologue) and MucP (a RseP homologue) and later in the cytoplasm by the ClpX and ClpP2 proteases (Qiu *et al.*, 2007, 2008). This leads to the full degradation of MucA and the release of active σ^{AlgT} into the cytoplasm, where it binds to the RNAPc.

Similar to the σ^{E} signalling cascade, envelope stress is also the signal inducing MucA proteolysis and σ^{AlgT} activation. A screening using a large variety of chemical agents showed that antibiotics that block various steps in peptidoglycan synthesis, including D-cycloserine, fosfomycin, tunicamycin, penicillin G and several cephalosporins, are strong activators of σ^{AlgT} activity (Wood *et al.*, 2006). Moreover, a transposon mutant screen has identified MucE as an activator of AlgW in a manner similar to the way how unfolded outer membrane proteins regulate σ^{E} activity in *E. coli* (Qiu *et al.*, 2007).

The σ^{AlgT} regulon has been determined by microarray analysis (Wood *et al.*, 2006; Wood and Ohman, 2009). Besides the alginate biosynthetic genes, about ~300 genes are part of this regulon. The σ^{AlgT} regulon includes mainly genes under the category of adaptation and protection, like genes involved in cell wall synthesis and the lipoproteins OsmC and OsmE. These last two proteins are induced upon osmotic shock and function to counteract stress due to defects in the cell wall. σ^{AlgT} also controls its own expression and that of the *mucABCD* genes. The last gene of this operon, *mucD*, encodes a serine protease similar to the periplasmic protease DegP of *E. coli*. The function of the DegP/MucD proteases is to eliminate misfolded proteins in the periplasm avoiding DegS/AlgW activation and consequently RseA/MucA proteolysis (Alba *et al.*, 2002; Kanehara *et al.*, 2003; Qiu *et al.*, 2007). The function of *mucC* is not clear at this point. In contrast to *mucA*, *mucB* and *mucD* mutants, inactivation of *mucC* does not cause any overt effects on alginate synthesis. However, it affects growth of *P. aeruginosa* under conditions of combined elevated temperature and increased ionic strength or osmolarity (Boucher *et al.*, 1997). Finally, the up-regulation of MucA and MucB by σ^{AlgT} ensures a negative feedback loop to restore σ^{AlgT} inactivation. In addition to σ^{AlgT}, there are several other transcriptional regulatory genes whose expression is elevated during the σ^{AlgT}-mediated envelope stress response (Wood and Ohman, 2009). In brief, the σ^{AlgT} response results in the expression of genes encoding products that can bring relief from stress and ultimately leads to reappearance of MucA and restoration of the σ^{AlgT} sequestration.

B. subtilis SigW (σ^{W}), SigX (σ^{X}) and SigM (σ^{M})

The *sigW*, *sigX* and *sigM* genes are three of the seven genes that encode ECF sigma factors in *B. subtilis* (Helmann, 2002). These sigma factors are known to play a crucial role in the response of *B. subtilis* to inhibitors of cell envelope integrity, mainly to antimicrobial compounds and cell wall antibiotics, although these responses have different features (Helmann, 2002). σ^{X} was the first ECF sigma factor investigated in *B. subtilis*, and, based on the genes controlled by this sigma factor, a role in the regulation of the composition and properties of the cell envelope was proposed (Huang and Helmann, 1998; Helmann, 2002; Cao and Helmann, 2004). However, since other *B. subtilis* sigma factors, mainly σ^{W}, also control the expression of most of the σ^{X}-dependent genes (Huang *et al.*, 1998; Helmann, 2002), it has been difficult to assign a specific function to σ^{X}. The overlap between the σ^{W} and σ^{X} regulons seems to depend on the −10 box of the promoter region. Whereas σ^{X} recognizes CGAC −10 sequences and σ^{W} recognizes CGTA sequences, both can recognize CGTC sequences (Qiu and Helmann, 2001). A consensus for σ^{M}-dependent promoters has also been proposed consisting of TGAAAC-N_{17}-CGTC (Jervis *et al.*, 2007), which is similar to σ^{X}- and σ^{W}-recognized promoters. This has made it difficult to define a specific and individual role also for σ^{M}. However, the assignment of a specific role for σ^{W} has been easier since of the ~60 genes that comprise its regulon only 15 are also regulated by σ^{X}. The σ^{W} regulon is often referred to as an 'antibiosis' regulon, since it is mainly involved in the production of, and resistance to, compounds that inhibit the growth of competing microorganisms (Helmann, 2002, 2006; Luo and Helmann, 2009). σ^{W} promotes the expression of genes involved in the defence to antibiotics and antimicrobial peptides, detoxification and cell envelope synthesis. In addition, σ^W also promotes the production of one or more antimicrobial compounds. Thus, exposure to antibiotics seems to induce both a defensive as well as an offensive response (Helmann, 2002, 2006). Despite the overlap of σ^W, σ^X and σ^M in promoter recognition, these three sigma factors differ in other aspects, as for example their activity at different phases of the cell growth and their induction by different stimuli (Cao *et al.*, 2002; Helmann, 2002; Thackray and Moir, 2003).

sigW, *sigX* and *sigM* genes are cotranscribed with their own membrane-spanning anti-sigma factor, *rsiW*, *rsiX* and yhdL respectively, from an autoregulated promoter (Huang *et al.*, 1998; Huang and Helmann, 1998; Horsburgh and Moir, 1999). Despite the presence of a specific anti-sigma factor little is known about the mechanisms activating σ^X and σ^M. However, the mechanism activating *B. subtilis* σ^W has been extensively studied and it is known to be similar to *E. coli* σ^E (Fig. 3.3). Upon receiving the stimulus, the σ^W anti-sigma factor RsiW is subjected to regulated intramembrane proteolysis (Fig. 3.2A), which is mediated by the PrsW (formerly YpdC) and RasP (YluC) proteases (Schöbel *et al.*, 2004; Heinrich and Wiegert, 2006). PrsW is responsible for the degradation of the periplasmic domain of RsiW, but in contrast to *E. coli* DegS and *P. aeruginosa* AlgW (Fig. 3.3) this protease does not belong to the HtrA family of proteases, but to a novel family of metalloproteases. RasP is a homologue of *E. coli* RseP and *P. aeruginosa* MucP (Fig. 3.3) and acts in a similar way by cleaving the transmembrane domain of RsiW after the degradation of its periplasmic domain. Complete degradation of RsiW is also mediated by cytoplasmic proteases such as ClpXP (Fig. 3.3; Zellmeier *et al.*, 2006).

Myxococcus xanthus CarQ (σ^{CarQ})

M. xanthus has a complex life cycle, involving a large degree of social interactions and signalling between cells. In response to starvation, it undergoes a developmental programme that culminates in the formation of multicellular fruiting bodies and conversion of the vegetative cells into myxospores (Kaiser, 2003). *M. xanthus* is also highly pigmented. Some of these pigments are made constitutively, but the synthesis of other pigments, such as the red-coloured carotenoids, are induced upon illumination with blue light. The synthesis of carotenoids is a protective response that prevents cellular damage caused by photo-excited molecules such as protoporphyrin, which absorbs light generating single oxygen ($^1O_2$), a potent antioxidant that damages the cells. Given their chemical structure, carotenoids can quench $^1O_2$ providing protection to the cell (Cogdell *et al.*, 2000). The induction of carotenoid synthesis by light in *M. xanthus* requires the expression of the *carQRS* operon. CarQ is an ECF sigma factor (σ^{CarQ}), whose activity is controlled by the transmembrane anti-sigma factor CarR (Browning *et al.*, 2003). Upon illumination, CarR is inactivated (probably by proteolysis) in a process that requires the action of CarF (Browning *et al.*, 2003). Free σ^{CarQ} activates expression of carotenoid synthesis genes and of the *carQRS* operon. This results in high expression of σ^{CarQ}, the anti-sigma CarR, and of CarS, the third component of the *carQRS* operon. CarS interacts with CarA, a repressor protein encoded by the carotenoid biosynthetic operon. CarA represses expression of the carotenoid biosynthetic operon in absence of light. Interaction of CarA with CarS allows σ^{CarQ}-RNAP to start transcription of the carotenoid biosynthetic genes (López-Rubio *et al.*, 2004; Pérez-Marín *et al.*, 2004).

Cytoplasmic stress sensing ECF sigma factors

A subgroup of ECF sigma factors has been recently classified as cytoplasmic-sensing ECF sigma factors (Staroń *et al.*, 2009). In contrast to the RpoE-like group, the ECF sigma factors belonging to this group do not share a high degree of sequence similarity. Instead, their unifying feature is the presence of a cytoplasmic anti-sigma factor (Staroń *et al.*, 2009). The anti-sigma factors of this group contain a zinc-binding motif (referred to ZAS domain from zinc-dependent anti-sigma) in the ASD, which binds Zn^{2+} and contacts the sigma factor (Paget *et al.*, 2001a). Most of these ECF sigma factors are involved in redox sensing and oxidative stress responses, and in some cases, like σ^E of *Mycobacterium tuberculosis*, respond to both cytoplasmic as well as periplasmic stresses. In addition to *M. tuberculosis* σ^E, the best studied examples within this group include σ^R of *Streptomyces coelicolor*, and σ^{EcfG} of α-proteobacteria (e.g. *Methylobacterium extorquens*).

M. tuberculosis SigE (σ^E)

σ^E of *M. tuberculosis* is one of the 10 ECF sigma factors encoded by this pathogen (Rodrigue *et al.*, 2006; Sachdeva *et al.*, 2010). The gene encoding σ^E, *sigE*, is transcribed from three different promoters: promoter P1, active during growth under normal conditions; promoter P2, induced under surface stress and alkaline pH by the MprAB two-component system; and promoter P3, induced under oxidative stress and heat shock by σ^H, another mycobacterial ECF sigma factor (see below) (Rodrigue *et al.*, 2006; Sachdeva *et al.*, 2010). σ^E activity is controlled by RseA, a ZAS anti-sigma factor encoded by a downstream gene. A recent study has shown that σ^E responds to both cytoplasmic as well as cell envelope stress, and that σ^E–RseA interaction is regulated *in vivo* by different mechanisms under these different conditions of stress (Barik *et al.*, 2010). Two cysteine residues within the anti-sigma domain (ASD) of RseA are required for the interaction of σ^E with RseA. This interaction is abrogated under oxidative stress, probably due to the oxidation of the cysteine residues and the concomitant release of the Zn^{2+} ion as described for other bacteria (Fig. 3.2D). No degradation of RseA was observed under these conditions (Barik *et al.*, 2010). In cells subjected to envelope stress, such as the presence of SDS or vancomycin, RseA undergoes a PknB-mediated phosphorylation that is a prerequisite for the proteolytic degradation of RseA. This proteolytic degradation is mediated by the ClpC1P2 complex (Barik *et al.*, 2010).

The *sigE rseA* operon also encodes a membrane-localized HtrA-like serine protease and a twin-arginine translocation (Tat) system protein (TatB). Interestingly the N-terminal region of RseA contains a twin arginine couple and it has been proposed that RseA is targeted to the bacterial membrane through the Tat system, where it is degraded by the HtrA-like protease in response to the stimuli releasing σ^E (Dona *et al.*, 2008). However, the involvement of these proteins in the activation of *M. tuberculosis* σ^E has not been demonstrated yet.

σ^E-dependent genes encode proteins belonging to different subclasses, including transcriptional regulators, enzymes involved in fatty acid degradation, heat-shock proteins, surface-exposed proteins, and proteins involved in maintaining cell envelope integrity (Rodrigue *et al.*, 2006; Fontán and Smith, 2009; Sachdeva *et al.*, 2010). Therefore it is not surprising that the *M. tuberculosis* σ^E regulon also includes genes required for cell envelope integrity during infection. In accordance with these observations, a *M. tuberculosis sigE* mutant has decreased viability in macrophages and is strongly attenuated in mice (Rodrigue *et al.*, 2006; Sachdeva *et al.*, 2010).

S. coelicolor SigR (σ^R)

In *S. coelicolor* σ^R plays a key role in the regulation of oxidative stress responses. Mutations in the *sigR* gene (encoding σ^R) make *S. coelicolor* sensitive to both superoxide generators and diamide, a chemical that oxidizes thiols in the cytoplasm, leading to the formation of disulphides bonds (so-called disulphide stress) (Paget *et al.*, 2001a; Newton and Fahey, 2008; Park and Roe, 2008). Disulphide bonds are normally not present in cytoplasmic proteins and their formation can inhibit protein activity. The *sigR* gene forms an operon with the *rsrA* gene. RsrA is a zinc-dependent anti-sigma factor (ZAS) that, in its reduced state, binds σ^R and by doing so prevents σ^R-mediated transcription (Fig. 3.2D). Upon oxidative stress, which changes the oxidation-reduction state of the cytoplasmic thiols, RsrA is inactivated by the reversible formation of an intramolecular disulphide bond, causing dissociation of the Zn^{2+} ion and the release of σ^R (Fig. 3.2D) (Li *et al.*, 2002; Zdanowski *et al.*, 2006). The ~30 known σ^R target genes include the *sigR rsrA* operon, those encoding thioredoxins (to reduce cytoplasmic disulphide bonds in proteins), enzymes involved in mycothiol biosynthesis, proteases, and other gene products involved in the disulphide stress response (Paget *et al.*, 2001b). Expression of thiol reducers and mycothiol systems result in the reduction of RsrA, which is then again able to sequester σ^R, thereby shutting off the response. However, there is also a positive feedback loop in σ^R-mediated regulation, which involves the synthesis of a higher molecular weight form of σ^R (referred to as $\sigma^{R'}$) (Kim *et al.*,

2009). This induced $\sigma^{R'}$ isoform is encoded by a *sigR* transcript derived from an upstream P2 promoter, which is σ^{R}-dependent (Kim *et al.*, 2009). $\sigma^{R'}$ is subjected to proteolysis by the ClpP1P2 proteases, which is part of the homeostatic pathway that enables an effective turn off stress response.

In other actinomycetales, such as *Corynebacterium glutamicum* and *M. tuberculosis*, the *S. coelicolor* σ^{R} orthologue σ^{H} controls the heat-shock response. The gene encoding σ^{H}, *sigH*, is induced after heat-shock and after exposure to diamide (Raman *et al.*, 2001; Manganelli *et al.*, 2002; Ehira *et al.*, 2009). In *M. tuberculosis*, *sigH* expression is also induced during macrophage infection, although σ^{H} is dispensable for growth in this condition (Manganelli *et al.*, 2002). A *sigH* mutant is able to reach the same bacterial load in infected mice as the wild-type strain, but produces fewer granulomas and kills the animal more slowly (Kaushal *et al.*, 2002).

EcfG-like sigma factors

The group of EcfG-like sigma factors is exclusively found in α-proteobacteria and seems to control the general stress response in this group (Staroń *et al.*, 2009). These ECF sigma factors share a conserved genomic context that includes an EcfG-like sigma factor (σ^{EcfG}), a gene encoding a NepR-like anti-sigma factor, a PhyR-like response regulator and, usually, a sensor histidine kinase (Fig. 3.2B) (Staroń *et al.*, 2009). The NepR-like anti-sigma factor is a soluble protein, usually smaller than 50 amino acids, that does not contain sequence similarity to other anti-sigma factors (Staroń *et al.*, 2009). The PhyR-like response regulator belongs to a new family of response regulators only found in α-proteobacteria (Francez-Charlot *et al.*, 2009; Herrou *et al.*, 2010). These proteins contain an N-terminal ECF sigma factor domain and a C-terminal receiver domain of a response regulator. PhyR does not seem to act as a transcriptional regulator itself, but instead controls gene expression through interaction with the anti-sigma factor (hence known as anti-anti-sigma factor). Accordingly, the sigma factor domain of PhyR does not contain conserved region 2.4 involved in −10 binding, and the −35 binding region 4.2 is degenerate (Francez-Charlot *et al.*, 2009). The best studied example within this group is PhyR-/NepR-/σ^{EcfG1}-mediated regulation in *Methylobacterium extorquens* (Francez-Charlot *et al.*, 2009). In unstressed cells, PhyR is inactive and σ^{EcfG1} is sequestered by the anti-sigma factor NepR (Fig. 3.2B). In response to stress, which is probably sensed by the associated histidine kinase, PhyR is phosphorylated and interacts with NepR, thereby releasing its interaction with σ^{EcfG1} (Fig. 3.2B). σ^{EcfG1} binds to the RNA polymerase and initiates transcription of the general stress response regulon (Francez-Charlot *et al.*, 2009). σ^{EcfG}-dependent gene expression is induced by heat, osmotic stress, carbon/nitrogen starvation and desiccation (Staroń and Mascher, 2010b).

ECF sigma factors in the regulation of iron uptake – cell-surface signalling (CSS)

The FecI-like sigma factors are one of the predominant groups of ECF sigma factors. This group has been classified as iron starvation sigma factors based on a common role in the regulation of iron uptake (Leoni *et al.*, 2000). Iron is essential for bacterial growth and survival, but in most environments iron acquisition is greatly complicated owing to its insolubility and extremely low concentration. To fulfil their requirements for iron bacteria produce and secrete high affinity iron chelating compounds called siderophores, or use siderophores produced by other organisms (referred as heterologous siderophores). In addition, pathogens can also use host iron complexes as iron sources, such as transferrin, lactoferrin, haem or haemoglobin (Ratledge and Dover, 2000; Wandersman and Delepelaire, 2004). However, although iron is crucial for bacterial growth, bacteria need to strictly regulate iron acquisition to prevent the accumulation of iron-induced toxic radicals inside the cell. Iron homeostasis is mainly controlled by the general iron regulator Fur that represses gene expression in iron-sufficient conditions (Escolar *et al.*, 1999; Hantke, 2001). Furthermore, the expression of iron uptake systems for heterologous siderophores and host molecules only makes sense if these molecules are present in the environment. In contrast to other ECF sigma factors, the expression of iron starvation ECF sigma factors, and of their cognate anti-sigma factors, is not autoregulated but induced by low iron concentrations through Fur. Once produced, these sigma factors mainly mediate expression of genes required for iron acquisition.

In Gram-negative bacteria the activity of iron starvation sigma factors is controlled by a signal transduction mechanism involving both the outer and the cytoplasmic membranes, the so-called cell-surface signalling (CSS) systems. In this system, the functional unit of ECF-dependent signalling is formed not only by the ECF sigma factor and its cognate anti-sigma factor, but also by an outer membrane receptor (Fig. 3.2C). The CSS outer membrane receptor belongs to the family of TonB-dependent receptors, which are usually involved in iron-siderophore uptake. The signal transduction pathway of CSS starts by binding of the inducing signal to the outer membrane receptor, which transduces the signal to the anti-sigma factor in the cytoplasmic membrane leading to the activation of the ECF sigma factor in the cytoplasm (Fig. 3.2C). The anti-sigma factors of CSS systems are transmembrane proteins that contain a conserved periplasmic domain involved in the interaction with the outer membrane receptor and a cytoplasmic N-terminal region that binds the sigma factor. Although the exact mode of action of CSS anti-sigma factors is not completely understood, some of these proteins seem to function as classical anti-sigma factors and inhibit the binding of the ECF sigma factor to the RNA polymerase. Deletion of these anti-sigma factors usually results in increased sigma-dependent transcription even in the absence of the inducing signal. However, other CSS anti-sigma factors are not only involved in inhibition of the sigma factor, but are also required for sigma factor activity. These proteins are therefore referred to as sigma factor regulators. In this part of the chapter we review the current knowledge on CSS regulation, with the main focus on the *E. coli* Fec system, which is the paradigm CSS system, and the *P. aeruginosa* Fpv system, which represents an important variation of these systems. Other CSS systems involved in the regulation of iron uptake together with CSS systems regulating virulence will be also discussed.

The *E. coli* Fec system: the paradigm CSS system

The *E. coli* iron citrate-mediated signalling pathway is one of the most studied and best understood CSS systems. It was recognized already years ago that this bacterium could employ citrate as a source of iron, and that the uptake of iron citrate was mediated by an iron citrate inducible transport system (Frost and Rosenberg, 1973; Hussein *et al.*, 1981). Although at first the general belief was that the whole iron citrate complex was taken up into the cell, it was later demonstrated that there is no co-transport of iron and citrate into the cytoplasm (Hussein *et al.*, 1981). This meant that iron citrate could not activate expression of its transport system from the cytoplasm and that signal perception took place outside the cell, which revealed the existence of a novel type of signal transduction mechanism in bacteria (Braun, 1997).

The *E. coli* iron citrate transport system is encoded by the *fecABCDE* operon and is regulated by the *fecIR* operon located upstream *fecABCDE* (Braun, 1997). The *fecI* gene encodes an ECF sigma factor and *fecR* a sigma factor regulator. Transcription of *fecIR* is iron-regulated through the Fur repressor protein while transcription of *fecABCDE* is negatively regulated by Fur and positively by σ^{FecI} (Enz *et al.*, 1995; Härle *et al.*, 1995). The *fecA* gene codes for a TonB-dependent outer membrane receptor and the *fecBCDE* genes for a cytoplasmic membrane ABC transport system that carries iron from the periplasm into the cytoplasm (Braun, 1997). TonB-dependent receptors are predominantly involved in the transport of iron–siderophore complexes across the outer membrane. In order to facilitate substrate passage, the receptor is energized by a cytoplasmic membrane protein complex composed of TonB-ExbB-ExbD. Of this complex the TonB protein is the one that physically interacts with the outer membrane receptor, hence the name TonB-dependent receptor (Fig. 3.2C) (Braun and Endriss, 2007; Krewulak and Vogel, 2008). Coupling with the cytoplasmic membrane is necessary as the iron–siderophore complex has to be actively transported across the outer membrane, where there is no other source of energy available. Crystal structures of several

TonB-dependent receptors have revealed, despite low sequence similarity, a common domain architecture (recently reviewed in Noinaj *et al.*, 2010). The C-terminal part of these proteins forms a large β-barrel in the outer membrane that consists of 22 antiparallel β-strands that are connected via short periplasmic turns and larger surface exposed loops. This β-barrel produces a big pore of approximately 35–40 Å (Fig. 3.4). This pore is occluded by a globular plug domain, also called cork or hatch, in absence of the substrate, which prevents the aspecific diffusion of large molecules across the outer membrane. In order for the substrate to pass, the plug must undergo conformational changes that temporarily allow opening of the pore. Whether the plug domain opens the pore by complete retraction or by creating an opening upon conformational rearrangement is still a matter of debate (Noinaj *et al.*, 2010). TonB-dependent receptors also contain the so-called TonB box, a β-strand of only six semi-conserved amino acids situated near the N-terminus of the protein. This region is located in the periplasm and is the domain interacting with the TonB protein (Noinaj *et al.*, 2010). How TonB exactly energizes the receptor is unknown, but simulation experiments have demonstrated that, despite the small interface, the interaction between TonB and the receptor is relatively stable, which would allow a strong mechanical coupling (Gumbart *et al.*, 2007).

Most TonB-dependent receptors have only a role in siderophore transport. However, the *E. coli* FecA receptor functions as both a transport and a signalling protein, mediating the uptake of iron citrate and regulating *fecABCDE* gene transcription (Härle *et al.*, 1995). These functions are independent and transport of iron citrate across the outer membrane is not required for transcription induction to occur and vice versa (Härle *et al.*, 1995; Kim *et al.*, 1997). This finding and the identification of a similar regulatory

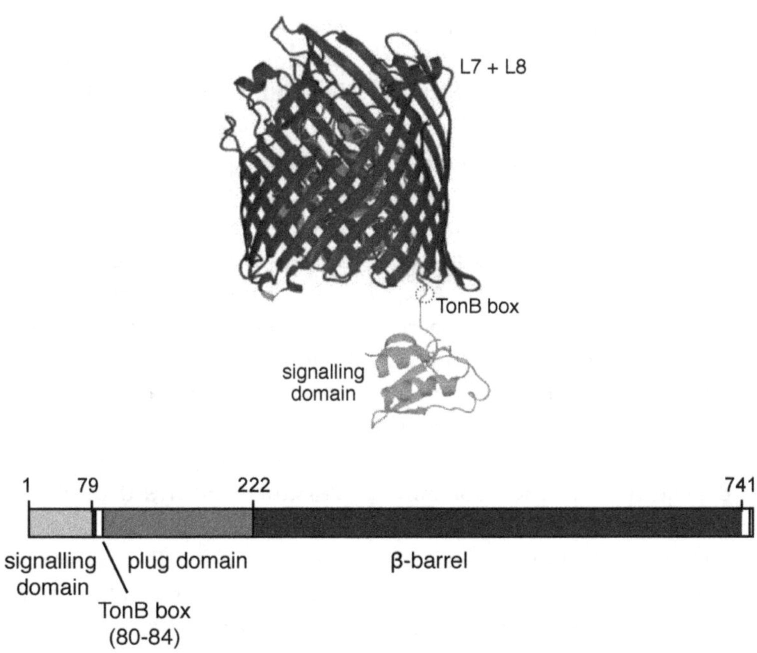

Figure 3.4 Structure of the *E. coli* FecA TonB-dependent transducer. The mature apo-FecA receptor (without iron citrate bound) is shown. The C-terminal β-barrel (dark) consists of 22 anti-parallel β-strands and forms a pore in the outer membrane. Extracellular loops 7 and 8, which play a key role in substrate recognition, are indicated. The plug domain occludes the β-barrel and contains the periplasm-located TonB-box. The N-terminal signalling domain comprises residues 1–79 and extends into the periplasm. Below is a schematic overview of the FecA structure, with amino acid positions of respective domains of the mature protein indicated. Structure was adapted from Ferguson *et al.* (2002) and Garcia-Herrero and Vogel (2005).

system in *Pseudomonas putida* (Koster *et al.*, 1994) resulted in the identification of a whole subfamily of TonB-dependent receptors involved in both transport and signalling. These receptors have been designated as TonB-dependent transducers (Koebnik, 2005). TonB-dependent transducers are easily recognizable, since they contain an additional N-terminal extension of approximately 70–80 amino acids that is not present in normal TonB-dependent receptors (Fig. 3.4). This extension contains the so-called signalling domain that interacts with the anti-sigma factor (or sigma factor regulator) in the cytoplasmic membrane (Fig. 3.2C). The signalling domain determines the specificity of the transduction pathway but has no effect on the binding and transport of the siderophore. FecA, as most TonB-dependent transducers, can bind both iron-free and iron-loaded siderophore (Ferguson *et al.*, 2002; Yue *et al.*, 2003). While interaction with iron-free citrate only induces small conformational changes in FecA, binding of iron-loaded citrate produces remarkable conformational changes in the receptor, mainly in the extracellular loops 7 and 8 of the β-barrel, in the plug domain and the TonB-box (Ferguson *et al.*, 2002; Yue *et al.*, 2003; Sauter and Braun, 2004). These conformational changes induce the interaction of the signalling domain of FecA with FecR, which is the starting point of the signalling pathway. FecA–FecR interaction has been demonstrated *in vitro* by binding assays. These assays showed that the N-terminal $FecA_{1-79}$ domain interacts with the C-terminal FecR periplasmic domain (Enz *et al.*, 2003a). Two solvent exposed residues, D_{43} and G_{39}, of the FecA signalling domain seem to be crucial for this interaction. These residues are not conserved in other members of the TonB-dependent transducer subfamily and are therefore probably involved in determining the specificity of the transduction pathway.

The FecR protein is a three-domain protein that contains an N-terminal cytoplasmic domain (residues 1–85), a transmembrane domain (residues 86–100) and a large periplasmic C-terminal domain (residues 101–317) (Braun *et al.*, 2006). Binding assays have shown that the region comprising residues 237–317 of the periplasmic domain is sufficient for interaction with FecA (Enz *et al.*, 2003a). The FecR N-terminal cytoplasmic region binds σ^{FecI}, mainly domain 4 of σ^{FecI} (Mahren *et al.*, 2002). Although full-length FecR is not able to activate σ^{FecI} in the absence of ferric citrate, the cytoplasmic tail of FecR ($FecR_{1-85}$) is required for σ^{FecI} activity (Ochs *et al.*, 1995; Stiefel *et al.*, 2001; Mahren and Braun, 2003). In fact, it was demonstrated that $FecR_{1-85}$ increases the affinity of σ^{FecI} for the RNA polymerase and protects σ^{FecI} from proteolytic degradation (Mahren and Braun, 2003). Interestingly, $FecR_{1-85}$ was co-purified with σ^{FecI} bound to the RNAP and it is therefore possible that the N-terminal domain of FecR is included in the σ^{FecI}–RNAP complex (Mahren and Braun, 2003). A model has been proposed in which, upon iron citrate induction, the cytoplasmic tail of FecR is cleaved, possibly by RseP, and together with σ^{FecI} released into the cytoplasm (Braun *et al.*, 2006). This complex will then bind the RNAPc and initiate *fecABCDE* transcription. σ^{FecI} has a very narrow promoter specificity, it seems to recognize only the *fecA* promoter (Maeda *et al.*, 2000b). σ^{FecI} recognizes the promoter sequence TGTCCT at –10 and GAAAAT at –35 (Enz *et al.*, 2003b). Insertion of point mutations in these promoter elements showed that, whereas the –35 box can be partially replaced, point mutations in the –10 box strongly reduce *fecA* expression (Enz *et al.*, 2003b). A comparable result was obtained with the *sigX* promoter of *B. subtilis* σ^{SigX} (Huang and Helmann, 1998). Since, as mentioned before, ECF sigma factors recognize similar –35 regions, it is not unexpected that nucleotide substitutions are tolerated in this region but not in the –10 region, which determines the specificity of the sigma factor promoter recognition.

The *P. aeruginosa* Fpv system: a variation on the theme

The *P. aeruginosa* Fpv system responds to the presence of the *P. aeruginosa*'s own siderophore pyoverdine and regulates both the uptake and the synthesis of this siderophore (Beare *et al.*, 2003). This CSS system resembles the *E. coli* Fec system in several aspects, but is also unique in several ways. One common feature is the structure and function of the outer membrane component FpvA. FpvA belongs to the TonB-dependent transducer family and its structure is similar to FecA (Fig. 3.4) (Cobessi *et al.*, 2005;

Brillet *et al.*, 2007; Wirth *et al.*, 2007). Pyoverdine binding to the receptor is mediated by the extracellular pocket of the receptor and the plug domain. Reconstitution experiments have very recently shown that the plug and the β-barrel domains of FpvA function independently and are able to form an active receptor even when they are expressed as two different proteins (Nader *et al.*, 2011). Moreover, iron-pyoverdine is not able to diffuse through the FpvA β-barrel if the plug domain is absent, which suggests that the plug is required for pyoverdine transport. This is consistent with a gated mechanism for pyoverdine uptake in which the plug domain does not exit the β-barrel. Therefore, the receptor does not form a complete open channel during pyoverdine transport (Nader *et al.*, 2011). FpvA can bind both iron-free and iron-loaded pyoverdine, but only upon interaction with iron-loaded pyoverdine a number of conformational changes are induced (Schalk *et al.*, 2001; Folschweiller *et al.*, 2002). Especially the signalling domain of FpvA is subjected to major changes upon iron-pyoverdine binding (Brillet *et al.*, 2007). In the apo-FpvA form (without pyoverdine), the signalling domain interacts with the TonB box. Binding of Fe-pyoverdine to FpvA destabilizes this interaction and these domains are then free to interact with their partners in the cytoplasmic membrane, the FpvR anti-sigma factor and the TonB protein, respectively. The interaction between the TonB box and the signalling domain in the absence of an inducing signal could explain why some point mutations within the signalling domain of FpvA block transport of pyoverdine (Breidenstein *et al.*, 2006), while this domain can be completely removed in FecA without affecting its transport function. It is possible that these point mutations stabilize the interaction of the signalling domain with the TonB box that, as a result, can no longer interact with the TonB protein upon signal perception, thereby abolishing pyoverdine uptake.

The pyoverdine CSS system of *P. aeruginosa* is in several aspects an unusual and unique CSS system. Whereas most CSS systems are induced by siderophores produced by other organisms or other iron sources (such as citrate or haem), here the inducing signal is its own siderophore. This situation could be more common in pseudomonads, since homologues to the *P. aeruginosa* pyoverdine TonB-dependent transducer are present in most *Pseudomonas* species (Llamas and Bitter, 2010). Secondly, most CSS systems, like the *E. coli* Fec system, activate only genes involved in the transport of the inducing signal (i.e. iron citrate). However, the pyoverdine CSS system regulates not only pyoverdine uptake, but also genes involved in pyoverdine biosynthesis (Shen *et al.*, 2002) and, most importantly, also the synthesis of two virulence factors, exotoxin A and the endoprotease PrpL (Lamont *et al.*, 2002; Beare *et al.*, 2003). Because the signal is produced by the bacterium itself and regulates its own production and that of several other genes, one could argue that pyoverdine is in fact an autoinducer. A third and final unusual characteristic of the pyoverdine CSS is that the sigma factor regulator FpvR does not interact with one, but with two ECF sigma factors: PvdS (σ^{PvdS}) and FpvI (σ^{FpvI}) (Lamont *et al.*, 2002; Beare *et al.*, 2003).

In contrast to *E. coli* FecR, FpvR functions solely as an anti-sigma factor (Mettrick and Lamont, 2009). The first 67 residues of FpvR are sufficient to bind both σ^{PvdS} and σ^{FpvI} (Rédly and Poole, 2005), and in absence of pyoverdine, FpvR promotes proteolytic degradation of σ^{PvdS} thereby reinforcing its role as anti-sigma factor (Spencer *et al.*, 2008). In agreement with a purely anti-sigma function, deletion of *fpvR* results in the activation of σ^{PvdS}- and σ^{FpvI}-dependent transcription in absence of pyoverdine. Furthermore, *fpvR* overexpression inhibits the activity of both sigma factors (Lamont *et al.*, 2002; Beare *et al.*, 2003). However, in contrast to σ^{PvdS}, σ^{FpvI} seems to be more unstable in absence of FpvR, suggesting that FpvR could protect this sigma factor from proteolytic degradation (Rédly and Poole, 2005). σ^{PvdS}-and σ^{FpvI} are quite different in sequence and apparently they bind FpvR in a similar but not identical manner (Rédly and Poole, 2005), which could explain this difference. The *fpvR* gene is located adjacent to *fpvI* but in the opposite orientation, which means that, in contrast to most ECF/anti–sigma pairs, they are not co-transcribed. The *pvdS* gene is not located near *fpvR*, but next to the pyoverdine biosynthetic locus as a single gene. The expression of the three genes is controlled by iron via Fur (Ochsner and Vasil, 1996). σ^{FpvI}

only controls the expression of the pyoverdine outer membrane receptor FpvA, whereas σ^{PvdS} regulates the expression of more than 20 genes. The σ^{PvdS} regulon mainly includes genes involved in pyoverdine biosynthesis, but also genes encoding the virulence factors exotoxin A and PrpL endoprotease, as well as the *regAB* and *ptxR* genes that positively regulate exotoxin A production (Vasil *et al.*, 1998; Leoni *et al.*, 2000; Wilderman *et al.*, 2001; Wilson *et al.*, 2001; Gaines *et al.*, 2007). A consensus binding promoter sequence has been predicted for *P. aeruginosa* σ^{PvdS}, consisting of TAAAT-N_{16}-CGT (Swingle *et al.*, 2008). In agreement with the main role of the −10 region in ECF sigma factors promoter specificity, point mutations in the −10 region of the σ^{PvdS} promoter resulted in a stronger effect on promoter activity as compared with mutations in the −35 region (Swingle *et al.*, 2008).

Other CSS systems regulating iron uptake

Iron–siderophore-mediated CSS systems

CSS systems involved in the regulation of iron uptake have been mainly studied in bacteria of the genus *Pseudomonas*. Fluorescent pseudomonads are entangled in a fierce competition for iron, which is reflected by their ability to use many heterologous siderophores of bacterial or fungal origin. For this they have an impressive amount of different TonB-dependent receptors (Stover *et al.*, 2000; Nelson *et al.*, 2002; Paulsen *et al.*, 2005). Most of these receptors are only produced in presence of their cognate siderophore, and CSS often controls this regulatory process. CSS systems in *Pseudomonas* have been recently reviewed in Llamas and Bitter (2010). The saprophytes *P. putida* and *Pseudomonas fluorescens* contain 13 and 20 CSS systems, respectively, although only a few of those have been characterized. For example, the PupIRB system of *P. putida* was, together with the *E. coli* FecIRA system, the first CSS system reported (Koster *et al.*, 1994). PupIRB regulates the uptake of the heterologous siderophores pseudobactin BN7 and BN8 produced by other *Pseudomonas* strains. PbrA, the *P. fluorescens* homologue of *P. aeruginosa* σ^{PvdS} is involved in the transcription of the pseudobactin M114 biosynthetic genes and in the synthesis of the major extracellular protease (Maunsell *et al.*, 2006). However, CSS regulation has been analysed in most detail for *P. aeruginosa*. In addition to the Fpv system described above, this bacterium contains another twelve CSS systems, most of them involved in the regulation of heterologous siderophore uptake systems. For example, the FoxIRA CSS system regulates the uptake of ferrioxamine, a siderophore produced by different bacteria species, the FiuIRA system regulates the uptake of ferrichrome, a siderophore produced by fungi, and the FemIRA system regulates the uptake of (carboxy)mycobactin that is the siderophore produced by *Mycobacterium* species (Llamas *et al.*, 2006, 2008). These systems have the normal CSS features and show mechanistically high similarities to the *E. coli* FecIRA CSS system (Llamas *et al.*, 2006, 2008; Mettrick and Lamont, 2009). Orthologues to these three *P. aeruginosa* CSS systems are also present in most other *Pseudomonas* species (Llamas and Bitter, 2010). Moreover, microarray analyses have shown that *P. aeruginosa* contains a CSS system involved not only in the regulation of iron uptake but also in the production of pyocins, specifically the R2, F2 and S5 pyocins (Llamas *et al.*, 2008). Pyocins are *P. aeruginosa* bacteriocins that are normally effective against the same or closely related species. The reason why *P. aeruginosa* has a CSS system involved in pyocin synthesis regulation is still unknown. However, since the outer membrane transducers of CSS are normally involved in both (siderophore) uptake and signalling, it is possible that *P. aeruginosa* uses this CSS system to 'steal' siderophores from other bacteria and simultaneously uses it to sense the presence of these bacteria and kill them. The identification of the signal activating this CSS would provide a clue to determine the function of this interesting CSS system.

Haem-mediated CSS systems

Many pathogens are able to use free haem or haem complexed to haemoproteins as a source of iron. In general, haem acquisition systems can be separated in three categories: the first involves a haem-binding outer-membrane lipoprotein; the second comprises an uptake system for free haem that involves a specific TonB-dependent receptor and an ABC transport system; and the third consists of an extracellular haem-binding protein (haemophore) (Wandersman and Delepelaire, 2004). In the third system, the haemophore is secreted via a type I secretion system and is, upon binding of haem, incorporated into the cell by a TonB-dependent receptor. Haemophores are produced only by Gram-negative bacteria. These proteins have high affinity for haem and can acquire haem from different sources, including haemoglobin, haemopexin, myoglobin and haem-loaded albumin (Wandersman and Delepelaire, 2004). In some bacteria, the TonB-dependent receptors involved in haem acquisition are also part of a CSS pathway. For example in most *Bordetella* species, including the causative agent of the human whooping cough *B. pertussis*, binding of free haem to its TonB-dependent receptor BhuR induces a CSS cascade via the σ^{RhuI} and RhuR regulatory proteins. RhuR functions as a sigma factor regulator, inhibiting σ^{RhuI} activity in the absence of haem and showing pro-sigma activity when haem is bound to BhuR. This CSS pathway leads to the expression of the *bhuRSTUV* operon that encodes the haem acquisition system (King *et al.*, 2005). *Serratia marcescens* and *P. aeruginosa* contain a haem-haemophore receptor that belongs to the family of TonB-dependent transducers (Ochsner *et al.*, 2000; Rossi *et al.*, 2003), which means that they have a dual role in both haem uptake and CSS-mediated regulation. This has been demonstrated for the HasR receptor of *S. marcescens*. This bacterium produces and secretes the HasA haemophore, which binds haem and delivers it to HasR. The structure of the HasR receptor resembles that of TonB-dependent receptors (Fig. 3.4), except for the extracellular loops that are exceptionally large (Krieg *et al.*, 2009). These long loops interact with the HasA haemophore. HasR can internalize both free and HasA-bound haem, although a much higher concentration of free haem is required for uptake. When haem-HasA binds to HasR, haem is spontaneously transferred to the HasR haem binding site (Izadi-Pruneyre *et al.*, 2006). Energy is not required for this process but it is required for the entry of haem in the periplasm and the dissociation of the HasA haemophore from the receptor (Paquelin *et al.*, 2001). There are no structural differences between HasR bound only to haem, to HasA or to the haem–HasA complex (Krieg *et al.*, 2009), but only binding of haem-HasA triggers the CSS cascade (Biville *et al.*, 2004). As usual for CSS receptors, the N-terminal periplasmic extension of HasR is required for signalling, but not for haem uptake (Biville *et al.*, 2004). HasR transduces the presence of the haem–HasA complex to the HasS protein, which functions solely as an anti-sigma factor, resulting in σ^{HasI} activation (Rossi *et al.*, 2003). σ^{HasI} directs transcription of the *hasRADEB* operon encoding the haem-acquisition system. Interestingly, σ^{HasI} also regulates the expression of *hasS*. Two transcripts are formed within the *hasIS* cluster, a Fur-regulated transcript comprising both *hasI* and *hasS* and a σ^{HasI}-dependent transcript comprising only *hasS* (Biville *et al.*, 2004). The σ^{HasI}-dependent regulation of *hasS* suggests that the Has system is both activated and repressed by the availability of external haem. When there is enough haem, the anti-sigma activity of HasS is turned off and σ^{HasI} induces *hasS* transcription, which leads to the storage of inactive HasS molecules that become active when HasR receptor is no longer occupied by haem-HasA. A similar σ^{HasI} motif has been identified upstream the *Erwinia carotovora hasR* and *hasS* homologues, suggesting a similar regulation for the haemophore-dependent haem acquisition in this bacterium (Biville *et al.*, 2004).

Regulation of virulence by CSS

CSS systems are involved in bacterial virulence by regulating the uptake of iron within the host during infection. One of the innate immune defences of the host is to reduce the amount of iron available for invading microorganisms through the production of iron-scavenging proteins such as lactoferrin (Valenti and Antonini, 2005). Another example is the production of lipocalin, a host molecule that specifically binds and neutralizes a number of bacterial siderophores (Flo *et al.*, 2004). As previously discussed, pathogenic bacteria have developed several iron acquisition systems to obtain iron during infection and their synthesis is often regulated by CSS. In *P. aeruginosa* CSS is even more intricately linked with virulence since, as mentioned before, the CSS Fpv system regulates the production of two virulence factors, exotoxin A and PrpL endoprotease, that are not involved in the direct acquisition of iron. In addition, another *P. aeruginosa* CSS was recently identified that seems to be dedicated to the regulation of several potential virulence factors (Llamas *et al.*, 2009). This CSS system is formed by the *vreA*, *vreI* and *vreR* genes, which encode a putative receptor, an ECF sigma factor (σ^{VreI}) and a sigma factor regulator, respectively. The first interesting characteristic of this system is the receptor component VreA, which is considerably smaller (23 kDa) than other CSS outer membrane receptors (75–85 kDa). Homology modelling shows that VreA has an unique domain architecture. The N-terminal domain resembles the signalling domain present in TonB-dependent transducers. However, the C-terminal domain resembles the periplasmic domain of TolA/TonB proteins, which are involved in the delivery of energy from the cytoplasmic membrane to the outer membrane (Llamas *et al.*, 2009). This prediction suggests that VreA is not the signal receptor itself, but interacts both with an anti-sigma factor (probably VreR) and with other outer membrane protein(s). Furthermore, since VreA does not have the C-terminal β-barrel domain typical of TonB-dependent receptors, this protein is probably only involved in signal transduction. A second important characteristic of the Vre system is that, although there is a functional Fur box in the promoter region upstream of the *vreA-vreI* genes (Ochsner and Vasil, 1996), this CSS system does not seem to be mainly involved in the regulation of iron uptake. Microarray analyses showed that σ^{VreI} regulates the expression of genes encoding secreted proteins and components of secretion systems (Llamas *et al.*, 2009). These include components of a type II secretion system (the Hxc system), involved in the secretion of a low-molecular-weight alkaline phosphatase (Ball *et al.*, 2002), and a putative two-partner secretion system (TPS). TPS systems mediate the secretion of high-molecular-weight surface-exposed proteins that are often involved in cell adhesion and pathogen dissemination (Jacob-Dubuisson *et al.*, 2001). Although the role of these σ^{VreI}-regulated genes in *P. aeruginosa* virulence have not been established yet, it has been shown that overexpression of σ^{VreI} increases the virulence of *P. aeruginosa* in zebrafish (*Danio rerio*) embryos (Llamas *et al.*, 2009). Moreover, the sera of most *P. aeruginosa*-infected patients contain antibodies directed against σ^{VreI}-regulated proteins (Llamas *et al.*, 2009), which indicates that the σ^{VreI} regulon is transcribed *in vivo* during infection. Apparently, the Vre system is activated by a host signal, which is consistent with previous reports showing that interaction of *P. aeruginosa* with human airway epithelial cells induces the expression of many VreI-regulated genes (Frisk *et al.*, 2004; Chugani and Greenberg, 2007). Homologues to the *vre* system are present in the genome of other bacteria, often associated with genes encoding Hxc-like type II and a TPS-like secretion system (Llamas *et al.*, 2009), which further suggests a functional interaction between Vre and these systems.

The Vre system is not the first example of a CSS system involved in the positive regulation of virulence upon interaction with the host. The plant pathogen *Ralstonia solanacearum*, the causal agent of bacterial wilt, uses a CSS pathway to regulate the expression of a type III protein secretion system in response to a non-diffusible plant signal (Aldon *et al.*, 2000). This system is formed by the TonB-dependent transducer PrhA, the anti-sigma factor PrhR and the ECF sigma factor σ^{PrhI}, and controls the expression of the *hrp* gene cluster encoding a type III secretion system essential for *R. solanacearum* virulence (Marenda *et al.*, 1998; Brito *et al.*, 2002). The *hrp* cluster is highly expressed upon co-culture with plant cell suspensions. Interestingly, induction of this system requires the presence of the plant cell wall and is not triggered by diffusible signals (Marenda *et al.*, 1998; Aldon *et al.*, 2000). This induction needs a functional PrhA receptor but not a functional type III secretion apparatus (Aldon *et al.*, 2000). The identification of the PrhA ligand would shed light on how the plant cell contact-dependent induction of *hrp* occurs, although unfortunately this signal is hitherto unknown.

Conclusions

Because ECF sigma factors are considerably smaller and more diverse as compared with other sigma factors, they have been discovered much later than primary sigma factors (Lonetto *et al.*, 1994). However, since then the analysis of these alternative sigma factors has blossomed; genome sequencing has revealed a plethora of ECF sigma factors in a diverse array of microbes and microarray experiments have been instrumental in identifying the ECF regulons. Identifying the regulon is relatively straightforward since overexpression of ECF sigma factors often leads to (moderate) induction of genes within the regulon. The regulated genes can be highly different, but in general there are two types of ECF sigma factors: those involved in sensing cell envelope stress conditions and those involved in sensing extracellular compounds. The stress-induced ECF sigma factors usually regulate genes that play a role in mitigating the effects of cell envelope stresses, such as chaperones and proteases for handling misfolded proteins. Within the second group, the extracellular compounds that induce these ECF sigma factors are usually heterologous siderophores or other alternative iron sources. These ECF sigma factors belong to a special regulatory system known as CSS. The regulons controlled by CSS systems are usually small and generally control genes encoding proteins involved in the uptake of these alternative iron sources.

Owing to their small size, ECF sigma factors miss several domains, especially (part of) domain 1 and 3 that are involved in the interaction with the RNAPc. Probably because of this the ECF sigma factors have a lower binding affinity for RNA polymerase. This in itself is advantageous, because activation of ECF sigma factors will therefore not result in decreased binding of the primary sigma factor to RNAP and a concomitant decreased transcription of household genes. This effect was also noticed in microarray experiments using bacteria with overexpression of ECF sigma factors. Even though high levels of overexpression were achieved, expression levels of household genes were not affected. In general ECF sigma factors are a highly efficient and productive way to regulate gene expression, which is used by many different bacteria, albeit in very different amounts.

Future trends

A general theme for the regulation of ECF sigma factors is the involvement of anti-sigma factors. Initially, these anti-sigma factors were considered as classical inhibiting proteins, but recently it has become clear that this view is oversimplified. For a number of anti-sigma factors it has been shown or suggested that their role is not only blocking activity in the absence of an inducing signal, but that upon activation the anti-sigma factor is in fact required for the activity of the ECF sigma factor. This is an important conceptual change that might prove more widespread for the activation of ECF sigma factors. Perhaps we should call all anti-sigma factors in the future sigma factor regulators to have a more unbiased name for their role in ECF sigma factor regulation.

Another emerging theme is the involvement of a proteolytic cascade in the activation of ECF sigma factors. These cascades turn out to be highly regulated and complex. Although actual details have only been analysed for a few of these systems, it is expected that in the near future more knowledge will be gained for proteolytic cascades in more of these systems. It will also be highly interesting to see if these cascades show interconnections, especially if more than one signal is perceived. Also the effect of competition of different activated sigma factors for binding to the RNAPc should be analysed in more detail to understand the interplay between these various systems.

A special subclass of ECF sigma factors are the ones that are involved in CSS. These systems depend on a specific outer membrane protein of the TonB-dependent receptor family that senses and transduces an environmental signal. Exactly how the signal is transduced to the cytoplasmic membrane sigma factor regulator (anti-sigma factor) is a field of active research and should become clear in the near future. As mentioned previously, most of these systems are involved in sensing heterologous siderophores. However, recently it has become clear that CSS systems are sometimes also used to sense other environmental signals, including host signals. In the near future we can expect the identification of more of these host-sensing CSS systems and hopefully also of the inducing signals that activate such systems.

References

Ades, S.E. (2008). Regulation by destruction: design of the σ^E envelope stress response. Curr. Opin. Microbiol. *11*, 535–540.

Alba, B.M., Leeds, J.A., Onufryk, C., Lu, C.Z., and Gross, C.A. (2002). DegS and YaeL participate sequentially in the cleavage of RseA to activate the σ^E-dependent extracytoplasmic stress response. Genes Dev. *16*, 2156–2168.

Aldon, D., Brito, B., Boucher, C., and Genin, S. (2000). A bacterial sensor of plant cell contact controls the transcriptional induction of *Ralstonia solanacearum* pathogenicity genes. EMBO J. *19*, 2304–2314.

Anthony, M., Rose, B., Pegler, M.B., Elkins, M., Service, H., Thamotharampillai, K., Watson, J., Robinson, M., Bye, P., Merlino, J., *et al.* (2002). Genetic analysis of *Pseudomonas aeruginosa* isolates from the sputa of Australian adult cystic fibrosis patients. J. Clin. Microbiol. *40*, 2772–2778.

Ball, G., Durand, E., Lazdunski, A., and Filloux, A. (2002). A novel type II secretion system in *Pseudomonas aeruginosa* Mol. Microbiol. *43*, 475–485.

Barik, S., Sureka, K., Mukherjee, P., Basu, J., and Kundu, M. (2010). RseA, the SigE specific anti-sigma factor of *Mycobacterium tuberculosis*, is inactivated by phosphorylation-dependent ClpC1P2 proteolysis. Mol. Microbiol. *75*, 592–606.

Beare, P.A., For, R.J., Martin, L.W., and Lamont, I.L. (2003). Siderophore-mediated cell signalling in *Pseudomonas aeruginosa*: divergent pathways regulate virulence factor production and siderophore receptor synthesis. Mol. Microbiol. *47*, 195–207.

Biville, F., Cwerman, H., Letoffe, S., Rossi, M.S., Drouet, V., Ghigo, J.M., and Wandersman, C. (2004). Haemophore-mediated signalling in *Serratia marcescens*: a new mode of regulation for an extra cytoplasmic function (ECF) sigma factor involved in haem acquisition. Mol. Microbiol. *53*, 1267–1277.

Boucher, J.C., Schurr, M.J., Yu, H., Rowen, D.W., and Deretic, V. (1997). *Pseudomonas aeruginosa* in cystic fibrosis: role of *mucC* in the regulation of alginate production and stress sensitivity. Microbiology *143*, 3473–3480.

Braun, V. (1997). Surface signaling: novel transcription initiation mechanism starting from the cell surface. Arch. Microbiol. *167*, 325–331.

Braun, V., and Endriss, F. (2007). Energy-coupled outer membrane transport proteins and regulatory proteins. Biometals *20*, 219–231.

Braun, V., Mahren, S., and Sauter, A. (2006). Gene regulation by transmembrane signaling. Biometals *19*, 103–113.

Breidenstein, E., Mahren, S., and Braun, V. (2006). Residues involved in FecR binding are localized on one side of the FecA signaling domain in *Escherichia coli*. J. Bacteriol. *188*, 6440–6442.

Brillet, K., Journet, L., Celia, H., Paulus, L., Stahl, A., Pattus, F., and Cobessi, D. (2007). A beta strand lock exchange for signal transduction in TonB-dependent transducers on the basis of a common structural motif. Structure *15*, 1383–1391.

Brito, B., Aldon, D., Barberis, P., Boucher, C., and Genin, S. (2002). A signal transfer system through three compartments transduces the plant cell contact-dependent signal controlling *Ralstonia solanacearum hrp* genes. Mol. Plant Microbe Interact. *15*, 109–119.

Brown, M.S., Ye, J., Rawson, R.B., and Goldstein, J.L. (2000). Regulated intramembrane proteolysis: a control mechanism conserved from bacteria to humans. Cell *100*, 391–398.

Browning, D.F., and Busby, S.J. (2004). The regulation of bacterial transcription initiation. Nat. Rev. Microbiol. *2*, 57–65.

Browning, D.F., Whitworth, D.E., and Hodgson, D.A. (2003). Light-induced carotenogenesis in *Myxococcus xanthus*: functional characterization of the ECF sigma factor CarQ and antisigma factor CarR. Mol. Microbiol. *48*, 237–251.

Butcher, B.G., Mascher, T., and Helmann, J.D. (2008). Environmental sensing and the role of extracytoplasmic function sigma factors. In Bacterial physiology – a molecular approach, El-Sharoud, W., ed. (Springer-Verlag , Berlin Heidelberg, Germany), pp. 233–261.

Campbell, E.A., Muzzin, O., Chlenov, M., Sun, J.L., Olson, C.A., Weinman, O., Trester-Zedlitz, M.L., and Darst, S.A. (2002). Structure of the bacterial RNA polymerase promoter specificity sigma subunit. Mol. Cell *9*, 527–539.

Campbell, E.A., Tupy, J.L., Gruber, T.M., Wang, S., Sharp, M.M., Gross, C.A., and Darst, S.A. (2003). Crystal structure of *Escherichia coli* σ^E with the cytoplasmic domain of its anti-sigma RseA. Mol. Cell *11*, 1067–1078.

Campbell, E.A., Greenwell, R., Anthony, J.R., Wang, S., Lim, L., Das, K., Sofia, H.J., Donohue, T.J., and Darst, S.A. (2007). A conserved structural module regulates transcriptional responses to diverse stress signals in bacteria. Mol. Cell *27*, 793–805.

Cao, M., and Helmann, J.D. (2004). The *Bacillus subtilis* extracytoplasmic-function σ^X factor regulates modification of the cell envelope and resistance to cationic antimicrobial peptides. J. Bacteriol. *186*, 1136–1146.

Cao, M., Wang, T., Ye, R., and Helmann, J.D. (2002). Antibiotics that inhibit cell wall biosynthesis induce expression of the *Bacillus subtilis* σ^W and σ^M regulons. Mol. Microbiol. *45*, 1267–1276.

Cobessi, D., Celia, H., Folschweiller, N., Schalk, I.J., Abdallah, M.A., and Pattus, F. (2005). The crystal structure of the pyoverdine outer membrane receptor FpvA from *Pseudomonas aeruginosa* at 3.6 Å resolution. J. Mol. Biol. *347*, 121–134.

Cogdell, R.J., Howard, T.D., Bittl, R., Schlodder, E., Geisenheimer, I., and Lubitz, W. (2000). How carotenoids protect bacterial photosynthesis. Philos. Trans. R. Soc. Lond. B. Biol. Sci. *355*, 1345–1349.

Costanzo, A., and Ades, S.E. (2006). Growth phase-dependent regulation of the extracytoplasmic stress factor, σ^E, by guanosine 3′,5′-bispyrophosphate (ppGpp). J. Bacteriol. *188*, 4627–4634.

Costanzo, A., Nicoloff, H., Barchinger, S.E., Banta, A.B., Gourse, R.L., and Ades, S.E. (2008). ppGpp and DksA likely regulate the activity of the extracytoplasmic stress factor σ^E in *Escherichia coli* by both direct and indirect mechanisms. Mol. Microbiol. *67*, 619–632.

Chugani, S., and Greenberg, E.P. (2007). The influence of human respiratory epithelia on *Pseudomonas aeruginosa* gene expression. Microb. Pathog. *42*, 29–35.

De Las Peñas, A., Connolly, L., and Gross, C.A. (1997). σ^E is an essential sigma factor in *Escherichia coli*. J. Bacteriol. *179*, 6862–6864.

Dombroski, A.J., Walter, W.A., and Gross, C.A. (1993). Amino-terminal amino acids modulate sigma-factor DNA-binding activity. Genes Dev. *7*, 2446–2455.

Dona, V., Rodrigue, S., Dainese, E., Palu, G., Gaudreau, L., Manganelli, R., and Provvedi, R. (2008). Evidence of complex transcriptional, translational, and posttranslational regulation of the extracytoplasmic function sigma factor σ^E in *Mycobacterium tuberculosis*. J. Bacteriol. *190*, 5963–5971.

Ehira, S., Teramoto, H., Inui, M., and Yukawa, H. (2009). Regulation of *Corynebacterium glutamicum* heat shock response by the extracytoplasmic-function sigma factor SigH and transcriptional regulators HspR and HrcA. J. Bacteriol. *191*, 2964–2972.

England, P., Westblade, L.F., Karimova, G., Robbe-Saule, V., Norel, F., and Kolb, A. (2008). Binding of the unorthodox transcription activator, Crl, to the components of the transcription machinery. J. Biol. Chem. *283*, 33455–33464.

Enz, S., Braun, V., and Crosa, J.H. (1995). Transcription of the region encoding the ferric dicitrate-transport system in *Escherichia coli*: similarity between promoters for *fecA* and for extracytoplasmic function sigma factors. Gene *163*, 13–18.

Enz, S., Brand, H., Orellana, C., Mahren, S., and Braun, V. (2003a). Sites of interaction between the FecA and FecR signal transduction proteins of ferric citrate transport in *Escherichia coli* K-12. J. Bacteriol. *185*, 3745–3752.

Enz, S., Mahren, S., Menzel, C., and Braun, V. (2003b). Analysis of the ferric citrate transport gene promoter of *Escherichia coli*. J. Bacteriol. *185*, 2387–2391.

Escolar, L., Perez-Martin, J., and de Lorenzo, V. (1999). Opening the iron box: transcriptional metalloregulation by the Fur protein. J. Bacteriol. *181*, 6223–6229.

Ferguson, A.D., Chakraborty, R., Smith, B.S., Esser, L., van der Helm, D., and Deisenhofer, J. (2002). Structural basis of gating by the outer membrane transporter FecA. Science *295*, 1715–1719.

Flo, T.H., Smith, K.D., Sato, S., Rodriguez, D.J., Holmes, M.A., Strong, R.K., Akira, S., and Aderem, A. (2004). Lipocalin 2 mediates an innate immune response to bacterial infection by sequestrating iron. Nature *432*, 917–921.

Flynn, J.M., Neher, S.B., Kim, Y.I., Sauer, R.T., and Baker, T.A. (2003). Proteomic discovery of cellular substrates of the ClpXP protease reveals five classes of ClpX-recognition signals. Mol. Cell *11*, 671–683.

Flynn, J.M., Levchenko, I., Sauer, R.T., and Baker, T.A. (2004). Modulating substrate choice: the SspB adaptor delivers a regulator of the extracytoplasmic-stress response to the AAA+ protease ClpXP for degradation. Genes Dev. *18*, 2292–2301.

Folschweiller, N., Gallay, J., Vincent, M., Abdallah, M.A., Pattus, F., and Schalk, I.J. (2002). The interaction between pyoverdin and its outer membrane receptor in *Pseudomonas aeruginosa* leads to different conformers: a time-resolved fluorescence study. Biochemistry *41*, 14591–14601.

Fontán, P.A., and Smith, I. (2009). *M. tuberculosis* σ^E Protects against environmental stress, immune responses. Microbe *4*, 119–123.

Francez-Charlot, A., Frunzke, J., Reichen, C., Ebneter, J.Z., Gourion, B., and Vorholt, J.A. (2009). Sigma factor mimicry involved in regulation of general stress response. Proc. Natl. Acad. Sci. U.S.A. *106*, 3467–3472.

Frisk, A., Schurr, J.R., Wang, G., Bertucci, D.C., Marrero, L., Hwang, S.H., Hassett, D.J., and Schurr, M.J. (2004). Transcriptome analysis of *Pseudomonas aeruginosa* after interaction with human airway epithelial cells. Infect. Immun. 72, 5433–5438.

Frost, G.E., and Rosenberg, H. (1973). The inducible citrate-dependent iron transport system in *Escherichia coli* K12. Biochim. Biophys. Acta *330*, 90–101.

Gaines, J.M., Carty, N.L., Tiburzi, F., Davinic, M., Visca, P., Colmer-Hamood, J.A., and Hamood, A.N. (2007). Regulation of the *Pseudomonas aeruginosa toxA*, *regA* and *ptxR* genes by the iron-starvation sigma factor PvdS under reduced levels of oxygen. Microbiology *153*, 4219–4233.

Garcia-Herrero, A., and Vogel, H.J. (2005). Nuclear magnetic resonance solution structure of the periplasmic signalling domain of the TonB-dependent outer membrane transporter FecA from *Escherichia coli*. Mol. Microbiol. *58*, 1226–1237.

Govan, J.R., and Deretic, V. (1996). Microbial pathogenesis in cystic fibrosis: mucoid *Pseudomonas aeruginosa* and *Burkholderia cepacia*. Microbiol. Rev. *60*, 539–574.

Grigorova, I.L., Phleger, N.J., Mutalik, V.K., and Gross, C.A. (2006). Insights into transcriptional regulation and sigma competition from an equilibrium model of RNA polymerase binding to DNA. Proc. Natl. Acad. Sci. U.S.A. *103*, 5332–5337.

Gruber, T.M., and Gross, C.A. (2003). Multiple sigma subunits and the partitioning of bacterial transcription space. Annu. Rev. Microbiol. *57*, 441–466.

Gumbart, J., Wiener, M.C., and Tajkhorshid, E. (2007). Mechanics of force propagation in TonB-dependent outer membrane transport. Biophys. J. *93*, 496–504.

Hantke, K. (2001). Iron and metal regulation in bacteria. Curr. Opin. Microbiol. *4*, 172–177.

Härle, C., Kim, I., Angerer, A., and Braun, V. (1995). Signal transfer through three compartments: transcription initiation of the *Escherichia coli* ferric citrate transport system from the cell surface. EMBO J. *14*, 1430–1438.

Haugen, S.P., Ross, W., and Gourse, R.L. (2008). Advances in bacterial promoter recognition and its control by factors that do not bind DNA. Nat. Rev. Microbiol. *6*, 507–519.

Hayden, J.D., and Ades, S.E. (2008). The extracytoplasmic stress factor, σ^E, is required to maintain cell envelope integrity in *Escherichia coli*. PLoS One 3, e1573.

Heinrich, J., and Wiegert, T. (2006). YpdC determines site-1 degradation in regulated intramembrane proteolysis of the RsiW anti-sigma factor of *Bacillus subtilis*. Mol. Microbiol. *62*, 566–579.

Heinrich, J., and Wiegert, T. (2009). Regulated intramembrane proteolysis in the control of extracytoplasmic function sigma factors. Res. Microbiol. *160*, 696–703.

Helmann, J.D. (2002). The extracytoplasmic function (ECF) sigma factors. Adv. Microb. Physiol. *46*, 47–110.

Helmann, J.D. (2006). Deciphering a complex genetic regulatory network: the *Bacillus subtilis* σ^W protein and intrinsic resistance to antimicrobial compounds. Sci. Prog. *89*, 243–266.

Helmann, J.D., and Chamberlin, M.J. (1988). Structure and function of bacterial sigma factors. Annu. Rev. Biochem. *57*, 839–872.

Hengge, R. (2009). Proteolysis of σ^S (RpoS) and the general stress response in *Escherichia coli*. Res. Microbiol. *160*, 667–676.

Herrou, J., Foreman, R., Fiebig, A., and Crosson, S. (2010). A structural model of anti-anti-sigma inhibition by a two-component receiver domain: the PhyR stress response regulator. Mol. Microbiol. *78*, 290–304.

Hershberger, C.D., Ye, R.W., Parsek, M.R., Xie, Z.D., and Chakrabarty, A.M. (1995). The *algT* (*algU*) gene of *Pseudomonas aeruginosa*, a key regulator involved in alginate biosynthesis, encodes an alternative sigma factor (σ^E). Proc. Natl. Acad. Sci. U.S.A. *92*, 7941–7945.

Hinton, D.M., Vuthoori, S., and Mulamba, R. (2006). The bacteriophage T4 inhibitor and coactivator AsiA inhibits *Escherichia coli* RNA Polymerase more rapidly in the absence of σ^{70} region 1.1: evidence that region 1.1 stabilizes the interaction between σ^{70} and core. J. Bacteriol. *188*, 1279–1285.

Hong, H.J., Paget, M.S., and Buttner, M.J. (2002). A signal transduction system in *Streptomyces coelicolor* that activates the expression of a putative cell wall glycan operon in response to vancomycin and other cell wall-specific antibiotics. Mol. Microbiol. *44*, 1199–1211.

Horsburgh, M.J., and Moir, A. (1999). σ^M, an ECF RNA polymerase sigma factor of *Bacillus subtilis* 168, is essential for growth and survival in high concentrations of salt. Mol. Microbiol. *32*, 41–50.

Huang, X., and Helmann, J.D. (1998). Identification of target promoters for the *Bacillus subtilis* σ^X factor using a consensus-directed search. J. Mol. Biol. *279*, 165–173.

Huang, X., Fredrick, K.L., and Helmann, J.D. (1998). Promoter recognition by *Bacillus subtilis* σ^W: autoregulation and partial overlap with the σ^X regulon. J. Bacteriol. *180*, 3765–3770.

Hussein, S., Hantke, K., and Braun, V. (1981). Citrate-dependent iron transport system in *Escherichia coli* K-12. Eur. J. Biochem. *117*, 431–437.

Ishihama, A. (2000). Functional modulation of *Escherichia coli* RNA polymerase. Annu. Rev. Microbiol. *54*, 499–518.

Izadi-Pruneyre, N., Huche, F., Lukat-Rodgers, G.S., Lecroisey, A., Gilli, R., Rodgers, K.R., Wandersman, C., and Delepelaire, P. (2006). The heme transfer from the soluble HasA hemophore to its membrane-bound receptor HasR is driven by protein–protein interaction from a high to a lower affinity binding site. J. Biol. Chem. *281*, 25541–25550.

Jacob-Dubuisson, F., Locht, C., and Antoine, R. (2001). Two-partner secretion in Gram-negative bacteria: a thrifty, specific pathway for large virulence proteins. Mol. Microbiol. *40*, 306–313.

Jervis, A.J., Thackray, P.D., Houston, C.W., Horsburgh, M.J., and Moir, A. (2007). SigM-responsive genes of *Bacillus subtilis* and their promoters. J. Bacteriol. *189*, 4534–4538.

Jishage, M., and Ishihama, A. (1998). A stationary phase protein in *Escherichia coli* with binding activity to the major sigma subunit of RNA polymerase. Proc. Natl. Acad. Sci. U.S.A. *95*, 4953–4958.

Kaiser, D. (2003). Coupling cell movement to multicellular development in myxobacteria. Nat. Rev. Microbiol. *1*, 45–54.

Kanehara, K., Ito, K., and Akiyama, Y. (2002). YaeL (EcfE) activates the σ^E pathway of stress response through a site-2 cleavage of anti-σ^E, RseA. Genes Dev. *16*, 2147–2155.

Kanehara, K., Ito, K., and Akiyama, Y. (2003). YaeL proteolysis of RseA is controlled by the PDZ domain of YaeL and a Gln-rich region of RseA. EMBO J. *22*, 6389–6398.

Kaushal, D., Schroeder, B.G., Tyagi, S., Yoshimatsu, T., Scott, C., Ko, C., Carpenter, L., Mehrotra, J., Manabe, Y.C., Fleischmann, R.D., *et al.* (2002). Reduced immunopathology and mortality despite tissue persistence in a *Mycobacterium tuberculosis* mutant lacking alternative sigma factor, SigH. Proc. Natl. Acad. Sci. U.S.A. *99*, 8330–8335.

Kazmierczak, M.J., Wiedmann, M., and Boor, K.J. (2005). Alternative sigma factors and their roles in bacterial virulence. Microbiol. Mol. Biol. Rev. *69*, 527–543.

Kim, I., Stiefel, A., Plantor, S., Angerer, A., and Braun, V. (1997). Transcription induction of the ferric citrate transport genes via the N-terminus of the FecA outer membrane protein, the Ton system and the electrochemical potential of the cytoplasmic membrane. Mol. Microbiol. *23*, 333–344.

Kim, M.S., Hahn, M.Y., Cho, Y., Cho, S.N., and Roe, J.H. (2009). Positive and negative feedback regulatory loops of thiol-oxidative stress response mediated by an unstable isoform of σ^R in actinomycetes. Mol. Microbiol. *73*, 815–825.

King, N.D., Kirby, A.E., and Connell, T.D. (2005). Transcriptional control of the *rhuIR-bhuRSTUV* heme acquisition locus in *Bordetella avium*. Infect. Immun. *73*, 1613–1624.

Koebnik, R. (2005). TonB-dependent trans-envelope signalling: the exception or the rule? Trends Microbiol. *13*, 343–347.

Koster, M., van Klompenburg, W., Bitter, W., Leong, J., and Weisbeek, P. (1994). Role for the outer membrane ferric siderophore receptor PupB in signal transduction across the bacterial cell envelope. EMBO J. *13*, 2805–2813.

Krewulak, K.D., and Vogel, H.J. (2008). Structural biology of bacterial iron uptake. Biochim. Biophys. Acta *1778*, 1781–1804.

Krieg, S., Huche, F., Diederichs, K., Izadi-Pruneyre, N., Lecroisey, A., Wandersman, C., Delepelaire, P., and Welte, W. (2009). Heme uptake across the outer membrane as revealed by crystal structures of the receptor–hemophore complex. Proc. Natl. Acad. Sci. U.S.A. *106*, 1045–1050.

Lamont, I.L., Beare, P.A., Ochsner, U., Vasil, A.I., and Vasil, M.L. (2002). Siderophore-mediated signaling regulates virulence factor production in *Pseudomonas aeruginosa*. Proc. Natl. Acad. Sci. U.S.A. *99*, 7072–7077.

Lane, W.J., and Darst, S.A. (2006). The structural basis for promoter −35 element recognition by the group IV sigma factors. PLoS Biol. *4*, e269.

Leoni, L., Orsi, N., de Lorenzo, V., and Visca, P. (2000). Functional analysis of PvdS, an iron starvation sigma factor of *Pseudomonas aeruginosa*. J. Bacteriol. *182*, 1481–1491.

Li, W., Stevenson, C.E., Burton, N., Jakimowicz, P., Paget, M.S., Buttner, M.J., Lawson, D.M., and Kleanthous, C. (2002). Identification and structure of the anti-sigma factor-binding domain of the disulphide-stress regulated sigma factor σ^R from *Streptomyces coelicolor*. J. Mol. Biol. *323*, 225–236.

Lonetto, M., Gribskov, M., and Gross, C.A. (1992). The σ^{70} family: sequence conservation and evolutionary relationships. J. Bacteriol. *174*, 3843–3849.

Lonetto, M.A., Brown, K.L., Rudd, K.E., and Buttner, M.J. (1994). Analysis of the *Streptomyces coelicolor sigE* gene reveals the existence of a subfamily of eubacterial RNA polymerase sigma factors involved in the regulation of extracytoplasmic functions. Proc. Natl. Acad. Sci. U.S.A. *91*, 7573–7577.

López-Rubio, J.J., Padmanabhan, S., Lázaro, J.M., Salas, M., Murillo, F.J., and Elías-Arnanz, M. (2004). Operator design and mechanism for CarA repressor-mediated down-regulation of the photoinducible *carB* operon in *Myxococcus xanthus*. J. Biol. Chem. *279*, 28945–28953.

Llamas, M.A., and Bitter, W. (2010). Cell-surface signalling in *Pseudomonas*. In Pseudomonas: Molecular Microbiology, Infection and Biodiversity, Ramos, J.L., and Filloux, A., eds (Springer Science, Dordrecht, Heidelberg, London, New York), pp. 59–95.

Llamas, M.A., Sparrius, M., Kloet, R., Jimenez, C.R., Vandenbroucke-Grauls, C., and Bitter, W. (2006). The heterologous siderophores ferrioxamine B and ferrichrome activate signaling pathways in *Pseudomonas aeruginosa*. J. Bacteriol. *188*, 1882–1891.

Llamas, M.A., Mooij, M.J., Sparrius, M., Vandenbroucke-Grauls, C.M., Ratledge, C., and Bitter, W. (2008). Characterization of five novel *Pseudomonas aeruginosa* cell-surface signalling systems. Mol. Microbiol. *67*, 458–472.

Llamas, M.A., van der Sar, A., Chu, B.C., Sparrius, M., Vogel, H.J., and Bitter, W. (2009). A Novel extracytoplasmic function (ECF) sigma factor regulates virulence in *Pseudomonas aeruginosa*. PLoS Pathog. *5*, e1000572.

Luo, Y., and Helmann, J.D. (2009). Extracytoplasmic function sigma factors with overlapping promoter specificity regulate sublancin production in *Bacillus subtilis*. J. Bacteriol. *191*, 4951–4958.

Maeda, H., Fujita, N., and Ishihama, A. (2000a). Competition among seven *Escherichia coli* sigma subunits: relative binding affinities to the core RNA polymerase. Nucleic Acids Res. *28*, 3497–3503.

Maeda, H., Jishage, M., Nomura, T., Fujita, N., and Ishihama, A. (2000b). Two extracytoplasmic function sigma subunits, σ^E and σ^{FecI}, of *Escherichia coli*: promoter selectivity and intracellular levels. J. Bacteriol. *182*, 1181–1184.

Mahren, S., and Braun, V. (2003). The FecI extracytoplasmic-function sigma factor of *Escherichia coli* interacts with the beta' subunit of RNA polymerase. J. Bacteriol. *185*, 1796–1802.

Mahren, S., Enz, S., and Braun, V. (2002). Functional interaction of region 4 of the extracytoplasmic function sigma factor FecI with the cytoplasmic portion of the FecR transmembrane protein of the *Escherichia coli* ferric citrate transport system. J. Bacteriol. *184*, 3704–3711.

Manganelli, R., Voskuil, M.I., Schoolnik, G.K., Dubnau, E., Gomez, M., and Smith, I. (2002). Role of the extracytoplasmic-function sigma factor σ^H in *Mycobacterium tuberculosis* global gene expression. Mol. Microbiol. *45*, 365–374.

Marenda, M., Brito, B., Callard, D., Genin, S., Barberis, P., Boucher, C., and Arlat, M. (1998). PrhA controls a novel regulatory pathway required for the specific induction of *Ralstonia solanacearum hrp* genes in the presence of plant cells. Mol. Microbiol. *27*, 437–453.

Maunsell, B., Adams, C., and O'Gara, F. (2006). Complex regulation of AprA metalloprotease in *Pseudomonas fluorescens* M114: evidence for the involvement of iron, the ECF sigma factor, PbrA and pseudobactin M114 siderophore. Microbiology *152*, 29–42.

Mekler, V., Kortkhonjia, E., Mukhopadhyay, J., Knight, J., Revyakin, A., Kapanidis, A.N., Niu, W., Ebright, Y.W., Levy, R., and Ebright, R.H. (2002). Structural organization of bacterial RNA polymerase holoenzyme and the RNA polymerase-promoter open complex. Cell *108*, 599–614.

Merrick, M.J. (1993). In a class of its own – the RNA polymerase sigma factor σ^{54} (σ^N). Mol. Microbiol. *10*, 903–909.

Mettrick, K.A., and Lamont, I.L. (2009). Different roles for anti-sigma factors in siderophore signalling pathways of *Pseudomonas aeruginosa*. Mol. Microbiol. *74*, 1257–1271

Missiakas, D., and Raina, S. (1998). The extracytoplasmic function sigma factors: role and regulation. Mol. Microbiol. *28*, 1059–1066.

Monteil, V., Kolb, A., Mayer, C., Hoos, S., England, P., and Norel, F. (2010). Crl binds to domain 2 of σ^S and confers a competitive advantage on a natural *rpoS* mutant of *Salmonella enterica* serovar Typhi. J. Bacteriol. *192*, 6401–6410.

Mooney, R.A., Darst, S.A., and Landick, R. (2005). Sigma and RNA polymerase: an on-again, off–again relationship? Mol. Cell *20*, 335–345.

Murakami, K.S., and Darst, S.A. (2003). Bacterial RNA polymerases: the wholo story. Curr. Opin. Struct. Biol. *13*, 31–39.

Murakami, K.S., Masuda, S., Campbell, E.A., Muzzin, O., and Darst, S.A. (2002a). Structural basis of transcription initiation: an RNA polymerase holoenzyme–DNA complex. Science *296*, 1285–1290.

Murakami, K.S., Masuda, S., and Darst, S.A. (2002b). Structural basis of transcription initiation: RNA polymerase holoenzyme at 4Å resolution. Science *296*, 1280–1284.

Nader, M., Journet, L., Meksem, A., Guillon, L., and Schalk, I.J. (2011). Mechanism of ferrisiderophore uptake by *Pseudomonas aeruginosa* outer membrane transporter FpvA: no diffusion channel formed at any time during ferrisiderophore uptake. Biochemistry *50*, 2530–2540.

Navarro Llorens, J.M., Tormo, A., and Martinez-Garcia, E. (2010). Stationary phase in Gram-negative bacteria. FEMS Microbiol. Rev. *34*, 476–495.

Nelson, K.E., Weinel, C., Paulsen, I.T., Dodson, R.J., Hilbert, H., Martins dos Santos, V.A., Fouts, D.E., Gill, S.R., Pop, M., Holmes, M., *et al.* (2002). Complete genome sequence and comparative analysis of the metabolically versatile *Pseudomonas putida* KT2440. Environ. Microbiol. *4*, 799–808.

Newton, G.L., and Fahey, R.C. (2008). Regulation of mycothiol metabolism by σ^R and the thiol redox sensor anti-sigma factor RsrA. Mol. Microbiol. *68*, 805–809.

Noinaj, N., Guillier, M., Barnard, T.J., and Buchanan, S.K. (2010). TonB-dependent transporters: regulation, structure, and function. Annu. Rev. Microbiol. *64*, 43–60.

Ochs, M., Veitinger, S., Kim, I., Welz, D., Angerer, A., and Braun, V. (1995). Regulation of citrate-dependent iron transport of *Escherichia coli*: *fecR* is required for transcription activation by FecI. Mol. Microbiol. *15*, 119–132.

Ochsner, U.A., and Vasil, M.L. (1996). Gene repression by the ferric uptake regulator in *Pseudomonas aeruginosa*: cycle selection of iron-regulated genes. Proc. Natl. Acad. Sci. U.S.A. *93*, 4409–4414.

Ochsner, U.A., Johnson, Z., and Vasil, M.L. (2000). Genetics and regulation of two distinct haem-uptake systems, *phu* and *has*, in *Pseudomonas aeruginosa*. Microbiology *146*, 185–198.

Österberg, S., Del Peso-Santos, T., and Shingler, V. (2011). Regulation of alternative sigma factor use. Annu. Rev. Microbiol. *65*, 37–55.

Paget, M.S., and Helmann, J.D. (2003). The σ^{70} family of sigma factors. Genome Biol *4*, 203.

Paget, M.S., Bae, J.B., Hahn, M.Y., Li, W., Kleanthous, C., Roe, J.H., and Buttner, M.J. (2001a). Mutational analysis of RsrA, a zinc-binding anti-sigma factor with a thiol-disulphide redox switch. Mol. Microbiol. *39*, 1036–1047.

Paget, M.S., Molle, V., Cohen, G., Aharonowitz, Y., and Buttner, M.J. (2001b). Defining the disulphide stress response in *Streptomyces coelicolor* A3(2): identification of the σ^R regulon. Mol. Microbiol. *42*, 1007–1020.

Paquelin, A., Ghigo, J.M., Bertin, S., and Wandersman, C. (2001). Characterization of HasB, a *Serratia marcescens* TonB-like protein specifically involved in the haemophore-dependent haem acquisition system. Mol. Microbiol. *42*, 995–1005.

Park, J.H., and Roe, J.H. (2008). Mycothiol regulates and is regulated by a thiol-specific antisigma factor RsrA and σ^R in *Streptomyces coelicolor*. Mol. Microbiol. *68*, 861–870.

Paulsen, I.T., Press, C.M., Ravel, J., Kobayashi, D.Y., Myers, G.S., Mavrodi, D.V., DeBoy, R.T., Seshadri, R., Ren, Q., Madupu, R., *et al.* (2005). Complete genome sequence of the plant commensal *Pseudomonas fluorescens* Pf-5. Nat Biotechnol *23*, 873–878.

Pérez-Marín, M.C., López-Rubio, J.J., Murillo, F.J., Elías-Arnanz, M., and Padmanabhan, S. (2004). The N terminus of *Myxococcus xanthus* CarA repressor is an autonomously folding domain that mediates physical and functional interactions with both operator DNA and antirepressor protein. J. Biol. Chem. *279*, 33093–33103.

Qiu, D., Eisinger, V.M., Rowen, D.W., and Yu, H.D. (2007). Regulated proteolysis controls mucoid conversion in *Pseudomonas aeruginosa*. Proc. Natl. Acad. Sci. U.S.A. *104*, 8107–8112.

Qiu, D., Eisinger, V.M., Head, N.E., Pier, G.B., and Yu, H.D. (2008). ClpXP proteases positively regulate alginate overexpression and mucoid conversion in *Pseudomonas aeruginosa*. Microbiology *154*, 2119–2130.

Qiu, J., and Helmann, J.D. (2001). The −10 region is a key promoter specificity determinant for the *Bacillus subtilis* extracytoplasmic-function sigma factors σ^X and σ^W. J. Bacteriol. *183*, 1921–1927.

Raivio, T.L., and Silhavy, T.J. (2001). Periplasmic stress and ECF sigma factors. Annu. Rev. Microbiol. *55*, 591–624.

Raman, S., Song, T., Puyang, X., Bardarov, S., Jacobs, W.R., Jr., and Husson, R.N. (2001). The alternative sigma factor SigH regulates major components of oxidative and heat stress responses in *Mycobacterium tuberculosis*. J. Bacteriol. *183*, 6119–6125.

Ratledge, C., and Dover, L.G. (2000). Iron metabolism in pathogenic bacteria. Annu. Rev. Microbiol. *54*, 881–941.

Rédly, G.A., and Poole, K. (2005). FpvIR control of *fpvA* ferric pyoverdine receptor gene expression in *Pseudomonas aeruginosa*: demonstration of an interaction between FpvI and FpvR and identification of mutations in each compromising this interaction. J. Bacteriol. *187*, 5648–5657.

Rhodius, V.A., Suh, W.C., Nonaka, G., West, J., and Gross, C.A. (2006). Conserved and variable functions of the σ^E stress response in related genomes. PLoS Biol *4*, e2.

Rodrigue, S., Provvedi, R., Jacques, P.E., Gaudreau, L., and Manganelli, R. (2006). The sigma factors of *Mycobacterium tuberculosis*. FEMS Microbiol. Rev. *30*, 926–941.

Rossi, M.S., Paquelin, A., Ghigo, J.M., and Wandersman, C. (2003). Haemophore-mediated signal transduction across the bacterial cell envelope in *Serratia marcescens*: the inducer and the transported substrate are different molecules. Mol. Microbiol. *48*, 1467–1480.

Ruiz, N., Kahne, D., and Silhavy, T.J. (2006). Advances in understanding bacterial outer-membrane biogenesis. Nat. Rev. Microbiol. *4*, 57–66.

Sachdeva, P., Misra, R., Tyagi, A.K., and Singh, Y. (2010). The sigma factors of *Mycobacterium tuberculosis*: regulation of the regulators. FEBS J. *277*, 605–626.

Sauter, A., and Braun, V. (2004). Defined inactive FecA derivatives mutated in functional domains of the outer membrane transport and signaling protein of *Escherichia coli* K-12. J. Bacteriol. *186*, 5303–5310.

Schalk, I.J., Hennard, C., Dugave, C., Poole, K., Abdallah, M.A., and Pattus, F. (2001). Iron-free pyoverdin binds to its outer membrane receptor FpvA in *Pseudomonas aeruginosa*: a new mechanism for membrane iron transport. Mol. Microbiol. *39*, 351–360.

Schöbel, S., Zellmeier, S., Schumann, W., and Wiegert, T. (2004). The *Bacillus subtilis* σ^W anti-sigma factor RsiW is degraded by intramembrane proteolysis through YluC. Mol. Microbiol. *52*, 1091–1105.

Schwartz, E.C., Shekhtman, A., Dutta, K., Pratt, M.R., Cowburn, D., Darst, S., and Muir, T.W. (2008). A full-length group 1 bacterial sigma factor adopts a compact structure incompatible with DNA binding. Chem. Biol. *15*, 1091–1103.

Sharp, M.M., Chan, C.L., Lu, C.Z., Marr, M.T., Nechaev, S., Merritt, E.W., Severinov, K., Roberts, J.W., and Gross, C.A. (1999). The interface of σ with core RNA polymerase is extensive, conserved, and functionally specialized. Genes Dev. *13*, 3015–3026.

Shen, J., Meldrum, A., and Poole, K. (2002). FpvA receptor involvement in pyoverdine biosynthesis in *Pseudomonas aeruginosa*. J. Bacteriol. *184*, 3268–3275.

Spencer, M.R., Beare, P.A., and Lamont, I.L. (2008). Role of cell surface signaling in proteolysis of an alternative sigma factor in *Pseudomonas aeruginosa*. J. Bacteriol. *190*, 4865–4869.

Staroń, A., and Mascher, T. (2010a). Extracytoplasmic function σ factors come of age. Microbe *5*, 164–170.

Staroń, A., and Mascher, T. (2010b). General stress response in α-proteobacteria: PhyR and beyond. Mol. Microbiol. *78*, 271–277.

Staroń, A., Sofia, H.J., Dietrich, S., Ulrich, L.E., Liesegang, H., and Mascher, T. (2009). The third pillar of bacterial signal transduction: classification of the extracytoplasmic function (ECF) sigma factor protein family. Mol. Microbiol. *74*, 557–581.

Stiefel, A., Mahren, S., Ochs, M., Schindler, P.T., Enz, S., and Braun, V. (2001). Control of the ferric citrate transport system of *Escherichia coli*: mutations in region 2.1 of the FecI extracytoplasmic-function sigma factor suppress mutations in the FecR transmembrane regulatory protein. J. Bacteriol. *183*, 162–170.

Stover, C.K., Pham, X.Q., Erwin, A.L., Mizoguchi, S.D., Warrener, P., Hickey, M.J., Brinkman, F.S., Hufnagle, W.O., Kowalik, D.J., Lagrou, M., *et al.* (2000). Complete genome sequence of *Pseudomonas aeruginosa* PAO1, an opportunistic pathogen. Nature *406*, 959–964.

Swingle, B., Thete, D., Moll, M., Myers, C.R., Schneider, D.J., and Cartinhour, S. (2008). Characterization of the PvdS-regulated promoter motif in *Pseudomonas syringae* pv.*tomato* DC3000 reveals regulon members and insights regarding PvdS function in other pseudomonads. Mol. Microbiol. *68*, 871–889.

Thackray, P.D., and Moir, A. (2003). SigM, an extracytoplasmic function sigma factor of *Bacillus subtilis*, is activated in response to cell wall antibiotics, ethanol, heat, acid, and superoxide stress. J. Bacteriol. *185*, 3491–3498.

Tiburzi, F., Imperi, F., and Visca, P. (2008). Intracellular levels and activity of PvdS, the major iron starvation sigma factor of *Pseudomonas aeruginosa*. Mol. Microbiol. *67*, 213–227.

Valenti, P., and Antonini, G. (2005). Lactoferrin: an important host defence against microbial and viral attack. Cell. Mol. Life Sci. *62*, 2576–2587.

Vasil, M.L., Ochsner, U.A., Johnson, Z., Colmer, J.A., and Hamood, A.N. (1998). The *fur*-regulated gene encoding the alternative sigma factor PvdS is required for iron-dependent expression of the LysR-type regulator *ptxR* in *Pseudomonas aeruginosa*. J. Bacteriol. *180*, 6784–6788.

Visca, P., Leoni, L., Wilson, M.J., and Lamont, I.L. (2002). Iron transport and regulation, cell signalling and genomics: lessons from *Escherichia coli* and *Pseudomonas*. Mol. Microbiol. *45*, 1177–1190.

Vuthoori, S., Bowers, C.W., McCracken, A., Dombroski, A.J., and Hinton, D.M. (2001). Domain 1.1 of the σ^{70} subunit of *Escherichia coli* RNA polymerase modulates the formation of stable polymerase/promoter complexes. J. Mol. Biol. *309*, 561–572.

Wandersman, C., and Delepelaire, P. (2004). Bacterial iron sources: from siderophores to hemophores. Annu. Rev. Microbiol. *58*, 611–647.

Wilderman, P.J., Vasil, A.I., Johnson, Z., Wilson, M.J., Cunliffe, H.E., Lamont, I.L., and Vasil, M.L. (2001). Characterization of an endoprotease (PrpL) encoded by a PvdS-regulated gene in *Pseudomonas aeruginosa*. Infect. Immun. *69*, 5385–5394.

Wilson, M.J., McMorran, B.J., and Lamont, I.L. (2001). Analysis of promoters recognized by PvdS, an extracytoplasmic-function sigma factor protein from *Pseudomonas aeruginosa*. J. Bacteriol. *183*, 2151–2155.

Wirth, C., Meyer-Klaucke, W., Pattus, F., and Cobessi, D. (2007). From the periplasmic signaling domain to the extracellular face of an outer membrane signal transducer of *Pseudomonas aeruginosa*: crystal structure of the ferric pyoverdine outer membrane receptor. J. Mol. Biol. *368*, 398–406.

Wood, L.F., and Ohman, D.E. (2009). Use of cell wall stress to characterize σ^{22} (AlgT/U) activation by regulated proteolysis and its regulon in *Pseudomonas aeruginosa*. Mol. Microbiol. *72*, 183–201.

Wood, L.F., Leech, A.J., and Ohman, D.E. (2006). Cell wall-inhibitory antibiotics activate the alginate biosynthesis operon in *Pseudomonas aeruginosa*: Roles of sigma (AlgT) and the AlgW and Prc proteases. Mol. Microbiol. *62*, 412–426.

Young, B.A., Anthony, L.C., Gruber, T.M., Arthur, T.M., Heyduk, E., Lu, C.Z., Sharp, M.M., Heyduk, T., Burgess, R.R., and Gross, C.A. (2001). A coiled-coil from the RNA polymerase beta' subunit allosterically induces selective nontemplate strand binding by σ^{70}. Cell *105*, 935–944.

Young, B.A., Gruber, T.M., and Gross, C.A. (2004). Minimal machinery of RNA polymerase holoenzyme sufficient for promoter melting. Science *303*, 1382–1384.

Yue, W.W., Grizot, S., and Buchanan, S.K. (2003). Structural evidence for iron-free citrate and ferric citrate binding to the TonB-dependent outer membrane transporter FecA. J. Mol. Biol. *332*, 353–368.

Zdanowski, K., Doughty, P., Jakimowicz, P., O'Hara, L., Buttner, M.J., Paget, M.S., and Kleanthous, C. (2006). Assignment of the zinc ligands in RsrA, a redox-sensing ZAS protein from *Streptomyces coelicolor*. Biochemistry *45*, 8294–8300.

Zellmeier, S., Schumann, W., and Wiegert, T. (2006). Involvement of Clp protease activity in modulating the *Bacillus subtilis* σ^{W} stress response. Mol. Microbiol. *61*, 1569–1582.

Quorum Sensing in Gram-negative Bacteria: Signals, Role and Networking

4

Zulma R. Suárez-Moreno, Juan F. González, Giulia Devescovi and Vittorio Venturi

Abstract

Bacteria regulate gene expression in a population dependent manner using a sophisticated mechanism based on the production and sensing of chemical signals, known as quorum sensing. Such synchronized response in bacterial populations constitutes a form of multicellularity and enables adaption and survival in challenging environments. Although current evidence shows that the predominant signalling molecules produced by Gram-negative bacteria are *N*-acyl derivatives of homoserine lactones (AHLs), bacteria use a wide variety of signals. In this chapter we provide an overview of quorum sensing in Gram-negative bacteria, and discuss current and future trends in this field of research.

Introduction

Bacterial cell–cell communication is the phenomenon by which a bacterium regulates its gene expression in response to the concentration of diffusible signals produced and released into the local environment by itself or neighbouring bacteria (Fuqua *et al.*, 1994; Platt and Fuqua, 2010). When the signal reaches a critical level, the population responds through the coordinated expression of specific target genes (Fuqua *et al.*, 1994).

The term 'quorum sensing' (QS) was proposed in mid-1990s (Fuqua *et al.*, 1994); however, it stemmed from the notion of 'autoinduction', which referred to amplified gene expression in response to self-produced signals as proposed in the late 1960s. Pioneering studies were performed on the fruiting body formation in *Myxococcus xanthus* (McVittie *et al.*, 1962), the genetic regulation of cellular competence in *Streptococcus pneumoniae* (Tomasz, 1965), and the study of bioluminescence in *Vibrio fisheri*, a symbiont of marine fishes and squids (Kempner and Hanson, 1968; Nealson *et al.*, 1970).

Studies in *V. fisheri* revealed that bioluminescence is produced during the middle of the logarithmic phase of growth (Kempner and Hanson, 1968). Importantly, in 1970 Nealson, Platt and Hastings noted that addition of spent supernatants from stationary-phase cultures induced production of light in early log-phase cultures (Nealson *et al.*, 1970). This observation suggested that bacteria were able to produce a diffusible compound or 'autoinducer', which was able to induce luciferase synthesis when its concentration reached a critical level as consequence of increasing cell-density (Nealson *et al.*, 1970). Further studies also revealed the autoinducer to be species-specific, as supernatants from other *Vibrio* species were not able to induce luciferase synthesis in *V. fisheri* (Eberhard, 1972). The structure of the compound was then identified as *N*-(3-oxohexanoyl) homoserine lactone (3-oxo-C6-HSL) (Eberhard *et al.*, 1981), and the regulatory circuit controlling bioluminescence in *V. fisheri* was later identified to comprise seven luminescence genes (*lux*) organized in two transcriptional units (Engebrecht *et al.*, 1983; Engebrecht and Silverman, 1984). One unit comprised the *luxR* gene, while the other was identified as an operon containing *luxI* followed by five genes required for light production *luxCDABEG*. In this system, *luxI* and *luxR* work as regulatory components, as the LuxI protein is responsible for the synthesis of the AHL, while LuxR binds the AHL at high cell density, allowing it to then activate transcription of the *luxICDABE* operon (Engebrecht *et al.*, 1983; Engebrecht and Silverman, 1984).

For many years it was thought that this regulatory mechanism was unique to certain marine species of *Vibrio;* however in the early 1990s, it was determined to be a more general phenomenon as many other Gram-negative bacteria were reported to produce AHLs with different phenotypes described to be regulated in a cell density manner. This then led to the proposal of the quorum-sensing (QS) concept of gene regulation (Fuqua *et al.*, 1994, 1996). Production of AHLs was detected in *Erwinia carotovora* (Bainton *et al.*, 1992a,b), *Pantoea stewartii* (Beck von Bodman and Farrand, 1995), *Agrobacterium tumefaciens* (Piper *et al.*, 1993) and *Pseudomonas aeruginosa* (Bainton *et al.*, 1992a) in some of the first studies related to QS. Since then, homologues of the *luxI* and *luxR* genes have been identified in a large number of bacterial genomes, and other important functions have been identified to be regulated by AHL-QS (Table 4.1). Additional complexity exists in many bacterial species as some of them possess several *luxI/luxR* systems, which are often organized in hierarchical networks, as for example in *P. aeruginosa, Sinorhizobium meliloti* and *Vibrio* species. AHL QS to date has been only associated with Gram-negative proteobacteria.

The existence of QS was also reported to regulate population and differentiation behaviours in Gram-positive bacteria such as *Lactococcus lactis* (van der Meer *et al.*, 1993), *Staphylococcus aureus* (Ji *et al.*, 1995) and *Bacillus subtilis* (Solomon *et al.*, 1996); however in these Gram-positive species, the autoinducers are short peptides processed from precursors. The peptides are then actively exported out of the cell and they interact with external domains of membrane-bound sensor proteins. Signal transduction occurs by a phosphorylation cascade that culminates in the activation of a DNA binding protein that controls transcription of target genes.

QS has revolutionized the thinking that bacteria exist in isolation as single-celled organisms, but rather reveal them as highly social organisms capable of cooperative behaviours. The adaptive significance of QS varies depending on its evolutionary and ecological context. In fact it is understood that although the ability to communicate is fundamental to bacteria, differences in the QS systems most likely exist as each system has been optimized to promote survival in a specialized niche in which a particular species of bacteria resides (Keller and Surette, 2006). In recent years, more efforts have been made to explore the role of QS in multispecies communities and in cross-kingdom communication. Other concepts such as diffusion and efficiency sensing have been proposed to refine the repercussions of signalling in terms of space and distribution (Redfield, 2002; Hense *et al.*, 2007). Studies have also focused on identifying possible mechanisms to interfere with QS to control pathogenicity and virulence, a strategy known as quorum quenching. Altogether, these research trends of bacterial socio-biology are beginning to create a more complete picture of the impact of bacterial communication in nature.

QS signals

The mechanisms of QS are varied but generally involve the production and detection of signal molecules that, at sufficiently high concentrations influence the expression of target genes. In a canonical system, the signal molecules must be synthesized inside the cell and passively released or actively secreted. As the number of cells in a population increases, the extracellular concentration likewise increases. When the signals accumulate above the minimal threshold level required for detection, cognate receptors bind them and trigger signal transduction cascades that result in population-wide changes in gene expression.

The signal compounds employed in QS regulation as mentioned previously are most often referred to as autoinducers or signal molecules or more recently as signal cues (Platt and Fuqua, 2010). Although many microbial metabolites may have signalling roles, it is important to differentiate between the signals involved in cell-to cell communication from other metabolites. In 2002, Winzer and colleagues proposed the following features, which a QS signal should have:

1 The production of the molecule should occur during specific stages of growth, under

Table 4.1 Examples of AHL QS systems

Species	QS signal	QS system	QS-regulated phenotypes	References
Acidithibacillus ferroxidans	3-OH-C8-HSL, 3-OH-C10-HSL, C12-HSL, 3-oxo-C12-HSL, 3-OH-C12-HSL, C14-HSL, 3-oxo-C14-HSL, 3-OH-C14-HSL, and 3-OH-C16-HSL	AfeI/AfeR	Unknown	Farah *et al*. (2005)
Aeromonas hydrophila	C4-HSL	AhyI/AhyR	Extracellular protease Biofilm formation T6SS, metalloprotease production and biofilm formation	Swift *et al*. (1997, 1999), Lynch *et al*. (2002), Khajanchi *et al*. (2009)
Aeromonas salmonicida	C4-HSL	AsaI-AsaR	Extracellular protease	Schwenteit *et al*. (2010), Swift *et al*. (1997)
Agrobacterium tumefaciens	C8–3-oxo-HSL	TraI/TraR	Virulence plasmid copy number and conjugal transfer	Piper *et al*. (1993), Zhang *et al*. (1993), Fuqua and Winans (1994)
Agrobacterium vitis	C14:1-HSL C16:1–3-oxo-HSL	AvsI/R	Virulence	Hao and Burr (2006)
Burkholderia cepacia	C6-HSL, C8-	CepI/CepR	Extracellular proteases, chitinases, biofilm formation, virulence factors	Huber *et al*. (2001), Aguilar *et al*. (2003)
Burkholderia cenocepacia	C6-HSL, C8-HSL	CepI/CepR	Virulence factors	Lewenza *et al*. (1999), Lewenza and Sokol (2001), Malott *et al*. (2005), O'Grady *et al*. (2009)
	C6-HSL, C8-HSL	CciI/CciR	Extracellular protease and biofilm formation	
Burkholderia mallei	C8-HSL	BmaI1/BmaR1	Unknown	Ulrich *et al*. (2004b) Duerkop *et al*. (2007, 2008)
	3-OH-HSL	BmaI3/BmaR3	Virulence in animal models	
Burkholderia pseudomallei	C10-HSL C8-HSL, C10-HSL, 3-OH-C8-HSL, 3-OH-C10-HSL,3-oxo-C14-HSL	PmiI/PmiR BpmI2/BpmR2 BpmI3/BpmR3	MprA metalloprotease, pathogenicity in mice	Valade *et al*. (2004), Ulrich *et al*. (2004a), Song *et al*. (2005), Kiratisin and Sanmee (2008)
	C8-HSL	BpsI1/BpsR1 BpsI1/BpsR1 BpsI2/BpsR2 BpsI3/BpsR3	Siderophore production, virulence, phospholipase	
Burkholderia vietnamiensis	C8-HSL, C6-HSL	CepI/CepR	Unknown	Conway and Greenberg (2002), Malott and Sokol (2007)
	C10-HSL,C12-HSL	BviI/BviR	Unknown	

Table 4.1 Continued

Species	QS signal	QS system	QS-regulated phenotypes	References
Burkholderia kururiensis	3-oxo-C14-HSL,3-oxo-C12-HSL, 3-oxo-C10-HSL, 3-oxo-C8-HSL, 3-oxo-C6-HSL	$BraI_{KUR}/BraR_{KUR}$	Exopolisaccharide production, biofilm formation, plant colonization	Suarez-Moreno *et al*. (2008)
Burkholderia xenovorans	3-oxo-C14-HSL,3-oxo-C12-HSL, 3-oxo-C10-HSL, 3-oxo-C8-HSL, 3-oxo-C6-HSL	$BraI_{XEN}/BraR_{XEN}$	Exopolysaccharide production	Suarez Moreno *et al*. (2008, 2010)
	3-OH-C8-HSL	XenI2/XenR2	Unknown	
Burkholderia unamae	3-oxo-C14-HSL,3-oxo-C12-HSL, 3-oxo-C10-HSL, 3-oxo-C8-HSL, 3-oxo-C6-HSL	$BraI_{UNA}/BraR_{UNA}$	Exopolysaccharide production, phenol degradation, biofilm formation	Suarez Moreno *et al*. (2008, 2010)
Burkholderia thailandiensis	C8-HSL	BtaI1/BtaR1	Aggregation	Ulrich *et al*. (2004c), Chandler *et al*, (2009), Duerkop *et al*. (2009)
	3-OH-C8-HSL,3-OH-C10HSL	BtaI2/BtaR2	Lipase, swarming, swimming, bactobolin production	
	3-OH-C8-HSL	BtaI3/BtaR3		
Chromobacterium violaceum	C6-HSL	CviI/CviR	Antibiotics,violacein, exoenzymes, cyanide	McClean *et al*. (1997)
Pectobacterium carotovorum (formerly *Erwinia carotovora*)	3-oxo-C6-HSL	CarI/CarR (ExpI/R)	Carbapenem antibiotics and exoenzymes	Bainton *et al*. (1992a,b), McGowan *et al*. (1995), Barnard and Salmond (2007)
Pantoea agglomerans pv. *Gypsophilae*	C4 –HSL,C6-HSL	PagI/PagR	Gall formation, plant virulence	Chalupowicz *et al*. (2008)
Pantoea stewartii	3-Oxo-C6-HSL	EsaI/EsaR	Exopolysaccharide	Beck von Bodman and Farrand (1995), von Bodman *et al*. (2003)
Pseudomonas aeruginosa	C12–3-oxo-HSL	LasI/LasR,	Virulence factors	Gambello and Iglewski (1991), Passador *et al*. (1993)
	C4-HSL	RhlI/RhlR	Rhamnolipid synthesis and virulence, elastase, protease,biofilm formation, RpoS, rhamnolipids	
Pseudomonas aureofaciens	C6-HSL	PhzI/PhzR	Phenazine biosynthesis	Pierson *et al*. (1994), Zhang and Pierson (2001)
		CsaI/CsaR	Biosynthesis of cell surface components	
Pseudomonas fuscovaginae	3-oxo-C12-HSL,3-oxo-C10-HSL C10-HSL, C12-HSL	PfvI/PfvR PfsI/PfsR	Rice pathogenicity	Mattiuzzo *et al*. (2010)

Pseudomonas putida	3-oxo-C12-HSL,3-oxo-C10-HSL,3-oxo-C8-HSL, 3-oxo-C6-HSL	PpuI/PpuR	Unknown	Steidle *et al*. (2002), Bertani and Venturi (2004)
Rhizobium leguminosarum bv. *viciae*	C14–7-*cis*-HSL C6-HSL C7-HSL C8-HSL C8–3-oxo-HSL C8–3-OH-HSL	CinI/CinR RhiI/RhiR RaiI/RaiR TraI/TraR	Functions required for symbiosis	Lithgow *et al*. (2000), Wisniewski-Dye and Downie, (2002), Wisniewski-Dye *et al*. (2002), Wilkinson *et al*. (2002)
Serratia liquefaciens	C4-HSL	SwrI/SwrR	Swarming motility	Eberl *et al*. (1999)
Serratia ATCC 39006	C4-HSL	SmaI/SmaR	Regulation of antibiotic prodigiosin, cellulolytic activity, prodigiosin production	Fineran *et al*. (2005)
Vibrio anguillarum	3-oxo-C10-HSL C6-HSL;3-OH-C6-HSL	VanI/VanR VanM/VanR	Pigment, EmpA metalloprotease, biofilm	Milton *et al*. (1997, 2002), Croxatto *et al*. (2002)
Vibrio fisheri	C6–3-oxo-HSL C8-HSL	LuxI/LuxR AinS/AinR	Bioluminescence Bioluminescence	Engebrecht and Silverman, (1984), Gilson *et al*. (1995)
Sinorhizobium melliloti	3-oxo-C14-HSL;C16:1-HSL, 3-oxo-C16:1-HSL,3-oxo-C16-HSL, C18-HSL, C12-HSL	SinI/SinR	EPS II production, nodulation, swarming	Marketon *et al*., (2002), Marketon and Gonzalez (2002)
Yersinia pestis	3-Oxo-C6-HSL, 3-oxo-C8-HSL	YspI/YspR	Unknow	Kirwan *et al*. (2006)
	3-Oxo-C6-HSL, 3-oxo-C8-HSL	YpeI/YpeR	Unknown	
Yersinia pseudotuberculosis	C6-HSL C6–3-oxo-HSL C8-HSL	YpsI/R YtbI/R	Motility and clumping	Atkinson *et al*. (1999)

physiological conditions or in response to changes in the environment.

2 The molecule should be secreted and accumulated in an extracellular environment.

3 The signal should be able to generate a concerted response after reaching a threshold via recognition by a specific receptor (located on the cell surface or in the cytoplasm).

4 The cellular response should extend beyond physiological changes required to metabolize or detoxify the autoinducer molecules (Winzer *et al.*, 2002).

As mentioned above, in Gram-negative bacterial species, the most common and studied signals thus far are the AHLs; importantly, several other signals with different chemical structures have been reported. Some examples are the unsaturated fatty acid diffusible signal factors (DSF) produced by *Xanthomonas* and *Stenotrophomonas* species (Barber *et al.*, 1997; Dow *et al.*, 2003; Ryan and Dow, 2011), esters such as the 3-OH palmitic acid methyl esters (3-OH-PAME) from *Ralstonia solanacearum* (Flavier *et al.*, 1997b), the PQS quinolone produced by *P. aeruginosa* (Pesci *et al.*, 1999), α-hydroketones produced by *Vibrio* and *Legionella* (Higgins *et al.*, 2007; Spirig *et al.*, 2008), and dicyclic peptides produced by *Pseudomonas* species (Holden *et al.*, 1999). In contrast, the predominant signal cues in Gram-positive bacteria are modified oligopeptides. Finally, there is a signal produced by both Gram-negative and Gram-positive bacteria belonging to a family of related molecules which is broadly termed as AI-2. This chapter will summarize the signals produced and roles played by QS in Gram-negative bacteria and how it is interconnected with other regulatory pathways. Fig. 4.1 presents the structures of the most known QS signals characterized in Gram-negative bacteria. As AHL QS has been by far the most studied to date, more attention will be dedicated to these systems. We will however focus more on new aspects of these signalling systems since they have been extensively studied. There is a large body of literature of the genetics, molecular biology and regulon studies of AHL QS systems in many different Gram-negative bacterial species (Table 4.1 and references therein).

AHLs

AHLs have in common a core structure composed of a fatty acyl chain ligated to a lactonized homoserine through an amide bond. An increasing number of Gram-negative species have been reported to produce AHLs, which can vary in the length of the acyl chain from C4 to C18 and in the degree of saturation at position C7 or C8. The acyl chains may also vary in the oxidation level at the C3, which might either be a carbonyl, a hydroxyl group or to be fully reduced.

In AHL-QS circuits, AHLs are synthesized by an AHL synthase (LuxI-type protein or LuxM-type synthase); AHLs then diffuse in and out of the cell. As the cell density increases, the concentration of the AHLs increases both inside as outside of the cell until reaching a critical point or threshold. At this stage, typically the AHL binds a transcriptional regulator belonging to the LuxR family, which is activated by exposing a DNA-binding domain that recognizes a palindromic *lux* box *cis*-element in the promoter region of the target genes (Egland and Greenberg, 1999). Different acyl chains ensure that different LuxR proteins will recognize different AHLs. Most bacterial species produce a restricted range of AHLs that is coupled to the AHL selectivity of their cognate LuxR proteins (Fig. 4.2). Examples of known LuxI/R circuits and the phenotypes regulated in a cell density manner are presented in Table 4.1.

AHL synthesis

AHL biosynthesis is primarily dependent on members of the LuxI family of synthases (Engebrecht and Silverman, 1984; More *et al.*, 1996), although AHLs may also be synthesized by LuxM/AinS type proteins (Bassler *et al.*, 1993). LuxI type proteins have been reported so far in over 70 genera and LuxM/AinS type proteins are predominant in *Vibrio* species (Case *et al.*, 2008).

From studies of the LuxI-type synthases in *V. fisheri*, *A. tumefaciens* and *P. aeruginosa* it was elucidated that these enzymes direct the formation of an amide bond between the *S*-adenosylmethionine (SAM) and the acyl moiety of a cognate acyl–acyl carrier protein (acyl ACP). The formation of acyl-SAM is then followed by lactone ring formation to yield the acyl homoserine lactone and 5′-methylthioadenosine (More *et*

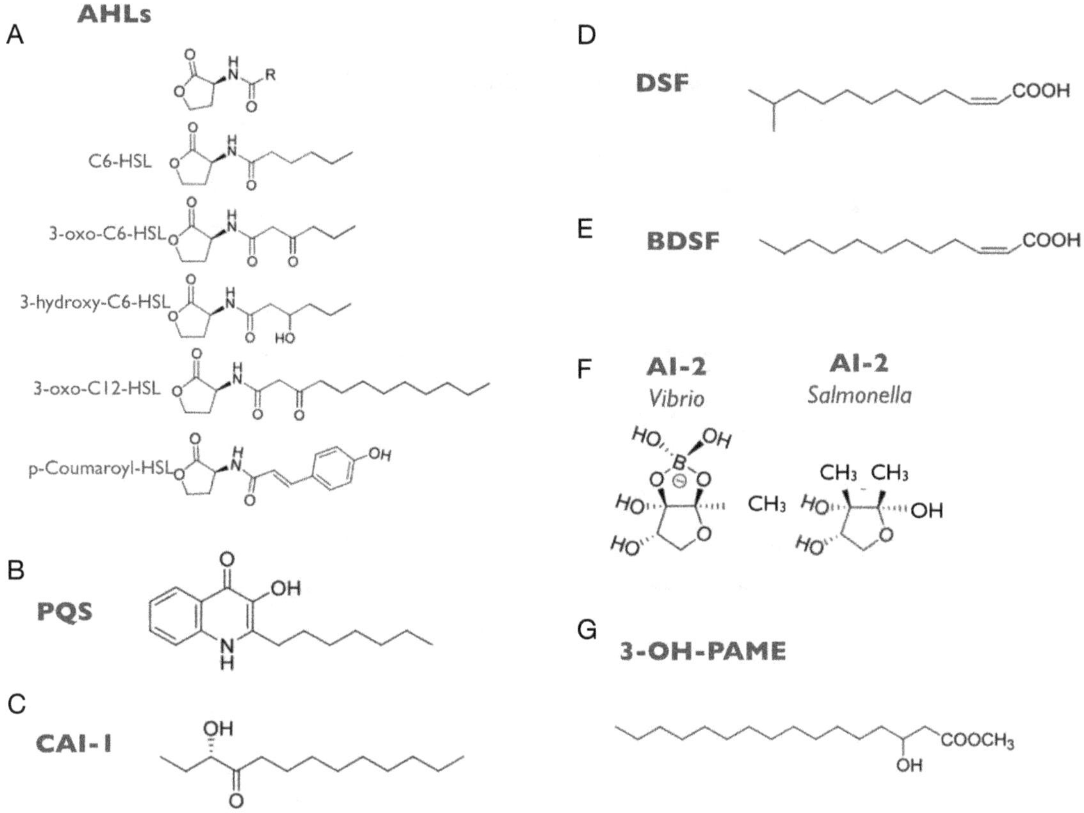

Figure 4.1 (a) AHLs produced by several Gram-negative bacteria (b) Structure of the PQS autoinducer of *Pseudomonas aeruginosa* (c) Structure of *V. cholerae* CAI-1, an alpha-hydroxyketone (d) DSF produced by *Xanthomonas* spp (e) BDSF produced by *Burkholderia cenocepacia* (f) Autoinducer 2 (AI-2), in the presence of boron, it exists as S-THMF borate, whereas in absence of boron it exists as R-THMF (g) 3-OH PAME from *Ralstonia solanacearum*.

al., 1996; Schaefer *et al.*, 1996; Parsek *et al.*, 1997, 1999). The homoserine lactone ring is derived from SAM, while the acyl portion of the AHL is derived from fatty-acid precursors conjugated to the acyl carrier protein (acyl-ACP) (More *et al.*, 1996; Parsek *et al.*, 1999). The choice of fatty acyl substrates is partly owing to substrate preference of the enzyme and to substrate availability. AHL synthases can accept several acyl-ACPs leading to the synthesis of a set of different AHLs, as demonstrated for YtbI from *Yersinia pseudotuberculosis*, which can synthesize up to 24 different AHLs (Ortori *et al.*, 2007). Recently, one exception was reported in the photosynthetic bacterium *Rhodopseudomonas palustris*, which synthesizes the signal *p*-coumaroyl-HSL from *p*-coumaric acid obtained from the extracellular environment (Schaefer *et al.*, 2008).

Characterization of many LuxI-type proteins shows that these proteins range from 190 to 226 amino acids in size and share 4 conserved sequence domains with an average of 37% identity with each other (Fuqua *et al.*, 1994; Watson *et al.*, 2002). Within these domains, 10 residues are conserved in the amino terminal portion of the protein, and it is believed that they are essential for the catalysis of the AHL synthesis since their mutation impaired AHL production by the LuxI, RhlI and EsaI synthases of *V. fisheri*, *P. aeruginosa* and *P. stewartii*, respectively (Hanzelka *et al.*, 1997; Parsek *et al.*, 1997; Watson *et al.*, 2002). These studies also determined that the identity of the residue at position 143 accounts for the different C3 substitutions in AHLs produced in different bacterial species. In fact synthases producing 3-oxo substituted

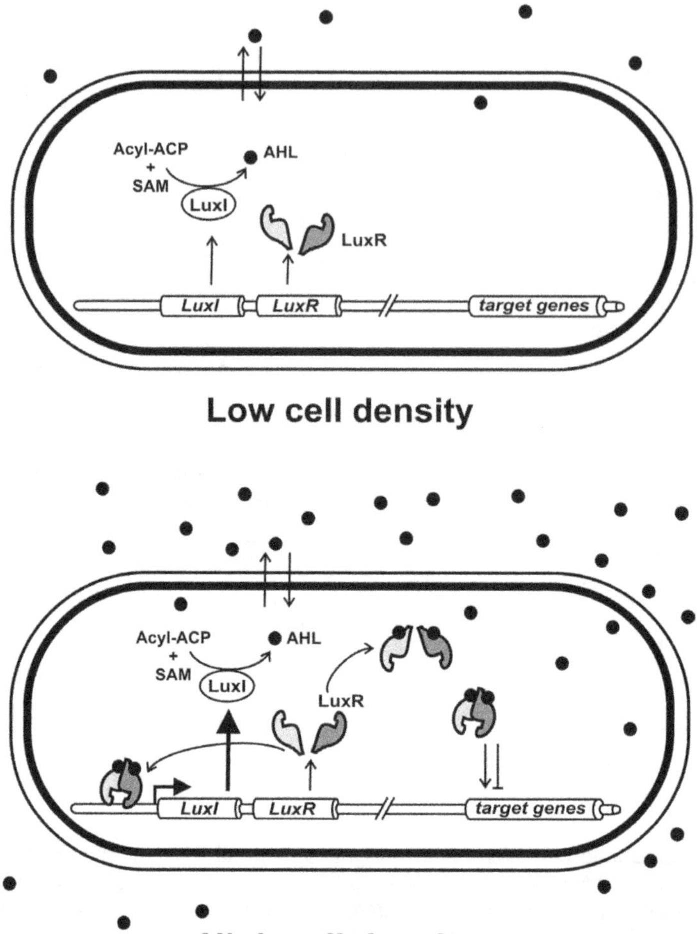

Figure 4.2 Schematic representation of an archetypical *N*-acyl homoserine lactone (AHL) dependent quorum sensing system in Gram-negative bacteria. The LuxI-type proteins are the main class of enzymes synthesizing AHLs and they use the cellular metabolites *S*-adenosyl-methionine and acetylated acyl carrier proteins to form AHLs. At low cell densities AHLs are being produced but their concentration is to low to affect LuxR protein activity. At high cell density, the AHL signal accumulates beyond a quorum concentration and interacts directly with the LuxR-type protein. This induces a conformational change (usually allowing homodimerization) and exposing the helix–turn–helix DNA-binding domain altering the affinity for specific DNA sequences (known as *lux*-boxes) at target gene promoters thus affecting gene expression.

acyl chains have a threonine at this position as in EsaI and LasI, while serine or alanine residues are characteristic of synthases producing unsubstituted AHLs, as is the case for RhlI, CerI and AsaI (Watson *et al.*, 2002; Gould *et al.*, 2006). An additional factor determining the specificity of the LuxI-type proteins is the acyl chain binding pocket, as suggested by studies with LasI from *P. aeruginosa* and EsaI from *P. stewartii*. LasI synthesizes long chain AHLs (3-oxo-C12-HSL), as their tunnel shape places no steric restriction on the length. In contrast, the acyl binding pocket of EsaI restricts access of acyl-ACPs to acyl chain lengths significantly longer than C6 to the active site. Therefore, even though the preferred donor of LasI is 3-oxo-C12-ACP, it is still able to synthesize a variety of shorter and longer acyl-chain-AHLs, when expressed in *E. coli* (Gould *et al.*, 2004, 2006).

LuxM proteins are another class of AHL synthases with distinct amino acid composition from LuxI homologues, but analogous in function

(Hanzelka *et al.*, 1999). AinS for *V. fisheri*, VanM from *V. anguillarum* and LuxM from *Vibrio harveyi* share 34% similarity at the amino terminal domain and are able to synthesize AHLs. Members of this family use the same substrates as LuxI-type proteins, although acyl-CoA may substitute the acyl-ACP protein (Bassler *et al.*, 1993; Hanzelka *et al.*, 1999; Milton *et al.*, 2001).

AHL sensing and response: the LuxR-type proteins

In general, AHL sensing is mediated by a LuxR-subfamily of transcriptional regulators, although in some *Vibrio* species the AHL receptor may also be a histidine kinase sensor protein, coupled to a two-component system, which upon activation induces a phosphorelay cascade leading to the transcriptional activation of the target genes.

LuxR-type proteins are modular and about 250 amino acids long consisting of two domains: the amino-terminal region containing an AHL-binding domain, and the carboxy-terminal region containing a helix–turn–helix (HTH) DNA-binding domain (Choi and Greenberg, 1991; Hanzelka and Greenberg, 1995). Protein sequence comparisons have revealed that LuxR type proteins share 18–25% identity, although their functional domains share much higher sequence conservation. Importantly, nine residues are identical in at least 95% of the LuxR proteins: six of them are located in the N terminal AHL-binding domain while the other three are located in the DNA-binding motif (Fuqua *et al.*, 2001; Nasser and Reverchon, 2007). Structure-function analysis indicated that the DNA binding domain is highly conserved while the AHL binding domain tends to vary in several LuxR type proteins, most likely to accommodate the variety of AHL activating signals (Vannini *et al.*, 2002; Schuster *et al.*, 2004; Urbanowski *et al.*, 2004).

Once bound to the AHL, LuxR-family proteins undergo a conformational change which most often results in homo-dimerization. While the N-terminal domain of each monomer is bound to the AHL ligand, the DNA binding domain recognizes half of a *cis* element known as *lux*-box (Vannini *et al.*, 2002; Schuster *et al.*, 2004). A *lux*-box is an 18–22 bp inverted repeat, usually centred between −40 and −80 relative to the transcriptional site of target genes (Fuqua and Greenberg, 2002). Although many target genes of the LuxR type proteins contain *lux*-boxes within their promoters, target genes lacking canonical *lux*-boxes in their promoters have also been described, as is the case for *traM* and *traR* in *A. tumefaciens* (Fuqua and Parsek, 2002).

LuxR-type protein folding normally requires the presence of the AHL ligand (Zhu and Winans, 2001). LuxR-type proteins bind their cognate AHL with a stoichiometric ratio of one molecule of protein to one molecule of AHL. In contrast, the affinity of the LuxR protein for its cognate AHL and the stability of the AHL–LuxR complex may vary. For example, the binding of LuxR to 3-oxo-C6-HSL is reversible, while in other LuxR proteins such as TraR or LasR the binding is irreversible (Zhu and Winans, 2001; Schuster *et al.*, 2004; Urbanowski *et al.*, 2004).

LuxR-proteins may work as transcriptional activators or as repressors. Transcriptional activation occurs when the AHL–LuxR complex recognizes the *lux* box in the promoter region of the target genes, and recruits RNA polymerase through direct contact, as is the case for LuxR and LasR. In the absence of AHL, some LuxR-type proteins may act as repressors by binding promoters of target genes where they prevent transcription, thus blocking access to the RNA polymerase (Tsai and Winans, 2010). Binding to the AHL causes conformation changes that release the protein from DNA and relieve repression, as described for EsaR of *P. stewartii* and ExpR of *Erwinia* (Von Bodman *et al.*, 1998; Minogue *et al.*, 2002).

Importantly, in most cases there is a positive feedback loop that amplifies the AHL synthesis, allowing an efficient coordination of regulation at the population level (Fuqua and Greenberg, 2002; Fuqua and Parsek, 2002). The amplification of the AHL signal ensures that all the target genes of the LuxR protein are regulated in response to the AHL signal. This population density dependent regulatory circuit therefore guarantees that maximal production of specific proteins occurs during the late logarithmic and stationary phases of growth when the population density is high.

AHL release and accumulation

Bacterial populations live in complex environments where biotic and abiotic parameters are constantly fluctuating. The AHL signalling process is influenced by those conditions that can modulate the generation, stability, diffusion and perception of AHL signals (Horswill *et al.*, 2007; Boyer and Wisniewski-Dye, 2009).

An important condition for the quorum response is the diffusion of the AHLs from inside to outside the cell. The diffusion of short chain AHL has been verified by using ^{3}H-labelled derivatives in 3-oxo-C6-HSL (Kaplan and Greenberg, 1985). It is believed that the hydrophobicity of AHLs may affect their diffusion, and therefore the length of the acyl chain, the polarity given by the substitution of the C3, and the level of unsaturations may affect their rate of release. Several studies in *P. aeruginosa* suggest that efflux of the 3-oxo-C12-HSL is mediated by a multidrug-efflux pump encoded by the *mexAB-oprM* operon, as mutants lacking this pump show less external 3-oxo-C12-AHL levels suggesting that this molecule is actively secreted (Evans *et al.*, 1998; Pearson *et al.*, 1999; Kohler *et al.*, 2001; Aendekerk *et al.*, 2002). Similarly, *B. pseudomallei* KHW uses the efflux pump BpeAB-OprB to release the AHLs, and thus efflux pump-deficient mutants display a reduction in virulence, siderophores production and biofilm formation (Chan and Chua, 2005; Chan *et al.*, 2007). Recently the role of BpeAB-OprB in the export of AHLs in *B. pseudomallei* has been questioned as AHL production remained unaffected in *bpeAB* mutants in other strains of this species (Mima and Schweizer, 2010). The role of efflux pumps in the delivery of the AHL molecules was also studied in *B. cenocepacia*, as mutation in two different efflux pumps resulted in 30% decrease in AHL production (Buroni *et al.*, 2009).

AHL molecules are stable at neutral and acidic pH, while basic conditions cause chemical hydrolysis of the lactone ring. Although in many environments this might be insignificant, extreme conditions of high pH could lead to spontaneous alterations of the AHL molecules. Other factors that affect AHL stability in the environment include the enzymatic degradation caused by bacterial acylase proteins (e.g. the AiiD protein family) and lactonases (e.g. the AiiA protein family), which influence the integrity of the acyl chain or the lactone ring, respectively (Dong *et al.*, 2000; Lin *et al.*, 2003). The enzymatic degradation of AHLs is a mechanism identified in many prokaryotic and eukaryotic organisms in natural environments as a strategy to regulate microbial activities by interfering with bacterial QS. In recent years, several scientists have explored ways to control bacterial infections and pathogenicity via blockage or interference with the AHL response; this is most often referred to as quorum quenching.

Examples of AHL-QS networks: the cases of *P. aeruginosa*, *A. tumefaciens* and *B. cenocepacia*

A very large number of data have been accumulated with respect to AHL QS in Gram-negative proteobacteria. It is not the scope here to exhaustively review all the AHL QS systems studied in bacteria as this would be very lengthy; Table 4.1 lists the most studied AHL QS systems in bacteria with specific references. It is noteworthy that some species harbour two or more circuits to coordinate their population-based responses. When several LuxI/R systems are present, they are often organized in hierarchical networks. In fact, it was recently estimated that as much as the 66% of the *luxI/R* hosting bacteria have more LuxR than LuxI homologues (Case *et al.*, 2008). In this chapter we will only very briefly review three well studied AHL QS networks of three bacterial species, one belonging to α-, one to β- and one to γ-proteobacteria.

The *P. aeruginosa* network

One of the most studied examples is the QS network of the opportunistic pathogen *P. aeruginosa*, which hosts two LuxI/R circuits; the LasI/LasR and RhlI/RhlR systems, producing and responding to 3-oxo-C-12-HSL and C4-HSL, respectively. These systems are interconnected with other regulatory systems and their epigenetic complexity may constitute the basis for the exceptional adaptability of *P. aeruginosa*. LasI/R and RhlI/R are arranged in a hierarchical fashion, since the complex 3-oxo-C12-HSL/LasR drives the expression of *lasI* (constituting a positive feedback loop) as well as *rhlR* and *rhlI* (Latifi *et al.*, 1996). A variation to this scheme was recently reported in an environmental *P. aeruginosa* strain PuPA3, in which the LasI/R and RhlI/R QS systems are not hierarchically arranged, but act independently to regulate the same functions (Steindler *et al.*, 2009). Transcriptome analyses of *P. aeruginosa* revealed that the regulon of these QS-elements are partially overlapping, as QS controls the expression of approx. 300 genes, which are widespread in the genome. This fact suggests that the QS circuitry constitutes a global regulatory system that affects many different cellular functions including many virulence factors (Schuster *et al.*, 2003; Wagner *et al.*, 2003).

Further levels of complexity are involved in the regulatory circuitry of the LasR/RhlR systems. For example the protein RsaL directly represses both *lasI* and *rsaL* transcription by binding to the bi-directional *rsaL-lasI* promoter, adjacent to the LasR binding site. RsaL binding generates a negative feedback loop that counteracts the positive feedback mediated by LasR/3-oxo-C12-HSL contributing to the homeostasis of the system (Rampioni *et al.*, 2007, 2009). Another transcriptional regulator which is interconnected with QS regulation in *P. aeruginosa* is VqsR. It was first described as a virulence and QS regulator (VqsR) (Juhas *et al.*, 2004). Its coding gene has a *las* box in its promoter and has been experimentally found to be under the direct control of the LasI/R quorum sensing system (Schuster *et al.*, 2003; Wagner *et al.*, 2003; Li *et al.*, 2007). Additionally, it is important for production of AHLs and quinolones and for the expression of various QS-controlled genes such as siderophore biosynthesis and membrane proteins associated to antimicrobial resistance (Juhas *et al.*, 2004, 2005). VqsR is a member of the FixJ-NarL superfamily of transcription factors it however lacks an AHL-binding domain making it a very distant relative of the LuxR-type family of transcriptional regulators (Schuster and Greenberg, 2006). *P. aeruginosa* also hosts a LuxR solo protein (these lack a cognate LuxI AHL synthase, see below) called QscR, which also detects 3-oxo-C12-HSL further extending the QS regulon (Ledgham *et al.*, 2003; Lequette *et al.*, 2006). *P. aeruginosa* also produces a quinolone signals, including PQS (*Pseudomonas* quinolone signal; see below), which acts as a link between the LasI/R and the RhlI/R systems (Diggle *et al.*, 2003, 2007b). Other global regulators have been reported to be linked to the AHL QS network, for example RpoS, RpoN, MvaT, RsmA, GacA and QteE, linking QS to many other responses in *P. aeruginosa* (Juhas *et al.*, 2006; Venturi *et al.*, 2006).

The AHL–QS network in the model plant pathogen *A. tumefaciens*

A. tumefaciens possesses the ability to transform plant cells by directly transferring DNA fragments (called T-DNA) to their genomes. This causes the infected cells to proliferate developing tumours (also called crown galls) and to produce novel compounds called opines which are sources of nutrients for the colonizing bacteria (White and Winans, 2007a). Nearly all the genes necessary for tumorigenesis are harboured in a large plasmid called Ti (tumour-inducing); this plasmid also harbours the functions for self-conjugation into other bacterial cells. The *A. tumefaciens* Ti plasmids carry an AHL system called TraI/R. TraI synthesizes *N*-(3-oxo-octanoyl)-L-homoserine lactone (3-oxo-C8-HSL) and at high cell-densities the TraR/3-oxo-C8-HSL complex activates transcription of several target genes on the Ti plasmid, most importantly the ones responsible for conjugation. But the expression of the *traR* regulator gene requires the presence of opines and the AHL system is functional only in the tumour environment. This means that plasmid conjugation actively takes place in crown gall tumours allowing consolidation of *A. tumefaciens* colonization and simultaneously keeping away other plant-associated bacteria. Experiments on the timing and regulation of the TraI/R system as well as extensive biochemical/structural studies have prompted important concepts and working models of AHL quorum-sensing systems (White and Winans, 2007b).

The role of AHL-QS in the multifarious *B. cenocepacia*

The existence of AHL-QS in species belonging to the *Burkholderia cepacia* complex (BCC) human opportunistic pathogenic species was first suggested in 1995, when it was shown that spent supernatants from *P. aeruginosa* were able to activate the production of siderophores, protease and lipase in a *B. cepacia* strain (McKenney *et al.*, 1995). Further studies determined the presence of an AHL-based QS system in a clinical isolate of *B. cenocepacia* strain (K56-2), which was designated *cepI/R* (Lewenza *et al.*, 1999). The *cepI/R* system produces and responds to C8-HSL, although minor amounts of C6-HSL have also been detected. CepR positively controls the transcription of *cepI* via binding to a 20 bp *lux*-box in the putative *cepI* promoter region, and negatively regulates its own transcription (Lewenza and Sokol, 2001).

A similar *cepI/R* scenario was also verified in *B. cenocepacia* strain H111 and in *B. cepacia* strain ATCC25416^T (Huber *et al.*, 2001; Aguilar *et al.*, 2003). The presence of *cepI/R* orthologue systems in other species of the BCC complex showed that the *cepI/R* system is conserved in all species of the BCC complex (Gotschlich *et al.*, 2001; Lutter *et al.*, 2001). Many different BCC virulence phenotypes have been shown to be positively regulated by the *cepI/R* system, including the extracellular enzymes such as proteases and polygalacturonases, swarming motility, and biofilm formation, while negative QS regulation was observed in the production of the siderophore ornibactin (Reviewed by Eberl, 2006; Venturi *et al.*, 2004). In addition, the *cepI/R* system is required for full virulence in both rodent and worm models of infection (Kothe *et al.*, 2003; Sokol *et al.*, 2003).

In epidemic strains of *B. cenocepacia* that possess the *B. cenocepacia* pathogenicity island (*cci*), CepR is required for the expression of an additional QS system called the *cciI/R* system (Malott *et al.*, 2005). The predominant AHL produced by the CciI synthase is C6-HSL, the *cciI* and *cciR* genes are cotranscribed, and CciR negatively regulates the expression of the *cciI/R* operon and *cepI*. CciI/R also regulates extracellular protease production, swarming motility and is important for virulence in murine hosts (Baldwin *et al.*, 2004). In summary, the two B. *cenocepacia* QS systems form a global regulatory system, in which both systems reciprocally regulate many genes (O'Grady *et al.*, 2009).

Unpaired LuxR solos

An emerging aspect of AHL QS, is that in addition to the canonical LuxI/LuxR quorum-sensing systems many bacteria also have AHL quorum sensing LuxR-type proteins that are not paired to a cognate LuxI. This is a common feature among proteobacteria, regardless whether the organism has a LuxI/R pair or not (Fig. 4.3). In fact, an analysis of 265 sequenced proteobacterial genomes revealed that 68 (26%) had a LuxI/LuxR system;

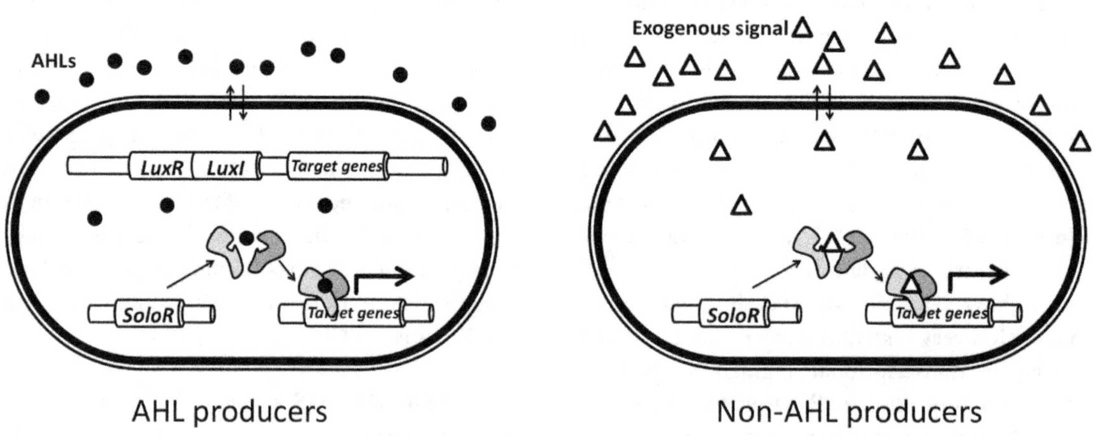

Figure 4.3 Current working models of LuxR solos in AHL producing and non-AHL producing proteobacteria. In AHL-producing bacteria, LuxR solos most often bind to endogenously produced AHLs and regulated gene expression; this can extend the AHL QS regulon (e.g. the *P. aeruginosa* QscR LuxR solo). In addition these solos can also bind exogenous AHLs produced by neighbouring bacteria (e.g. the *P. aeruginosa* QscR LuxR solo). In non-AHL-producing bacteria, LuxR solos regulate gene expression upon binding to exogenous AHLs produced by neighbouring bacteria (e.g. the *E. coli* and *S. enterica* SdiA LuxR solo) or to eukaryotic signals (e.g. XccR and OryR of *X. campestris* and *X. oryzae* respectively which bind plant compounds).

a subset of 45 (66%) of these contained more LuxRs than LuxIs. Furthermore, of the 265 fully sequenced proteobacterial genomes analysed, 45 (17%) contained only QS LuxRs (Case *et al.*, 2008). These unpaired proteins have been initially referred to as 'orphans' (Fuqua, 2006) although more recently the term 'solos' was introduced, to better describe their unpaired nature (Subramoni and Venturi, 2009a). These solos are homologous to classical LuxR-type proteins having an autoinducer binding domain in the N-terminus and a DNA binding helix–turn–helix domain (HTH) in the C-terminus. Clearly, evolutionary and environmental pressures have favoured the presence of these proteins, which can give the bacteria various advantages (Subramoni and Venturi, 2009a). In bacteria which harbour a complete AHL QS system, the presence of a solo(s) can expand the regulatory repertoire of the LuxI/R QS systems by responding to endogenous or exogenous autoinducers independently or in conjunction with the I/R pair (Subramoni and Venturi, 2009a; Patankar and Gonzalez, 2009). Furthermore, they can regulate target genes or discern the presence of favourable environmental conditions by 'eavesdropping' on external prokaryotic signals (Ahmer, 2004) or via interkingdom signalling when responding to eukaryotic signals (Zhang *et al.*, 2007; Ferluga *et al.*, 2007). The most studied examples of LuxR solos are described below and it is expected that many more studies on LuxR solos of diverse bacterial species will soon follow.

TrlR

As previously mentioned, the plant pathogen *A. tumefaciens* is the causal agent of crown gall tumours in higher plants. The disease is initiated by the transfer from the tumour-inducing (Ti) plasmid of a DNA fragment (T-DNA) into the plant genome. The movement of this plasmid is achieved through conjugation and is controlled by an elaborate QS cascade; the QS encoding genes are all located in the Ti plasmid (White and Winans, 2007a). The process is triggered once the transcriptional regulator OccR senses the tumour-procured compound octopine and eventually resulting in the activation of the transcription of the QS LuxR-family *traR* (Fuqua and Winans, 1996). TraR forms a stable complex with 3-oxo-C8-HSL, the autoinducer produced by its cognate synthase TraI. At high cell densities the autoinducer accumulates and the TraR/3-oxo-C8-HSL complex starts off the expression of the genes necessary for conjugation. A gene in the Ti plasmid called *trlR* is highly similar to the *traR* gene (Fuqua and Winans, 1994; Oger *et al.*, 1998). *trlR* resides within the *mot* operon which is involved in the catabolism of mannopines. Owing to a frameshift mutation, the region corresponding to the DNA binding HTH domain of TrlR is incapable of binding DNA, even though its N-terminus can bind AHLs. TrlR interacts with the QS LuxR protein TraR, resulting in TraR–TrlR-3OC8 heterodimers with only one functional DNA-binding domain and hence being unable to regulate target genes (Chai *et al.*, 2001). This negative regulation is exerted indirectly through mannopines, which activate expression of the *mot* operon that includes the *trlR* gene. When richer sources of carbon (for example succinate) are present, mannopine catabolism is repressed, reducing TrlR levels thus guaranteeing that conjugation is only carried out in favourable conditions (Chai *et al.*, 2001).

QscR

QS in the opportunistic pathogen *P. aeruginosa* has been extensively studied (Lazdunski *et al.*, 2004). As previously mentioned, it consists of two I/R protein pairs; the LasI/R pair synthesizes and responds to 3-oxo-C12-HSL, while the RhlI/R pair, which is under the regulation of LasI/R, uses C4-HSL as its cognate inducer. *P. aeruginosa* also has an unpaired LuxR-type solo called QscR (quorum sensing control repressor). Initial studies showed that QscR mutants were hypervirulent, killing the animal model *Drosophila melanogaster* more rapidly than the wild type. Additionally, the *qscR* mutant synthesized the LasI autoinducer (3-oxo-C12-HSL) prematurely at low cell densities and formed blue colonies because of the overproduction of the pigment phenazine (Chugani *et al.*, 2001). It was proposed that the mechanism, by which QscR repressed these and other QS regulated virulence factors, was that at low AHL concentrations QscR can form heterodimers with both LasR and RhlR and inhibiting their function (Ledgham *et al.*, 2003). Interestingly, QscR has its own regulon that includes some genes shared with both LasR and RhlR regulons (Lequette *et al.*, 2006). QscR binding to DNA requires 3-oxo-C12-HSL, the cognate AHL of the LasR/I QS system, but it possesses a more relaxed AHL specificity compared with LasR, being able to activate target promoters in the presence other of long-chain AHLs including C10, 3-oxo-C10-HSL, C12-HSL or 3-oxo-C12-HSL. This relaxed specificity could allow it to respond to exogenous AHLs (Lazdunski *et al.*, 2004; Lee *et al.*, 2006).

SdiA

Salmonella enterica and *Escherichia coli* do not produce AHLs and possess a LuxR solo called SdiA, which responds to exogenous AHLs resulting in signal interception or 'eavesdropping' on AHLs produced by neighbouring bacteria. SdiA has been more extensively studied in *S. enterica*. It was demonstrated that overexpression of SdiA can bypass the need of a signal molecule; this allowed the identification of several SdiA target genes (Ahmer *et al.*, 1998; Michael *et al.*, 2001) that include the *rck* operon and *srgE*. The *rck* (resistance to complement killing) is a six-gene operon believed to be involved in adhesion and resistance to complement killing, and *srgE* is a gene of unknown function which has apparently been acquired horizontally (Smith and Ahmer, 2003). *S. enterica* SdiA responds to a wide range of exogenous AHLs, albeit with different sensitivity: 3-oxo-C8-HSL at 1 nM, 3-oxo-C6-HSL at 5nM and 3-oxo-C10-HSL, C6-HSL and C8-HSL at 50 nM (Ahmer, 2004). Interestingly, temperature also affects the SdiA regulon in *S. enterica*, since lowering the temperature to 30°C triggers SdiA-dependent activation in the absence of AHLs (Smith and Ahmer, 2003). The working model that SdiA eavesdrop in *S. enterica* has been strengthened recently by findings that SdiA detects AHLs during transit through turtles (Smith *et al.*, 2008). In addition, SdiA is also activated by AHLs produced by *Yersinia enterocolitica* in infected mice and not activated in mice infected with a non-AHL-producing *Y. enterocolitica yenI* mutant (Dyszel *et al.*, 2010).

E. coli SdiA, when overexpressed, confers the ability to bypass inhibition of cell division, thus it was named SdiA (suppressor of division inhibition). SdiA in *E. coli* responds to AHLs and up-regulates the transcription of the cell division operon *ftsQAZ* (Wang *et al.*, 1991). *E. coli* SdiA has also been reported to increase resistance to some antibiotics and quinolones, apparently through the activation of efflux pumps (Wei *et al.*, 2001; Rahmati *et al.*, 2002; Ahmer, 2004). In addition, microarray studies when *sdiA* is overexpressed have indicated that the gene *uvrY* is regulated by SdiA in *E. coli* (Gyaneshwar *et al.*, 2001; Suzuki *et al.*, 2002). A genetic screen further determined that six up-regulated and nine down-regulated promoters were under SdiA regulation; this screen also identified *uvrY* as being regulated by SdiA (Van Houdt *et al.*, 2006). Another microarray study identified 40 down-regulated and 42 up-regulated genes to be under *sdiA* regulation, however only flagellar genes were found in common with previous regulon studies (Lee *et al.*, 2008). A recent study also highlighted that many of the phenotypes previously observed to be under the regulation of *sdiA* when overexpressed in a plasmid, did not display regulation by the chromosomal *sdiA* and exogenous addition of AHLs (Dyszel *et al.*, 2010). Future studies are therefore needed to precisely establish the induction conditions required for SdiA regulation.

XccR and OryR

Interestingly, there is a subgroup of LuxR-type solos that have substitutions in one or two of the highly conserved amino acids in the AHL autoinducer-binding domain. These changes give the proteins the ability to bind and respond to plant low molecular weight molecules thus this subfamily is involved in interkingdom signalling (Zhang *et al.*, 2007; Ferluga *et al.*, 2007; Ferluga and Venturi, 2009;). This group includes XccR of *Xanthomonas campestris*, OryR of *Xanthomonas oryzae* and NesR in *Sinorhizobium meliloti* (Patankar and Gonzalez, 2009a).

X. campestris pv. campestris (Xcc) and *X. oryzae* pv. *oryzae* (*Xoo*) are plant pathogens which do not have a LuxI/R pair and do not produce AHLs but have QS systems based on the DSF (*cis*-11-methyl-2-dodecenoic acid) and DF (butyrolactone) signal molecules (He and Zhang, 2008). Both pathogens possess an orthologous LuxR solo which responds to plant compounds; namely XccR in *X. campestris* and OryR in *X. oryzae*. XccR and OryR have a conserved AHL autoinducer-binding domain in their N-terminus, however they have substitutions in two of the six key amino acids; the tryptophan residue at position 57 and the tyrosine at position 61 are switched to a methionine and a tryptophan respectively. These amino acids are known to be important for AHL interactions because they form hydrogen bonds with the autoinducers (Nasser and Reverchon, 2007). XccR and OryR do not bind AHLs but are involved in interkingdom signalling, responding to an unidentified plant low molecular weight compound and are both important for virulence (Zhang *et al.*, 2007; Ferluga *et al.*, 2007; Ferluga and Venturi, 2009; Subramoni and Venturi, 2009a). Interestingly, both XccR and OryR regulate the proline iminopeptidase (*pip*) gene, which is located adjacent to *xccR* and *oryR* and both *pip* genes contain a well conserved *lux* box in their promoter region. Importantly several XccR and OryR orthologues are present but only in chromosomes of plant-associated bacteria indicating that this protein is widespread and most probably involved in interkingdom signalling between plants and bacteria (Subramoni and Venturi, 2009).

Other LuxR Solos

As mentioned previously, approximately one-third of the proteobacterial sequenced genomes contain LuxR solos, hence this family of proteins is very poorly studied and their role is far from clear. A few other solos are beginning to be studied especially with respect to their role in the AHL QS response and network; these are discussed below.

The plant pathogens *Pectobacterium carotovorum* subsp. *carotovora* (formerly *Erwinia carotovora* (*Ecc*)) and *Pectobacterium carotovorum* subsp. *atroseptica* (*Eca*) carry a LuxR solo called VirR or ExpR2. These solos act as repressors in the absence of AHLs; however, when the threshold concentration of AHL is reached, the binding of the signal molecule relieves the transcriptional repression. Moreover, at low population density VirR/ExpR2 is able to promote the transcription of the global repressor RsmA that suppresses the production of plant cell wall degrading enzymes (Barnard and Salmond, 2007). In *Ecc* Strain GS101 there is a second LuxR solo denominated CarR that regulates the production of the antibiotic carbapenem when bound to 3-oxo-C6-HSL, which is produced by the endogenous CarI/ExpR system (Barnard and Salmond, 2007; Patankar and Gonzalez, 2009b). Interestingly, in the opportunistic pathogen *Serratia marcescens* CarR also regulates the production of carbapenem but in a ligand-independent manner (Cox *et al.*, 1998).

The nitrogen-fixing plant symbiont *Rhizobium leguminosarum* bv. *viciae* has a complex QS system consisting of three LuxI/R pairs, CinI/R, RhI/R and TraI/R. The latter two are located on the symbiotic plasmid pRL1J1 and the LuxR solo BisR is also found on the symbiotic plasmid. BisR induces the expression of *traR* upon sensing the CinI product 3OH-C14-HSL and also represses the production of this AHL by binding to *cinI* promoter. Additionally, BisR mutants showed reduced plasmid transfer (Wilkinson *et al.*, 2002; Gonzalez and Marketon, 2003).

Transcriptomic studies in *S. meliloti*, a soil bacterium that can establish a symbiotic relationship with the legume *Medicago sativa*, have demonstrated that the solo ExpR can modulate a wide range of genes involved in motility, chemotaxis and symbiotically relevant exopolysaccharides (Hoang *et al.*, 2004, 2008; Glenn *et al.*, 2007). Interestingly, the *S. meliloti* Rm1021 genome codes for four additional LuxR solos. One of them, NesR, was shown to be involved in the active methyl cycle within the methionine biosynthesis pathway and to affect the nutritional and stress responses. Furthermore, the *nesR* mutant resulted to be less efficient in plant nodulation (Patankar and Gonzalez, 2009a).

In *A. vitis*, the causative agent of crown gall and necrosis of grape, two solo LuxR-type regulators, AviR and AvhR were found. A mutation in the *aviR* locus results in the loss of the ability to induce necrosis on grape and hypersensitive response in non-host plants. Moreover these mutants produce fewer long-chain AHLs (Zheng *et al.*, 2003). On the other hand, AvhR is necessary for the hypersensitive response, but when mutated, it only partially affects grape necrosis (Hao *et al.*, 2005).

Brucella melitensis, an intracellular pathogen and the aetiological agent of brucellosis, possesses two LuxR solos: BlxR (or BabR) and VjbR. Microarray studies on a BlxR mutant indicated that this solo regulates a wide variety of virulence related genes, including those encoding for the type IV secretion system and flagella. Mutants in *blxR* showed attenuated virulence in mice and a decreased ability to grow in macrophages (Rambow-Larsen *et al.*, 2008). These results are not in line with those reported by another study (Weeks *et al.*, 2010), which showed that intracellular survival in macrophages is reduced in *vjbR* mutants but not in *blxR* mutants. Furthermore *vjbR* mutants are fully attenuated in mice virulence. Microarray analysis of a *vjbR* deletion mutant showed that this regulator controls some metabolic genes as well as a large number of genes involved in virulence; including adhesins, proteases, lipoproteins, haemolysin secretion system components and effectors (Weeks *et al.*, 2010). Interestingly, BlxR and VjbR appear to be positively autoregulated and also cross-regulate the expression of each other (Rambow-Larsen *et al.*, 2008). *B. melitensis* has been reported to produce C12-HSL (Taminiau *et al.*, 2002) however it does not possess a LuxI homologue hence it has been suggested that it harbours a different yet unknown AHL synthase. Interestingly, both BlxR and VjbR respond to C12-HSL and regulate gene expression (Weeks *et al.*, 2010), hence it cannot be excluded that these two LuxR regulators are not solos.

In the opportunistic pathogen *B. cenocepacia* two QS systems, CepI/R and CciI/R and a LuxR solo, CepR2, were described. Microarray analysis in *B cenocepacia* strain K56-2 showed that CepR2 negatively regulates its own expression but has no effect on cep*I/R* and *cciI/R* expression. On the other hand, CepR2 regulates a wide range of known QS-regulated genes involved in virulence, chemotaxis and motility, thus fine-tuning the quorum sensing network (Malott *et al.*, 2009).

The causal agent of melioidosis, *B. pseudomallei* possesses a complex QS system consisting of three LuxI/R pairs and two LuxR solos; in *B. pseudomallei* K96243 the LuxI/R pairs are designated BpsI1-R1, BpsI2/R2, BpsI3/R3 and the solos BpsR4 and BpsR5. Expression studies showed that both BpsR4 and BpsR5 are involved in the regulation of the QS components. It was demonstrated that BpsR5 partially activates *bpsI1* expression, both in the presence and absence of AHLs. The constitutive transcription of *bpsI2* is partially enhanced by BpsR4 and BpsR5 in the presence of different AHLs. The situation gets even more complicated owing to the fact that besides the formation of the typical homodimers, these LuxR homologues are able to interact with each other through the formation of a wide combination of heterodimers. Whether these heterodimers can play a role in *in vivo* gene expression has still to be confirmed (Kiratisin and Sanmee, 2008).

Only about one-third of the soil-borne or plant-associated *P. putida* strains possess a complete AHL QS system. Nevertheless, a genomic database search revealed that all the sequenced strains possess a LuxR solo called PpoR. In *P. putida* WCS358 PpoR binds to 3-oxo-C6 HSL and its expression is growth phase dependent. Microarray analysis revealed that PpoR down regulates a series of genes involved in amino acid catabolism, while it up-regulates genes involved in protein synthesis, sulfur metabolism, inorganic ion utilization and oxidative stress (Subramoni and Venturi, 2009b).

Quinolones

Besides the two AHL-based QS systems, *P. aeruginosa* also utilizes the signal, 2-heptyl-3-hydroxy quinolone, known as PQS (*<u>P</u>seudomonas* <u>q</u>uinolone <u>s</u>ignal) (Fig. 4.1). This system also relies upon the PQS precursor molecule HHQ (2-heptyl-4-quinolone) to control global gene expression and important virulence factors (Pesci *et al.*, 1999; Heeb *et al.*, 2011).

Although quinolone production by *P. aeruginosa* was reported in the 1950s (Cornforth and James, 1956), the realization that PQS was a QS signal was not until 1999, when it was demonstrated that addition of spent culture medium extract from a *P. aeruginosa* wild type strain (PAO1) caused induction of the *lasB* gene in a *lasR*-deficient mutant; the LasB elastase is a virulence factor transcriptionally controlled both by the LasI/R. These data suggested that a non-AHL signal produced by PAO1 was still capable of activating *lasB* expression. The molecule responsible for the non-AHL mediated QS was purified and identified as 2-heptyl-3 hydroxy-4 quinolone (Pesci *et al.*, 1999).

PQS biosynthesis and genetic regulation

The alkyl quinolones in *P. aeruginosa* are derived via the condensation of anthranilate and β-keto fatty acids (Cornforth and James, 1956). Anthranilate is converted into the intermediate precursor of PQS, HHQ (2-heptyl-4-quinolone) by the proteins PqsA, PqsB, PqsC and PqsD (Calfee *et al.*, 2001). HHQ can then be passed between cells and converted intracellularly into PQS through the action of PqsH (Gallagher *et al.*, 2002). Quinolone synthesis in *P. aeruginosa* requires multiple genes, which were initially identified by screening transposon mutant libraries for clones displaying reduced pyocyanin production, and were denoted as *pqsABCDE*, *pqsR* (*mfvR*), *pqsH* and *pqsL*. The *pqsABCDE* genes are arranged in an operon, and adjacent to these are the anthranilate synthase genes *phnAB* and *pqsR* (Gallagher *et al.*, 2002). Two other genes, *pqsH* and *pqsL* are also involved in alkyl-quinolones synthesis, but both of these are located separately elsewhere on the chromosome (Fig. 4.4).

PqsA is involved in the activation of anthranilate, whereas PqsB and PqsC are similar to β-keto-acyl-acyl carrier protein synthases, which are involved in the fatty acid metabolism and thus in the elongation of the acyl side chains of the quinolones (Diggle *et al.*, 2003; Deziel *et al.*, 2004). In contrast, PqsD acts as a condensing enzyme, which may either catalyse the condensation of anthranoyl-CoA with a 3-oxo-acid or may be involved in the formation of a 3-oxo-acid precursor. For this reason, mutations in the *pqsD*, as well as in *pqsA* and *pqsB*, completely abolish the PQS synthesis (Zhang *et al.*, 2008). Recent studies have provided new insights regarding other *pqs* genes, as *pqsE*, which is not involved in the PQS synthesis but is an effector of its response. PqsE has similarities to the metallo-β-hydrolase superfamily, and its crystal structure has been determined, although its exact function is still unknown (Yu *et al.*, 2009). Mutations in *pqsE* reduce the expression of important virulence factors (as for example pyocyanin, lectin and HCN), while its overexpression produces the opposite effect (Farrow *et al.*, 2008). Interestingly a recent transcriptomic analysis has revealed that PqsE regulates a major subset of the PQS controlled genes, as well as the virulence in animal and plant models and therefore may be considered as a global regulator within the QS network. Furthermore it was verified that it is required for swarming motility and biofilm formation (Rampioni *et al.*, 2010). PqsE can repress the transcription of the *pqsABCDE* operon and therefore its own transcription. PqsE is not involved in the alkyl-quinolone biosynthesis and is a crucial element mediating the cellular response to PQS (Diggle *et al.*, 2003; Deziel *et al.*, 2005).

Other members of the PQS synthetic machinery are PqsH and PqsL. The *pqsH* gene encodes a NADH-dependent flavin monooxygenase that hydroxylates the 3′ carbon atom of the HHQ in the final step of the PQS biosynthesis, and therefore deficient mutants in *pqsH* are not able to produce PQS, but may still produce HHQ (Deziel *et al.*, 2004; Schertzer *et al.*, 2010). The *pqsL* gene encodes a second, distinct monooxygenase that is required for the synthesis of HQNO and other quinolone N oxides. Although the function of HQNO molecules is still poorly understood, it is known that *pqsL* mutants overproduce PQS, which suggest that PqsL interacts with and diverts a fraction of the HHQ precursor towards HQNO and away from HHQ and PQS biosynthesis (D'Argenio *et al.*, 2002).

Unlike many other QS signalling molecules, PQS is extremely hydrophobic, but it is found associated with the cell envelope as well as in the supernatant after solubilization by rhamnolipids. It was demonstrated that PQS is trafficked within the *P. aeruginosa* population via membrane vesicles (MVs) (Mashburn and Whiteley, 2005). Importantly the addition of PQS containing MVs was able to restore the production of pyocyanin in a PQS-deficient mutant, which suggest that PQS in this state is bioactive and bioavailable (Mashburn and Whiteley, 2005). Recent studies suggested that PQS induces MV formation by interacting strongly with the lipid A component of the LPS. Similarly it was observed that HHQ, which differs from PQS by the absence of the third position hydroxyl, exhibits significantly less interaction with LPS, suggesting a critical role of this hydroxyl in mediating interactions with lipid A (Mashburn-Warren *et*

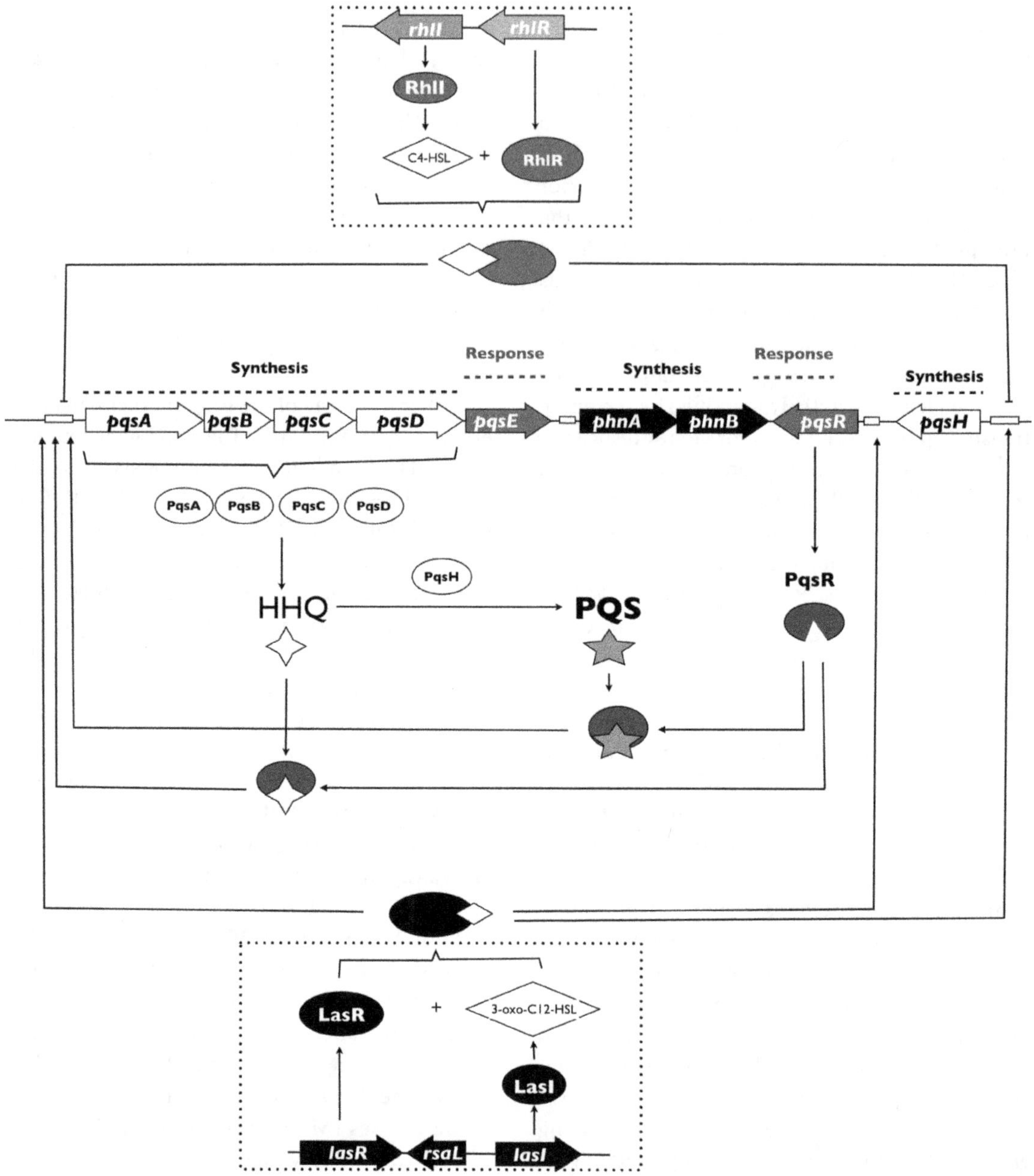

Figure 4.4 The PQS biosynthesis and regulation in *Pseudomonas aeruginosa.* The synthesis of HHQ and PQS is catalysed by the products of the two operons *pqsABCDE* and *phnAB* which are independently regulated by the LysR type regulator PqsR. PqsH catalyses the last step in the conversion of HHQ and PQS. These two molecules are able to bind PqsR to positively regulate the transcription of *pqsABCDE.* The dot –squares represent the AHL-QS systems RhlIR and LasIR. The complex RhlR-C4-HSL negatively regulate both *pqsABCDE* and *pqsH.* In contrast, the complex LasR-(3-oxo-C12-HSL) positively regulates the transcription of these two elements and thus the production of PQS.

al., 2008, 2009). The use of MVs to coordinate group behaviour in a *P. aeruginosa* draws parallels with eukaryote vesicle trafficking systems and illustrates that bacteria posses communication systems analogous to those commonly used by multicelullar organisms.

Regulation of PQS biosynthesis

One of the key factors in the PQS biosynthesis pathway is the LysR-type transcriptional regulator PqsR, which has a helix–turn–helix motif at the N-terminus. This protein is the transcriptional regulator of both the *pqsABCDE* and the *phnAB* operons, so mutation in this regulator results in abolition of the PQS biosynthesis, which affects other virulence determinants as the pyocyanin, elastase, exoprotein as well as reduced ability to cause disease in plants and animals (Cao *et al.*, 2001). PqsR binds the promoter of *pqsABCDE*, but this binding increases dramatically in the presence of PQS, suggesting that PQS acts as a PqsR co-inducer. *In vitro*, PqsR binds at two different locations upstream of *pqsA*, and the strength and position of the binding depends on the presence of PQS (Wade *et al.*, 2005). Similarly, the transcriptional start site for *pqsR* was mapped, which revealed that the transcription of *pqsR* is positively regulated by LasR and negatively regulated by RhlR. These results suggested that a regulatory chain occurs where *pqsR* is under the control of LasR and RhlR (Fig. 4.4; Wade *et al.*, 2005; Xiao *et al.*, 2006b). HHQ potentiates PqsR binding to the *pqsABCDE* promoter, as suggested by the fact that *pqsH* mutants producing HHQ but no PQS, display normal PqsR-dependent gene expression, which demonstrates that HHQ may also work as a QS signal. (Xiao *et al.*, 2006a)

Both the LasIR and the RhlIR systems are linked to the PQS production (Fig. 4.4). Firstly, LasI/R positively regulates PQS production, while RhlI/R acts as negative modulator of their regulatory effects. The LasI/R system positively controls PQS production by inducing the *pqsR* and *pqsA* promoters. The LasR-3-oxo-C12-HSL complex induces *pqsR* transcription by binding to a conserved *lasR* box located 517–536 bp upstream of its transcription start site (McGrath *et al.*, 2004; Gilbert *et al.*, 2009). However, a functional LasI/R system is not required for alkyl quinolone biosynthesis, as *lasR*-deficient mutants still produce PQS in the late stationary phase (Diggle *et al.*, 2003) and the expression of *pqsH* and *pqsR* genes is delayed but not abolished (Pesci *et al.*, 1999; McKnight *et al.*, 2000; Wade *et al.*, 2005). Importantly, *lasR*-deficient mutants accumulate HHQ, but produce very low levels PQS in early phases of the growth, as LasR positively regulates *pqsH*, encoding for the monooxgenase required to convert HHQ in PQS (Gallagher *et al.*, 2002; Deziel *et al.*, 2004).

In contrast to the LasI/R scenario, RhlR has a repressive effect on *pqsR* transcription, as *rhlR*-deficient mutants exhibit at least 50% higher transcription than wild type strains (Wade *et al.*, 2005). Similarly, transcription from the *pqsA* promoter is enhanced in *rhlI*-deficient mutants (McGrath *et al.*, 2004), which supports the negative regulation exerted by the RhlIR QS system. It has also been established that PQS regulates the RhlI/R system (Pesci *et al.*, 1999) and it positively regulates *rhlI* expression, influencing C4-HSL accumulation, generating a negative autoregulatory feedback loop (Diggle *et al.*, 2003).

PQS and iron

In addition to its role as a cell-to-cell signalling molecule, PQS is also able to chelate ferric iron. Many genes regulated via PQS are also regulated by iron and oxidative stress, which has prompted efforts to determine the role of PQS in the cell's ability to perceive environmental iron (Bredenbruch *et al.*, 2006). Multiple independent reports have now concluded that PQS is an iron chelator (Diggle *et al.*, 2007b; Schertzer *et al.*, 2009). Furthermore, compounds similar to PQS, also possess iron-binding capabilities, but molecules lacking the 3-hydroxy group such as HHQ, are unable to do so (Diggle *et al.*, 2007b). Iron availability also influences the levels at which PQS induces the activity of PqsR as transcriptional activator, and thus iron may also act as a modulator of the PQS signalling in *P. aeruginosa* (Hazan *et al.*, 2010). Interestingly, iron has also been found bound to PqsE, although the biological significance of this finding remains poorly understood (Yu *et al.*, 2009).

Quinolones in other bacteria

Comparative analyses have revealed that the alkyl quinolone biosynthesis genes are also present in other bacterial species. The presence of an operon homologous to *pqsABCDE* was first identified in *B. pseudomallei* and *B. thailandensis*, although other *Burkholderia* species were also screened for the production of alkyl quinolones by liquid chromatography/mass spectrometry (Diggle *et al.*, 2006). Such studies also demonstrated that homologous operons were also present in *B. ambifaria*, a member of the *Burkholderia cepacia* complex. In contrast with those of *P. aeruginosa*, the quinolones produced by *Burkholderia* typically bear a methyl group, hence their designation as 4-hydroxy-3-methyl-2-alkylquinolines (HMAQs) (Vial *et al.*, 2008). Three families of HMAQs with a saturated or unsaturated alkyl chain at the 2′ position were also identified. Importantly, the operon in these species contains two more genes downstream of the *pqsE* homologue, resulting in the *hmqABCDEFG* operon, although none of these species harbour *pqsH* orthologues, nor produce PQS. Mutational analysis also showed that the inactivation of *hmqA* inhibits the production of HMAQs, while the methylation of the quinoline ring requires a putative methyltransferase encoded by *hmqG*. Interestingly, *hmqA* or *hmqG* mutations increase the production of acyl homoserine lactones and, consequently, phenotypes under the control of AHL QS in *B. ambifaria* including antifungal activity, siderophore production, and proteolytic activity. These results indicate that only HAQs bearing a methyl group (HMAQs) are involved in QS regulation, indicating high levels of complexity in the QS network in these species (Vial *et al.*, 2008; Heeb *et al.*, 2011).

The AI-2/LuxS system

The QS systems of Gram-negative bacteria described thus far are believed to allow mostly intraspecies communication; these could play an important role in bacteria that successfully colonize specific niches enabling them to monitor their numbers and coordinate behaviours. The species specificity of the QS systems could also be a way for bacteria to recognize self from other signals in the environment. It is very likely that bacteria also need a mechanism to detect and/or communicate with other bacterial species or even to be 'informed' of its ratio to that of other species in a mixed bacterial community.

Initial evidence of the presence of signals involved in inter-species communication came from studies in the bioluminescent marine pathogen *V. harveyi*. This bacterium produces three signals, one is an α-hydroxyketone (AHK; see below), another one is an AHL (3-OH-C4-HSL) which is used for intra-species signalling, and the third one is most often called AI-2, and is believed to be used for interspecies signalling (reviewed by Federle and Bassler, 2003). The AHL signal of *V. harveyi* is synthesized by the protein LuxM (see above) and is detected by the hybrid sensor kinase-response regulator LuxN. The *V. harveyi* AI-2 signal is a furanosyl borate diester (Fig. 4.1) which is synthesized by an enzyme called LuxS. In common with the AHLs, AI-2 is derived from SAM; however, it is synthesized in three enzymatic steps (Schauder *et al.*, 2001). The detection of AI-2 is initially carried out by the LuxP protein forming a LuxP–AI-2 complex, which then interacts with a second protein called LuxQ, which is a hybrid sensor kinase-response regulator just like LuxN. The LuxN and LuxPQ then pass on the signal to the phosphotransferase protein, LuxU, which finally transmits the signal to the response regulator LuxO. The result is that both AHL and AI-2 acts as QS signals transmitting the signal to a common protein LuxU, which then finally signals other proteins leading to the control of gene expression. This cascade in *V. harveyi* occurs via phosphorylation typical of two-component bacterial systems in the following way: at low signal concentration (i.e. at low cell density), LuxN and LuxQ autophosphorylate their conserved aspartate residue in the response regulator domain. This phosphate is then transferred to the histidine residue of the phosphotransferase LuxU protein, which then finally transfers it to the conserved aspartate residue of LuxO. Phosphorylated-LuxO results in indirect repression of the luciferase operon. At high AHL or AI-2 signal concentrations (i.e. at high cell density) interactions of the signals with LuxN and LuxPQ respectively, induces the proteins to switch to the phosphatase mode. This results in the dephosphorylation and inactivation of LuxO leading to de-repression of the luciferase target operon (Fig. 4.5). The de-repression in *V. harveyi* is indirect and associated with transcriptional activation by the LuxR master regulator (this is not related to AHL-LuxR proteins), which controls genes of the QS regulon (Neiditch *et al.*, 2005).

The LuxS synthase is present in many genomes of both Gram-negative and Gram-positive

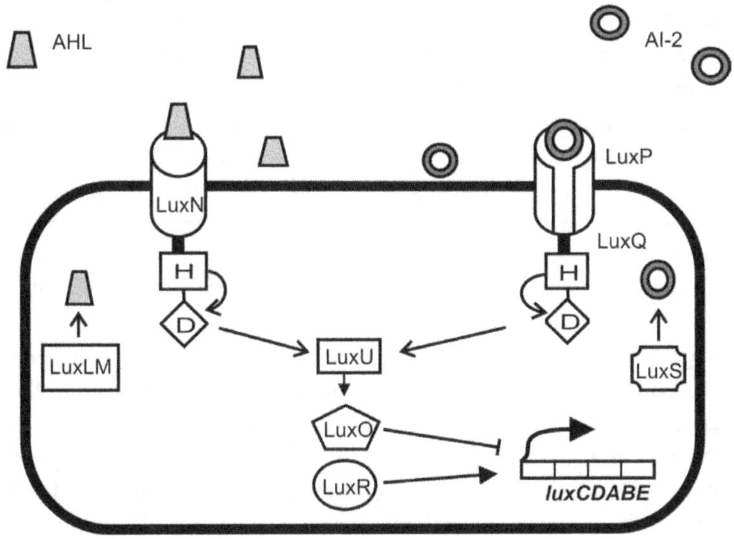

Figure 4.5 The quorum sensing circuitry of *V. harveyi* which integrates in the same pathway the AHL and AI-2 response. The AHLs are synthesized by LuxLM and AI-2 (furanosyl borate diester) is synthesized by LuxS. In the absence of AHLs and AI-2 the LuxN and LuxPQ sensors act as kinases and the phosphorylation cascade finally leads to phospho-LuxU which indirectly represses the luciferase *luxCDABE* operon. Binding of the signals to LuxN and LuxPQ on the other hand, leads to dephosphorylation of LuxU and LuxO; this relieves repression of the luciferase operon. Expression of the luciferase operon can then be activated by LuxR (not related to AHL quorum sensing LuxRs of Fig. 4.2). This picture has been extrapolated from Federle and Bassler, 2003.

bacteria hence a wide variety of bacteria are capable of synthesizing AI-2. This observation has led scientists to postulate that AI-2 is a signal involved in interspecies bacterial communication. Importantly, the LuxS enzyme has also been proposed to have a metabolic role in cells in addition to being responsible for AI-2 biosynthesis (Vendeville *et al.*, 2005). This could also be an alternative reason for the widespread conservation and presence of LuxS in bacteria. In spite of this controversy, a large number of data have been accumulated that AI-2 is in fact a universal language of interspecies communication (Federle and Bassler, 2003; Vendeville *et al.*, 2005; Galloway *et al.*, 2011). AI-2 signalling has been associated with pathogenicity in a variety of bacteria regulating many different behaviours in pathogenic bacteria including virulence and biofilm formation (Xavier and Bassler, 2003). Biofilm formation and/or production of virulence-associated factors have been shown to be regulated by AI-2/LuxS in *V. cholerae, Staphylococcus epidermidis, Streptococcus pyogenes, Clostridium perfringens* and *Porphyromonas gingivalis.*

The LuxS enzyme synthesizes 4,5-dihydroxy-2,3-pentanedione (DPD), which undergoes spontaneous rearrangements to form a series of DPD derivatives that exist in equilibrium forming an AI-2 pool. Different DPDs are recognized by different bacteria; however, it is believed that constant conversion and equilibrium in the AI-2 pool allows bacteria to respond to their own AI-2 as well as to AI-2 produced by other neighbouring bacteria. For this reason AI-2 can be regarded as a universal cell–cell signalling system among bacteria. Owing to the different chemical nature of the active signal molecule in the AI-2 signal pool for different bacteria, it is not surprising that several AI-2 receptors have been identified. Apart from LuxP of *V. harveyi* as described above, in *E. coli* and *S. typhimurium* the protein detecting the AI-2 signal is called LsrB (Miller *et al.*, 2004) and in *Actinobacillus actinomycetemcomitans* the RbsB receptor for AI-2 has been found (O'Sullivan *et al.*, 2007).

AI-2 signalling has attracted considerable interest in the scientific community in recent years. The nature of the AI-2 signal pool being produced by such a large number of bacteria probably means the broad-spectrum modulation of gene expression in a bacterial community. In addition the important role in regulating virulence in a wide variety of pathogens, both Gram-negative and Gram-positive, will push scientists to further precisely decipher the response in several bacteria since to date only the response to AI-2 in *V. harveyi* and *S. typhimurium* is understood in some detail.

DSF (diffusible signal factor)

The signal molecules collectively known as diffusible signal factors (DSFs) are unsaturated fatty acids (Fig. 4.1). The first was described in the plant pathogen *X. campestris* pv. *campestris* (*Xcc*). Structurally related molecules were later identified in *Burkholderia, Stenotrophomonas* and *Xylella.* In all cases these molecules have been shown to be involved in the regulation of virulence, biofilm formation and other phenotypes in a cell density manner. In addition, DSF molecules have also been linked to interspecies signalling and modulation of the bacterial behaviour (reviewed by Ryan and Dow, 2011).

The understanding that DSF was a QS signal molecule originally emerged from studies intended to identify regulatory mechanisms of virulence factors in *Xcc* (Tang *et al.*, 1991; Barber *et al.*, 1997). These identified a cluster of genes termed *rpf* (regulation of pathogenicity factors), which positively regulated EPS production, exoenzymes biosynthesis and plant virulence. Furthermore, the phenotype of *rpfF*-deficient mutants could be restored to wild type levels either by growing them adjacent to *rpfF* non-deficient mutants, or in the presence of extracts from their spent supernatants. This suggested that a diffusible molecule was most likely involved in the regulation of the virulence-associated factors.

The DSF family of signals

Structural analyses of several DSF-like molecules demonstrated that the *cis*-unsaturated double bond at the 2-position is a key structural feature for DSF activity. DSF is structurally and functionally related to farnesoic acid (FA), which is a signal molecule that regulates morphological transition and virulence on the fungal pathogen *Candida albicans*. Interestingly, similar to FA, which is also an unsaturated fatty acid, DSF inhibits the dimorphic transition of *C. albicans* (Wang *et al.*, 2004). DSF activity in *Xcc* relies in the methyl branch and the length of the chain, since unbranched *cis* unsaturated fatty acids with chains longer or shorter than 12 or 13 carbons have very little or no activity when compared with DSF (Wang *et al.*, 2004).

DSF-like molecules have also been identified in other *Xanthomonas*, such as the rice pathogen *X. oryzae* pv. *oryzae* (*Xoo*), in *Stenotrophomonas maltophilia* and *Xylella fastidiosa*. *Xoo* and *Xcc* share the conserved mechanisms for DSF biosynthesis and regulation, although *Xoo* produces DSF, BDSF (*cis*-2-dodecenoic acid) and a third signal denoted as CDSF (*cis,cis*-11-methyl dodeca-2-5-dienoic acid). All the three DSF-family signals promote EPS production and xylanase activity in *Xoo*, but CDSF is less active than its analogues DSF and BDSF. Importantly, the composition and ratio of the three DSF-family signals produced by *Xoo* are influenced by the culture medium (He *et al.*, 2010).

S. maltophilia encodes a cell–cell signalling system highly related to the DSF encoding system of *Xcc*. This involves the synthesis of DSF and seven other related molecules, and the disruption of their biosynthesis results in pleiotrophic effects in *S. maltophilia*. In fact, this signalling controls factors contributing to the virulence and antibiotic resistance of this emerging nosocomial pathogen (Fouhy *et al.*, 2007). Homologous genes to the *rpf* cluster are also present in the genome of the plant pathogen *X. fastidiosa* and the DSF molecule produced was identified as 12-methyl-tetradecanoic acid (Colnaghi Simionato *et al.*, 2007). Furthermore, plant inoculation experiments indicated that DSF negatively regulates plant virulence in *X. fastidiosa*, as DSF-deficient strains were hypervirulent, although they exhibited reduced capacity to colonize the insect vector (Chatterjee *et al.*, 2008a,b). Multiple DSF-like molecules have also been identified in bacteria belonging to the *Burkholderia cepacia* complex. A *cis*-2-dodecenoic acid or BDSF (*Burkholderia* DSF) was isolated in spent supernatants of *B. cenocepacia* (Boon *et al.*, 2008), and more recently the same molecule was also found to be produced by *B. lata, B. vietnamiensis, B. dolosa* and *B. ambifaria*, whereas *B. multivorans, B.*

stabilis, *B. anthina* and *B. pyrrocinia* also synthesize *cis,cis*-11-methyldodeca-2,5-dienoic acid, a new structure among the DSF-like molecules (Deng *et al.*, 2010). A DSF-like molecule (*cis*-2-decenoic acid) was also characterized in *P. aeruginosa*. This signal was found capable of inducing the dispersion of established biofilms and of inhibiting biofilm development at nanomolar concentrations (Davies and Marques, 2009).

DSF biosynthesis and perception

DSF synthesis and perception is thus far best understood in *Xcc*. Three genes, *rpfF*, *rpfB* and *rpfC* are involved in DSF biosynthesis. The synthesis of DSF is dependent on RpfF, which has amino acid sequence relatedness to enoyl CoA hydratase and is also partially dependent on RpfB, which is a long-chain fatty acyl CoA ligase. The *rpfC* gene on the other hand encodes a membrane-associated two component sensor kinase (Barber *et al.*, 1997). RpfC positively regulates the synthesis of virulence factors, but negatively regulates the DSF synthesis, and thus *rpfC*-deficient mutants exhibit increased DSF production (Slater *et al.*, 2000). The *rpfB* and *rpfF* genes are genetically linked in the *Xcc* chromosome and are transcribed as an operon, although *rpfF* also has its own gene promoter (Fig. 4.6).

The sensing and transduction of the DSF signal in bacteria involves two-component phosphorelay signal transduction systems (Fig. 4.6). In *Xcc*, DSF sensing depends upon the two-component system encoded by the *rpfGHC* operon, which is adjacent to *rpfF* but convergently transcribed (Slater *et al.*, 2000). RpfC is a complex sensor kinase comprising a sensory input domain, a histidine kinase (HisKA) domain, a Che-Y-like two-component receiver domain (REC), and a C-terminal histidine phosphotransfer domain (HPT). RpfG contains a REC domain and an HD-GYP domain. DSF binding to the sensory input domain of RpfC activates autophosphorylation at the HisKA acceptor domain. This event is followed by phosphorelay involving an aspartic acid residue of the REC domain and a histidine residue of the HPT domain and finally phosphotransfer to the REC domain of the RpfG regulator (Slater *et al.*, 2000; He *et al.*, 2006). Phosphorylation of RpfG leads to its activation as a cyclic di-GMP phosphodiesterase and consequent alterations in the level of cyclic di-GMP in the cell, which influences the synthesis of virulence factors by diverse mechanisms (Ryan

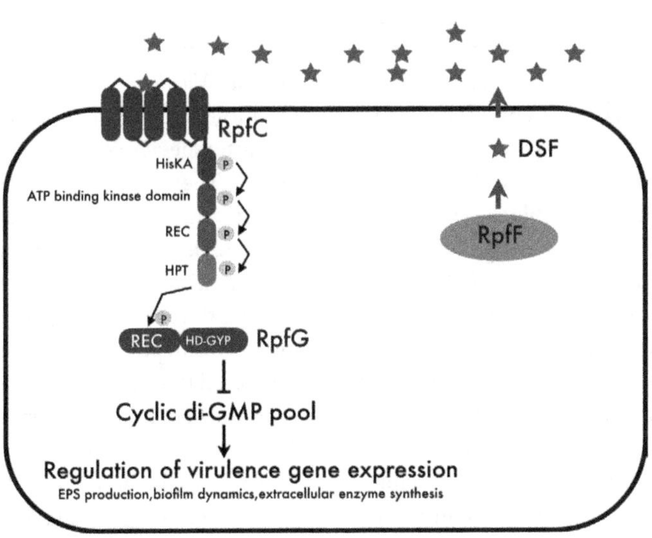

Figure 4.6 Schematic representation of the DSF signalling network in Xcc. DSF is synthesized by RpfF, and freely diffuses to the extracellular milieu. Perception of DSF by RpfC leads to autophosphorylation and phosphotransfer to the REC domain of RpfG. This leads to activation of RpfG as a cyclic di-GMP phosphodiesterase, an activity associated with the HD-GYP domain. The consequent reduction of cyclic di-GMP levels promotes synthesis of extracellular enzymes and EPS but leads to inhibition of biofilm formation. This picture has been extrapolated from He *et al.* (2008) and Ryan and Dow (2011).

et al., 2006). The effect of Rpf on cyclic di-GMP levels is reviewed in detail in another chapter of this book thus it will not be covered here. RpfH is a novel protein with four transmembrane helices that has amino acid sequence similarity to the sensory input domain of RpfC; however, thus far no function for RpfH in DSF synthesis or perception is evident (Fig. 4.6). It is worth mentioning that, in contrast to the *rpfF-deficient* mutant, strains lacking RpfC or RpfG do not respond to addition of exogenous DSF, consistent with a role in DSF sensing and signal transduction.

DSF signalling contributes to a range of bacterial functions including virulence, biofilm formation, motility and antibiotic tolerance (He and Zhang, 2008). Recent transcriptomic and proteomic studies in *Xcc*, *X. fastidiosa* and *B. cenocepacia* have additionally included other cell functions such as the stress tolerance, synthesis of extracellular enzymes, chemotaxis, iron uptake and EPS synthesis, among others (reviewed by Ryan and Dow, 2011). The production of DSF is carefully regulated. As previously mentioned, RpfC acts as a positive regulator of virulence factor synthesis but as a negative regulator of DSF synthesis. This effect of RpfC on DSF synthesis is not at the transcriptional level of the *rpfF* gene (Slater *et al.*, 2000). Current evidence suggests that DSF repression occurs via protein–protein interactions between the REC domain of RpfC and RpfF (He *et al.*, 2006; Cheng *et al.*, 2010). Structural analysis of a co-crystal of the REC domain and RpfF suggests that the REC domain may restrict DSF synthesis by blocking substrate access to the RpfF active site (Cheng *et al.*, 2010). It has been proposed that release of RpfF as a result of conformational changes in RpfC that occur upon DSF binding may act as a mechanism to allow autoinduction of DSF synthesis (He *et al.*, 2006; Cheng *et al.*, 2010). Such protein–protein interaction represents both a novel mechanism for control of synthesis of a QS signal and a novel role for the regulatory action of two-component sensory proteins.

Other QS signals

It is widely accepted that most bacterial species have at least one mechanism of cell–cell communication. This is highlighted by the discovery of different signals from various bacterial species with varied chemical structures. Above we have reviewed the currently most studied QS signals in Gram-negative bacteria, below we list and briefly describe other QS signals. It is most likely that some of these are more widespread in the bacterial world and that many more new signals will be discovered as this research field progresses.

3-OH-PAME

In the plant pathogen *R. solanacearum* cell–cell communication is integrated in a complex regulatory network, allowing the induction of virulence genes in response to plant signals and nutritional starvation. In addition to AHLs, *R. solanacearum* produces the volatile fatty acid 3-hydroxypalmitic acid (3-OH-PAME; Fig. 4.3) to regulate traits involved in pathogenicity in a cell density-dependant manner.(Flavier *et al.*, 1997a,c). At a high cell density, 3-OH PAME allows the expression of the transcriptional regulator PhcA, which induces the expression of virulence factors and modulates the synthesis of AHLs (Flavier *et al.*, 1997c). In spite of the evidence of cross talk between both QS systems, the 3-OH PAME regulon remains to be characterized, as well as its precise role in plant virulence.

α-Hydroxyketones

Several *Vibrio* species produce more than one QS molecule, which allow them to promote intra and interspecies communication (Ng and Bassler, 2009). These signals include AI-2, AHLs and the recently discovered α-hydroxyketones (AHK) (Fig. 4.3). This latter group of molecules were first identified in *V. cholerae* (Miller *et al.*, 2002), but most recently they were also characterized in *Legionella pneumophila* and *V. harveyi*, and it is predicted that they may also be produced in other bacterial species (Spirig *et al.*, 2008; Tiaden *et al.*, 2010; Ng *et al.*, 2011).

So far, the best characterized AHK systems are the *L. pneumophila lqs* (*Legionella* quorum sensing) and the *V. cholerae cqs* (cholera quorum sensing), which produce the signalling molecules (3-hydroxypentadecan-4-one) or cholera autoinducer-1 (3-hydroxytridecan-4-one) (CAI-1), respectively. The *lqs/cqs* gene clusters encode the homologous autoinducer synthases LqsA/CqsA and the sensor kinases LqsS/CqsS. In addition, the *lqs* cluster also harbours a prototypic response regulator LqsR and a homologue to HdeD, which is involved in acid resistance in *E. coli* (Higgins *et al.*, 2007; Tiaden *et al.*, 2008b). AHK signalling regulates pathogen–host cell interactions, bacterial virulence, formation of biofilms or extracellular filaments, and expression of a genomic island (Tiaden *et al.*, 2008a,b).

Recently, two AHKs produced by *V. harveyi* were identified as (Z)-3 aminoundec-2-en-4-ona (Ea-C-8-CAI-1), and a CAI-1 analogue carrying an eight-carbon tail (C-8-CAI-1). Although the *V. harveyi* CqsS kinase only detects Ea-C-8-CAI-1, it has been proposed that this molecule fulfils the requirements for being an inter-*Vibrio* QS signal because it is produced and detected by multiple *Vibrio* species (Ng *et al.*, 2011). Importantly, the interaction between the signals and CqsS two-component histidine sensor kinase (Cqs possesses both kinase and phosphatase activities representing the active or inactive state respectively) represents an elegant model to study this class of proteins as the environmental signal of this family of sensor proteins is in most cases unknown. Among different CqsS proteins, the transmembrane regions located at the N-terminus are most conserved and are involved in the interaction of chemical moieties conserved among all CAI-1-type molecules. By contrast, the transmembrane domains at the C-terminus are less conserved and thus interact with distinct moieties of CAI-1-type molecules (Ng *et al.*, 2011). Interestingly, the effect of a given CAI-1-type molecule on a given CqsS depends on the interaction with the ligand binding pocket; if a ligand prefers the inactive state the molecules act as agonist, whereas if the CAI-1type molecule prefers the active state they are antagonists (Ng *et al.*, 2011). Furthermore, orthologues to *lqsA*, *lqsS* and *lqsR* have been identified in all *Legionella* species sequenced, as well as in *Nitrococcus mobilis*, *Burkholderia xenovorans* and *Polaromonas* spp., among others (reviewed by Tiaden *et al.*, 2010).

The AarA rhomboid system of *Providencia stuartii*

Some bacterial and eukaryotic signalling mechanisms seem to have a common evolutionary origin, as suggested by the particular QS system identified en *Providencia stuartii*. In this species the rhomboid intramembrane serine protease protein AarA is required for the release of a QS signal molecule, whose structure has not yet been elucidated (Rather *et al.*, 1999). The AarA protein is homologous to the Rho protein from *Drosophila melanogaster*, which is a serine protease required for the intramembrane cleavage, release and activation of epidermal growth factor receptor ligands (Klambt, 2002). Orthologues to AarA have been identified in species from the three kingdoms, which suggest that this mechanism may be useful for interkingdom communication. Recently, an additional component of the twin-arginine translocase (Tat) protein secretion pathway was identified as involved in the production of secreted signalling molecules in *P. stuartii*, although the chemical structure of the signal still needs to be determined (Stevenson *et al.*, 2007).

Cyclic dipeptides

A range of cyclic dipeptides (or diketopiperazines, DKPs) have been isolated from supernatants of several Gram-negative bacteria and have been reported to modulate the activity of QS LuxR-family proteins, albeit at much higher concentrations than AHLs (Holden *et al.*, 1999; Degrassi *et al.*, 2002; Park *et al.*, 2006). Since DKPs have been isolated (very often as mixtures) from supernatants of a wide range of Gram-negative including *P. putida*, *P. fluorescens*, *P. aeruginosa*, *P. mirabilis*, *Enterobacter agglomerans* and *V. vulnificus*, they have been proposed to be involved in interspecies signalling. Importantly, a recent report has highlighted that modulation of AHL QS by DKPs does not occur through direct interactions with the LuxR-type proteins (Campbell *et al.*, 2009). Future studies will need to determine the precise mode of action of DKPs and whether they can be regarded as signals and/or modulators of QS systems.

Future trends in QS research

The aim of this chapter was to present an overview of the most studied QS signals and their mode of action. It is likely that an enormous number of data on QS in Gram-negative has been generated in the last 15 years. It is most likely however that many more signals will be discovered in the near future and/or that many more species will be found to produce already known ones. Recent studies have shown that apart from producing and sensing endogenous signals, we are beginning to understand that bacteria sense and respond to exogenous QS signals produced by others and that QS-related proteins also respond to eukaryotic signals. Many questions have therefore been raised and this research field will definitively continue to be very active in the future. For example, what is the role of the QS signals in inter-species and interkingdom signalling? In natural environments bacteria in most cases exist as multispecies communities and in close association with eukaryotes; thus, QS signalling must now be studied more in this context. In fact, most studies to date have involved bacterial laboratory growth in a bacterial monospecies (strain) set up. In addition many open questions remain regarding the evolutionary forces that shape QS and how QS behaviours of individual cells translate into collective properties of bacterial groups. Evolutionary studies are being performed in *P. aeruginosa* (Sandoz *et al.*, 2007; Diggle *et al.*, 2007a), but how the QS networks in other bacterial species may evolve is an issue, which also deserves further studies. It must also be noted that several species harbour more than one QS system, and often these systems are hierarchically organized. How multiple quorum-sensing signals may be channelled into a single circuit, and how different signals lead to differential gene expression profiles in bacteria is an issue that merits further attention. The study of QS regulated virulence factors has been pivotal to understand host–pathogen interactions in both animals and plants. Significant efforts have been invested in developing therapeutic strategies based in the quenching or interfering with QS systems; this topic of research needs to continue and hopefully result in an effective and practical application in the control of bacterial infections. In addition, the possible exploitation of QS in beneficial bacterial species remains to be evaluated and studied.

Acknowledgements

We thank Julieta Manrique for her assistance in preparing the figures. We thank Sujatha Subramoni for reading and helpful discussions. ZRSM and JFG were funded by an ICGEB fellowship.

References

Aendekerk, S., Ghysels, B., Cornelis, P., and Baysse, C. (2002). Characterization of a new efflux pump, MexGHI-OpmD, from *Pseudomonas aeruginosa* that confers resistance to vanadium. Microbiology *148*, 2371–2381.

Aguilar, C., Bertani, I., and Venturi, V. (2003). Quorum-sensing system and stationary-phase sigma factor (rpoS) of the onion pathogen *Burkholderia cepacia* genomovar I type strain, ATCC 25416. Appl. Environ. Microbiol. *69*, 1739–1747.

Ahmer, B.M. (2004). Cell-to-cell signalling in *Escherichia coli* and *Salmonella enterica*. Mol. Microbiol. *52*, 933–945.

Ahmer, B.M., van Reeuwijk, J., Timmers, C.D., Valentine, P.J., and Heffron, F. (1998). *Salmonella typhimurium* encodes an SdiA homolog, a putative quorum sensor of the LuxR family, that regulates genes on the virulence plasmid. J. Bacteriol. *180*, 1185–1193.

Atkinson, S., Throup, J.P., Stewart, G.S., and Williams, P. (1999). A hierarchical quorum-sensing system in *Yersinia pseudotuberculosis* is involved in the regulation of motility and clumping. Mol. Microbiol. *33*, 1267–1277.

Bainton, N.J., Bycroft, B.W., Chhabra, S.R., Stead, P., Gledhill, L., Hill, P.J., Rees, C.E., Winson, M.K., Salmond, G.P., Stewart, G.S., *et al.* (1992a). A general role for the *lux* autoinducer in bacterial cell signalling: control of antibiotic biosynthesis in *Erwinia*. Gene *116*, 87–91.

Bainton, N.J., Stead, P., Chhabra, S.R., Bycroft, B.W., Salmond, G.P., Stewart, G.S., and Williams, P. (1992b). N-(3-oxohexanoyl)-L-homoserine lactone regulates carbapenem antibiotic production in *Erwinia carotovora*. Biochem. J. *288*, 997–1004.

Baldwin, A., Sokol, P.A., Parkhill, J., and Mahenthiralingam, E. (2004). The *Burkholderia cepacia* epidemic strain marker is part of a novel genomic island encoding both virulence and metabolism-associated genes in *Burkholderia cenocepacia*. Infect. Immun. *72*, 1537–1547.

Barber, C.E., Tang, J.L., Feng, J.X., Pan, M.Q., Wilson, T.J., Slater, H., Dow, J.M., Williams, P., and Daniels, M.J. (1997). A novel regulatory system required for pathogenicity of *Xanthomonas campestris* is mediated by a small diffusible signal molecule. Mol. Microbiol. *24*, 555–566.

Barnard, A.M., and Salmond, G.P. (2007). Quorum sensing in *Erwinia* species. Anal. Bioanal. Chem. *387*, 415–423.

Bassler, B.L., Wright, M., Showalter, R.E., and Silverman, M.R. (1993). Intercellular signalling in *Vibrio harveyi*: sequence and function of genes regulating expression of luminescence. Mol. Microbiol. *9*, 773–786.

Beck von Bodman, S., and Farrand, S.K. (1995). Capsular polysaccharide biosynthesis and pathogenicity in *Erwinia stewartii* require induction by an

N-acylhomoserine lactone autoinducer. J. Bacteriol. *177*, 5000–5008.

Bertani, I., and Venturi, V. (2004). Regulation of the N-Acyl homoserine lactone-dependent quorum-sensing system in rhizosphere *Pseudomonas putida* WCS358 and cross-talk with the stationary-phase RpoS sigma factor and the global regulator GacA. Appl. Environ. Microbiol. *70*, 5493 – 5502.

von Bodman, S.B., Majerczak, D.R., and Coplin, D.L. (1998). A negative regulator mediates quorum-sensing control of exopolysaccharide production in *Pantoea stewartii* subsp. *stewartii*. Proc. Natl. Acad. Sci. U.S.A. *95*, 7687–7692.

von Bodman, S.B., Ball, J.K., Faini, M.A., Herrera, C.M., Minogue, T.D., Urbanowski, M.L., and Stevens, A.M. (2003). The quorum sensing negative regulators EsaR and ExpR(Ecc), homologues within the LuxR family, retain the ability to function as activators of transcription. J. Bacteriol. *185*, 7001–7007.

Boon, C., Deng, Y., Wang, L.H., He, Y., Xu, J.L., Fan, Y., Pan, S.Q., and Zhang, L.H. (2008). A novel DSF-like signal from *Burkholderia cenocepacia* interferes with *Candida albicans* morphological transition. ISME J. *2*, 27–36.

Boyer, M., and Wisniewski-Dye, F. (2009). Cell–cell signalling in bacteria: not simply a matter of quorum. FEMS Microbiol. Ecol. *70*, 1–19.

Bredenbruch, F., Geffers, R., Nimtz, M., Buer, J., and Haussler, S. (2006). The *Pseudomonas aeruginosa* quinolone signal (PQS) has an iron-chelating activity. Environ. Microbiol. *8*, 1318–1329.

Buroni, S., Pasca, M.R., Flannagan, R.S., Bazzini, S., Milano, A., Bertani, I., Venturi, V., Valvano, M.A., and Riccardi, G. (2009). Assessment of three Resistance-Nodulation-Cell Division drug efflux transporters of *Burkholderia cenocepacia* in intrinsic antibiotic resistance. BMC Microbiol. *9*, 200.

Calfee, M.W., Coleman, J.P., and Pesci, E.C. (2001). Interference with Pseudomonas quinolone signal synthesis inhibits virulence factor expression by *Pseudomonas aeruginosa*. Proc. Natl. Acad. Sci. U.S.A. *98*, 11633–11637.

Campbell, J., Lin, Q., Geske, G.D., and Blackwell, H.E. (2009). New and unexpected insights into the modulation of LuxR-type quorum sensing by cyclic dipeptides. ACS Chem. Biol. *4*, 1051–1059.

Cao, H., Krishnan, G., Goumnerov, B., Tsongalis, J., Tompkins, R., and Rahme, L.G. (2001). A quorum sensing-associated virulence gene of *Pseudomonas aeruginosa* encodes a LysR-like transcription regulator with a unique self-regulatory mechanism. Proc. Natl. Acad. Sci. U.S.A. *98*, 14613–14618.

Case, R.J., Labbate, M., and Kjelleberg, S. (2008). AHL-driven quorum-sensing circuits: their frequency and function among the Proteobacteria. ISME J. *2*, 345–349.

Chai, Y., Zhu, J., and Winans, S.C. (2001). TrlR, a defective TraR-like protein of *Agrobacterium tumefaciens*, blocks TraR function *in vitro* by forming inactive TrlR:TraR dimers. Mol. Microbiol. *40*, 414–421.

Chalupowicz, L., Manulis-Sasson, S., Itkin, M., Sacher, A., Sessa, G., and Barash, I. (2008). Quorum-sensing system affects gall development incited by *Pantoea agglomerans* pv. *gypsophilae*. Mol. Plant Microbe Interact. *21*, 1094–1105.

Chan, Y.Y., and Chua, K.L. (2005). The *Burkholderia pseudomallei* BpeAB-OprB efflux pump: expression and impact on quorum sensing and virulence. J. Bacteriol. *187*, 4707–4719.

Chan, Y.Y., Bian, H.S., Tan, T.M., Mattmann, M.E., Geske, G.D., Igarashi, J., Hatano, T., Suga, H., Blackwell, H.E., and Chua, K.L. (2007). Control of quorum sensing by a *Burkholderia pseudomallei* multidrug efflux pump. J. Bacteriol. *189*, 4320–4324.

Chandler, J.R., Duerkop, B.A., Hinz, A., West, T.E., Herman, J.P., Churchill, M.E., Skerrett, S.J., and Greenberg, E.P. (2009). Mutational analysis of *Burkholderia thailandensis* quorum sensing and self-aggregation. J. Bacteriol. *191*, 5901–5909.

Chatterjee, S., Newman, K.L., and Lindow, S.E. (2008a). Cell-to-cell signaling in *Xylella fastidiosa* suppresses movement and xylem vessel colonization in grape. Mol. Plant Microbe Interact. *21*, 1309–1315.

Chatterjee, S., Wistrom, C., and Lindow, S.E. (2008b). A cell–cell signaling sensor is required for virulence and insect transmission of Xylella fastidiosa. Proc. Natl. Acad. Sci. U.S.A. *105*, 2670–2675.

Cheng, Z., He, Y.W., Lim, S.C., Qamra, R., Walsh, M.A., Zhang, L.H., and Song, H. (2010). Structural basis of the sensor–synthase interaction in autoinduction of the quorum sensing signal DSF biosynthesis. Structure *18*, 1199–1209.

Choi, S.H., and Greenberg, E.P. (1991). The C-terminal region of the *Vibrio fischeri* LuxR protein contains an inducer-independent lux gene activating domain. Proc. Natl. Acad. Sci. U.S.A. *88*, 11115–11119.

Chugani, S.A., Whiteley, M., Lee, K.M., D'Argenio, D., Manoil, C., and Greenberg, E.P. (2001). QscR, a modulator of quorum-sensing signal synthesis and virulence in *Pseudomonas aeruginosa*. Proc. Natl. Acad. Sci. U.S.A. *98*, 2752–2757.

Colnaghi Simionato, A.V., da Silva, D.S., Lambais, M.R., and Carrilho, E. (2007). Characterization of a putative *Xylella fastidiosa* diffusible signal factor by HRGC-EI-MS. J. Mass Spectrom. *42*, 1375–1381.

Conway, B.A., and Greenberg, E.P. (2002). Quorum-sensing signals and quorum-sensing genes in *Burkholderia vietnamiensis*. J. Bacteriol. *184*, 1187–1191.

Cornforth, J.W., and James, A.T. (1956). Structure of a naturally occurring antagonist of dihydrostreptomycin. Biochem. J. *63*, 124–130.

Cox, A.R., Thomson, N.R., Bycroft, B., Stewart, G.S., Williams, P., and Salmond, G.P. (1998). A pheromone-independent CarR protein controls carbapenem antibiotic synthesis in the opportunistic human pathogen *Serratia marcescens*. Microbiology *144*, 201–209.

Croxatto, A., Chalker, V.J., Lauritz, J., Jass, J., Hardman, A., Williams, P., Camara, M., and Milton, D.L. (2002). VanT, a homologue of *Vibrio harveyi* LuxR, regulates serine, metalloprotease, pigment, and biofilm production in *Vibrio anguillarum*. J. Bacteriol. *184*, 1617–1629.

D'Argenio, D.A., Calfee, M.W., Rainey, P.B., and Pesci, E.C. (2002). Autolysis and autoaggregation in *Pseudomonas aeruginosa* colony morphology mutants. J. Bacteriol. *184*, 6481–6489.

Davies, D.G., and Marques, C.N. (2009). A fatty acid messenger is responsible for inducing dispersion in microbial biofilms. J. Bacteriol. *191*, 1393–1403.

Degrassi, G., Aguilar, C., Bosco, M., Zahariev, S., Pongor, S., and Venturi, V. (2002). Plant growth-promoting *Pseudomonas putida* WCS358 produces and secretes four cyclic dipeptides: cross-talk with quorum sensing bacterial sensors. Curr. Microbiol. *45*, 250–254.

Deng, Y., Wu, J., Eberl, L., and Zhang, L.H. (2010). Structural and functional characterization of diffusible signal factor family quorum-sensing signals produced by members of the *Burkholderia cepacia* complex. Appl. Environ. Microbiol. *76*, 4675–4683.

Deziel, E., Lepine, F., Milot, S., He, J., Mindrinos, M.N., Tompkins, R.G., and Rahme, L.G. (2004). Analysis of *Pseudomonas aeruginosa* 4-hydroxy-2-alkylquinolines (HAQs) reveals a role for 4-hydroxy-2-heptylquinoline in cell-to-cell communication. Proc. Natl. Acad. Sci. U.S.A. *101*, 1339–1344.

Deziel, E., Gopalan, S., Tampakaki, A.P., Lepine, F., Padfield, K.E., Saucier, M., Xiao, G., and Rahme, L.G. (2005). The contribution of MvfR to *Pseudomonas aeruginosa* pathogenesis and quorum sensing circuitry regulation: multiple quorum sensing-regulated genes are modulated without affecting *lasRI*, *rhlRI* or the production of N-acyl-L-homoserine lactones. Mol. Microbiol. *55*, 998–1014.

Diggle, S.P., Lumjiaktase, P., Dipilato, F., Winzer, K., Kunakorn, M., Barrett, D.A., Chhabra, S.R., Camara, M., and Williams, P. (2006). Functional genetic analysis reveals a 2-Alkyl-4-quinolone signaling system in the human pathogen *Burkholderia pseudomallei* and related bacteria. Chem. Biol. *13*, 701–710.

Diggle, S.P., Winzer, K., Chhabra, S.R., Worrall, K.E., Camara, M., and Williams, P. (2003). The *Pseudomonas aeruginosa* quinolone signal molecule overcomes the cell density-dependency of the quorum sensing hierarchy, regulates *rhl*-dependent genes at the onset of stationary phase and can be produced in the absence of LasR. Mol. Microbiol. *50*, 29–43.

Diggle, S.P., Gardner, A., West, S.A., and Griffin, A.S. (2007a). Evolutionary theory of bacterial quorum sensing: when is a signal not a signal? Philos. Trans. R. Soc. Lond. B. Biol. Sci. *362*, 1241–1249.

Diggle, S.P., Matthijs, S., Wright, V.J., Fletcher, M.P., Chhabra, S.R., Lamont, I.L., Kong, X., Hider, R.C., Cornelis, P., Camara, M., *et al.* (2007b). The *Pseudomonas aeruginosa* 4-quinolone signal molecules HHQ and PQS play multifunctional roles in quorum sensing and iron entrapment. Chem. Biol. *14*, 87–96.

Dong, Y.H., Xu, J.L., Li, X.Z., and Zhang, L.H. (2000). AiiA, an enzyme that inactivates the acylhomoserine lactone quorum-sensing signal and attenuates the virulence of *Erwinia carotovora*. Proc. Natl. Acad. Sci. U.S.A. *97*, 3526–3531.

Dow, J.M., Crossman, L., Findlay, K., He, Y.Q., Feng, J.X., and Tang, J.L. (2003). Biofilm dispersal in *Xanthomonas campestris* is controlled by cell–cell signaling and is required for full virulence to plants. Proc. Natl. Acad. Sci. U.S.A. *100*, 10995–11000.

Duerkop, B.A., Ulrich, R.L., and Greenberg, E.P. (2007). Octanoyl-homoserine lactone is the cognate signal for *Burkholderia mallei* BmaR1-BmaI1 quorum sensing. J. Bacteriol. *189*, 5034–5040.

Duerkop, B.A., Herman, J.P., Ulrich, R.L., Churchill, M.E., and Greenberg, E.P. (2008). The *Burkholderia mallei* BmaR3-BmaI3 quorum-sensing system produces and responds to N-3-hydroxy-octanoyl homoserine lactone. J. Bacteriol. *190*, 5137–5141.

Duerkop, B.A., Varga, J., Chandler, J.R., Peterson, S.B., Herman, J.P., Churchill, M.E., Parsek, M.R., Nierman, W.C., and Greenberg, E.P. (2009). Quorum-sensing control of antibiotic synthesis in *Burkholderia thailandensis*. J. Bacteriol. *191*, 3909–3918.

Dyszel, J.L., Smith, J.N., Lucas, D.E., Soares, J.A., Swearingen, M.C., Vross, M.A., Young, G.M., and Ahmer, B.M. (2010). *Salmonella enterica* serovar Typhimurium can detect acyl homoserine lactone production by *Yersinia enterocolitica* in mice. J. Bacteriol. *192*, 29–37.

Eberhard, A. (1972). Inhibition and activation of bacterial luciferase synthesis. J. Bacteriol. *109*, 1101–1105.

Eberhard, A., Burlingame, A.L., Eberhard, C., Kenyon, G.L., Nealson, K.H., and Oppenheimer, N.J. (1981). Structural identification of autoinducer of *Photobacterium fischeri* luciferase. Biochemistry *20*, 2444–2449.

Eberl, L. (2006). Quorum sensing in the genus *Burkholderia*. Int. J. Med. Microbiol. *296*, 103–110.

Eberl, L., Molin, S., and Givskov, M. (1999). Surface motility of *Serratia liquefaciens* MG1. J. Bacteriol. *181*, 1703–1712.

Egland, K.A., and Greenberg, E.P. (1999). Quorum sensing in *Vibrio fischeri*: elements of the *luxl* promoter. Mol. Microbiol. *31*, 1197–1204.

Engebrecht, J., and Silverman, M. (1984). Identification of genes and gene products necessary for bacterial bioluminescence. Proc. Natl. Acad. Sci. U.S.A. *81*, 4154–4158.

Engebrecht, J., Nealson, K., and Silverman, M. (1983). Bacterial bioluminescence: isolation and genetic analysis of functions from *Vibrio fischeri*. Cell *32*, 773–781.

Evans, K., Passador, L., Srikumar, R., Tsang, E., Nezezon, J., and Poole, K. (1998). Influence of the MexAB-OprM multidrug efflux system on quorum sensing in *Pseudomonas aeruginosa*. J. Bacteriol. *180*, 5443–5447.

Farah, C., Vera, M., Morin, D., Haras, D., Jerez, C.A., and Guiliani, N. (2005). Evidence for a functional quorum-sensing type AI-1 system in the extremophilic bacterium *Acidithiobacillus ferrooxidans*. Appl. Environ. Microbiol. *71*, 7033–7040.

Farrow, J.M., 3rd, Sund, Z.M., Ellison, M.L., Wade, D.S., Coleman, J.P., and Pesci, E.C. (2008). PqsE functions independently of PqsR-*Pseudomonas* quinolone signal and enhances the rhl quorum-sensing system. J. Bacteriol. *190*, 7043–7051.

Federle, M.J., and Bassler, B.L. (2003). Interspecies communication in bacteria. J. Clin. Invest. *112*, 1291–1299.

Ferluga, S., and Venturi, V. (2009). OryR Is a LuxR-family protein involved in interkingdom signaling between pathogenic *Xanthomonas oryzae* pv. *oryzae* and rice. J. Bacteriol. *191*, 890–897.

Ferluga, S., Bigiriman, J., Höfte, M., and Venturi, V. (2007). A LuxR homologue of *Xanthomonas oryzae* pv. *oryzae* is required for optimal rice virulence. Mol. Plant Pathol. *8*, 529–538.

Fineran, P.C., Slater, H., Everson, L., Hughes, K., and Salmond, G.P. (2005). Biosynthesis of tripyrrole and beta-lactam secondary metabolites in *Serratia*: integration of quorum sensing with multiple new regulatory components in the control of prodigiosin and carbapenem antibiotic production. Mol. Microbiol. *56*, 1495–1517.

Flavier, A.B., Clough, S.J., Schell, M.A., and Denny, T.P. (1997a). Identification of 3-hydroxypalmitic acid methyl ester as a novel autoregulator controlling virulence in *Ralstonia solanacearum*. Mol. Microbiol. *26*, 251–259.

Flavier, A.B., Clough, S.J., Schell, M.A., and Denny, T.P. (1997b). Identification of 3-hydroxypalmitic acid methyl ester as a novel autoregulator controlling virulence in *Ralstonia solanacearum*. Mol. Microbiol. *26*, 251–259.

Flavier, A.B., Ganova-Raeva, L.M., Schell, M.A., and Denny, T.P. (1997c). Hierarchical autoinduction in Ralstonia solanacearum: control of acyl-homoserine lactone production by a novel autoregulatory system responsive to 3-hydroxypalmitic acid methyl ester. J. Bacteriol. *179*, 7089–7097.

Fouhy, Y., Scanlon, K., Schouest, K., Spillane, C., Crossman, L., Avison, M.B., Ryan, R.P., and Dow, J.M. (2007). Diffusible signal factor-dependent cell–cell signaling and virulence in the nosocomial pathogen *Stenotrophomonas maltophilia*. J. Bacteriol. *189*, 4964–4968.

Fuqua, C. (2006). The QscR quorum-sensing regulon of *Pseudomonas aeruginosa*: an orphan claims its identity. J. Bacteriol. *188*, 3169–3171.

Fuqua, C., and Greenberg, E.P. (2002). Listening in on bacteria: acyl-homoserine lactone signalling. Nat. Rev. Mol. Cell Biol. *3*, 685–695.

Fuqua, C., and Parsek, M.R. (2002). Molecular mechanisms of quorum sensing. In Modern Microbial Genetics, Streips, U.R.Y., ed (Wiley-Lis, Inc, New York), pp. 362–379.

Fuqua, C., and Winans, S.C. (1996). Conserved cis-acting promoter elements are required for density-dependent transcription of *Agrobacterium tumefaciens* conjugal transfer genes. J. Bacteriol. *178*, 435–440.

Fuqua, C., Winans, S.C., and Greenberg, E.P. (1996). Census and consensus in bacterial ecosystems: the LuxR-LuxI family of quorum-sensing transcriptional regulators. Annu. Rev. Microbiol. *50*, 727–751.

Fuqua, C., Parsek, M., and Greenberg, E. (2001). Regulation of gene expression by cell-to-cell communication: acyl-homoserine lactone quorum sensing. Annu. Rev. Genet. *35*, 439–468.

Fuqua, W.C., and Winans, S.C. (1994). A LuxR-LuxI type regulatory system activates *Agrobacterium* Ti plasmid conjugal transfer in the presence of a plant tumor metabolite. J. Bacteriol. *176*, 2796–2806.

Fuqua, W.C., Winans, S.C., and Greenberg, E.P. (1994). Quorum sensing in bacteria: the LuxR-LuxI family of cell density-responsive transcriptional regulators. J. Bacteriol. *176*, 269–275.

Gallagher, L.A., McKnight, S.L., Kuznetsova, M.S., Pesci, E.C., and Manoil, C. (2002). Functions required for extracellular quinolone signaling by *Pseudomonas aeruginosa*. J. Bacteriol. *184*, 6472–6480.

Galloway, W.R., Hodgkinson, J.T., Bowden, S.D., Welch, M., and Spring, D.R. (2011). Quorum sensing in Gram-negative bacteria: small-molecule modulation of AHL and AI-2 quorum sensing pathways. Chem. Rev. *111*, 28–67.

Gambello, M.J., and Iglewski, B.H. (1991). Cloning and characterization of the *Pseudomonas aeruginosa lasR* gene, a transcriptional activator of elastase expression. J. Bacteriol. *173*, 3000–3009.

Gilbert, K.B., Kim, T.H., Gupta, R., Greenberg, E.P., and Schuster, M. (2009). Global position analysis of the *Pseudomonas aeruginosa* quorum-sensing transcription factor LasR. Mol. Microbiol. 73, 1072–1085.

Gilson, L., Kuo, A., and Dunlap, P.V. (1995). AinS and a new family of autoinducer synthesis proteins. J. Bacteriol. *177*, 6946–6951.

Glenn, S.A., Gurich, N., Feeney, M.A., and Gonzalez, J.E. (2007). The ExpR/Sin quorum-sensing system controls succinoglycan production in *Sinorhizobium meliloti*. J. Bacteriol. *189*, 7077–7088.

Gonzalez, J.E., and Marketon, M.M. (2003). Quorum sensing in nitrogen-fixing rhizobia. Microbiol. Mol. Biol. Rev. *67*, 574–592.

Gotschlich, A., Huber, B., Geisenberger, O., Togl, A., Steidle, A., Riedel, K., Hill, P., Tummler, B., Vandamme, P., Middleton, B., *et al.* (2001). Synthesis of multiple N-acylhomoserine lactones is wide-spread among the members of the *Burkholderia cepacia* complex. Syst. Appl. Microbiol. *24*, 1–14.

Gould, T.A., Herman, J., Krank, J., Murphy, R.C., and Churchill, M.E. (2006). Specificity of acyl-homoserine lactone synthases examined by mass spectrometry. J. Bacteriol. *188*, 773–783.

Gould, T.A., Schweizer, H.P., and Churchill, M.E. (2004). Structure of the *Pseudomonas aeruginosa* acyl-homoserinelactone synthase LasI. Mol. Microbiol. *53*, 1135–1146.

Gyaneshwar, P., James, E.K., Mathan, N., Reddy, P.M., Reinhold-Hurek, B., and Ladha, J.K. (2001). Endophytic Colonization of Rice by a Diazotrophic Strain of *Serratia marcescens*. J. Bacteriol. *183*, 2634–2645.

Hanzelka, B.L., and Greenberg, E.P. (1995). Evidence that the N-terminal region of the *Vibrio fischeri* LuxR protein constitutes an autoinducer-binding domain. J. Bacteriol. *177*, 815–817.

Hanzelka, B.L., Stevens, A.M., Parsek, M.R., Crone, T.J., and Greenberg, E.P. (1997). Mutational analysis of the *Vibrio fischeri* LuxI polypeptide: critical regions of an autoinducer synthase. J. Bacteriol. *179*, 4882–4887.

Hanzelka, B.L., Parsek, M.R., Val, D.L., Dunlap, P.V., Cronan, J.E., Jr., and Greenberg, E.P. (1999).

Acylhomoserine lactone synthase activity of the *Vibrio fischeri* AinS protein. J. Bacteriol. *181*, 5766–5770.

Hao, G., and Burr, T.J. (2006). Regulation of long-chain N-acyl-homoserine lactones in *Agrobacterium vitis*. J. Bacteriol. *188*, 2173–2183.

Hao, G., Zhang, H., Zheng, D., and Burr, T.J. (2005). *luxR* homolog *avhR* in *Agrobacterium vitis* affects the development of a grape-specific necrosis and a tobacco hypersensitive response. J. Bacteriol. *187*, 185–192.

Hazan, R., He, J., Xiao, G., Dekimpe, V., Apidianakis, Y., Lesic, B., Astrakas, C., Deziel, E., Lepine, F., and Rahme, L.G. (2010). Homeostatic interplay between bacterial cell–cell signaling and iron in virulence. PLoS Pathog *6*, e1000810.

He, Y.W., and Zhang, L.H. (2008). Quorum sensing and virulence regulation in *Xanthomonas campestris*. FEMS Microbiol. Rev. *32*, 842–857.

He, Y.W., Wang, C., Zhou, L., Song, H., Dow, J.M., and Zhang, L.H. (2006). Dual signaling functions of the hybrid sensor kinase RpfC of *Xanthomonas campestris* involve either phosphorelay or receiver domain–protein interaction. J. Biol. Chem. *281*, 33414–33421.

He, Y.W., Wu, J., Cha, J.S., and Zhang, L.H. (2010). Rice bacterial blight pathogen *Xanthomonas oryzae* pv. *oryzae* produces multiple DSF-family signals in regulation of virulence factor production. BMC Microbiol. *10*, 187.

Heeb, S., Fletcher, M.P., Chhabra, S.R., Diggle, S.P., Williams, P., and Camara, M. (2011). Quinolones: from antibiotics to autoinducers. FEMS Microbiol. Rev. *35*, 247–274.

Hense, B.A., Kuttler, C., Muller, J., Rothballer, M., Hartmann, A., and Kreft, J.U. (2007). Does efficiency sensing unify diffusion and quorum sensing? Nat. Rev. Microbiol. *5*, 230–239.

Higgins, D.A., Pomianek, M.E., Kraml, C.M., Taylor, R.K., Semmelhack, M.F., and Bassler, B.L. (2007). The major *Vibrio cholerae* autoinducer and its role in virulence factor production. Nature *450*, 883–886.

Hoang, H.H., Becker, A., and Gonzalez, J.E. (2004). The LuxR homolog ExpR, in combination with the Sin quorum sensing system, plays a central role in *Sinorhizobium meliloti* gene expression. J. Bacteriol. *186*, 5460–5472.

Hoang, H.H., Gurich, N., and Gonzalez, J.E. (2008). Regulation of motility by the ExpR/Sin quorum-sensing system in *Sinorhizobium meliloti*. J. Bacteriol. *190*, 861–871.

Holden, M.T., Ram Chhabra, S., de Nys, R., Stead, P., Bainton, N.J., Hill, P.J., Manefield, M., Kumar, N., Labatte, M., England, D., *et al.* (1999). Quorum–sensing cross talk: isolation and chemical characterization of cyclic dipeptides from *Pseudomonas aeruginosa* and other Gram-negative bacteria. Mol. Microbiol. *33*, 1254–1266.

Horswill, A.R., Stoodley, P., Stewart, P.S., and Parsek, M.R. (2007). The effect of the chemical, biological, and physical environment on quorum sensing in structured microbial communities. Anal. Bioanal. Chem. *387*, 371–380.

Huber, B., Riedel, K., Hentzer, M., Heydorn, A., Gotschlich, A., Givskov, M., Molin, S., and Eberl, L. (2001). The cep quorum-sensing system of *Burkholderia cepacia* H111 controls biofilm formation and swarming motility. Microbiology *147*, 2517–2528.

Ji, G., Beavis, R.C., and Novick, R.P. (1995). Cell density control of staphylococcal virulence mediated by an octapeptide pheromone. Proc. Natl. Acad. Sci. U.S.A. *92*, 12055–12059.

Juhas, M., Eberl, L., and Tummler, B. (2005). Quorum sensing: the power of cooperation in the world of *Pseudomonas*. Environ. Microbiol. *7*, 459–471.

Juhas, M., Wiehlmann, L., Huber, B., Jordan, D., Lauber, J., Salunkhe, P., Limpert, A.S., von Gotz, F., Steinmetz, I., Eberl, L., *et al.* (2004). Global regulation of quorum sensing and virulence by VqsR in *Pseudomonas aeruginosa*. Microbiology *150*, 831–841.

Kaplan, H.B., and Greenberg, E.P. (1985). Diffusion of autoinducer is involved in regulation of the *Vibrio fischeri* luminescence system. J. Bacteriol. *163*, 1210–1214.

Keller, L., and Surette, M.G. (2006). Communication in bacteria: an ecological and evolutionary perspective. Nat. Rev. Microbiol. *4*, 249–258.

Kempner, E.S., and Hanson, F.E. (1968). Aspects of light production by *Photobacterium fischeri*. J. Bacteriol. *95*, 975–979.

Khajanchi, B.K., Sha, J., Kozlova, E.V., Erova, T.E., Suarez, G., Sierra, J.C., Popov, V.L., Horneman, A.J., and Chopra, A.K. (2009). N-acylhomoserine lactones involved in quorum sensing control the type VI secretion system, biofilm formation, protease production, and *in vivo* virulence in a clinical isolate of *Aeromonas hydrophila*. Microbiology *155*, 3518–3531.

Kiratisin, P., and Sanmee, S. (2008). Roles and interactions of *Burkholderia pseudomallei* BpsIR quorum-sensing system determinants. J. Bacteriol. *190*, 7291–7297.

Kirwan, J.P., Gould, T.A., Schweizer, H.P., Bearden, S.W., Murphy, R.C., and Churchill, M.E. (2006). Quorum-sensing signal synthesis by the *Yersinia pestis* acyl-homoserine lactone synthase YspI. J. Bacteriol. *188*, 784–788.

Klambt, C. (2002). EGF receptor signalling: roles of star and rhomboid revealed. Curr. Biol. *12*, R21-R23.

Kohler, T., van Delden, C., Curty, L.K., Hamzehpour, M.M., and Pechere, J.C. (2001). Overexpression of the MexEF-OprN multidrug efflux system affects cell-to-cell signaling in *Pseudomonas aeruginosa*. J. Bacteriol. *183*, 5213–5222.

Kothe, M., Antl, M., Huber, B., Stoecker, K., Ebrecht, D., Steinmetz, I., and Eberl, L. (2003). Killing of *Caenorhabditis elegans* by *Burkholderia cepacia* is controlled by the cep quorum-sensing system. Cell. Microbiol. *5*, 343–351.

Latifi, A., Foglino, M., Tanaka, K., Williams, P., and Lazdunski, A. (1996). A hierarchical quorum-sensing cascade in *Pseudomonas aeruginosa* links the transcriptional activators LasR and RhIR (VsmR) to expression of the stationary-phase sigma factor RpoS. Mol. Microbiol. *21*, 1137–1146.

Lazdunski, A.M., Ventre, I., and Sturgis, J.N. (2004). Regulatory circuits and communication in Gram-negative bacteria. Nat. Rev. Microbiol. *2*, 581–592.

Ledgham, F., Ventre, I., Soscia, C., Foglino, M., Sturgis, J.N., and Lazdunski, A. (2003). Interactions of the quorum sensing regulator QscR: interaction with itself and the other regulators of *Pseudomonas aeruginosa* LasR and RhlR. Mol. Microbiol. *48*, 199–210.

Lee, J., Zhang, X.S., Hegde, M., Bentley, W.E., Jayaraman, A., and Wood, T.K. (2008). Indole cell signaling occurs primarily at low temperatures in *Escherichia coli*. ISME J. *2*, 1007–1023.

Lee, J.H., Lequette, Y., and Greenberg, E.P. (2006). Activity of purified QscR, a *Pseudomonas aeruginosa* orphan quorum-sensing transcription factor. Mol. Microbiol. *59*, 602–609.

Lequette, Y., Lee, J.H., Ledgham, F., Lazdunski, A., and Greenberg, E.P. (2006). A distinct QscR regulon in the *Pseudomonas aeruginosa* quorum-sensing circuit. J. Bacteriol. *188*, 3365–3370.

Lewenza, S., Conway, B., Greenberg, E.P., and Sokol, P.A. (1999). Quorum sensing in *Burkholderia cepacia*: identification of the LuxRI homologs CepRI. J. Bacteriol. *181*, 748–756.

Lewenza, S., and Sokol, P.A. (2001). Regulation of ornibactin biosynthesis and N-acyl-L-homoserine lactone production by CepR in *Burkholderia cepacia*. J. Bacteriol. *183*, 2212–2218.

Li, L.L., Malone, J.E., and Iglewski, B.H. (2007). Regulation of the *Pseudomonas aeruginosa* quorum-sensing regulator VqsR. J. Bacteriol. *189*, 4367–4374.

Lin, Y.H., Xu, J.L., Hu, J., Wang, L.H., Ong, S.L., Leadbetter, J.R., and Zhang, L.H. (2003). Acyl-homoserine lactone acylase from *Ralstonia* strain XJ12B represents a novel and potent class of quorum-quenching enzymes. Mol. Microbiol. *47*, 849–860.

Lithgow, J.K., Wilkinson, A., Hardman, A., Rodelas, B., Wisniewski-Dye, F., Williams, P., and Downie, J.A. (2000). The regulatory locus cinRI in *Rhizobium leguminosarum* controls a network of quorum-sensing loci. Mol. Microbiol. *37*, 81–97.

Lutter, E., Lewenza, S., Dennis, J.J., Visser, M.B., and Sokol, P.A. (2001). Distribution of quorum-sensing genes in the *Burkholderia cepacia* complex. Infect. Immun. *69*, 4661–4666.

Lynch, M.J., Swift, S., Kirke, D.F., Keevil, C.W., Dodd, C.E., and Williams, P. (2002). The regulation of biofilm development by quorum sensing in *Aeromonas hydrophila*. Environ. Microbiol. *4*, 18–28.

McClean, K.H., Winson, M.K., Fish, L., Taylor, A., Chhabra, S.R., Camara, M., Daykin, M., Lamb, J.H., Swift, S., Bycroft, B.W., *et al.* (1997). Quorum sensing and *Chromobacterium violaceum*: exploitation of violacein production and inhibition for the detection of N-acylhomoserine lactones. Microbiology *143*, 3703–3711.

McGowan, S., Sebaihia, M., Jones, S., Yu, B., Bainton, N., Chan, P.F., Bycroft, B., Stewart, G.S., Williams, P., and Salmond, G.P. (1995). Carbapenem antibiotic production in *Erwinia carotovora* is regulated by CarR, a homologue of the LuxR transcriptional activator. Microbiology *141*, 541–550.

McGrath, S., Wade, D.S., and Pesci, E.C. (2004). Dueling quorum sensing systems in *Pseudomonas aeruginosa* control the production of the *Pseudomonas* quinolone signal (PQS). FEMS Microbiol. Lett. *230*, 27–34.

McKenney, D., Brown, K.E., and Allison, D.G. (1995). Influence of *Pseudomonas aeruginosa* exoproducts on virulence factor production in *Burkholderia cepacia*: evidence of interspecies communication. J. Bacteriol. *177*, 6989–6992.

McKnight, S.L., Iglewski, B.H., and Pesci, E.C. (2000). The *Pseudomonas* quinolone signal regulates rhl quorum sensing in *Pseudomonas aeruginosa*. J. Bacteriol. *182*, 2702–2708.

McVittie, A., Messik, F., and Zahler, S.A. (1962). Developmental Biology of *Mixococcus*. J. Bacteriol. *84*, 546–551.

Malott, R.J., and Sokol, P.A. (2007). Expression of the *bviIR* and *cepIR* quorum-sensing systems of *Burkholderia vietnamiensis*. J. Bacteriol. *189*, 3006–3016.

Malott, R.J., Baldwin, A., Mahenthiralingam, E., and Sokol, P.A. (2005). Characterization of the *cciIR* quorum-sensing system in *Burkholderia cenocepacia*. Infect. Immun. 73, 4982–4992.

Malott, R.J., O'Grady, E.P., Toller, J., Inhulsen, S., Eberl, L., and Sokol, P.A. (2009). A *Burkholderia cenocepacia* orphan LuxR homolog is involved in quorum-sensing regulation. J. Bacteriol. *191*, 2447–2460.

Marketon, M.M., and Gonzalez, J.E. (2002). Identification of two quorum-sensing systems in *Sinorhizobium meliloti*. J. Bacteriol. *184*, 3466–3475.

Marketon, M.M., Gronquist, M.R., Eberhard, A., and Gonzalez, J.E. (2002). Characterization of the *Sinorhizobium meliloti sinR/sinI* locus and the production of novel N-acyl homoserine lactones. J. Bacteriol. *184*, 5686–5695.

Mashburn, L.M., and Whiteley, M. (2005). Membrane vesicles traffic signals and facilitate group activities in a prokaryote. Nature *437*, 422–425.

Mashburn-Warren, L., Howe, J., Garidel, P., Richter, W., Steiniger, F., Roessle, M., Brandenburg, K., and Whiteley, M. (2008). Interaction of quorum signals with outer membrane lipids: insights into prokaryotic membrane vesicle formation. Mol. Microbiol. *69*, 491–502.

Mashburn-Warren, L., Howe, J., Brandenburg, K., and Whiteley, M. (2009). Structural requirements of the Pseudomonas quinolone signal for membrane vesicle stimulation. J. Bacteriol. *191*, 3411–3414.

Mattiuzzo, M., Bertani, I., Ferluga, S., Cabrio, L., Bigirimana, J., Guarnaccia, C., Pongor, S., Maraite, H., and Venturi, V. (2010). The plant pathogen *Pseudomonas fuscovaginae* contains two conserved quorum sensing systems involved in virulence and negatively regulated by RsaL and the novel regulator RsaM. Environ Microbiol. *13*, 145–162.

van der Meer, J.R., Polman, J., Beerthuyzen, M.M., Siezen, R.J., Kuipers, O.P., and De Vos, W.M. (1993). Characterization of the *Lactococcus lactis* nisin A operon genes *nisP*, encoding a subtilisin-like serine protease involved in precursor processing, and *nisR*, encoding a regulatory protein involved in nisin biosynthesis. J. Bacteriol. *175*, 2578–2588.

Michael, B., Smith, J.N., Swift, S., Heffron, F., and Ahmer, B.M. (2001). SdiA of *Salmonella enterica* is a LuxR homolog that detects mixed microbial communities. J. Bacteriol. *183*, 5733–5742.

Miller, M.B., Skorupski, K., Lenz, D.H., Taylor, R.K., and Bassler, B. (2002). Parallel quorum sensing systems converge to regulate virulence in *Vibrio cholerae*. Cell *110*, 303–314.

Miller, S.T., Xavier, K.B., Campagna, S.R., Taga, M.E., Semmelhack, M.F., Bassler, B.L., and Hughson, F.M. (2004). *Salmonella typhimurium* recognizes a chemically distinct form of the bacterial quorum-sensing signal AI-2. Mol. Cell *15*, 677–687.

Milton, D.L., Hardman, A., Camara, M., Chhabra, S.R., Bycroft, B.W., Stewart, G.S., and Williams, P. (1997). Quorum sensing in *Vibrio anguillarum*: characterization of the *vanI/vanR* locus and identification of the autoinducer N-(3-oxodecanoyl)-L-homoserine lactone. J. Bacteriol. *179*, 3004–3012.

Milton, D.L., Chalker, V.J., Kirke, D., Hardman, A., Camara, M., and Williams, P. (2001). The LuxM homologue VanM from *Vibrio anguillarum* directs the synthesis of N-(3-hydroxyhexanoyl)homoserine lactone and N-hexanoylhomoserine lactone. J. Bacteriol. *183*, 3537–3547.

Mima, T., and Schweizer, H.P. (2010). The BpeAB-OprB efflux pump of *Burkholderia pseudomallei* 1026b does not play a role in quorum sensing, virulence factor production, or extrusion of aminoglycosides but is a broad-spectrum drug efflux system. Antimicrob. Agents Chemother. *54*, 3113–3120.

Minogue, T.D., Wehland-von Trebra, M., Bernhard, F., and von Bodman, S.B. (2002). The autoregulatory role of EsaR, a quorum-sensing regulator in *Pantoea stewartii* ssp. stewartii: evidence for a repressor function. Mol. Microbiol. *44*, 1625–1635.

More, M.I., Finger, L.D., Stryker, J.L., Fuqua, C., Eberhard, A., and Winans, S.C. (1996). Enzymatic synthesis of a quorum-sensing autoinducer through use of defined substrates. Science *272*, 1655–1658.

Nasser, W., and Reverchon, S. (2007). New insights into the regulatory mechanisms of the LuxR family of quorum sensing regulators. Anal. Bioanal. Chem. *387*, 381–390.

Nealson, K.H., Platt, T., and Hastings, J.W. (1970). Cellular control of the synthesis and activity of the bacterial luminescent system. J. Bacteriol. *104*, 313–322.

Neiditch, M.B., Federle, M.J., Miller, S.T., Bassler, B.L., and Hughson, F.M. (2005). Regulation of LuxPQ receptor activity by the quorum-sensing signal autoinducer-2. Mol. Cell *18*, 507–518.

Ng, W.L., and Bassler, B.L. (2009). Bacterial quorum-sensing network architectures. Annu. Rev. Genet. *43*, 197–222.

Ng, W.L., Wei, Y., Perez, L.J., Cong, J., Long, T., Koch, M., Semmelhack, M.F., Wingreen, N.S., and Bassler, B.L. (2010). Probing bacterial transmembrane histidine kinase receptor–ligand interactions with natural and synthetic molecules. Proc. Natl. Acad. Sci. U.S.A. *107*, 5575–5580.

Ng, W.L., Perez, L.J., Wei, Y., Kraml, C.M., Semmelhack, M., and Bassler, B. (2011). Signal production and detection specificity in *Vibrio* CqsA/CqsS quorum-sensing systems. Mol. Microbiol. *79*, 1407–1417

O'Grady, E.P., Viteri, D.F., Malott, R.J., and Sokol, P.A. (2009). Reciprocal regulation by the CepIR and CciIR quorum sensing systems in *Burkholderia cenocepacia*. BMC Genomics *10*, 441.

O'Sullivan, L., Andrew, J.W., Jones, T.H., Angela, M.M., James, M.T., and Eshwar, M. (2007). Identifying the genetic basis of ecologically and biotechnologically useful functions of the bacterium *Burkholderia vietnamiensis*. Environ. Microbiol. *9*, 1017–1034.

Oger, P., Kim, K.S., Sackett, R.L., Piper, K.R., and Farrand, S.K. (1998). Octopine-type Ti plasmids code for a mannopine-inducible dominant-negative allele of traR, the quorum-sensing activator that regulates Ti plasmid conjugal transfer. Mol. Microbiol. *27*, 277–288.

Ortori, C.A., Atkinson, S., Chhabra, S.R., Camara, M., Williams, P., and Barrett, D.A. (2007). Comprehensive profiling of N-acylhomoserine lactones produced by *Yersinia pseudotuberculosis* using liquid chromatography coupled to hybrid quadrupole-linear ion trap mass spectrometry. Anal. Bioanal. Chem. *387*, 497–511.

Park, D.K., Lee, K.E., Baek, C.H., Kim, I.H., Kwon, J.H., Lee, W.K., Lee, K.H., Kim, B.S., Choi, S.H., and Kim, K.S. (2006). Cyclo(Phe-Pro) modulates the expression of *ompU* in *Vibrio* spp. J. Bacteriol. *188*, 2214–2221.

Passador, L., Cook, J.M., Gambello, M.J., Rust, L., and Iglewski, B.H. (1993). Expression of *Pseudomonas aeruginosa* virulence genes requires cell-to-cell communication. Science *260*, 1127–1130.

Parsek, M.R., Schaefer, A.L., and Greenberg, E.P. (1997). Analysis of random and site-directed mutations in rhII, a *Pseudomonas aeruginosa* gene encoding an acylhomoserine lactone synthase. Mol. Microbiol. *26*, 301–310.

Parsek, M.R., Val, D.L., Hanzelka, B.L., Cronan, J.E., Jr., and Greenberg, E.P. (1999). Acyl homoserine-lactone quorum-sensing signal generation. Proc. Natl. Acad. Sci. U.S.A. *96*, 4360–4365.

Patankar, A.V., and Gonzalez, J.E. (2009a). An orphan LuxR homolog of *Sinorhizobium meliloti* affects stress adaptation and competition for nodulation. Appl. Environ. Microbiol. *75*, 946–955.

Patankar, A.V., and Gonzalez, J.E. (2009b). Orphan LuxR regulators of quorum sensing. FEMS Microbiol. Rev. *33*, 739–756.

Pearson, J.P., Van Delden, C., and Iglewski, B.H. (1999). Active efflux and diffusion are involved in transport of *Pseudomonas aeruginosa* cell-to-cell signals. J. Bacteriol. *181*, 1203–1210.

Pesci, E.C., Milbank, J.B., Pearson, J.P., McKnight, S., Kende, A.S., Greenberg, E.P., and Iglewski, B.H. (1999). Quinolone signaling in the cell-to-cell communication system of *Pseudomonas aeruginosa*. Proc. Natl. Acad. Sci. U.S.A. *96*, 11229–11234.

Pierson, L.S., 3rd, Keppenne, V.D., and Wood, D.W. (1994). Phenazine antibiotic biosynthesis in *Pseudomonas aureofaciens* 30–84 is regulated by PhzR in response to cell density. J. Bacteriol. *176*, 3966–3974.

Piper, K.R., Beck von Bodman, S., and Farrand, S.K. (1993). Conjugation factor of *Agrobacterium tumefaciens* regulates Ti plasmid transfer by autoinduction. Nature *362*, 448–450.

Platt, T.G., and Fuqua, C. (2010). What's in a name? The semantics of quorum sensing. Trends Microbiol. *18*, 383–387.

Rahmati, S., Yang, S., Davidson, A.L., and Zechiedrich, E.L. (2002). Control of the AcrAB multidrug efflux pump by quorum-sensing regulator SdiA. Mol. Microbiol. *43*, 677–685.

Rambow-Larsen, A.A., Rajashekara, G., Petersen, E., and Splitter, G. (2008). Putative quorum-sensing regulator BlxR of *Brucella melitensis* regulates virulence factors including the type IV secretion system and flagella. J. Bacteriol. *190*, 3274–3282.

Rampioni, G., Schuster, M., Greenberg, E.P., Bertani, I., Grasso, M., Venturi, V., Zennaro, E., and Leoni, L. (2007). RsaL provides quorum sensing homeostasis and functions as a global regulator of gene expression in *Pseudomonas aeruginosa*. Mol. Microbiol. *66*, 1557–1565.

Rampioni, G., Schuster, M., Greenberg, E.P., Zennaro, E., and Leoni, L. (2009). Contribution of the RsaL global regulator to *Pseudomonas aeruginosa* virulence and biofilm formation. FEMS Microbiol. Lett. *301*, 210–217.

Rampioni, G., Pustelny, C., Fletcher, M.P., Wright, V.J., Bruce, M., Rumbaugh, K.P., Heeb, S., Camara, M., and Williams, P. (2010). Transcriptomic analysis reveals a global alkyl-quinolone-independent regulatory role for PqsE in facilitating the environmental adaptation of *Pseudomonas aeruginosa* to plant and animal hosts. Environ. Microbiol. *12*, 1659–1673.

Rather, P.N., Ding, X., Baca-De Lancey, R.R., and Siddiqui, S. (1999). *Providencia stuartii* genes activated by cell-to-cell signaling and identification of a gene required for production or activity of an extracellular factor. J. Bacteriol. *181*, 7185–7191.

Redfield, R.J. (2002). Is quorum sensing a side effect of diffusion sensing? Trends Microbiol. *10*, 365–370.

Ryan, R.P., and Dow, J.M. (2011). Communication with a growing family: diffusible signal factor (DSF) signaling in bacteria. Trends Microbiol. *19*, 145–152.

Ryan, R.P., Fouhy, Y., Lucey, J.F., Crossman, L.C., Spiro, S., He, Y.W., Zhang, L.H., Heeb, S., Camara, M., Williams, P., *et al.* (2006). Cell–cell signaling in *Xanthomonas campestris* involves an HD-GYP domain protein that functions in cyclic di-GMP turnover. Proc. Natl. Acad. Sci. U.S.A. *103*, 6712–6717.

Sandoz, K.M., Mitzimberg, S.M., and Schuster, M. (2007). Social cheating in *Pseudomonas aeruginosa* quorum sensing. Proc. Natl. Acad. Sci. U.S.A. *104*, 15876–15881.

Schaefer, A., Val, D.L., Hanzelka, B.L., Cronan, J.E., Jr., and Greenberg, E.P. (1996). Generation of cell-to-cell signals in quorum sensing: acyl homoserine lactone synthase activity of a purified *Vibrio fischeri* LuxI protein. Proc. Natl. Acad. Sci. U.S.A. *93*, 9505–9509.

Schaefer, A.L., Greenberg, E.P., Oliver, C.M., Oda, Y., Huang, J.J., Bittan-Banin, G., Peres, C.M., Schmidt, S., Juhaszova, K., Sufrin, J.R., *et al.* (2008). A new class of homoserine lactone quorum-sensing signals. Nature *454*, 595–599.

Schauder, S., Shokat, K., Surette, M.G., and Bassler, B.L. (2001). The LuxS family of bacterial autoinducers: biosynthesis of a novel quorum-sensing signal molecule. Mol. Microbiol. *41*, 463–476.

Schertzer, J.W., Boulette, M.L., and Whiteley, M. (2009). More than a signal: non-signaling properties of quorum sensing molecules. Trends Microbiol. *17*, 189–195.

Schertzer, J.W., Brown, S.A., and Whiteley, M. (2010). Oxygen levels rapidly modulate *Pseudomonas aeruginosa* social behaviours via substrate limitation of PqsH. Mol. Microbiol. *77*, 1527–1538.

Schuster, M., and Greenberg, E.P. (2006). A network of networks: quorum-sensing gene regulation in *Pseudomonas aeruginosa*. Int. J. Med. Microbiol. *296*, 73–81.

Schuster, M., Lostroh, C.P., Ogi, T., and Greenberg, E.P. (2003). Identification, timing, and signal specificity of *Pseudomonas aeruginosa* quorum-controlled genes: a transcriptome analysis. J. Bacteriol. *185*, 2066–2079.

Schuster, M., Urbanowski, M.L., and Greenberg, E.P. (2004). Promoter specificity in *Pseudomonas aeruginosa* quorum sensing revealed by DNA binding of purified LasR. Proc. Natl. Acad. Sci. U.S.A. *101*, 15833–15839.

Schwenteit, J., Gram, L., Nielsen, K.F., Fridjonsson, O.H., Bornscheuer, U.T., Givskov, M., and Gudmundsdottir, B.K. (2010). Quorum sensing in *Aeromonas salmonicida* subsp. *achromogenes* and the effect of the autoinducer synthase AsaI on bacterial virulence. Vet. Microbiol. *147*, 389–397

Slater, H., Alvarez-Morales, A., Barber, C.E., Daniels, M.J., and Dow, J.M. (2000). A two-component system involving an HD-GYP domain protein links cell–cell signalling to pathogenicity gene expression in *Xanthomonas campestris*. Mol. Microbiol. *38*, 986–1003.

Smith, J.N., and Ahmer, B.M. (2003). Detection of other microbial species by *Salmonella*: expression of the SdiA regulon. J. Bacteriol. *185*, 1357–1366.

Smith, J.N., Dyszel, J.L., Soares, J.A., Ellermeier, C.D., Altier, C., Lawhon, S.D., Adams, L.G., Konjufca, V., Curtiss, R., 3rd, Slauch, J.M., *et al.* (2008). SdiA, an N-acylhomoserine lactone receptor, becomes active during the transit of *Salmonella enterica* through the gastrointestinal tract of turtles. PLoS ONE 3, e2826.

Sokol, P.A., Sajjan, U., Visser, M.B., Gingues, S., Forstner, J., and Kooi, C. (2003). The CepIR quorum-sensing system contributes to the virulence of *Burkholderia cenocepacia* respiratory infections. Microbiology *149*, 3649–3658.

Solomon, J.M., Lazazzera, B.A., and Grossman, A.D. (1996). Purification and characterization of an extracellular peptide factor that affects two different developmental pathways in *Bacillus subtilis*. Genes Dev. *10*, 2014–2024.

Song, Y., Xie, C., Ong, Y.M., Gan, Y.H., and Chua, K.L. (2005). The BpsIR quorum-sensing system of *Burkholderia pseudomallei*. J. Bacteriol. *187*, 785–790.

Spirig, T., Tiaden, A., Kiefer, P., Buchrieser, C., Vorholt, J.A., and Hilbi, H. (2008). The *Legionella* autoinducer synthase LqsA produces an alpha-hydroxyketone signaling molecule. J. Biol. Chem. *283*, 18113–18123.

Steidle, A., Allesen-Holm, M., Riedel, K., Berg, G., Givskov, M., Molin, S., and Eberl, L. (2002). Identification and characterization of an N-acylhomoserine lactone-dependent quorum-sensing system in *Pseudomonas putida* strain IsoF. Appl. Environ. Microbiol. *68*, 6371–6382.

Steindler, L., Bertani, I., De Sordi, L., Schwager, S., Eberl, L., and Venturi, V. (2009). LasI/R and RhlI/R quorum sensing in a strain of *Pseudomonas aeruginosa* beneficial to plants. Appl. Environ. Microbiol. *75*, 5131–5140.

Stevenson, L.G., Strisovsky, K., K.M., C., Bhatt, S., Freeman, M., and Rather, P.N. (2007). Rhomboid protease AarA mediates quorum-sensing in *Providencia stuartii* by activating TatA of the twin-arginine translocase. Proc. Natl. Acad. Sci. U.S.A. *4*, 1003–1008.

Suarez-Moreno, Z.R., Caballero-Mellado, J., and Venturi, V. (2008). The new group of non-pathogenic plant-associated nitrogen-fixing *Burkholderia* spp. shares a conserved quorum-sensing system, which is tightly regulated by the RsaL repressor. Microbiology *154*, 2048–2059.

Suarez-Moreno, Z.R., Devescovi, G., Myers, M., Hallack, L., Mendonca-Previato, L., Caballero-Mellado, J., and Venturi, V. (2010). Commonalities and differences in N-acyl homoserine lactone quorum sensing regulation in the species cluster of beneficial plant-associated *Burkholderia*. Appl. Environ. Microbiol. *76*, 4302–4317.

Subramoni, S., and Venturi, V. (2009a). LuxR-family 'solos': bachelor sensors/regulators of signalling molecules. Microbiology *155*, 1377–1385.

Subramoni, S., and Venturi, V. (2009b). PpoR is a conserved unpaired LuxR solo of *Pseudomonas putida* which binds N-acyl homoserine lactones. BMC Microbiology *9*, 125.

Suzuki, K., Wang, X., Weilbacher, T., Pernestig, A.K., Melefors, O., Georgellis, D., Babitzke, P., and Romeo, T. (2002). Regulatory circuitry of the CsrA/CsrB and BarA/UvrY systems of *Escherichia coli*. J. Bacteriol. *184*, 5130–5140.

Swift, S., Karlyshev, A.V., Fish, L., Durant, E.L., Winson, M.K., Chhabra, S.R., Williams, P., Macintyre, S., and Stewart, G.S. (1997). Quorum sensing in *Aeromonas hydrophila* and *Aeromonas salmonicida*: identification of the LuxRI homologs AhyRI and AsaRI and their cognate N-acylhomoserine lactone signal molecules. J. Bacteriol. *179*, 5271–5281.

Swift, S., Lynch, M.J., Fish, L., Kirke, D.F., Tomas, J.M., Stewart, G.S., and Williams, P. (1999). Quorum sensing-dependent regulation and blockade of exoprotease production in *Aeromonas hydrophila*. Infect. Immun. *67*, 5192–5199.

Taminiau, B., Daykin, M., Swift, S., Boschiroli, M.L., Tibor, A., Lestrate, P., De Bolle, X., O'Callaghan, D., Williams, P., and Letesson, J.J. (2002). Identification of a quorum-sensing signal molecule in the facultative intracellular pathogen *Brucella melitensis*. Infect. Immun. *70*, 3004–3011.

Tang, J.L., Liu, Y.N., Barber, C.E., Dow, J.M., Wootton, J.C., and Daniels, M.J. (1991). Genetic and molecular analysis of a cluster of rpf genes involved in positive regulation of synthesis of extracellular enzymes and polysaccharide in *Xanthomonas campestris* pathovar *campestris*. Mol. Gen. Genet. *226*, 409–417.

Tiaden, A., Spirig, T., Carranza, P., Bruggemann, H., Riedel, K., Eberl, L., Buchrieser, C., and Hilbi, H. (2008a). Synergistic contribution of the *Legionella pneumophila lqs* genes to pathogen–host interactions. J. Bacteriol. *190*, 7532–7547.

Tiaden, A., Spirig, T., Sahr, T., Walti, M.A., Boucke, K., Buchrieser, C., and Hilbi, H. (2008b). The autoinducer synthase LqsA and putative sensor kinase LqsS regulate phagocyte interactions, extracellular filaments and a genomic island of *Legionella pneumophila*. Environ. Microbiol. *12*, 1243–1259.

Tiaden, A., Spirig, T., and Hilbi, H. (2010). Bacterial gene regulation by alpha-hydroxyketone signaling. Trends Microbiol. *18*, 288–297.

Tomasz, A. (1965). Control of the competent state in *Pneumococcus* by a hormone-like cell product: an example for a new type of regulatory mechanism in bacteria. Nature *208*, 155–159.

Ulrich, R.L., Deshazer, D., Brueggemann, E.E., Hines, H.B., Oyston, P.C., and Jeddeloh, J.A. (2004a). Role of quorum sensing in the pathogenicity of *Burkholderia pseudomallei*. J. Med. Microbiol. *53*, 1053–1064.

Ulrich, R.L., Deshazer, D., Hines, H.B., and Jeddeloh, J.A. (2004b). Quorum sensing: a transcriptional regulatory system involved in the pathogenicity of *Burkholderia mallei*. Infect. Immun. *72*, 6589–6596.

Ulrich, R.L., Hines, H.B., Parthasarathy, N., and Jeddeloh, J.A. (2004c). Mutational analysis and biochemical characterization of the *Burkholderia thailandensis* DW503 quorum-sensing network. J. Bacteriol. *186*, 4350–4360.

Urbanowski, M.L., Lostroh, C.P., and Greenberg, E.P. (2004). Reversible acyl-homoserine lactone binding to purified *Vibrio fischeri* LuxR protein. J. Bacteriol. *186*, 631–637.

Valade, E., Thibault, F.M., Gauthier, Y.P., Palencia, M., Popoff, M.Y., and Vidal, D.R. (2004). The PmlI-PmlR quorum-sensing system in *Burkholderia pseudomallei* plays a key role in virulence and modulates production of the MprA protease. J. Bacteriol. *186*, 2288–2294.

Van Houdt, R., Aertsen, A., Moons, P., Vanoirbeek, K., and Michiels, C.W. (2006). N-acyl-L-homoserine lactone signal interception by *Escherichia coli*. FEMS Microbiol. Lett. *256*, 83–89.

Vannini, A., Volpari, C., Gargioli, C., Muraglia, E., Cortese, R., De Francesco, R., Neddermann, P., and Marco, S.D. (2002). The crystal structure of the quorum sensing protein TraR bound to its autoinducer and target DNA. EMBO J. *21*, 4393–4401.

Vendeville, A., Winzer, K., Heurlier, K., Tang, C.M., and Hardie, K.R. (2005). Making 'sense' of metabolism: autoinducer-2, LuxS and pathogenic bacteria. Nat. Rev. Microbiol. *3*, 383–396.

Venturi, V., Friscina, A., Bertani, I., Devescovi, G., and Aguilar, C. (2004). Quorum sensing in the *Burkholderia cepacia* complex. Res. Microbiol. *155*, 238–244.

Vial, L., Lepine, F., Milot, S., Groleau, M.C., Dekimpe, V., Woods, D.E., and Deziel, E. (2008). *Burkholderia pseudomallei, B. thailandensis*, and *B. ambifaria* produce

4-hydroxy-2-alkylquinoline analogues with a methyl group at the 3 position that is required for quorum-sensing regulation. J. Bacteriol. *190*, 5339–5352.

Wade, D.S., Calfee, M.W., Rocha, E.R., Ling, E.A., Engstrom, E., Coleman, J.P., and Pesci, E.C. (2005). Regulation of Pseudomonas quinolone signal synthesis in *Pseudomonas aeruginosa*. J. Bacteriol. *187*, 4372–4380.

Wagner, V.E., Bushnell, D., Passador, L., Brooks, A.I., and Iglewski, B.H. (2003). Microarray analysis of *Pseudomonas aeruginosa* quorum-sensing regulons: effects of growth phase and environment. J. Bacteriol. *185*, 2080–2095.

Wang, X.D., de Boer, P.A., and Rothfield, L.I. (1991). A factor that positively regulates cell division by activating transcription of the major cluster of essential cell division genes of *Escherichia coli*. EMBO J. *10*, 3363–3372.

Wang, L.H., He, Y., Gao, Y., Wu, J.E., Dong, Y.H., He, C., Wang, S.X., Weng, L.X., Xu, J.L., Tay, L., *et al.* (2004). A bacterial cell–cell communication signal with cross-kingdom structural analogues. Mol. Microbiol. *51*, 903–912.

Watson, W.T., Minogue, T.D., Val, D.L., von Bodman, S.B., and Churchill, M.E. (2002). Structural basis and specificity of acyl-homoserine lactone signal production in bacterial quorum sensing. Mol. Cell *9*, 685–694.

Weeks, J.N., Galindo, C.L., Drake, K.L., Adams, G.L., Garner, H.R., and Ficht, T.A. (2010). *Brucella melitensis* VjbR and C12-HSL regulons: contributions of the N-dodecanoyl homoserine lactone signaling molecule and LuxR homologue VjbR to gene expression. BMC Microbiol. *10*, 167.

Wei, Y., Lee, J.M., Smulski, D.R., and LaRossa, R.A. (2001). Global impact of sdiA amplification revealed by comprehensive gene expression profiling of *Escherichia coli*. J. Bacteriol. *183*, 2265–2272.

White, C.E., and Winans, S.C. (2007a). Cell–cell communication in the plant pathogen Agrobacterium tumefaciens. Philos. Trans. R. Soc. Lond. B. Biol. Sci. *362*, 1135–1148.

White, C.E., and Winans, S.C. (2007b). The quorum-sensing transcription factor TraR decodes its DNA binding site by direct contacts with DNA bases and by detection of DNA flexibility. Mol. Microbiol. *64*, 245–256.

Wilkinson, A., Danino, V., Wisniewski-Dye, F., Lithgow, J.K., and Downie, J.A. (2002). N-acyl-homoserine lactone inhibition of rhizobial growth is mediated by two quorum-sensing genes that regulate plasmid transfer. J. Bacteriol. *184*, 4510–4519.

Winzer, K., Hardie, K.R., and Williams, P. (2002). Bacterial cell-to-cell communication: sorry, cannot talk now – gone to lunch! Curr. Opin. Microbiol. *5*, 216–222.

Wisniewski-Dye, F., and Downie, J.A. (2002). Quorum-sensing in Rhizobium. Antonie Van Leeuwenhoek *81*, 397–407.

Wisniewski-Dye, F., Jones, J., Chhabra, S.R., and Downie, J.A. (2002). raiIR genes are part of a quorum-sensing network controlled by cinI and cinR in *Rhizobium leguminosarum*. J. Bacteriol. *184*, 1597–1606.

Xavier, K.B., and Bassler, B.L. (2003). LuxS quorum sensing: more than just a numbers game. Curr. Opin. Microbiol. *6*, 191–197.

Xiao, G., Deziel, E., He, J., Lepine, F., Lesic, B., Castonguay, M.H., Milot, S., Tampakaki, A.P., Stachel, S.E., and Rahme, L.G. (2006a). MvfR, a key *Pseudomonas aeruginosa* pathogenicity LTTR-class regulatory protein, has dual ligands. Mol. Microbiol. *62*, 1689–1699.

Xiao, G., He, J., and Rahme, L.G. (2006b). Mutation analysis of the *Pseudomonas aeruginosa* mvfR and pqsABCDE gene promoters demonstrates complex quorum-sensing circuitry. Microbiology *152*, 1679–1686.

Yu, S., Jensen, V., Seeliger, J., Feldmann, I., Weber, S., Schleicher, E., Haussler, S., and Blankenfeldt, W. (2009). Structure elucidation and preliminary assessment of hydrolase activity of PqsE, the *Pseudomonas* quinolone signal (PQS) response protein. Biochemistry *48*, 10298–10307.

Zhang, L., Murphy, P.J., Kerr, A., and Tate, M.E. (1993). *Agrobacterium* conjugation and gene regulation by N-acyl-L-homoserine lactones. Nature *362*, 446–448.

Zhang, L., Jia, Y., Wang, L., and Fang, R. (2007). A proline iminopeptidase gene up-regulated *in planta* by a LuxR homologue is essential for pathogenicity of *Xanthomonas campestris* pv. *campestris*. Mol. Microbiol. *65*, 121–136.

Zhang, Y.M., Frank, M.W., Zhu, K., Mayasundari, A., and Rock, C.O. (2008). PqsD is responsible for the synthesis of 2,4-dihydroxyquinoline, an extracellular metabolite produced by *Pseudomonas aeruginosa*. J. Biol. Chem. *283*, 28788–28794.

Zhang, Z., and Pierson, L.S., 3rd (2001). A second quorum-sensing system regulates cell surface properties but not phenazine antibiotic production in *Pseudomonas aureofaciens*. Appl. Environ. Microbiol. *67*, 4305–4315.

Zheng, D., Zhang, H., Carle, S., Hao, G., Holden, M.R., and Burr, T.J. (2003). A luxR homolog, *aviR*, in *Agrobacterium vitis* is associated with induction of necrosis on grape and a hypersensitive response on tobacco. Mol. Plant Microbe Interact. *16*, 650–658.

Zhu, J., and Winans, S.C. (2001). The quorum-sensing transcriptional regulator TraR requires its cognate signaling ligand for protein folding, protease resistance, and dimerization. Proc. Natl. Acad. Sci. U.S.A. *98*, 1507–1512.

Cyclic di-GMP Signalling and Regulation in Bacteria

5

J. Maxwell Dow, Yvonne McCarthy, Karen O'Donovan, Delphine Caly and Robert P. Ryan

Abstract

Cyclic di-GMP is now recognized as an almost universal second messenger in eubacteria that acts to regulate a wide range of functions including developmental transitions, adhesion, biofilm formation, motility and the synthesis of virulence factors. Cyclic di-GMP is synthesized from two GTP molecules by diguanylate cyclases that have a GGDEF domain and degraded by phosphodiesterases with either an EAL or HD-GYP domain. These proteins often have associated signal input domains, suggesting that their enzymatic activity may be modulated by different environmental or cellular cues. Cyclic di-GMP exerts a regulatory action through binding to diverse receptors that include a small protein domain called PilZ, transcription factors, enzymatically inactive GGDEF, EAL or HD-GYP domains and riboswitches. The multiplicity of GGDEF, EAL and HD-GYP proteins together with a range of receptors within the same bacterial cell indicates the considerable complexity of cyclic di-GMP signalling. This has led to the concept of discrete pools of the nucleotide that are generated locally and act to influence intimately associated targets. A number of signalling proteins may be organized in a regulatory network to control a common function(s). Understanding cyclic di-GMP signalling may afford strategies for inhibition of biofilm formation and virulence factor synthesis in bacterial pathogens.

Introduction

Cyclic di-GMP was originally described in 1987 as an allosteric regulator of cellulose synthesis in *Acetobacter xylinum* (now *Gluconacetobacter xylinus*) by the laboratory of Moshe Benziman (Ross *et al.*, 1987). It is now appreciated that this nucleotide occurs in the majority of bacterial species (both Gram-positive and Gram-negative) where it acts to regulate a wide range of functions (Jenal and Malone, 2006). A balance of synthesis and degradation controls the level of cyclic di-GMP in the bacterial cell. Cyclic di-GMP is synthesized from two molecules of GTP by GGDEF domain diguanylate cyclases (DGC) whereas cyclic di-GMP degradation is catalysed by phosphodiesterases (PDE) that have either an EAL or HD-GYP domain. These domains are often found in association with other signalling domains. This suggests that the activity of these proteins in cyclic di-GMP turnover is regulated either directly or indirectly by environmental cues. Cyclic di-GMP thus acts as a second messenger linking the perception of different environmental cues to changes in bacterial behaviour. The mechanisms by which cyclic di-GMP exerts these regulatory influences are diverse and include binding to 'adaptor' proteins with a PilZ domain, binding to transcription factors, interaction with riboswitches and binding to enzymatically inactive GGDEF/EAL/HD-GYP domains.

Many bacterial genomes encode multiple proteins with GGDEF, EAL and HD-GYP domains (for example some *Vibrio* species have over 60 proteins). This raises the issue of how these multiple signalling systems are functionally organized. A related question is whether cyclic di-GMP exists within the cell as a single general pool, several discrete pools or both. The available evidence indicates that within the same cell, several signalling systems can act together to regulate a common function whereas other systems have a discrete regulatory action. Such 'insulated' regulatory effects may involve spatial separation of signalling systems or temporal separation, where the structural genes may be expressed only in certain phases of growth.

In this review, we will begin with a brief overview of the proteins involved in cyclic di-GMP synthesis and degradation before going on to discuss the different mechanisms by which cyclic di-GMP exerts its action in the cell. Following this we will illustrate the diversity of regulatory networks involving cyclic di-GMP using specific examples from different bacteria, before concluding with a discussion of outstanding research questions.

The generation and degradation of cyclic di-GMP

Studies of the regulation of cellulose in *G. xylinus* showed that cyclic di-GMP synthesis and degradation are controlled by enzymes containing two protein domains that were named GGDEF and EAL after conserved amino acid motifs (Tal *et al.*, 1998). The GGDEF domain was first identified in PleD, a regulatory protein controlling swarmer-to-stalked-cell transition in *Caulobacter crescentus* (Hecht and Newton, 1985). The EAL domain was first described in BvgR, a repressor of virulence gene expression in *Bordetella pertussis* (Merkel *et al.*, 1998). The findings in *G. xylinus* were of seminal importance. By linking GGDEF and EAL domains to cyclic di-GMP turnover they gave the first indication that cyclic di-GMP signalling might be widespread in bacteria and that it might control functions other than cellulose synthesis (Tal *et al.*, 1998). The proteins identified in *G. xylinus* had both GGDEF and EAL domains, so it was not possible to ascribe a specific function to either domain.

The GGDEF domain is a diguanylate cyclase

Direct evidence for the role of the GGDEF domain was obtained by biochemical studies of purified GGDEF domain proteins including PleD (Paul *et al.*, 2004; Ryjenkov *et al.*, 2005) and of isolated GGDEF domains from proteins from different bacterial phyla (Ryjenkov *et al.*, 2005). Each of these GGDEF domains/proteins converted two molecules of GTP to cyclic di-GMP with Mg^{2+} as a cofactor, but had no activity with other nucleotides. PleD comprises tandem CheY-like receiver (REC) domains with a C-terminal GGDEF domain (Aldridge *et al.*, 2003). Activation of PleD occurs by phosphorylation on an aspartic acid residue in the first REC domain; the second REC domain is inactive. Studies of PleD in solution and determination of the crystal structure support a model in which the protein dimerizes upon phosphorylation to achieve a form that is active in cyclic di-GMP synthesis (Chan *et al.*, 2004). The GGDEF signature motif forms part of the active site or A-site where GTP is bound (one molecule of GTP substrate per monomer). Intriguingly, the structural analysis showed that PleD bound dimeric cyclic di-GMP at an allosteric site, the I site, which is characterized by the motif RxxD (Chan *et al.*, 2004). Binding at this site may account for the strong non-competitive product inhibition of PleD, which may act in the cell to set a limit on the concentration of cyclic di-GMP to be attained, preventing 'runaway' synthesis and needless consumption of GTP (Chan *et al.*, 2004).

There are a number of other conserved residues within the GGDEF domain in addition to the signature residues, although not all GGDEF domain proteins have a recognisable I site. Variations within the GGDEF signature motif are also found. A common variant is GGEEF, which is found in many GGDEF domain proteins that do not also have an EAL domain (Kulesekara *et al.*, 2006). Experimental evidence indicates that some of these proteins have cyclic di-GMP synthase activity (Kulesekara *et al.*, 2006). Other variations however suggest that the enzymatic activity is likely to have been lost. As we will discuss later, enzymatically inactive GGDEF domains may have a regulatory role, influencing the activity of an attached EAL domain in a GGDEF-EAL domain protein or through physical interaction with other proteins.

The EAL domain is a phosphodiesterase

Experimental evidence for the role of the EAL domain as a cyclic di-GMP phosphodiesterase

(PDE) has been provided by studies of intact proteins as well as isolated EAL domains (Christen *et al*, 2005; Schmidt *et al.*, 2005; Tamayo *et al.*, 2005; Ryan *et al.*, 2006a) In most of these cases the major product of the enzymatic action of the EAL domain was the linear nucleotide pGpG, which was more slowly converted to GMP. This activity of the EAL domain was absolutely dependent on the presence of Mg^{2+} or Mn^{2+} (Schmidt *et al.*, 2005). As with the GGDEF domain, mutational analysis indicates the essential role of the conserved EAL motif in enzymatic activity. Nevertheless, several variations within the EAL motif (specifically ETL and EVL), still allow enzymatic activity in cyclic di-GMP hydrolysis (Schmidt *et al.*, 2005; Kazmierczak *et al.*, 2006; Kulesekara *et al.*, 2006). Other conserved residues within the EAL domain are also required for activity. These include residues that are directly or indirectly involved in positioning the Mg^{2+} for catalysis, residues of the extended conserved motif DDFG(T/A)GYSS that form a loop and a E residue (not in the EAL motif) which is the general base catalyst (Rao *et al.*, 2008; 2009; Römling, 2009). A number of proteins with variations in some of these conserved residues lack PDE activity but may retain a regulatory role.

The activities of GGDEF –EAL domain fusions

As noted above, the proteins that were first implicated in cyclic di-GMP turnover have both GGDEF and EAL domains. This domain organization, which is also found in a substantial number of other proteins, raises the question of what determines the overall activity of such proteins. One possible answer is that one of the two domains is not enzymatically functional. The GGDEF-EAL domain protein CC3396 from *C. crescentus* contains an altered active site motif (GEDEF) and lacks DGC activity. However, this altered GGDEF domain is able to bind GTP leading to activation of the attached EAL domain as a PDE (Christen *et al.*, 2005). This activation involves a lowering of the Km of the EAL domain for cyclic di-GMP and is specific for GTP. A similar activation mechanism has been proposed for FimX from *P. aeruginosa*, which has the variant sequence GDSIF in the GGDEF domain (Kazmierczak *et al.*, 2006). In a similar fashion, it is plausible that enzymatically inactive EAL domains may be capable of binding cyclic di-GMP to affect the DGC activity of a linked GGDEF domain through an allosteric mechanism (Qi *et al.*, 2011)

Although for the majority of GGDEF-EAL domain proteins either one or other of the domains are active, a few bifunctional enzymes have been described (see, for example, Tarutina *et al.*, 2006). In this case the proteins may be able to switch between states able to synthesize and hydrolyse cyclic di-GMP. This may be associated with the oligomerization state of the protein. Structural analyses indicate that the GGDEF domain acts as a dimer in cyclic di-GMP synthesis whereas the PDE activity of the EAL domain is apparently independent of protein oligomerization (Chan *et al.*, 2004; Schmidt *et al.*, 2005). Regulation of the oligomerization state of the GGDEF-EAL proteins may then serve to determine which activity is expressed. Associated sensory input domains may influence the state of oligomerization in response to environmental cues. A second possibility is that the DGC and PDE activities may be independently set to ensure a very precise concentration emanating from a point source (Ryan *et al.*, 2006b).

The HD-GYP domain is a second PDE

The suggestion that the HD-GYP domain is a second class of PDE came from bioinformatic studies of multiple genome sequences (Galperin *et al.*, 1999, 2001). HD-GYP is a subgroup of the HD superfamily of metal-dependent phosphohydrolases. The HD-GYP nomenclature reflects some of the most conserved amino acid residues. In many bacterial proteomes the HD-GYP domain is associated with the two-component receiver (REC) domain, thus indicating a role in signalling. A role for HD-GYP in cyclic di-GMP hydrolysis was proposed based on an examination of the distribution and numbers of GGDEF, EAL and HD-GYP domains encoded by different bacterial genomes, where several genomes encode proteins with the GGDEF and HD-GYP domains but no EAL domain (Galperin *et al.*, 1999, 2001, 2005). Although it is less abundant than GGDEF and EAL domains, the HD-GYP domain is also widely distributed in bacteria. In a very limited number of proteins the HD-GYP domain is found in association with the GGDEF domain; no proteins with both EAL and HD–GYP associations are known.

Biochemical studies of the HD-GYP domain have largely focussed on RpfG from the plant pathogen *Xanthomonas campestris*. RpfG is a positive regulator of synthesis of extracellular enzyme virulence factors and negative regulator of biofilm formation (Slater *et al.*, 2000; Dow *et al.*, 2003). The isolated HD-GYP domain from RpfG can hydrolyse cyclic di-GMP to GMP via the linear intermediate pGpG (Ryan *et al.*, 2006a). This reaction depends upon Mn^{2+}, for which Mg^{2+} cannot substitute. The signature HD residues constitute the presumed catalytic diad of the HD-GYP domain and alanine substitution of these residues abolishes both the enzymatic activity against cyclic di-GMP and the regulatory activity of the protein. Multiple sequence alignments suggest that the signature HD and GYP motifs might be more usefully considered as part of the larger motifs HDxGK and HHExxDGxGYP which are present in subdomains separated by a region of high sequence diversity (Ryan *et al.*, 2010). There are a number of other well-conserved charged and hydrophobic residues. Other conserved H and D residues (i.e. those not in the HD diad) may have a role in metal binding. It was originally proposed that the GYP motif has a role in determining substrate specificity (Galperin, 2001). More recent work however has indicated that these residues have a role in the protein–protein interaction between the HD-GYP domain of RpfG and GGDEF domain proteins that control motility in *X. campestris* (Andrade *et al.*, 2006; Ryan *et al.*, 2010; Fig. 5.1). Furthermore, alanine substitutions within the GYP motif do not influence the enzymatic activity of RpfG as a PDE (Ryan *et al.*, 2010).

As with GGDEF and EAL domains, enzymatically inactive variants of the HD-GYP domain are found. In some cases the proteins with these variant domains have been shown to play a regulatory role. For example, *Pseudomonas aeruginosa* has three HD-GYP domain proteins; PA4108 and PA4781 have the signature HD-GYP residues and are active as PDEs whereas PA2572 has the YN-GYP variation and is inactive. Nevertheless, mutation of *PA2572* leads to alteration in a number of phenotypes including biofilm formation and rhamnolipid production (Ryan *et al.*, 2009).

Diverse environmental signals are transduced through cyclic di-GMP signalling

Many GGDEF, EAL and HD-GYP domain proteins have additional domains that may directly sense environmental cues. These domains include various membrane-associated or periplasmic domains that may be involved in sensing small molecules, PAS, which binds flavin or haem and may sense molecular oxygen or redox potential, GAF, which binds cyclic mononucleotides and other small molecular weight effectors, and BLUF, a FAD-dependent blue light sensor (Römling *et al.*, 2005). In some cases multiple sensory input domains are found.

Binding of effectors to the sensory input domain is believed to affect the enzyme activity of the associated domains as DGCs or PDEs. Examples of this come from studies of PdeA1 and AxDGC2 from *G. xylinus* (Chang *et al.*, 2001; Qi *et al.*, 2009). PdeA1 is a PDE that has a PAS-GAF-GGDEF-EAL domain structure. The PAS domain of this protein contains a haem moiety that binds molecular oxygen, leading to a decrease in PDE activity (Chang *et al.*, 2001). Conversely, molecular oxygen positively influences the DGC activity of AxDGC2 (Qi *et al.*, 2009). This protein has a PAS-GGDEF-EAL domain structure with an FAD cofactor non-covalently bound to the PAS domain. The redox status of this FAD cofactor modulates the DGC activity of the GGDEF domain such that the oxidized form of FAD gives a higher catalytic activity (Qi *et al.*, 2009). These findings suggest that PdeA1 and AxDGC2 act in concert to regulate the cellular cyclic di-GMP level in response to a change in oxygen concentration or cellular redox status. Both PdeA1 and AxDGC2 are associated with the regulation of cellulose synthesis and pellicle formation, which are promoted at higher oxygen levels through elevation in the cellular levels of cyclic di-GMP.

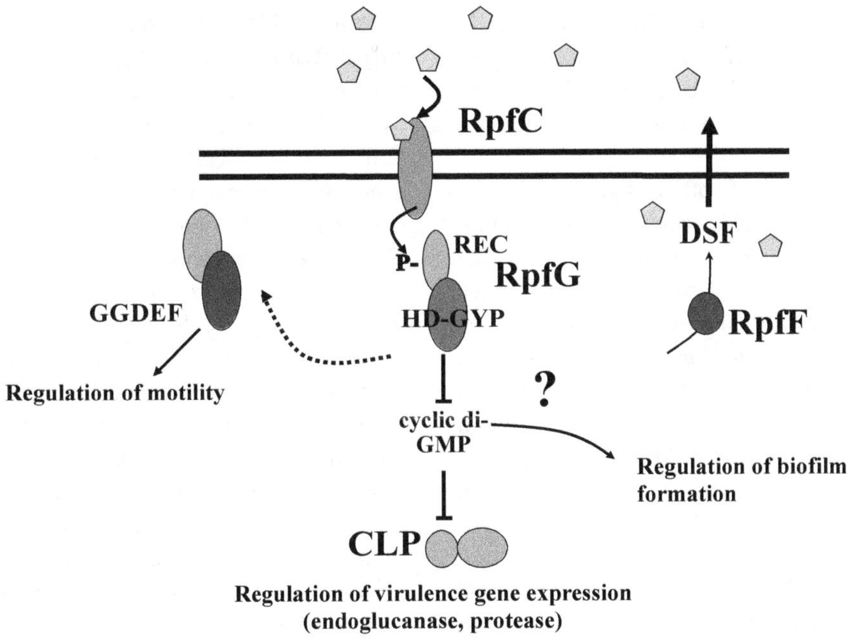

Figure 5.1 Cyclic di-GMP as a second messenger in the regulation of virulence in *Xanthomonas campestris* by cell–cell signalling: role of the HD-GYP domain regulator RpfG. The HD-GYP domain regulator RpfG plays multiple roles in transduction of the cell–cell signal DSF in *Xanthomonas campestris*. The synthesis of the diffusible signal factor DSF (shown as pentagons) depends upon RpfF. Perception of DSF by the sensor kinase RpfC leads to phosphorylation of RpfG, thus activating it as a cyclic di-GMP phosphodiesterase and reducing the cellular levels of cyclic di-GMP. At high levels, cyclic di-GMP acts to inhibit synthesis of extracellular enzymes through binding to the transcriptional activator Clp, preventing promoter binding. Degradation of cyclic di-GMP relieves this inhibition. Phosphorylation of RpfG, also promotes its physical interaction with two GGDEF domain DGCs. This interaction, which serves to regulate motility, depends upon the GYP motif. By contrast, effects of RpfG on extracellular enzyme synthesis do not require the GYP motif. Additional complexity is evident in that the influence of the Rpf/DSF signalling system on biofilm formation is apparently independent of both Clp and the two DGCs.

Many cyclic di-GMP signalling proteins have transmembrane domains and are probably localized at the cytoplasmic membrane. In some cases these proteins have substantial periplasmic domains, and both the membrane-associated and periplasmic domains could be involved in sensing environmental signals. The nature of these signals remains obscure, although the structural relationship of two of these domains, 7TMR-DISMED2 and 7TMR-DISM-7TM, to sugar binding domains of glucosidases and glucuronidases suggest a likely function as carbohydrate sensors (Anantharaman and Aravind, 2003).

In addition to those proteins with the sensory input domains described above, a number of GGDEF, EAL and HD-GYP domain proteins contain a CheY-like receiver (REC) domain. These proteins can be part of two-component signal transduction systems and their enzymatic activity may be altered by phosphorylation. Examples include WspR (REC-GGDEF) that is part of a modified chemotaxis system controlling the wrinkly spreader phenotype of *Pseudomonas* species, RpfG (REC-HD-GYP) that is involved in transduction of the cell–cell signalling molecule DSF in *X. campestris*, PvrR, a REC-EAL domain protein controlling phenotypic variation in *Pseudomonas aeruginosa* and RocR, a REC-EAL domain protein controlling expression of fimbrial cup genes in *P. aeruginosa* (Drenkard and Ausubel, 2002; Hickman *et al.*, 2005; Kulasekara *et al.*, 2005; Ryan *et al.*, 2006a; Malone *et al.*, 2007). RpfG works in concert with the histidine kinase RpfC, which is the sensor for the cell–cell signal DSF (Slater *et al.*, 2000; Fig. 5.1). The environmental signals recognized by

WspA, the methyl-accepting chemotaxis protein of the Wsp system, PvrS, the sensor kinase presumed to be associated with PvrR, or RocS1, the sensor associated with RocR are as yet unknown however.

The perception of other signals may be linked to cyclic di-GMP turnover by different mechanisms involving the interaction of a sensing protein with a cyclic di-GMP signalling partner. The polyamine norspermidine promotes the formation of biofilms in *Vibrio cholerae*, an action that depends upon the GGDEF-EAL domain protein MbaA and the periplasmic binding protein NspS (Karatan *et al.*, 2005). MbaA was originally characterized as a repressor of *V. cholerae* biofilm formation (Bomchil *et al.*, 2003). NspS is thought to act as a sensor for norspermidine and upon binding the ligand to interact with the periplasmic portion of MbaA, thereby reducing its ability to inhibit biofilm formation (Karatan *et al.*, 2005). The *nspS* and *mbaA* genes are co-transcribed in an operon with a third gene encoding a predicted cytoplasmic protein.

One of two haem nitric oxide/oxygen (H-NOX) binding domain proteins in *Legionella pneumophila* has been shown to regulate biofilm formation in response to NO (Carlson *et al.*, 2010). Genetic and biochemical studies suggest that this protein, Hnox1, upon binding the ligand NO, inhibits the DGC activity of the GGDEF-EAL domain protein LPG1057, thus reducing cyclic di-GMP levels and biofilm formation. The *hnox1* and *lpg1057* genes are adjacent in the *L. pneumophila* chromosome.

Subinhibitory concentrations of the antibiotic tobramycin have been shown to trigger biofilm formation in *P. aeruginosa* (Hoffman *et al.*, 2005). This effect depends upon the aminoglycoside response regulator, PA2818 (Arr), a membrane-associated EAL domain protein with a periplasmic domain. This action of Arr required the PDE activity, as indicated by mutational analysis affecting the conserved EAL motif. It is proposed that tobramycin enhances the PDE activity of Arr, although whether this occurs through binding to the protein or by an indirect mechanism is not known.

Cyclic di-GMP exerts a regulatory influence by different mechanisms

By 2006, the enzymatic activities of the GGDEF, EAL and HD-GYP domains had been established and it was appreciated that cyclic di-GMP regulates a wide range of functions in bacteria. However the mechanisms by which the nucleotide exerted these diverse regulatory influences and the nature of cyclic di-GMP receptors or effector proteins were unknown. Bioinformatics provided the first insight into such mechanisms by defining a small protein domain that was called PilZ as a putative cyclic di-GMP receptor, a prediction that was subsequently confirmed experimentally. It is now evident that cyclic di-GMP has a range of cellular targets of different structure that include proteins that contain the PilZ domain, unrelated transcription factors, enzymatically inactive GGDEF, EAL or HD-GYP domain proteins and RNA riboswitches. Thus, cyclic di-GMP can exert a regulatory influence at the level of transcription, post-transcriptionally and post-translationally (for example as an allosteric activator of cellulose synthase). In the following sections we discuss the different classes of cyclic di-GMP receptor or effector.

The PilZ domain

As noted above, the discovery of cyclic di-GMP came about through its description as an allosteric activator of cellulose synthesis in *Gluconobacter xylinus*. Cyclic di-GMP was originally described to bind the BcsB subunit of cellulose synthase of *G. xylinus*, although later reports suggested it bound to an alternative target that co-purified with the cellulose synthase. Bioinformatic analysis of proteins constituting the cellulose synthase complex identified a small domain (of approximately 100 amino acids) in the C-terminus of the BcsA subunit that was also found in the single domain protein PilZ of *P. aeruginosa* (118 amino acids; Alm *et al.*, 1996) as well as in other proteins in many bacteria (Amikam and Galperin, 2006). This domain, which was called PilZ, was proposed to be a cyclic di-GMP receptor, and binding of the nucleotide was demonstrated experimentally for both the PilZ domain of BcsA and for

YcgR, an protein that contains a C-terminal PilZ domain and a N-terminal domain with a similar fold (Ryjenkov *et al.*, 2006). Bacterial genomes generally encode multiple PilZ domain proteins, both those comprising a single PilZ domain and those in which other domains are present. PilZ domain proteins within a single organism show very limited sequence similarity although bioinformatic predictions and structural studies indicate that they all show the same fold (Benach *et al.*, 2007; Ramelot *et al.*, 2007; Li *et al.*, 2008; Guzzo *et al.*, 2009). The most highly conserved residues include a motif in the N-terminus of the protein, (q/e)rRxxxR) that contains two invariant arginines (indicated in capitals) essential for cyclic di-GMP binding and a second motif, dzSxxG, (where z is any hydrophobic residue) which also participates in nucleotide binding.

Cyclic di-GMP binding to the PilZ domain may exert a regulatory action by intramolecular allosteric effects or by promoting protein–protein interactions. This is illustrated by the role of YcgR in controlling swimming speed in *E. coli* in response to the level of cyclic di-GMP (Fig. 5.2). YcgR binds cyclic di-GMP through its C-terminal PilZ domain to act as a brake to regulate the speed of flagellum rotation (Ko and Park, 2000; Ryenkov *et al.*, 2006). During exponential growth, when the PDE YjhH is active, cyclic-di-GMP levels remain low. This promotes motility through

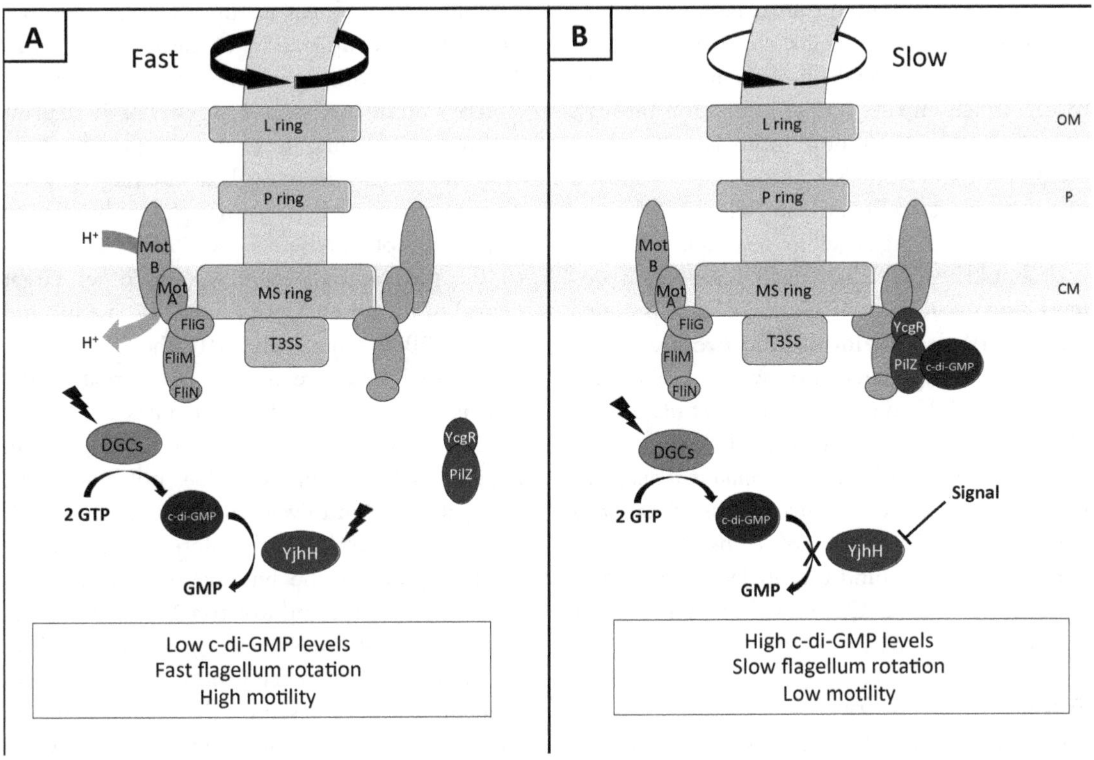

Figure 5.2 Cyclic di-GMP influences swimming velocity in *Escherichia coli* through the PilZ domain protein YcgR. YcgR binds cyclic-di-GMP through its C-terminal PilZ domain and regulates the speed of flagellum rotation through action as a brake. During exponential growth, cyclic-di-GMP levels remain low because the PDE YjhH is active and motility is promoted through 'normal' flagellum rotation speed. During starvation, cyclic di-GMP concentrations increase, in part owing to inactivation of YjhH. The YcgR protein interacts with the consequently available cyclic di-GMP, which activates YcgR and causes a reduction of the speed of flagellum rotation. Different binding targets within the flagellar motor proteins have been proposed and the precise mode of action of the cyclic-di-GMP-bound YcgR protein remains unclear (see text for details). OM, outer membrane; CM, cytoplasmic membrane.

'normal' flagellum rotation speed. During starvation, cyclic di-GMP concentrations increase, in part owing to inactivation of YjhH. The YcgR protein interacts with the consequently available cyclic di-GMP, which activates YcgR and causes a reduction of the speed of flagellum rotation (Wolfe and Visick, 2008). Different binding targets within the flagellar motor proteins have been proposed and the precise mode of action of the cyclic-di-GMP-bound YcgR protein remains unclear (Boehm *et al.*, 2010; Fang and Gomelsky, 2010; Paul *et al.*, 2010). Paul and colleagues have proposed that YcgR normally interacts with the rotor proteins FliM and FliG, but that upon interaction with c-di-GMP, YcgR undergoes a conformational change, which prevents the interaction of FliG with the stator protein MotA, an interaction required for torque generation (Paul *et al.*, 2010). Boehm and colleagues have proposed that the activated YcgR protein interacts with MotA, which impairs torque generation through the reduction of the proton influx (Boehm *et al.*, 2010). Fang and Gomelsky (2010) have proposed that YcgR binds to FliG to alter FliG–FliM interactions, thus biasing flagellum rotation.

It should be noted that not all PilZ domain proteins bind cyclic di-GMP. Some of these proteins may nevertheless be involved in cyclic di-GMP signalling through interaction with proteins with GGDEF or EAL domains (Guzzo *et al.*, 2009). This is illustrated by the proposed mechanism of control of pilus action in *Xanthomonas axonopodis* by XAC2398, a protein encoding enzymatically inactive GGDEF and EAL domains. A PilZ protein that does not bind the nucleotide mediates the interaction of XAC2398 with the pilus motor protein PilB (Fig. 5.3).

Transcription factors

The first evidence that cyclic di-GMP may directly influence transcription came from studies of FleQ, a transcription factor from *P. aeruginosa* which is a master regulator of flagellar biosynthesis and chemotaxis gene expression (Hickman and Harwood, 2008). FleQ also represses expression of genes of the *pel* and *psl* operons, which are responsible for the synthesis of different EPS polymers. The action of FleQ on the *pel* operon is exerted through binding to the promoter of the *pelA* gene, which is the most upstream gene. This repression is partially relieved by a second protein, FleN, in the presence of ATP or ADP. Binding of cyclic di-GMP to the FleQ–FleN complex causes FleQ to dissociate from the DNA, thus further relieving the repression of *pel* expression (Hickman and Harwood, 2008).

A second example of a cyclic di-GMP-responsive transcription factor is the cyclic AMP receptor-like protein Clp from *Xanthomonas* species (Leduc and Roberts 2009; Chin *et al.*, 2010; Tao *et al.*, 2010). Clp exhibits high sequence similarity to the cAMP receptor protein (Crp) of *E. coli* (also known as CAP because of its function as a catabolite activation protein). Both proteins have a N-terminal cNMP binding domain and a C-terminal DNA-binding domain. However Clp exhibits very different features to Crp. Clp does not affect the utilization of carbon sources in *Xanthomonas*, but significantly alters the expression of a number of virulence genes of these plant pathogens including those encoding extracellular enzymes such as the endoglucanase EngXCA and polygalacturonate lyase. Although Crp binding to target promoters requires cyclic AMP, Clp binds to the promoter of *engXCA* (and other target genes) in the absence of cyclic AMP (Leduc and Roberts 2009; Chin *et al.*, 2010; Tao *et al.*, 2010). The crystal structure of Clp shows that in the absence of any ligand, the protein has an intrinsic conformation adapted for DNA binding (Chin *et al.*, 2010). Although Clp does not bind cyclic AMP, it does bind cyclic di-GMP at physiologically relevant levels. Furthermore, cyclic di-GMP binding prevents the binding of Clp to target promoters (Leduc and Roberts 2009; Chin *et al.*, 2010; Tao *et al.*, 2010). The precise binding site for cyclic di-GMP to Clp is not known, although modelling studies based on the crystal structure suggest that the nucleotide binds at a site between the cNMP and DNA binding domains, rather than at the cNMP-binding domain. Clp may serve as a checkpoint for controlling expression of virulence factors in response to environmental cues or intracellular conditions that act to modulate the cellular level of cyclic di-GMP (Fig. 5.1).

A third example of a cyclic di-GMP-responsive transcriptional regulator is VpsT from *Vibrio cholerae*. VpsT controls extracellular matrix

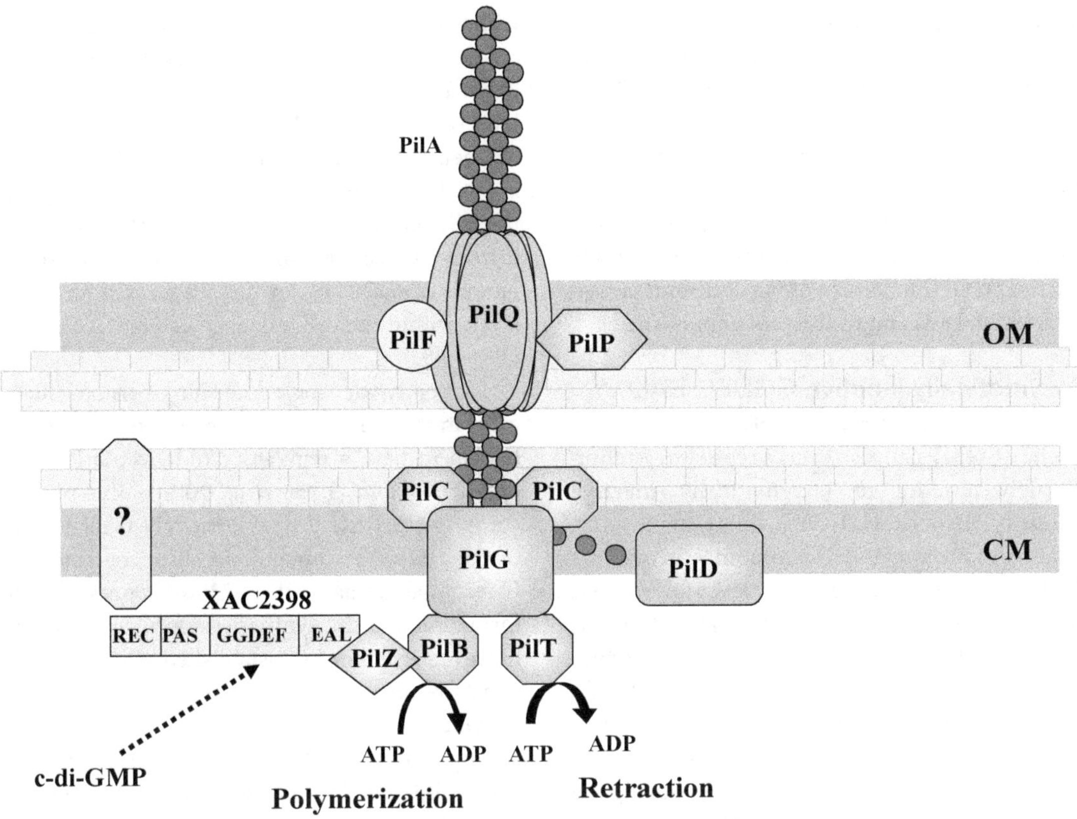

Figure 5.3 Regulation of type IV pilus action in *Xanthomonas axonopodis*: protein–protein interactions involving an enzymatically inactive GGDEF-EAL domain protein and a PilZ domain protein. The filament of type IV pilus is a polymer of pilin subunits (mostly PilA). After being processed by the prepilin peptidase, pilin subunits associated with the inner membrane are incorporated into the base of the pilus filament in a process that requires the inner membrane protein PilC and the ATPase activity of PilB, which forms a hexameric ring associated with the cytoplasmic face of the inner membrane. The growing pilin polymer passes through pores formed by PilQ in the outer membrane. Pili are depolymerized in a process that requires the hexameric PilT ATPase. PilZ interacts with PilB and with XAC2398, a protein encoding both GGDEF and EAL domains, but which is enzymatically inactive. Note that this PilZ protein does not bind cyclic di-GMP. XAC2398 itself may be tethered to the inner membrane via interactions between its REC domain and an unknown sensory protein. These interactions point to a means by which specific signals that impinge on XAC2398 could regulate PilB function. In addition the system may be responsive to the cellular level of cyclic di-GMP that has been shown to bind to XAC2398. OM, outer membrane; CM, cytoplasmic membrane.

production, motility and biofilm formation and acts as a positive regulator of expression of *Vibrio* polysaccharide (*vps*) genes and genes encoding matrix proteins. VpsT comprises an N-terminal receiver (REC) domain and a C-terminal helix–turn–helix DNA binding domain. Although possessing a REC domain, VpsT is not regulated by phosphorylation. The binding of cyclic di-GMP, however, promotes oligomerization, activating VpsT for DNA binding and transcription (Krasteva *et al.*, 2010).

PelD

In *P. aeruginosa*, the *pel* operon (comprising *pelA–G*) encodes enzymes and proteins involved in the synthesis and export of an EPS. The levels of this PEL polysaccharide increase when cyclic di-GMP levels are high. This effect of cyclic di-GMP is exerted in part at the transcriptional level through modulation of FleQ action on expression of the *pel* operon (see above). However, cyclic di-GMP also influences polysaccharide synthesis through binding to PelD, a membrane bound protein that may interact with other *pel* gene products (Lee *et al.*, 2007). PelD has the RXXD motif characteristic of the I site found in some DGCs and the predicted secondary structure in this region of the protein is similar to the I site (Lee *et al.*, 2007). Although there is no recognisable GGDEF motif in PelD, bioinformatic analysis predicts a structure for the domain that is related to that of the (enzymatically inactive) GGDEF domain of FimX from *P. aeruginosa*. (J. M. Dow and R. P. Ryan, unpublished).

Enzymatically inactive GGDEF, EAL or HD-GYP domain proteins

Cyclic di-GMP can also exert a regulatory action through binding to enzymatically inactive GGDEF, EAL or HD-GYP domain proteins. Binding to variant GGDEF domains can occur at the I site rather than the catalytic A site whereas binding to variant EAL or HD-GYP domains most probably occurs at the substrate binding site. PopA from *C. crescentus* is an example of a variant GGDEF domain protein that mediates regulation by cyclic di-GMP (Duerig *et al.*, 2009). PopA localizes to the cell poles during the cell cycle where it acts to control chromosome replication. This occurs through recruitment of CtrA, the master regulator of the cell cycle and its degradation at the cell pole by the protease ClpXP. The localization of PopA depends upon the binding of cyclic di-GMP to the I site (Duerig *et al.*, 2009).

LapD of *Pseudomonas fluorescens* provides an example of a regulatory role for an enzymatically inactive EAL domain protein (Newell *et al.*, 2009). LapD is a membrane-associated GGDEF-EAL domain protein in which both domains appear to lack enzymatic activity. This protein influences the secretion and retention at the cell surface of LapA, a large adhesin protein required for surface attachment and biofilm formation, in a manner controlled by cyclic di-GMP (Monds *et al.*, 2007). This effect is exerted through binding of cyclic di-GMP to the EAL domain of LapD. Mutagenesis studies indicate that nucleotide binding is probably at a site equivalent to the substrate-binding site of an active EAL domain PDE. Gjermansen and colleagues (2010) have provided genetic evidence that the link between LapD and LapA is a protein designated LapG, which is a periplasmic proteinase. More recent studies have shown that LapD regulates the activity of LapG, repressing it when the intracellular level of cyclic di-GMP is high and de-repressing it when the intracellular level of cyclic di-GMP is low. LapG determines the presence of LapA at the cell surface by cleaving LapA, thus releasing it from surface attachment (Navarro *et al*, 2011; Newell *et al*, 2011). In this way, cyclic di-GMP levels within the cell influence cell-surface events (Fig. 5.4).

Currently the single example of an enzymatically inactive HD-GYP domain protein that is known to have a regulatory function is PA2572 of *P. aeruginosa* (Ryan *et al*, 2009). This protein comprises a REC domain, the YN-GYP variant of the HD-GYP domain and an uncharacterized C-terminal domain. PA2572 influences biofilm formation and the synthesis of virulence factors; however, it is not yet established whether these effects require binding of cyclic di-GMP to the protein.

PopA and LapD provide examples of how cyclic di-GMP binding to degenerate GGDEF and EAL domain proteins can influence their interaction with other proteins. Intramolecular regulation may also occur, for example where cyclic di-GMP binding to a degenerate EAL domain in a GGDEF–EAL fusion protein influences the activity of an associated GGDEF domain (or vice versa) (see Qi *et al.*, 2011). It should also be noted that some proteins with degenerate GGDEF and/or EAL domains may have functions that are not responsive to cyclic di-GMP but may instead reflect the capacity of these domains to participate in protein–protein interactions or interactions with other cellular macromolecules. Examples include YcgF of *E. coli*, an EAL domain protein that binds the transcriptional repressor YcgE to antagonize its action (Tschowri *et al.*, 2009) and CsrD, a GGDEF-EAL domain also from *E. coli* that is involved in the turnover of small RNAs (Suzuki *et al.*, 2006).

Riboswitches

Riboswitches are receptors for small molecules that are part of the 5′ untranslated region (UTR) of messenger RNAs. These RNA sequences, which are called aptamers, bind different

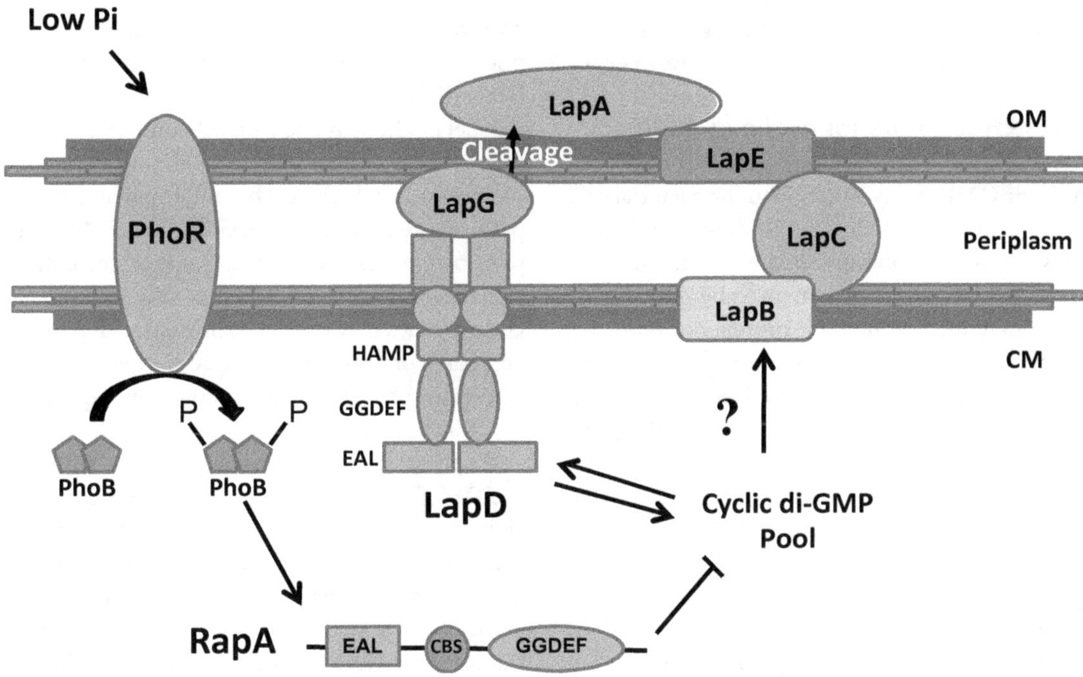

Figure 5.4 Regulation of cell surface properties by cyclic di-GMP: the Lap system and 'inside-out' signalling in *Pseudomonas fluorescens.* The *lap* locus (for large adhesin protein) is absolutely required for biofilm formation by *P. fluorescens* on abiotic surfaces. LapA is an adhesin that associates with the surface of cells, and is required for their transition from reversible to irreversible attachment to a surface. LapBCE is an ABC transporter required for LapA secretion. LapD is a GGDEF-EAL domain protein that localizes to the inner membrane but controls retention of LapA at the cell surface through modulation of the activity of the periplasmic protease LapG. Although LapD does not have DGC or PDE activity, it binds cyclic di-GMP to promote retention of LapA. When cyclic di-GMP levels are high, the activity of LapG is repressed. Under low levels of cyclic di-GMP, LapG is de-repressed and acts to cleave LapA, thus releasing it from the cell surface. The cellular levels of cyclic di-GMP are influenced by RapA, which is a PDE whose expression is controlled by the levels of environmental phosphate. Conditions of low environmental phosphate are sensed by PhoR, which activates PhoB through phosphorylation. Phosphorylated PhoB activates the transcription of *rapA* as well as other genes within the *pho* regulon. In this way, low phosphate levels inhibit bacterial attachment and biofilm formation. OM, outer membrane; CM, cytoplasmic membrane; HAMP, a domain that is found in bacterial sensor kinases and chemotaxis proteins; CBS, a small domain which usually occurs in tandem repeats and may play a regulatory role, although its exact function is unknown.

metabolites to repress or activate gene expression at the transcriptional or translational levels. The regulated genes typically encode protein products that are involved in the synthesis or uptake of the metabolite. Bioinformatic approaches identified the GEMM element as a candidate for a riboswitch aptamer. GEMM elements (of which there are two classes) are frequently located in the UTR of genes encoding DGCs or PDEs, suggesting an association with cyclic di-GMP signalling. Experimental work with a particular GEMM element, the Vc2 RNA of *Vibrio cholerae,* demonstrated that this element binds cyclic di-GMP (Sudarsan *et al.,* 2008). The Vc2 element is upstream of the *tfoX* gene, which regulates bacterial competence in the presence of chitin. Binding of cyclic di-GMP to the Vc2 element is thought to activate expression of the *tfoX* gene.

In the Gram-positive pathogen, *Clostridium difficile,* cyclic di-GMP binding to a second type of riboswitch regulates the activity of a self-splicing ribozyme that controls expression of a gene that is putatively involved in virulence (Lee *et al.,* 2010). Precursor mRNA processed in the presence of GTP alone lacks a ribosome-binding site. Cyclic di-GMP binding in the presence of GTP promotes alternative RNA processing that creates a ribosome-binding site for the downstream gene.

Further receptors for cyclic di-GMP

The body of work outlined above shows that cyclic di-GMP exerts an action in the bacterial cell through diverse mechanisms, binding to both proteins and RNA molecules. It is likely that further cyclic di-GMP effectors remain to be discovered. These may include binding proteins of different structural classes as well as further types of interaction with RNA molecules. Indeed, very recent work has described cyclic di-GMP binding for BdcA, (formerly YjgI), a protein from *Escherichia coli* with a Rossmann-fold NAD(P)H/NAD(P)(+) binding domain (Ma *et al.*, 2010).

Functional organisation of cyclic di-GMP signalling; the concept of discrete pools of the second messenger

Most bacterial genomes encode a number of proteins with GGDEF, EAL or HD-GYP domains, for example *P. aeruginosa* at ~6.3 Mb encodes 40 such proteins and *Vibrio cholerae* El Tor at ~4.03 Mb has 62 (Galperin, 2005). As we have discussed in the previous sections, cyclic di-GMP exerts a regulatory action on different cellular functions by a range of different mechanisms. The large numbers of proteins involved in generating or degrading the second messenger together with the range of targets indicate that there must be considerable complexity in the organization of cyclic di-GMP signalling within a single organism.

One possible organizational motif is that a number of signalling systems are configured in a network that integrates sensing of different environmental cues to determine the cellular level of cyclic di-GMP. Differences in affinity constants of different receptors for cyclic di-GMP would determine which processes were activated or repressed by such a general pool of the second messenger. An alternative but not mutually exclusive arrangement is that certain cyclic di-GMP signalling systems are dedicated to specific cellular tasks associated with the generation and degradation of localized pools of the nucleotide. These systems may activate processes that are determined by co-localizing proteins. The concept of discrete pools of a second messenger that is ostensibly freely diffusible is controversial. Perhaps a useful parallel might be made with substrate channelling in multienzyme complexes catalysing metabolic pathways, in which soluble intermediates in the pathway remain in association with the protein complexes. As well as spatial separation, signalling systems may be temporally separated as certain genes may only be expressed with particular phases of bacterial growth (e.g. exponential vs. stationary phase) (Hengge, 2009; Sommerfeldt *et al.*, 2009).

Experimental observations that different GGDEF, EAL or HD-GYP domain proteins have significant roles in regulation of the same bacterial processes but under different environmental conditions are consistent with the notion of the existence of networks. This has been reported for the role of GGDEF domain proteins in cellulose synthesis in *Salmonella* (Garcia *et al.*, 2004; Simm *et al.*, 2004; Römling, 2005) and GGDEF/EAL/HD-GYP domain proteins influencing the synthesis of extracellular enzymes in *X. campestris* (Ryan *et al.*, 2007; He *et al.*, 2009). In addition, the interplay between the actions of three cyclic di-GMP signalling proteins, CdgC, RocS and MbaA, has been shown to regulate rugosity and polysaccharide production in *Vibrio cholerae* (Lim *et al.*, 2006). In contrast, the findings that some signalling proteins have particular localization in the bacterial cell provide support for the notion of discrete pools of cyclic di-GMP. This has been shown for several proteins including PleD and TipF of *C. crescentus* and FimX, of *P. aeruginosa*. FimX, a GGDEF-EAL domain protein, has been shown to locate to a single pole of the cell, an effect that depends upon both the EAL and GGDEF domains (Huang *et al.*, 2003; Kazmierczak *et al.*, 2006). As noted above, PleD is a regulator of swarmer to stalk cell transitions and pole development in *C. crescentus* and has a REC-REC-GGDEF domain structure (Aldridge *et al.*, 2003). Upon phosphorylation, PleD locates to the pole of the cell where the new stalk will be formed (Paul *et al.*, 2004). Phosphorylation of the protein, which occurs on the N-terminal REC domain, also activates the protein for cyclic di-GMP synthesis. TipF is an EAL domain protein that is a flagellum assembly factor and localizes to the division septum and the newborn pole after division (Huitema *et al.*, 2006). TipF positioning relies on a second protein, TipN. In the absence of TipN, flagella are assembled at ectopic locations where TipF is mislocalized.

A second strand of evidence for the 'insulated' action of particular cyclic di-GMP signalling systems comes from mutational analyses that indicate that different systems control different phenotypic outcomes. In *X. campestris*, a number of signalling proteins have significant regulatory effects on the synthesis of extracellular enzymes whereas others do not influence extracellular enzyme synthesis but control other functions such as motility (Ryan *et al.* 2007, 2010). In *P. aeruginosa*, multiple proteins with GGDEF and/or EAL domains regulate the formation of biofilms and cytotoxicity, but a number of proteins with an influence on biofilm formation have no effects on cytotoxicity, indicating a specificity of action (Kulesekara *et al.*, 2006).

An emerging theme from many studies on different bacteria is that high cellular levels of cyclic di-GMP promote sessility and aggregative behaviours such as biofilm formation whereas low cyclic di-GMP levels promote motility. However recent analyses in *P. aeruginosa* show phenotypic effects of mutation of some genes encoding cyclic di-GMP signalling systems that do not correspond to this generalization; biofilm formation is enhanced by mutation of certain genes encoding DGCs and prevented by mutation of other genes encoding EAL domain proteins (Hoffman *et al.*, 2005; Kulesekara *et al.*, 2006). These apparently anomalous roles of particular signalling systems are difficult to reconcile with the concept that all components influence a single pool of cyclic di-GMP and have been ascribed to highly localized effects on discrete pools of the nucleotide. By inference, some functions contributing to biofilm formation in *P. aeruginosa* could be activated by low levels of the nucleotide. One caveat is that if cyclic di-GMP signalling proteins form complexes with other proteins that influence biofilm formation, loss of this signalling component through mutation may adversely affect complex assembly and hence function in a manner that would not necessarily reflect the enzymatic activity against cyclic di-GMP.

Evidence for the co-existence of networks of signalling systems and dedicated signalling systems in the same cell has come from studies of the regulation of cellulose synthesis in *Salmonella*. A hierarchical arrangement of cyclic di-GMP signalling systems regulates cellulose synthesis in *Salmonella* (Kader *et al.*, 2006). Two GGDEF-EAL domain proteins (STM3388 and STM2123) additively contribute to the expression of the transcriptional regulator CsgD. This protein regulates transcription of *adrA*, which encodes a GGDEF domain protein that is directly implicated in regulation of cellulose synthesis, but not in the expression of *csgD* (Kader *et al.*, 2006) (Fig. 5.5). These findings indicate that AdrA and STM3388/STM2123 operate independently, perhaps influencing discrete pools of cyclic di-GMP

Interplay of cyclic di-GMP signalling with other regulatory networks

A body of work demonstrates that cyclic di-GMP signalling systems can have considerable interplay with a number of other regulatory networks in the bacterial cell, to include growth phase-dependent gene expression, quorum sensing, and the Csr post-transcriptional regulatory network. Genes encoding cyclic di-GMP signalling proteins can be differentially expressed as a function of bacterial growth phase (Sommerfeldt *et al.*, 2009). In *E. coli*, the stationary phase sigma factor sigma S positively controls expression of a number of genes encoding GGDEF/EAL domain proteins that can control the synthesis of cellulose and curli fimbriae through an influence on cyclic di-GMP levels (Sommerfeldt *et al.*, 2009; Fig. 5.6).

Cyclic di-GMP signalling regulates a range of similar phenotypes to quorum sensing to include biofilm formation, other multicellular behaviours and virulence. This knowledge has led to the suggestion of interplay between the two regulatory systems (Camilli and Bassler, 2006). Work on DSF signal transduction in xanthomonads (Fig. 5.1) suggests that a direct link between cell–cell signalling and cyclic di-GMP signalling can occur. For quorum sensing systems in other bacteria such as *Vibrio* spp., regulatory interplay with cyclic

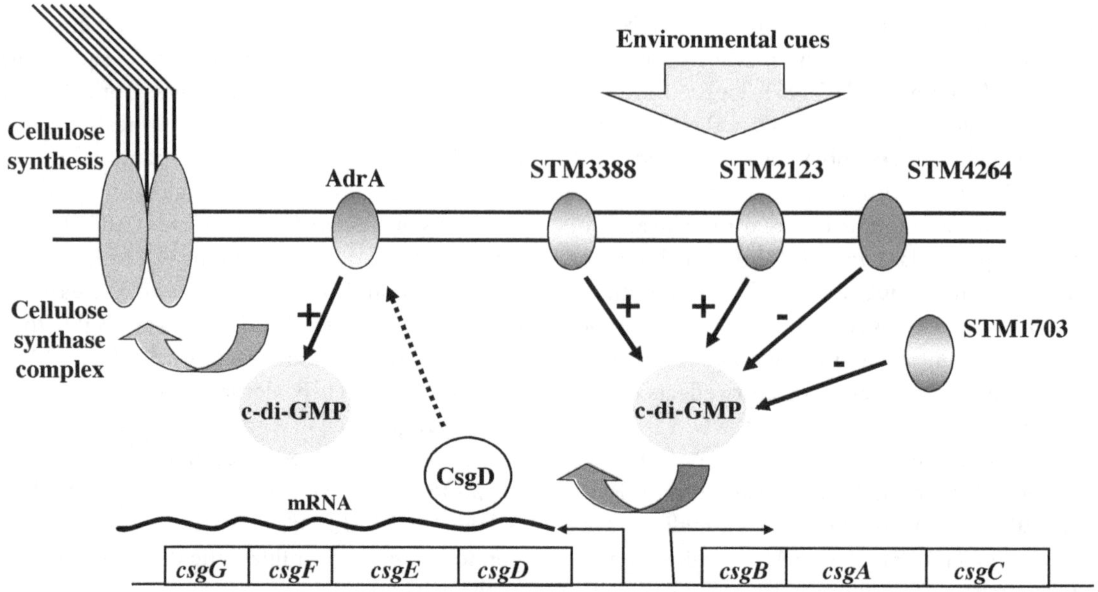

Figure 5.5 Hierarchical organization of cyclic di-GMP signalling systems in regulation of cellulose synthesis in *Salmonella.* Two GGDEF-EAL domain proteins (STM3388 and STM2123) additively contribute to a pool of cyclic di-GMP that activates the expression of the transcriptional regulator CsgD. The PDEs STM1703 and STM4264 negatively influence this pool. The CsgD protein regulates transcription of *adrA*, (indicated by the dotted arrow), which encodes a GGDEF domain protein that is directly implicated in regulation of cellulose synthesis, but not in the expression of *csgD*. This system can operate independently from that involving STM3388 and STM2123, perhaps by influencing a discrete pool of cyclic di-GMP. The activities of all of these proteins may also be influenced by environmental cues. Activation of cellulose synthesis involves binding of cyclic di-GMP to the PilZ domain in the C-terminus of the BcsA subunit of cellulose synthetase.

di-GMP signalling has been shown, but appears to be more indirect (Hammer and Bassler, 2008).

Interplay of cyclic di-GMP signalling with post-transcriptional regulation has been described for the Csr system in *E. coli* (Suzuki *et al.*, 2006; Jonas *et al.*, 2008). CsrA is a post-transcriptional regulator that binds to the mRNA of target genes to prevent ribosome binding and inhibit translation. CsrA inhibits biosynthesis and catabolism of glycogen, gluconeogenesis and biofilm formation. Binding of CsrA to two small RNAs, CsrB and CsrC, reverses this inhibition. CsrD, a GGDEF-EAL domain protein, is involved in targeting CsrB and CsrC for turnover by RNaseE and thus acts to modulate the inhibitory activity of CsrA (Suzuki *et al.*, 2006). The mode of action of CsrD is unknown; the protein has degenerate GGDEF and EAL domains, has no detectable enzymatic activity in cyclic di-GMP metabolism or as a ribonuclease and does not appear to bind cyclic di-GMP. More recently it has been shown that CsrA influences the expression of *ycdT* and *ydeH,* genes that encode proteins with a GGDEF domain (Jonas *et al.*, 2008; Fig. 5.6). CsrA binds to the mRNA transcripts of these genes, suggesting regulation at the post-transcriptional level (Jonas *et al.*, 2008).

Discussion

It is evident that there is a considerable diversity in regulatory systems mediated by cyclic di-GMP in bacteria. The identification of an expanding repertoire of receptors or effectors for the nucleotide poses the balancing question of whether alternative enzymes for the synthesis and degradation occur. A related question is why there is only one class of DGC (the GGDEF domain) but two classes of PDE (the EAL and HD-GYP domains).

The issue of alternative synthases is brought into focus by recent work describing the occurrence

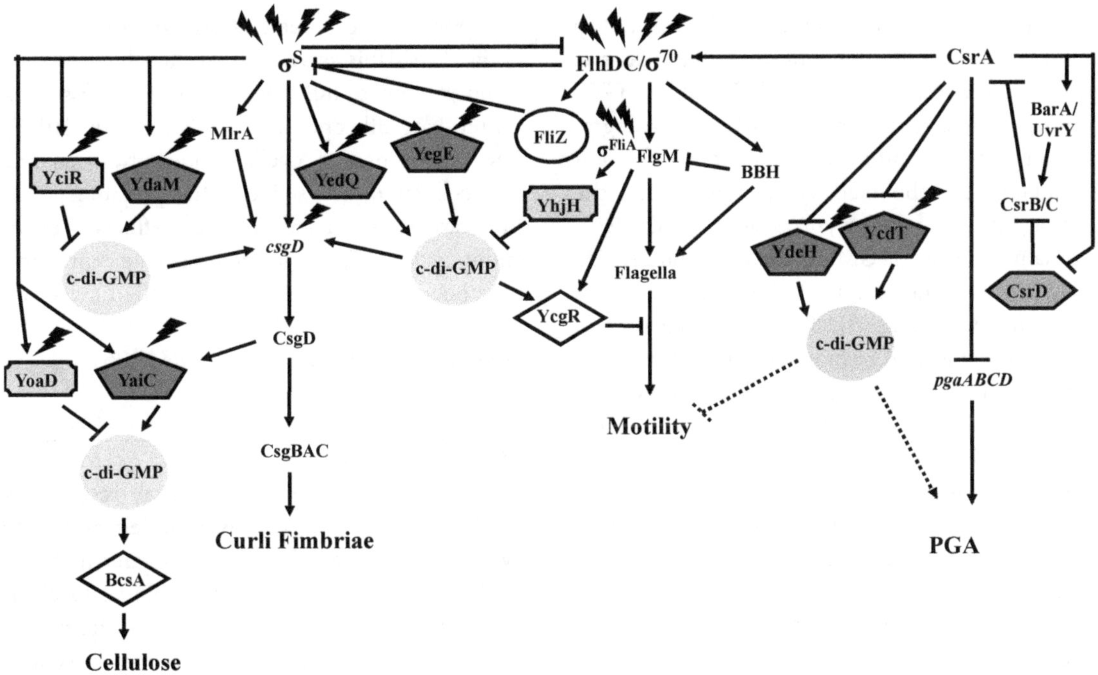

Figure 5.6 Interplay between cyclic di-GMP signalling and other regulatory systems in regulation of motility and biofilm formation in *Escherichia coli*. Interplay between the Csr and FliZ/σS systems and cyclic di-GMP signalling regulates motility and biofilm-associated functions in an inverse fashion in *Escherichia coli*. The figure summarizes the various levels of communication between the σS-controlled curli cascade and the FlhDC/σ70-controlled motility cascade. Another cyclic di-GMP control module (YciR/YdaM) acts in parallel to control curli expression, while the CsgD-dependent GGDEF protein YaiC and the σS-dependent EAL protein YoaD control cellulose synthesis at later time points in stationary phase. The direct and indirect effects of the Csr system on motility and synthesis of PGA (poly beta (1,6)-*N*-acetyl glucosamine), which is required for biofilm formation, are also illustrated. GGDEF and EAL domain proteins are indicated by pentagons and rectangles respectively, PilZ domain proteins by diamonds and CsrD, which has inactive GGDEF and EAL domains, as a hexagon. Lightning bolts indicate further potential signalling and regulatory inputs. Dashed lines indicate regulatory connections requiring further characterization. BBH is basal body hook. BarA/UvrY is a two-component system regulating expression of the small RNAs CsrB and CsrC.

of cyclic di-AMP as a new second messenger in bacteria (Witte *et al.*, 2008). Studies of the DNA integrity scanning protein (DisA) of *Bacillus subtilis* have shown that a domain previously of unknown function has diadenylate cyclase activity and was renamed DAC. The DAC activity of DisA is suppressed by binding to branched DNA, but not to duplex or single-stranded DNA. This suggests that DisA has role as a monitor of DNA integrity and in particular the presence of stalled replication forks or recombination intermediates, for which the output is alteration in the level of cyclic di-AMP (Witte *et al.*, 2008). The DAC domain is unrelated to the GGDEF domain and DisA only has a low activity in cyclic di-GMP synthesis. In parallel, studies of the *B. subtilis* protein YybT have identified a different domain, DHH/DHHA1 that is involved in cyclic di-AMP and cyclic di-GMP hydrolysis (Rao *et al.*, 2010). Measurements of affinity constants together with the distribution of related proteins in bacterial genomes suggest that the physiological substrate for YybT is cyclic di-AMP. Intriguingly YybT also contains a modified GGDEF domain that has ATPase rather than DGC activity (Rao *et al.*, 2010). Nothing is currently known of whether there is any interplay between cyclic di-AMP and cyclic di-GMP regulation. Nevertheless the demonstration of alternative synthases and hydrolases for cyclic di-nucleotides indicate that the existence of further classes of DGC and PDE should not be discounted.

The occurrence of a second PDE unrelated to the EAL domain has been rationalized in several ways. Differences between the EAL and HD-GYP domains are seen in the relative activities in degradation of pGpG and in requirements for divalent metal ions. For the majority of cases studied, the major product of the enzymatic action of the EAL domain on cyclic di-GMP is the linear nucleotide pGpG; this product is then slowly converted to GMP. In contrast the HD-GYP domain generates pGpG in a transient fashion, and the major product is GMP. It has been suggested that pGpG may have a regulatory role or may itself act as a signal molecule (Römling *et al.*, 2005). In this case, the HD-GYP domain could serve to regulate pGpG levels, although this task may be performed by other, non-specific, nucleases. The enzymatic activity of the HD-GYP domain depends upon Mn^{2+} and is not supported by Mg^{2+}, although the significance of these differences is unclear. In *X. campestris*, the physical interaction of the HD-GYP domain protein RpfG with GGDEF domain proteins acts to regulate motility (Ryan *et al.*, 2010). This may suggest that the HD-GYP domain evolved to modulate the activity of GGDEF domain proteins that are not (covalently) associated with an EAL domain. This may allow additional regulation of specific cellular processes in response to particular environmental cues.

Identification of the environmental signals to which different cyclic di-GMP signalling systems respond has only been made in a limited number of cases, and other candidates remain to be tested. As pointed out above, cyclic di-GMP signalling regulates many of the same phenotypes as quorum sensing. In xanthomonads, the perception of the cell–cell signal DSF is linked to alterations in cyclic di-GMP through the RpfC/RpfG two-component system. The possibility that quorum-sensing molecules of other classes may bind to and activate (or inhibit) a GGDEF/EAL domain protein cannot be overlooked.

An interesting possibility is that cyclic di-GMP itself may be an environmental signal. Both cyclic di-GMP and cyclic di-AMP can have effects on eukaryotic cells, influencing host innate immunity (Karaolis *et al.*, 2007; Woodward *et al.*, 2010). These molecules may be released from bacteria that are undergoing lysis, but specific mechanisms of export have also been described. For example, cyclic di-AMP is secreted by intracellular *Listeria monocytogenes* to influence host cell processes via a multidrug efflux pump (Woodward *et al.*, 2010). The occurrence of cyclic dinucleotides in the extracellular milieu suggests the possibility that they can act as a signal. Although effects of exogenous cyclic di-GMP on bacterial gene expression have been described, the levels of the nucleotide used were high since the aim was to elevate the internal level (Méndez-Ortiz *et al.*, 2006).

Future trends

It is evident that a great deal more research is needed to gain a comprehensive understanding of cyclic di-GMP signalling in any organism. These studies should address issues of cellular localization, interactions with other proteins, the nature of the associated effectors or receptors as well as the effects of gene disruption and stacking of mutations on expression of specific cellular activities. Such research should resolve not only the role of individual signalling systems but also their potential interplay within the bacterial cell.

Determination of structures for further signalling proteins and receptors allied to site-directed mutagenesis studies should provide insight into their functions at the molecular level. For example, at the time of writing, no structure of a full-length GGDEF-EAL domain protein has been reported although a model for a complete FimX of *P. aeruginosa* has been suggested based upon structural analysis of the isolated (degenerate) GGDEF and EAL domains (Navarro *et al.*, 2009). Structural studies of proteins crystallized in the presence of cyclic di-GMP should identify new binding sites for the nucleotide in the expanding range of receptors that include transcription factors such as FleQ and Clp.

Although techniques have been developed to extract and quantify cyclic di-GMP levels from cells, methods to examine the levels of the nucleotide at the cellular or subcellular level need to be established. A recent advance in this area has been the development of fluorescence resonance energy transfer (FRET)-based biosensor proteins to monitor cyclic di-GMP concentrations within individual bacterial cells using microscopy

(Christen *et al.*, 2010). A surprising finding using this technique in *P. aeruginosa* and *C. crescentus* was the asymmetrical distribution of the nucleotide in the progeny at cell division. Cyclic di-GMP-responsive riboswitches engineered to control expression of fluorescent proteins could also prove useful to detect levels of the messenger in single cells. Whether such techniques can be refined for examination of subcellular localization of cyclic di-GMP is an important question for the field.

Acknowledgements

The work of the authors has been supported in part by grants awarded by the Science Foundation of Ireland (SFI 07/IN.1/B955 to J.M.D. and SFI 09/SIRG/B1654 to R.P.R.), ESCMID through a Research Grant (to R.R.P) and the Wellcome Trust.

References

Aldridge, P., Paul, R., Goymer, P., Rainey, P., and Jenal U. (2003). Role of the GGDEF regulator PleD in polar development of *Caulobacter crescentus*. Mol. Microbiol. *47*, 1695–1708.

Alm, R.A., Bodero, A.J., Free, P.D., and Mattick, J.S. (1996). Identification of a novel gene, *pilZ*, essential for type 4 fimbrial biogenesis in *Pseudomonas aeruginosa*. J. Bacteriol. *178*, 46–53.

Amikam, D., and Galperin, M.Y. (2006). PilZ is part of the bacterial c-di-GMP binding protein. Bioinformatics *22*, 3–6.

Anantharaman, V., and Aravind, L. (2003). Application of comparative genomics in the identification and analysis of novel families of membrane-associated receptors in bacteria. BMC Genomics *4*, 34

Andrade, M.O., Alegria, M.C., Guzzo, C.R., Docena, C., Rosa, M.C.P., Ramos, C.H.I., and Farah, C.S. (2006). The HD-GYP domain of RpfG mediates a direct linkage between the Rpf quorum-sensing pathway and a subset of diguanylate cyclase proteins in the phytopathogen *Xanthomonas axonopodis* pv citri. Mol. Microbiol. *62*, 537–551

Benach, J, Swaminathan, S.S., Tamayo, R., Handelman, S.K., Folta-Stogniew, E., Ramos, J.E., Forouhar, F., Neely, H., Seetharaman, J., Camilli, A., *et al.* (2007). The structural basis of cyclic diguanylate signal transduction by PilZ domains. EMBO J. *26*, 5153–5166.

Boehm, A., Kaiser, M., Li, H., Spangler, C., Kasper, C.A., Ackermann, M., Kaever, V., Sourjik, V., Roth, V., and Jenal, U. (2010). Second messenger-mediated adjustment of bacterial swimming velocity. Cell *141*, 107–116.

Bomchil, N., Watnick, P., and Kolter, R. (2003). Identification and characterization of a *Vibrio cholerae* gene, *mbaA*, involved in maintenance of biofilm architecture. J. Bacteriol. *185*, 1384–1390.

Camilli, A., and Bassler, B.L. (2006). Bacterial small-molecule signaling pathways. Science *311*, 1113–1116.

Carlson, H.K., Vance, R.E., and Marletta, M.A. (2010). H-NOX regulation of c-di-GMP metabolism and biofilm formation in *Legionella pneumophila*. Mol. Microbiol. *77*, 930–942.

Chan, C., Paul, R., Samoray, D., Amiot, N.C., Giese, B., Jenal, U., and Schirmer, T. (2004). Structural basis of activity and allosteric control of diguanylate cyclase. Proc. Natl. Acad. Sci. U.S.A. *101*, 17084–17089.

Chang, A.L., Tuckerman, J.R., Gonzalez, G., Mayer, R., Weinhouse, H., Volman, G., Amikam, D., Benziman, M., and Gilles-Gonzalez, M. A. (2001). Phosphodiesterase A1, a regulator of cellulose synthesis in *Acetobacter xylinum*, is a heme-based sensor. Biochemistry *40*, 3420–3426.

Chin, K.-H., Lee, Y.-C., Tu, Z.-L., Chen, C.-H., Tseng, Y.-H., Yang, J.-M., Ryan, R.P., McCarthy, Y., Dow, J.M., Wang, A. H.-J., and Chou, S.-H. (2010). The cAMP receptor-like protein CLP is a novel c-di-GMP receptor linking cell–cell signaling to virulence gene expression in *Xanthomonas campestris*. J. Mol. Biol. *396*, 646–662.

Christen, M., Christen, B., Folcher, M., Schauerte, A., and Jenal, U. (2005). Identification and characterization of a cyclic di-GMP-specific phosphodiesterase and its allosteric control by GTP. J. Biol. Chem. *280*, 30829–30837.

Christen, M., Kulasekara, H.D., Christen, B., Kulasekara, B.R., Hoffman, L.R., and Miller, S.I. (2010). Asymmetrical distribution of the second messenger c-di-GMP upon bacterial cell division. Science. *328*, 1295–1297.

Dow, J.M., Crossman, L., Findlay, K., He, Y.Q., Feng, J.X., and Tang, J.L. (2003). Biofilm dispersal in *Xanthomonas campestris* is controlled by cell–cell signaling and is required for full virulence to plants. Proc. Natl. Acad. Sci. U.S.A. *100*, 10995–11000.

Drenkard, E., and Ausubel, F. M. (2002). *Pseudomonas* biofilm formation and antibiotic resistance are linked to phenotypic variation. Nature *416*, 740–743.

Duerig, A., Abel, S., Folcher, M., Nicollier, M., Schwede, T., Amiot, N., Giese, B., and Jenal, U. (2009). Second messenger-mediated spatiotemporal control of protein degradation regulates bacterial cell cycle progression. Genes Dev. *23*, 93–104.

Fang, X., and Gomelsky, M. (2010). A post-translational, c-di-GMP-dependent mechanism regulating flagellar motility. Mol. Microbiol. *76*, 1295–1305.

Galperin, M.Y. (2005). A census of membrane-bound and intracellular signal transduction proteins in bacteria: Bacterial IQ, extroverts and introverts. BMC Microbiol. *5*, 35

Galperin, M.Y., Natale, D.A., Aravind, L., and. Koonin, E.V. (1999). A specialized version of the HD hydrolase domain implicated in signal transduction. J. Mol. Microbiol. Biotechnol. *1*, 303–305.

Galperin, M.Y., Nikolskaya, A.N., and Koonin, E.V. (2001). Novel domains of the prokaryotic two-component signal transduction systems. FEMS Microbiol. Lett. *203*, 11–21.

Garcia, B., Latasa, C., Solano, C., Portillo, F.G., Gamazo, C., and Lasa, I. (2004). Role of the GGDEF protein family in Salmonella cellulose biosynthesis and biofilm formation. Mol. Microbiol. *54*, 264–277.

Gjermansen, M., Nilsson, M., Yang, L., and Tolker-Nielsen, T. 2010. Characterization of starvation-induced dispersion in *Pseudomonas putida* biofilms: genetic elements and molecular mechanisms. Mol. Microbiol. *75*, 815–826.

Guzzo, C.R., Salinas, R.K., Andrade, M.O., and Farah, C.S. (2009). PILZ protein structure and interactions with PILB and the FIMX EAL domain: implications for control of type IV pilus biogenesis. J. Mol. Biol. *393*, 848–866.

Hammer, B.K., and Bassler, B.L. (2009). Distinct sensory pathways in *Vibrio cholerae* El Tor and classical biotypes modulate cyclic dimeric GMP levels to control biofilm formation. J. Bacteriol. *191*, 169–177

He, Y.W., Boon, C., Zhou, L., and Zhang, L.H. (2009). Co-regulation of *Xanthomonas campestris* virulence by quorum sensing and a novel two-component regulatory system RavS/RavR. Mol. Microbiol. *71*, 1464–1476.

Hecht, G.B., and Newton, A. (1995). Identification of a novel response regulator required for the swarmer-to-stalked-cell transition in *Caulobacter crescentus*. J. Bacteriol. *177*, 6223–6229.

Hengge, R. (2009). Principles of c-di-GMP signalling in bacteria. Nat. Rev. Microbiol. *7*, 263–273.

Hickman, J.W., and Harwood, C.S. (2008). Identification of FleQ from *Pseudomonas aeruginosa* as a c-di-GMP-responsive transcription factor. Mol. Microbiol. *69*, 376–389.

Hickman, J.W., Tifrea, D.F., and Harwood, C.S. (2005). A chemosensory system that regulates biofilm formation through modulation of cyclic diguanylate levels. Proc. Natl. Acad. Sci. U.S.A. *102*, 14422–14427.

Hoffman, L.R., D'Argenio, D.A., MacCoss, M.J., Zhang, Z., Jones, R.A., and Miller, S.I.. (2005). Aminoglycoside antibiotics induce bacterial biofilm formation. Nature *436*, 1171–1175.

Huang, B., Whitchurch, C.B., and Mattick, J.S. (2003). FimX, a multidomain protein connecting environmental signals to twitching motility in *Pseudomonas aeruginosa*. J. Bacteriol. *185*, 7068–7076.

Huitema, E., Pritchard, S., Matteson, D., Radhakrishnan, S.K., and Viollier, P.H. (2006). Bacterial birth scar proteins mark future flagellum assembly site. Cell *124*, 1025–1037.

Jenal, U and Malone, J. (2006). Mechanisms of Cyclic-di-GMP signaling in bacteria. Annu. Rev. Genet. *40*, 385–407.

Jonas, K., Edwards, A.N., Simm, R., Romeo, T., Römling, U., and Melefors, O. (2008). The RNA binding protein CsrA controls cyclic di-GMP metabolism by directly regulating the expression of GGDEF proteins. Mol. Microbiol. *70*, 236–257.

Kader, A., Simm, R., Gerstel, U., Morr, M., and Römling, U. (2006). Hierarchical involvement of various GGDEF domain proteins in rdar morphotype development of *Salmonella enterica* serovar *Typhimurium*. Mol. Microbiol. *60*, 602–616.

Karaolis, D.K., Means, T.K., Yang, D., Takahashi, M., Yoshimura, T., Muraille, E., Philpott, D., Schroeder, J.T., Hyodo, M., Hayakawa, Y., *et al.* (2007). Bacterial c-di-GMP is an immunostimulatory molecule. J. Immunol. *178*, 2171–2181.

Karatan, E., Duncan, T.R., and Watnick, P.I. (2005). NspS, a predicted polyamine sensor, mediates activation of *Vibrio cholerae* biofilm formation by norspermidine. J. Bacteriol. *187*, 7434–7443.

Kazmierczak, B.I., Lebron, M.B., and Murray, T.S. (2006). Analysis of FimX, a phosphodiesterase that governs twitching motility in *Pseudomonas aeruginosa*. Mol. Microbiol. *60*, 1026–1043.

Ko, M., and Park, C. (2000). Two novel flagellar components and H-NS are involved in the motor function of *Escherichia coli*. J. Mol. Biol. *303*, 371–382.

Krasteva, P.V, Fong, J.C., Shikuma, N.J., Beyhan, S., Navarro, M.V., Yildiz, F.H., and Sondermann, H. (2010). *Vibrio cholerae* VpsT regulates matrix production and motility by directly sensing cyclic di-GMP. Science. *327*, 866–868.

Kulasekara, H.D., Ventre, I., Kulasekara, B.R., Lazdunski, A., Filloux, A., and Lory, S. (2005). A novel two-component system controls the expression of *Pseudomonas aeruginosa* fimbrial *cup* genes. Mol. Microbiol. *55*, 368–380.

Kulesekara, H., Lee, V., Brencic, A., Liberati, N., Urbach, J., Miyata, S., Lee, D.G., Neely, A.N., Hyodo, M., Hayakawa, Y., *et al.* (2006). Analysis of *Pseudomonas aeruginosa* diguanylate cyclases and phosphodiesterases reveals a role for bis-(3′-5′)-cyclic-GMP in virulence. Proc. Natl. Acad. Sci. U.S.A. *103*, 2839–2844.

Leduc, J.L., and Roberts, G.P. (2009). Cyclic di-GMP Allosterically Inhibits the CRP-Like Protein (Clp) of *Xanthomonas axonopodis* pv. citri. J. Bacteriol. *191*, 7121–7122.

Lee, E.R., Baker, J.L., Weinberg, Z., Sudarsan, N., and Breaker, R.R. (2010). An allosteric self-splicing ribozyme triggered by a bacterial second messenger. Science *329*, 845–848.

Lee, V.T., Matewish, J.M., Kessler, J.L., Hyodo, M., Hayakawa, Y., and Lory, S. (2007). A cyclic-di-GMP receptor required for bacterial exopolysaccharide production. Mol. Microbiol. *65*, 1474–1484.

Li, T.N., Chin, K.H., Liu, J.H., Wang, A.H., and Chou, S.H. (2009). XC1028 from adopts a PilZ domain-like structure without a c-di-GMP switch. Proteins *75*, 282–288.

Lim, B., Beyhan, S., Meir, J., and Yildiz, F.H. (2006). Cyclic-diGMP signal transduction systems in *Vibrio cholerae*: modulation of rugosity and biofilm formation. Mol. Microbiol. *60*, 331–348.

Ma, Q., Yang, Z., Pu, M., Peti, W., and Wood, T. K. (2010). Engineering a novel c-di-GMP-binding protein for biofilm dispersal. Environ. Microbiol. *13*, 631–642.

Malone, J.G., Williams, R., Christen, M., Jenal, U., Spiers, A.J., and Rainey, P.B. (2007). The structure–function relationship of WspR, a *Pseudomonas fluorescens* response regulator with a GGDEF output domain. Microbiology *153*, 980–994.

Méndez-Ortiz, M.M., Hyodo, M., Hayakawa, Y., and Membrillo-Hernández, J. (2006). Genome-wide transcriptional profile of *Escherichia coli* in response to high levels of the second messenger 3′,5′-cyclic diguanylic acid. J. Biol. Chem. *281*, 8090–8099

Merkel, T.J., Barros, C., and Stibitz, S. (1998). Characterization of the *bvgR* locus of *Bordetella pertussis*. J. Bacteriol. *180*, 1682–1690.

Monds, R.D., Newell, P.D., Gross, R.H., and O'Toole, G.A. (2007). Phosphate-dependent modulation of c-di-GMP levels regulates *Pseudomonas fluorescens* Pf0-1 biofilm formation by controlling secretion of the adhesin LapA. Mol. Microbiol. *63*, 656–679.

Navarro, M.V., De, N., Bae, N., Wang, Q., and Sondermann, H. (2009). Structural analysis of the GGDEF-EAL domain-containing c-di-GMP receptor FimX. Structure *17*, 1104–1116.

Navarro, M.V., Newell, P.D., Krasteva, P.V., Chatterjee, D., Madden, D.R., O'Toole, G.A., and Sondermann, H. (2011). Structural basis for c-di-GMP-mediated inside-out signaling controlling periplasmic proteolysis. PLoS Biol. *9*, e1000588.

Newell, P.D., Monds, R.D., and O'Toole, G.A. (2009). LapD is a bis-(3′,5′)-cyclic dimeric GMP-binding protein that regulates surface attachment by *Pseudomonas fluorescens* Pf0-1. Proc. Natl. Acad. Sci. U.S.A. *106*, 3461–2466.

Newell, P.D., Boyd, C.D., Sondermann, H., and O'Toole, G.A. (2011). A c-di-GMP effector system controls cell adhesion by inside-out signaling and surface protein cleavage. PLoS Biol. *9*, e1000587.

Paul, K. Nieto V., Carlquist W.C., Blair D.F. and Harshey, R.M. (2010). The c-di-GMP binding protein YcgR controls flagellar motor direction and speed to affect chemotaxis by a "backstop brake" mechanism. Mol. Cell. *38*, 128–139.

Paul, R., Weiser, S., Amiot, N.C., Chan, C., Schirmer, T., Giese, B., and Jenal, U. (2004). Cell cycle-dependent dynamic localization of a bacterial response regulator with a novel di-guanylate cyclase output domain. Genes Dev. *18*, 715–727.

Qi, Y., Rao, F., Luo, Z., and Liang, Z.X. (2009). A flavin cofactor-binding PAS domain regulates c-di-GMP synthesis in AxDGC2 from *Acetobacter xylinum*. Biochemistry *48*, 10275–10285

Qi, Y., Chuah, M.L., Dong, X., Xie, K., Luo, Z., Tang, K., and Liang, Z.X. (2011). Binding of C-di-GMP in the non-catalytic EAL domain of FimX induces a long-range conformational change. J. Biol. Chem. *19*, 1378–1384.

Ramelot, T.A., Yee, A., Cort, J.R., Semesi, A., Arrowsmith, C.H., and Kennedy, M.A. (2007). NMR structure and binding studies confirm that PA4608 from *Pseudomonas aeruginosa* is a PilZ domain and a c-di-GMP binding protein. Proteins *66*, 266–271.

Rao, F., Yang, Y., Qi, Y., and Liang, Z.X. (2008). Catalytic mechanism of cyclic di-GMP-specific phosphodiesterase: a study of the EAL domain-containing RocR from *Pseudomonas aeruginosa*. J. Bacteriol. *190*, 3622–3631.

Rao, F., Qi, Y., Chong, H.S., Kotaka, M., Li, B., Li, J., Lescar, J., Tang, K., and Liang, Z.X. (2009). The functional role of a conserved loop in EAL domain-based cyclic di-GMP-specific phosphodiesterase. J. Bacteriol. *191*, 4722–4731.

Rao, F., See, R.Y., Zhang, D., Toh, D.C., Ji, Q., and Liang, Z.X. (2010). YybT is a signaling protein that contains a cyclic dinucleotide phosphodiesterase domain and a GGDEF domain with ATPase activity. J. Biol. Chem. *285*, 473–482.

Römling, U. (2005). Characterization of the rdar morphotype, a multicellular behaviour in Enterobacteriaceae. Cell. Mol. Life Sci. *62*, 1–13.

Römling U. (2009). Rationalizing the evolution of EAL domain-based cyclic di-GMP-specific phosphodiesterases. J. Bacteriol. *194*, 4697–700.

Römling, U., Gomelsky M., and Galperin M.Y. (2005). C-di-GMP: The dawning of a novel bacterial signalling system. Mol. Microbiol. *57*, 629–639.

Ross, P., Weinhouse, H., Aloni, Y., Michaeli, D., Weinberger-Ohana, P., Mayer, R., Braun, S., de Vroom, E., van der Marel, G.A., van Boom, J.H., *et al.* (1987). Regulation of cellulose synthesis in *Acetobacter xylinum* by cyclic diguanylate. Nature *325*, 279–281.

Ryan, R.P., Fouhy, Y., Lucey, J.F., Crossman, L.C., Spiro, S., He, Y.-W., Zhang, L.-H., Heeb, S., Williams, P., and Dow, J. M. (2006a). Cell–cell signaling in *Xanthomonas campestris* involves an HD-GYP domain protein that functions in cyclic di-GMP turnover. Proc. Natl. Acad. Sci. U.S.A. *103*, 6712–6717.

Ryan, R.P., Fouhy, Y., Lucey, J.F., and Dow, J.M. (2006b). Cyclic di-GMP signaling in bacteria: recent advances and new puzzles. J. Bacteriol. *188*, 8327–8333

Ryan, R.P., Fouhy, Y., Lucey, J.F., Jiang, B.L., He, Y.Q., Feng, J.X., Tang, J.L., and Dow, J.M. (2007). Cyclic di-GMP signalling in the virulence and environmental adaptation of *Xanthomonas campestris*. Mol. Microbiol. *63*, 429–442.

Ryan, R.P., Lucey, J., O'Donovan, K., McCarthy, Y., Yang, L., Tolker-Nielsen, T., and Dow, J.M. (2009). HD-GYP domain proteins regulate biofilm formation and virulence in *Pseudomonas aeruginosa*. Environ. Microbiol. *11*, 1126–36.

Ryan, R.P., McCarthy, Y., Andrade, M., Farah, C.S., Armitage, J.P., and Dow, J.M. (2010). Cell–cell signal-dependent dynamic interactions between HD-GYP and GGDEF domain proteins mediate virulence in *Xanthomonas campestris*. Proc. Natl. Acad. Sci. U.S.A. *107*, 5989–5994.

Ryjenkov, D.A., Tarutina, M., Moskvin, O.M., and Gomelsky, M. (2005). Cyclic diguanylate is a ubiquitous signaling molecule in Bacteria: insights into biochemistry of the GGDEF protein domain. J. Bacteriol. *187*, 1792–1798.

Ryjenkov, D.A., Simm, R., Römling, U., and Gomelsky, M. (2006). The PilZ domain is a receptor for the second messenger c-di-GMP. The PilZ domain protein YcgR controls motility in enterobacteria. J. Biol. Chem. *281*, 30310–30314.

Schmidt, A.J., Ryjenkov, D.A., and Gomelsky, M. (2005). Ubiquitous protein domain EAL encodes cyclic diguanylate-specific phosphodiesterase: enzymatically active and inactive EAL domains. J. Bacteriol. *187*, 4774–4781.

Simm, R., Morr, M., Kader, A., Nimtz, M., and Römling, U. (2004). GGDEF and EAL domains inversely regulate cyclic di-GMP levels and transition from sessility to motility. Mol. Microbiol. *53*, 1123–1134.

Slater, H., Alvarez-Morales, A., Barber, C.E., Daniels, M.J., and Dow, J.M. (2000). A two-component system involving an HD-GYP domain protein links cell–cell signalling to pathogenicity gene expression in *Xanthomonas campestris*. Mol. Microbiol. *38*, 986–1003.

Sommerfeldt, N., Possling, A., Becker, G., Pesavento, C., Tschowri, N., and Hengge, R. (2009). Gene expression patterns and differential input into curli fimbriae regulation of all GGDEF/EAL domain proteins in *Escherichia coli*. Microbiology *155*, 1318–1331.

Sudarsan, N., Lee, E.R., Weinberg, Z., Moy, R.H., Kim, J.N., Link, K.H., and Breaker, R.R. (2008). Riboswitches in eubacteria sense the second messenger cyclic di-GMP. Science. *321*, 411–413.

Suzuki, K., Babitzke, P., Kushner, S.R., and Romeo, T. (2006). Identification of a novel regulatory protein (CsrD) that targets the global regulatory RNAs CsrB and CsrC for degradation by RNase E. Genes Dev. *20*, 2605–2617.

Tal, R., Wong, H.C., Calhoon, R., Gelfand, D., Fear, A.L., Volman, G., Mayer, R., Ross, P., Amikam, D., Weinhouse, H., *et al.* (1998). Three *cdg* operons control cellular turnover of cyclic di-GMP in *Acetobacter xylinum*: genetic organization and occurrence of conserved domains in isoenzymes. J. Bacteriol. *180*, 4416–4425.

Tamayo, R., Tischler, A.D., and Camilli, A. (2005). The EAL domain protein VieA is a cyclic diguanylate phosphodiesterase. J. Biol. Chem. *280*, 33324–33330.

Tao, F., He, Y.-W., Wu, D.-H., Swarup, S., and Zhang, L.-H. (2010). The cyclic nucleotide monophosphate domain of *Xanthomonas campestris* global regulator Clp defines a new class of cyclic di-GMP effectors. J. Bacteriol. *192*, 1020–1029.

Tarutina, M., Ryjenkov, D.A., and Gomelsky, M. (2006). An unorthodox bacteriophytochrome from *Rhodobacter sphaeroides* involved in turnover of the second messenger c-di-GMP. J. Biol. Chem. *281*, 34751–34758.

Tschowri, N., Busse, S., and Hengge, R. 2009. The BLUF-EAL protein YcgF acts as a direct anti-repressor in a blue-light response of *Escherichia coli*. Genes Dev. *23*, 522–534.

Witte, G., Hartung, S., Büttner, K., and Hopfner, K.P. (2008). Structural biochemistry of a bacterial checkpoint protein reveals diadenylate cyclase activity regulated by DNA recombination intermediates. Mol. Cell. *30*, 167–178.

Wolfe, A.J., and Visick, K. L. (2008). Get the message out: cyclic-Di-GMP regulates multiple levels of flagellum-based motility. J. Bacteriol. *190*, 463–475

Woodward, J.J., Iavarone, A.T., and Portnoy, D.A. (2010). c-di-AMP secreted by intracellular *Listeria monocytogenes* activates a host type I interferon response. Science *328*, 1703–1705.

RNA-mediated Regulation of Virulence Gene Expression: Another Layer of Complexity

6

Efthimia Lioliou, Cédric Romilly, Thomas Geissmann, François Vandenesch and Pascale Romby

Abstract

Many pathogenic bacteria cause serious diseases in humans, animals, and plants. Owing to the appearance of resistance to multiple antibiotics, it has become important to fully understand the regulatory networks that lead to the production of virulence factors that help the bacteria combat the host defence machinery, acquire nutrients, and survive and/or proliferate within the host. In recent years, complex interplays between transcriptional regulatory proteins, two-component systems, and regulatory RNAs have been described, establishing the gene expression patterns in pathogenic bacteria. In this review, several examples will illustrate the diversity of regulatory RNAs and how they are integrated into the regulatory circuits required for virulence gene expression, with special emphasis on the mechanisms of regulation at the molecular level.

Introduction

Pathogenic bacteria have evolved a plethora of sensory systems that coordinate the expression of virulence genes in response to both environmental and host signals. Protein-mediated transcriptional regulation is central in many of these regulatory networks. In addition to proteins, RNAs have recently emerged as major regulators of numerous physiological and adaptive responses in bacteria (Waters and Storz, 2009; Papenfort and Vogel, 2010). Using systematic computational searches for conserved intergenic regions, cloning-based techniques and microarrays, several groups have discovered that conserved intergenic regions express functional small non-coding RNAs (sRNA) in *Escherichia coli* (reviewed in Altuvia, 2007). With the recent technical advances, such as high throughput sequencing (e.g. Sittka *et al.*, 2008; Beaume *et al.*, 2010; Sharma *et al.*, 2010; Deltcheva *et al.*, 2011), and tiled microarrays with full genome coverage (e.g. Toledo-Arana *et al.*, 2009), hundreds of candidate regulatory RNA genes as well as genome-wide antisense transcription have been discovered in various bacteria including pathogens. RNA molecules have several features that make them ideal regulators for adaptive responses. Their structures are versatile and can undergo conformational changes upon environmental cues or ligand binding as it was demonstrated for several 5′ untranslated regions (5′-UTR) of mRNAs (Ramesh and Winkler, 2010). Functional and mechanistic studies show that RNAs are some of the missing links in the regulatory pathways that allow bacteria to sense population density and various environmental changes, modulate and modify cell surface properties, adjust their metabolism during cell growth, and regulate virulence gene expression. Regulatory RNAs are diverse and comprise mRNA leaders that affect expression *in cis*, non-coding RNAs that sequester proteins or base-pair with target mRNAs, and CRISPR RNAs that direct specific degradation of foreign DNA. The present review will illustrate several representative mechanisms exploited by regulatory RNAs in pathogenic bacteria to control gene expression. In all the cases discussed, the outcome of RNA-mediated regulation on virulence and host–pathogen interactions has been elucidated.

Cis-acting regions of mRNAs and regulation of virulence gene expression

Large 5′ untranslated regions (5′-UTR) of mRNAs have evolved dedicated regulatory sites, which are usually structured. These regions of mRNAs respond to a wide variety of *trans*-acting factors [e.g. metabolites (riboswitch), divalent ions, uncharged tRNAs (T-boxes), and RNA-binding proteins] and to environmental cues [e.g. temperature (thermosensor) and pH]. They usually act as direct sensors of the physical and metabolic states of the cell (Narberhaus, 2010; Ramesh and Winkler, 2010; Smith *et al.*, 2010). Among these *cis*-acting regulatory elements, thermosensors detect and respond to temperature changes primarily through regulating the translation of downstream genes (Narberhaus, 2010), while riboswitches exert their regulatory effect through conformational changes of the RNA induced by the metabolite binding and promoting termination/anti-termination of transcription, translational inhibition/activation or RNA degradation (e.g. Barrick and Breaker, 2007; Zhang *et al.*, 2010). Riboswitches respond to changes in metabolite concentration by binding the respective metabolite (e.g. SAM, TPP, FMN, purine, pre-Q1, c-di-GMP, lysine, Mg^{2+}) and regulating the synthesis of proteins involved in amino acid transport and biosynthesis, and in co-factors and nucleotide biosynthesis (see below).

The thermosensors described thus far are involved in the regulation of heat/cold-shock and virulence genes. In most cases, the Shine–Dalgarno (SD) sequence is sequestered into a secondary structure at low temperatures (< 30°C). Following a temperature shift, the SD becomes exposed, allowing the ribosome access to promote translation. Although these regulatory elements do not exhibit sequence conservation, structural signatures such as the ROSE or the four-U element that are characterized by non-canonical base-pairings have been described (Narberhaus, 2010). An NMR study of the ROSE temperature-responsive element revealed that the SD sequence forms several non-canonical base-pairs, which are the first to melt at 37°C, and make it accessible to the 30S ribosome subunit (Chowdhury *et al.*, 2006). Thus, melting of non-canonical base-pairs within a hairpin structure might be the initial event that senses the temperature upshift and starts the gradual melting of the hairpin. Another well-studied case of a thermosensor is *prfA* mRNA, which encodes a major transcriptional regulator of virulence genes in *Listeria monocytogenes*. The PrfA coding sequence is preceded by a 5′UTR which folds into a stable hairpin structure that sequesters the SD at 30°C (Johansson *et al.*, 2002). At 37°C, this structure is destabilized allowing PrfA synthesis and consequently, the virulence factors are made (Johansson *et al.*, 2002). Regulation of PrfA synthesis is unique because the expression of *prfA* mRNA appears to be additionally regulated by a *trans*-acting riboswitch leading to the fine tuning of expression levels of the regulator under specific metabolic conditions (Loh *et al.*, 2009). Indeed, under rich medium conditions and in mid-logarithmic phase, the SAM riboswitch element (SreA) of *L. monocytogenes* was shown to act *in trans* through binding to *prfA* mRNA 80 nucleotides upstream the SD sequence and acts to repress translation by an unknown mechanism (Fig. 6.1). This additional level of regulation also occurs in the absence of binding of SAM. However, the *trans*-acting regulatory effect only occurred at 37°C, when the complementary sequences of PrfA are available for SreA binding. Moreover, the expression of SreA is PrfA-dependent, thereby forming a feedback regulatory loop (Fig. 6.1). This elegant mechanism highlights the importance of mRNA regulatory elements in the expression of virulence factors and reveals a link between virulence and the metabolic state of the cell. Whether other *cis*-acting riboswitch elements regulate *in trans* the translation of other mRNAs remains to be determined.

Cis-acting riboswitches as tractable drug targets

As mentioned above, *cis*-acting riboswitches are found upstream of mRNAs whose products are involved in the biosynthesis or transport of the metabolite they respond to. When the intracellular concentration of the metabolite reaches a critical threshold, the metabolite binds to a specific riboswitch, usually leading to repression of the downstream gene(s), but cases of activation

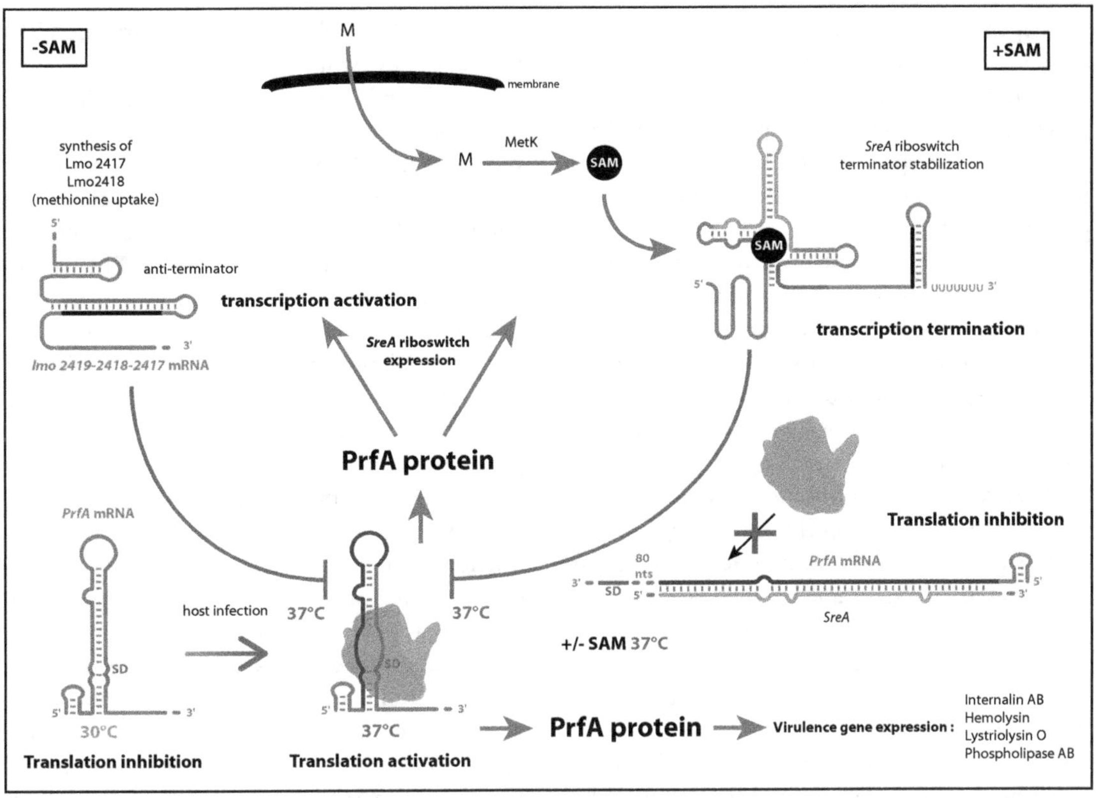

Figure 6.1 Dual regulation of *prfA* mRNA in *L. monocytogenes*. Translation of *prfA* mRNA is inhibited at low temperature (<30°C) because the Shine and Dalgarno sequence (SD) is sequestered into an inhibitory hairpin. When temperature increases to 37°C, the secondary structure 'opens', allowing translation of PrfA. Moreover, PrfA synthesis is regulated *in trans* by a riboswitch responding to the intracellular concentration of S-adenosylmethionine (SAM), SreA, which regulates the expression of the downstream genes (lmo2417–2418) involved in methionine uptake. When nutrients are limiting (low levels of SAM), transcription of the downstream genes takes place owing to the formation of an anti-terminator structure. Conversely, under high levels of SAM, SAM binds to SreA to induce a premature transcription and SreA acts as a sRNA to repress the translation of *prfA* mRNA by binding to the 5′UTR 80 bp upstream of the SD sequence. This interaction is only allowed at high temperature (37°C) where the regulatory sequences of PrfA are exposed for interaction. The complementary regions of SreA and *prfA* mRNA are coloured in blue and purple, respectively. A colour version of this figure is available in the plate section at the back of the book.

also exist. Riboswitches typically consist of a highly conserved aptamer domain (sensor), responsible for metabolite recognition and the expression platform (effector), which mediates the genetic response. In many cases, riboswitches can adopt alternative conformations, and, upon binding to the metabolite, one of the conformers is stabilized to favour termination of transcription and/or translation attenuation, or degradation of the downstream gene(s). Crystallographic studies performed on most of the riboswitch classes reveal that, when the metabolite is bound, the relatively disordered aptamer domain adopts a compact structure that engages in multiple interactions with the metabolite thus providing specificity to the interaction (Serganov, 2010). The riboswitch function can depend on the concentration of the ligand or can be kinetically driven, depending on the elongation speed of RNA polymerase, and the speed at which the metabolite binds to the sensor (Barrick and Breaker, 2007; Zhang *et al.*, 2010). Recently, another riboswitch that responds to alkaline pH conditions was reported in *Salmonella typhimurium* (Nechooshtan *et al.*, 2009). Interestingly this riboswitch must be transcribed *de novo* to allow mRNA translation at alkaline pH. This pH-dependent translational regulation relies on specific transcriptional pausing, which facilitates the active conformational state of *alx* mRNA to promote ribosome binding and translation. Influencing the speed of RNA polymerase thus has direct consequences on RNA folding and the genetic response.

Although a large number of riboswitches control the synthesis of enzymes involved in essential metabolic pathways, some of them as the second messenger cyclic di-guanosine monophosphate (c-di-GMP) trigger a wide-range of physiological changes such as the conversion between motile and biofilm lifestyles, and virulence (Sudarsan *et al.*, 2008). In *Vibrio cholerae*, c-di-GMP binds to several 5′-UTR of mRNAs and regulates the expression of proteins known to be required for bacterial infection and mammalian intestine colonization (Sudarsan *et al.*, 2008). The crystal structure of the aptamer bound to c-di-GMP from *V. cholerae* has been solved revealing that the symmetric ligand is recognized asymmetrically with both the guanines and the backbone. The c-di-GMP binds into a cleft formed by a three-helix junction that involves basepairing and extensive base stacking (Kulshina *et al.*, 2009; Smith *et al.*, 2009). Surprisingly, in *Clostridium difficile*, it was demonstrated that a GTP-dependent self-splicing ribozyme and a riboswitch aptamer responding to c-di-GMP cooperate to function as an allosteric RNA to promote splicing (Lee *et al.*, 2010). Binding of the c-di-GMP to the aptamer domain of the 5′UTR induces folding changes at a splice site junction to yield a processed form of the mRNA. Consequently, a typical Shine and Dalgarno sequence became positioned at an optimal distance of the start codon to induce translation of the downstream gene encoding a putative virulence factor (Lee *et al.*, 2010). Obviously these tandem riboswitches generate more complex gene control elements leading to a variety of responses according to the signalling molecules (Sudarsan *et al.*, 2006; Lee *et al.*, 2010).

As the need for novel drug development became apparent, several efforts have focused on designing antibiotics with restricted specificity to minimize the emergence of antibiotic resistance. Bacterial riboswitches appear to be tractable drug targets, as they often regulate functions that are essential for cell growth or for virulence. The specificity by which aptamers recognize their cognate ligands made it possible to design ligand analogues, non-metabolizable by the cell, that can recognize the riboswitch and repress growth even in the absence of the natural ligand. Several recent studies explored the potential of riboswitches as drug targets. Namely, analogues to TPP, lysine, FMN and guanine-dependent riboswitches have been designed and tested for their antibacterial potential (Blount and Breaker, 2006; Mulhbacher *et al.*, 2010b). A well-known antibiotic, pyrithiamine, was found to act by inhibiting expression of genes regulated by the TPP riboswitch (Sudarsan *et al.*, 2005). In a more recent study, a pyrimidine analogue was found to inhibit bacterial growth due to binding to a guanine riboswitch upstream of the *Staphylococcus aureus guaA* gene, which encodes a GMP synthetase, and thereby affecting cellular nucleotide synthesis. The outcome was specifically mediated through the riboswitch and was not due to off-side effects. The same compound was shown to reduce the bacterial load in a murine mastitis model, and *S. aureus* did not develop mutations after successive passages in the presence of the compound (Mulhbacher *et al.*, 2010a). As this compound was highly effective against *S. aureus* and *Clostridium difficile*, two pathogenic bacteria that have evolved multiple drug resistance, this strategy holds promise for the design of future antimicrobial drugs. Because the purine riboswitch exists in few bacteria, it minimizes any selective pressure for resistance in non-targeted bacteria.

Trans-acting sRNAs targeting mRNAs and their impact on virulence

Most regulatory RNAs act by an antisense mechanism, which requires base-pairings to target mRNAs and effect up- or down-regulation of gene expression at the post-transcriptional level (Figs. 6.2 and 6.3). Binding to a target mRNA often inhibits translation and/or facilitates mRNA decay, but it can also activate translation. Two classes of RNAs acting through base-pairings with mRNA targets can be distinguished based on their genomic localization and the extent of base-pairings with mRNA targets. The first class involves the so-called antisense RNAs (asRNA), which form extended perfect matches to mRNAs because these asRNAs are expressed from the gene strand complementary to the mRNA target. These asRNAs, in general, regulate the expression of a single mRNA target. The second class of regulatory RNAs forms imperfect base-pairings with mRNA targets because the two RNAs are encoded by different gene loci. Many of these RNAs behave as key effectors of adaptive responses, including adjustment to environmental cues, stress responses and virulence by coordinating the expression of numerous mRNAs encoding functionally related proteins. Examples of sRNAs regulating the synthesis of virulence factors, cell invasion or colonization in an animal model are given in Table 6.1.

Fully complementary asRNAs: mobile elements, virulence

Small asRNAs regulate essential functions of genetic mobile elements, such as plasmids, transposons, pathogenicity islands, and bacteriophages (reviewed in Wagner *et al.*, 2002; Felden *et al.*, 2011; Brantl, 2007). These elements are important motors of genetic variability but some can also compromise genome integrity. This is why transposition is tightly regulated, and asRNAs have been shown to control transposase synthesis in various bacteria (Brantl, 2007). Another class of small asRNAs represses the synthesis of potentially small hydrophobic and toxic peptides, the so-called type I toxin–antitoxin system (reviewed in Fozo *et al.*, 2008). The small asRNAs are very efficient repressors and recognize their target RNAs in an Hfq-independent manner (Brantl, 2007). In this case, the structures of the sense RNA and asRNAs are critical to promote fast binding, which has a direct consequence on the repression efficiency (Wagner *et al.*, 2002). Recent high-throughput sequencing and tiling array approaches uncovered a hitherto unappreciated amount of large antisense transcription in several pathogens including *Helicobacter pylori* (Sharma *et al.*, 2010), *Vibrio cholerae* (Liu *et al.*, 2009), *L. monocytogenes* (Toledo-Arana *et al.*, 2009) and *S. aureus* (e.g. Beaume *et al.*, 2010). Although one could speculate that some of these asRNAs originate from transcriptional interference or from an RNA quality control mechanism, the conditions that govern their synthesis, their physiological role and mechanism of action remain largely unknown. A recent report in *S. enterica* identified a novel 1.2-kb-long asRNA, called AmgR, which is complementary to the *mgtC* mRNA encoding the MgtC protein, which is necessary for survival within macrophages, virulence in mice and growth in low Mg^{2+} (Lee and Groisman, 2010). The expression of AmgR decreased MgtC protein levels in an RNase E-dependent process. Strikingly, inactivation of the chromosomal *amgR* promoter enhanced *Salmonella* virulence, and *amgR* transcription was shown to be governed by PhoP, which also directs transcription of the sense mRNA. This novel mechanism would indicate that AmgR functions as a timing device, controlling virulence depending on the metabolic state of the bacteria, i.e. Mg^{2+} concentration. Finally, several overlapping 3′-UTRs or 5′-UTRs of convergent or divergent mRNAs were detected in *S. aureus* (Felden *et al.*, 2011) and *L. monocytogenes* (Toledo-Arana *et al.*, 2009) in which the UTR of one mRNA overlaps the mRNA on the opposite strand. Whether these overlapping UTRs affect gene expression is not yet known.

Dual functions of sRNAs targeting mRNAs regulate quorum sensing and virulence

Many pathogenic bacteria extensively control the expression of virulence genes at the transcriptional level, e.g. by two-component systems. However, many of these two component systems also activate the synthesis of sRNAs, which in turn regulate virulence gene expression at the post-transcriptional level. The involvement of regulatory proteins and sRNAs in complex regulatory networks was particularly well demonstrated in the control of the quorum sensing system in *V. cholerae* and *S. aureus*. The quorum sensing system engages the bacteria in cell-to-cell communication through the production, release, and detection of extracellular signalling molecules that in turn drive gene expression programs in a collective manner, such as virulence factor production and biofilm formation. In *S. aureus*, this cascade of events is

Table 6.1 Selected examples of *trans*-acting regulatory RNAs and their respective targets with an established role in virulence

Bacterial species/sRNA	Target	Target function	Phenotype in cell culture or animal model	Regulatory mechanisms	Regulation of sRNA expression or characteristics	References
C. perfringens VR RNA	*colA*	Kappa-toxin		Antisense regulation, stabilization of mRNA	Regulated by the ViR/VirS system	Shimizu *et al*. (2002), Ohtani *et al*. (2003), Obana *et al*. (2010)
	cpd	2′,3′-Cyclic nucleotide phosphodiesterase Alpha-toxin		ND		
L. monocytogenes						
rilB	*lmo2104*	Ferrous iron transport protein FeoA	Mutant showed increased colonization of liver in mice	Antisense regulation?, activation of translation	Carries five long sequence repeats	Mandin *et al*. (2007), Toledo-Arana *et al*. (2009
	lmo2105	Ferrous iron transport protein FeoB		Antisense regulation?, activation of translation		
rliE[1]	*comEA*	Competence factors		Antisense regulation[1], mRNA degradation		
	comFA	Competence factors		Antisense regulation, mRNA degradation		
	lmo0945	Competence factors		Antisense regulation, mRNA degradation		
rliI	*lmo1035*	Phosphotransferase system		Antisense regulation, mRNA degradation		
rli23, rli25, rli35[1]	*Lmo0172, 0330, 0828*	Transposase		Antisense regulation[1]		
rli29	*lmo0471*	Unknown		sRNA complementary to the 5′-UTR	ND	
S. aureus						
RNAIII	*hla*	Alpha-haemolysin	Mutant showed attenuated virulence in endocarditis rabbit model of infection	Antisense regulation, activation of translation	Regulated by the *agr* system	Cheung *et al*. (1994), Novick and Muir (1999), Huntzinger *et al*. (2005), Geisinger *et al*. (2006), Boisset *et al*. (2007), Geissmann *et al*. (2009), Chevalier *et al*. (2010)
	rot	Repressor of toxin rot		Antisense regulation, repression of translation and RNase III-dependent degradation	UCCC sequence motif of RNAIII used as a mRNA recognition signal	
	spa	Protein A				

	sa1000	Fibrinogen-binding protein				
	lytM	Peptidoglycan hydrolase				
	coa	Coagulase				
SprD	*sbi*	Immune-evasion molecule	Virulence of the mutant strain abolished in mice	Antisense regulation, inhibition of translation	Pathogenicity island encoded sRNA	Chabelskaya *et al*. (2010)
S. pyogenes						
fasX	*fbp54*	Fibronectin-binding protein	Mutant showed decreased adherence and internalization to human epithelial cells	ND	Regulated by amino-acid starvation, the FasBAC and *luxS* systems	Kreikemeyer *et al*. (2001), Steiner and Malke (2001), Klenk *et al*. (2005) Siller *et al*. (2008), Ramirez-Pena *et al*. (2010)
	mrp	Fibrinogen-binding protein		ND		
	Ska	Streptokinase		Antisense regulation, stabilization of mRNA		
	SLS	Streptolysin S		ND		
pel	*emm*	M protein		ND	Regulated by cell-density mechanism	Mangold *et al*. (2004)
	sic	Inhibitor of complement		ND		
	SpeB	Cysteine protease		ND		
rivX	*emm*	M protein		Dependent on RivR, ND	Controlled by the regulator CovR	Roberts and Scott (2007)
	fbpA	Fibronectin-binding protein				
	grm	Gene-regulated by Mga				
	Mga	Response regulator				
	scpA	C5a peptidase				
	scrA	Phosphotransferase system				
	speB	Cysteine protease				
S. Typhimurium						
mgrR	*eptB*	LPS modification		Antisense regulation, repression of translation and degradation	Regulated by PhoP	Moon and Gottesmann (2009)
invR	*ompD*	Porin synthesis		Antisense regulation, repression of translation and degradation	Regulated by HilD, encoded by SPI-1	Pfeiffer *et al*. (2007)
amgR[1]	*mgtBC*	Magnesium homeostasis/ virulence	Mutant showed enhanced virulence in mice	Antisense regulation[1], mRNA degradation	Regulated by PhoP	Lee and Groisman (2010)

Table 6.1 Continued

Bacterial species/sRNA	Target	Target function	Phenotype in cell culture or animal model	Regulatory mechanisms	Regulation of sRNA expression or characteristics	References
isrC	*msgA*	Virulence		Antisense regulation, repression of translation and degradation		Padalon-Brauch *et al*. (2008)
csrB, csrC	CsrA		Double mutant showed defect in epithelial cell invasion	Sequestration of protein	Regulated by SirA	Fortune *et al*. (2006)
V. cholerae						
Qrr1–4	*hapR, vca0939, aphA*	Quorum sensing		Antisense regulation, repression of translation and degradation	Regulated by LuxO, s54, AphA, LuxR	Lenz *et al*. (2004), Hammer and Bassler (2007), Steven *et al*., (2011)
csrB, csrC, csrD	CsrA	Quorum sensing		Sequestration of protein	Regulated by VarS-VarA	Lenz *et al*. (2005)
E. carotovora						
rsmB	RsmA	Exopolysaccharide production		Sequestration of protein	Regulated by GacS-GacA	Cui *et al*. (2008)
L. pneumophila						
rsmY, rsmZ	RsmA		Double mutant showed defect in growth in macrophages	Sequestration of protein	Regulated by LetA	Rasis and Segal (2009), Sahr *et al*. (2009)
P. aeruginosa						
PhrS	*pqsR*	Quorum sensing regulator, virulence		Antisense regulation, activation of translation	Regulated by ANR	Sonnleitner *et al*. (2011)
crcZ	Crc	Catabolite repression		Sequestration of protein	Regulated by CbrA-CbrB	Sonnleitner *et al*. (2009)
rsmY, rsmZ	RsmA	Type VI secretion		Sequestration of protein	Regulated by GacS-GacA	Brencic *et al.* (2009), Moreno *et al.* (2011) for *P. putida*
Y. pseudotuberculosis						
csrB, csrC	CsrA	Virulence		Sequestration of protein	Regulated by BarA-UvrY	Heroven *et al*. (2008)

[1]cis-encoded asRNAs, i.e. sRNAs encoded in the same locus as their target mRNA but on the opposite strand.
ND, not determined.

orchestrated by the accessory gene regulator *agr* system, primarily via its two main intracellular effectors, the response regulator AgrA and the multifunctional RNAIII (reviewed in Novick and Geisinger, 2008; Queck *et al.*, 2008; Felden *et al.*, 2011). The expression of RNAIII accumulates at high cell density and coincides with a drastic change in gene expression. Because the *agr* system and RNAIII have impact on virulence factor expression, it was thought to be of importance for virulence, notably during acute infection (Felden *et al.*, 2011). Indeed an *agr* mutant of *S. aureus* causes an attenuated virulence in the rabbit model of endocarditis (Cheung *et al.*, 1994). RNAIII is stable and has the unique property of acting as an mRNA encoding the 26 aa delta-haemolysin peptide, as a repressor of early expressed virulence factors and as an activator of a large number of exotoxins (Fig. 6.2A). The elucidation of the RNAIII secondary structure has largely contributed to discovering its mechanism of action (Benito *et al.*, 2000). This structure reveals independent structural domains and three hairpin domains carrying a redundant CCCU sequence motif. The 5′-UTR of RNAIII positively regulates *hla* translation (encoding alpha haemolysin) by competing directly with an intramolecular RNA secondary structure that sequesters the *hla* ribosomal binding site (RBS) (Fig. 6.3A). The large 3′-UTR coordinates the repression of the synthesis of early expressed virulence factors (i.e. coagulase, protein A, fibrinogen-binding protein SA1000) and the repressor of toxins Rot at the post-transcriptional level by a shared mechanism that does not require the RNA chaperone protein Hfq (Fig. 6.2A): RNAIII anneals to target mRNAs, represses translation initiation and induces mRNA degradation that is triggered by the double-strand-specific RNase III (Boisset *et al.*, 2007). Depending on the location and the structure of the mRNA regulatory sites, the structures of the mRNA–sRNA complexes vary considerably. For instance, RNAIII forms an imperfect duplex with *coa* mRNA that sequesters the RBS stabilized by a loop–loop interaction involving the coding region of the mRNA, whereas it forms two loop–loop interactions with *rot* mRNA (Fig. 6.2A). With the exception of *hla* translational activation, all the direct effects of RNAIII lead to the repression of mRNA targets. However, as Rot is a transcriptional regulatory protein, its repression by RNAIII results in indirect effects on the transcription of many genes, including activation of many exotoxins and repression of adhesins. These complex regulatory circuits involve several coherent feed-forward loops in which RNAIII and Rot jointly regulate the activation and the repression of several virulence factors (protein A, coagulase, alpha-haemolysin, etc.) at both the transcriptional and post-transcriptional levels. These double controls promote dynamic complexity and suggest that the amount and timing of production of virulence factors is precisely controlled during the course of infection. However, it will be interesting to use computational approaches to analyse how the RNAIII-based regulation determines the dynamics of the feed-forward loops and how these dynamics shape the bacterial responses at high cell density (Beisel and Storz, 2010). We predict that the list of RNAIII targets will probably grow because the *agr* system is known to regulate an impressive list of other processes that affect biofilm formation, peptidoglycan metabolism, and transport pathways. With the discovery of novel *S. aureus* sRNAs, we speculate that other sRNAs could participate in these regulatory networks to counteract or to act in a synergistic manner with RNAIII (reviewed in Felden *et al.*, 2011).

In contrast to RNAIII, four and five redundant Qrr sRNAs from *V. cholerae* and *V. harveyi*, respectively, were shown to play major functions at low cell density by regulating the synthesis of two master quorum-sensing regulatory proteins, which operate either at the low cell density (AphA) or at high cell density (LuxR in *V. harveyi*/HapR in *V. cholerae*). These are perhaps the most complex examples of sRNA-mediated regulatory circuits that regulate quorum sensing and that are characterized by several feedback loops (Lenz *et al.*, 2004). At low cell density, phosphorylation of LuxO activates the expression of the four Qrr RNAs, which in turn promotes the translation of *aphA* mRNA and represses the translation of *luxR* mRNA (Fig. 6.2B). In addition, AphA represses the transcription of *luxR* and the Qrr RNAs. Conversely, at high cell density, high concentrations of the autoinducer reverse the phosphate flow leading to the dephosphorylation of LuxO, and

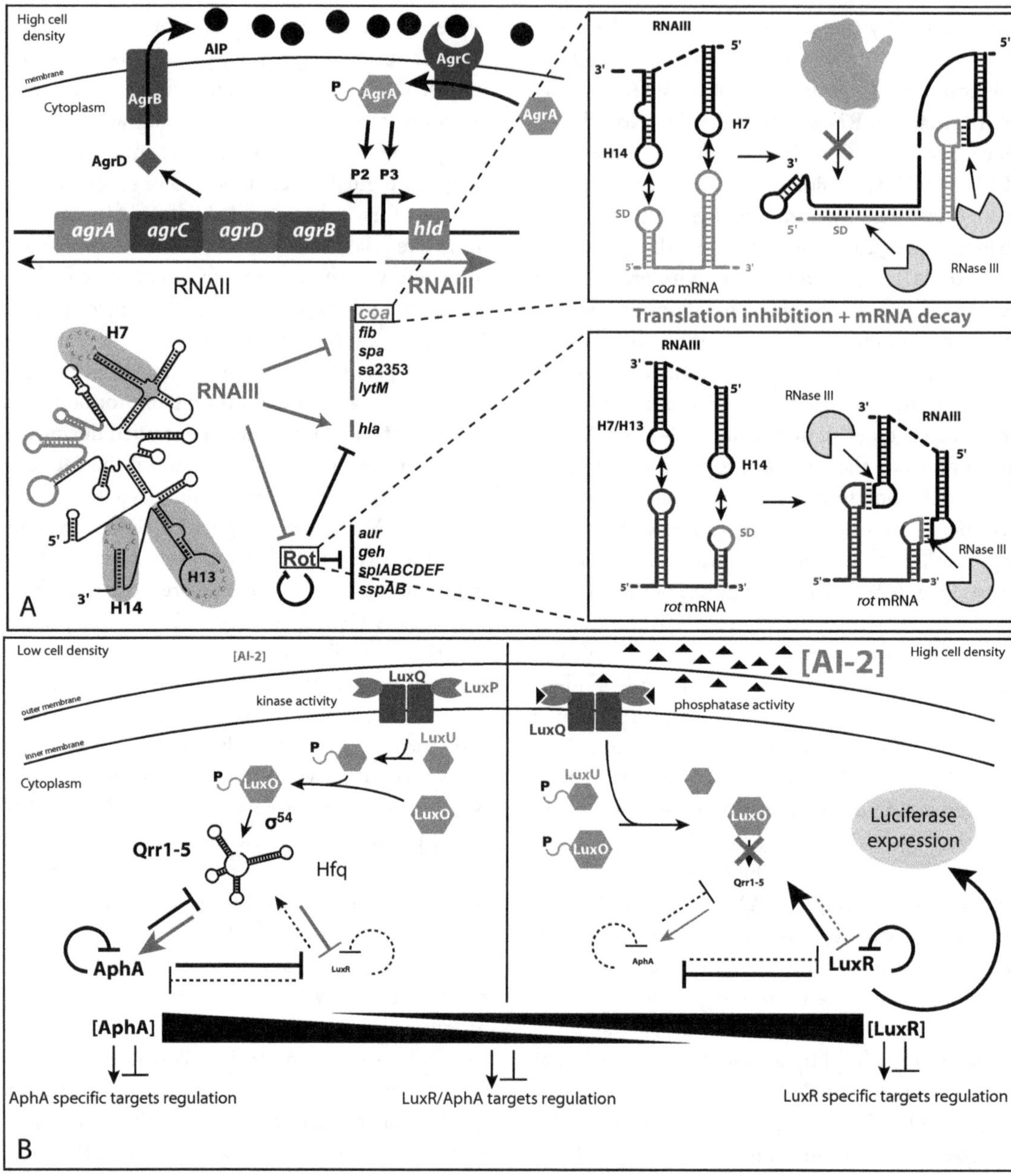

Figure 6.2 Regulatory networks in virulence gene expression. (A) Quorum sensing regulates virulence gene expression in *S. aureus*. Cell population density is sensed by the *agr* system. When population reaches a critical threshold, AgrB (red), a membrane peptidase matures the auto-inducing peptide AIP encoded by AgrD (grey), which is then secreted in the extracellular environment. Upon threshold concentration of AIP, autophosphorylation of the histidine kinase AgrC (purple) takes place and through a phosphorelay the response regulator, AgrA (light blue) is activated (Geisinger and Novick, 2008). AgrA autoactivates the transcription of its own operon and of RNAIII, which encodes delta-haemolysin (in green). The secondary structure of RNAIII is shown. The 3′UTR (in black) of RNAIII acts as an antisense RNA to repress the synthesis of adhesins (coagulase, protein A, SA1000), of amidases and hydrolases (SA2353, LytM) and of the repressor of toxins Rot, and the 5′UTR (red) activates the translation of *hla* mRNA (see Fig. 6.3). For repression, RNAIII uses two hairpin domains to bind *coa* and *rot* mRNAs. The hairpins H7 and H14 of RNAIII bind to *coa* mRNA to form a duplex stabilized by a loop–loop interaction (upper insert) while they form two loop–loop interactions with the 5′UTR of *rot* mRNA (bottom insert). In both cases, the duplexes prevent

consequently, the synthesis of the four Qrr sRNAs is stopped (Fig. 6.2B). This in turn blocks AphA translation and activates the synthesis of LuxR/HapR. The interesting feature is that LuxR and AphA mutually repress their expression, and, in addition, the LuxR protein activates the synthesis of Qrr sRNAs. Furthermore, AphA and LuxR/HapR are transcriptional regulators that affect the expression of hundreds of genes involved in virulence, biofilm formation, and bioluminescence, and both regulons overlap (Rutherford *et al.*, 2011). Based on these studies, an elegant model was proposed, in which interplay between the transcriptional regulatory proteins and Qrr sRNAs generates a reciprocal gradient of AphA and LuxR expression that establishes the quorum sensing gene expression patterns at low and high density levels. This also provides more possibilities for the cell to generate sophisticated regulatory programs in response to various cell densities. In contrast to *S. aureus* RNAIII, the same region of the Qrr sRNAs binds to the 5′-UTR far upstream the SD sequence of *aphA* mRNA to activate translation, and binds to the ribosome binding site of *luxR* mRNA to repress translation. Thus, activation and repression depend, in this case, on the location of the regulatory regions on the target mRNAs. Repression requires the RNA chaperone protein Hfq to facilitate the binding of Qrr RNAs to *luxR* mRNA, to occlude ribosome binding, and to promote mRNA degradation. For translation activation of *aphA* mRNA, it was predicted that the Qrr sRNA would prevent the formation of an inhibitory hairpin structure that would sequester the RBS of *aphA* mRNA. Thus, binding of Qrr sRNAs to the 5′-UTR of *aphA* mRNA would favour the recruitment of the 30S subunit to its RBS, which is a mechanism that is referred to as anti-antisense RNA regulation and is similar to the mechanism by which *S. aureus* RNAIII activates the translation of *hla* mRNA (Fig. 6.3A). It is intriguing that in *V. cholerae* and *V. harveyi*, the number of copies of the Qrr sRNAs generates different effects on the regulatory responses. In *V. harveyi*, the five sRNAs are not redundant; instead, they act in an additive manner to translate the signal and thus contribute to the generation of a precise gradient for the repression of *luxR* mRNA. In contrast, in *V. cholerae*, the four Qrr sRNAs are fully redundant because all four sRNAs must be deleted to observe an effect on *hapR* mRNA stability, and this effect is directly linked to a gene dosage compensation mechanism (Lenz *et al.*, 2004). These behaviours might be linked to the adaptation of the two bacteria to different niches. For instance, *V. cholerae* colonizes the human intestinal tract, while *V. harveyi* does not.

In Gram-negative bacteria, networks of sRNAs regulate the expression of outer membrane proteins (OMP), some of which are highly immunogenic and play key roles in host–pathogen interactions. In *E. coli* and *S. typhimurium*, a network of sRNAs associated with Hfq represses the translation of several *omp* mRNAs that are subsequently degraded via endoribonuclease E (Guillier *et al.*, 2006; Papenfort and Vogel, 2009). In addition, another Hfq-dependent sRNA in *E. coli* (MgrR) represses the synthesis of the enzyme EptB, which modifies lipopolysaccharides anchored to the outer membrane (Moon and Gottesman, 2009). This sRNA is induced by the

ribosome binding and recruit the RNase III for mRNA degradation. The drawing was adapted from Boisset *et al.* (2007). (B) Several sRNAs regulate QS in *V. harveyi* and *V. cholerae*. At low cell density (LCD), in the absence of the autoinducer (AI-2), the histidine kinase-receptor LuxPQ phosphorylates LuxU (green), which in turn phosphorylates LuxO (light blue). LuxO-P together with the RNAP-σ54 holoenzyme up-regulate expression of 5 sRNAs, Qrr1–5. These sRNAs activate the synthesis of the master regulator AphA through an anti-antisense mechanism and repress the translation of *luxR* mRNA. At high cell density (HCD), and high concentrations of AI-2, the LuxPQ receptor acts as a phosphatase and de-phosphorylates LuxO, thus repressing synthesis of Qrr sRNAs. Under these conditions, LuxR is de-repressed. Maximal levels of AphA or LuxR are attained in LCD or HCD, respectively. A gradient of the two intracellular regulator concentrations is established to provide a large panel of responses. Black and red bars denote transcriptional and post-transcriptional repression, respectively; black and red arrows denote transcriptional and post-transcriptional activation, respectively. The drawing is adapted from Rutherford *et al.* (2011). A colour version of this figure is available in the plate section at the back of the book.

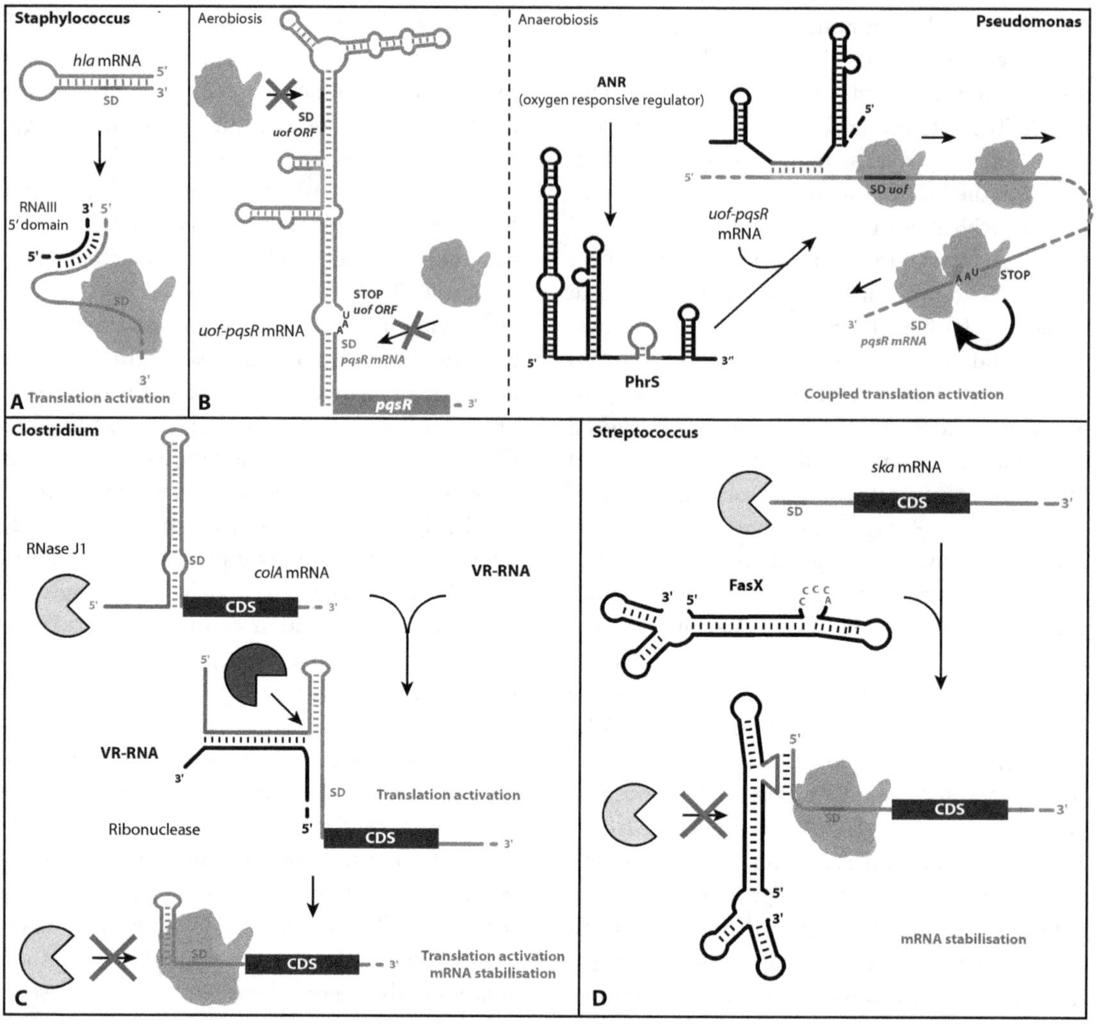

Figure 6.3 sRNA-dependent activation of translation. (A, B) Activation by an anti-antisense mechanism. (A) *S. aureus* RNAIII binds to the 5′UTR of *hla* mRNA thereby preventing the formation of an inhibitory structure. SD is for Shine and Dalgarno sequence. (B) A similar mechanism exploited by PhrS sRNA in *P. aeruginosa*. PhrS sRNA binding induces translation of the quorum-sensing regulator PqsR through activation of translation of an upstream ORF (uof) which is translationally coupled to PqsR. Expression of PhrS is induced by ANR, an oxygen responsive regulator, at high cell density population when oxygen becomes limiting. The drawing is adapted from Sonnleitner *et al.* (2011). (C, D) Activation by stabilization of mRNA structure: (C) In *C. perfringens* VR-RNA binds to the 5′UTR of *colA* mRNA creating a specific site for an unknown RNase (red circle). As a result, a short hairpin structure is generated at the 5′ end, and the SD sequence becomes accessible to the ribosome (Obana *et al.*, 2010). (D) In *S. pyogenes*, FasX binds at the 5′ end of *ska* mRNA encoding streptokinase 30 nts upstream of the SD sequence (Ramirez-Peña *et al.*, 2010). In both cases, the mRNAs are stabilized by preventing access of the RNase J1 (yellow circle). A colour version of this figure is available in the plate section at the back of the book.

two-component system PhoP/PhoQ, which regulates Mg^{2+} homeostasis, resistance to antimicrobial peptides and LPS synthesis (Kato and Groisman, 2008). OMP proteins and LPS determine the cell envelope integrity and represent a selective barrier that protects the bacteria against effectors of the immune system of the host, such as antimicrobial peptides. Although the composition of the membrane is different in Gram-positive bacteria, recent observations suggest that *S. aureus* RNAIII also contributes to cell wall integrity by regulating peptidoglycan metabolism at high cell densities. Deregulation of the amidases and hydrolases changes the properties of the cell wall (Lioliou, Fechter *et al.*, personal communication). Finally, other sRNAs were shown to mediate cross-talks between the core genome and horizontally acquired pathogenicity islands (reviewed in Papenfort and Vogel, 2010; Felden *et al.*, 2011). For instance, *S. aureus* SprD RNA, which is expressed from the pathogenicity island φ, represses translation initiation of the *sbi* mRNA, which encodes an immune-evasion molecule located in the core genomic locus (Chabelskaya *et al.*, 2010). In *S. typhimurium*, the SPI-1 gene, encoding InvR RNA, controls the synthesis of *ompD* mRNA (Pfeiffer *et al.*, 2007). Thus, it is tempting to propose that such regulatory RNAs, which bring a benefit for the genome, might also contribute to the control of the acquisition and mobility of horizontally acquired virulence factors.

Novel mechanism for sRNA-dependent activation of gene expression

Conceptually, there is no major difference between the ways the sRNAs bind to mRNAs to repress or to activate translation. In general, activator sRNAs bind to mRNA in the 5′-UTR omit far of the RBS while repressor sRNAs interact within the RBS to occlude ribosome binding. However, several recent examples have shown that repressor sRNAs can also bind far upstream of the RBS to inhibit translation. The mRNA signals that lead to repression or activation are not yet fully understood (Darfeuille *et al.*, 2007; Holmqvist *et al.*, 2010). One of these factors concerns mRNA structure. Indeed, the action of many activator sRNAs relies on a competition between the formation of a duplex with the mRNA and an intramolecular structure of the mRNA that sequesters the RBS. Hence, in this so-called anti-antisense mechanism, the sRNA liberates the RBS to facilitate ribosome binding and translation, as *S. aureus* RNAIII does for activating of the translation of the *hla* mRNA encoding α-haemolysin (Fig. 6.3A). Even more complex, the sRNA PhrS from *Pseudomonas aeruginosa* activates the translation of the quorum sensing regulatory protein PqsR via a leader peptide. PhrS belongs to the bifunctional sRNAs (Wadler and Vanderpool, 2007), which act through an antisense mechanism and simultaneously encode a peptide, although the function of this peptide is still not known. PhrS binds to the 5′-UTR of a short open reading frame to which the *pqsR* gene is translationally coupled (Fig. 6.3B). The short sRNA–mRNA interaction activates the translation of the short leader peptide most likely by preventing the formation of an inhibitory hairpin structure, which sequesters the translational coupling signal (Fig. 6.3B). Interestingly, expression of PqsR is under the control of the oxygen-responsive regulator ANR, thereby linking oxygen availability and cell density. Such a control might be of importance for biofilm formation (Sonnleitner *et al.*, 2011).

Another mechanism, that has recently emerged, is the fact that sRNAs activates virulence gene expression primarily through mRNA stabilization (Fröhlich and Vogel, 2009). In *Clostridium perfringens*, the RBS of *colA* mRNA, which encodes collagenase, is sequestered into a large and stable hairpin structure. In this mechanism, VR-RNA utilizes its 3′ end to bind to the 5′ end of *colA* mRNA and form a 35-bp irregular helix (Fig. 6.3C). This helix directs a specific endoribonuclease cleavage site just downstream of the RNA duplex, which generates a shorter mRNA starting with a smaller but stable stem–loop structure and followed by an accessible RBS (Obana *et al.*, 2010). The absence of unpaired residues at the 5′ end most probably prevents degradation from the 5′–3′ exoribonuclease RNase J1. In addition, binding of the ribosome to the shorter mRNA contributes to mRNA stabilization (Fig. 6.3C). The dramatic increase of the levels of *colA* mRNA correlates nicely with the synthesis of ColA protein, suggesting that mRNA stabilization is the main cause of the protein synthesis activation. In the second example, the sRNA FasX, from the group A *Streptococcus pyogenes* (GAS), activates the expression of the virulence factor streptokinase (SKA) by increasing the level of *ska* mRNA (Kreikemeyer *et al.*, 2001; Ramirez-Peña *et al.*, 2010). SKA, the secreted streptokinase is a key factor of *Streptococcus* and aids the dissemination of the pathogen by activating the host plasminogen. FasX binds to *ska* mRNA at its 5′-end at a UCCC sequence motif,

30 nucleotides upstream of the RBS (Fig. 6.3D). Interestingly, this short interaction is similar to the sequence motif used by *S. aureus* RNAIII and other sRNAs to bind target mRNAs (Fig. 6.2A; Boisset *et al.*, 2007; Geissmann *et al.*, 2009). This interaction causes a dramatic increase of SKA and correlated with enhanced mRNA steady state levels (Ramirez-Peña *et al.*, 2010). The discovery of the dual-functioning RNase J1 in Gram-positive bacteria, which specifies both an endoribonuclease activity and the 5′ to 3′ exoribonuclease activity, might explain why perturbation of the 5′ end of mRNA has direct consequences on mRNA stability (Condon and Bechhofer, 2011). In contrast, some sRNAs have been shown to work in a reciprocal manner to FasX by competing with the formation of such a stabilizing hairpin structure. This is, for instance, true in the case of RNAIII from *S. aureus*, which represses the translation and induces rapid degradation of the stable *spa* mRNA triggered by RNase III (Huntzinger *et al.*, 2005). Thus, regulation of mRNA stability through sRNA–mRNA pairings is more common than expected for both activation and repression mechanisms (Condon and Bechhofer, 2011). Whether such mechanisms exist in Gram-negative bacteria, which have evolved different RNases, remains to be addressed. These examples illustrate the intimately connections between regulatory RNAs and ribonucleases in RNA regulation.

Regulatory circuits through sRNA targeting of regulatory proteins

Although the 'traditional' mode of action of a sRNA is regulation of expression via base-pairing with an mRNA, certain sRNAs exert their function by sequestering RNA binding proteins (Fig. 6.4). A well-characterized example involves members of the CsrB family of sRNAs of *E. coli* antagonizing the activity of the major carbon storage regulator protein CsrA. *E. coli* CsrA and its homologue RsmA (repressor of secondary metabolites) in *Pseudomonas aeruginosa* regulate the expression of proteins involved in carbon metabolism, peptide transport, biofilm formation, quorum sensing and virulence (e.g. Babitzke and Romeo, 2007; Bejerano-Sagie and Xavier, 2007; Brencic and Lory, 2009; Table 6.1). The proteins are homodimeric and act at the post-transcriptional level by binding the RBS of mRNAs and preventing translation. The solution structure of a complex formed between *Pseudomonas fluorescens* RsmE (homologue of CsrA/RsmA), and a target RNA, encompassing the ribosome-binding site of a target mRNA, has been solved by NMR (Schubert *et al.*, 2007) and revealed the molecular basis of the recognition and a minimal RNA binding site characterized by the sequence A/UCANGGANGU/A. The homodimeric protein makes base-specific contacts with the GGA sequence, forming a loop, and with the phosphate backbone of the small helix (Fig. 6.4). This structure of the complex explains how, by sequestering the Shine–Dalgarno sequence, the RsmA/CsrA proteins repress translation. In most cases, the inhibition of translation is subsequently followed by the degradation of the mRNA. This is reminiscent of the mechanism by which many sRNAs, which bind to the SD sequence of target mRNAs, repress their translation. Hence, sequestration of the SD sequence can be directed through specific RNA-protein or RNA–RNA interaction. CsrA (RsmA) can also act as an activator, but the mechanism of this outcome has not been elucidated yet. The *E. coli* sRNAs of the CsrB family (RsmY-Z of *P. aeruginosa*) contain 22 repeats of the GGA motif, which is essential for recognition by CsrA and is located in the loop of a hairpin. Each sRNA can bind approximately 9 CsrA dimers, thus titrating out CsrA from its mRNA targets and inactivating it. CsrA has been shown to up-regulate the expression of CsrB sRNAs through a two-component system (BarA/UvrY or the homologous GacS/GacA), thereby producing an autoregulatory circuit (Babitzke and Romeo, 2007; Timmermans and Van Melderen, 2010). Recently it has been shown that this two-component system responds to Krebs cycle intermediates and weak acids, such as formic and acetic acid (Chavez *et al.*, 2010). In the human pathogen *P. aeruginosa*, the expression of the two homologous sRNAs RsmY and RsmZ is regulated by a complex regulatory cascade involving multiple signalling pathways (Goodman *et al.*, 2004; Ventre *et al.*, 2006; Bordi *et al.*, 2010). Interestingly, the sensor RetS and the histidine phosphotransfer protein HptB belong to two

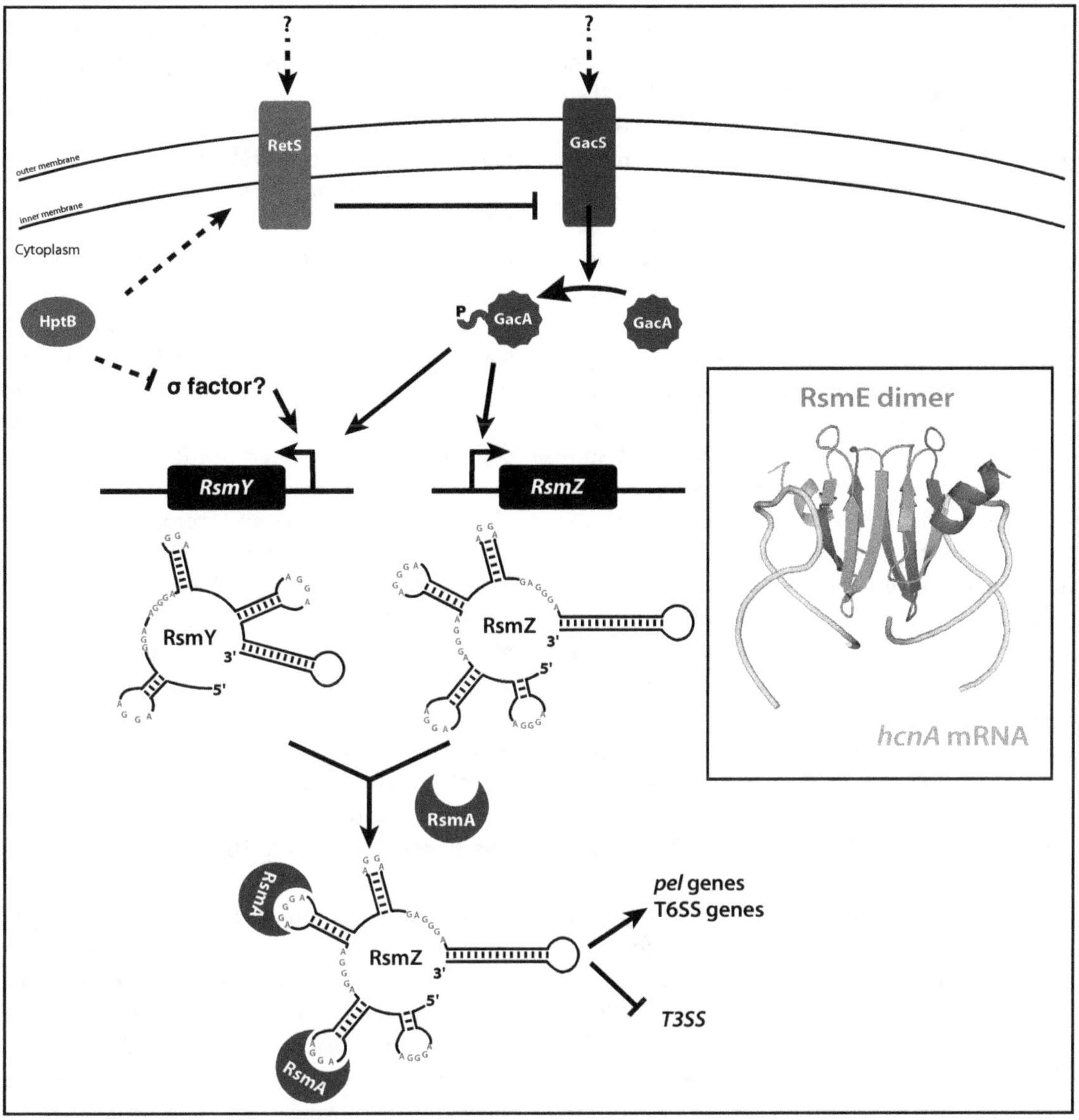

Figure 6.4 Complex regulatory networks controlling virulence gene expression in *P. aeruginosa*. Regulation of expression of RsmY/RsmZ sRNAs is mediated through a two-component system characterized by the histidine kinase GacS (violin) and the response regulator GacA (red). The autophosphorylation of GacS (purple) is inhibited by the sensor protein, RetS (green). The histidine phosphotransfer protein HptB (blue) up-regulates the synthesis of RsmY (Bordi *et al.*, 2010). The synthesis of RsmY/RsmZ out titrates RsmA protein (violin) from its mRNA targets, resulting in activation of *pel* and T6SS genes and repression of T3SS genes. The sRNAs carry a redundant AGGA sequence mimicking the Shine and Dalgarno sequence of the target mRNAs. Inner panel: The solution structure of the complex between RsmE (RsmA homologue) of *P. fluorescens* and a 20-nt fragment of *hcnA* mRNA encompassing the RBS is shown (Schubert *et al.*, 2007). A colour version of this figure is available in the plate section at the back of the book.

different signalling pathways that converge to the two-component system GacS-GacA, known to activate the synthesis of RsmY/Z (Fig. 6.4). RetS interferes directly with the autophosphorylation of GacS and in turn represses the synthesis of the two RNAs, whereas HptB exclusively controls the synthesis of RsmY (Bordi *et al.*, 2010). Thus, the final output of these two signalling pathways leads to a differential expression of the two RNAs. Surprisingly, the two RNAs exert a redundant action on the repression of the T3SS genes, while they have an additive effect for the activation of the *pel* genes (Bordi *et al.*, 2010). Although the reason for these different behaviours is not known, this example clearly illustrates the complexity of these sRNA-dependent circuits that fine-tune the regulation of biofilm formation and virulence gene expression.

A similar system operates in *Pseudomonas* species to regulate Crc protein, the master regulator of catabolite repression. Crc is an RNA-binding protein that, in the presence of preferred substrates in the environment, represses catabolic genes for substrates that are not optimal for bacterial growth. Recently, it was shown that in *P. aeruginosa* a small regulatory RNA, CsrZ, could relieve the regulation exerted by Crc by sequestering the protein in a similar manner to what was described for RsmA/CsrA (Sonnleitner *et al.*, 2009). The expression of CrsZ itself is under the transcriptional control of a carbon source-sensitive two-component system, CbrA/CbrB. A motif (U/CAAU/CAA) was shown to be important for recognition of mRNA targets by Crc and was also found to be present in several copies on CrsZ, thus enabling sequestration of the protein (Sonnleitner *et al.*, 2009). The same motif was found to be essential for recognition of numerous mRNA targets by Crc in *P. putida* (Moreno *et al.*, 2009). Recently, two *P. putida* sRNAs (CrcZ and CrcY), characterized by six Crc binding motifs, were shown to antagonize the regulatory activity of Crc (Moreno *et al.*, 2012). As the links between bacterial metabolism and virulence become more and more appreciated and because Crc has recently been reported to affect antibiotic resistance and virulence (Linares *et al.*, 2010), the importance of this type of RNA regulation in bacterial pathogenesis becomes increasingly evident.

Finally, 6S RNA is a well-conserved ncRNA and is reported to interact with the housekeeping form of RNA polymerase in complex with σ70 to inhibit expression from certain promoters in the stationary phase of growth (Wassarman, 2007). A current study reveals that a 6S RNA deletion mutant in *Legionella pneumophila* shows decreased multiplication in protists and mammalian hosts (Faucher *et al.*, 2010), highlighting a significant contribution of 6S RNA to pathogenicity.

RNA acting at the DNA level and defence mechanism: CRISPR

CRISPR elements are clustered regularly interspaced short palindromic repeats that encode an RNA-based immunity system to protect bacteria from invading bacteriophages and plasmids (e.g. Barrangou *et al.*, 2007; Marraffini and Sontheimer, 2008). These elements have been compared with the eukaryotic RNA interference defence system against viruses. CRISPR are ubiquitous in archaea and numerous bacteria, including pathogenic bacteria such as *Shigella dysenteriae*, *S. typhimurium* and *S. pyogenes* (Mojica *et al.*, 2000; Grissa *et al.*, 2007, Makarova *et al.*, 2011). CRISPR elements are highly variable DNA regions that consist of a long leader sequence followed by a series of repeats-spacer units. In general, these units consist of repeats of 23–47 nucleotides that are separated by a spacer of similar size but of unique sequences (Horvath and Barrangou, 2010). The repeat sequence is usually identical in a given CRISPR and can be repeated 2 to 249 times. These repeats diverged among bacteria and often contain short palindromic stem-loop structures. In contrast, the spacer sequences are not conserved but are derived from bacteriophages and plasmids, suggesting that they can mediate a defence mechanism (Makarova *et al.*, 2006). One year later, Barrangou *et al.* (2007) demonstrated that the addition of a new spacer sequence derived from a phage upon infection in *S. thermophilus* made the bacteria resistant to that specific phage. In *S. epidermidis*, CRISPR elements interfere with plasmid conjugation and transformation (Marraffini and Sontheimer, 2008). Adjacent to the CRISPR DNA were between two and six CRISPR-associated proteins genes (CAS) genes (Deveau *et al.*, 2008). The functions of the *cas* genes are not fully understood, and their number varies between species. These proteins often contain RNA and DNA binding motifs as well as endo- and exonuclease domains. In *E. coli*, six proteins CasA-CasE associate to form a complex named cascade, which is required for antiviral defence (Brouns *et al.*, 2008).

What is the mode of action of CRISPR locus? The current model is that CRISPR elements are transcribed as long precursor RNAs and are processed into small RNA fragments containing a single repeat-spacer unit (crRNA). In various systems, the maturation process is governed by a specific Cas endoribonuclease protein (CasE, Cas6, Csy4), which is missing in many CRISPR/Cas subtypes. Interestingly, it was recently discovered that in *S. pyogenes*, a small non-coding RNA (tracrRNA) encoded on the CRISPR locus forms 22 base-pairings with the long precursor RNA and guides the host factor RNase III to generate the mature crRNA (Deltcheva *et al.*, 2011). This novel pathway shows, for the first time, the cross-talk between the CRISPR locus and the core genome, which is mediated through an sRNA (Fig. 6.5). The spacers of the crRNA serve as a guide for targeting the invader nucleic acids (Brouns *et al.*, 2008; Carte *et al.*, 2008; Hale *et al.*, 2009; Deltcheva *et al.*, 2011). It was shown that, *in vivo*, the crRNA targets and cleaves bacteriophage and plasmid DNAs (Marraffini and Sontheimer, 2008; Garneau *et al.*, 2010), although homology-dependent cleavage of invader RNAs has also been reported (Hale *et al.*, 2009). Interestingly, the crRNA can discriminate between self and non-self sequences because of additional pairings with the repeat sequences flanking the spacers (Marraffini and Sontheimer, 2010).

Up until now, there has been no direct evidence that CRISPR elements are required for virulence. However, they must be relevant for evolution because CRISPRs can interfere with the exchange of genetic material, such as plasmids and phages, thereby avoiding the acquisition of new genetic information. It is tempting to propose that CRISPR might control mobile elements such as prophage-encoded virulence factors. For instance, it was shown that lysogenic infection of *P. aeruginosa* inhibits biofilm formation and swarming mobility, and these effects are dependent on the presence of the intact CRISPR system (Andersson and Banfield, 2008; Zegans *et al.*, 2009). Further investigations will surely yield more evidence on the gain of functions due

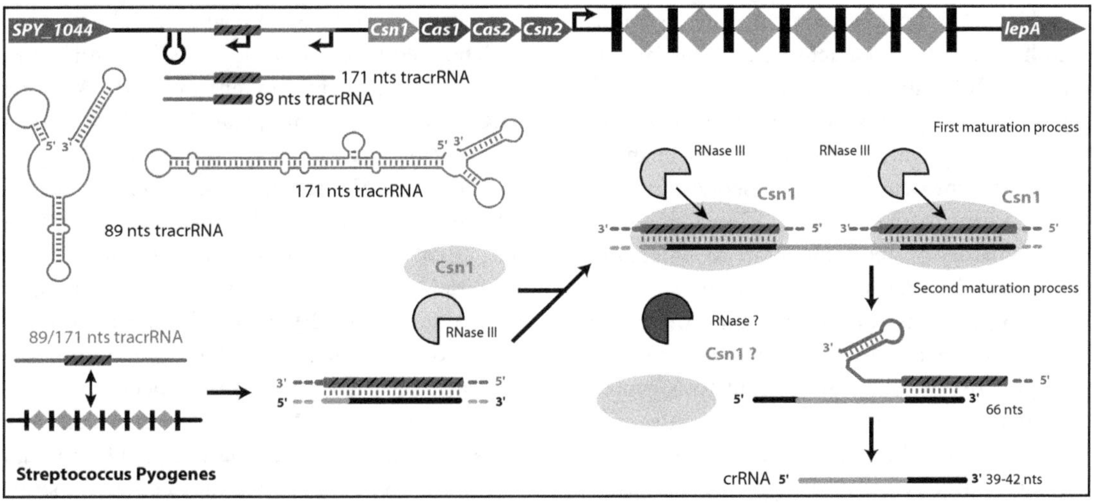

Figure 6.5 An example of CRISPR interference mechanism in *Streptococcus pyogenes*. The regularly interspaced short palindromic repeat locus is shown with the four protein encoding genes (csn1–2, cas1–2) and the tracrRNA gene located in an inverted orientation. The repeats (black rectangles) and the spacers (orange) are transcribed into a long precursor that is processed by the binding of the small sRNA tcacrRNA (red) to the repeat motifs to form short duplexes of 22 nts. These duplexes are then cleaved by the endoribonuclease III (RNase III, yellow circle) in the presence of Csn1 protein (light blue) to generate 66 nts long RNA fragments. A second cleavage induced by an unknown RNase generates a 39–42 nts long RNA fragments. This short RNA fragment is the active form that targets the foreign DNA (Deltcheva *et al.*, 2011). A colour version of this figure is available in the plate section at the back of the book.

to CRISPR elements on adaptation to environments and virulence.

Perspectives

This review shows that tremendous efforts have been made recently to identify the set of sRNAs that contribute to gene regulation in several pathogenic bacteria. However, the primary targets of sRNAs and their direct regulators have been identified in only few cases, and this finding is particularly true for the sRNAs regulating the expression of mRNA through base-pairings. Target identification and mechanistic studies will be necessary to build a precise map of the connections between sRNAs, their targets, and any associated regulatory proteins. Nevertheless, several general rules can already be drawn. Signal transduction in bacteria is mainly conducted through phosphorelays mediated by two-component systems, and in many cases, these systems are responsible for the transcriptional regulation of sRNAs involved in pathogenesis or in stress responses that can be encountered during the infection process. However, the signalling molecules that are responsible for the establishment of the signal transduction and the bacterial responses are not always known. Another common characteristic is the fact that many sRNAs induce downstream effects by regulating the synthesis of pleiotropic regulatory proteins using a variety of molecular mechanisms, such as thermosensor, protein sequestration, or base-pairings with mRNAs. Therefore, sRNA and regulatory proteins cooperate to affect global changes in the expression patterns in response to signalling pathways and to fine-tune the expression of virulence factors. The relative importance of regulatory RNAs versus regulatory proteins and whether RNAs confer more advantages to the cells has been questioned. Many of the regulatory RNAs act at the post-transcriptional levels and therefore achieve faster regulation than transcriptional regulators. This finding might be of importance for adaptive processes, such as the stress response or environmental changes – including the sudden exposure to an antibiotic regimen – encountered during the infection process. The versatility of the RNA structure and the capacity of RNAs to adopt alternative structures in response to environmental cues or ligand binding may also help to integrate more signals. By controlling individual genes, the sRNAs add another layer of complexity. Moreover, many pathogens often use several copies of sRNAs for regulation, inducing additive or redundant effects on gene regulation, or even both at once (Lenz *et al.*, 2004; Bordi *et al.*, 2010).

The mechanisms of the sRNA-dependent regulation are in general very similar in pathogenic and commensal bacteria. They are usually conserved in closely related pathogenic bacteria while different solutions have evolved in evolutionary distant bacteria. For instance in low GC Gram-positive bacteria, *glmS* expression is regulated by a *cis*-acting regulatory element that catalyses cleavage in the mRNA while in enterobacteria, a regulatory cascade involving two *trans*-acting sRNAs leads to the activation of *glmS* mRNA translation (Görke and Vogel, 2008). Interestingly, novel mechanisms have been found in representative pathogens of Gram-positive bacteria where sRNAs bind at the 5′ end of the target mRNA to protect the mRNAs against degradation, which is a mechanism that has not yet been found in Gram-negative bacteria. In enteric bacteria, many sRNAs regulate gene expression with the RNA chaperone Hfq protein, which is required for virulence (Chao and Vogel, 2010). The protein can enhance sRNA–target mRNA association rates, stabilize sRNAs against degradation, act as a chaperone by unfolding an sRNA or mRNA structure, and help to recruit RNase E for the mutual degradation of the sRNA and target mRNAs. However, several pathogenic bacteria do not carry an obvious Hfq homologue (Chao and Vogel, 2010). Furthermore, Hfq was shown dispensable for sRNA-mediated riboregulation in several Gram-positive species as for *S. aureus* (Felden *et al.*, 2011). However, recent work shows that in several *S. aureus* strains, a *hfq* mutant was found to be more resistant to oxidative stress and its pathogenicity was reduced (Liu *et al.*, 2010). In *Listeria monocytogenes*, Hfq is required for virulence (Christiansen *et al.*, 2004) and has been recently shown to facilitate the binding of the sRNA LhrA to its target mRNAs. This LhrA-mRNA, in turn, represses the translation and induces rapid degradation of the mRNA target (Nielsen *et al.*, 2009). However, as in *S. aureus*, the majority of *L. monocytogenes* sRNAs do not require Hfq for their stability and for the interaction with mRNAs (Toledo-Arana *et al.*, 2009). These findings indicate that modulation of translation by trans-encoded sRNAs may occur by both Hfq-dependent and -independent mechanisms, adding another layer of complexity in Gram-positive bacteria. Therefore, identification of proteins that bind to Hfq-independent sRNAs *in vivo* might provide leads for proteins with similar functions to Hfq.

This review illustrates the great diversity in sizes, structures and mechanisms of sRNAs and shows that RNA structure and sequence determinants are essential for regulation. For instance, non-canonical base-pairs play critical functions for thermosensor activity, and other regulatory RNAs carry a C-rich motif located in hairpin loops or in accessible single strands, which is a specific recognition signature for mRNA binding (Fig. 6.2). Other sRNAs carry repeated sequences mimicking the binding sites present on mRNAs that are recognized by post-transcriptional regulatory proteins (CsrA, Crc). Such a decoy confers to the sRNA the ability to bind tightly to the protein. One aspect, which is often neglected, is the importance of the structures of the sRNA and the targets in regulation. In bacteria co-transcriptional RNA folding has direct impact on gene regulation and the speed of transcription, and the efficiency of translation is known to affect the folding pathway of RNAs and thus the genetic response (Zemora and Waldsich, 2010). This aspect is critical for large regulatory RNAs, such as the *cis*-acting riboswitches.

Most of the studies of the mechanism of action of the regulatory RNAs provide a static picture of gene regulation in a population of cells. However, identical genotype and growth environments are insufficient to ensure that individual cells within clonal microbial cultures will exhibit the same phenotype (Gefen and Balaban, 2009). This heterogeneity and diversity are thought to be fundamental to fitness and give the microbial population the ability to persist in the face of external changes and to exploit new niches. Obvious links have been shown between sRNAs, quorum sensing, metabolism and virulence. However, little is known about the expression pattern of sRNAs and their roles during infection. Is there a specific sRNA pattern of expression in acute or chronic infections? We also do not know how the host microbial flora influences sRNA expression and regulatory networks in pathogenic bacteria. What is the contribution of the *cis*-acting sRNAs versus *trans*-acting sRNAs in virulence? Regulatory RNAs activating virulence factors, affecting mobility of pathogenicity islands, expression of host genes or controlling cell-to-cell communication might also be important players in virulence expression. Continued studies of sRNA functions in many pathogens will certainly reveal novel strategies that will modify the known pattern of gene expression.

Acknowledgements

We thank I. Caldelari and P. Fechter for stimulating discussions. This work was supported by the Centre National de la Recherche Scientifique (CNRS; P.R.), the Université de Strasbourg (UdS, P.R.), Institut National de la Recherche Médicale (INSERM, F.V.), the Université de Lyon (F.V.), and the Agence Nationale pour la Recherche (ANR09-BLAN-0024-01; ANR2010-PATH-005-01; P.R., F.V.). E.L. is holder of a long-term fellowship from the FEBS, and CR received support from the Région Alsace and from Délégation Générale pour l'Armement (D.G.A.).

References

Altuvia, S. (2007). Identification of bacterial small non-coding RNAs: experimental approaches. Curr. Opin. Microbiol. *10*, 257–261.

Andersson, A.F., and Banfield, J.F. (2008). Virus population dynamics and acquired virus resistance in natural microbial communities. Science *320*, 1047–1050.

Babitzke, P., and Romeo, T. (2007). CsrB sRNA family: sequestration of RNA-binding regulatory proteins. Curr. Opin. Microbiol. *10*, 156–163.

Barrangou, R., Fremaux, C., Deveau, H., Richards, M., Boyaval, P., Moineau, S., Romero, D.A., and Horvath, P. (2007). CRISPR provides acquired resistance against viruses in prokaryotes. Science *315*, 1709–1712.

Barrick, J.E., and Breaker, R.R. (2007). The distributions, mechanisms, and structures of metabolite-binding riboswitches. Genome Biol. *8*, R239.

Beaume, M., Hernandez, D., Farinelli, L., Deluen, C., Linder, P., Gaspin, C., Romby, P., Schrenzel, J., and Francois, P. (2010). Cartography of methicillin-resistant *S. aureus* transcripts: detection, orientation and temporal expression during growth phase and stress conditions. PLoS One *5*, e10725.

Beisel, C.L., and Storz, G. (2010). Basepairing small RNAs and their roles in global regulatory networks. FEMS Microbiol. Rev. *34*, 866–882.

Bejerano-Sagie, M., and Xavier, K.B. (2007). The role of small RNAs in quorum sensing. Curr. Opin. Microbiol. *10*, 189–198.

Benito, Y., Kolb, F.A., Romby, P., Lina, G., Etienne, J., and Vandenesch, F. (2000). Probing the structure of RNAIII, the *Staphylococcus aureus agr* regulatory RNA, and identification of the RNA domain involved in repression of protein A expression. RNA *6*, 668–679.

Blount, K.F., and Breaker, R.R. (2006). Riboswitches as antibacterial drug targets. Nat. Biotechnol. *24*, 1558–1564.

Boisset, S., Geissmann, T., Huntzinger, E., Fechter, P., Bendridi, N., Possedko, M., Chevalier, C., Helfer, A.C., Benito, Y., Jacquier, A., *et al.* (2007). *Staphylococcus aureus* RNAIII coordinately represses the synthesis of virulence factors and the transcription regulator Rot by an antisense mechanism. Genes Dev. *21*, 1353–1366.

Bordi, C., Lamy, M.C., Ventre, I., Termine, E., Hachani, A., Fillet, S., Roche, B., Bleves, S., Mejean, V., Lazdunski, A., *et al.* (2010). Regulatory RNAs and the HptB/RetS signalling pathways fine-tune *Pseudomonas aeruginosa* pathogenesis. Mol. Microbiol. *76*, 1427–1443.

Brantl, S. (2007). Regulatory mechanisms employed by *cis*-encoded antisense RNAs. Curr. Opin. Microbiol. *10*, 102–109.

Brencic, A., and Lory, S. (2009). Determination of the regulon and identification of novel mRNA targets of *Pseudomonas aeruginosa* RsmA. Mol. Microbiol. *72*, 612–632.

Brencic, A., McFarland, K.A., McManus, H.R., Castang, S., Mogno, I., Dove, S.L., and Lory, S. (2009). The GacS/GacA signal transduction system of *Pseudomonas aeruginosa* acts exclusively through its control over the transcription of the RsmY and RsmZ regulatory small RNAs. Mol. Microbiol. *73*, 434–445.

Brouns, S.J., Jore, M.M., Lundgren, M., Westra, E.R., Slijkhuis, R.J., Snijders, A.P., Dickman, M.J., Makarova, K.S., Koonin, E.V., and van der Oost, J. (2008). Small CRISPR RNAs guide antiviral defense in prokaryotes. Science *321*, 960–964.

Carte, J., Wang, R., Li, H., Terns, R.M., and Terns, M.P. (2008). Cas6 is an endoribonuclease that generates guide RNAs for invader defense in prokaryotes. Genes Dev. *22*, 3489–3496.

Chabelskaya, S., Gaillot, O., and Felden, B. (2010). A *Staphylococcus aureus* small RNA Is required for bacterial virulence and regulates the expression of an immune-evasion molecule. PLoS Pathog. *6*, e1000927.

Chao, Y., and Vogel, J. (2010). The role of Hfq in bacterial pathogens. Curr. Opin. Microbiol. *13*, 24–33.

Chavez, R.G., Alvarez, A.F., Romeo, T., and Georgellis, D. (2010). The physiological stimulus for the BarA sensor kinase. J. Bacteriol. *192*, 2009–2012.

Cheung, A.L., Eberhardt, K.J., Chung, E., Yeaman, M.R., Sullam, P.M., Ramos, M., and Bayer, A. S. (1994). Diminished virulence of a sar-/agr- mutant of *Staphylococcus aureus* in the rabbit model of endocarditis. J. Clin. Invest. *94*, 1815–1822.

Chowdhury, S., Maris, C., Allain, F.H., and Narberhaus, F. (2006). Molecular basis for temperature sensing by an RNA thermometer. EMBO J. *25*, 2487–2497.

Christiansen, J.K., Larsen, M.H., Ingmer, H., Sogaard-Andersen, L., and Kallipolitis, B. H. (2004). The RNA-binding protein Hfq of *Listeria monocytogenes*: role in stress tolerance and virulence. J. Bacteriol. *186*, 3355–3362.

Condon, C., and Bechhofer, D.H. (2011). Regulated RNA stability in the Gram-positives. Curr. Opin. Microbiol. *14*, 148–154.

Cui, Y., Chatterjee, A., Yang, H., and Chatterjee, A. K. (2008). Regulatory network controlling extracellular proteins in *Erwinia carotovora* subsp. carotovora: FlhDC, the master regulator of flagellar genes, activates rsmB regulatory RNA production by affecting *gacA* and *hexA* (lrhA) expression. J. Bacteriol. *190*, 4610–4623.

Darfeuille, F., Unoson, C., Vogel, J., and Wagner, E.G.H. (2007). An antisense RNA inhibits translation by competing with standby ribosomes. Mol. Cell *26*, 381 – 392.

Deltcheva, E., Chylinski, K., Sharma, C.M., Gonzales, K., Chao, Y., Pirzada, Z.A., Eckert, M.R., Vogel, J., and Charpentier, E. (2011). CRISPR RNA maturation by trans-encoded small RNA and host factor RNase III. Nature *471*, 602–607.

Deveau, H., Barrangou, R., Garneau, J.E., Labonte, J., Fremaux, C., Boyaval, P., Romero, D.A., Horvath, P., and Moineau, S. (2008). Phage response to CRISPR-encoded resistance in *Streptococcus thermophilus*. J. Bacteriol. *190*, 1390–1400.

Faucher, S.P., Friedlander, G., Livny, J., Margalit, H., and Shuman, H.A. (2010). *Legionella pneumophila* 6S RNA optimizes intracellular multiplication. Proc. Natl. Acad. Sci. U.S.A. *107*, 7533–7538.

Felden, B., Vandenesch, F., Bouloc, P., and Romby, P. (2011). The *Staphylococcus aureus* RNome and its commitment to virulence. PLoS Pathog *7*, e1002006.

Fortune, D.R., Suyemoto, M., and Altier, C. (2006). Identification of CsrC and characterization of its role in epithelial cell invasion in *Salmonella enterica serovar Typhimurium*. Infect. Immun. *74*, 331–339.

Fozo, E.M., Hemm, M.R., and Storz, G. (2008). Small toxic proteins and the antisense RNAs that repress them. Microbiol. Mol. Biol. Rev. *72*, 579–589.

Fröhlich, K.S., and Vogel, J. (2009). Activation of gene expression by small RNA. Curr. Opin. Microbiol. *12*, 674–682.

Garneau, J.E., Dupuis, M.-, Villion, M., Romero, D.A., Barrangou, R., Boyaval, P., Fremaux, C., Horvath, P., Magadán, A.H., and Moineau, S. (2010). The CRISPR/Cas bacterial immune system cleaves bacteriophage and plasmid DNA. Nature *468*, 67–71.

Gefen, O., and Balaban, N.Q. (2009). The importance of being persistent: heterogeneity of bacterial populations under antibiotic stress. FEMS Microbiol. Rev. *33*, 704–717.

Geisinger, E., Adhikari, R.P., Jin, R., Ross, H.F., and Novick, R. P. (2006). Inhibition of rot translation by RNAIII, a key feature of agr function. Mol. Microbiol. *61*, 1038–1048.

Geissmann, T., Chevalier, C., Cros, M.J., Boisset, S., Fechter, P., Noirot, C., Schrenzel, J., Francois, P., Vandenesch, F., Gaspin, C., *et al.* (2009). A search for small noncoding RNAs in *Staphylococcus aureus* reveals a conserved sequence motif for regulation. Nucleic Acids Res. *37*, 7239–7257.

Goodman, A.L., Kulasekara, B., Rietsch, A., Boyd, D., Smith, R.S., Lory, S. (2004). A signaling network reciprocally regulates genes associated with acute infection and chronic persistence in *Pseudomonas aeruginosa.* Dev. Cell 7, 745–54.

Görke, B., and Vogel, J. (2008). Noncoding RNA control of the making and breaking of sugars. Genes Dev. *22*, 2914–2925.

Grissa, I., Vergnaud, G., and Pourcel, C. (2007). The CRISPRdb database and tools to display CRISPRs and to generate dictionaries of spacers and repeats. BMC Bioinformatics *8*, 172–182.

Guillier, M., Gottesman, S., and Storz, G. (2006). Modulating the outer membrane with small RNAs. Genes Dev. *20*, 2338–2348.

Hale, C.R., Zhao, P., Olson, S., Duff, M.O., Graveley, B.R., Wells, L., Terns, R.M., and Terns, M.P. (2009). RNA-guided RNA cleavage by a CRISPR RNA-Cas protein complex. Cell *139*, 945–956.

Hammer, B.K., and Bassler, B. L. (2007). Regulatory small RNAs circumvent the conventional quorum sensing pathway in pandemic *Vibrio cholerae.* Proc. Natl. Acad. Sci. U.S.A. *104*, 11145–11149.

Heroven, A.K., Bohme, K., Rohde, M., and Dersch, P. (2008). A Csr-type regulatory system, including small non-coding RNAs, regulates the global virulence regulator RovA of *Yersinia pseudotuberculosis* through RovM. Mol. Microbiol. *68*, 1179–1195.

Holmqvist, E., Reimegard, J., Sterk, M., Grantcharova, N., Romling, U., and Wagner, E.G. (2010). Two antisense RNAs target the transcriptional regulator CsgD to inhibit curli synthesis. EMBO J. *29*, 1840–1850.

Horvath, P., and Barrangou, R. (2010). CRISPR/Cas, the immune system of Bacteria and Archaea. Science *327*, 167–170.

Huntzinger, E., Boisset, S., Saveanu, C., Benito, Y., Geissmann, T., Namane, A., Lina, G., Etienne, J., Ehresmann, B., Ehresmann, C., *et al.* (2005). *Staphylococcus aureus* RNAIII and the endoribonuclease III coordinately regulate *spa* gene expression. EMBO J. *24*, 824–835.

Johansson, J., Mandin, P., Renzoni, A., Chiaruttini, C., Springer, M., and Cossart, P. (2002). An RNA thermosensor controls expression of virulence genes in *Listeria monocytogenes.* Cell *110*, 551–561.

Kato, A., and Groisman, E.A. (2008). The PhoQ/PhoP regulatory network of *Salmonella enterica.* Adv. Exp. Med. Biol. *631*, 7–21.

Klenk, M., Koczan, D., Guthke, R., Nakata, M., Thiesen, H.J., Podbielski, A., and Kreikemeyer, B. (2005). Global epithelial cell transcriptional responses reveal *Streptococcus pyogenes* Fas regulator activity association with bacterial aggressiveness. Cell. Microbiol. *7*, 1237–1250.

Kreikemeyer, B., Boyle, M.D., Buttaro, B.A., Heinemann, M., and Podbielski, A. (2001). Group A streptococcal growth phase-associated virulence factor regulation by a novel operon (Fas) with homologies to two-component-type regulators requires a small RNA molecule. Mol. Microbiol. *39*, 392–406.

Kulshina, N., Baird, N.J., and Ferre-D'Amare, A. R. (2009). Recognition of the bacterial second messenger cyclic diguanylate by its cognate riboswitch. Nat. Struct. Mol. Biol. *16*, 1212–1217.

Lee, E.J., and Groisman, E.A. (2010). An antisense RNA that governs the expression kinetics of a multifunctional virulence gene. Mol. Microbiol. *76*, 1020–1033.

Lee, E.R., Baker, J.L., Weinberg, Z., Sudarsan, N., and Breaker, R. R. (2010). An allosteric self-splicing ribozyme triggered by a bacterial second messenger. Science *329*, 845–848.

Lenz, D.H., Mok, K.C., Lilley, B.N., Kulkarni, R.V., Wingreen, N.S., and Bassler, B.L. (2004). The small RNA chaperone Hfq and multiple small RNAs control quorum sensing in *Vibrio harveyi* and *Vibrio cholerae.* Cell *118*, 69–82.

Lenz, D.H., Miller, M.B., Zhu, J., Kulkarni, R.V., and Bassler, B. L. (2005). CsrA and three redundant small RNAs regulate quorum sensing in *Vibrio cholerae.* Mol. Microbiol. *58*, 1186–1202.

Linares, J.F., Moreno, R., Fajardo, A., Martinez-Solano, L., Escalante, R., Rojo, F., and Martinez, J.L. (2010). The global regulator Crc modulates metabolism, susceptibility to antibiotics and virulence in *Pseudomonas aeruginosa.* Environ. Microbiol. *12*, 3196–3212.

Liu, J.M., Livny, J., Lawrence, M.S., Kimball, M.D., Waldor, M.K., and Camilli, A. (2009). Experimental discovery of sRNAs in *Vibrio cholerae* by direct cloning, 5S/tRNA depletion and parallel sequencing. Nucleic Acids Res. *37*, e46.

Liu, Y., Wu, N., Dong, J., Gao, Y., Zhang, X., Mu, C., Shao, N., and Yang, G. (2010). Hfq is a global regulator that controls the pathogenicity of *Staphylococcus aureus.* PLoS One *5*, e13069.

Loh, E., Dussurget, O., Gripenland, J., Vaitkevicius, K., Tiensuu, T., Mandin, P., Repoila, F., Buchrieser, C., Cossart, P., and Johansson, J. (2009). A trans-acting riboswitch controls expression of the virulence regulator PrfA in *Listeria monocytogenes.* Cell *139*, 770–779.

Makarova, K.S., Grishin, N.V., Shabalina, S.A., Wolf, Y.I., and Koonin, E.V. (2006). A putative RNA-interference-based immune system in prokaryotes: computational analysis of the predicted enzymatic machinery, functional analogies with eukaryotic RNAi, and hypothetical mechanisms of action. Biol. Direct *1*, 7.

Makarova, K.S., Haft, D.H., Barrangov, R., Brouns, S.J., Charpentier, E., Horvath, P., Moineau, S., Mojica, F.J., Wolf, Y.I., Yakunin, A.F., *et al.* (2011). Evolution and classification of the CRISPR-Cas systems. Nat. Rev. Microbiol. *9*, 467–477.

Mandin, P., Repoila, F., Vergassola, M., Geissmann, T., and Cossart, P. (2007). Identification of new noncoding RNAs in *Listeria monocytogenes* and prediction of mRNA targets. Nucleic Acids Res. *35,* 962–974.

Mangold, M., Siller, M., Roppenser, B., Vlaminckx, B.J., Penfound, T.A., Klein, R., Novak, R., Novick, R.P., and Charpentier, E. (2004). Synthesis of group A streptococcal virulence factors is controlled by a regulatory RNA molecule. Mol. Microbiol. *53,* 1515–1527.

Marraffini, L.A., and Sontheimer, E.J. (2008). CRISPR interference limits horizontal gene transfer in Staphylococci by targeting DNA. Science *322,* 1843–1845.

Marraffini, L.A., and Sontheimer, E.J. (2010). Self versus non-self discrimination during CRISPR RNA-directed immunity. Nature *463,* 568–571.

Mojica, F.J., Diez-Villasenor, C., Soria, E., and Juez, G. (2000). Biological significance of a family of regularly spaced repeats in the genomes of Archaea, Bacteria and mitochondria. Mol. Microbiol. *36,* 244–246.

Moon, K., and Gottesman, S. (2009). A PhoQ/P-regulated small RNA regulates sensitivity of *Escherichia coli* to antimicrobial peptides. Mol. Microbiol. *74,* 1314–1330.

Moreno, R., Marzi, S., Romby, P., and Rojo, F. (2009). The Crc global regulator binds to an unpaired A-rich motif at the *Pseudomonas putida alkS* mRNA coding sequence and inhibits translation initiation. Nucleic Acids Res. *37,* 7678–7690.

Moreno, R., Fonseca, P., and Rojo, F. (2012). Two small RNAs, CrcY and CrcZ, act in concert to sequester the Crc global regulator in *Pseudomonas putida,* modulating catabolite repression. Mol. Microbiol. *83,* 24–40.

Mulhbacher, J., Brouillette, E., Allard, M., Fortier, L.C., Malouin, F., and Lafontaine, D.A. (2010a). Novel riboswitch ligand analogs as selective inhibitors of guanine-related metabolic pathways. PLoS Pathog. *6,* e1000865.

Mulhbacher, J., St-Pierre, P., and Lafontaine, D.A. (2010b). Therapeutic applications of ribozymes and riboswitches. Curr. Opin. Pharmacol. *10,* 551–556.

Narberhaus, F. (2010). Translational control of bacterial heat shock and virulence genes by temperature-sensing mRNAs. RNA Biol. *7,* 84–89.

Nechooshtan, G., Elgrably-Weiss, M., Sheaffer, A., Westhof, E., and Altuvia, S. (2009). A pH-responsive riboregulator. Genes Dev. *23,* 2650–2662.

Nielsen, J.S., Lei, L.K., Ebersbach, T., Olsen, A.S., Klitgaard, J.K., Valentin-Hansen, P., and Kallipolitis, B. H. (2010). Defining a role for Hfq in Gram-positive bacteria: evidence for Hfq-dependent antisense regulation in *Listeria monocytogenes.* Nucleic Acids Res. *38,* 907–919.

Novick, R.P., and Geisinger, E. (2008). Quorum sensing in staphylococci. Annu. Rev. Genet. *42,* 541–564.

Novick, R.P., and Muir, T. W. (1999). Virulence gene regulation by peptides in staphylococci and other Gram-positive bacteria. Curr. Opin. Microbiol. *2,* 40–45.

Obana, N., Shirahama, Y., Abe, K., and Nakamura, K. (2010). Stabilization of *Clostridium perfringens* collagenase mRNA by VR-RNA-dependent cleavage in 5′ leader sequence. Mol. Microbiol. *77,* 1416–1428.

Ohtani, K., Kawsar, H.I., Okumura, K., Hayashi, H., and Shimizu, T. (2003). The VirR/VirS regulatory cascade affects transcription of plasmid-encoded putative virulence genes in *Clostridium perfringens* strain 13. FEMS Microbiol. Lett. *222,* 137–141.

Padalon-Brauch, G., Hershberg, R., Elgrably-Weiss, M., Baruch, K., Rosenshine, I., Margalit, H., and Altuvia, S. (2008). Small RNAs encoded within genetic islands of *Salmonella typhimurium* show host-induced expression and role in virulence. Nucleic Acids Res. *36,* 1913–1927.

Papenfort, K., and Vogel, J. (2009). Multiple target regulation by small noncoding RNAs rewires gene expression at the post-transcriptional level. Res. Microbiol. *160,* 278–287.

Papenfort, K., and Vogel, J. (2010). Regulatory RNA in bacterial pathogens. Cell Host Microbe *8,* 116–127.

Pfeiffer, V., Sittka, A., Tomer, R., Tedin, K., Brinkmann, V., and Vogel, J. (2007). A small non-coding RNA of the invasion gene island (SPI-1) represses outer membrane protein synthesis from the *Salmonella* core genome. Mol. Microbiol. *66,* 1174–1191.

Queck, S.Y., Jameson-Lee, M., Villaruz, A.E., Bach, T.-H.L., Khan, B.A., Sturdevant, D.E., Ricklefs, S.M., Li, M., and Otto, M. (2008). RNAIII-independent target gene control by the *agr* quorum-sensing system: insight into the evolution of virulence regulation in *Staphylococcus aureus.* Mol. Cell *32,* 150–158.

Ramesh, A., and Winkler, W.C. (2010). Magnesium-sensing riboswitches in bacteria. RNA Biol. *7,* 77–83.

Ramirez-Peña, E., Treviño, J., Liu, Z., Perez, N., and Sumby, P. (2010). The group A *Streptococcus* small regulatory RNA FasX enhances streptokinase activity by increasing the stability of the *ska* mRNA transcript. Mol. Microbiol. *78,* 1332–1347.

Rasis, M., and Segal, G. (2009). The LetA-RsmYZ-CsrA regulatory cascade, together with RpoS and PmrA, post-transcriptionally regulates stationary phase activation of *Legionella pneumophila* Icm/Dot effectors. Mol. Microbiol. *72,* 995–1010.

Roberts, S.A., and Scott, J. R. (2007). RivR and the small RNA RivX: the missing links between the CovR regulatory cascade and the Mga regulon. Mol. Microbiol. *66,* 1506–1522.

Rutherford, S.T., van Kessel, J.C., Shao, Y., and Bassler, B.L. (2011). AphA and LuxR/HapR reciprocally control quorum sensing in vibrios. Genes Dev. *25,* 397–408.

Sahr, T., Bruggemann, H., Jules, M., Lomma, M., Albert-Weissenberger, C., Cazalet, C., and Buchrieser, C. (2009). Two small ncRNAs jointly govern virulence and transmission in *Legionella pneumophila.* Mol. Microbiol. *72,* 741–762.

Schubert, M., Lapouge, K., Duss, O., Oberstrass, F.C., Jelesarov, I., Haas, D., and Allain, F.H.-T. (2007). Molecular basis of messenger RNA recognition by the specific bacterial repressing clamp RsmA/CsrA. Nat. Struct. Mol. Biol. *14,* 807–813.

Serganov, A. (2010). Determination of riboswitch structures: light at the end of the tunnel? RNA Biol. 7, 98–103.

Sharma, C.M., Hoffmann, S., Darfeuille, F., Reignier, J., Findeiss, S., Sittka, A., Chabas, S., Reiche, K., Hackermuller, J., Reinhardt, R., *et al.* (2010). The primary transcriptome of the major human pathogen *Helicobacter pylori*. Nature *464*, 250–255.

Shimizu, T., Yaguchi, H., Ohtani, K., Banu, S., and Hayashi, H. (2002). Clostridial VirR/VirS regulon involves a regulatory RNA molecule for expression of toxins. Mol. Microbiol. *43*, 257–265.

Siller, M., Janapatla, R.P., Pirzada, Z.A., Hassler, C., Zinkl, D., and Charpentier, E. (2008). Functional analysis of the group A streptococcal luxS/AI-2 system in metabolism, adaptation to stress and interaction with host cells. BMC Microbiol. *8*, 188.

Sittka, A., Lucchini, S., Papenfort, K., Sharma, C.M., Rolle, K., Binnewies, T.T., Hinton, J.C., and Vogel, J. (2008). Deep sequencing analysis of small noncoding RNA and mRNA targets of the global post-transcriptional regulator, Hfq. PLoS Genet. *4*, e1000163.

Smith, A.M., Fuchs, R.T., Grundy, F.J., and Henkin, T.M. (2010). Riboswitch RNAs: regulation of gene expression by direct monitoring of a physiological signal. RNA Biol. 7, 104–110.

Smith, K.D., Lipchock, S.V., Ames, T.D., Wang, J., Breaker, R.R., and Strobel, S. A. (2009). Structural basis of ligand binding by a c-di-GMP riboswitch. Nat. Struct. Mol. Biol. *16*, 1218–1223.

Sonnleitner, E., Abdou, L., and Haas, D. (2009). Small RNA as global regulator of carbon catabolite repression in *Pseudomonas aeruginosa*. Proc. Natl. Acad. Sci. U.S.A. *106*, 21866–21871.

Sonnleitner, E., Gonzalez, N., Sorger-Domenigg, T., Heeb, S., Richter, A.S., Backofen, R., Williams, P., Huttenhofer, A., Haas, D., and Bläsi, U. (2011). The small RNA PhrS stimulates synthesis of the *Pseudomonas aeruginosa* quinolone signal. Mol. Microbiol. *80*, 868–885.

Steiner, K., and Malke, H. (2001). relA-Independent amino acid starvation response network of *Streptococcus pyogenes*. J. Bacteriol. *183*, 7354–7364.

Sudarsan, N., Cohen-Chalamish, S., Nakamura, S., Emilsson, G.M., and Breaker, R.R. (2005). Thiamine pyrophosphate riboswitches are targets for the antimicrobial compound pyrithiamine. Chem. Biol. *12*, 1325–1335.

Sudarsan, N., Hammond, M.C., Block, K.F., Welz, R., Barrick, J.E., Roth, A., and Breaker, R. R. (2006). Tandem riboswitch architectures exhibit complex gene control functions. Science *314*, 300–304.

Sudarsan, N., Lee, E.R., Weinberg, Z., Moy, R.H., Kim, J.N., Link, K.H., and Breaker, R. R. (2008). Riboswitches in eubacteria sense the second messenger cyclic di-GMP. Science *321*, 411–413.

Timmermans, J., and Van Melderen, L. (2010). Post-transcriptional global regulation by CsrA in bacteria. Cell. Mol. Life Sci. *67*, 2897–2908.

Toledo-Arana, A., and Solano, C. (2010). Deciphering the physiological blueprint of a bacterial cell: revelations of unanticipated complexity in transcriptome and proteome. BioEssays *32*, 461–467.

Toledo-Arana, A., Dussurget, O., Nikitas, G., Sesto, N., Guet-Revillet, H., Balestrino, D., Loh, E., Gripenland, J., Tiensuu, T., Vaitkevicius, K., *et al.* (2009). The *Listeria* transcriptional landscape from saprophytism to virulence. Nature *459*, 950–956.

Ventre, I., Goodman, A.L., Vallet-Gely, I., Vasseur, P., Soscia, C., Molin, S., Bleves, S., Lazdunski, A., Lory, S., and Filloux, A. (2006). Multiple sensors control reciprocal expression of *Pseudomonas aeruginosa* regulatory RNA and virulence genes. Proc. Natl. Acad. Sci. U.S.A. *103*, 171–176.

Vogel, J. (2009). A rough guide to the non-coding RNA world of *Salmonella*. Mol. Microbiol. *71*, 1–11.

Wadler, C.S., and Vanderpool, C.K. (2007). A dual function for a bacterial small RNA: SgrS performs basepairing-dependent regulation and encodes a functional polypeptide. Proc. Natl. Acad. Sci. U.S.A. *104*, 20454–20459.

Wagner, E.G.H., Altuvia, S., and Romby, P. (2002). Antisense RNAs in bacteria and their genetic elements. Adv. Genet. *46*, 361–398.

Wassarman, K.M. (2007). 6S RNA: a small RNA regulator of transcription. Curr. Opin. Microbiol. *10*, 164–168.

Waters, L.S., and Storz, G. (2009). Regulatory RNAs in Bacteria. Cell *136*, 615–628.

Zegans, M.E., Wagner, J.C., Cady, K.C., Murphy, D.M., Hammond, J.H., and O'Toole, G.A. (2009). Interaction between bacteriophage DMS3 and host CRISPR region inhibits group behaviors of *Pseudomonas aeruginosa*. J. Bacteriol. *191*, 210–219.

Zemora, G., and Waldsich, C. (2010). RNA folding in living cells. RNA Biol. 7, 634–641.

Zhang, J., Lau, M.W., and Ferre-D'Amare, A.R. (2010). Ribozymes and riboswitches: modulation of RNA function by small molecules. Biochemistry *49*, 9123–9131.

H-NS, Global Regulator of Gene Expression and Organizer of the Bacterial Nucleoid

7

Charles J Dorman

Abstract

H-NS is an abundant DNA binding protein that has been found to influence the expression of hundreds of genes in those Gram-negative bacteria, chiefly *Escherichia coli* and *Salmonella enterica*, where its regulatory effects have been investigated. It also has the potential to organize the structure of the nucleoid. H-NS has a preference for binding to A+T-rich DNA and this preference underlies its targeting of genes that have been acquired by horizontal transfer. H-NS usually acts as a transcriptional silencer by binding to a nucleation site followed by lateral spreading with or without the creation of DNA-protein-DNA bridges; it may also act as an architectural component in the nucleoid. Bacteria use a multitude of mechanisms to displace H-NS or to attenuate its negative influence on gene expression. A paralogue of H-NS, called StpA, is an efficient RNA chaperone and controls a regulon of genes in *S. enterica* by influencing expression of the RpoS sigma factor. Orthologues of H-NS have been discovered on large self-transmissible plasmids, introducing a new dimension in considerations of the roles of H-NS-like proteins in horizontal gene transfer. Importantly, analogues of H-NS are now being discovered and characterized in Gram-negative bacteria such as *Pseudomonas* that are only distantly related to *E. coli*, in the medically important actinomycete *Mycobacterium tuberculosis* and even in Gram-positive organisms such as *Bacillus subtilis*.

Introduction

The application of techniques for genome-wide study of transcription regulation in bacteria is transforming H-NS research from a specialist interest of those investigating nucleoid structure or the regulation of a specific gene to something that is of interest to scientists studying genome architecture, genome evolution, DNA recombination, horizontal gene transfer, global gene regulation and the details of individual gene regulatory switches (Corcoran and Dorman, 2009; Browning *et al.*, 2010; Dame and Dorman, 2010; Dillon and Dorman, 2010; Navarre, 2010). The H-NS protein was recognized early on as a factor capable of modulating the activity of RNA polymerase during *in vitro* transcription (Jacquet *et al.*, 1971; Cukier-Kahn *et al.*, 1972) and a little later as a component of the bacterial nucleoid (Varshavsky *et al.*, 1977; Bakaev, 1981). Its name derives from its description as a heat-stable nucleoid structuring (H-NS) protein and it was classified at first as a bacterial 'histone-like' protein (Lammi *et al.*, 1984). In the past, the same protein has been named 16K (Laine *et al.*, 1984), B1 (Varshavsky *et al.*, 1977), H1 (Cukier-Kahn *et al.*, 1972) and H1a (Spassky *et al.*, 1984). The term 'histone-like' protein is going out of use, as it is clear that H-NS shares no amino acid sequence similarities to eukaryotic histones. Instead, H-NS is now grouped with the bacterial nucleoid-associated proteins or NAPs (Dorman, 2004, 2009; Browning *et al.*, 2010; Dillon and Dorman, 2010). This assignment is consistent with the location of the H-NS protein in the cell and its roles in nucleoid structuring and gene regulation.

Different research groups discovered the *hns* gene independently over a period of several years, with each group naming the gene according to the particular genetic system in which they found a regulatory role for this locus. For example, *hns* was named *pilG* by those studying the phase-variable expression of type 1 fimbriae/pili in *E. coli* (Spears *et al.*, 1986) and was named *bglY* by research groups studying derepression of the cryptic *bgl* (beta-glucoside utilization) operon (Defez and De Felice, 1981). The *hns* gene was linked to virulence in *Shigella flexneri* where it was known as the *virR* locus (Maurelli and Sansonetti, 1988) and in uropathogenic *E. coli* where it was named *drdX* (Goransson *et al.*, 1990). Genetic studies showed that these different loci were allelic with *osmZ*, a regulator of the response to osmotic upshock, particularly the osmoprotectant transporter operon, *proU* (Higgins *et al.*, 1988; Dorman *et al.*, 1990) and subsequently DNA sequencing revealed that this very pleiotropic regulatory gene encoded the H-NS protein, leading to a unification of nomenclature and the use of the name *hns* (Hromockyj *et al.*, 1990; Hulton *et al.*, 1990; May *et al.*, 1990) as proposed by Pon *et al.* (1988).

The structure of H-NS

The 15-kDa, 137-amino-acid H-NS protein consists of two functional domains that are connected by a flexible linker. An oligomerization domain (amino acids 1 to 83) occupies the amino terminal portion of H-NS while a nucleic acid binding domain (amino acids 91 to 137) is located in its C-terminus (Dorman *et al.*, 1999; Rimsky, 2004; Sete *et al.*, 2009). The residues in the nucleic acid binding domain adopt a winged helix–turn–helix conformation (Shino *et al.*, 1999). Structural analyses have been performed on the proteins from *E. coli* (Bloch *et al.*, 2003), *Salmonella enterica* (Renzoni *et al.*, 2001; Esposito *et al.*, 2002) and *Vibrio cholerae* (Cerdan *et al.*, 2003) and these show a role for the alpha helix-1 and helix-2 subdomains of the amino terminus of H-NS in promoting dimerization. Higher-level oligomerization requires two more alpha helices (3 and 4) composed of amino acids 67 to 83 of the amino-terminal domain (Leonard *et al.*, 2009). Crystallographic analysis of the isolated amino-terminal domain suggests that head-to-head interactions between pairs of helices 1 and 2 create H-NS dimers and that tail-to-tail interactions between pairs of helices 3 and 4 create oligomers from pairs of dimers (Fig. 7.1). These interactions promote the formation of an H-NS superhelix with six molecules per superhelical turn, a superstructure that lends itself to bridging two adjacent DNA superhelices (Arold *et al.*, 2010). This is consistent with bridging by H-NS across plectomenically interwound supercoiled DNA observed *in vitro* using atomic force microscopy (Maurer *et al.*, 2009). However, in the absence of crystallographic data for the full-length protein, it remains unclear how the presence of the nucleic-acid-binding domain might affect interactions at the amino terminus of H-NS. It is also important to consider that the isolated oligomerization domains of H-NS have been described as interacting both in parallel and anti-parallel and the issue of which arrangement is dominant *in vivo* remains unresolved (Esposito *et al.*, 2002; Bloch *et al.*, 2003; Dorman, 2004).

DNA-binding activity of H-NS

Early work suggested that the H-NS protein had a preference for binding to intrinsically curved DNA sequences (Yamada *et al.*, 1990; Dame *et al.*, 2001) and planar curvature was shown subsequently to contribute to targeting of particular DNA sequences by H-NS (Rimsky *et al.*, 2001). Curved DNA is often found in association with bacterial transcription promoters (Nov Klaiman *et al.*, 2009), so a preference for binding to such structures might make H-NS an ideal protein for targeting the transcriptional machinery (Dorman, 2007a). Recent investigations using chromatin immunoprecipitation on chip (ChIP-on-chip) have shown a strong correspondence between H-NS binding sites and high average A+T content in DNA (Grainger *et al.*, 2006; Lucchini *et al.*, 2006; Navarre *et al.*, 2006, 2007; Oshima *et al.*, 2006; Wade *et al.*, 2007; Dorman, 2007a; Fang and Rimsky, 2008; Dillon *et al.*, 2010). DNase I footprinting of H-NS bound to DNA has shown that a typical target promoter has more than one site protected from DNase I digestion by H-NS and that the sites often extend for scores of base-pairs along the DNA (Dattananda *et al.*, 1991; Overdier and Csonka, 1992; Schnetz, 1995; Beloin and Dorman, 2003; Yang *et al.*, 2005; Olekhnovich and Kadner, 2006; Wolf *et al.*, 2006; Corbett *et al.*, 2007; Lithgow *et al.*, 2007; Pul *et al.*, 2008; Sette *et al.*, 2009). Transcriptomic comparisons

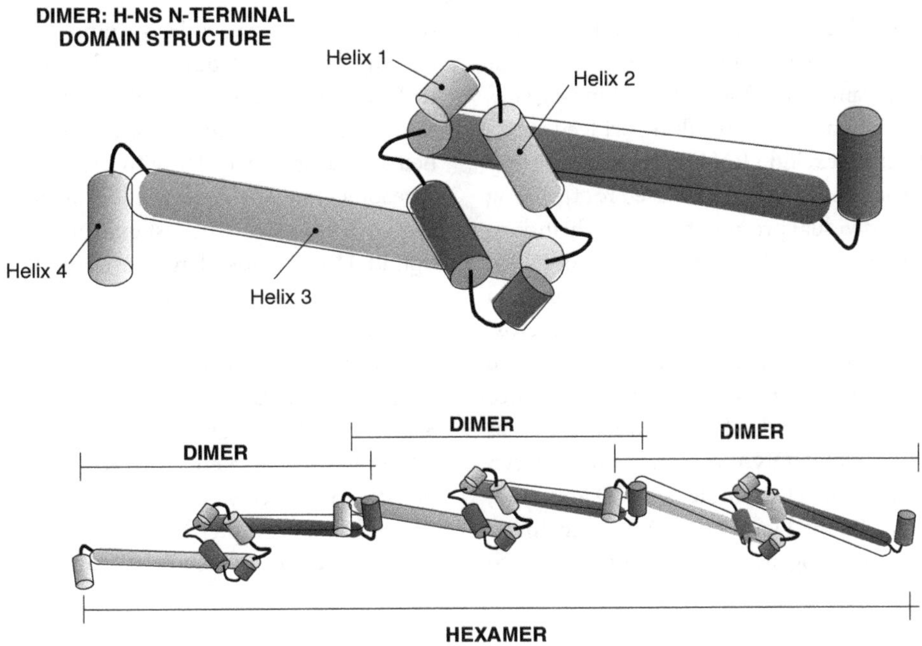

Figure 7.1 Dimerization and oligomerization interactions of the amino-terminal domains of H-NS. The four alpha helices formed by amino acids 1 to 82 of H-NS are shown in white and in grey. The helices for one monomer are numbered; helix 1 is the most N-terminal. The primary dimer interfaces reside in helices 1 and 2 of the paired monomers while the oligomerization is provided by the secondary dimer interfaces in helices 4. These primary and secondary dimerization regions are designated as sites 1 and 2, respectively (Arold *et al.*, 2010).

of wild-type bacteria with mutants deficient in H-NS show that loss of this protein can affect gene expression positively as well as negatively. However, in cases where the influence of H-NS has been dissected in detail at the molecular level, H-NS has been found to act as a transcription repressor or gene silencer. Presumably, H-NS influences indirectly the expression of those genes that show a positive response to its absence.

The *proU* promoter was one of the first to be studied in detail from the perspective of H-NS-mediated repression (Higgins *et al.*, 1988; Lucht *et al.*, 1994) and it was shown recently that a high-affinity binding site for the protein exists at two locations within the *proU* negative regulatory element that lies immediately downstream from the promoters (Bouffartigues *et al.*, 2007). When this DNA sequence (5′-TCGATATATT-3′) is excised and transferred to a location that is not otherwise attractive for H-NS binding, the new site is converted into a preferred H-NS binding site (Bouffartigues *et al.*, 2007). This sequence is present at a frequency that is higher than would be expected if it were to arise by chance alone among A+T-rich DNA regions that bind H-NS (Lang *et al.*, 2007). It is thought likely that this sequence can serve as an initial binding (or recruitment) site for H-NS from which the protein can then spread along the DNA (Bouffartigues *et al.*, 2007; Lang *et al.*, 2007; Navarre, 2010). Interestingly, this high affinity motif consists of A-T or T-A base-pairs at 8 of its 10 positions, in keeping with other observations that H-NS targets A+T-rich DNA.

DNA bridging and DNA-dependent polymerization

At least two mechanisms of transcription silencing by H-NS have been described. One involves the binding of the protein to a nucleation site in the vicinity of the target promoter followed by spreading of H-NS along the DNA to mask the binding sites for RNA polymerase; the other involves the creation of a DNA-protein-DNA bridge in which each H-NS dimer simultaneously contacts two distinct binding sites in DNA (Amit *et al.*, 2003; Dame *et al.*, 2006). If the binding sites bracket the promoter, the resulting bridged structure may either exclude RNA polymerase or trap it at the promoter (Fig. 7.2). DNase I footprinting studies of individual promoters gave early indications that H-NS binds cooperatively to DNA, protecting long stretches of the double helix from enzyme degradation. Later, atomic force microscopy and single-molecule methods based on the application of optical tweezers provided evidence that H-NS can form bridges between different parts of the same DNA molecule or between different, adjacent DNA molecules (Dame *et al.*, 2000, 2005, 2006; Dorman, 2007b; Wiggins *et al.*, 2009). Bridging activity is consistent with the possession by each H-NS dimer of two DNA binding domains. Cooperative binding of H-NS to DNA, beginning at a high-affinity site, allows the protein to spread laterally along the double helix, coating the DNA with H-NS. This lateral spreading may or may not be associated with bridge formation. The establishment of DNA-bound H-NS across large sections of a gene is consistent with transcription silencing and creates nucleoprotein structures that are analogous to chromatin in eukaryotic cells (Schröder and Wagner, 2000; Dame *et al.*, 2002; Dorman and Kane, 2009). Bridge formation lends itself to a more sophisticated form of transcriptional blockade in which RNA polymerase may be quarantined at a promoter lying within a DNA loop that is closed by a DNA-H-NS-DNA bridge (Fig.

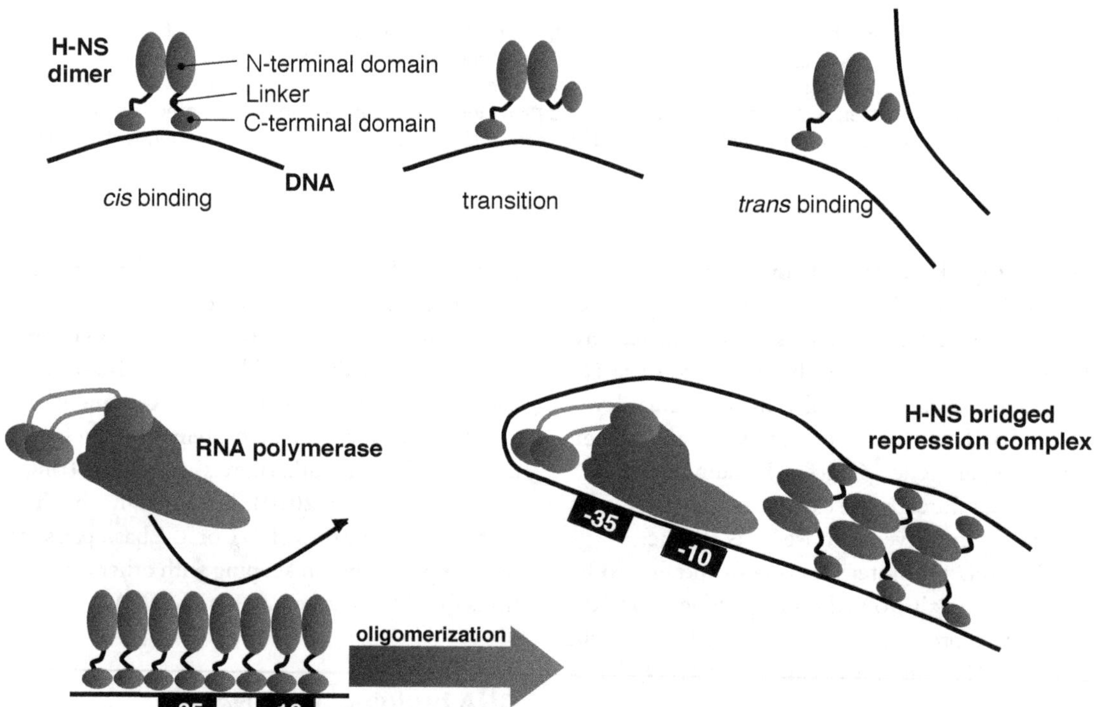

Figure 7.2 H-NS interactions with DNA. The H-NS dimer is a bi-dentate DNA binding protein that can interact with a DNA molecule through both DNA binding domains (*cis* interaction) or with two DNA molecules (or two separate parts of the same DNA duplex) though its *trans* binding mode. These are illustrated at the top of the figure, together with a transition state in which the H-NS dimer is attached to DNA via just one of its DNA binding domains. These interactions are developments of the H-NS-DNA relationships proposed by Wiggins *et al.* (2009). The H-NS protein can bind and polymerize along DNA, sequestering transcription promoters such that RNA polymerase is prevented from forming a closed transcription complex (bottom right). It may also bind and bridge DNA such that RNA polymerase bound to a promoter is unable to isomerize the closed transcription complex to an open one (bottom right).

7.2). The efficiency of H-NS-mediated promoter repression seems to depend on the nature of the sigma factor of RNA polymerase that is in use: H-NS can block the activity of RNA polymerase containing sigma-70 but not sigma-38 (Shin *et al.*, 2005; Grainger *et al.*, 2008). Recent *in vitro* studies have shown that H-NS switches between bridging and spreading modes in response to divalent cations, especially magnesium (Liu *et al.*, 2010). The physiological significance of this ability to toggle between two binding modes is currently unclear. It has been proposed that each H-NS dimer may form three types of relationship with DNA (Wiggins *et al.*, 2009). In the bridging mode, each DNA binding domain is connected *in trans* to a DNA duplex. The protein can also adopt a *cis*-binding mode in which both DNA binding domains contact adjacent parts of the same duplex. In the third mode, the protein is attached to a DNA molecule via one of its two DNA binding domains while the other domain is free and available for DNA interaction, either *in cis* or *in trans*. Perhaps the magnesium-dependent toggling effect drives a transition between these modes, with high magnesium favouring the trans-binding or bridging mode (Fig. 7.2).

Antagonism of H-NS-mediated transcription silencing

Promoter inactivation by the H-NS protein can be reversed by a variety of mechanisms. Indeed, one of the themes of the H-NS field concerns the largely *ad hoc* nature of the multitude of H-NS antagonism mechanisms that have been described (Stoebel *et al.*, 2008). This may reflect the fact that H-NS has a rather weak DNA binding activity (Shin *et al.*, 2005), perhaps allowing competing forces to displace it from DNA with relative ease. The *proU* operon was introduced earlier in the context of H-NS binding to high-affinity sites within the negative regulatory element. This operon encodes a transport system by osmoprotectants that are accumulated by the bacterium in response to the influx of potassium that immediately follows osmotic upshock (Sutherland *et al.*, 1986). Shifting the microbe to a high-osmolarity growth medium causes a strong stimulation of *proU* transcription, partly due to the displacement of H-NS (Lucht *et al.*, 1994; Gowrishankar and Manna, 1996; Nagarajavel *et al.*, 2007). No second protein has been found to mediate this displacement. Instead, it appears to result from a remodelling of the local topology of the *proU* promoter, a promoter that is sensitive to the changes in DNA supercoiling that accompany osmotic stress (Higgins *et al.*, 1988). Crystallographic analysis of the oligomerization domain of H-NS suggests a link between the pathway followed by this protein in its polymerized state and the trajectory of DNA in a plectonemic supercoil (Arold *et al.*, 2010). In addition, DNA-H-NS-DNA bridges may be weakened if the underlying DNA adopts an alternative course in three-dimensional space due to a resetting of its superhelicity in response to osmotic stress. Upshifts in growth medium osmolarity result in more negative supercoiling of DNA (Higgins *et al.*, 1988; McClellan *et al.*, 1990; Hsieh *et al.*, 1991; Cameron *et al.*, 2011) and also reduce the degree of intrinsic curvature of DNA (Sinden *et al.*, 1998). The resulting displacement of H-NS is thought to permit RNA polymerase to transcribe the *proU* operon.

Like osmotic upshock, increases in growth temperature also modify intrinsic curvature of DNA and its degree of negative supercoiling (Goldstein and Drlica, 1984; Sinden *et al.*, 1998). Derepression of the *virF* master control gene in the bacillary dysentery pathogen *Shigella flexneri* results in the transcriptional activation of the rest of the virulence cascade in that organism (Adler *et al.*, 1989; Hale, 1991; Dorman *et al.*, 2001). The chief environmental signal is an increase in the temperature to that of the human body, 37°C (Hale, 1991; Porter and Dorman, 1994; Durand *et al.*, 2000). The *virF* gene is silenced by H-NS and its promoter is bracketed by H-NS binding sites (Falconi *et al.*, 1998; Prosseda *et al.*, 1998). It has been proposed that the bridging by H-NS in its *trans*-acting DNA-binding mode is responsible for blockading the *virF* promoter and that elevated temperature alters both the degree of curvature and the location of the apex of the curve at the promoter DNA such that maintenance of the bridge becomes unsustainable. The result is H-NS displacement with concomitant activation of *virF* transcription (Prosseda *et al.*, 2004). The *proU* and *virF* promoters represent examples where the relief of H-NS-mediated transcription silencing occurs through modifications to local DNA structure without the need for input from other regulatory proteins. However, there are far more examples of the latter type of H-NS antagonism mechanism in the literature. In fact, the *virF* derepression example, while reliant on changes in local DNA architecture, is also one where a second protein, FIS, is known to play a positive role in transcription regulation (Falconi *et al.*, 1996).

One of the first examples of protein-mediated antagonism of transcription repression by H-NS to be discovered was also described in *S. flexneri*. This concerns VirB, an intermediate regulator within the virulence cascade, whose gene, *virB*, is silenced by H-NS and activated by the AraC-like DNA binding protein, VirF (Adler *et al.*, 1989; Porter and Dorman, 2002). VirF and H-NS compete for access to a common set of DNA binding sites at the *virB* promoter and displacement of H-NS results in *virB* transcription activation provided that the promoter DNA is appropriately negatively supercoiled (Tobe *et al.*, 1993, 1995). Another AraC-like protein that displaces H-NS at a virulence gene promoter is ToxT, a protein that controls the transcription of genes in the virulence cascade of the human pathogen *Vibrio cholerae*. The ToxT does not act simply as an antirepressor that displaces H-NS, it also makes protein–protein contact with RNA polymerase at the promoter, a role that is more typical of conventional transcription factors (Hulbert and Taylor, 2002; Yu and DiRita, 2002). Other H-NS antagonists have also been described as acting in this way, for example the SsrB DNA binding response regulator of the SsrA-SsrB two-component regulatory system governing the expression of virulence genes in the SPI-2 pathogenicity island of *Salmonella enterica* that allow the bacterium to survive inside macrophage (Walthers *et al.*, 2011).

The *S. flexneri* VirB DNA binding protein controls most of the genes involved in expression of the virulence phenotype of this intracellular pathogen (Adler *et al.*, 1989; Hale, 1991; Dorman and Porter, 1998; Dorman *et al.*, 2001). This protein is of particular interest because it appears to be a reprogrammed plasmid partition factor rather than a dedicated transcription factor (Buchrieser *et al.*, 2000). VirB influences the expression of its target genes by binding (at least initially) at a distance (Taniya *et al.*, 2003; Turner and Dorman, 2007). It displaces the H-NS protein, acting as a classic anti-repressor rather than as a transcription activator (Turner and Dorman, 2007). The modern large virulence plasmid in *S. flexneri/S. sonnei*/enteroinvasive *E. coli* has two functional protein-dependent plasmid partition systems (Buchrieser *et al.*, 2000; Dorman, 2008). It is thought that the *virB* gene is a vestige of a third, now redundant plasmid partition system of the ParA/ParB/*parS* type, in which the VirB protein is homologous to DNA-binding ParB-like proteins and its box1/box2 binding site in DNA is similar to the *parS* sequence (Turner and Dorman, 2007). No counterpart of ParA has been found in association with VirB (Buchrieser *et al.*, 2000). VirB binds at a distance from its target promoter and exerts its influence by interacting with DNA through a mechanism that is incompatible with maintenance of an H-NS–DNA complex. Specifically, the VirB protein has a DNA binding and wrapping activity which undermines the DNA bridging activity of H-NS leading to disruption of the H-NS–DNA complex that silences transcription of the target promoter at the *icsB* operon (Stoebel *et al.*, 2008; Turner *et al.*, 2007). Similarly, the H-NS-like Ler protein antagonist of H-NS, encoded by the LEE (locus for enterocyte effacement) pathogenicity island of enteropathogenic *E. coli* appears to form protein–DNA structures that may be incompatible with sustained binding of H-NS to the same DNA (Mellies *et al.*, 2011).

Antagonism of transcription repression due to H-NS does not result necessarily in the complete removal of the protein from DNA. Instead, the H-NS–DNA complex may be remodelled by the intervention of the antagonist to create a structure capable of supporting transcription initiation. The SlyA DNA-binding protein has been reported to remodel transcriptionally silenced promoters such as the *hlyE* (haemolysin E) promoter in *E. coli*. Here, SlyA binds to two regions that encompass the promoter and which overlap with two regions bound by H-NS. The resulting protein–DNA complex can present a spectrum of transitions from H-NS-DNA to SlyA-DNA-H-NS to SlyA-DNA that seem to reflect the relative concentrations of the two DNA binding proteins.

Although SlyA at high concentration can completely displace H-NS, it also becomes inhibitory to transcription when it binds in high amounts. Moreover, when H-NS concentration increases, it can displace SlyA. The mechanism provides for the re-establishment of H-NS-mediated repression of the *hlyE* promoter and makes the SlyA protein both an anti-repressor of the promoter and a repressor, depending on the local SlyA concentration (Lithgow *et al.*, 2007). At other promoters, SlyA acts to antagonise transcription repression by H-NS while another protein carries out the functions of a dedicated transcription factor, activating the promoter through interaction with RNA polymerase. For example, the PhoP response regulator up-regulates transcription of the *ugtL* and *pagC* genes in *S. enterica* while SlyA interferes with H-NS mediated repression of the same promoters (Perez *et al.*, 2008). The orthologue of SlyA in *Yersinia*, RovA, is also an antagonist of H-NS that positively affects the expression of virulence genes (Ellison and Miller, 2006a).

Natural and artificial truncates of H-NS

The ability of H-NS to repress transcription is also counteracted by hetero-oligomer formation. The H-NST$_{EPEC}$ and H-NST$_{UPEC}$ proteins, respectively from enteropathogenic *E. coli* (EPEC) and from uropathogenic *E. coli* (UPEC), resemble a truncated H-NS molecule that lacks a DNA binding domain (Williamson and Free, 2005) (Table 7.1). The presence of H-NS–H-NST heterodimers with their single DNA binding domain may result in weaker protein–DNA complexes at H-NS targets in the genome, leading to relief of transcriptional repression. The gene for the H-NST protein is part of a pathogenicity island in enteropathogenic strains of *E. coli* whose genes are subject to H-NS repression. It is tempting to speculate that the arrival of the H-NST gene in EPEC and UPEC along with the rest of the island via horizontal DNA transfer has modified the transcription regulatory activity of H-NS to facilitate the expression of the virulence genes. Not all H-NS orthologues are equally sensitive to interference by H-NST (Baños *et al.*, 2011).

Hha-like proteins and H-NS

The Hha family of proteins shows limited amino acid sequence and protein structural similarity to the oligomerization domain of H-NS (McFeeters *et al.*, 2007; Madrid *et al.*, 2007) and these proteins can bind to H-NS, perhaps altering its binding site preferences (Baños *et al.*, 2009) (Table 7.1). In *E. coli*, the 8 kDa Hha protein and its paralogue YdgT (Cnu) are present throughout growth at concentrations lower than that of H-NS so they are unlikely to form one-to-one complexes with every member of the H-NS population. It seems more probable that Hha-like proteins interact with a subpopulation of H-NS, perhaps influencing just some of the promoters to which H-NS binds, a proposition that is supported by data from *S. enterica* (Vivero *et al.*, 2008). The mechanism of Hha-like protein modulation of H-NS activity is unknown but is likely to be distinct from that exhibited by H-NST. H-NST, like genetically engineered truncates consisting of just the oligomerization domain of H-NS, are dominant negative when over-expressed and simply reduce H-NS transcription repression activity globally (Ueguchi *et al.*, 1997; Baños *et al.*, 2008; Navarre, 2010). In contrast, the effects of the Hha-like proteins seem to be more subtle and targeted (Baños *et al.*, 2009).

H-NS and FIS

H-NS has a complicated relationship with other NAPs in governing gene expression. For example, H-NS and FIS (the factor for inversion stimulation) sometimes work in partnership to repress the same promoter while in other cases FIS acts to antagonize H-NS repression. This includes a role for FIS as an activator of *hns*, the autorepressed gene that encodes H-NS (Falconi *et al.*, 1996), a supporting role for FIS in activating the *virF* promoter (Falconi *et al.*, 2001) and a positive role for FIS at the H-NS-repressed rRNA operon *rrnB* (Tippner *et al.*, 1994). Like H-NS, FIS prefers to bind to an A+T-rich DNA sequence (Kahramanoglou *et al.*, 2011) so it is possible that binding sites for the two proteins overlap frequently simply by chance. Given that A+T-rich DNA is a feature of many bacterial promoters, FIS-H-NS antagonism might provide the basis of a simple genetic

Table 7.1 H-NS-like proteins discussed in this chapter.

Protein	Organism	Gene location	Binding preferences	Native protomer
H-NS	*Escherichia coli*, *Salmonella* spp. and many others	Chromosome (the *hns* gene is part of the large chromosomal inversion differentiating *E. coli* from *Salmonella*)	A+T-rich DNA; A+U-rich RNA (*E. coli*); High-affinity site in DNA: 5′-TCGATATATT-3′	Homodimer, heterodimer (with StpA, Sfh, Hfp)
Bph3	*Bordetella* spp.	Chromosome	A+T-rich DNA	ND
Cnu	*E. coli*	Chromosome	None: acts via H-NS	Heteromer formed with H-NS (Homodimers?)
Hfp	*E. coli* 536	Pathogenicity island	Similar to H-NS	Homodimer; heterodimer (with H-NS)
Hha	*E. coli, Salmonella* spp. and many others	Chromosome and plasmid	None: acts via H-NS	Heteromers formed with H-NS (Homodimers?)
H-NST	*E. coli* (EPEC and UPEC)	Pathogenicity islands	None: acts via H-NS interaction	Heteromers formed with H-NS (homodimers?)
HvrA	*Rhodococcus capsulatus*	Chromosome	ND	ND
Lsr2	*Mycobacterium* spp.	Chromosome (plasmid in *Rhodococcus equi*)	A+T-rich DNA	Homodimers
MvaT	*Pseudomonas* spp.	Chromosome	A+T-rich DNA	Homodimer; cooperates with MvaU
Pmr	*Pseudomonas* spp.	IncP-7 plasmids	A+T-rich DNA	Homodimer?
Rok	*Bacillus subtilis*	Chromosome	A+T-rich DNA	Homodimer
Sfh	*Salmonella* spp., *Shigella flexneri*	IncHI1 plasmids	DNA binding; similar preferences to H-NS and StpA	Homodimer (possibly sensitive to proteolysis); Heterodimers formed with H-NS, Sfh
SPB	*Acinetobacter* spp., *Psychrobacter* spp., *Rhodobacter sphaeroides*	Chromosome	ND	ND
StpA	*Escherichia coli, Salmonella* spp. and many others	Chromosome	RNA binding Binds similar sites to H-NS in DNA	Proteolytically sensitive homodimer; proteolysis-resistant heterodimer formed with H-NS
TurA, TurB, TurC, TurD, TurE	*Pseudomonas putida*	Chromosome	ND	TurA–TurB heterodimers; orthoglogues of MvaT
VicH	*Vibrio chloerae*	Chromosome	A+T-rich DNA	Homodimer?
XrvA	*Xanthomonas oryzae*	Chromosome	ND	ND
YdgT	*Salmonella* spp.	Chromosome	None: acts via H-NS	Heteromers formed with H-NS (homodimers?)

ND, not determined.

switch, modulating transcription activity. In this way, antagonism of H-NS DNA binding activity by FIS may be commonplace, at least under growth conditions where FIS is plentiful. FIS is at its most abundant in bacteria entering the exponential phase of growth and declines rapidly thereafter (Ball *et al.*, 1992; Keane and Dorman, 2003) although an exception occurs in bacteria that are maintained in conditions of poor aeration where FIS levels are sustained into later stages of the growth cycle (Ó Cróinín and Dorman, 2007). Unfortunately, investigations using whole-genome methods based on chromatin immunoprecipitation have produced conflicting results. One survey of FIS and H-NS binding sites across the *E. coli* chromosome revealed a striking correspondence in the regions that are bound by these proteins (Grainger *et al.*, 2006) while a second study found that the sites bound by the two proteins were mutually exclusive (Kahramanoglou *et al.*, 2011). It should be noted that these experiments were performed under different conditions, including the use of minimal medium containing fructose to grow the bacteria in the 2006 study and rich LB medium without sugar supplements in the later work. These different growth regimes are likely to result in physiologically distinct populations of bacteria that are difficult to compare.

H-NS targets horizontally acquired foreign DNA

The genome of the facultative intracellular pathogen *S. enterica* serovar Typhimurium (*S.* Typhimurium) contains several pathogenicity islands that consist of contiguous regions of A+T-rich DNA that are thought to have been acquired by horizontal transfer after the *Salmonella* and *E. coli* lineages diverged from their common ancestor about 100 million years ago (Ochman and Groisman, 1996). Genes in these *Salmonella* pathogenicity islands (SPI) confer on *S.* Typhimurium the ability to invade host cells (SPI-1) and to survive in host macrophage (SPI-2) (Valdez *et al.*, 2009). SPI gene expression is subject to very complex regulation (Hensel, 2000; Rhen and Dorman, 2005; Ellermeier and Slauch, 2007) and a role for H-NS in *S.* Typhimurium virulence was established early on (Schechter *et al.*, 2003; Harrison *et al.*, 1994; O'Byrne and Dorman, 1994). Studies using DNA microarrays and chromatin immunoprecipitation-on-chip methods revealed that H-NS preferentially silences transcription of these A+T-rich genes (Lucchini *et al.*, 2006; Navarre *et al.*, 2006, 2007). It has been suggested that this represents a form of self/non-self recognition in which H-NS preferentially targets DNA of foreign origin, silencing any transcription units that this DNA contains (Navarre, 2010). Of course, H-NS also targets parts of the core genome, so it is probably not appropriate to think of it exclusively in terms of a defence against foreign DNA (Dorman, 2007a). Nevertheless, it is attractive to think of H-NS as contributing to a process in which genes acquired laterally are incorporated into the genome of a bacterium without causing undue disturbance to its physiology through inappropriate expression. Although many bacteria are exposed constantly to lateral transfer of DNA, much of the work on the role of H-NS in this process has focused on long-established DNA sequences of foreign origin, such as pathogenicity islands. There is some evidence that genetic elements that are currently being transferred horizontally have evolved mechanisms for dealing with H-NS. For example, bacteriophage T7 expresses from among its 'middle' genes the protein 5.5 which inhibits the transcription silencing activity of H-NS in *E. coli* (Liu and Richardson, 1993), suggesting that eliminating this DNA binding protein is part of the strategy used by the phage to proliferate.

Regulation of the *hns* gene

H-NS is an abundant nucleoid-associated protein with approximately 20,000 copies per cell in *E. coli* (Williams and Rimsky, 1996; Atlung and Ingmer, 1997; Dorman *et al.*, 1999). The overall quantity of H-NS protein per cell in a bacterial population only varies within a very narrow window during growth in batch culture and *hns* transcription declines as the bacteria enter stationary phase (Free and Dorman, 1995). Transcription of the *hns* gene is driven from a single promoter and the *hns* mRNA is monocistronic; the *hns* promoter does not respond to global changes in DNA supercoiling (Free and Dorman, 1994). Expression of the *hns* gene is controlled at both the transcriptional and posttranscriptional levels;

there is no evidence that H-NS is subject to allosteric control or to posttranslational modification (Dorman, 2004). The *hns* promoter is repressed by H-NS and is activated by the FIS protein and by the cold shock protein CspA (Brandi *et al.*, 1994; Giangrossi *et al.*, 2001). H-NS is itself a cold shock protein and its level increases in bacteria that experience a sudden fall in temperature (La Teana *et al.*, 1991). The positive role of FIS involves antagonism of H-NS-mediated repression of the *hns* promoter. The iron response regulator FUR is a repressor of the *hns* gene in *S. enterica* (Troxell *et al.*, 2011) and H-NS is involved in controlling iron metabolism by repressing the *ftnA* gene coding for ferritin in *E. coli* (Nandal *et al.*, 2010) (Fig. 7.3). At a posttranscriptional level, translation of the *hns* message is inhibited by the untranslated small regulatory RNA DsrA (Fig. 7.3). This molecule forms a double-stranded RNA with the 5′ end of the *hns* message, sequestering its translation signals. Efficient interaction between DsrA and the *hns* message depends on the highly pleiotropic RNA chaperone protein Hfq (Lease *et al.*, 1998; Chao and Vogel, 2010). Interestingly, DsrA links *hns* expression to that of *rpoS*, the gene coding for the stress and stationary phase sigma factor RpoS. Expression of H-NS and RpoS is affected in a reciprocal manner by DsrA (Lease and Belfort, 2000). A further level of complexity is revealed by the discovery that H-NS can specifically bind and alter the stability of the RpoS mRNA and the DsrA sRNA (Brescia *et al.*, 2004). As was mentioned above, RNA polymerase containing RpoS is not constrained in transcription initiation by H-NS in the same way as the same polymerase containing RpoD. Expression of the *dsrA* gene is repressed by the LysR-like transcription factor, LeuO, a protein that is encoded by the H-NS-repressed gene *leuO* (Klauck *et al.*, 1997). These connections point to the presence of a complicated regulatory network

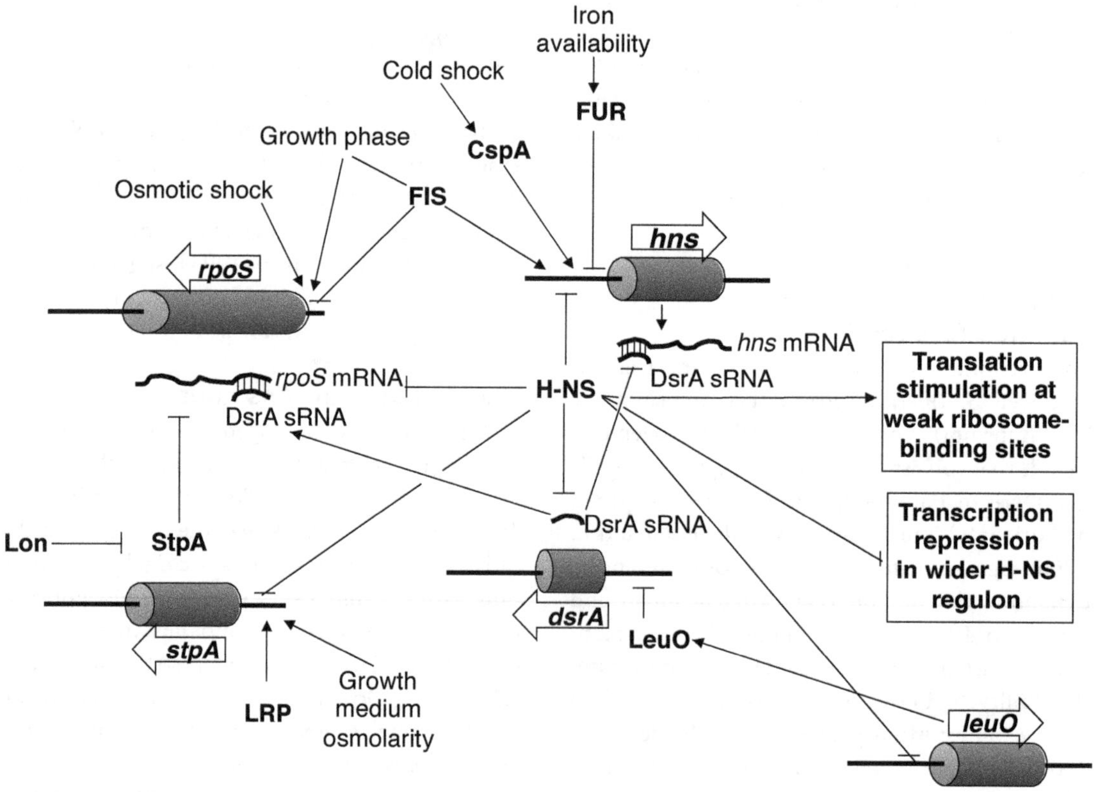

Figure 7.3 H-NS as a nexus of global regulation. The principal genes are represented by cylinders with block arrows to indicate the direction of transcription. Positive regulatory inputs are shown by thin arrows and negative ones by 'T' symbols. The genes are oriented within the diagram for convenience and not to suggest their relative orientations in the chromosome.

acting at both the transcriptional and posttranscriptional levels whose details are still being worked out (Fig. 7.3).

A link exists between *hns* transcription repression and the replication of the chromosome in *E. coli* (Free and Dorman, 1995). If chromosome replication fork movement is arrested, the H-NS protein represses transcription of the *hns* gene. In mutants unable to express functional H-NS, transcription of *hns* is not affected by fork arrest. The result is a rough matching of the H-NS population in the cell to chromosome copy number. This raises the possibility that the cell may require a minimal amount of the protein to meet its needs, either in terms of nucleoid structure, transcription control or both. Artificial overexpression of H-NS or elimination of its expression by mutation alters the fitness of the bacterium, which is consistent with the need for an optimal level of H-NS (McGovern *et al.*, 1994; Lucchini *et al.*, 2006; Navare *et al.*, 2006). Mutants expressing H-NS derivatives unable to bind to DNA display an altered cell cycle. At 30°C the mutants have a chromosome replication time that is halved compared with the wild type. The molecular basis of this phenotype is unclear (Atlung and Hansen, 2002) as is the reported link between DnaA and H-NS at the origin of chromosome replication (Katayama *et al.*, 1996).

The StpA protein

StpA, the 'suppressor of the td$^-$ phenotype', is an RNA chaperone that was discovered during a search in *E. coli* for genetic suppressors of bacteriophage T4 mutants that are deficient in splicing the group I intron of the T4 thymidylate synthase gene (Zhang and Belfort, 1992; Zhang *et al.*, 1995). StpA was immediately recognized as being a paralogue of H-NS and the two proteins are 58% identical at the amino acid sequence level, with many overlapping functions, including DNA binding activity (Dorman *et al.*, 1999; Doetsch *et al.*, 2010) (Table 7.1). StpA has a similar domain organization to H-NS and has its RNA/DNA binding activity in the carboxy-terminal domain (Cusick and Belfort, 1998). It stimulates group I intron slicing by accelerating intron RNA folding (Zhang *et al.*, 1995; Rajkowitsch and Schroeder, 2007) as well as loosening non-specific RNA–RNA interactions as part of its chaperone activity (Waldsich *et al.*, 2002). StpA displays no specific RNA sequence preferences for binding (Mayer *et al.*, 2007).

StpA forms heterodimers with H-NS (Sonden and Uhlin, 1996; Williams *et al.*, 1996; Zhang *et al.*, 1996) but it is unclear if these have properties that are distinct from those of H-NS or StpA homodimers. The StpA protein is sensitive to proteolytic degradation by the Lon protease whereas H-NS is not; formation of heterodimers with H-NS protects StpA from proteolysis (Johansson and Uhlin, 1999; Johansson *et al.*, 2001).

The *stpA* gene is monocistronic and is transcribed from a single promoter that is repressed by H-NS and activated by the leucine-responsive regulatory protein, LRP (Free and Dorman, 1997) (Fig. 7.3). The positive role of LRP is especially important in bacteria growing in minimal medium. Transcription of *stpA* is enhanced in response to osmotic shock, a finding that is especially interesting in the light of the recent finding that StpA influences the expression of RpoS, the stress and stationary phase sigma factor (Lucchini *et al.*, 2009). Despite strong repression by H-NS, the *stpA* gene is transiently expressed in bacteria growing in rich medium during the exponential phase of growth (Free and Dorman, 1997). It is not clear if this expansion of the StpA protein population leads to changes in the transcription profile of the cell. In *S. enterica*, StpA contributes to the suppression of RpoS sigma factor expression in exponentially growing cells and this should bring benefits by preventing the inappropriate expression of genes whose products are not required before stationary phase (Fig. 7.3). StpA does not control a separate regulon of genes in *S. enterica*, instead it targets a subset of the H-NS regulon, many of which are also controlled by RpoS, cAMP-CRP and PhoP (Lucchini *et al.*, 2009). In *E. coli*, inactivation of *stpA* has no effect on the global transcription pattern; experiments with *hns* and *hns stpA* mutants of *E. coli* reveal changes in similar groups of genes but the detail of the responses differ between the single and double mutants (Müller *et al.*, 2006). These data suggest that StpA can modulate the influence of H-NS at some promoters but do not necessarily mean that these effects require the formation of StpA-H-NS heteromers. Loss of both StpA and H-NS expression results in a reduction in growth rate in *E. coli* but this defect can be suppressed by a specific point mutation (A404E) in the *spoT* gene that reduces the level of the alarmone ppGpp in the cell. This reduction has consequences for the expression of the cAMP-CRP regulon because the *crp* gene is regulated negatively by ppGpp (Johansson *et al.*, 2000). This reveals a complex set of connections between the H-NS and StpA paralogues, the stringent response and the cAMP-CRP regulon. It further complicates the already-complex set of links between H-NS, StpA and the RpoS regulon (Fig. 7.3).

Although StpA is the H-NS-like protein that has been examined in most detail in terms of RNA binding, it is becoming clear that H-NS itself has significant RNA binding activity and that this is physiologically very relevant. This protein can reposition the small ribosomal subunit on the mRNA that specifies the MalT protein in *E. coli* to compensate for the presence of a poor ribosome-binding site in the message (Park *et al.*, 2010). These interactions occur within a region of A+U-rich RNA, calling to mind the preference shown by H-NS for binding to A+T-rich regions in DNA. It seems likely that this ribosome repositioning activity is a general property of H-NS and is not confined to the message specified by the *malT* gene (Park *et al.*, 2010) (Fig. 7.3).

Hetero-oligomers with other H-NS-like proteins

The StpA–H-NS heterodimer relationship is not unique. For example, the uropathogenic strain 536 of *E. coli* expresses the Hfp H-NS-like protein in addition to StpA and Hfp can form heteromers with H-NS (Müller *et al.*, 2010) (Table 7.1). Mutants unable to express H-NS and Hfp displayed severe growth defects at low temperature whereas mutants deficient in H-NS and StpA had growth defects at high temperatures, suggesting that the Hfp and StpA proteins exert distinct influences within strain 536 physiology. Expression of the *hfp* gene is dependent on the FIS protein and repressed by H-NS, properties that it shares with the *hns* gene, and *hfp* may be a component of the RpoS regulon. This shows that the *hfp* gene, which is part of a horizontally acquired genetic element, is now embedded in the regulatory circuitry of the bacterium. Interestingly, the Hfp protein concentration peaks at a later period of the cell cycle than does the concentration of the StpA protein; in this respect Hfp mimics the expression pattern of yet another horizontally acquired H-NS-like protein, Sfh (Dorman, 2010).

Unlike Hfp and related proteins that are found within pathogenicity islands in the chromosomes of pathogenic bacteria (Williamson and Free, 2005; Müller *et al.*, 2010), the gene that encodes the Sfh protein is carried on a broad-host-range self-transmissible plasmid known as R27 (Sherburne *et al.*, 2000; Beloin *et al.*, 2003; Deighan *et al.*, 2003) (Table 7.1). Like Hfp, the Sfh protein is expressed late in growth although transcription of the *sfh* gene is maximal at earlier stages of the growth cycle. This indicates that *sfh* gene expression is subject to a posttranscriptional step that delays maximal build-up of the protein until the transition from the exponential to the stationary phase of growth (Doyle and Dorman, 2006). The role of the Sfh protein, which can form heteromers with either H-NS or StpA (Deighan *et al.*, 2003), extends beyond the plasmid that encodes it. For example, the protein binds to a subset of the targets in the *Salmonella* chromosome to which H-NS binds and when H-NS is removed from the cell, Sfh can expand its target range to bind to most of the locations previously occupied by H-NS (Dillon *et al.*, 2010). While Sfh may modulate the activity of H-NS, perhaps by forming heteromers (Baños *et al.*, 2009), it can also substitute for it (Beloin *et al.*, 2003). The DNA of the R27 plasmid is A+T-rich and it contains many matches to the high-affinity binding sites for H-NS (Lang *et al.*, 2007). It has been suggested that the carriage of a gene coding for a protein with the ability to 'top up' the cellular content of H-NS-like proteins may obviate any undesirable disturbances to the H-NS to DNA ratio in the cell because of the presence of the R27 plasmid (Doyle *et al.*, 2007).

Other H-NS-like proteins

Orthologues of H-NS have been identified in many Gram-negative bacteria (Bertin *et al.*, 1999) and functional analogues are being discovered in Gram-positives. For example, the Rok protein of *Bacillus subtilis* was characterized originally as a regulator of competence for DNA uptake; it has recently been found to target A+T-rich DNA sequences throughout the *B. subtilis* genome, including sites known to be occupied by mobile genetic elements that are capable of horizontal transmission (Smits and Grossman, 2010) (Table 7.1). The Lsr2 protein represents a class of H-NS-like proteins in the pathogen *Mycobacterium tuberculosis* and related actinomycetes. It has DNA bridging activity and can substitute for H-NS, complementing many of the phenotypes of an *hns* mutant; in its turn, the *hns* gene has been shown to complement an *lsr2* mutant of the genetically tractable *M. smegmatis* (Gordon *et al.*, 2008). Although Lsr2 is not a homologue of H-NS, it shares a similar functional domain structure and mode of interaction with DNA, indicating that it is a true functional analogue that has evolved to fulfil a similar role in actinomycetes related to *Mycobacterium* species (Gordon *et al.*, 2010).

In *Pseudomonas aeruginosa*, MvaT is the founding member of a family of H-NS-like proteins that influence the expression of genes involved in both core metabolic functions and virulence (Tendeng *et al.*, 2003a; Li *et al.*, 2009) (Table 7.1). As with H-NS, oligomerization is critical for the biological activities of MvaT (Castang and Dove, 2010) and this proterin has been shown to participate in DNA–protein–DNA bridging (Dame *et al.*, 2005). It works in combination with its paralogue MvaU to regulate the transcription of genes involved in adhesion and biofilm formation (Vallet-Gely *et al.*, 2007) arginine metabolism, prophage induction, pigmentation (Li *et al.*, 2009) and the expression of small regulatory RNA (Brencic *et al.*, 2009) by binding to A+T-rich regions of the genome (Castang *et al.*, 2008). Orthologues of MvaT/MvaU have been identified in *Ps. putida* (Renzi *et al.*, 2010) and *Ps. fluorescens* (Baehler *et al.*, 2006) where they also act as global regulators of gene expression. The Pmr protein is an IncP-7 plasmid-encoded H-NS-like protein that has been described in *Pseudomonas* spp.; it seems to perform a similar role to the Sfh protein from IncHI.1 R27-like plasmids in *Shigella/Salmonella* (Takeda *et al.*, 2010; Yun *et al.*, 2011).

H-NS has been investigated in the plant pathogens *Erwinia amylovora* (Hildebrand *et al.*, 2006) and in *E. chrysanthemi* (Nasser *et al.*, 2001; Nasser and Reverchon, 2002) where it is a virulence factor. It is also a negative regulator of virulence gene expression in *Yersinia enterocolitica* (Ellison and Miller, 2006b), *Y. pestis* (Lawrenz and Miller, 2007) and *Y. pseudotuberculosis* (Heroven *et al.*, 2007). The *Vibrio* counterpart of H-NS, VicH, regulates negatively the transcription of the major virulence determinants of this important human pathogen (Stonehouse *et al.*, 2011). VicH is structurally similar to H-NS (Cerdan *et al.*, 2003) and is 50% identical in amino acid sequence (Tendeng *et al.*, 2000). VicH-mediated repression of key promoters in the *V. cholerae* virulence regulon is antagonised by the binding of transcription factors such as ToxT, an AraC-like protein (Yu and DiRita, 2002; Stonehouse *et al.*, 2011). Thus the pattern of regulation by H-NS (targeting A+T-rich DNA, silencing virulence gene expression and antagonism by other DNA binding proteins) is similar in *V. cholerae* to that described in bacteria such as *Salmonella* where the role of H-NS in gene regulation has been studied in more detail. Other cases of H-NS-like proteins controlling gene expression in organisms away from the most intensively studied examples include the H-NS-like protein HvrA that regulates nitrogen fixation genes in *Rhodobacter capsulatus* (Raabe *et al.*, 2002), the Bph3 protein in the whooping cough pathogen *Bordetella pertussis* (Goyard and Bertin, 1997), the SPB protein of *Rhodobacter sphaeroides* (Bertin *et al.*, 1999), the H-NS and SPB-like proteins in the low-temperature specialists *Acinetobacter* spp. and *Psychrobacter* spp. (Tendeng *et al.*, 2003b) and XrvA in the rice pathogen *Xanthomonas oryzae* (Feng *et al.*, 2009) (Table 7.1). The available information is once again indicative of a negative role for these proteins in influencing the transcription of their target genes.

Nucleoid structure

The family of nucleoid-associated proteins is so-named to reflect the location of these molecules in the bacterium, but the name also implies that they contribute to the higher organization of the nucleoid. Set against this is observation that nucleoid structure has been reported to be preserved in bacteria devoid of nucleoid-associated proteins, including H-NS (Zimmerman, 2001). This probably reflects the fact that the factors involved in the maintenance of nucleoid structure are multifactorial and complex (Dillon and Dorman, 2010).

H-NS certainly has the potential to fulfil a structural role in the nucleoid (Luijsterburg *et al.*, 2006, 2008). It is abundant and it is known to bind widely within the genome. It does not seem to be excluded from any large sections of those chromosomes where its distribution has been studied (Lucchini *et al.*, 2006; Navarre *et al.*, 2006; Oshima *et al.*, 2006; Dillon *et al.*, 2010). This is in contrast to the MatP, SeqA and SlmA proteins that organize the macrodomain structure of the genome and which show a macro-domain-associated restricted distribution (Bernardt and de Boer, 2005; Mercier *et al.*, 2008; Sanchez-Romero *et al.*, 2010; Cho *et al.*, 2011; Tonthat *et al.*, 2011). This may indicate that H-NS is not involved uniquely in macrodomain organization but it may instead play an important structural role at a more local level. Genetic evidence has identified H-NS as a possible domainin, that is a protein that contributes to the formation of the boundaries at the ends of the 10- to 14-kbp chromosomal domains that form during exponential phase of the growth cycle and then become dispersed as the bacterium enters stationary phase (Hardy and Cozzarelli, 2005). The bridging activity of H-NS might lend itself to domainin activity by permitting H-NS to tie together otherwise distant segments of DNA, causing the intervening sequences to form the domain loop (Noom *et al.*, 2007). This proposal is also consistent with the observation form chromatin immunoprecipitation studies that indicate that H-NS has a preference for intergenic regions since one might expect complete genes rather than just segments of genes to be encompassed within domains. The bridging activity of the MvaT protein in *Pseudomonas* (Dame *et al.*, 2005) may also influence nucleoid structure in that organism. Fluorescence resonance energy transfer experiments have revealed RNA bridging activity of the H-NS paralogue StpA that involves the same domains known to bind DNA (Rajkowitsch and Schroeder, 2007); this might also be indicative of a contribution by StpA to nucleoid structure. Moreover, StpA may also contribute to the structuring of the nucleoid through its oligomeric relationship with H-NS and/or through its RNA binding activity because transcripts remain adjacent to the genes that specify them in *E. coli* (Montero Llopis *et al.*, 2010). Recent physical analysis of chromosome domains has implicated H-NS (but not StpA) as an architectural element in the maintenance of nucleoid superstructure (Wang *et al.*, 2011).

Conclusions

The H-NS protein exerts a very wide-ranging influence over the gene expression profiles of those bacteria that harbour it. This influence is overwhelmingly negative, at least at those genes were H-NS has a direct influence on transcription. There is a striking correspondence between H-NS-mediated repression and the control of expression of genes involved in bacterial virulence (Dorman, 2004). In many cases this can be explained by the fact that such genes consist of DNA that has a higher average A+T content than the rest of the genome, making them ideal targets for H-NS binding and associated transcription silencing. Despite the breadth of its influence over transcription, H-NS is not an essential protein in *E. coli*; in *S. enterica* compensatory mutations in loci such as *rpoS* and *phoP* allow the mutant to remain viable (Lucchini *et al.*, 2006; Navarre *et al.*, 2006). Nevertheless, the loss of H-NS with the associated disorganization of the transcription pattern of the cell is likely to make the mutant uncompetitive with the wild type. The impressive range of H-NS antagonists that has been described indicates that H-NS-mediated repression is a significant force to be reckoned with, and may have created multiple opportunities for gene regulatory switches to have evolved. Because the proteins that counter H-NS-mediated repression are themselves responsive to a wide variety of environmental signals, H-NS has indirectly brought the members of its regulon under the positive influence of multiple aspects of the environment. This simple pattern of repression alleviated by environmentally sensitive response also lends itself to ever-increasing degrees of refinement, opening the way for the evolution of more and more sophisticated mechanisms of transcription regulation and transcription coordination. The link between H-NS and nucleoid structure is not so straightforward as one might have predicted based on our knowledge of the protein and our expectations of the properties of nucleoid structuring elements. H-NS continues to have the potential to influence nucleoid structure and there is already some evidence that it does so, but there is still some distance to go in our investigations of this aspect of this fascinating protein.

Future trends

One may predict that in the immediate future the number of genes described as being under regulation by H-NS will continue to increase and that the number of bacterial species described as having H-NS or an H-NS analogue will also expand – this is merely to extrapolate from what has been happening in the immediate past. Similarly we may anticipate the discovery of more examples of genes coding for H-NS orthologues on mobile genetic elements (Takeda *et al.*, 2011). We may also see more detailed descriptions of H-NS antagonism mechanisms based on investigations of individual promoters, possibly supported by ever more sophisticated studies performed using single-molecule techniques. Such methods have already revolutionized our understanding of H-NS biology and have the potential to teach us much more in the future. The application of high-resolution imaging techniques to the study of H-NS in relation to the structure of the nucleoid is now under way and will surely assist in advancing this part of the field. While H-NS–DNA interactions will continue to be the focus of research for most in the field, the interaction of H-NS with RNA is also likely to receive more attention as its potential to affect gene expression post-transcriptionally is studied more intensively (Park *et al.*, 2010). There is a need to examine the behaviour of H-NS under a wider range of growth conditions than has been attempted so far, both at a whole genome level and at the level of individual genes. Ideally, these experiments should include work that studies the activities of H-NS in pathogenic bacteria during infection. Work of this type is likely to be especially valuable in developing new approaches to treat or prevent disease, especially in the case of strategies that do not rely merely on killing the pathogen but on making it less competitive owing to disruption of its gene expression programme and/or nucleoid structure.

Web resources

http://cellobservatory.leidenuniv.nl/research/chromatin-structure-and-epigenetics-bacteria-and-archaea
http://www.epidna.com/showabstract.php?pmid=17108966
http://www.nat.vu.nl/en/research/physics-life-health/single-molecule-cell-level-biophysics/dna-enzymes/
http://commons.wikimedia.org/wiki/File:DNA-H-NS-Bridge.jpg
http://sciblogs.co.nz/code-for-life/2010/08/24/coiling-bacterial-dna/
http://2010.igem.org/Team:Cambridge/Bioluminescence/Background
http://www.weizmann.ac.il/complex/stavans/elasticity

Acknowledgements

This work was supported by grants from Science Foundation Ireland.

References

Adler, B., Sasakawa, C., Tobe, T., Makino, S., Komatsu, K., and Yoshikawa, M. (1989). A dual transcriptional activation system for the 230 kb plasmid genes coding for virulence-associated antigens of *Shigella flexneri*. Mol. Microbiol. 3, 627–635.

Amit, R., Oppenheim, A.B., and Stavans, J. (2003). Increased bending rigidity of single DNA molecules by H-NS, a temperature and osmolarity sensor. Biophys. J. *84*, 2467–2473.

Arold, S.T., Leomard, P.G., Parkinson, G.N., and Ladbury, J.E. (2010). H-NS forms a superhelical protein scafold for DNA condensation. Proc. Natl. Acad. Sci. U.S.A. *107*, 15728–15732.

Atlung, T., and Hansen, F.G. (2002). Effect of different concentrations of H-NS protein on chromosome replication and the cell cycle in *Escherichia coli*. J. Bacteriol. *184*, 1843–1850.

Atlung, T., and Ingmer, H. (1997). H-NS: a modulator of environmentally regulated gene expression. Mol. Microbiol. *24*, 7–17.

Baehler, E., de Werra, P., Wick, L.Y., Péchy-Tarr, M., Mathys, S., Maurhofer, M., and Keel, C. (2006). Two novel MvaT-like global regulators control exoproduct formation and biocontrol activity in root-associated

Pseudomonas fluorescens CHA0. Mol. Plant Microbe Interact. *19*, 313–329.

Bakaev, V.V. (1981). Structure of chromosomal deoxyribonucleoproteins. XI. Organization of deoxyribonucleoprotein complex in bacterial cells. Mol. Biol. (Mosk.) *15*, 1350–1363.

Ball, C.A., Osuna, R., Ferguson, K.C., and Johnson, R.C. (1992). Dramatic changes in Fis levels upon nutrient upshift in *Escherichia coli*. J. Bacteriol. *174*, 8043–8056.

Baños, R.C., Pons, J.I., Madrid, C., and Juárez, A. (2008). A global modulatory role for the *Yersinia enterocolitica* H-NS protein. Microbiology *154*, 1281–1289.

Baños, R.C., Vivero, A., Aznar, S., García, J., Pons, M., Madrid, C., and Juárez, A. (2009). Differential regulation of horizontally acquired and core genome genes by the bacterial modulator H-NS. PLoS Genet. *5*, e1000513.

Baños, R.C., Aznar, S., Madrid, C., and Juárez, A. (2011). Differential functional properties of chromosomal- and plasmid-encoded H-NS proteins. Res. Microbiol. *162*, 382–385.

Beloin, C., and Dorman, C.J. (2003). An extended role for the nucleoid structuring protein H-NS in the virulence gene regulatory cascade of *Shigella flexneri*. Mol. Microbiol. *47*, 825–838.

Beloin, C., Deighan, P., Doyle, M., and Dorman, C.J. (2003). *Shigella flexneri* 2a strain 2457T expresses three members of the H-NS-like protein family: characterization of the Sfh protein. Mol. Genet. Genomics *270*, 66–77.

Bernhardt, T.G., and de Boer, P.A. (2005). SlmA, a nucleoid-associated, FtsZ binding protein required for blocking septal ring assembly over chromosomes in *E. coli*. Mol. Cell *18*, 555–564.

Bertin, P., Benhabiles, N., Krin, E., Laurent-Winter, C., Tendeng, C., Turlin, E., Thomas, A., Danchin, A., and Brasseur, R. (1999). The structural and functional organization of H-NS-like proteins is evolutionarily conserved in Gram-negative bacteria. Mol. Microbiol. *31*, 319–329.

Bloch, V., Yang, Y., Margeat, E., Chavanieu, A., Augé, M.T., Robert, B., Arold, S., Rimsky, S., and Kochoyan, M. (2003). The H-NS dimerization domain defines a new fold contributing to DNA recognition. Nat. Struct. Biol. *10*, 212–218.

Bouffartigues, E., Buckle, M., Badaut, C., Travers, A., and Rimsky, S. (2007). H-NS cooperative binding to high-affinity sites in a regulatory element results in transcriptional silencing. Nat. Struct. Mol. Biol. *14*, 441–448.

Brandi, A., Pon, C.L., and Gualerzi, C.O. (1994). Interaction of the main cold shock protein CS7.4 (CspA) of *Escherichia coli* with the promoter region of *hns*. Biochimie *76*, 1090–1098.

Brencic, A., McFarland, K.A., McManus, H.R., Castang, S., Mogno, I., Dove, S.L., and Lory, S. (2009). The GacS/GacA signal transduction system of *Pseudomonas aeruginosa* acts exclusively through its control over the transcription of the RsmY and RsmZ regulatory small RNAs. Mol. Microbiol. 73, 434–445.

Brescia, C.C., Kaw, M.K., and Sledjeski, D.D. (2004). The DNA binding protein H-NS binds to and alters the stability of RNA *in vitro* and *in vivo*. J. Mol. Biol. *339*, 505–514.

Browning, D.F., Grainger, D.C., and Busby, S.J.W. (2010). Effects of nucleoid-associated proteins on bacterial chromosome structure and gene expression. Curr. Opin. Microbiol. *13*, 773–780.

Buchrieser, C., Glaser, P., Rusniok, C., Nedjari, H., D'Hauteville, H., Kunst, F., Sansonetti, P., and Parsot, C. (2000). The virulence plasmid pWR100 and the repertoire of proteins secreted by the type III secretion apparatus of *Shigella flexneri*. Mol. Microbiol. *38*, 760–771.

Cameron, A.D., Stoebel, D.M., and Dorman, C.J. (2011). DNA supercoiling is differentially regulated by environmental factors and FIS in *Escherichia coli* and *Salmonella* enterica. Mol. Microbiol. *80*, 85–101.

Castang, S., and Dove, S.L. (2010). High-order oligomerization is required for the function of the H-NS family member MvaT in *Pseudomonas aeruginosa*. Mol. Microbiol. *78*, 916–931.

Castang, S., McManus, H.R., Turner, K.H., and Dove, S.L. (2008). H-NS family members function coordinately in an opportunistic pathogen. Proc. Natl. Acad. Sci. U.S.A. *105*, 18947–18952.

Castellanos M.I., Harrison, D.J., Smith, J.M., Labahn, S.K., Levy, K.M., and Wing, H.J. (2009). VirB alleviates H-NS repression of the *icsP* promoter in *Shigella flexneri* from sites more than one kilobase upstream of the transcription start site. J. Bacteriol. *191*, 4047–4050.

Cerdan, R., Bloch, V., Yang, Y., Bertin, P., Dumas, C., Rimsky, S., Kochoyan, M., and Arold, S.T. (2003). Crystal structure of the N-terminal dimerization domain of VicH, the H-NS-like protein of *Vibrio cholerae*. J. Mol. Biol. *334*, 179–185.

Chao, Y., and Vogel, J. (2010). The role of Hfq in bacterial pathogens. Curr. Opin. Microbiol. *13*, 24–33.

Cho, H., McManus, H.R., Dove, S.L., and Bernardt, T.G. (2011). Nucleoid occlusion factor SlmA is a DNA-activated FtsZ polymerization antagonist. Proc. Natl. Acad. Sci. U.S.A. *108*, 3773–3778.

Corbett, D., Bennett, H.J., Askar, H., Green, J., and Roberts, I.S. (2007). SlyA and H-NS regulate transcription of the *Escherichia coli* K5 capsule gene cluster, and expression of slyA in *Escherichia coli* is temperature-dependent, positively autoregulated, and independent of H-NS. J. Biol. Chem. *282*, 33326–33335.

Corcoran, C.P., and Dorman, C.J. (2009). DNA relaxation dependent phase biasing of the *fim* genetic switch in *Escherichia coli* depends on the interplay of H-NS, IHF and LRP. Mol. Microbiol. *74*, 1071–1082.

Cukier-Kahn, R., Jacquet, M., and Gros, F. (1972). Two heat-resistant, low molecular weight proteins from *Escherichia coli* that stimulate DNA directed RNA synthesis. Proc. Natl. Acad. Sci. U.S.A. *69*, 3643–3647.

Cusick, M.E., and Belfort, M. (1998). Domain structure and RNA annealing activity of the *Escherichia coli* regulatory protein StpA. Mol. Microbiol. *28*, 847–857.

Dame, R.T., and Dorman, C.J. (2010). Bacterial Chromatin. (Springer , Dordrecht, Netherlands).

Dame, R.T., Wyman, C., and Goosen, N. (2000). H-NS mediated compaction of DNA visualised by atomic force microscopy. Nucleic Acids Res. *28*, 3504–3510.

Dame, R.T., Wyman, C., and Goosen, N. (2001). Structural basis for preferential binding of H-NS to curved DNA. Biochimie *83*, 231–234.

Dame, R.T., Wyman, C., Wurm, R., Wagner, R., and Goosen, N. (2002). Structural basis for H-NS-mediated trapping of RNA polymerase in the open initiation complex at the *rrnB* P1. J. Biol. Chem. *277*, 2146–2150.

Dame, R.T., Luijsterburg, M.S., Krin, E., Bertin, P.N., Wagner, R., and Wuite, G.J. (2005). DNA bridging: a property shared among H-NS-like proteins. J. Bacteriol. *187*, 1845–1848.

Dame, R.T., Noom, M.C., and Wuite, G.J. (2006). Bacterial chromatin organization by H-NS protein unravelled using dual DNA manipulation. Nature *444*, 387–390.

Dattananda, C.S., Rajkumari, K., and Gowrishankar, J. (1991). Multiple mechanisms contribute to osmotic inducibility of proU operon expression in *Escherichia coli*: demonstration of two osmoresponsive promoters and of a negative regulatory element within the first structural gene. J. Bacteriol. *173*, 7481–7490.

Defez, R., and De Felice, M. (1981). Cryptic operon for beta-glucoside metabolism in *Escherichia coli* K12: genetic evidence for a regulatory protein. Genetics *97*, 11–25.

Deighan, P., Beloin, C., and Dorman, C.J. (2003). Three-way interactions among the Sfh, StpA and H-NS nucleoid-structuring proteins of *Shigella flexneri* 2a strain 2457T. Mol. Microbiol. *48*, 1401–1416.

Dillon, S.C., and Dorman, C.J. (2010). Bacterial nucleoid-associated proteins, nucleoid structure and gene expression. Nature Rev. Microbiol. *8*, 949–959.

Dillon, S.C., Cameron, A.D., Hokamp, K., Lucchini, S., Hinton, J.C., and Dorman, C.J. (2010). Genome-wide analysis of the H-NS and Sfh regulatory networks in *Salmonella* Typhimurium identifies a plasmid-encoded transcription silencing mechanism. Mol. Microbiol. *76*, 1250–1265.

Doetsch, M., Gstrein, T., Schroeder, R., and Förtig, B. (2010). Mechanisms of StpA-mediated RNA remodeling. RNA Biology, *7*, 735–743.

Dorman, C.J. (2004). H-NS: a universal regulator for a dynamic genome. Nature Rev. Microbiol. *2*, 391–400.

Dorman, C.J. (2007a). H-NS, the genome sentinel. Nature Rev. Microbiol. *5*, 157–161.

Dorman, C.J. (2007b). Probing bacterial nucleoid structure with optical tweezers. Bioessays *29*, 212–216.

Dorman, C.J. (2008). The virulence plasmids of *Shigella flexneri*. In Microbial Megaplasmids, Schwartz, E., ed (Springer, Heidelberg, Germany), pp 151–170.

Dorman, C.J. (2009). Nucleoid-associated proteins and bacterial physiology. Adv. App. Microbiol. *67*, 47–64.

Dorman, C.J. (2010). Horizontally acquired homologues of the nucleoid-associated protein H-NS: implications for gene regulation. Mol. Microbiol. *75*, 264–267.

Dorman, C.J., and Kane, K.A. (2009). DNA bridging and antibridging: a role for bacterial nucleoid-associated proteins in regulating the expression of laterally acquired genes. FEMS Microbiol. Rev. *33*, 587–592.

Dorman, C.J., and Porter, M.E. (1998). The Shigella virulence gene regulatory cascade: a paradigm of bacterial gene control mechanisms. Mol. Microbiol. *29*, 677–684.

Dorman, C.J., Ní Bhriain, N., and Higgins, C.F. (1990). DNA supercoiling and the environmental regulation of virulence gene expression in *Shigella flexneri*. Nature *344*, 789–792.

Dorman CJ, Hinton JC, Free A. (1999). Domain organization and oligomerization among H-NS-like nucleoid-associated proteins in bacteria. Trends Microbiol. *7*, 124–128.

Dorman, C.J., McKenna, S., and Beloin, C. (2001). Regulation of virulence gene expression in *Shigella flexneri*, a facultative intracellular pathogen. Int. J. Med. Microbiol. *291*, 89–96.

Doyle, M., and Dorman, C.J. (2006). Reciprocal transcriptional and posttranscriptional growth-phase-dependent expression of *sfh*, a gene that encodes a paralogue of the nucleoid-associated protein H-NS. J. Bacteriol. *188*, 7581–7591.

Doyle, M., Fookes, M., Ivens, A., Mangan, M.W., Wain, J., and Dorman, C.J. (2007). An H-NS-like stealth protein aids horizontal DNA transmission in bacteria. Science *315*, 251–252.

Durand, J.M, Dagberg, B., Uhlin, B.E., and Björk, G.R. (2000). Transfer RNA modification, temperature and DNA superhelicity have a common target in the regulatory network of the virulence of *Shigella flexneri*: the expression of the *virF* gene. Mol. Microbiol. *35*, 924–935.

Ellermeier, J.R., and Slauch, J.M. (2007). Adaptation to the host environment: regulation of the SPI1 type III secretion system in *Salmonella* enterica serovar Typhimurium. Curr. Opin. Microbiol. *10*, 24–29.

Ellison, D.W., and Miller, V.L. (2006a). Regulation of virulence by members of the MarR/SlyA family. Curr. Opin. Microbiol. *9*, 153–159.

Ellison, D.W., and Miller, V.L. (2006b). H-NS represses inv transcription in *Yersinia enterocolitica* through competition with RovA and interaction with YmoA. J. Bacteriol. *188*, 5101–5112.

Esposito, D., Petrovic, A., Harris, R., Ono, S., Eccleston, J.F., Mbabaali, A., Haq, I., Higgins, C.F., Hinton, J.C., Driscoll, P.C., and Ladbury, J.E. (2002). H-NS oligomerization domain structure reveals the mechanism for high order self-association of the intact protein. J. Mol. Biol. *324*, 841–850.

Falconi, M., Colonna, B., Prosseda, G., Michaeli, G., and Gualerzi, C.O. (1996). Antagonistic involvement of FIS and H-NS proteins in the transcriptional control of hns expression. Mol. Microbiol. *19*, 965–975.

Falconi, M., Colonna, B., Prosseda, G., Micheli, G., and Gualerzi, C.O. (1998). Thermoregulation of Shigella and *Escherichia coli* EIEC pathogenicity. A temperature-dependent structural transition of DNA modulates accessibility of *virF* promoter to transcriptional repressor H-NS. EMBO J. *17*, 7033–7043.

Falconi, M., Prosseda, G., Giangrossi, M., Beghetto, E., and Colonna, B. (2001). Involvement of FIS in the H-NS-mediated regulation of *virF* gene of *Shigella* and enteroinvasive *Escherichia coli*. Mol. Microbiol. *42*, 439–452.

Fang, F.C., and Rimsky, S. (2008). New insights into transcriptional regulation by H-NS. Curr. Opin. Microbiol. *11*, 113–120.

Feng, J.X., Song, Z.Z., Duan, C.J., Zhao, S., Wu, Y.Q., Wang, C., Dow, J.M., and Tang, J.L. (2009). The *xrvA* gene of *Xanthomonas oryzae* pv. *oryzae*, encoding an H-NS-like protein, regulates virulence in rice. Microbiology *155*, 3033–3044.

Free, A., and Dorman, C.J. (1994). *Escherichia coli tyrT* gene transcription is sensitive to DNA supercoiling in its native chromosomal context: effect of DNA topoisomerase IV overexpression on *tyrT* promoter function. Mol. Microbiol. *14*, 151–161.

Free, A., and Dorman, C.J. (1995). Coupling of *Escherichia coli hns* mRNA levels to DNA synthesis by autoregulation: implications for growth phase control. Mol. Microbiol. *18*, 101–113.

Free, A., and Dorman, C.J. (1997). The *Escherichia coli stpA* gene is transiently expressed during growth in rich medium and is induced in minimal medium and by stress conditions. J. Bacteriol. *179*, 909–918.

Giangrossi, M., Gualerzi, C.O., and Pon, C.L. (2001). Mutagenesis of the downstream region of the *Escherichia coli hns* promoter. Biochimie *83*, 251–259.

Goldstein, E., and Drlica, K. (1984). Regulation of bacterial DNA supercoiling: plasmid linking numbers vary with growth temperature. Proc. Natl. Acad. Sci. U.S.A. *81*, 4046–4050.

Göransson, M., Sondén, B., Nilsson, P., Dagberg, B., Forsman, K., Emanuelsson, K., and Uhlin, B.E. (1990). Transcriptional silencing and thermoregulation of gene expression in *Escherichia coli*. Nature *344*, 682–685.

Gordon, B.R., Imperial, R., Wang, L., Navarre, W.W., and Liu, J. (2008). Lsr2 of *Mycobacterium* represents a novel class of H-NS-like proteins. J. Bacteriol. *190*, 7052–7059.

Gordon, B.R., Li, Y., Wang, L., Sintsova, A., van Bakel, H., Tian, S., Navarre, W.W., Xia, B., and Liu, J. (2010). Lsr2 is a nucleoid-associated protein that targets AT-rich sequences and virulence genes in *Mycobacterium tuberculosis*. Proc. Natl. Acad. Sci. U.S.A. *107*, 5154–5159.

Gowrishankar, J., and Manna, D. (1996). How is osmotic regulation of transcription of the *Escherichia coli proU* operon achieved? A review and a model. Genetica *97*, 363–378.

Goyard, S., and Bertin, P. (1997). Characterization of BpH3, an H-NS-like protein in *Bordetella pertussis*. Mol. Microbiol. *24*, 815–823.

Grainger, D.C., Hurd, D., Goldberg, M.D., and Busby, S.J.W. (2006). Association of nucleoid proteins with coding and non-coding segments of the *Escherichia coli* genome. Nucleic Acids Res. *34*, 4642–4652.

Grainger, D.C., Goldberg, M.D., Lee, D.J., and Busby, S.J.W. (2008). Selective repression by Fis and H-NS at the *Escherichia coli dps* promoter. Mol. Microbiol. *68*, 1366–1377.

Hale, T.L. (1991). Genetic basis of virulence in *Shigella* species. Microbiol Rev. *55*, 206–224.

Hardy, C.D., and Cozzerelli, N.R. (2005). A genetic selection for supercoiling mutants of *Escherichia coli* reveals proteins implicated in chromosome structure. Mol. Microbiol. *57*, 1636–1652.

Harrison, J.A., Pickard, D., Higgins, C.F., Khan, A., Chatfield, S.N., Ali, T., Dorman, C.J., Hormaeche, C.E., and Dougan, G. (1994). Role of *hns* in the virulence phenotype of pathogenic *Salmonellae*. Mol. Microbiol. *13*, 133–140.

Hensel, M. (2000). *Salmonella* pathogenicity island 2. Mol. Microbiol. *36*, 1015–1023.

Heroven, A.K., Böhme, K., Tran-Winkler, H., and Dersch, P. (2007). Regulatory elements implicated in the environmental control of invasin expression in enteropathogenic *Yersinia*. Adv. Exp. Med. Biol. *603*, 156–166.

Higgins, C.F., Dorman, C.J., Stirling, D.A., Waddell, L., Booth, I.R., May, G., and Bremer, E. (1988). A physiological role for DNA supercoiling in the osmotic regulation of gene expression in *S. typhimurium* and *E. coli*. Cell *52*, 569–584.

Hildebrand, M., Aldridge, P., and Geider, K. (2006). Characterization of *hns* genes from *Erwinia amylovora*. Mol. Genet. Genomics *275*, 310–319.

Hromockyj, A.E., Tucker, S.C., and Maurelli, A.T. (1990). Temperature regulation of *Shigella* virulence: identification of the repressor gene *virR*, an analogue of *hns*, and partial complementation by tyrosyl transfer RNA ($tRNA1^{Tyr}$). Mol. Microbiol. *6*, 2113–2124.

Hsieh, L.S., Rouviére-Yaniv, J., and Drlica, K. (1991). Bacterial DNA supercoiling and [ATP]/[ADP] ratio: changes associated with salt shock. J. Bacteriol. *173*, 3914–3917.

Hulbert, R.R., and Taylor, R.K. (2002). Mechanism of ToxT-dependent transcriptional activation at the *Vibrio cholerae tcpA* promoter. J. Bacteriol. *184*, 5533–554.

Hulton, C.S., Seirafi, A., Hinton, J.C.D., Sidebotham, J.M., Waddell, L., Pavitt, G.D., Owen-Hughes, T., Spassky, A., Buc, H., and Higgins, C.F. (1990). Histone-like protein H1 (H-NS), DNA supercoiling, and gene expression in bacteria. Cell *63*, 631–642.

Jacquet, M., Cukier-Kahn, R., Pla, J., and Gros, F. (1971). A thermostable protein factor acting on *in vitro* DNA transcription. Biochem. Biophys. Res. Comm. *45*, 1597–1607.

Johansson, J., and Uhlin, B.E. (1999). Differential protease-mediated turnover of H-NS and StpA revealed by a mutation altering protein stability and stationary-phase survival of *Escherichia coli*. Proc. Natl. Acad. Sci. U.S.A. *96*, 10776–10781.

Johansson, J., Balsalobre, C., Wang, S.Y., Urbonaviciene, J., Jin, D.J., Sondén, B., and Uhlin, B.E. (2000). Nucleoid proteins stimulate stringently controlled bacterial promoters: a link between the cAMP-CRP and the (p)ppGpp regulons in *Escherichia coli*. Cell *102*, 475–485.

Johansson, J., Eriksson, S., Sondén, B., Wai, S.N., and Uhlin, B.E. (2001). Heteromeric interactions among nucleoid-associated bacterial proteins: localization of StpA-stabilizing regions in H-NS of *Escherichia coli*. J. Bacteriol. *183*, 2343–2347.

Kahramanoglou, C., Seshasayee, A.S., Prieto, A.I., Ibberson, D., Schmidt, S., Zimmermann, J., Benes, V., Fraser, G.M., and Luscombe, N.M. (2011). Direct and indirect effects of H-NS and Fis on global gene expression control in *Escherichia coli*. Nucleic Acids Res. *39*, 2073–2091.

Katayama, T., Takata, M., and Sekimizu, K. (1996). The nucleoid protein H-NS facilitates chromosome DNA replication in *Escherichia coli dnaA* mutants. J. Bacteriol. *178*, 5790–5792.

Keane, O.M., and Dorman, C.J. (2003). The gyr genes of *Salmonella* enterica serovar Typhimurium are repressed by the factor for inversion stimulation, Fis. Mol. Genet. Genomics *270*, 56–65.

Klauck, E., Böhringer, J., and Hengge-Aronis, R. (1997). The LysR-like regulator LeuO in *Escherichia coli* is involved in the translational regulation of *rpoS* by affecting the expression of the small regulatory DsrA-RNA. Mol. Microbiol. *25*, 559–569.

Laine, B., Sautiere, P., Spassky, A., and Rimsky, S. (1984). A DNA binding protein from *E. coli*: isolation, characterization and its relationship with proteins H1 and B1. Biochem. Biophys. Res. Comm. *119*, 1147–1153.

Lammi, M., Paci, M., Pon, C.L., Losso, M.A., Miano, A., Pawlik, R.T., Gianfranceschi, G.L., and Gualerzi, C.O. (1984). Proteins from the prokaryotic nucleoid: biochemical and 1H NMR studies on three bacterial histone-like proteins. In Proteins Involved in DNA Replication, vol 179, Hubscher, U., and Spadari, S., eds (Plenum, NY), pp 467–477.

La Teana, A., Brandi, A., Falconi, M., Spurio, R., Pon, C.L., and Gualerzi, C.O. (1991). Identification of a cold shock transcriptional enhancer of the *Escherichia coli* gene encoding nucleoid protein H-NS. Proc. Natl. Acad. Sci. U.S.A. *88*, 10907–10911.

Lang, B., Blot, N., Bouffartigues, E., Buckle, M., Geertz, M., Gualerzi, C.O., Mavathur, R., Muskhelishvili, G., Pon, C.L., Rimsky, S., *et al.* (2007). High-affinity DNA binding sites for H-NS provide a molecular basis for selective silencing within proteobacterial genomes. Nucleic Acids Res. *35*, 6330–6337.

Lawrenz, M.B., and Miller, V.L. (2007). Comparative analysis of the regulation of *rovA* from the pathogenic yersiniae. J. Bacteriol. *189*, 5963–5975.

Lease, R.A., and Belfort, M. (2000). A trans-acting RNA as a control switch in *Escherichia coli*: DsrA modulates function by forming alternative structures. Proc. Natl. Acad. Sci. U.S.A. *97*, 9919–9924.

Lease, R.A., Cusick, M.E., and Belfort, M. (1998). Riboregulation in *Escherichia coli*: DsrA RNA acts by RNA:RNA interactions at multiple loci. Proc. Natl. Acad. Sci. U.S.A. *95*, 12456–12461.

Leonard, P.G., Ono, S., Gor, J., Perkins, S.J., and Ladbury, J.E. (2009). Investigation of the self-association and hetero-association interactions of H-NS and StpA from Enterobacteria. Mol. Microbiol. *73*, 165–179.

Li, C., Wally, H., Miller, S.J., and Lu, C.D. (2009). The multifaceted proteins MvaT and MvaU, members of the H-NS family, control arginine metabolism, pyocyanin synthesis, and prophage activation in *Pseudomonas aeruginosa* PAO1. J. Bacteriol. *191*, 6211–6218.

Lithgow JK, Haider F, Roberts IS, Green J. (2007). Alternate SlyA and H-NS nucleoprotein complexes control *hlyE* expression in *Escherichia coli* K-12. Mol. Microbiol. *66*, 685–698.

Liu, Q., and Richardson, C.C. (1993). Gene 5.5 protein of bacteriophage T7 inhibits the nucleoid protein H-NS of *Escherichia coli*. Proc. Natl. Acad. U.S.A. *90*, 1761–1765.

Liu, Y., Chen, H., Kenney, L.J., and Yan, J. (2010). A divalent switch drives H-NS/DNA-binding conformations between stiffening and bridging modes. Genes Dev. *24*, 339–344.

Lucchini, S., Rowley, G., Goldberg, M.D., Hurd, D., Harrison, M., and Hinton, J.C. (2006). H-NS mediates the silencing of laterally acquired genes in bacteria. PLoS Pathog. *2*, e81.

Lucchini, S., McDermott, P., Thompson, A., and Hinton, J.C.D. (2009). The H-NS-like protein StpA represses the RpoS (sigma 38) regulon during exponential growth of *Salmonella* Typhimurium. Mol. Microbiol. *74*, 1169–1186.

Lucht, J.M., Dersch, P., Kempf, B., and Bremer, E. (1994). Interactions of the nucleoid-associated DNA-binding protein H-NS with the regulatory region of the osmotically controlled *proU* operon of *Escherichia coli*. J. Biol. Chem. *269*, 6578–6586.

Luijsterburg, M.S., Noom, M.C., Wuite, G.J., and Dame, R.T. (2006). The architectural role of nucleoid-associated proteins in the organization of bacterial chromatin: a molecular perspective. J. Struct. Biol. *156*, 262–272.

Luijsterburg, M.S., White, M.F., van Driel, R., and Dame, R.T. (2008). The major architects of chromatin: architectural proteins in bacteria, archaea and eukaryotes. Crit. Rev. Biochem. Mol. Biol. *43*, 393–418.

McClellan, J.A., Boublíková, P., Palecek, E., and Lilley, D.M.J. (1990). Superhelical torsion in cellular DNA responds directly to environmental and genetic factors. Proc. Natl. Acad. Sci. U.S.A. *87*, 8373–8377.

McFeeters, R.L., Altieri, A.S., Cherry, S., Tropea, J.E., Waugh, D.S., and Byrd, R.A. (2007). The high-precision solution structure of *Yersinia* modulating protein YmoA provides insight into interaction with H-NS. Biochemistry *46*, 13975–13982.

McGovern V, Higgins NP, Chiz RS, Jaworski A. (1994). H-NS over-expression induces an artificial stationary phase by silencing global transcription. Biochimie *76*, 1019–1029.

Madrid, C., Balsalobre, C., García, J., and Juárez, A. (2007). The novel Hha/YmoA family of nucleoid-associated proteins: use of structural mimicry to modulate the activity of the H-NS family of proteins. Mol. Microbiol. *63*, 7–14.

Maurelli, A.T., and Sansonetti, P.J. (1988). Identification of a chromosomal gene controlling temperature-regulated expression of *Shigella* virulence. Proc. Natl. Acad. Sci. U.S.A. *85*, 2820–2824.

Maurer, G., Fritz, J., and Mushelishvili, G. (2009). A systematic *in vitro* study of nucleoprotein complexes formed by bacterial nucleoid-associated proteins revealing novel types of DNA organization. J. Mol. Biol. *387*, 1261–1276.

May, G., Dersch, P., Haardt, M., Middendorf, A., and Bremer, E. (1990). The *osmZ* (*bglY*) gene encodes the DNA-binding protein H-NS (H1a), a component of the *Escherichia coli* K12 nucleoid. Mol. Gen. Genet. *224*, 81–90.

Mayer, O., Rajkowitsch, L., Lorenz, C., Konrat, R., Schroeder, R. (2007). RNA chaperone activity and RNA-binding properties of the *E. coli* StpA protein. Nucleic Acid Res. *35*, 1257–1269.

Mellies, J.L., Benison, G., McNitt, W., Mavor, D., Boniface, C., and Larabee, F.J. (2011). Ler of Pathogenic *Escherichia coli* forms toroidal protein–DNA complexes. Microbiology *157*, 1123–1133.

Mercier, R., Petit, M.A., Schbath, S., Robin, S., El Karoui, M., Boccard, F., and Espéli, O. (2008). The MatP/*matS* site-specific system organizes the terminus region of the *E. coli* chromosome into a macrodomain. Cell *135*, 475–485.

Montero Llopis, P., Jackson, A.F., Sliusarenko, O., Surovtsev, I., Heinritz, J., Emonet, T., and Jacobs-Wagner, C. (2010). Spatial organization of the flow of genetic information in bacteria. Nature *466*, 77–81.

Müller, C.M., Dobrindt, U., Nagy, G., Emödy, L., Uhlin, B.E., and Hacker, J. (2006). Role of histone-like proteins H-NS and StpA in expression of virulence determinants of uropathogenic *Escherichia coli*. J. Bacteriol. *188*, 5428–5438.

Müller, C.M., Schneider, G., Dobrindt, U., Emödy, L., Hacker, J., and Uhlin, B.E. (2010). Differential effects and interactions of endogenous and horizontally acquired H-NS-like proteins in pathogenic *Escherichia coli*. Mol. Microbiol. *75*, 280–293.

Nagarajavel, V., Madhusudan, S., Dole, S., Rahmouni, A.R., and Schnetz, K. (2007). Repression by binding of H-NS within the transcription unit. J. Biol. Chem. *282*, 23622–23630.

Nandal, A., Huggins, C.C., Woodhall, M.R., McHugh, J., Rodríguez-Quiñones, F., Quail, M.A., Guest, J.R., and Andrews S.C. (2010). Induction of the ferritin gene (*ftnA*) of *Escherichia coli* by $Fe(^{2+})$-Fur is mediated by reversal of H-NS silencing and is RyhB independent. Mol. Microbiol. *75*, 637–657.

Nasser, W., and Reverchon, S. (2002). H-NS-dependent activation of pectate lyases synthesis in the phytopathogenic bacterium *Erwinia chrysanthemi* is mediated by the PecT repressor. Mol. Microbiol. *43*, 733–748.

Nasser, W., Faelen, M., Hugouvieux-Cotte-Pattat, N., and Reverchon, S. (2001). Role of the nucleoid-associated protein H-NS in the synthesis of virulence factors in the phytopathogenic bacterium *Erwinia chrysanthemi*. Mol. Plant Microbe Interact. *14*, 10–20.

Navarre, W.W. (2010). H-NS as a defense system. In Bacterial Chromatin, Dame, R.T., and Dorman, C.J., eds (Springer, Dordrecht, Netherlands), pp. 245–322.

Navarre, W.W., Porwollik, S., Wang, Y., McClelland, M., Rosen, H., Libby, S.J., and Fang, F.C. (2006). Selective silencing of foreign DNA with low GC content by the H-NS protein in *Salmonella*. Science *313*, 236–238.

Navarre, W.W., McClelland, M., Libby, S.J., and Fang, F.C. (2007). Silencing of xenogeneic DNA by H-NS-facilitation of lateral gene transfer in bacteria by a defense system that recognizes foreign DNA. Genes Dev. *21*, 1456–1471.

Noom, M.C., Navarre, W.W., Oshima, T., Wuite, G.J., and Dame, R.T. (2007). H-NS promotes looped domain formation in the bacterial chromosome. Curr. Biol. *17*, R913-R914.

Nov Klaiman, T., Hosid, S., and Bolshoy, A. (2009). Upstream curved sequences in *E. coli* are related to the regulation of transcription initiation. Comput. Biol. Chem. 33, 275–282.

O'Byrne, C.P., and Dorman, C.J. (1994). Transcription of the *Salmonella typhimurium spv* virulence locus is regulated negatively by the nucleoid-associated protein H-NS. FEMS Microbiol. Lett. *121*, 99–105.

Ochman, H., and Groisman, E.A. (1996). Distribution of pathogenicity islands in *Salmonella* spp. Infect. Immun. *64*, 5410–5412.

Ó Cróinín, T., and Dorman, C.J. (2007). Expression of the Fis protein is sustained in late exponential and stationary phase cultures of *Salmonella enterica* serovar Typhimurium grown in the absence of aeration. Mol. Microbiol. *66*, 237–251.

Olekhnovich, I.N., and Kadner, R.J. (2006). Crucial roles of both flanking sequences in silencing of the *hilA* promoter in *Salmonella enterica*. J. Mol. Biol. *357*, 373–386.

Oshima, T., Ishikawa, S., Kurokawa, K., Aiba, H., and Ogasawara, N. (2006). *Escherichia coli* histone-like protein H-NS preferentially binds to horizontally acquired DNA in association with RNA polymerase. DNA Res. *13*, 141–153.

Overdier, D.G., and Csonka, L.N. (1992). A transcriptional silencer downstream of the promoter in the osmotically controlled *proU* operon of *Salmonella typhimurium*. Proc. Natl. Acad. Sci. U.S.A. *89*, 3140–3144.

Park, H.S., Ostberg, Y., Johansson, J., Wagner, E.G. and Uhlin, B.E. (2010). Novel role for a bacterial nucleoid protein in translation of mRNAs with suboptimal ribosome-binding sites. Genes Dev. *24*, 1345–1350.

Perez, J.C., Latifi, T., and Groisman, E.A. (2008). Overcoming H-NS-mediated transcriptional silencing of horizontally acquired genes by the PhoP and SlyA proteins in *Salmonella enterica*. J. Biol. Chem. *283*, 10773–10783.

Pon, C.L., Calogero, R.A., and Gualerzi, C.O. (1988). Identification, cloning, nucleotide sequence and chromosomal map location of *hns*, the structural gene for *Escherichia coli* DNA-binding protein H-NS. Mol. Gen. Genet. *212*, 199–202.

Porter, M.E., and Dorman, C.J. (1994). A role for H-NS in the thermo-osmotic regulation of virulence gene expression in *Shigella flexneri*. J. Bacteriol. *176*, 4187–4191.

Porter, M.E., and Dorman, C.J. (2002). *In vivo* DNA-binding and oligomerization properties of the *Shigella flexneri* AraC-like transcriptional regulator VirF as identified by random and site-specific mutagenesis. J. Bacteriol. *184*, 531–539.

Prosseda, G., Fradiani, P.A., Di Lorenzo, M., Falconi, M., Micheli, G., Casalino, M., Nicoletti, M., and Colonna, B. (1998). A role for H-NS in the regulation of the *virF* gene of *Shigella* and enteroinvasive *Escherichia coli*. Res. Microbiol. *149*, 15–25.

Prosseda, G., Falconi, M., Giangrossi, M., Gualerzi, C.O., Micheli, G., and Colonna, B. (2004). The *virF* promoter in *Shigella*: more than just a curved DNA stretch. Mol. Microbiol. *51*, 523–537.

Pul, U., Lux, B., Wurm, R., and Wagner, R. (2008). Effect of upstream curvature and transcription factors H-NS and LRP on the efficiency of *Escherichia coli* rRNA promoters P1 and P2 – a phasing analysis. Microbiology *154*, 2546–2558.

Raabe, K., Drepper, T., Riedel, K.U., Masepohl, B., and Klipp, W. (2002). The H-NS-like protein HvrA modulates expression of nitrogen fixation genes in the phototrophic purple bacterium *Rhodobacter capsulatus* by binding to selected *nif* promoters. FEMS Microbiol. Lett. *216*, 151–158.

Rajkowitsch, L., and Schroeder, R. (2007). Dissecting RNA chaperone activity. RNA *13*, 2053–2060.

Renzi, F., Rescalli, E., Galli, E., and Bertoni, G. (2010). Identification of genes regulated by the MvaT-like paralogues TurA and TurB of *Pseudomonas putida* KT2440. Environ. Microbiol. *12*, 254–263.

Renzoni, D., Esposito, D., Pfuhl, M., Hinton, J.C., Higgins, C.F., Driscoll, P.C., and Ladbury, J.E. (2001). Structural characterization of the N-terminal oligomerization domain of the bacterial chromatin-structuring protein, H-NS. J. Mol. Biol. *306*, 1127–1137.

Rhen, M., and Dorman, C.J. (2005). Hierarchical gene regulators adapt *Salmonella enterica* to its host milieus. Int. J. Med. Microbiol. *294*, 487–502.

Rimsky, S. (2004). Structure of the histone-like protein H-NS and its role in regulation and genome superstructure. Curr. Opin. Microbiol. *7*, 109–114.

Rimsky, S., Zuber, F., Buckle, M., and Buc, H. (2001). A molecular mechanism for the repression of transcription by the H-NS protein. Mol. Microbiol. *42*, 1311–1323.

Sánchez-Romero, M.A., Busby, S.J., Dyer, N.P., Ott, S., Millard, A.D., and Grainger, D.C. (2010). Dynamic distribution of SeqA protein across the chromosome of *Escherichia coli* K-12. Mbio *1*, pii, e00012-10.

Schechter, L.M., Jain, S., Akbar, S., and Lee, C.A. (2003). The small nucleoid-binding proteins H-NS, HU, and Fis affect hilA expression in *Salmonella enterica* serovar Typhimurium. Infect. Immun. *71*, 5432–5435.

Schnetz, K. (1995). Silencing of *Escherichia coli bgl* promoter by flanking sequence elements. EMBO J. *14*, 2545–2550.

Schröder, O., and Wagner, R. (2000). The bacterial DNA-binding protein H-NS represses ribosomal RNA transcription by trapping RNA polymerase in the initiation complex. J. Mol. Biol. *298*, 737–748.

Sette, M., Spurio, R., Trotta, E., Brandizi, C., Brandi, A., Pon, C.L., Barbato, G., Boelens, R., and Gualerzi, C.O. (2009). Sequence-specific recognition of DNA by the C-terminal domain of nucleoid-associated protein H-NS. J. Biol. Chem. *284*, 30453–30462.

Sherburne, C.K., Lawley, T.D., Gilmour, M.W., Blattner, F.R., Burland, V., Grotbeck, E., Rose, D.J., and Taylor, D.E. (2000). The complete DNA sequence and analysis of R27, a large IncHI plasmid from *Salmonella typhi* that is temperature sensitive for transfer. Nucleic Acids Res. *28*, 2177–2186.

Shin, M., Song, M., Rhee, J.H., Hong, Y., Kim, Y.J., Seok, Y.J., Ha, K.S., Jung, S.H., and Choy, H.E. (2005). DNA looping-mediated repression by histonelike protein H-NS: specific requirement of E^{70} as a cofactor for looping. Genes Dev. *19*, 2388–2398.

Shindo, H., Ohnuki, A., Ginba, H., Katoh, E., Ueguchi, C., Mizuno, T., and Yamazaki, T. (1999). Identification of the DNA binding surface of H-NS protein from *Escherichia coli* by heteronuclear NMR spectroscopy. FEBS Lett. *455*, 63–69.

Sinden, R.R., Pearson, C.E., Potaman, V.N., and Ussery, D.W. (1998). DNA: structure and function. Adv. Genome Biol. *5A*, 1–141.

Smits, W.K., and Grossman, A.D. (2010). The transcriptional regulator Rok binds A+T-rich DNA and is involved in repression of a mobile genetic element in *Bacilus subtilis*. PLoS Genet. *6*, e1001207.

Sonden, B., and Uhlin, B.E. (1996). Coordinated and differential expression of histone-like proteins in *Escherichia coli*: regulation and function of the H-NS analog StpA. EMBO J. *15*, 4970–4980.

Spassky, A., Rimsky, S., Garreau, H., and Buc, H. (1984). H1a, an *E. coli* DNA binding protein which accumulates in stationary phase strongly compacts DNA *in vitro*. Nucleic Acids Res. *12*, 5321–5340.

Spears, P.A., Schauer, D., and Orndorff, P.E. (1986). Metastable regulation of type 1 piliation in *Escherichia coli* and isolation and characterization of a phenotypically stable mutant. J. Bacteriol. *168*, 179–185.

Stoebel, D.M., Free, A., and Dorman, C.J. (2008). Anti-silencing: overcoming H-NS-mediated repression of transcription in Gram-negative enteric bacteria. Microbiology *154*, 2533–2545.

Stonehouse, E.A., Hulbert, R.R., Nye, M.B., Skorupski, K., and Taylor, R.K. (2011). H-NS binding and repression of the *ctx* promoter in *Vibrio cholerae*. J. Bacteriol. *193*, 979–988.

Sutherland, L., Cairney, J., Elmore, M.J., Booth, I.R., and Higgins, C.F. (1986). Osmotic regulation of transcription: induction of the *proU* betaine transport gene is dependent on accumulation of intracellular potassium. J. Bacteriol. *168*, 805–814.

Takeda, T., Yun, C.S., Shintani, M., Yamane, H., and Nojiri, H. (2011). Distribution of genes encoding nucleoid-associated protein homologs in plasmids. Int. J. Evol. Biol. 2011, 685015.

Taniya, T., Mitobe, J., Nakayama, S., Mingshan, Q., Okuda, K., and Watanabe, H. (2003). Determination of the InvE binding site required for expression of IpaB of the *Shigella sonnei* virulence plasmid: involvement of a ParB BoxA-like sequence. J. Bacteriol. *185*, 5158–516.

Tendeng, C., Badaut, C., Krin, E., Gounon, P., Ngo, S., Danchin, A., Rimsky, S., and Bertin, P. (2000). Isolation and characterization of *vicH*, encoding a new pleiotropic regulator in *Vibrio cholerae*. J. Bacteriol. *182*, 2026–2032.

Tendeng, C., Soutourina, O.A., Danchin, A., and Bertin, P.N. (2003a). MvaT proteins in *Pseudomonas* spp.: a novel class of H-NS-like proteins. Microbiology *149*, 3047–3050.

Tendeng, C., Krin, E., Soutourina, O.A., Marin, A., Danchin, A., and Bertin, P.N. (2003b). A novel H-NS-like protein from an Antarctic psychrophilic bacterium reveals a crucial role for the N-terminal domain in thermal stability. J. Biol. Chem. *278*, 18754–18760.

Tippner, D., Afflerbach, H., Bradaczek, C., and Wagner, R. (1994). Evidence for a regulatory function of the histone-like *Escherichia coli* protein H-NS in ribosomal RNA synthesis. Mol. Microbiol. *11*, 589–604.

Tobe, T., Yoshikawa, M., Mizuno, T., and Sasakawa, C. (1993). Transcriptional control of the invasion regulatory gene *virB* of *Shigella flexneri*: activation by *virF* and repression by H-NS. J. Bacteriol. *175*, 6142–6149.

Tobe, T., Yoshikawa, M., and Sasakawa, C. (1995). Thermoregulation of *virB* transcription in *Shigella flexneri* by sensing of changes in local DNA superhelicity. J. Bacteriol. *177*, 1094–1097.

Tonthat, N.K., Arold, S.T., Pickering, B.F., van Dyke, M.W., Liang, S., Lu, Y., Beuria, T.K., Margolin, W., and Schumacher, M.A. (2011). Molecular mechanism by which the nucleoid occlusion factor, SlmA, keeps cytokenesis in check. EMBO J. *30*, 154–164.

Troxell, B., Sikes, M.L., Fink, R.C., Vazquez-Torres, A., Jones-Carson, J., and Hassan, H.M. (2011). Fur negatively regulates hns and is required for the expression of HilA and virulence in *Salmonella enterica* serovar Typhimurium. J. Bacteriol. *193*, 497–505.

Turner, E.C., and Dorman, C.J. (2007). H-NS antagonism in *Shigella flexneri* by VirB, a virulence gene transcription regulator that is closely related to plasmid partition factors. J. Bacteriol. *189*, 3403–3413.

Ueguchi, C., Seto, C., Suzuki, T., and Mizuno, T. (1997). Clarification of the dimerization domain and its functional significance for the *Escherichia coli* nucleoid protein H-NS. J. Mol. Biol. *274*, 145–151.

Valdez, Y., Ferreira, R.B., and Finlay, B.B. (2009). Molecular mechanisms of *Salmonella* virulence and host resistance. Curr. Top. Microbiol. Immunol. *337*, 93–127.

Vallet-Gely, I., Sharp, J.S., and Dove, S.L. (2007). Local and global regulators linking anaerobiosis to *cupA* fimbrial gene expression in *Pseudomonas aeruginosa*. J. Bacteriol. *189*, 8667–8676.

Varshavsky, A.J., Nedospasov, S.A., Bakayev, V.V., Bakayeva, T.G., and Georgiev, G.P. (1977). Histone-like proteins in the purified *Escherichia coli* deoxyribonucleoprotein. Nucleic Acids Res. *4*, 2725–2745.

Vivero, A., Baños, R.C., Mariscotti, J.F., Oliveros, J.C., García-del Portillo, F., Juárez, A., and Madrid, C. (2008). Modulation of horizontally acquired genes by the Hha-YdgT proteins in *Salmonella enterica* serovar Typhimurium. J. Bacteriol. *190*, 1152–1156.

Wade, J.T., Struhl, K., Busby, S.J.W., and Grainger, D.C. (2007). Genomic analysis of protein–DNA interactions in bacteria: insights into transcription and chromosome organization. Mol. Microbiol. *65*, 21–26.

Waldsich, C., Grossberger, R., and Schroeder, R. (2002). RNA chaperone StpA loosens interactions of the tertiary structure in the group I intron *in vivo*. Genes Dev. *16*, 2300–2312.

Walthers, D., Li, Y., Liu, Y., Anand, G., Yan, J., and Kenney, L.J. (2011). *Salmonella enterica* response regulator SsrB relieves H-NS silencing by displacing H-NS bound in polymerization mode and directly activates transcription. J. Biol. Chem. *286*, 1895–1902.

Wang, W., Li, G.W., Chen, C., Xie, X.S., and Zhuang, X. (2011). Chromosome organization by a nucleoid-associated protein in live bacteria. Science *333*, 1445–1449.

Wiggins, P.A., Dame, R.T., Noom, M.C., and Wuite, G.J. (2009). Protein-mediated molecular bridging: a key mechanism in biopolymer organization. Biophys J. *97*, 1997–2003.

Williams, R.M., and Rimsky, S. (1997). Molecular aspects of the *E. coli* nucleoid protein, H-NS: a central controller of gene regulatory networks. FEMS Microbiol. Lett. *156*, 175–185.

Williams, R.M., Rimsky, S., and Buc, H. (1996). Probing the structure, function, and interactions of the *Escherichia coli* H-NS and StpA proteins by using dominant negative derivatives. J. Bacteriol. *178*, 4335–4343.

Williamson, H.S., and Free, A. (2005). A truncated H-NS-like protein from enteropathogenic *Escherichia coli* acts as an H-NS antagonist. Mol. Microbiol. *55*, 808–827.

Wolf, T., Janzen, W., Blum, C., and Schnetz, K. (2006). Differential dependence of StpA on H-NS in autoregulation of StpA and in regulation of *bgl*. J. Bacteriol. *188*, 6728–6738.

Yamada, H., Muramatsu, S., and Mizuno, T. (1990). An *Escherichia coli* protein that preferentially binds to sharply curved DNA. J. Biochem. *108*, 420–425.

Yang, J., Tauschek, M., Strugnell, R., and Robins-Browne, R.M. (2005). The H-NS protein represses transcription of the *eltAB* operon, which encodes heat-labile enterotoxin in enterotoxigenic *Escherichia coli*, by binding to regions downstream of the promoter. Microbiology *151*, 1199–1208.

Yu, R.R., and DiRita, V.J. (2002). Regulation of gene expression in *Vibrio cholerae* by ToxT involves both antirepression and RNA polymerase stimulation. Mol. Microbiol. *43*, 119–134.

Yun, C.S., Suzuki, C., Naito, K., Takeda, T., Takahashi, Y., Sai, F., Terabayashi, T., Miyakoshi, M., Shintani, M., Nishida, H., *et al.* (2010). Pmr, a histone-like protein H1 (H-NS) family protein encoded by the IncP-7 plasmid pCAR1, is a key global regulator that alters host function. J. Bacteriol. *192*, 4720–4731.

Zhang, A., and Belfort, M. (1992). Nucleotide sequence of a newly-identified *Escherichia coli* gene, *stpA*, encoding an H-NS-like protein. Nucleic Acids Res. *20*, 6735.

Zhang, A., Derbyshire, V., Salvo, J.L., and Belfort, M. (1995). *Escherichia coli* protein StpA stimulates self-splicing by promoting RNA assembly *in vitro*. RNA *1*, 783–793.

Zhang, A., Rimsky, S., Reaban, M.E., Buc, H., and Belfort, M. (1996). *Escherichia coli* protein analogs StpA and H-NS: regulatory loops, similar and disparate effects on nucleic acid dynamics. EMBO J. *15*, 1340–1349.

Zimmerman, S.B. (2001). Cooperative transitions of isolated *Escherichia coli* nucleoids: implications for the nucleoid as a cellular phase. J. Struct. Biol. *153*, 160–175.

Two-component Regulatory Systems in Prokaryotes

8

David E. Whitworth

Abstract

Two-component systems (TCSs) are signalling pathways found abundantly in prokaryotes, and they are the dominant mechanism for stimulus-responsive adaptation in such organisms. An ever-increasing number of physiological phenomena are known to be regulated by TCSs, including cell cycle progression, pathogenesis, motility, and biofilm formation. The basic TCS comprises a receptor protein (sensor kinase) which autophosphorylates in response to a stimulus. The phosphoryl group is then directly transferred to a response regulator protein (the second component) that has a phosphorylation-dependent effector function. While the most basic TCSs are relatively well understood, there are many 'atypical' systems, which exhibit additional mechanistic features (for instance, regulation of subcellular location, intrinsic and extrinsic phosphatase activities, and cross-communication between TCSs), adding complexity to their signalling properties. The relatively recent availability of complete prokaryotic genome sequences has also provided new opportunities to appreciate global features of TCS function. For example, analyses have provided insights into TCS evolution, which in turn have yielded computational methods for evaluating TCS protein partnerships.

This chapter provides an overview of the common features of TCSs from a historical perspective, and then describes current understanding regarding the mechanisms of TCS function. Finally, outstanding questions regarding TCS function are discussed.

Introduction

Two-component systems (TCSs) are a conserved family of signal transduction pathways found in large numbers in prokaryotes. The latest version of the P2CS database of TCSs contains 64,960 TCS proteins encoded within 1125 completely sequenced prokaryotic genomes (Barakat *et al.*, 2011), with the average bacterium possessing around 30 TCSs. The large number of TCSs is a testament to the flexibility and robustness of the basic TCS scheme, which has been successfully elaborated upon by evolution in a variety of different ways. There has also accumulated an extensive literature regarding TCS biology. ISI Web of Knowledge^{SM} (Thomson Reuters) currently references nearly 5500 biochemistry/molecular biology articles on TCS, with 4400 of those being microbiological. Rather than providing an exhaustive survey of the literature, this chapter aims to provide an overview of the common features of TCS function. Further sections address a variety of aspects of TCS biology – their functions, genomics, evolution, mechanisms and structures. Finally, trends in TCS research are discussed and current gaps in knowledge highlighted.

The term 'two-component system' was first coined 25 years ago in a paper by Nixon *et al.* (1986) in which it was discovered that a variety of stimulus-responsive pathways were encoded by a pair of genes, each containing a conserved domain. One gene was suggested to encode a stimulus-responsive receptor protein, while the partner gene encoded an effector protein, or 'response regulator' (RR). Two years later Kofoid and Parkinson (1988) defined 'transmitter' domains found at the C-termini of diverse sensory proteins, and 'receiver' domains at the N-termini of partner RRs. The N-termini of sensors and the C-termini of RRs were relatively variable, suggesting roles in stimulus sensation and effector function respectively. A model was proposed wherein sensation of a stimulus or 'signal' led the transmitter domain of the sensory protein to interact specifically with the receiver domain of the RR, somehow propagating the signal. By the end of the 1980s it was accepted that signal propagation between sensor and RR occurred as a consequence of the direct transfer of a phosphoryl group between the two proteins (Stock *et al.*, 1989; Parkinson and Kofoid, 1992). The transmitter domains of sensors were recognized as autokinases, phosphorylating themselves upon stimulation, with the phosphoryl group coming from ATP. Phosphorylation of the sensor occurred on a conserved histidine residue, leading to the description of sensors as (variously) sensor kinases, histidine kinases (HKs), or histidine protein kinases. RRs were known to become phosphorylated on a conserved aspartate residue, with phosphorylation of an RR altering its effector activity (Stock *et al.*, 1989; Parkinson and Kofoid, 1992), and TCSs are thus often described as His-Asp phosphotransfer (or HAP) pathways.

Analysis of the fine structure of transmitter domains led to the identification of conserved sequence motifs, notably an H-box which housed the phosphorylatable histidine residue, and N, G1, F and G2 boxes which were subsequently found to bind to ATP and to be responsible for catalysis (Stock *et al.*, 1989). While the N, G1, F and G2 regions were quite close together, the N-terminal H-box was at a distance from the catalytic boxes (typically ~80 to 100 residues), joined by a linker of variable sequence (Stock *et al.*, 1989). Modern databases of domains group the N, G1, F and G2 boxes together within an HATPase_c domain (pfam entry PF02518), while the H-box containing region is referred to as a HisKA domain (PF00512). In a subset of transmitter domains the phosphorylatable histidine residue is found within a Hpt domain (PF01627) rather than a HisKA, and such systems typically (but not always) regulate chemotactic responses.

The basic TCS paradigm (illustrated in Fig. 8.1A) has survived intact for nearly two decades, and describes the vast majority of TCSs. However continuing research has shown that many TCSs do not conform to the basic paradigm, involving additional signalling or scaffolding domains/proteins. For instance, a large number of TCSs do not operate through a single His-Asp phosphotransfer, but rather involve a series of phosphotransfer reactions, typically along a series of His-Asp-His-Asp residues. These systems are usually referred to as 'phosphorelays' with the phosphoryl group transferred between successive HisKA, receiver^1, Hpt, and receiver^2 domains (Fig. 8.1B). It is thought that phosphorelays have evolved as they possess more points for regulation than the simpler TCSs. A good example of this is the sporulation phosphorelay of *Bacillus subtilis* (Perego, 1998), in which HKs contribute to phosphorylation of the RR Spo0A through a relay involving Spo0F (a RR) and Spo0B (a Hpt/HisKA domain analogue). The phosphorylation states of Spo0F and Spo0A are regulated not only by kinases, but also by specific phospho-aspartyl phosphatases, resulting in an extremely sophisticated signalling network (Perego, 1998).

The advent of whole genome sequencing has provided a more holistic appreciation of TCS biology. Originally it was thought that TCSs were exclusively found in eubacteria; however, archaebacterial and eukaryotic examples were soon identified (Chang and Meyerowitz, 1994; Kennelly and Potts, 1996). TCSs are far more numerous in eubacterial genomes than in archaebacteria or eukaryotes (Wuichet *et al.*, 2010), yet nevertheless fulfil important roles in such organisms (Santos and Shiozaki, 2001; Mizuno, 2005). TCSs have not yet been found in metazoans (Wuichet *et al.*, 2010).

Sequencing the *Synechocystis* sp. PCC6803 genome led to the discovery of at least 80 TCS proteins encoded by that organism's genome (Mizuno *et al.*, 1996). Similar analyses showed the presence of 62 TCS genes in *Escherichia coli* (Mizuno, 1997) and 70 TCS genes in *B. subtilis* (Fabret *et al.*, 1999). The core mechanism of every TCS involves a protein–protein interaction between conserved HisKA/Hpt and receiver domains, and thus there is the potential for any HK to signal to any RR. In the earliest studies

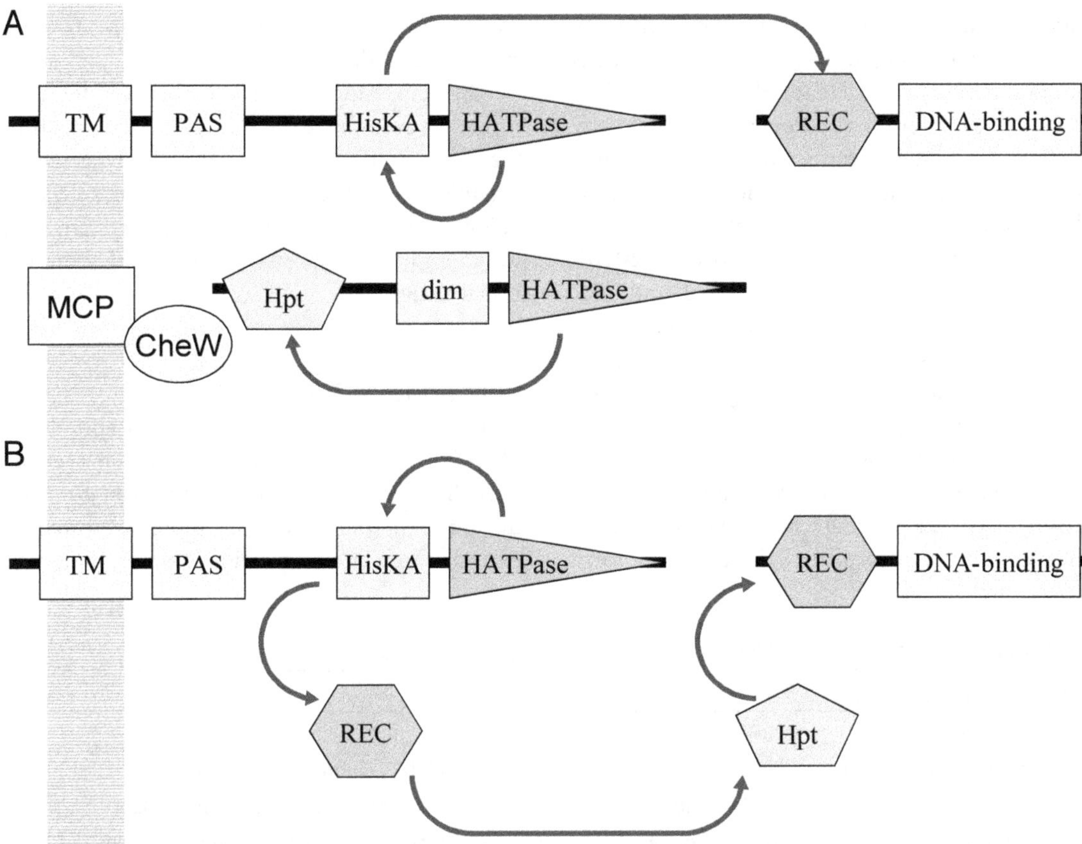

Figure 8.1 TCS architecture and phosphoryl group transfer (arrows). (A) A simple TCS. An environmental change is detected by an input domain of the HK (a PAS domain in this case), leading to autophosphorylation of the transmitter domain. The phosphoryl group is transferred to the receiver domain of the partner response regulator. In most HKs the transmitter domain comprises a HisKA domain and a HATPase domain. In chemosensory systems the transmitter domain possesses a phosphorylatable Hpt domain [and sometimes a vestigial HisKA dimerization (dim) domain lacking the critical His residue]. Stimuli are sensed by a separate receptor protein (MCP) complexed with an 'adaptor' protein, CheW. (B) A simple phosphorelay. The basic TCS is expanded by the inclusion of a second receiver domain and a phosphorelay domain (normally a Hpt domain, but Spo0B in the *Bacillus subtilis* sporulation pathway).

using purified TCSs *in vitro*, it seemed that HKs were generally able to transfer phosphoryl groups to any RR (Stock *et al.*, 1989). However *in vivo* the situation was less clear, as 'cross-regulation' between two discretely encoded TCSs was only seen in strains lacking a HK gene, or with overexpression of a RR (Stock *et al.*, 1989). It was known from studies in *E. coli* that a single HK CheA, could phosphorylate multiple RRs – namely CheY and CheB (Simon *et al.*, 1989). It was soon also shown that some RRs were phosphorylated by multiple HKs, for instance Spo0F of *B. subtilis* can be phosphorylated by KinA and KinB (Perego *et al.*, 1989; Spiegelman *et al.*, 1990). It became clear that for some TCSs cross-regulation was an important feature of their *in vivo* behaviour, while in other cases it was noise and actively suppressed. However for the majority of TCSs the 'one HK for one RR' rule seems to hold true, though examples are common where TCSs are combined into networks (Laub and Goulian, 2007).

In addition to showing the abundance of TCS genes in the average prokaryote, genome sequencing has also revealed a remarkable variability in most features of TCS biology. For instance, the output domains of RRs are diverse (Galperin,

2010), while there is an even greater diversity in the number and nature of the input domains of HKs (Cheung and Hendrickson, 2010; Krell *et al.*, 2010). There is also considerable diversity both in the number of transmitter and/or receiver domains encoded at a TCS gene locus, and in how those domains are distributed between different gene products (Williams and Whitworth, 2010). If all TCSs were encoded as paired HK and RR genes, then it would be trivial to suggest entire signalling pathways directly from genome sequences. However, around 40% of all TCS genes are found on their own in genome sequences (orphaned), with a further 5–7% found in complex loci of more than two TCS genes (Williams and Whitworth, 2010). Identifying which other TCS genes interact with such orphan TCS genes to form an entire signalling pathway, is often far from straightforward.

The remaining sections of this chapter each tackle a common feature of TCS biology. Thus the chapter discusses the functions, genetics, evolution, structure, and mechanisms of TCSs, culminating with a survey of recent system-level insights gained from experimental and computational analyses.

Functions of TCSs

As might be expected from the sheer abundance of TCS genes, examples of TCSs that regulate virtually any microbial phenotype can be readily found in the literature. Table 8.1 provides examples of TCS regulating diverse biological

Table 8.1 Examples of TCSs which regulate diverse biological processes

Process regulated by TCS	TCS proteins	References
DNA replication	MtrAB	Fol *et al.* (2006)
Peptidoglycan metabolism	VraSR, WalKR	Kuroda *et al.* (2003), Dubrac *et al.* (2007)
Nutrient uptake	UhpBA, PhoRP	Verhamme *et al.* (2001), Whitworth *et al.* (2008a)
Inorganic ion transport	ManSR, KdpDE	Walderhaug *et al.* (1992), Yamaguchi *et al.* (2002)
Fermentation/respiration	DcuSR, ResED	Sun *et al.* (1996), Zientz *et al.* (1998)
Gliding motility	PilSR, AglZ, FrzS	Hobbs *et al.* (1993), Yang *et al.* (2004), Leonardy *et al.* (2007)
Virulence/persistence	PrrAB, SaeRS	Ewann *et al.* (2002), Liang *et al.* (2006)
Chemotaxis	CheABY	Simon *et al.* (1989), Djordjevic and Stock (1998)
Circadian timing	SasA/RpaA	Takai *et al.* (2006)
Competence development	ComPA, ComDE	Weinrauch *et al.* (1990), Pestova *et al.* (1996)
EPS production	VsrBC, SivSR	Huang *et al.* (1995), Bolotin *et al.* (2007)
Photosynthesis	RegBA, NblSR	Mosley *et al.* (1994), Salinas *et al.* (2007)
Quorum sensing	LuxNQ, LuxU, LuxO	Ulrich *et al.* (2005), Milton (2006)
Multicellular development	FruA, RedCDEF	Ellehauge *et al.* (1998), Higgs *et al.* (2005), Whitworth and Cock (2008a)
Proteolysis	CpxAR, CssRS	Danese *et al.* (1995), Darmon *et al.* (2002)
Drug resistance	BaeSR, EvgAS	Eguchi *et al.* (2003), Nishino *et al.* (2007)
Flagella biosynthesis	QseBC	Sperandio *et al.* (2002)
Osmotic balancing	EnvZ/OmpR	Comeau *et al.* (1985)
Sporulation	KinAB, Spo0FBA	Perego *et al.* (1989), Trach and Hoch (1993)
Heavy metal tolerance	CusSR, HydHG	Munson *et al.* (2000), Leonhartsberger *et al.* (2001)
Intercellular signalling	ActABCD, AsgD	Cho and Zusman (1999), Gronewald and Kaiser (2001)
Heat shock response	HsfBA, CssRS	Darmon *et al.* (2002), Ueki and Inouye (2002)
Cell cycle progression	DivJK, PleCD	Hecht *et al.* (1995), Wheeler and Shapiro (1999), Aldridge *et al.* (2003)

phenomena. The vast majority of TCSs regulate physiological adaptation as a direct response to changing environmental conditions. Responses can be apparently specific to one environmental parameter, for instance the Pho TCSs of many organisms respond to the availability of phosphate (Watson *et al.*, 1996), while CusSR responds to copper by inducing expression of copper-resistance genes (Munson *et al.*, 2000).

However, examples abound where individual TCSs are regulated by multiple stimuli, and/or regulate multiple adaptive responses. For example the BaeSR TCS responds to envelope stress (indole-induced), and metal ions (copper and zinc), giving rise to a drug-resistant phenotype (Nishino *et al.*, 2007) as well as a general envelope-stress response (Raffa and Raivio, 2002). The chemotactic TCS of *E. coli* regulates changes in directional motility in response to the perception of multiple chemoattractants/chemorepellants, by integrating signals from a set of receptor proteins known as methyl-accepting chemotaxis proteins, or MCPs (Sourjik and Berg, 2004). In addition, many phenotypes are regulated by multiple TCSs, either acting separately, or as networks of interacting TCSs. An intriguing example where changes in cellular behaviour arise as a consequence of the integration of regulation by multiple independent TCSs, is provided by the EnvZ/OmpR, CpxAR and RcsCDB systems of *E. coli*, which have recently been shown to act together as a combinatorial sensor of membrane properties (Clarke and Voigt, 2011). Behaviour can also arise as the consequence of interactions between TCSs, as is the case for the cell fate-determining system of *Caulobacter crescentus*, which relies upon communication between the DivJK and PleCD TCSs (Paul *et al.*, 2008). Similarly, regulation of the Pho regulon of *B. subtilis* involves at least three interacting TCS pathways (Hulett, 1996).

Many TCS-regulated processes are of profound and direct relevance to mankind. The pathogenesis of infectious organisms is often under the control of TCSs, and for most stages of a typical infection cycle, examples are plentiful where regulation involves TCSs (Beier and Gross, 2006; Calva and Oropeza, 2006; Perego and Hoch, 2008). IrlRS of *Burkholderia pseudomallei* regulates invasion of eukaryotic cells (Jones *et al.*, 1997), while early proliferation of *Mycobacterium tuberculosis* during intracellular growth is regulated by the PrrAB system (Ewann *et al.*, 2002). Survival of *Yersinia pestis* in human neutrophils is promoted by the PhoPQ TCS (O'Loughlin *et al.*, 2010), and NetB toxin production by *Clostridium perfringens* is regulated by VirSR (Cheung *et al.*, 2010). In *Pseudomonas aeruginosa* an array of TCS proteins regulate virulence, including PhoPQ, GacAS, RetS, LadS and AlgR amongst others (Gooderham and Hancock, 2009). ParSR governs multidrug resistance of *P. aeruginosa*, and in chronic lung infections a major virulence factor is the secretion of alginate, which is regulated by the KinB/AlgB TCS (Ma *et al.*, 1997). TCSs are attractive targets for drug discovery, not just because of their involvement in pathogenesis, but also because metazoans do not possess any endogenous TCSs – potentially reducing the risks of side-effects (Matsushita and Janda, 2002; Stephenson and Hoch, 2004; Cai *et al.*, 2005; Watanabe *et al.*, 2008; Li *et al.*, 2009; Gotoh *et al.*, 2010).

Thankfully, TCSs also regulate microbial behaviours that can be exploited by mankind. The RapA1/RapA2, AfsQ1/AfsQ2, PhoRP and AbsA1/AbsA2 TCSs all regulate secondary metabolite production in *Streptomyces* (Sola-Landa *et al.*, 2003; Lu *et al.*, 2007; McKenzie and Nodwell, 2007; Shu *et al.*, 2009), and TCSs have also been shown to regulate secondary metabolite production in Myxobacteria (Sandmann *et al.*, 2008), *Serratia* (Fineran *et al.*, 2005), *Erwinia* (Hyytiäinen *et al.*, 2003), and *Pseudomonas* (Corbell and Loper, 1995). Degradation of lignocellulose is of increasing importance to the biorenewables industry, and this process is regulated by the AbfSR TCS in *Cellvibrio japonicus* (Emami *et al.*, 2009). TCS regulation can also be harnessed for bioremediation – for instance *Pseudomonas putida* F1 exhibits chemotaxis towards, and degradation of, toxic aromatic hydrocarbons and trichloroethylene – behaviour dependent on the TodST TCS (Parales *et al.*, 2000). With the emergence of 'synthetic biology', TCSs have also been shown to be useful in the rational design of behaviour in micro-organisms. For instance, *E. coli* has recently been engineered to exhibit multichromatic regulation of gene expression as a tool for systems/synthetic biology research, using two TCS signalling pathways (Tabor *et al.*, 2011).

Genetics and genomics of TCSs

The first genome to be completely sequenced (*Haemophilus influenzae* Rd) was published in 1995 and contained nine TCS genes (Fleischmann *et al.*, 1995). Since then genome sequences have become available at an ever-increasing rate, with around 1400 publicly accessible at the beginning of 2011. Identification of TCSs in genomes is relatively trivial, with early approaches using algorithms such as BLAST, to identify putative TCS genes through their sequence similarity to known TCSs. This allowed the identification and quantification of TCSs in individual genomes, and it soon became apparent that most genomes encoded rather more TCS proteins than *H. influenzae*. Since 2002, the perceived genomic abundance of TCS has not changed significantly – presently, the mean number of TCS genes per genome seems to be between 50 and 60 for Eubacteria, ~15 for Archaebacteria, and < 5 for Eukaryotes, although considerable variability exists between genomes (Ashby, 2004; Whitworth, 2008; Wuichet *et al.*, 2010). Most eukaryotic genomes (beyond plants and fungi) possess no TCS genes at all, and while many prokaryotic genomes also encode no TCS genes, others, such as Deltaproteobacteria, Cyanobacteria and the occasional Firmicute, can have more than 250 (Wang *et al.*, 2002; Schneiker *et al.*, 2007; Whitworth and Cock, 2008b; Sirota-Madi *et al.*, 2010). There is a clear relationship between the number of TCS genes and genome size, with approximately 22 genes per Mbp after the first 1 Mbp (Whitworth and Cock, 2008b). Unsurprisingly there is also a correlation between life-cycle and TCS gene complement, with more TCS genes found in organisms which live in relatively variable environments, and also in organisms with more sophisticated lifestyles (Ashby, 2004; Galperin, 2005).

As the number of available genomes began to increase, problems of 'annotation creep' in TCS genes started to arise, as a consequence of misannotation, overspecific annotation, and erroneous assumptions made during the annotation process. For instance, early annotation pipelines assigned annotation purely on the basis of significant sequence similarity, therefore in some genomes multiple TCSs were annotated (wrongly) as regulators of phosphate uptake, purely because they were most similar in sequence to the Pho TCSs of *E. coli*. As many non-TCS proteins contain domains also found as input/output domains in TCSs (and HATPase_c domains are also found commonly beyond TCS proteins, for example in DNA gyrases/topoisomerases), many of those proteins have been erroneously annotated as TCS proteins. As with regulatory proteins in general, sequence similarity between TCSs does not necessarily imply a similar functional role, therefore annotation of TCSs by sequence similarity, has been superseded by categorization on the basis of domain architecture (Galperin, 2010), where TCS proteins are divided into families, and family membership correlates with mechanism of action (for example, DNA-binding, c-di-GMP synthesis, receptor methylation, etc.). Typically, RRs are categorized by the nature of their output domains, while HKs are divided into families according to a phylogenetic analysis of the transmitter domain, and/or the sequence characteristics of the H-box region (see for example Fabret *et al.*, 1999; Hutchings *et al.*, 2004; Galperin and Nikolskaya, 2007). While early studies of TCS genomics used similar domain-based approaches to identify and categorize TCS genes, there were inevitably differences in approach, some subtle but others more fundamental. The emergence of specialized but comprehensive TCS databases such as MiST and P2CS (see 'Web resources' at end of chapter), will hopefully result in the harmonization of categorization and annotation schemes across the TCS literature.

There have now been many publications cataloguing the TCS gene sets of individual organisms (Table 8.2), and it has become clear that there are few overarching laws governing the nature of the TCS gene set in different genomes. But, while there are no hard-and-fast 'rules', there may instead be some 'guiding principles' which are adhered to more or less rigidly, depending on the organism. For instance it is generally true that most TCS genes within any genome are found

Table 8.2 Publications cataloguing genomic TCS gene sets

Organism	Date	TCS genes	Publication	Comments
Synechocystis PCC 6803	1996	80	Mizuno *et al.* (1996)	Categorized domain composition. Several complex hybrid HKs
Escherichia coli	1997	62	Mizuno (1997)	Grouped RRs and HK into families by domain structure
Bacillus subtilis	1999	72	Fabret *et al.* (1999)	Suggested co-evolution of H-box motif, RR output domain and gene order
Streptococcus pneumoniae	2000	27	Throup *et al.* (2000)	Virtually all genes paired, with most loci affecting pathogenesis
Pseudomonas aeruginosa	2000	127	Rodrigue *et al.* (2000)	Large number of independent Hpt domains
Anabaena/Nostoc PCC 7120	2002	211	Wang *et al.* (2002)	Several proteins also have Ser/Thr (de) phosphorylation domains
Rhizobia	2004	99–174	Hagiwara *et al.* (2004)	Definition of orthologous sets of TCS in three species
Streptomyces coelicolor	2004	164	Hutchings *et al.* (2004)	Several atypical RRs. Abundance of Nar and Pho family TCSs
Clostridia	2005	48–79	Doß *et al.* (2005)	Predominantly Pho family TCSs with HAMP domains. Orphans have shared heritage
Desulfovibrio vulgaris	2006	114	Zhang *et al.* (2006)	Computational prediction of partnerships between orphan proteins (correlated expression)
Bacillus species	2006	70–110	de Been *et al.* (2006)	Computational prediction of partnerships between orphan proteins (orthology-, synteny- and phylogeny-based)
Euryarchaeota	2006	3–87	Ashby (2006)	TCS divided into orthology groups
Cyanobacteria	2006	12–261	Ashby and Houmard (2006)	Diverse modes of evolution. Catalogued TCS into orthology groups
Streptomyces avermitilis	2007	135	Wei *et al.* (2007)	Orthology analysis allowed proposals for cognate partnerships between orphans
Pseudomonas syringae	2007	18–186	Lavín *et al.* (2007)	Differences in TCS complement associated with different pathogenic properties
Myxobacteria	2008	175–336	Whitworth and Cock (2008b)	High proportion of Ntr family TCSs, orphans and complex hybrid HKs. Many RRs lacking output domains
Erwinia	2009	46	Zhao *et al.* (2009)	Identification of TCS followed by systematic gene knockout analysis
Enterobacteria	2010	25–65	Qi *et al.* (2010)	Comparative evolutionary analysis of 53 genomes from 16 species
Brucella	2010	36	Lavín *et al.* (2010b)	Highly conserved TCS gene set, however remarkable variation in TCS pseudogenes

in HK-RR gene pairs, however the proportion of TCS genes found in pairs can vary between 11.5% and 82.3%, even for genomes with > 100 TCS genes (Williams and Whitworth, 2010). Similarly, within a genome, there tends to be conservation of gene order within a TCS family (for example in *B. subtilis*, all nine Nar family TCS are encoded with the HK upstream of the RR, while the 14 Pho family TCS are all encoded with the RR upstream), however some genomes exhibit significant proportions of family members with the opposite gene order (for example 30% of Pho family TCSs in *Myxococcus xanthus* have the HK gene upstream of the RR gene).

Comparisons of TCS gene sets from different organisms are made possible through the use of metrics, against which each gene set can be measured (Galperin and Kolker, 2006, Galperin *et al.*, 2010). Most commonly used metrics are simple counts (for instance, the number of TCS genes, the number of TCS domains, the number of different input domains in the gene set, etc.). Other metrics assess the relative numbers of subsets of TCSs (for instance, the proportion of sensor kinases with transmembrane helices in their sensory domains, the ratio of transmitter:receiver domains, the proportion of TCS genes found in pairs, etc.). As would be expected, for all TCS gene metrics there is a significant phylogenetic component, with closely related organisms tending to have similar TCS gene/domain organizations, and distantly related organisms exhibiting very different organizations (Galperin *et al.*, 2010). Therefore we must be careful to acknowledge that our perception of what constitutes 'normal' TCS gene sets, may well be biased by which genomes have been sequenced to date (i.e. the culturable minority and those of the greatest financial/medical importance to mankind), and at lower taxonomic levels, by the relative paucity of sequenced genomes (Whitworth, 2008).

Of the many generalities or 'guiding principles' regarding TCS gene organization which have emerged from comparative genomics analyses, most can be trivially explained. For instance, the number of TCS genes increases with increasing genome size, and with increasing complexity of lifestyle/environment, each of which would require an increased regulatory capacity. Similarly, most TCS are encoded as paired HK and RR genes, which allows co-regulation of gene expression, co-evolution and co-transfer of the entire TCS.

Another intriguing feature of TCS gene organization is the prevalence of 'hybrid kinases'. Hybrid kinases contain both transmitter and receiver domains, and thus resemble the presumed product of a fusion between paired HK and RR genes. Nearly a quarter (23%) of all HKs are actually hybrid kinases (Williams and Whitworth, 2010), however it is far from clear why some systems are encoded by partnered HK and RR genes, while others are encoded as a single hybrid kinase gene. The situation is further confused by considering genes encoding phosphorelays. Phosphorelays transfer phosphoryl groups between four successive TCS domains (Transmitter, Receiver[1], Hpt, Receiver[2]). In some systems the four domains are encoded by four separate genes. However in other cases some genes encode multiple TCS domains (Appleby *et al.*, 1996). For instance in *E. coli* three HKs (TorS, ArcB, and EvgS) have (from N- to C-terminus) successive transmitter, receiver and Hpt domains (Mizuno, 1993), and are found immediately upstream of their partner RRs (TorR, ArcA and EvgA respectively), together constituting intact phosphorelays. The reasons why some patterns of covalent connectivities are observed in phosphorelays, while other patterns are not observed remains obscure (Appleby *et al.*, 1996; Williams and Whitworth, 2010). In addition, many hybrid kinases are observed with very complex TCS domain architectures (Williams and Whitworth, 2010; Figs. 8.2 and 8.3); for example, the RodK hybrid kinase of *M. xanthus* contains a transmitter domain, with three C-terminal receiver domains, all of which are required for activity (Rasmussen *et al.*, 2006).

Evolution of TCSs

The ability to catalogue the entire TCS gene sets of completely sequenced genomes has enabled diverse analyses to probe the evolution of TCSs (reviewed recently by Whitworth and Cock, 2009). In addition to identifying global trends in TCS evolution, studies have focused on the evolution of contemporary gene sets in specific taxa, while other studies have provided insights into the relationship between selective pressures and mechanistic features of TCS action.

As more genome sequences became available, and the abundance of HK and RR sequences increased, it became possible to perform phylogenetic analyses on TCSs. It soon became clear that such analyses could group receiver and transmitter domain sequences into distinct clades or families (Grebe and Stock, 1999; Koretke *et al.*, 2000; Kim and Forst, 2001). Initially there were thought to be between 5 and 11 major TCS families, however it now seems that there are a plethora of minor families, with four particularly large families (the

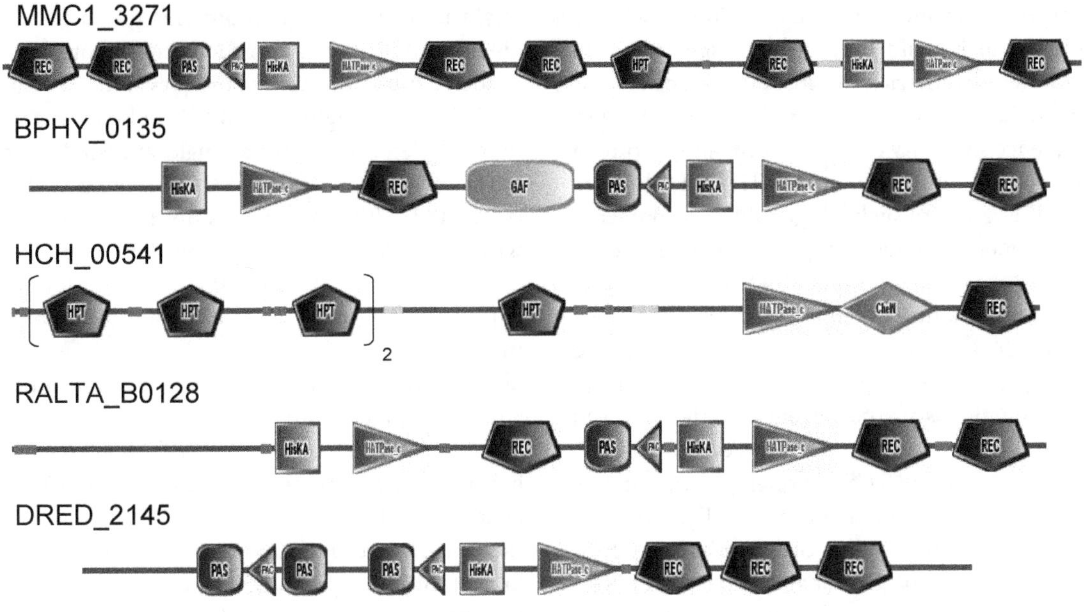

Figure 8.2 Domain architectures of complex TCS proteins. A selection of particularly complicated hybrid kinases. Each protein contains at least four transmitter, receiver and/or Hpt domains. Individual images generated by SMART (http://smart.embl-heidelberg.de/).

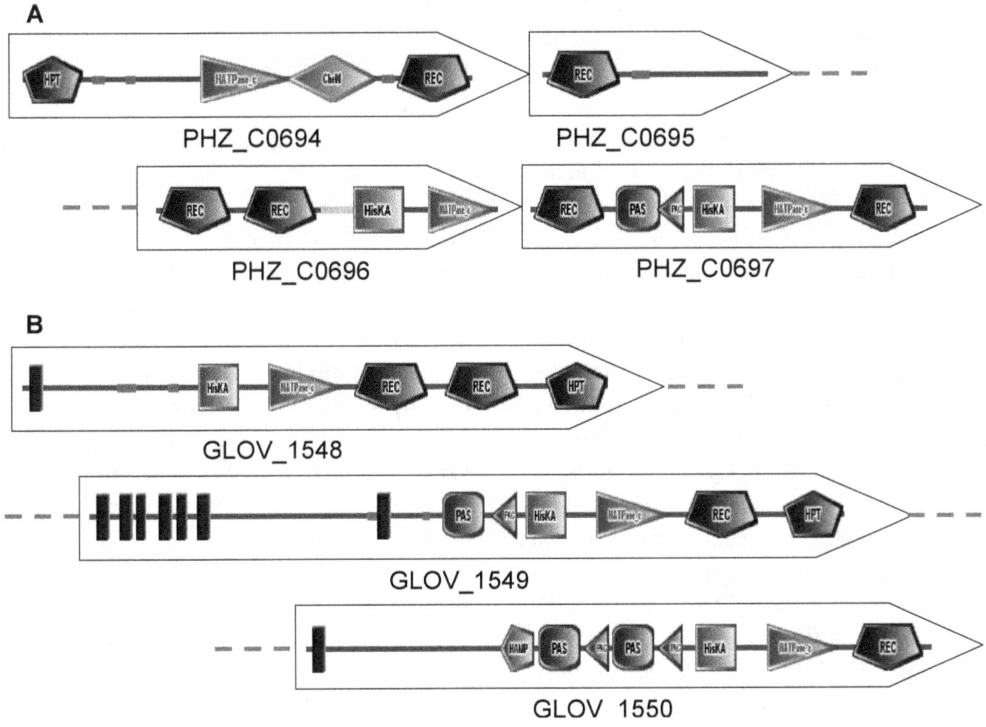

Figure 8.3 Organization of complex TCS gene loci. Two unusually complex TCS gene loci, each encoding nine receiver, transmitter and Hpt domains, distributed between four (A) and three (B) contiguous TCS genes, the majority of which are hybrid kinases. Arrows denote the direction of transcription. Individual images generated by SMART (http://smart.embl-heidelberg.de/).

Che, Ntr, Pho and Nar families). Grouping TCSs into families could be achieved using a variety of different criteria, for instance according to the sequence similarity of transmitter domains, sequence similarity of receiver domains, nature of RR output domain, nature of HK input domain, etc. The results of such categorization approaches were generally congruent, providing evidence for the co-evolution of input, transmitter, receiver and output domains – in turn implying that each family had derived from the duplication and divergence of a single ancestral TCS. Such an interpretation is also consistent with the observed conservation of gene order within each family.

The numbers of TCSs belonging to each family differs widely between organisms. For instance, the TCSs of *Streptomyces coelicolor* are predominantly members of the Nar and Pho families, with no members of the Ntr family (Hutchings *et al.*, 2004), however *M. xanthus* has few Nar family TCSs, but a large number of Ntr-family TCSs (Whitworth and Cock, 2008b). Thus in individual taxa, there has been lineage-specific amplification of particular families of TCSs, but in other cases deletion of entire TCS families seems to have occurred (Whitworth and Cock, 2008b; Qi *et al.*, 2010). While this appears to be generally the case, incongruence between the heritage of many adjacently encoded transmitter and receiver domains has been observed, suggesting non-linear descent, probably by recombination between TCS of different families (Koretke *et al.*, 2000; Chen *et al.*, 2004; Alm *et al.*, 2006). There is also evidence of extensive domain shuffling during evolution, with many minor families of TCS emerging from within one of the main families through the acquisition of novel input/output functions (Pao and Saier, 1995; Alm *et al.*, 2006; Qian *et al.*, 2008).

It is well documented that adjacent co-directional genes in prokaryotic genomes can overlap, resulting in translational coupling between pairs of genes (Rogozin *et al.*, 2002; Cock and Whitworth, 2010), and this is indeed the case for paired TCS genes (Cock and Whitworth, 2007a). The phenomenon of translational coupling means that such genes will generally be expressed together at a fixed ratio, independently of 'noise' in the transcription/translation machinery. An alternative strategy to achieve the same goal, might be to fuse the two proteins into a hybrid kinase, and it appears that such gene fusions have occurred numerous times during evolution. Originally it was thought that all hybrid kinases were monophyletic, implying a single ancestral gene fusion event (Grebe and Stock, 1999; Koretke *et al.*, 2000). This now seems not to be the case, with hybrid kinases found in all the major TCS families (Zhang and Shi, 2005). However it does appear that only a subset of TCSs are able to form functional hybrid kinases (Cock and Whitworth, 2007b), with constraints imposed by gene order and domain architecture.

More than 80% of all HKs (both true HKs and hybrid kinases) possess transmembrane (TM) helices (Whitworth and Cock, 2007b), implying they operate through the sensation of extracellular stimuli (Galperin, 2005). In addition, more than half of all RRs possess DNA-binding output domains (www.P2CS.org). TM helices and DNA-binding domains each need to be positioned in appropriate (and separate) subcellular locations in order to function, which largely precludes the formation of a functional hybrid kinase. In addition, in order to maintain appropriate membrane insertion and/or DNA-binding activity, fusion will generally only result in a functional hybrid kinase if the HK gene is upstream of the RR. As the order of paired TCS genes correlates with the TCS family, as does the nature of the HK input domain and whether the output domain is DNA-binding, fusion rate is highly dependent upon TCS family membership (Cock and Whitworth, 2007b). One-component systems (OCSs) are signalling proteins with an input domain linked directly to an output domain (Ulrich *et al.*, 2005). In a comparative genomic analysis, Ulrich *et al.*, (2005) found that OCSs evolved prior to the evolution of TCSs, are both more widespread than TCSs, and more diverse than TCSs. They postulated that one of the reasons that TCSs evolved in the first place, was to allow the physical separation of stimulus sensation from response elicitation, enabling extra-cytoplasmic sensing. Why then are hybrid kinases found at such a high frequency in genomes, when they represent TCSs that have lost their primary raison d'être? One possible explanation is that TCSs (even cytosolic hybrid kinases) are able to integrate multiple signals, through a variety of mechanisms, which OCSs are not, increasing their adaptive potential and evolutionary fitness (Krell *et al.*, 2009).

TCS protein structure

The structures of several TCS components have been determined, and there have now been many reviews on this topic (Robinson *et al.*, 2000; Stock *et al.*, 2001; West and Stock, 2001), including a trio of recent publications (Gao and Stock, 2009; Bourret, 2010; Stewart, 2010).

By the turn of the last century, the structures of signalling domains from several TCSs had been solved, providing insights into the mechanisms of signal transduction. The deduced structures of receiver domains are found to generally exhibit a common fold ($\alpha\beta_5$), with alternating alpha-helices and beta-strands (Fig. 8.4). The beta-strands form a five-stranded parallel beta-pleated sheet within the core of the domain, with the alpha-helices packed around the outside of the sheet (Stock *et al.*, 2001; Bourret, 2010). Receiver domains typically contain three highly conserved Asp residues in their primary structure, two towards the N-terminus of the domain, and one (the phosphorylatable Asp), towards the centre of the domain. In the receiver domain three-dimensional structure these residues lie close together, forming an acidic pocket which coordinates a divalent metal ion (typically Mg^{2+}) required for catalysis of phosphorylation (Lukat *et al.*, 1990; Bourret, 2010). *In vitro*, receiver domains can catalyse their own phosphorylation using a range of small phospho-donor molecules (e.g. acetyl phosphate, phophoramidate, carbamoyl phosphate, etc.), some of which may have physiological relevance (Wolfe, 2010). An apparent lack of substrate specificity and unsaturatable reaction kinetics (S.S. Da Re *et al.*, 1999), suggests that binding between receiver domains and their substrates is extremely weak (Bourret, 2010).

In 1998, NMR studies provided the solution structure of the HATPase domain of EnvZ, which also adopts a globular fold (Tanaka *et al.*, 1998). The HATPase domain has a two-layer sandwich structure, with one layer formed from three alpha-helices, and the other from five beta-strands (Fig. 8.5). As for receiver domains, highly conserved residues required for HATPase function are co-located in the structure, forming a binding site for Mg^{2+} and ATP, and an ATP lid (reviewed by Stewart, 2010). A year later, the NMR solution structure of the HisKA domain of EnvZ was published (Tomomori *et al.*, 1999). The HisKA domain is dimeric, and entirely helical, with two alpha-helices from each monomer coming together to form a four-helix bundle (Fig. 8.5). The two phosphorylatable His residues

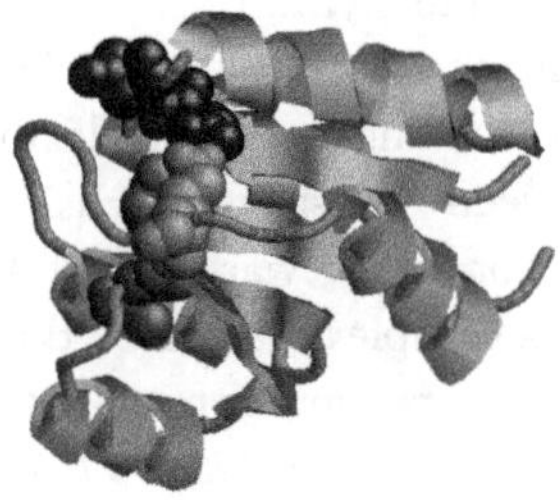

Figure 8.4 Receiver domain structure. Spo0F is a typical receiver domain, composed of five alternating sets of alpha-helices and beta-strands. A catalytic pocket is formed by a trio of Asp residue, a Ser/Thr residue, and a conserved Lys (side chains shown as space-filled spheres). A Phe/Tyr is often required for the propagation of conformational change upon phosphorylation, but is absent from Spo0F. Image created from PDB file 1NAT by Heather Currinn.

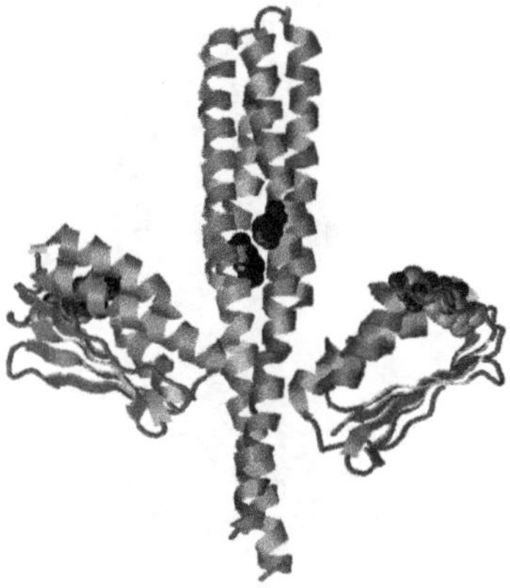

Figure 8.5 Transmitter domain structure. The transmitter domain dimer of DesK. Globular HATPase domains (in complex with the ATP analogue AMP-PCP) are found on either side of a central four-helix bundle containing a phosphorylatable His residue (highlighted side chains). Image created from PDB file 3GIE by Stuart Thomas.

extend into the solvent, making them easily accessible to HATPase domains (during autophosphorylation), or to receivers domains (during phosphotransfer).

The structures of some other phosphoaccepting TCS domains show remarkable structural similarity to the HisKA four-helix bundle. The Hpt domain structure of CheA from *E. coli* was solved in 1995 (Zhou *et al.*, 1995). This domain is monomeric, and possesses a single phosphorylatable His residue, however it is composed of four helices in a bundle that can be readily superimposed onto that of a HisKA (Fig. 8.6). Spo0B is a phosphotransfer protein unique to the *Bacillus* sporulation phosphorelay, where it fulfils the role of the Hpt domain found in more typical phosphorelays. Spo0B monomers have an N-terminal helix–loop–helix, through which dimerization occurs, resulting in a four-helix bundle akin to those of HisKAs (Fig. 8.7). Spo0B also possesses a C-terminal domain reminiscent of an HATPase domain (Varughese *et al.*, 1998). While there is considerable structural similarity between the HisKA, Hpt and Spo0B folds, and their structures can be readily superimposed on one another (for an example see Berger and Nimwegen, 2008), they exhibit no significant sequence similarity to one another, suggesting they have evolved convergently.

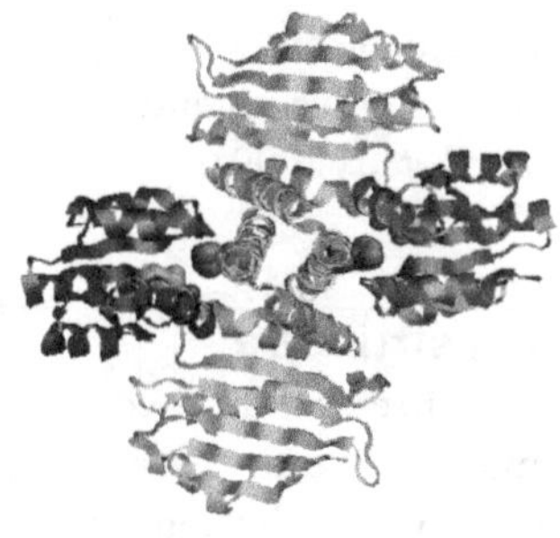

Figure 8.7 Structure of the complex between receiver and phosphotransfer domains. The complex formed between two Spo0F RRs (top left and bottom right darker domains) and a Spo0B dimer, looking down the four-helix bundle of Spo0B. The phosphorylatable His residues of Spo0B (dark space-filled spheres) are positioned next to the Asp residues of Spo0F (pale space-filled spheres). Image created from PDB file 1F51 by Simon Cameron.

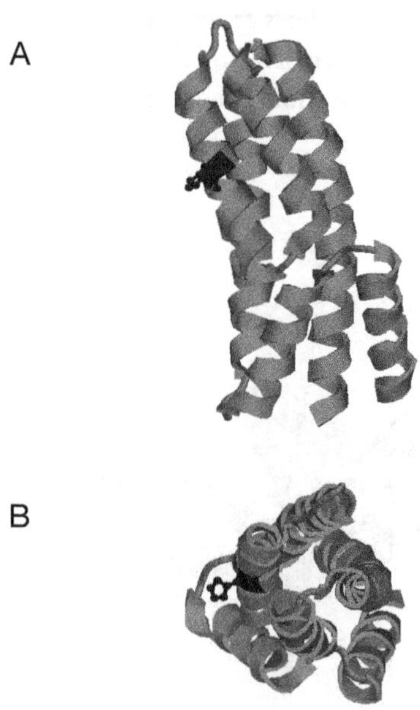

Figure 8.6 Hpt domain structure. Two views of the Hpt domain of ZmHP2 from maize. The four helix bundle is monomeric and contains a single phosphorylatable His residue (highlighted). Image created from PDB file 1WN0 by Stuart Thomas.

Today, the structures of over 200 receiver domains are known (including several solved as part of whole response regulators), along with the structures of several HATPases and phosphotransfer domains (Bourret, 2010; Stewart, 2010). However, it was only relatively recently that structures were first obtained for intact transmitter domain units (Marina *et al.*, 2005; Bick *et al.*, 2009; Albanesi *et al.*, 2009), probably because of difficulties arising from the flexibility of the linkers between constituent HisKA/Hpt and HATPase domains (Stewart, 2010). In 1995, the structure of a complex formed by the RR Spo0F, and the phosphotransfer protein Spo0B of *Bacillus subtilis*, was solved (Zapf *et al.*, 1995). For the first time the geometry of a complex formed between a four-helix bundle and a receiver domain was described. As expected the receiver domain bound to one side of the helical bundle, with the phosphorylatable Asp and His residues facing each other across the Spo0B–Spo0F interface (Fig. 8.7). A similar picture emerged from studies of the structure of the YPD1 (Hpt protein) complex with SLN1 (RR), with the phosphorylatable

His and Asp residues opposing each other across the protein–protein interface (Xu *et al.*, 2003; Zhao *et al.*, 2008). However, while structurally analogous to HisKA domains, Spo0B is an atypical phosphotransfer protein, and the lack of sequence similarity between Spo0B, Hpt and HisKA domains means that we cannot be certain that the binding between Spo0F and Spo0B, or between YPD1 and SLN1, are good proxies for receiver-HisKA binding. We will have to keep waiting for the structure of such a complex to be determined, which seems to be a non-trivial task, presumably due to the transience and low affinity of HisKA-receiver binding (Bourret, 2010; Schug *et al.*, 2010). Thankfully a low resolution structure of a HisKA–receiver complex has become available recently (Yamada *et al.*, 2009), which conforms to expectations based on the Spo0B–Spo0F interaction, and presumably higher resolution structures will become available in due course.

Mechanisms of TCS signal transduction

Signal perception by input domains

TCSs respond to a huge range of stimuli, and it is therefore of little surprise that they also possess a phenomenal diversity of sensory or input domains (Mascher *et al.*, 2006; Cheung and Hendrickson, 2010; Krell *et al.*, 2010). Initial attempts to classify HK input domains used the number and nature of any TM helices (Fabret *et al.*, 1999; Mascher *et al.*, 2006) as classification criteria, and while TM topology provided a guide to the cellular compartment being sensed, it soon became apparent that TM topology did not correlate with TCS family membership, or the nature of the perceived stimulus. In addition, a number of conserved domains are often found as HK input domains (typically N-terminal to the transmitter domain), including CACHE, CHASE, HAMP, GAF, PAS and PAC domains (Galperin *et al.*, 2001; Mascher *et al.*, 2006). These domains are often highly variable in sequence, frequently duplicated, and present in diverse combinations within a HK, suggesting either that the HK can sense multiple signals (and possibly integrate them in a sophisticated fashion), or that some 'sensory' domains are capable of acting as signal-transducing 'linker' domains (Fabret *et al.*, 1999; Mascher *et al.*, 2006) – particularly HAMP, PAS/PAC and GAF domains (Gao and Stock, 2009).

In most cases, the mechanism of signal transduction by input domains to the HK core is not understood, but recently structures have been deduced for a few signalling domains with bound ligands, providing insights into mechanisms of stimulus perception (Krell *et al.*, 2010). Many extracytoplasmic-sensing HKs share a common topology, with two TM helices separated by an extra-cytoplasmic loop. In the HKs EnvZ and NarX (and the Tar aspartate receptor which stimulates signalling through the Che TCS), the TM helices of the sensor can dimerize forming a four-helix bundle. Studies using Tar-EnvZ and Tar-NarX chimeras suggest that two helices of the four-helix bundle undergo a piston-like sliding motion upon ligand binding (Ward *et al.*, 2002; Yoshida *et al.*, 2007; Cheung and Hendrickson, 2009). Similar arrangements of helices, with piston-like motions upon stimulation are also observed for CitA and similar HKs (Sevvana *et al.*, 2008).

HAMP domains also possess a four-helix structure; however, the bundle is twisted into a coiled-coil, and it is proposed that signal transduction occurs through relative rotation of the helices within the coil (Hulko *et al.*, 2006). It is not clear how diverse signalling mechanisms (piston sliding or helix rotation) can bring about the same effect on HK autokinase activity, nor how in some HKs, TM signalling via a piston-sliding mechanism is transduced to a HAMP domain as a helix rotation (Gao and Stock, 2009; Cheung and Hendrickson, 2010). Most sensory domains are not prone to dimerization when present in isolation, and are thought to form dimers only as a consequence of their transmembrane localization, or because the HKs to which they are attached form dimers (Gao and Stock, 2009; Cheung and Hendrickson, 2010). As such the binding between dimers of sensory domains are possibly quite weak, and therefore the monomer–monomer interface easily rearranged – potentially transducing signals to the HK core through changes in symmetry across the HisKA dimerization interface (Gao and Stock, 2009; Cheung and Hendrickson, 2010).

Autophosphorylation

While relatively little is known about how stimulus perception activates HK autokinase activity, most transmitter domains become active when separated from their sensory domains, implying that in the absence of signal, sensory domains inhibit kinase activity (Skerker *et al.*, 2005; Yamamoto *et al.*, 2005). Consequently most transmitter domains in isolation exhibit autokinase activity and the kinetic parameters of autophosphorylation have been measured in several such systems (summarized by Krell *et al.*, 2010). Autophosphorylation typically occurs *in trans* within a HK dimer, with the HATPase of one subunit phosphorylating the HisKA of the other monomer (Ninfa *et al.*, 1993; Cai and Inouye, 2003). However, examples exist where phosphorylation occurs *in cis* (Casino *et al.*, 2009). The phosphoryl group for autophosphorylation usually comes from the γ-phosphoryl group of ATP, but a case where the phosphoryl group derives from GTP has been described (Scaramozzino *et al.*, 2008). Whilst most HKs are thought to act as homodimers, in at least one case a heterodimer formed from two different HK monomers has been shown to have physiological relevance. In *P. aeruginosa* the hybrid kinase GacS causes phosphorylation of the RR GacA during chronic infections, however during acute infection another hybrid kinase, RetS, forms heterodimers with GacS, preventing phosphorylation of GacA (Goodman *et al.*, 2009).

Phosphotransfer reactions

Binding of a RR receiver domain to a HK HisKA domain, allows phosphoryl groups to be transferred from the His residue of the HisKA to the conserved Asp residue of the receiver (for a discussion of the energetic properties of phospho-His and phospho-Asp residues see Stock *et al.*, 1989 and Gao and Stock, 2009). The K_D of RR-HK binding is typically around 1 μM, much higher than that observed generally between proteins, implying only loose association between HKs and RRs (Krell *et al.*, 2010). Phosphotransfer can also occur in reverse, from phosphorylated RR to HK (retrophosphorylation, Jiménez-Pearson *et al.*, 2005), or from RR to Hpt domains within phosphorelays.

The specificity of HK-RR binding is of paramount concern *in vivo*, as most organisms possess multiple HKs and RRs, all of which are homologous and therefore potentially capable of interacting with one another relatively indiscriminately (Bijlsma and Groisman, 2003). Conserved mechanisms seem to have evolved to ensure the specificity of phosphotransfer, or to cope with inappropriate phosphotransfer (Bijlsma and Groisman, 2003; Laub and Goulian, 2007). In some cases TCS proteins have multiple partners, for instance the HK CheA of *E. coli* phosphorylates both CheB and CheY, while the RR Spo0F of *B. subtilis* is phosphorylated by multiple HKs (Stock *et al.*, 1989; Fabret *et al.*, 1999). By virtue of this property complex signalling networks can be constructed, which are able to generate sophisticated behaviour, and/or assimilate multiple signals (Clarke and Voigt, 2011). However, the general rule seems to be that each HK has one cognate RR and vice-versa (Skerker *et al.*, 2005). In such cases RRs can still be phosphorylated by non-cognate HKs, however this represents 'noise' rather than signalling, and is referred to as 'cross-talk'. A common mechanism to reduce unwanted cross-talk is to endow the HK with phosphatase activity towards its cognate phospho-RR, such that when not actively phosphorylating the RR, the HK actively de-phosphorylates any RR that becomes phosphorylated (Alves and Savageau, 2003). Such phosphatase activity is not necessarily a consequence of retrophosphorylation, but can be a distinct activity of the HisKA domain, modulated by interaction with the HATPase (reviewed by Gao and Stock, 2009). Another potential mechanism to reduce cross-talk is RR competition, where an abundance of cognate RR outcompetes non-cognate RRs from HK-binding (Laub and Goulian, 2007).

Activation of effector domains

A plethora of output domains are found associated with receiver domains in RRs (Galperin, 2010). The most prevalent output domains are DNA-binding, consistent with roles in altering gene expression, however significant proportions of RRs have output domains with enzymatic activity, or lack an output domain altogether (Jenal and Galperin, 2009). Particularly common enzymatic output domains include methyl-esterase/transferase, di-guanylate cyclase, Ser/Thr protein kinase/phosphatase, and adenylate cyclase domains.

Despite the range of output domains in RRs, there seem to be common mechanisms for receiver-mediated regulation of effector domains/proteins (Gao and Stock, 2010; Bourret, 2010). The structures of several phosphorylated and unphosphorylated receiver domains and RRs have been solved, providing insights into the mechanisms of regulation (reviewed by Gao and Stock, 2010). Phosphorylation of receiver domains leads to the displacement of highly conserved residues around the phosphorylation site, which in turn leads to a conformational change that is propagated through the domain. The most dramatic changes in conformation all lie on one face of the receiver domain, defined by the secondary structure elements α4-β5-α5 (Gao and Stock, 2010; Bourret, 2010). In essence, it is thought that each receiver domain can adopt two possible conformations, with one stabilized by phosphorylation (the 'active' conformation).

Standalone receivers (RRs lacking an output domain), are thought to modulate the activity of other proteins through phosphorylation-dependent intermolecular protein–protein interactions. For example, phosphorylated CheY of *E. coli* interacts with the flagella protein FliM (Lee *et al.*, 2001). The separation of receiver domain from effector seems particularly important in TCSs that regulate chemotaxis, or spatial behaviours such as the cell cycle and polar morphogenesis (Jenal and Galperin, 2009). Receiver domains in RRs with output domains modulate the activity of their output domains either by phosphorylation-dependent alterations in intramolecular domain–domain interactions, or through changes in oligomerization state induced by phosphorylation. For example in NarL, the unphosphorylated receiver domain binds to the DNA-binding domain preventing its interaction with DNA, with phosphorylation relieving binding (Baikalov *et al.*, 1996). Similarly, phosphorylation of FixJ promotes the receiver domain to release (and activate) the DNA-binding output domain, whilst simultaneously promoting receiver domain dimerization (S. Da Re *et al.*, 1999). The RR PleD dimerizes in response to phosphorylation, bringing together two guanylate-binding domains, a requirement to produce the cyclic di-guanylate second messenger c-di-GMP (Paul *et al.*, 2007). Such cases, where the change in conformation affects the ability of a receiver domain to dimerize, are commonplace (Gao *et al.*, 2008), and are particularly important for RRs with DNA-binding output domains, whose operator sequences are often tandem or inverted repeats. Regardless of whether hetero/homo-domain, or inter-domain/protein, the changes in protein–protein interaction caused by phosphorylation are almost universally mediated through the α4-β5-α5 surface of the receiver domain (Fig. 8.8).

Modulation of TCS signalling

As described above, the basic TCS scheme is a relatively simple conserved mechanism for coupling stimulus sensation to response elicitation. Complexity in signalling pathway structure can emerge as a consequence of HKs and/or RRs having multiple partners, or through the inclusion of extra receiver and Hpt domains to create a phosphorelay. However, many TCSs are embellished versions of the basic scheme, with additional protein components, or unusual activities.

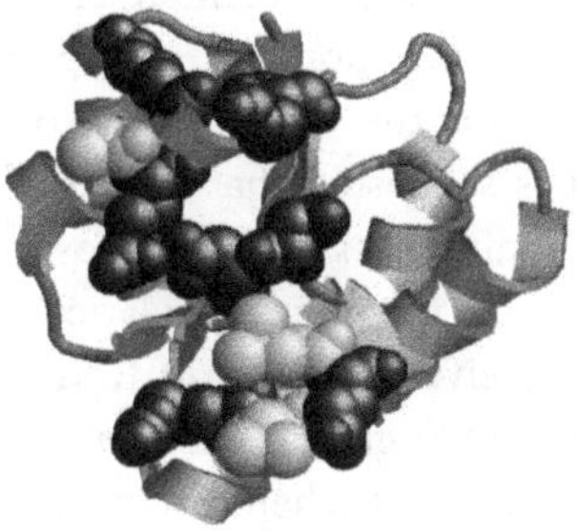

Figure 8.8 The Pho-family receiver domain intermolecular–interaction interface. Residues of the receiver domain α4-β5-α5 region are highlighted as space-filled spheres. Dark residues have charged side-chains, while pale residues are hydrophobic. The interaction interface has pseudo 2-fold rotational symmetry, which allows homodimerization. Image created from PDB file 1ZES by Stuart Thomas, inspired by Fig. 1C of Gao *et al.* (2008).

Accessory receiver domains

An extremely common TCS embellishment is the presence of additional receiver domains (Williams and Whitworth, 2010). Over recent years it has become clear that while the purpose of a typical (core) receiver domain is to receive phosphoryl groups from a partner HK, change conformation and then stimulate/inhibit some cellular activity, a large number of receiver domains do not perform that role, instead presumably having an 'accessory' function (Zhang and Shi, 2005). For example the RedCDEF phosphorelay of *M. xanthus* has one more receiver domain than necessary (Higgs *et al.*, 2005), while the CbbRRS phosphorelay of *Rhodopseudomonas palustris* possesses two extra receiver domains (Romagnoli and Tabita, 2006). In the CbbRRS system, a receiver domain within the hybrid kinase CbbSR dictates which of the two RRs, CbbRR1 or CbbRR2 becomes phosphorylated (Romagnoli and Tabita, 2007). In the Che TCS of *Helicobacter pylori*, phosphoryl groups are transferred from CheA to either the RR CheY1, or to a receiver domain at the C-terminus of CheA (CheY2). Phosphoryl groups can also flow from CheY1 to CheA to CheY2, and it is proposed that CheY2 acts as a phosphate sink to modulate the half-life of CheY1 phosphorylation (Jiménez-Pearson *et al.*, 2005). The unusual hybrid kinase RodK, again of *M. xanthus*, possesses three receiver domains each of which are important for function, and a cognate RR RokA (Wegener-Feldbrügge and Søgaard-Andersen, 2009). The RodK transmitter domain can phosphorylate both RokA and the C-terminal RodK receiver domain. It is suggested that which of these two receiver domains does become phosphorylated is dependent upon the phosphorylation status of the other two receiver domains of RodK (Wegener-Feldbrügge and Søgaard-Andersen, 2009). There are also some cases where a core RR possesses additional accessory functions. For instance, DivK of *C. crescentus* is the cognate RR of the HK DivJ, but it also acts as an allosteric activator of two HKs, DivJ and PleC (Paul *et al.*, 2008).

Phosphatase activities

As the phosphorylation levels of both HKs and RRs are critically important for appropriate signalling, it is not surprising that phospho-aspartyl phosphatases and phospho-histidinyl phosphatases are important regulators of TCS activity. RRs and some HKs have intrinsic phosphatase activity, while RRs also act as specific phosphatases of phospho-HKs, and most HKs appear to also act as phospho-RR phosphatases when not being stimulated. Characterized phospho-HKs have similar half-lives of phosphorylation (between 30 and 90 minutes), however phospho-RRs have a wide range of stabilities, with half-lives of phosphorylation varying between two seconds and days (Krell *et al.*, 2010).

There are also extrinsic or 'auxiliary' phosphatases that act upon phosphorylated TCS proteins. Four families of phosphatases that act upon phospho-RRs have been described, the CheZ, CheC/CheX/FliY, Spo0E and Rap families (Silversmith, 2010). Phosphatase activity can be regulated in various ways. CheZ phosphatase is activated upon binding to CheA (O'Connor and Matsumura, 2004), which regulates the subcellular localization of phosphatase activity, while CheC is stimulated by binding CheD, a co-translated deamidase of chemotactic receptors (Chao *et al.*, 2006). Spo0E activity is potentially regulated by proteolysis (Le and Schumann, 2009), while Rap family phosphatase activities can be regulated by mature Phr pentapeptides that are co-expressed with their target phosphatases (Perego, 1997). An auxiliary phospho-histidinyl phosphatase has also been identified (SixA) which dephosphorylates the Hpt domain of the ArcB phosphorelay hybrid kinase of *E. coli* (Ogino *et al.*, 1998). Proteins showing homology to SixA are found in many genomes, suggesting that regulated dephosphorylation of phospho-His TCS proteins may be more common than currently appreciated (Krell *et al.*, 2010).

HK inhibitors and activators

Several inhibitors of HKs have been described (reviewed by Krell *et al.*, 2010). KinA is a HK of the sporulation phosphorelay of *B. subtilis*, and its ability to autophosphorylate is inhibited by two proteins, Sda and KipI (Wang *et al.*, 1997; Rowland *et al.*, 2004). Both proteins are thought to bind to the HisKA domain, but at a distance from the phosphorylatable His residue. On the other hand, CpxP is a periplasmic protein of *E.*

coli, which acts an inhibitor of the HK CpxA, by binding to its sensory domain (Zhou *et al.*, 2011). A general stress response activates the protease DegP, which degrades CpxP, enabling signalling by CpxA (Buelow and Raivio, 2005). The TorSR TCS regulates production of proteins involved in anaerobic respiration of trimethylamine N-oxide (TMAO), including a cytochrome TorC. The immature cytochrome ApoTorC inhibits TorS autokinase activity (Ansaldi *et al.*, 1999), presumably preventing the accumulation of immature TorC. The FixLJ TCS of *Sinorhizobium meliloti* induces nitrogen fixation and respiration genes under anoxic conditions. Activity of the FixL HK is inhibited by FixT (Garnerone *et al.*, 1999), which exhibits sequence similarity to RRs, suggesting a mechanism of FixL down-regulation involving phosphatase activity (Crosson *et al.*, 2005).

In addition to inhibitors, a variety of auxiliary activators of HKs are known (Buelow and Raivio, 2009; Krell *et al.*, 2010). The most common examples are where the sensory function of HKs have been displaced onto a separate protein, resulting in TCSs described as three-component systems. Auxiliary sensors can be cytoplasmic (e.g. the PII protein of *E. coli*), periplasmic, or transmembrane (e.g. the MCPs of *E. coli*). As mentioned earlier, the mechanism of transmembrane signalling by MCPs (such as Tar) is thought to be similar to transmembrane signalling within HKs (Gao and Stock, 2009; Krell *et al.*, 2010), thus the only significant difference between an MCP-stimulated HK and a conventional HK is the devolution of sensory activity onto a separate polypeptide. In chemotactic systems, changes in the stoichiometry of receptor complexes are thought to be important for signalling (Kentner *et al.*, 2006), and this might therefore preclude the fusion of CheA and MCP functionalities into a single polypeptide. The cytoplasmic protein PII affects the activity of the HK NtrB, which regulates expression of nitrogen fixation genes. PII is sensitive to cellular levels of N and C, such that under conditions of nitrogen abundance PII is de-uridylylated, enabling it to bind to the transmitter domain of NtrB, inhibiting its autokinase activity (Atkinson *et al.*, 1994). Periplasmic binding proteins (PBPs) can also act as auxiliary regulators of HK activity. For instance the PBP BctC binds to the HK BctE upon binding citrate (Antoine *et al.*, 2005), the TorT PBP stimulates TorS activity upon binding trimethylamine N-oxide (Baraquet *et al.*, 2006), and the LuxP PBP modulates activity of the LuxQ quorum signalling HK upon binding the quorum signal AI-2 (Neiditch *et al.*, 2006). In short, auxiliary sensors enable TCSs to respond to an increased number of ligands/signals, facilitate communication between TCSs, and allow the integration of TCS regulation with other cellular regulatory mechanisms (Buelow and Raivio, 2009)

Spatial segregation and regulation

Subcellular localization is an important feature for the function of many TCS. In some cases HKs and their RRs are located at distinct places within the cell, with signalling through the TCS controlled by the regulation of their co-location (reviewed by García Véscovi *et al.*, 2010). The best characterized example of such a phenomenon is the TCS circuitry controlling the cell cycle and polar morphogenesis in *C. crescentus* (reviewed by Jenal and Galperin, 2009; García Véscovi *et al.*, 2010). Stalked cells are attached to a substrate through a polar holdfast structure. Cell division by a stalked cell produces one flagellated 'swarmer' daughter cell and one stalked cell. Prior to cell division, the two HKs DivJ and PleC are positioned at opposing cell poles, such that PleC is localized to the swarmer cell upon cell division, whereas DivJ localizes to the stalked pole. DivJ activity phosphorylates the RR DivK throughout the predivisional cell; however, post septation levels of phospho-DivK drop in the swarmer daughter cell (but not in the stalked cell), owing to the absence of DivJ, and the phosphatase activity of PleC, which actively dephosphorylates phospho-DivK (Paul *et al.*, 2008). In another asymmetry-generating pathway, the standalone receiver protein CpdR localizes itself and the ClpXP protease to the stalked pole in a phosphorylation-dependent fashion. Meanwhile the RR PopA causes the localization of CtrA to the stalked pole, but in a c-di-GMP-dependent, rather than phosphorylation-dependent, fashion (Deurig *et al.*, 2009). Co-localization of ClpXP leads to stalked cell-specific degradation of the RR master cell cycle regulator CtrA, and CpdR (both of which can be phosphorylated by a phosphorelay), thus post-division, CtrA is only found in the swarmer cell (Iniesta and Shapiro, 2008). The phosphorylation and dephosphorylation of CtrA is also regulated in a spatial fashion, generating asymmetry prior to cell division (Chen *et al.*, 2011).

In *E. coli* the CheA HK is positioned at a polar chemosensory cluster, mediated through its interaction with chemosensors and accessory proteins (Maddock and Shapiro, 1993; Kentner *et al.*, 2009). The cognate RR CheY is free to move between peritrichously arranged flagella motors (its sites of action), and the polar sensory cluster. *Rhodobacter sphaeroides* cells possess a single medially located flagellum, however they have multiple CheAs and CheYs, several of which have distinct subcellular locations (Wadhams *et al.*, 2005). CheA(2) is located at a polar chemosensory cluster, while CheA(3) and CheA(4) are located in a cytoplasmic chemosensory cluster. While CheA2 is able to phosphorylate all chemotaxis CheYs and CheBs of *R. sphaeroides*, CheA(3) is only able to phosphorylate one CheY and one CheB (Scott *et al.*, 2010). This suggests that chemotactic behaviour occurs as the consequence of the integration of signals from multiple chemotactic pathways, and somehow the localization of chemosensory machinery is important for appropriate signal integration.

M. xanthus moves by gliding in the direction of the long axis of the cell, with periodic reversals of direction, controlled by chemosensory TCSs. Motility is not powered by flagella rotation, but rather by two alternative mechanisms, a social movement driven by the extension/retraction of type-four pili (S-motility), and an adventurous movement of uncertain mechanism (A-motility). The activity of both the A- and S-motility engines are reciprocally regulated at opposing poles, and the frequency of periodic reversals in cellular motion is modulated by the Frz chemosensory pathway (Zusman *et al.*, 2007). The HK of the Frz pathway is the hybrid kinase FrzE, which acts upstream of FrzZ, a RR that consists solely of two tandem receiver domains (Inclán *et al.*, 2007). The receiver domain of FrzE is not required for phosphotransfer to downstream elements, but rather regulates the activity of the FrzE autokinase domain, in an allosteric manner (Inclán *et al.*, 2008). Each motility system of *M. xanthus* possesses an unusual RR with a coiled-coil output domain, but lacking the phosphorylatable Asp in the receiver domain (FrzS and AglZ of S- and A-motility respectively). FrzS has been shown to localize at the leading cell pole and to dynamically relocalize during a cellular reversal cycle in a phosphorylation-independent fashion (Mignot *et al.*, 2005; Fraser *et al.*, 2007), while AglZ localizes to proposed 'A-motility clusters' located at foci along the cell length (Nan *et al.*, 2010). RomR is a third motility RR, which possesses a novel output domain. RomR localizes predominantly to the lagging cell pole, with dynamic relocalization after a cellular reversal in a fashion regulated by the Frz pathway (Leonardy *et al.*, 2007).

Another important aspect of the spatial organization of TCSs is the covalent linkage between TCS domains. Standalone receivers, three-component systems, and hybrid kinases all exhibit relatively unusual patterns of covalent linkages, which are typically related to requirements for co-location or relative mobility of individual domains, and these issues have all been discussed briefly above. A further important consideration relates to the physicochemical consequences of covalent linkage. Tethering a transmitter and receiver domain together within a single polypeptide raises the effective concentration of one domain in the vicinity of the other, which can impact on signal transduction. For instance, the RodK HK of *M. xanthus* is a hybrid kinase with three C-terminal receiver domains, and a separate cognate RR RokA. Using isolated domains, the RodK transmitter preferentially phosphorylates RokA *in vitro*, however full length RodK preferentially phosphorylates its third receiver domain rather than RokA (Wegener-Feldbrügge and Søgaard-Andersen, 2009), presumably because its effective concentration at the RodK kinase is higher, by virtue of its covalent tethering.

Systems biology of TCS signalling

Recently, a variety of computational, synthetic and systems biology approaches have been applied to aspects of TCS behaviour. The advent of high-throughput methodologies have enabled organism-wide analyses of TCS gene complements, biochemical properties and protein–protein interactions, while advances in computational modelling have provided insights into the evolution and dynamics of TCSs.

One area where understanding has dramatically increased as a consequence of such advances is that of HK-RR specificity. In 1999, Fabret *et al.* grouped the HKs and RRs of *B. subtilis* into cognate families and observed that some amino acid residues thought to mediate the HK–RR interaction were globally conserved, others were conserved within families, while others yet were variable within a family. It was proposed that family specific residues restricted interactions to partners of the same family, while variable residues conferred within-family specificity (Fabret *et al.*, 1999). As such, it seemed likely that an understanding of the evolutionary forces acting on particular amino acids at the HK–RR interface would provide insights into the specificity of the HK–RR interaction (Hoch and Varughese, 2001). As more and more complete genomes were sequenced, it became possible to assemble datasets of HK and RR sequences that were assumed to be cognate partners by virtue of the adjacency of their genes, and thus to have co-evolved. Correlated variations in their amino acid residues allowed the identification of specificity-conferring amino acid positions (Fig. 8.9), and computational models which would predict whether a particular HK–RR combination were likely to interact (Burger and van Nimwegen, 2008; Skerker *et al.*, 2008; Szurmant *et al.*, 2008; Cock, 2009; Kolesov and Mirny, 2009; Weight *et al.*, 2009; Lunt *et al.*, 2010; Schug *et al.*, 2010). Such understanding has also allowed the rational engineering of TCS specificity, providing insights into the evolution of specificity (Capra *et al.*, 2010; Siryaporn *et al.*, 2010). Similar approaches have now also been applied to chemosensory TCSs, allowing their interaction specificities to be re-wired (Bell *et al.*, 2010). The success of such modelling approaches implies that the geometry of the receiver–transmitter complex is highly conserved. Interestingly, the predictive powers of one such model diminishes if trained on paired transmitter and receiver domain sequences from within hybrid kinases rather than from separate

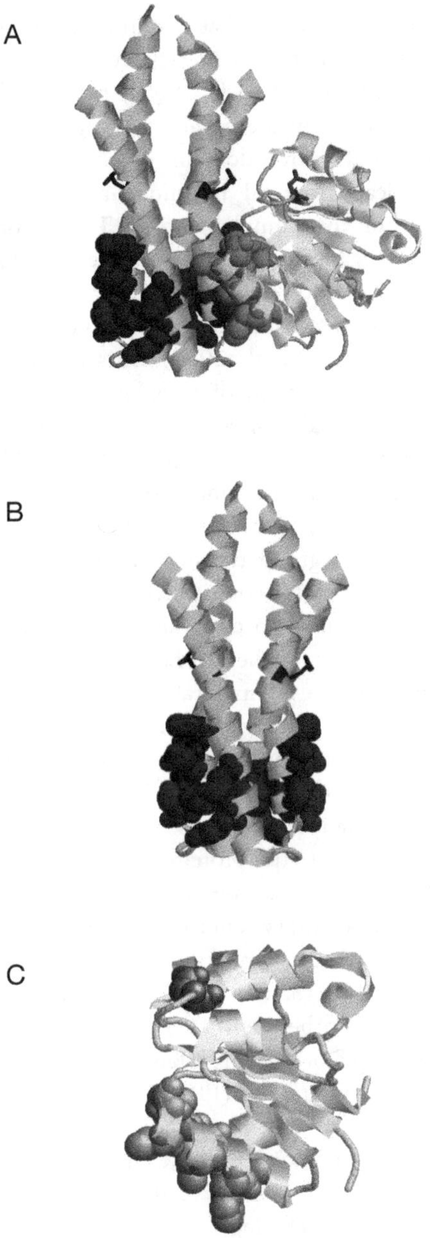

Figure 8.9 Computational deduction of the interaction between HisKA and receiver domains. (A) The four-helix bundle HisKA dimer of HK853, with the phosphorylatable His residues shown as dark sticks. Residues suggested to be involved in interactions with receiver domains are shown as dark spheres. (B) The receiver domain of RR468, with the phosphorylatable Asp residue (top) shown as dark spheres, and suggested HisKA-binding residues as pale spheres. (C) The complex of HK853 with RR468, coloured as in A and B. Images created from PDB file 3DGE by Stuart Thomas. Interacting residues suggested by Capra *et al.* (2010).

HKs and RRs, implying that within-polypeptide interactions are not as strong/specific as inter-protein interactions (Cock, 2009), presumably compensated for by the covalent linkage of the transmitter and receiver domains.

Major leaps forward in understanding the behaviour of the complement of TCSs from within an organism have been made possible courtesy of a method called phosphotransfer profiling, developed by Laub and co-workers (Skerker *et al.*, 2005; Laub *et al.*, 2007). This approach involves the production of purified transmitter and receiver domains *in vitro*, and radiolabelling of each transmitter domain with γ-^{32}P-ATP. Aliquots of each labelled transmitter domain are then mixed individually with every receiver domain, and phosphoryl group transfer to the receiver domains monitored – cognate receiver domains being preferentially phosphorylated *in vitro*. This approach has been applied to *E. coli* and *C. crescentus*, providing experimental evidence of the degree of specificity between components of the TCS network, and mapping HK–RR interactions (Skerker *et al.*, 2005; Biondi *et al.*, 2006).

High throughput methods have also been developed to assess the protein–protein interactions of TCS proteins. Studies have made use of the yeast two-hybrid system to probe the interactions between large numbers of transmitter and receiver domains (Sato *et al.*, 2007; Whitworth *et al.*, 2008) and between TCS and the rest of the proteome (Dortay *et al.*, 2008), while Förster resonance energy transfer (FRET) has been used to probe receiver domain dimerization (Gao *et al.*, 2008). Such studies indicate that weak protein–protein interactions between components of different TCSs (including receiver–receiver and transmitter–transmitter domain interactions), are relatively common – potentially indicating cross-regulation (Gao *et al.*, 2008; Whitworth *et al.*, 2008)

One approach within systems biology makes use of mathematical modelling to understanding the dynamic properties and robustness of systems. Several studies have addressed the properties of generic and specific TCSs (reviewed recently by Goulian, 2010). Models have shed light on the biological significance of the relative concentrations of HKs and RRs, linking observed values and regulatory mechanisms with robustness against fluctuations in the concentrations of the HK and RR (Batchelor and Goulian, 2003; Shinar *et al.*, 2007). Many TCS are autoregulatory, and positive autoregulation has been suggested to provide a short-term memory in some systems (Hoffer *et al.*, 2001), while negative autoregulation would give rise to an oscillating output from the TCS. It has been proposed that systems tailored to low strength signals respond fast and display effectively negative feedback, while systems that exhibit positive feedback have evolved to respond to noisy signals with a persistent output (Ray and Igoshin, 2010). Invoking common TCS mechanisms, such as a HK-independent dephosphorylation of phospho-RR, it has proven possible to demonstrate that TCS can exhibit bi-stability (Igoshin *et al.*, 2008), as seen in experiments (Zhou *et al.*, 2005), while a study considering phosphorelays suggests that such pathways are likely to filter noise from the sensory stimulus (Kim and Cho, 2006). As mentioned previously, models also suggest that having phosphatase activity resident within a HK, is an effective mechanism to suppress cross-talk between TCS (Alves and Savageau, 2003; Siryaporn and Goulian, 2008; Groban *et al.*, 2009). A very recent study has undertaken an analysis of the sensory logic of 14 TCS from *E. coli*, which could well guide the further development of synthetic TCSs (Olivera *et al.*, 2010).

A related discipline to systems biology is synthetic biology, which is the rational design of biological systems, often employing principles adapted from engineering. In TCS research, synthetic approaches have focused to date on creating novel couplings between input and output function (Ninfa *et al.*, 2007, 2010). Several engineered circuits conferring novel abilities to an organism have involved components derived from TCS (Atkinson *et al.*, 2003; Fung *et al.*, 2005; Levskaya *et al.*, 2005; Tabor *et al.*, 2011). As mentioned above, the grafting of different input functions onto TCSs have been undertaken to shed light on mechanisms of signal transduction in HKs (Ward *et al.*, 2002; Yoshida *et al.*, 2007; Cheung and Hendrickson, 2009; Nistala *et al.*, 2010). Potentially, the diverse range of input and output functions, the conserved mechanisms of signal transduction, and the different possibilities for extrinsic and intrinsic regulation inherent in TCSs, makes them versatile components for use in the engineering of synthetic regulatory circuits (Ninfa, 2010).

Discussion

TCSs are common prokaryotic signalling pathways, found in large numbers in most genomes. Many prokaryotic behaviours, including pathogenesis, are often regulated by TCSs, making them attractive targets for pharmaceutical development. The mechanisms of signal transduction within a TCS seem to be highly conserved, yet able to assimilate diverse input stimuli, generating a wide range of physiological responses. Their typically two-protein organization allows the separation of sensation from response elicitation, making them effective regulators of spatial processes, while differences in their enzymatic activities allows them to generate outputs exhibiting a wide range of dynamics.

Many TCSs, if not the majority, are elaborations on the TCS basic scheme, often functioning within large and sophisticated networks. Transmitter domains can signal to more than one receiver domain, while some receiver domains are phosphorylated by more than one transmitter domain, generating complex branched networks of interactions. In addition, the signalling activities of many TCSs are modulated by extrinsic factors – for example phosphatases, inhibitors, allosteric regulators, additional receiver domains etc. – allowing TCS networks to produce an integrated response to multiple stimuli. Basic TCSs can be expanded into phosphorelays, providing more opportunities for extrinsic regulation, while the presence of domains from other types of signalling pathways implies that in the cell TCSs act as components of higher-order regulatory networks. While the last two decades have proved a fruitful time for research into TCSs with many general principles emerging, there are still many gaps in our knowledge, with studies of individual TCSs frequently seeming to add something novel to our appreciation of TCSs. Hopefully the next two decades of TCS research will be equally rewarding – it is an exciting time to be involved in such a large and important field of microbiological research.

Future trends

Over recent years great strides have been made towards understanding the biology of TCSs. Comparative genomics has provided insights into the evolution of TCS gene sets, while structural information has shed light on the mechanisms of signal transduction by TCSs, particularly regarding the association between transmitter and receiver domains. Assisted by computational models of the transmitter–receiver interaction, it is now possible to rationally engineer TCS partnerships, while the construction of novel connectivities between input and output functions are possible through grafting experiments. Systems biology approaches are able to assess the importance of TCS signalling architectures and dynamics, opening a window onto phenomena common in TCSs, such as bi-stability and robustness, and in turn providing further avenues for the rational manipulation of TCSs.

However there are still many areas of TCS biology where fundamental questions remain unanswered. Structural studies are beginning to provide insights into the mechanisms of transmitter domain activation by input domains, but there are currently too few structures to claim a general understanding of the mechanisms involved. Similarly, while the phosphorylation-induced conformational switch of receiver domains is relatively well understood, it is not yet possible to routinely predict the substrates and/or targets of individual RR output domains, nor whether a particular receiver domain can dimerize. Properties of TCS proteins that are important in governing the dynamics of a TCS include the intrinsic phosphatase activity of phospho-RRs, and whether the partner HK also participates in phospho-RR dephosphorylation. The structural bases of these properties are currently obscure, however research is active in these areas, and it will hopefully not be long before the structural, sequence, mechanistic and evolutionary bases of such activities are generally understood.

Another area where further research is needed is the integration of TCS signalling with other modes of signalling in prokaryotes. Many TCSs have input/output domains characteristic of other types of signalling pathways, for example Ser/Thr kinase/phosphatase domains, anti-sigma factor domains, and anti-anti-sigma factor domains are all found fairly frequently in RRs (Galperin, 2010). To date there is a virtual absence of research into the mechanisms by which TCS signalling is transduced through these other types of signalling pathway, or how TCSs respond to sensory inputs from other types of signalling pathways.

Web resources

The following list of web resources is not at all exhaustive, as TCSs are so abundant that examples of TCSs are included in nearly all other databases. Rather, the list below highlights the few databases that focus exclusively or largely on TCSs. In addition, web-servers that allow analyses that are particularly useful for studying TCS are described. The three main TCS-focused databases are P2CS, MiST2 and the Census of prokaryotic response regulators, while KEGG provides an organism-specific view of the presence/absence of TCS orthologues:

P2CS

Prokaryotic 2-Component Systems (P2CS) is a TCS-centred database aimed at biologists (Barakat *et al.*, 2009, 2011). TCSs can be accessed by organism, by metagenome, by sequence cluster, or by domain architecture, and data can be downloaded in a variety of formats. Each TCS is fully described and links are provided for querying external servers. The genome context of each TCS is also characterized and TCS loci are viewable in an interactive map or genome browser. Two notable strengths of P2CS are that it searches an ORFeome for TCS domains to find TCS genes missed during genome annotation, and that it also builds and uses a sequence profile of an organism's HK H-box motifs to identify possible phosphorylation sites of HKs that apparently lack such motifs. P2CS currently contains data from 1125 prokaryotic genomes and 39 metagenomes, and can be accessed at http://www.p2cs.org.

MiST2

The Microbial Signal Transduction (MiST) database (Ulrich and Zhulin, 2007) has been updated recently to MiST2 (Ulrich and Zhulin, 2010). The database includes all prokaryotic genomes (including draft genomes) and classifies TCS proteins, one-component system proteins, chemotaxis proteins and ECF sigma factors. Summary information and sequence data can be obtained for each gene/protein, TCS gene organization is classified and there is a cart feature for genes of interest to the user. MiST2 can be found at http://mistdb.com/

Census of prokaryotic response regulators

This database provides a per-genome classification of RRs based on their output domains (Galperin, 2010). Individual entries are linked to analysis servers and other more general databases. The census can be accessed at http://www.ncbi.nlm.nih.gov/Complete_Genomes/RRcensus.html

KEGG

KEGG is a general database of genes, enzymes and metabolism; however, it has pages devoted to TCSs. The pages provide useful guides to the topology of HKs, their phosphotransfer to RRs, and the presence/absence of TCS orthologues in individual genomes (Nikitin *et al.*, 2004). The TCS pages of KEGG can be entered at http://www.genome.jp/kegg/pathway/ko/ko02020.html

SMART, CDART and TMHMM

Owing to the variety of input and output domains, and the diversity of domain architectures, normal sequence similarity-based methods of gene classification are inappropriate for many TCSs (Galperin, 2010). Instead, domain architecture analysis provides a more robust way of categorizing TCS proteins. Two webtools allowing such analyses are SMART and CDART, while the presence of transmembrane helices (suggestive of intra- or extracellular signal perception) is provided by TMHMM:

SMART

The Simple Modular Architecture Research Tool (SMART) is provided at http://smart.embl-heidelberg.de/. SMART allows an analysis of the domain architecture of inputted protein sequences with an elegant graphical representation of the results (including transmembrane helices), and allows the retrieval of proteins containing user–specified combinations of domains (Letunic *et al.*, 2009). See Figs. 8.2 and 8.3 for examples of SMART graphics.

CDART

The Conserved Domain Architecture Retrieval Tool (CDART) takes a protein sequence and identifies its conserved domains (Geer *et al.*, 2002). It also displays protein domain architectures that share domains with the query, with links to individual examples (results can also be filtered by domain composition). CDART is available at http://www.ncbi.nlm.nih.gov/Structure/lexington/lexington.cgi.

TMHMM Server v.2.0

This webtool based on the model created by Sonnhammer *et al.* (1998) predicts the presence of transmembrane helices in a protein sequence, providing an output of predicted protein topology. It is useful for defining whether HKs are likely to be responsive to cytoplasmic or extracytoplasmic stimuli. TMHMM Server v.2.0 can be accessed at http://www.cbs.dtu.dk/services/TMHMM/.

Acknowledgements

Images within Figs. 8.4 to 8.9 were kindly prepared by Heather Currinn, Stuart Thomas and Simon Cameron while undergraduate students at Aberystwyth University.

References

Albanesi, D., Martín, M., Trajtenberg, F., Mansilla, M.C., Haouz, A., Alzari, P.M., de Mendoza, D., and Buschiazzo, A. (2009). Structural plasticity and catalysis regulation of a thermosensor histidine kinase. Proc. Natl. Acad. Sci. U.S.A. *106*, 16185–16190.

Aldridge, P., Paul, R., Goymer, P., Rainey, P., and Jenal, U. (2003). Role of the GGDEF regulator PleD in polar development of *Caulobacter crescentus*. Mol. Microbiol. *47*, 1695–1708.

Alm, E., Huang, K., and Arkin, A. (2006). The evolution of two-component systems in bacteria reveals different strategies for niche adaptation. PLoS Comp. Biol. *2*, e143

Alves, R., and Savageau, M.A. (2003). Comparative analysis of prototype two-component systems with either bifunctional or monofunctional sensors: differences in molecular structure and physiological function. Mol. Microbiol. *48*, 25–51

Ansaldi, M., Bordi, C., Lepelletier, M., and Méjean, V. (1999). TorC apocytochrome negatively autoregulates the trimethylamine N-oxide (TMAO) reductase operon in *Escherichia coli*. Mol. Microbiol. *33*, 284–295.

Antoine, R., Huvent, I., Chemlal, K., Deray, I., Raze, D., Locht, C., and Jacob-Dubuisson, F. (2005). The periplasmic binding protein of a tripartite tricarboxylate transporter is involved in signal transduction. J. Mol. Biol. *351*, 799–809.

Appleby JL, Parkinson JS, and Bourret RB. (1996). Signal transduction via the multi-step phosphorelay: not necessarily a road less traveled. Cell *86*, 845–848.

Ashby, M.K. (2004). Survey of the number of two-component response regulator genes in the complete and annotated genome sequences of prokaryotes. FEMS Microbiol. Lett. *231*, 277–281.

Ashby, M.K. (2006). Distribution, structure and diversity of "bacterial" genes encoding two-component proteins in the Euryarchaeota. Archaea. *2*, 11–30.

Ashby, M.K., and Houmard, J. (2006). Cyanobacterial two-component proteins: structure, diversity, distribution, and evolution. Microbiol. Mol. Biol. Rev. *70*, 472–509.

Atkinson, M.R., Kamberov, E.S., Weiss, R.L., and Ninfa, A.J. (1994). Reversible uridylylation of the *Escherichia coli* PII signal transduction protein regulates its ability to stimulate the dephosphorylation of the transcription factor nitrogen regulator I (NRI or NtrC). J. Biol. Chem. *269*, 28288–28293.

Atkinson, M.R., Savageau, M.A., Myers, J.T., and Ninfa, A.J. (2003). Development of genetic circuitry exhibiting toggle switch or oscillatory behavior in *Escherichia coli*. Cell *113*, 597–607.

Baikalov, I., Schröder, I., Kaczor-Grzeskowiak, M., Grzeskowiak, K., Gunsalus, R.P., and Dickerson, R.E. (1996). Structure of the *Escherichia coli* response regulator NarL. Biochemistry *35*, 11053–11061.

Barakat, M., Ortet, P., Jourlin-Castelli, C., Mejean, V., and Whitworth, D.E. (2009). P2CS: a two-component system resource for prokaryotic signal transduction research. BMC Genomics *10*, 315.

Barakat, M., Ortet, P., and Whitworth, D.E. (2011). P2CS: a database of prokaryotic two-component systems. Nucleic Acids Res. *39(Database issue)*, D771–776.

Baraquet, C., Théraulaz, L., Guiral, M., Lafitte, D., Méjean, V., and Jourlin-Castelli, C. (2006). TorT, a member of a new periplasmic binding protein family, triggers induction of the Tor respiratory system upon trimethylamine N-oxide electron-acceptor binding in *Escherichia coli*. J. Biol. Chem. *281*, 38189–38199.

Batchelor, E., and Goulian, M. (2003). Robustness and the cycle of phosphorylation and dephosphorylation in a two-component regulatory system. Proc. Natl. Acad. Sci. U.S.A. *100*, 691–696.

de Been, M., Francke, C., Moezelaar, R., Abee, T., and Siezen, R.J. (2006). Comparative analysis of two-component signal transduction systems of *Bacillus cereus*, *Bacillus thuringiensis* and *Bacillus anthracis*. Microbiol. *152*, 3035–3048.

Beier, D., and Gross, R. (2006). Regulation of bacterial virulence by two-component systems. Curr. Opin. Microbiol. *9*, 143–152.

Bell, C.H., Porter, S.L., Strawson, A., Stuart, D.I., and Armitage, J.P. (2010). Using structural information to change the phosphotransfer specificity of a two-component chemotaxis signalling complex. PLoS Biol. *8*, e1000306.

Berger, L., and van Nimwegen, E. (2008). Accurate prediction of protein–protein interactions from sequence alignments using a Bayesian method. Mol. Syst. Bio. *4*, 165.

Bick, M.J., Lamour, V., Rajashankar, K.R., Gordiyenko, Y., Robinson, C.V., and Darst, S.A. (2009). How to switch off a histidine kinase: crystal structure of *Geobacillus stearothermophilus* KinB with the inhibitor Sda. J. Mol. Biol. *386*, 163–177

Bijlsma, J.J.E., and Groisman, E.A. (2003). Making informed decisions: regulatory interactions between two-component systems. Trends Microbiol. *11*, 359–366.

Biondi, E.G., Skerker, J.M., Arif, M., Prasol, M.S., Perchuk, B.S., and Laub, M.T. (2006). A phosphorelay system controls stalk biogenesis during cell cycle progression in *Caulobacter crescentus*. Mol. Microbiol. *59*, 386–401.

Bolotin, S., Fuller, J.D., Bast, D.J., Beveridge T.J., and de Azavedo, J.C. (2007). Capsule expression regulated by a two-component signal transduction system in *Streptococcus iniae*. FEMS Immunol. Med. Microbiol. *50*, 366–374.

Bourret, R.B. (2010). Receiver domain structure and function in response regulator proteins. Curr. Opin. Microbiol. *13*, 142–149.

Buelow, D.R., and Raivio, T.L. (2005). Cpx signal transduction is influenced by a conserved N-terminal domain in the novel inhibitor CpxP and the periplasmic protease DegP. J. Bacteriol. *187*, 6622–6630.

Buelow, D.R., and Raivio, T.L. (2009). Three (and more) component regulatory systems – auxiliary regulators of bacterial histidine kinases. Mol. Microbiol. *75*, 547–566.

Burger, L., and van Nimwegen, E. (2008). Accurate prediction of protein–protein interactions from sequence alignments using a Bayesian method. Mol. Syst. Biol. *4*, 165.

Cai, S.J., and Inouye, M. (2003). Spontaneous subunit exchange and biochemical evidence for trans-autophosphorylation in a dimer of *Escherichia coli* histidine kinase (EnvZ). J. Mol. Biol. *329*, 495–503.

Cai, X.H., Zhang, Q., Shi, S.Y., and Ding, D.F. (2005). Searching for potential drug targets in two-component and phosphorelay signal-transduction systems using three–dimensional cluster analysis. Acta Biochim. Biophys. Sin. (Shanghai). *37*, 293–302.

Calva, E., and Oropeza, R. (2006). Two-component signal transduction systems, environmental signals, and virulence. Microb. Ecol. *51*, 166–176.

Capra, E.J., Perchuk, B.S., Lubin, E.A., Ashenberg, O., Skerker, J.M., and Laub, M.T. (2010). Systematic dissection and trajectory-scanning mutagenesis of the molecular interface that ensures specificity of two-component signalling pathways. PLoS Genet. *6*, e1001220.

Casino, P., Rubio, V., and Marina, A. (2009). Structural insight into partner specificity and phosphoryl transfer in two-component signal transduction. Cell *139*, 325–336.

Chang, C., and Meyerowitz, E.M. (1994). Eukaryotes have "two-component" signal transducers. Res. Microbiol. *145*, 481–486.

Chao, X., Muff, T.J., Park, S.Y., Zhang, S., Pollard, A.M., Ordal, G.W., Bilwes, A.M., and Crane, B.R. (2006). A receptor-modifying deamidase in complex with a signalling phosphatase reveals reciprocal regulation. Cell *124*, 561–571.

Chen, Y.E., Tropini, C., Jonas, K., Tsokos, C.G., Huang, K.C., and Laub, M.T. (2011). Spatial gradient of protein phosphorylation underlies replicative asymmetry in a bacterium. Proc. Natl. Acad. Sci. U.S.A. *108*, 1052–1057.

Chen, Y.T., Chang, H.Y., Lu, C.L., and Peng, H.L. (2004). Evolutionary analysis of the two-component systems in *Pseudomonas aeruginosa* PAO1. J. Mol. Evol. *59*, 725–737.

Cheung, J., and Hendrickson, W.A. (2009). Structural analysis of ligand stimulation of the histidine kinase NarX. Structure *17*, 190–201.

Cheung, J., and Hendrickson, W.A. (2010). Sensor domains of two-component regulatory systems. Curr. Opin. Microbiol. *13*, 116–123.

Cheung, J.K., Keyburn, A.L., Carter, G.P., Lanckriet, A.L., Van Immerseel, F., Moore, R.J., and Rood, J.I. (2010). The VirSR two-component signal transduction system regulates NetB toxin production in *Clostridium perfringens*. Infect. Immun. *78*, 3064–3072.

Cho, K., and Zusman, D.R. (1999). AsgD, a new two-component regulator required for A-signalling and nutrient sensing during early development of *Myxococcus xanthus*. Mol. Microbiol. *34*, 268–281.

Clarke, E.J., and Voigt, C.A. (2011). Characterization of combinatorial patterns generated by multiple two-component sensors in *E. coli* that respond to many stimuli. Biotechnol. Bioeng. *108*, 666–675.

Cock, P.J.A. (2009). Two-component regulation: modelling, predicting and identifying protein–protein interactions and assessing signalling networks of bacteria. PhD Thesis, University of Warwick, UK.

Cock, P.J.A., and Whitworth, D.E. (2007a) Evolution of gene overlaps: relative reading frame bias in prokaryotic two-component system genes. J. Mol. Evol. *64*, 457–462.

Cock, P.J.A., and Whitworth, D.E. (2007b) Evolution of prokaryotic two-component system signalling pathways: gene fusions and fissions. Mol. Biol. Evol. *24*, 2355–2357.

Cock, P.J.A., and Whitworth, D.E. (2010). Evolution of relative reading frame bias in unidirectional prokaryotic gene overlaps. Mol. Biol. Evol. *27*, 753–756.

Comeau, D.E., Ikenaka, K., Tsung, K.L., and Inouye, M. (1985). Primary characterization of the protein products of the *Escherichia coli ompB* locus: structure and regulation of synthesis of the OmpR and EnvZ proteins. J. Bacteriol. 164, 578–584.

Corbell, N., and Loper, J.E. (1995). A global regulator of secondary metabolite production in *Pseudomonas fluorescens* Pf-5. J. Bacteriol. *177*, 6230–6236.

Crosson, S., McGrath, P.T., Stephens, C., McAdams, H.H., and Shapiro, L. (2005). Conserved modular design of an oxygen sensory/signaling network with species-specific output. Proc. Natl. Acad. Sci. U.S.A. *102*, 8018–8023.

Danese, P.N., Snyder, W.B., Cosma, C.L., Davis, L.J., and Silhavy, T.J. (1995). The Cpx two-component signal transduction pathway of *Escherichia coli* regulates transcription of the gene specifying the stress-inducible periplasmic protease DegP. Genes Dev. *9*, 387–398.

Da Re, S., Schumacher, J., Rousseau, P., Fourment, J., Ebel, C., and Kahn, D. (1999). Phosphorylation-induced dimerization of the FixJ receiver domain. Mol. Microbiol. *34*, 504–511.

Da Re, S.S., Deville-Bonne, D., Tolstykh, T., Veron, M., and Stock, J.B. (1999). Kinetics of CheY phosphorylation by small molecule phosphodonors. FEBS Lett. *457*, 323–326.

Darmon, E., Noone, D., Masson, A., Bron, S., Kuipers, O.P., Devine, K.M., and van Dijl, J.M (2002). A novel class of heat and secretion stress-responsive genes is controlled by the CssRS two-component system of *Bacillus subtilis*. J. Bacteriol. *184*, 5661–5671.

Djordjevic, S., and Stock, A.M. (1996). Structural analysis of bacterial chemotaxis proteins: components of a dynamic signalling system. J. Struct. Biol. *124*, 189–200.

Doß, S., Gröger, C., Knauber, T., Whitworth, D.E., and Treuner-Lange, A. (2005). Comparative genomic analysis of signal transduction proteins in clostridia. In Handbook on Clostridia, Dürre, P., ed (CRC Press , Boac Raton, FL), pp. 561–583.

Dortay, H., Gruhn, N., Pfeifer, A., Schwerdtner, M., Schmülling, T., and Heyl, A. (2008). Toward an interaction map of the two-component signaling pathway of *Arabidopsis thaliana*. J. Proteome Res. *7*, 3649–3660.

Dubrac, S., Boneca, I.G., Poupel, O., and Msadek, T. (2007). new insights into the WalK/WalR (YycG/YycF) essential signal transduction system pathway reveal a major role in controlling cell wall metabolism and biofilm formation in *Staphylococcus aureus*. J. Bacteriol. *189*, 8257–8269.

Duerig, A., Abel, S., Folcher, M., Nicollier, M., Schwede, T., Amiot, N., Giese, B., and Jenal, U. (2009). Second messenger-mediated spatiotemporal control of protein degradation regulates bacterial cell cycle progression. Genes Dev. *23*, 93–104.

Eguchi, Y., Oshima, T., Mori, H., Aono, R., Yamamoto, K., Ishihama, A., and Utsumi, R. (2003). Transcriptional regulation of drug efflux genes by EvgAS, a two-component system in *Escherichia coli*. Microbiol. *149*, 2819–2828.

Ellehauge, E., Nørregaard-Madsen, M., and Søgaard-Andersen, L. (1998). The FruA signal transduction protein provides a checkpoint for the temporal co-ordination of intercellular signals in *Myxococcus xanthus* development. Mol. Microbiol. 30, 807–817.

Emami, K., Topakas, E., Nagy, T., Henshaw, J., Jackson, K.A., Nelson, K.E., Mongodin, E.F., Murray, J.W., Lewis, R.J., and Gilbert, H.J. (2009). Regulation of the xylan-degrading apparatus of *Cellvibrio japonicus* by a novel two-component system. J. Biol. Chem. *284*, 1086–1096.

Ewann, F., Jackson, M., Pethe, K., Cooper, A., Mielcarek, N., Ensergueix, D., Gicquel, B., Locht, C., and Supply, P. (2002). Transient requirement of the PrrA-PrrB two-component system for early intracellular multiplication of *Mycobacterium tuberculosis*. Infect. Immun. *70*, 2256–2263.

Fabret, C., Feher, V.A., and Hoch, J.A. (1999). Two-component signal transduction in *Bacillus subtilis*: how one organism sees its world. J. Bacteriol. *181*, 1975–1983.

Fineran, P.C., Slater, H., Everson, L., Hughes, K., and Salmond, G.P. (2005). Biosynthesis of tripyrrole and beta-lactam secondary metabolites in *Serratia*: integration of quorum sensing with multiple new regulatory components in the control of prodigiosin and carbapenem antibiotic production. Mol. Microbiol. *56*, 1495–1517.

Fleischmann, R.D., Adams, M.D., White, O., Clayton, R.A., Kirkness, E.F., Kerlavage, A.R., Bult, C.J., Tomb, J.F., Dougherty, B.A., Merrick, J.M., *et al.* (1995). Whole-genome random sequencing and assembly of *Haemophilus influenzae* Rd. Science *269*, 496–512.

Fol, M., Chauhan, A., Nair, N.K., Maloney, E., Moomey, M., Jagannath, C., Madiraju, M.V., and Rajagopalan, M. (2006). Modulation of *Mycobacterium tuberculosis* proliferation by MtrA, an essential two-component response regulator. Mol. Microbiol. *60*, 643–657.

Fraser, J.S., Merlie, J.P. Jr,, Echols, N., Weisfield, S.R., Mignot, T., Wemmer, D.E., Zusman, D.R., and Alber, T. (2007). An atypical receiver domain controls the dynamic polar localization of the *Myxococcus xanthus* social motility protein FrzS. Mol. Microbiol. *65*, 319–332.

Fung, E., Wong, W.W., Suen, J.K., Bulter, T., Lee, S.G., and Liao, J.C. (2005). A synthetic gene-metabolic oscillator. Nature *435*, 118–122.

Galperin, M.Y. (2005). A census of membrane-bound and intracellular signal transduction proteins in bacteria: bacterial IQ, extroverts and introverts. BMC Microbiol. *5*, 35.

Galperin, M.Y. (2010). Diversity of structure and function of response regulator output domains. Curr. Opin. Microbiol. *13*, 150–159.

Galperin, M.Y., and Kolker, E. (2006). New metrics for comparative genomics. Curr. Opin. Biotechnol. *17*, 440–447.

Galperin, M.Y., and Nikolskaya, A.N. (2007). Identification of sensory and signal-transducing domains in two-component signalling systems. Methods Enzymol. *422*, 47–74.

Galperin, M.Y., Nikolskaya, A.N., and Koonin, E.V. (2001). Novel domains of the prokaryotic two-component signal transduction systems. FEMMS Microbiol. Lett. *203*, 11–21.

Galperin, M.Y., Higdon, R., and Kolker, E. (2010). Interplay of heritage and habitat in the distribution of bacterial signal transduction systems. Mol. Biosyst. *6*, 721–728.

Gao, R., and Stock, A.M. (2009). Biological insights from structures of two-component proteins. Annu. Rev. Microbiol. *63*, 133–154.

Gao, R., and Stock, A.M. (2010). Molecular strategies for phosphorylation-mediated regulation of response regulator activity. Curr. Opin. Microbiol. *13*, 160–167.

Gao, R., Tao, Y., and Stock, A.M. (2008). System-level mapping of *Escherichia coli* response regulator dimerization with FRET hybrids. Mol. Microbiol. *69*, 1358–1372.

García Véscovi, E., Sciara, M.I., and Castelli, M.E. (2010). Two component systems in the spatial program of bacteria. Curr. Opin. Microbiol. *13*, 210–218.

Garnerone, A.M., Cabanes, D., Foussard, M., Boistard, P., and Batut, J. (1999). Inhibition of the FixL sensor kinase by the FixT protein in *Sinorhizobium meliloti*. J. Biol. Chem. *274*, 32500–32506.

Geer, L.Y., Domrachev, M., Lipman, D.J., and Bryant, S.H. (2002). CDART: protein homology by domain architecture. Genome Res. *12*, 1619–1623.

Gooderham, W.J., and Hancock, R.E. (2009). Regulation of virulence and antibiotic resistance by two-component regulatory systems in *Pseudomonas aeruginosa*. FEMS Microbiol. Rev. *33*, 279–294.

Goodman, A.L., Merighi, M., Hyodo, M., Ventre, I., Filloux, A., and Lory, S. (2009). Direct interaction between sensor kinase proteins mediates acute and chronic disease phenotypes in a bacterial pathogen. Genes Dev. *23*, 249–259

Gotoh, Y., Doi, A., Furuta, E., Dubrac, S., Ishizaki, Y., Okada, M., Igarashi, M., Misawa, N., Yoshikawa, H., Okajima, T., Msadek, T., and Utsumi, R. (2010). Novel antibacterial compounds specifically targeting the essential WalR response regulator. J. Antibiot (Tokyo). *63*, 127–134.

Goulian, M. (2010). Two-component signaling circuit structure and properties. Curr. Opin. Microbiol. *13*, 184–189.

Grebe, T.W., and Stock, J.B. (1999). The histidine protein kinase superfamily. Adv. Microb. Physiol. *41*, 139–227.

Groban, E.S., Clarke, E.J., Salis, H.M., Miller, S.M., and Voigt, C.A. (2009). Kinetic buffering of cross talk between bacterial two-component sensors. J. Mol. Biol. *390*, 380–93.

Gronewold, T.M., and Kaiser, D. (2001). The *act* operon controls the level and time of C-signal production for *Myxococcus xanthus* development. Mol. Microbiol. *40*, 744–756.

Hagiwara, D., Yamashino, T., and Mizuno, T. Genome-wide comparison of the His-to-Asp phosphorelay signalling components of three symbiotic genera of *Rhizobia*. DNA Res. *11*, 57–65.

Hecht, G.B., Lane, T., Ohta, N., Sommer, J.M., and Newton, A. (1995). An essential single domain response regulator required for normal cell division and differentiation in *Caulobacter crescentus*. EMBO J. 14, 3915–3924.

Higgs, P.I., Cho, K., Evans, L.S., Whitworth, D.E., and Zusman, D.R. (2005). Four unusual two-component signal transduction homologs, RedC to RedF, are necessary for timely development in *Myxococcus xanthus*. J. Bacteriol. *187*, 8191–8195.

Hobbs, M., Collie, E.S., Free, P.D., Livingston, S.P., and Mattick, J.S. (1993). PilS and PilR, a two-component transcriptional regulatory system controlling expression of type 4 fimbriae in *Pseudomonas aeruginosa*. Mol. Microbiol. *7*, 669–682.

Hoch, J.A., and Varughese, K.I. (2001). Keeping signals straight in phosphorelay signal transduction. J. Bacteriol. *183*, 4941–4949.

Huang, J., Carney, B.F., Denny, T.P., Weissinger, A.K., and Schell, M.A. (1995). A complex network regulates expression of *eps* and other virulence genes of *Pseudomonas solanacearum*. J. Bacteriol. *177*, 1259–1267.

Hulett, F.M. (1996). The signal transduction network for Pho regulation in *Bacillus subtilis*. Mol. Microbiol. *19*, 933–999.

Hulko, M., Berndt, F., Gruber, M., Linder, J.U., Truffault, V., Schultz, A., Martin, J., Schultz, J.E., Lupas, A.N., and Coles, M. (2006). The HAMP domain structure implies helix rotation in transmembrane signalling. Cell. 126, 929–940.

Hutchings, M.I., Hoskisson, P.A., Chandra, G., and Buttner, M.J. (2004). Sensing and responding to diverse extracellular signals? Analysis of the sensor kinases and response regulators of *Streptomyces coelicolor* A3(2). Microbiol. *150*, 2795–2806.

Hyytiäinen, H., Sjöblom, S., Palomäki, T., Tuikkala, A., and Tapio Palva, E. (2003). The PmrA-PmrB two-component system responding to acidic pH and iron controls virulence in the plant pathogen *Erwinia carotovora* ssp. *carotovora*. Mol. Microbiol. *50*, 795–807.

Igoshin, O.A., Alves, R., and Savageau, M.A. (2008). Hysteretic and graded responses in bacterial two-component signal transduction. Mol. Microbiol. *68*, 1196–1215.

Inclán, Y.F., Vlamakis, H.C., and Zusman, D.R. (2007). FrzZ, a dual CheY-like response regulator, functions as an output for the Frz chemosensory pathway of *Myxococcus xanthus*. Mol. Microbiol. *65*, 90–102.

Inclán, Y.F., Laurent, S., and Zusman, D.R. (2008). The receiver domain of FrzE, a CheA–CheY fusion protein, regulates the CheA histidine kinase activity and downstream signalling to the A- and S-motility systems of *Myxococcus xanthus*. Mol. Microbiol. *68*, 1328–1339.

Iniesta, A.A., and Shapiro, L. (2008). A bacterial control circuit integrates polar localization and proteolysis of key regulatory proteins with a phospho-signalling cascade. Proc. Natl. Acad. Sci. U.S.A. *105*, 16602–16607.

Jenal, U., and Galperin, M.Y. (2009). Single domain response regulators: molecular switches with emerging roles in cell organization and dynamics. Curr. Opin. Microbiol. *12*, 152–160.

Jiménez-Pearson, M.A., Delany, I., Scarlato, V., and Beier, D. (2005). Phosphate flow in the chemotactic response system of *Helicobacter pylori*. Microbiology. *151*, 3299–3311.

Jones, A.L., DeShazer, D., and Woods, D.E. (1997). Identification and characterization of a two-component regulatory system involved in invasion of eukaryotic cells and heavy-metal resistance in *Burkholderia pseudomallei*. Infect. Immun. *65*, 4972–4977.

Kennelly, P.J., and Potts, M. (1996). Fancy meeting you here! A fresh look at "prokaryotic" protein phosphorylation. J. Bacteriol. *178*, 4759–4764.

Kentner, D., Thiem, S., Hildenbeutel, M., and Sourjik, V. (2006). Determinants of chemoreceptor cluster formation in *Escherichia coli*. Mol. Microbiol. *61*, 407–417.

Kim, J.R., and Cho, K.H. (2006). The multi-step phosphorelay mechanism of unorthodox two-component systems in *E. coli* realizes ultrasensitivity to stimuli while maintaining robustness to noises. Comput. Biol. Chem. *30*, 438–444.

Kim, D., and Forst, S. (2001). Genomic analysis of the histidine kinase family in bacteria and archaea. Microbiol. *147*, 1197–1212.

Kofoid, E.C., and Parkinson, J.S. (1988). Transmitter and receiver modules in bacterial signalling proteins. Proc. Natl. Acad. Sci. U.S.A. *85*, 4981–4985.

Kolesov, G., and Mirny, L.A. (2009). Using evolutionary information to find specificity-determining and co-evolving residues. Methods Mol. Biol. *541*, 421–448

Koretke, K.K., Lupas, A.N., Warren, P.V., Rosenberg, M., and Brown, J.R. (2000). Evolution of two-component signal transduction. Mol. Biol. Evol. *17*, 1956–1970

Krell, T., Busch, A., Lacal, J., Silva-Jiménez, H., and Ramos, J.L. (2009). The enigma of cytosolic two-component systems: a hypothesis. Environ. Microbiol. Rep. *1*, 171–176.

Krell, T., Lacal, J., Busch, A., Silva-Jiménez, H., Guazzaroni, M.E., and Ramos, J.L. (2010). Bacterial sensor kinases: diversity in the recognition of environmental signals. Annu. Rev. Microbiol. *64*, 539–559.

Kuroda, M., Kuroda, H., Oshima, T., Takeuchi, F., Mori, H., and Hiramatsu, K. (2003). Two-component system VraSR positively modulates the regulation of cell-wall biosynthesis pathway in *Staphylococcus aureus*. Mol. Microbiol. *3*, 807–821.

Laub, M.T., and Goulian, M. (2007). Specificity in two-component signal transduction pathways. Annu. Rev. Genet. *41*, 121–145.

Laub, M.T., Biondi, E.G. and Skerker, J.M. (2007). Phosphotransfer profiling: systematic mapping of two-component signal transduction pathways and phosphorelays. Methods Enzymol. *423*, 531–548.

Lavín, J.L., Kiil, K., Resano, O., Ussery, D.W., and Oguiza, J.A. (2007). Comparative genomic analysis of two-component regulatory proteins in *Pseudomonas syringae*. BMC Genomics *8*, 397.

Lavín, J.L., Binnewies, T.T., Pisabarro, A.G., Ussery, D.W., García-Lobo, J.M., and Oguiza, J.A. (2010b) Differences in two-component signal transduction proteins among the genus *Brucella*: implications for host preference and pathogenesis. Vet. Microbiol. *144*, 478–483.

Le, A.T., and Schumann, W. (2009). The Spo0E phosphatase of *Bacillus subtilis* is a substrate of the FtsH metalloprotease. Microbiology *155*, 1122–1132.

Lee, S.Y., Cho, H.S., Pelton, J.G., Yan, D., Henderson, R.K., King, D.S., Huang, L., Kustu, S., Berry, E.A., and Wemmer, D.E. (2001). Crystal structure of an activated response regulator bound to its target. Nat. Struct. Biol. *8*, 52–56.

Leonardy, S., Freymark, G., Hebener, S., Ellehauge, E., and Søgaard-Andersen, L. (2007). Coupling of protein localization and cell movements by a dynamically localized response regulator in *Myxococcus xanthus*. EMBO J. *26*, 4433–4444.

Leonhartsberger, S., Huber, A., Lottspeich, F., and Böck, A. (2001). The *hydH/G* genes from *Escherichia coli* code for a zinc and lead responsive two-component regulatory system. J. Mol. Biol. *307*, 93–105.

Letunic, I., Doerks, T., and Bork, P. (2009). SMART6: recent updates and new developments. Nucleic Acids Res. *37(Database issue)*, D229–232.

Levskaya, A,. Chevalier, A.A., Tabor, J.J., Simpson, Z.B., Lavery, L.A., Levy, M., Davidson, E.A., Scouras, A., Ellington, A.D., Marcotte, E.M., *et al.* (2005). Synthetic biology: engineering *Escherichia coli* to see light. Nature *438*, 441–442.

Li, N., Wang, F., Niu, S., Cao, J., Wu, K., Li, Y., Yin, N., Zhang, X., Zhu, W., and Yin Y. (2009). Discovery of novel inhibitors of *Streptococcus pneumoniae* based on the virtual screening with the homology-modeled structure of histidine kinase (VicK). BMC Microbiol. *9*, 129.

Liang, X., Yu, C., Sun, J., Liu, H., Landwehr, C., Holmes, D., and Ji, Y. (2006). Inactivation of a two-component signal transduction system, SaeRS, eliminates adherence and attenuates virulence of *Staphylococcus aureus*. Infect. Immun. *74*, 4655–4665.

Lu, Y., Wang, W., Shu, D., Zhang, W., Chen, L., Qin, Z., Yang, S., and Jiang, W. (2007). Characterization of a novel two-component regulatory system involved in the regulation of both actinorhodin and a type I polyketide in *Streptomyces coelicolor*. Appl. Microbiol. Biotechnol. *77*, 625–635.

Lukat, G.S., Stock, A.M., and Stock, J.B. (1990). Divalent metal ion binding to the CheY protein and its significance to phosphotransfer in bacterial chemotaxis. Biochemistry *29*, 5436–5442.

Lunt, B., Szurmant, H., Procaccini, A., Hoch, J.A., Hwa, T., and Weigt, M. (2010). Inference of direct residue contacts in two-component signalling. Methods Enzymol. *471*, 17–41.

Ma, S., Wozniak, D.J., and Ohman, D.E. (1997). Identification of the histidine protein kinase KinB in *Pseudomonas aeruginosa* and its phosphorylation of the alginate regulator *algB*. J. Biol. Chem. *272*, 17952–17960.

McKenzie, N.L., and Nodwell, J.R. (2007). Phosphorylated AbsA2 negatively regulates antibiotic production in *Streptomyces coelicolor* through interactions with pathway-specific regulatory gene promoters. J. Bacteriol. *189*, 5284–5292.

Maddock, J.R., and Shapiro, L. (1993). Polar location of the chemoreceptor complex in the *Escherichia coli* cell. Science *259*, 1717–1723.

Marina, A., Waldburger, C.D., and Hendrickson, W.A. (2005). Structure of the entire cytoplasmic portion of a sensor histidine-kinase protein. EMBO J. *24*, 4247–4259

Mascher, T., Helmann, J.D., and Unden, G. (2006). Stimulus perception in bacterial signal-transducing histidine kinases. Microbiol. Mol. Biol. Rev. *70*, 910–938.

Matsushita, M., and Janda, K.D. (2002). Histidine kinases as targets for new antimicrobial agents. Bioorg. Med. Chem. *10*, 855–867.

Mignot, T., Merlie, J.P. Jr., and Zusman, D.R. (2005). Regulated pole-to-pole oscillations of a bacterial gliding motility protein. Science *310*, 855–857.

Milton, D.L. (2006). Quorum sensing in vibrios: complexity for diversification. Int. J. Med. Microbiol. *296*, 61–71.

Mizuno, T. (1997). Compilation of all genes encoding two-component phosphotransfer signal transducers in the genome of *Escherichia coli*. DNA Res. *4*, 161–168.

Mizuno, T. (2005). Two-component phosphorelay signal transduction systems in plants: from hormone responses to circadian rhythms. Biosci. Biotechnol. Biochem. *69*, 2263–2276.

Mizuno, T., Kaneko, T., and Tabata, S. (1996). Compilation of all genes encoding bacterial two-component signal transducers in the genome of the cyanobacterium, *Synechocystis* sp. strain PCC 6803. DNA Res. *3*, 407–414.

Mosley, C.S., Suzuki, J.Y and Bauer, C.E. (1994). Identification and molecular genetic characterization of a sensor kinase responsible for coordinately regulating light harvesting and reaction center gene expression in response to anaerobiosis. J. Bacteriol. *176*, 7566–7573.

Munson, G.P., Lam, D.L., Outten, F.W., and O'Halloran, T.V. (2000). Identification of a copper-responsive two-component system on the chromosome of *Escherichia coli* K-12. J. Bacteriol. *182*, 5864–5871.

Nan, B., Mauriello, E.M., Sun, I.H., Wong, A., and Zusman, D.R. (2010). A multi–protein complex from *Myxococcus xanthus* required for bacterial gliding motility. Mol. Microbiol. *76*, 1539–1554.

Neiditch, M.B., Federle, M.J., Pompeani, A.J., Kelly, R.C., Swem, D.L., Jeffrey, P.D., Bassler, B.L., and Hughson, F.M. (2006). Ligand-induced asymmetry in histidine sensor kinase complex regulates quorum sensing. Cell. *126*, 1095–1108.

Nikitin, F., Rance, B., Itoh, M., Kanehisa, M., and Lisacek, F. (2004). Using protein motif combinations to update KEGG pathway maps and orthologue tables. Genome Inform. *15*, 266–275.

Ninfa, A.J. (2010). Use of two-component signal transduction systems in the construction of synthetic genetic networks. Curr. Opin. Microbiol. *13*, 240–245.

Ninfa, A.J., Selinsky, S., Perry, N., Atkins, S., Xiu Song, Q., Mayo, A., Arps, D., Woolf, P., and Atkinson, M.R. (2007). Using two-component systems and other bacterial regulatory factors for the fabrication of synthetic genetic devices. Methods Enzymol. *422*, 488–512.

Ninfa, E.G. Atkinson, M.R., Kamberov, E.S., and Ninfa, A.J. (1993). Mechanism of autophosphorylation of *Escherichia coli* nitrogen regulator II (NRII or NtrB): trans-phosphorylation between subunits. J. Bacteriol. *175*, 7024–7032.

Nishino, K., Nikaido, E., and Yamaguchi, A. (2007). Regulation of multidrug efflux systems involved in multidrug and metal resistance of *Salmonella enterica* serovar Typhimurium. J. Bacteriol. *189*, 9066–9075.

Nistala, G.J., Wu, K., Rao, C.V., and Bhalerao, K.D. (2010). A modular positive feedback-based gene amplifier. J. Biol. Eng. *4*, 4.

Nixon, B.T., Ronson, C.W., and Ausubel, F.M. (1986). Two-component regulatory systems responsive to environmental stimuli share strongly conserved domains with the nitrogen assimilation regulatory genes *ntrB* and *ntrC*. Proc. Natl. Acad. Sci. U.S.A. *83*, 7850–7854.

O'Connor, C., and Matsumura, P. (2004). The accessibility of Cys-120 in $CheA_S$ is important for the binding of CheZ and enhancement of CheZ phosphatase activity. Biochemistry *43*, 6909–6916.

Ogino, T., Matsubara, M., Kato, N., Nakamura, Y., and Mizuno, T. (1998). An *Escherichia coli* protein that exhibits phosphohistidine phosphatase activity towards the HPt domain of the ArcB sensor involved in the multistep His-Asp phosphorelay. Mol. Microbiol. *27*, 573–585.

Olivera, B.C., Ugalde, E., and Martínez-Antonio, A. (2010). Regulatory dynamics of standard two-component systems in bacteria. J. Theor. Biol. *264*, 560–569.

O'Loughlin, J.L., Spinner, J.L., Minnich, S.A., and Kobayashi, S.D. (2010). *Yersinia pestis* two-component gene regulatory systems promote survival in human neutrophils. Infect. Immun. *78*, 773–782.

Pao, G.M., and Saier, M.H., Jr. (1995). Response regulators of bacterial signal transduction systems: selective domain shuffling during evolution. J. Mol. Evol. *40*, 136–154.

Parales, R.E., Ditty, J.L., and Harwood, C.S. (2000). Toluene-degrading bacteria are chemotactic towards the environmental pollutants benzene, toluene, and trichloroethylene. Appl. Environ. Microbiol. *66*, 4098–4104.

Parkinson, J.S., and Kofoid, E.C. (1992). Communication modules in bacterial signalling proteins. Annu. Rev. Genet. *26*, 71–112.

Paul, R., Abel, S., Wassmann, P., Beck, A., Heerklotz, H., and Jenal, U. (2007). Activation of the diguanylate cyclase PleD by phosphorylation-mediated dimerization. J. Biol. Chem. *282*, 29170–29177.

Paul, R., Jaeger, T., Abel, S., Wiederkehr, I., Folcher, M., Biondi, E.G. Laub, M.T., and Jenal, U. (2008). Allosteric regulation of histidine kinases by their cognate response regulator determines cell fate. Cell *133*, 452–461.

Perego, M. (1997). A peptide export-import control circuit modulating bacterial development regulates protein phosphatases of the phosphorelay. Proc. Natl. Acad. Sci. U.S.A. *94*, 8612–8617.

Perego, M. (1998). Kinase-phosphatase competition regulates *Bacillus subtilis* development. Trends Microbiol. *6*, 366–370.

Perego, M., and Hoch, J.A. (2008). Commingling regulatory systems following acquisition of virulence plasmids by *Bacillus anthracis*. Trends Microbiol. *16*, 215–221.

Perego, M., Cole, S.P., Burbulys, D., Trach, K., and Hoch, J.A. (1989). Characterization of the gene for a protein kinase which phosphorylates the sporulation-regulatory proteins SpoOA and SpoOF of *Bacillus subtilis*. J. Bacteriol. *171*, 6187–6196.

Pestova, E.V., Håvarstein, L.S., and Morrison, D.A. (1996). Regulation of competence for genetic transformation in *Streptococcus pneumoniae* by an auto-induced peptide pheromone and a two-component regulatory system. Mol. Microbiol. *21*, 853–862.

Qi, M., Sun, F.J., Caetano-Anollés, G., and Zhao, Y. (2010). Comparative genomic and phylogenetic analyses reveal the evolution of the core two-component signal transduction systems in Enterobacteria. J. Mol. Evol. *70*, 167–180.

Qian, W., Han, Z.J., and He, C. (2008). Two-component signal transduction systems of Xanthomonas spp.: a lesson from genomics. Mol. Plant–Microbe Interact. *21*, 151–161.

Raffa, R.G., and Raivio, T.L. (2002). A third envelope stress signal transduction pathway in *Escherichia coli*. Mol. Microbiol. *45*, 1599–1611.

Rasmussen, A.Aa., Wegener-Feldbrügge, S., Porter, S.L., Armitage, J.P., and Søgaard-Andersen, L. (2006). Four signalling domains in the hybrid histidine protein kinase RodK of *Myxococcus xanthus* are required for activity. Mol. Microbiol. *60*, 525–534.

Ray, J.C., and Igoshin, O.A. (2010). Adaptable functionality of transcriptional feedback in bacterial two-component systems. PLoS Comput. Biol. *6*, e1000676.

Robinson, V.L., Buckler, D.R., and Stock, A.M. (2000). A tale of two components: a novel kinase and a regulatory switch. Nat. Struct. Biol. *7*, 626–633.

Rodrigue, A., Quentin, Y., Lazdunski, A., Méjean, V., and Foglino, M. (2000). Two-component systems in *Pseudomonas aeruginosa*: why so many? Trends Microbiol. *8*, 498–504.

Rogozin, I.B., Spiridonov, A.N., Sorokin, A.V., Wolf, Y.I., Jordan, I.K., Tatusov, R.L., and Koonin, E.V. (2002). Purifying and directional selection in overlapping prokaryotic genes. Trends Genet. *18*, 228–232

Romagnoli, S., and Tabita, F.R. (2006). A novel three-protein two-component system provides a regulatory twist on an established circuit to modulate expression of the *cbbI* region of *Rhodopseudomonas palustris* CGA010. J. Bacteriol. *188*, 2780–2791.

Romagnoli, S., and Tabita, F.R. (2007). Phosphotransfer reactions of the CbbRRS three-protein two-component system from *Rhodopseudomonas palustris* CGA010 appear to be controlled by an internal molecular switch on the sensor kinase. J. Bacteriol. *189*, 325–335.

Rowland, S.L., Burkholder, W.F., Cunningham, K.A., Maciejewski, M.W., Grossman, A.D., and King, G.F. (2004). Structure and mechanism of action of Sda, an inhibitor of the histidine kinases that regulate initiation of sporulation in *Bacillus subtilis*. Mol. Cell *13*, 689–701.

Salinas, P., Ruiz, D., Cantos, R., Lopez-Redondo, M.L., Marina, A., and Contreras, A. (2007). The regulatory factor SipA provides a link between NblS and NblR signal transduction pathways in the cyanobacterium *Synechococcus* sp. PCC 7942. Mol. Microbiol. *66*, 1607–1619.

Sandmann, A., Frank, B., and Müller, R. (2008). A transposon-based strategy to scale up myxothiazol production in myxobacterial cell factories. J. Biotechnol. *135*, 255–261.

Santos, J.L., and Shiozaki, K. (2001). Fungal histidine kinases. Sci STKE, re1

Sato, S., Shimoda, Y., Muraki, A., Kohara, M., Nakamura, Y., and Tabata, S. (2007). A large-scale protein protein interaction analysis in *Synechocystis* sp. PCC6803. DNA Res. *14*, 207–216.

Scaramozzino, F., White, A., Perego, M., and Hoch, J.A. (2009). A unique GTP-dependent sporulation sensor histidine kinase in *Bacillus anthracis*. J. Bacteriol. *191*, 687–692.

Schneiker, S., Perlova, O., Kaiser, O., *et al.* (2007). Complete sequence of the largest known bacterial genome from the myxobacterium *Sorangium cellulosum*. Nat. Biotechnol. *25*, 1281–1289.

Schug, A., Weigt, M., Hoch, J.A., Onuchic, J.N., Hwa, T., and Szurmant, H. (2010). Computational modeling of phosphotransfer complexes in two-component signalling. Methods Enzymol. *471*, 43–58.

Scott, K.A., Porter, S.L., Bagg, E.A., Hamer, R., Hill, J.L., Wilkinson, D.A., and Armitage, J.P. (2010). Specificity of localization and phosphotransfer in the CheA proteins of *Rhodobacter sphaeroides*. Mol. Microbiol. *76*, 318–330.

Sevvana, M., Vijayan, V., Zweckstetter, M., Reinelt, S., Madden, D.R., Herbst-Irmer, R., Sheldrick, G.M., Bott, M., Griesinger, C., and Becker, S. (2008). A ligand-induced switch in the periplasmic domain of sensor histidine kinase CitA. J. Mol. Biol. 377, 512–523.

Shinar, G., Milo, R., Martínez, M.R., and Alon, U. (2007). Input output robustness in simple bacterial signaling systems. Proc. Natl. Acad. Sci. U.S.A. *104*, 19931–19935.

Shu, D., Chen, L., Wang, W., Yu, Z., Ren, C., Zhang, W., Yang, S., Lu, Y., and Jiang, W. *afsQ1-Q2-sigQ* is a pleiotropic but conditionally required signal transduction system for both secondary metabolism and morphological development in *Streptomyces coelicolor*. Appl. Microbiol. Biotechnol. *81*, 1149–1160.

Silversmith, R.E. (2010). Auxiliary phosphatases in two-component signal transduction. Curr. Opin. Microbiol. *13*, 177–183.

Simon, M.I., Borkovich, K.A., Bourret, R.B., and Hess, J.F. (1989). Protein phosphorylation in the bacterial chemotaxis system. Biochimie *71*, 1013–1019.

Sirota-Madi, A., Olender, T., Helman, Y., Ingham, C., Brainis, I., Roth, D., Hagi, E., Brodsky, L., Leshkowitz, D., Galatenko, V., *et al.* (2010). Genome sequence of the pattern forming *Paenibacillus vortex* bacterium reveals potential for thriving in complex environments. BMC Genomics *11*, 710.

Siryaporn, A., and Goulian, M. (2008). Cross-talk suppression between the CpxA-CpxR and EnvZ-OmpR two-component systems in *E. coli*. Mol. Microbiol. *70*, 494–506.

Siryaporn, A., Perchuk, B.S., Laub, M.T., and Goulian, M. (2010). Evolving a robust signal transduction pathway from weak cross-talk. Mol. Syst. Biol. *6*, 452.

Skerker, J.M., Prasol, M.S., Perchuk, B.S., Biondi, E.G., and Laub, M.T. (2005). Two-component signal transduction pathways regulating growth and cell cycle progression in a bacterium: a system-level analysis. PLoS Biol. *3*, e334.

Skerker, J.M., Perchuk, B.S., Siryaporn, A., Lubin, E.A., Ashenberg, O., Goulian, M., and Laub, M.T. (2008). Rewiring the specificity of two-component signal transduction systems. Cell *133*, 1043–1054.

Sola-Landa, A., Moura, R.S., and Martín, J.F. (2003). The two-component PhoR-PhoP system controls both primary metabolism and secondary metabolite biosynthesis in *Streptomyces lividans*. Proc. Natl. Acad. Sci. U.S.A. *100*, 6133–6138.

Sonnhammer, E.L.L., von Heijne, G., and Krogh, A. (1998). A hidden Markov model for predicting transmembrane helices in protein sequences. In Proc. of Sixth Int. Conf. on Intelligent Systems for Molecular Biology. Glasgow, J., Littlejohn, T., Major, F., Lathrop, R.,. Sankoff, D., and Sensen, S, eds. (AAAI Press , Menlo Park, CA), pp. 175–182.

Sourjik, V., and Berg, H.C. (2004). Functional interactions between receptors in bacterial chemotaxis. Nature *428*, 437–441.

Sperandio, V., Torres, A.G., and Kaper, J.B. (2002). Quorum sensing *Escherichia coli* regulators B and C (QseBC): a novel two-component regulatory system involved in the regulation of flagella and motility by quorum sensing in *E. coli*. Mol. Microbiol. *43*, 809–821.

Spiegelman, G., Van Hoy, B., Perego, M., Day, J., Trach, K., and Hoch, J.A.(1990). Structural alterations in the *Bacillus subtilis* SpoOA regulatory protein which suppress mutations at several *spoO* loci. J. Bacteriol. *172*, 5011–5019.

Stephenson, K., and Hoch, J.A. (2004). Developing inhibitors to selectively target two-component and phosphorelay signal transduction systems of pathogenic microorganisms. Curr. Med. Chem. *11*, 765–773.

Stewart, R.C. (2010). Protein histidine kinases: assembly of active sites and their regulation in signalling pathways. Curr. Opin. Microbiol. *13*, 133–141.

Stock, A.M., Robinson, V.L., and Goudreau, P.N. (2000). Two-component signal transduction. Annu. Rev. Biochem. *69*, 183–215.

Stock, J.B., Ninfa, A.J., and Stock, A.M. (1989). Protein phosphorylation and regulation of adaptive responses in bacteria. Microbiol. Rev. *53*, 450–490.

Sun, G., Sharkova, E., Chesnut, R., Birkey, S., Duggan, M.F., Sorokin, A., Pujic, P., Ehrlich, S.D., and Hulett, F.M. (1996). Regulators of aerobic and anaerobic respiration in *Bacillus subtilis*. J. Bacteriol. *178*, 1374–1385.

Szurmant, H., Bobay, B.G., White, R.A., Sullivan, D.M., Thompson, R.J., Hwa, T., Hoch, J.A., and Cavanagh, J. (2008). Co-evolving motions at protein–protein interfaces of two-component signalling systems identified by covariance analysis. Biochemistry *47*, 7782–7784.

Tabor, J.J., Levskaya, A., and Voigt, C.A. (2011). Multichromatic Control of gene expression in *Escherichia coli*. J. Mol. Biol. *405*, 315–324.

Takei, N., Nakajima, M., Oyama, T., Kito, R., Sugita, C., Sugita, M., Kondo, T., and Iwasaki, H. (2006). A KaiC-associating SasA-RpaA two-component regulatory system as a major circadian timing mediator in cyanobacteria. Proc. Natl. Acad. Sci. U.S.A. *103*, 12109–12114.

Tanaka, T., Saha, S.K., Tomomori, C., Ishima, R., Liu, D., Tong, K.I., Park, H., Dutta, R., Qin, L., Swindells, M.B., *et al.*, (1998). NMR structure of the histidine kinase domain of the *E. coli* osmosensor EnvZ. Nature *396*, 88–92.

Throup, J.P., Koretke, K.K., Bryant, A.P., Ingraham, K.A., Chalker, A.F., Ge, Y., Marra, A., Wallis, N.G., Brown, J.R., Holmes, D.J., *et al.* (2000). A genomic analysis of two-component signal transduction in *Streptococcus pneumoniae*. Mol. Microbiol. *35*, 566–576.

Tomomori, C., Tanaka, T., Dutta, R., Park, H., and Saha, S. (1999). Solution structure of the homodimeric core domain of *Escherichia coli* histidine kinase EnvZ. Nat. Struct. Biol. *6*, 729–734.

Trach, K.A., and Hoch, J.A. (1993). Multisensory activation of the phosphorelay initiating sporulation in *Bacillus subtilis*: identification and sequence of the protein kinase of the alternate pathway. Mol. Microbiol. *8*, 69–79.

Ueki, T., and Inouye, S. (2002). Transcriptional activation of a heat-shock gene, *lonD*, of *Myxococcus xanthus* by a two component histidine-aspartate phosphorelay system. J. Biol. Chem. *277*, 6170–6177.

Ulrich, D.L., Kojetin, D., Bassler, B.L., Cavanagh, J., and Loria, J.P. (2005). Solution structure and dynamics of LuxU from *Vibrio harveyi*, a phosphotransferase protein involved in bacterial quorum sensing. J. Mol. Biol. *347*, 297–307.

Ulrich, L.E., and Zhulin, I.B. (2007). MIST: a microbial signal transduction database. Nucleic Acids Res. *35*(Database issue), D386-D390.

Ulrich, L.E., and Zhulin, I.B. (2010). The MiST2 database: a comprehensive genomics resource on microbial signal transduction. Nucleic Acids Res. *38*(Database issue), D401-D407.

Ulrich, L.E., Koonin, E.V., and Zhulin, I.B. (2005). One-component systems dominate signal transduction in prokaryotes. Trends Microbiol. *31*, 52–56

Varughese, K.I., Madhusudan, Zhou, X.Z., Whiteley, J.M., and Hoch, J.A. (1998). Formation of a novel four-helix bundle and molecular recognition sites by dimerization of a response regulator phosphotransferase. Mol. Cell *2*, 485–493.

Verhamme, D.T., Arents, J.C., Postma, P.W., Crielaard, W., and Hellingwerf K.J. (2001). Glucose-6-phosphate-dependent phosphoryl flow through the Uhp two-component regulatory system. Microbiol. *147*, 3345–3352.

Wadhams, G.H., Martin, A.C., Warren, A.V., and Armitage, J.P. (2005). Requirements for chemotaxis protein localization in *Rhodobacter sphaeroides*. Mol. Microbiol. *58*, 895–902.

Walderhaug, M.O., Polarek, J.W., Voelkner, P., Daniel, J.M., Hesse, J.E., Altendorf, K., and Epstein, W. (1992). KdpD and KdpE, proteins that control expression of the *kdpABC* operon, are members of the two-component sensor-effector class of regulators. J. Bacteriol. *174*, 2152–2159.

Wang, L., Grau, R., Perego, M., and Hoch, J.A. (1997). A novel histidine kinase inhibitor regulating development in *Bacillus subtilis*. Genes Dev. *11*, 2569–2579.

Wang, L., Sun, Y.P., Chen, W.L., Li, J.H., and Zhang, C.C. (2002). Genomic analysis of protein kinases, protein phosphatases and two-component regulatory systems of the cyanobacterium *Anabaena* sp. strain PCC 7120. FEMS Microbiol. Lett. *217*, 155–165

Ward, S.M., Delgado, A., Gunsalus, R.P., and Manson, M.D. (2002). A NarX-Tar chimera mediates repellent chemotaxis to nitrate and nitrite. Mol. Microbiol. *44*, 709–719.

Watanabe, T., Okada, A., Gotoh, Y., and Utsumi, R. (2008). Inhibitors targeting two-component signal transduction. Adv. Exp. Med. Biol. *631*, 229–236

Watson, G.M., Scanlan, D.J., and Mann, N.H. (1996). Characterization of the genes encoding a phosphate-regulated two component sensory system in the marine cyanobacterium *Synechococcus* sp. WH7803. FEMS Microbiol. Lett. *142*, 105–109.

Wegener-Feldbrügge, S., and Søgaard-Andersen, L. (2009). The atypical hybrid histidine protein kinase RodK in *Myxococcus xanthus*: spatial proximity supersedes kinetic preference in phosphotransfer reactions. J. Bacteriol. 191, 1765–1776.

Wei, W., Wang, W., Cao, Z., Yu, H., Wang, X., Zhao, J., Tan, H., Xu, H., Jiang, W., and Li, Y. (2007). Comparative analysis of two-component signal transduction system in two streptomycete genomes. Acta Biochim Biophys Sin (Shanghai). *39*, 317–325.

Weigt, M., White, R.A., Szurmant, H., Hoch, J.A., and Hwa, T. (2009). Identification of direct residue contacts in protein–protein interaction by message passing. Proc. Natl. Acad. Sci. U.S.A. *106*, 67–72.

Weinrauch, Y., Panchev, R, Dubnau, E., Smith, I., and Dubnau, D. (1990). A *Bacillus subtilis* regulatory gene product for genetic competence and sporulation resembles sensor protein members of the bacterial two-component signal-transduction systems. Genes Dev. *4*, 860–872.

West, A.H., and Stock, A.M. (2001). Histidine kinases and response regulator proteins in two-component signalling systems. Trends Biochem. Sci. *26*, 369–376.

Wheeler, R.T., and Shapiro, L. (1999). Differential localization of two histidine kinases controlling bacterial cell differentiation. Mol. Cell. *4*, 683–694.

Whitworth, D.E. (2008). Genomes and knowledge – a questionable relationship? Trends Microbiol. *16*, 512–519.

Whitworth, D.E., and Cock, P.J.A. (2008a). Myxobacterial Two-Component Systems. In Myxobacteria: Multicellularity and Differentiation, D.E. Whitworth, ed. (ASM Press, Washington, DC), pp. 169–189.

Whitworth, D.E., and Cock, P.J.A. (2008b). Two-component systems of the myxobacteria: structure, diversity and evolutionary relationships. Microbiol. *154*, 360–372.

Whitworth, D.E., and Cock, P.J.A. (2009). Evolution of prokaryotic two-component systems: insights from comparative genomics. *Amino Acids*. *47*, 459–466.

Whitworth, D.E., Holmes, A.B., Irvine, A.G., Hodgson, D.A., and Scanlan, D.J. (2008a). P-acquisition components of the *Myxococcus xanthus* Pho regulon are regulated by both P-availability and development. J. Bacteriol. *190*, 1997–2003.

Whitworth, D.E., Millard, A., Hodgson, D.A., and Hawkins, P.F. (2008b). Protein–protein interactions between two-component system transmitter and receiver domains of *Myxococcus xanthus*. Proteomics *8*, 1839–1842.

Williams, R.H.N., and Whitworth, D.E. (2010). Genetic organisation of prokaryotic two-component system signalling pathways. BMC Genomics *11*, 720.

Wolfe, A.J. (2010). Physiologically relevant small phosphodonors link metabolism to signal transduction. Curr. Opin. Microbiol. 13, 204–209.

Wuichet, K., Cantwell, B.J., and Zhulin, I.B. (2010). Evolution and phyletic distribution of two-component signal transduction systems. Curr. Opin. Microbiol. *13*, 219–225.

Xu, Q., Porter, S.W., and West, A.H. (2003). The yeast YPD1/SLN1 complex: insights into molecular recognition in two-component signalling systems. Structure *11*, 1569–1581.

Yamada, S., Sugimoto, H., Kobayashi, M., Ohno, A., Nakamura, H., and Shiro, Y. (2009). Structure of PAS-linked histidine kinase and the response regulator complex. Structure *17*, 1333–1344.

Yamaguchi, K., Suzuki, I, Yamamoto, H., Lyukevich, A., Bodrova, I., Los, D.A., Piven, I., Zinchenko, V., Kanehisa, M., and Murata, N. (2002). A two-component Mg^{2+}-sensing system negatively regulates expression of the mntCAB operon in *Synechocystis*. Plant Cell *14*, 2901–2913.

Yamamoto, K., Hirao, K., Oshima, T., Aiba, H., Utsumi, R., and Ishihama, A. (2005). Functional characterization *in vitro* of all two-component signal transduction systems from *Escherichia coli*. J. Biol. Chem. *280*, 1448–1456.

Yang, R., Bartle, S., Otto, R., Stassinopoulos, A., Rogers, M., Plamann, L., and Hartzell, P. (2004). AglZ is a filament-forming coiled-coil protein required for adventurous gliding motility of *Myxococcus xanthus*. J. Bacteriol. *186*, 6168–6178.

Yoshida, T., Phadtare, S., and Inouye, M. (2007). The design and development of Tar-EnvZ chimeric receptors. Methods Enzymol. *423*, 166–183.

Zapf, J., Sen, U., Madhusudan, Hoch, J.A., and Varughese, K.I. (2000). A transient interaction between two phosphorelay proteins trapped in a crystal lattice reveals the mechanism of molecular recognition and phosphotransfer in signal transduction. Struct. Fold Des. *8*, 851–862

Zhang, W., and Shi, L. (2005). Distribution and evolution of multiple-step phosphorelay in prokaryotes: lateral domain recruitment involved in the formation of hybrid-type histidine kinases. Microbiology *151*, 2159–2173

Zhang, W., Culley, D.E., Wu, G., and Brockman, F.J. (2006). Two-component signal transduction systems of *Desulfovibrio vulgaris*: structural and phylogenetic analysis and deduction of putative cognate pairs. J. Mol. Evol. *62*, 473–487.

Zhao, X., Copeland, D.M., Soares, A.S., and West, A.H. (2008). Crystal structure of a complex between the phosphorelay protein YPD1 and the response regulator domain of SLN1 bound to a phosphoryl analog. J. Mol. Biol. *375*, 1141–1151.

Zhao, Y., Wang, D., Nakka, S., Sundin, G.W., and Korban, S.S. (2009). Systems level analysis of two-component signal transduction systems in *Erwinia amylovora*: role in virulence, regulation of amylovoran biosynthesis and swarming motility. BMC Genomics *10*, 245.

Zhou, H., Lowry, D.F., Swanson, R.V., Simon, M.I., and Dahlquist, F.W. (1995). NMR studies of the phosphotransfer domain of the histidine kinase CheA from *Escherichia coli*: assignments, secondary structure, general fold, and backbone dynamics. Biochemistry *34*, 13858–13870.

Zhou, L., Grégori, G., Blackman, J.M., Robinson, J.P., and Wanner, B.L. (2005). Stochastic activation of the response regulator PhoB by noncognate histidine kinases. J. Integr. Bioinform. *2*, 11.

Zhou, X., Keller, R., Volkmer, R., Krauss, N., Scheerer, P., and Hunke, S. (2011). Structural basis for two-component system inhibition and pilus sensing by the auxiliary CpxP protein. J. Biol. Chem. *286*, 9805–9814.

Zientz, E., Bongaerts, J., and Unden, G. (1998). Fumarate regulation of gene expression in *Escherichia coli* by the DcuSR (*dcuSR* genes) two-component regulatory system. J. Bacteriol. *180*, 5421–5425.

Zusman, D.R., Scott, A.E., Yang, Z., and Kirby, J.R. (2007). Chemosensory pathways, motility and development in *Myxococcus xanthus*. Nat. Rev. Microbiol. *5*, 862–872.

Bacterial Chemotaxis

9

Kathryn A. Scott, Elizabeth E. Jefferys, Benjamin A. Hall, Mark A.J. Roberts and Judith P. Armitage

Abstract

Chemotaxis is the process by which bacteria migrate towards environments that are favourable for growth. Changes in the concentration of attractants or repellents are detected by receptors, which are usually transmembrane proteins. These receptors transduce the signal to the interior of the cell where a two-component system ultimately leads to changes in motile behaviour. Chemotaxis emerged as a beneficial trait for survival early in the evolution of bacteria and archaea. A core set of proteins is common to the chemosensory networks in many different species. During the evolution of bacteria this core network has diversified and expanded. Here we describe the conserved apparatus in the steps necessary for chemotaxis; sensing of chemoeffectors, signalling to the motility apparatus, rapid signal termination, and adaptation. We then highlight examples from species with complex chemosensory networks to illustrate the variations in chemotactic apparatus that have arisen from the common core.

Introduction

Peering through the lens of his hand-crafted microscope Antony van Leeuwenhoek found resident within the plaque from his teeth a cornucopia of 'very little living animalcules, very prettily a-moving'. In a 1683 letter to the Royal Society van Leeuwenhoek described how these living entities swam nimbly and 'of-times spun round like a top', thus presenting one of the first reports of living bacteria (the discovery of microorganisms by van Leeuwenhoek and Hooke is reviewed in Gest, 2004). Unbeknownst to van Leeuwenhoek were the complex and integrated decision making processes underlying these various behaviours. Two centuries later Wilhelm Pfeffer reported the first insights into the factors affecting bacterial motility. Pfeffer demonstrated the directed movement of bacteria into capillary tubes loaded with nutrients and the aversion of bacteria to capillaries filled with toxic substances (Pfeffer, 1888). Further, he showed that if a repulsive chemical was mixed in appropriate proportions with an attractive substance, bacteria would opt to travel into the tube despite the threat of the noxious compound. What van Leeuwenhoek, Pfeffer and their contemporaries who found interest in the activities of these tiny living entities were among the first to observe was what is now one of the best-understood of bacterial behaviours – chemotaxis.

Chemotaxis is the mechanism by which bacteria sense and migrate along concentration gradients of specific chemical substances in their environment. The movement of an organism towards an environment optimal for metabolic activity is referred to as energy taxis, whilst phototaxis is movement in response to changing light intensity. The ability to respond and adapt to the external conditions through movement is a key survival trait that emerged early in the evolution of bacteria and archaea (Ulrich *et al.*, 2005). Since the pioneering work by Pfeffer bacterial chemosensory pathways have been examined using a wide variety of techniques including mutational studies, x-ray crystallography, genomic/bioinformatic analysis and mathematical modelling. The underlying features fundamental to chemotactic behaviour have emerged from the detailed characterization of chemotaxis in a number of model species. As for many other prokaryotic signalling processes, bacterial chemotaxis appears to be based on a conserved set of molecular components that have been elaborated over the course of evolution (Wuichet and Zhulin, 2010). Here we provide an overview of the fundamental features of bacterial chemotaxis, followed by a more in-depth discussion of the biochemical steps involved and how these steps differ between species. Chemosensory-like networks have also been adapted in many different organisms to the control of cellular processes other than swimming motility; we briefly discuss the importance of these networks in the control of gliding motility and in the regulation of gene expression.

Navigation on the microscale

Successful tactic behaviour clearly relies upon the ability of an organism to accurately perceive which direction is favourable and then to move in that direction. Some motile bacteria move by swimming through liquid media whilst others move along surfaces. Free-swimming bacteria are propelled through liquid media by the rotation of one or more helical filaments, known as flagella. The flagella are attached at their base to a rotary motor embedded in the cell envelope. This motor, powered by the movement of H^+ or Na^+ ions across the cytoplasmic membrane down their electrochemical gradient, is capable of rotating the flagella at up to 1000 Hz (Sowa and Berry, 2008). There are two important components in any rotary motor: the rotor and the stator. The structures of the flagellar motors from a number of species have been determined using cryoelectron microscopy (Chen *et al.*, 2011). These structures have demonstrated that the bacterial flagellar motors analysed to-date are comprised of a common core consisting of five rings spanning the cell envelope, but that there is significant species variation outside this core. The core rings are termed the L-, P-, S-, M- and C-rings based on their location within the cell; lipopolysaccharide, peptidoglycan layer, submembranous, membranous, and cytoplasmic respectively. It should be noted that although originally thought to be composed of distinct protein species, the M- and S-rings were subsequently shown to consist multiple copies of a single protein (Ueno *et al.*, 1992). Torque generation is achieved through interactions between the C-ring, also known as the switch complex, and the stator proteins (MotA/B in H^+ driven motors, PomA/B in Na^+ driven motors), however the precise molecular mechanism for this process is not yet fully understood (Sowa and Berry, 2008).

Bacteria swim in a characteristically erratic pattern and are unable to move directly to a food source. Unlike most chemotactic eukaryotic cells which appear to effect seeking and escape movements, for a bacterial cell such directed linear motion is probably an impossibility (see Swaney *et al.*, 2010, for a recent review of eukaryotic chemotaxis). Gradual rotation of the cell body as it moves through the highly viscous environment causes a bacterium to travel in a curved trajectory from which it is readily displaced owing to Brownian motion. In addition the small size of a bacterium relative to its environment means that it experiences very high viscosity such that in the absence of a propulsive force it stops moving very rapidly. These factors, combined with the frequent switching of the flagellar motor, results in, over time, the cell performing a three-dimensional random walk (Berg and Brown, 1972). Bacteria differ in the mechanisms used to change direction. *Escherichia coli* has a bidirectional motor; clockwise rotation of the motor leads to random tumbling whilst counterclockwise rotation gives rise to smooth swimming or runs. In contrast, *Rhodobacter sphaeroides* has a unidirectional, stopping motor and relies on Brownian motion during pauses in flagellar rotation to effect reorientation (Armitage and Macnab, 1987). *Sinorhizobium meliloti* also has a unidirectional motor, but in this case reorientation occurs through slowing of flagellar rotation (Gotz and Schmitt, 1987; Platzer *et al.*, 1997; Scharf *et al.*, 1998). In the absence of chemoeffector molecules the period of time between redirection is largely invariant. Under such conditions the unsystematic journey a cell undertakes may allow effective exploration of its environment. Upon encountering gradients of either positive or negative stimuli a chemotactic bacterial cell will exhibit biased motility with the chemotactic pathway controlling the frequency of reorientation events (Fig. 9.1) (Berg and Brown, 1972). In the presence of an increased concentration

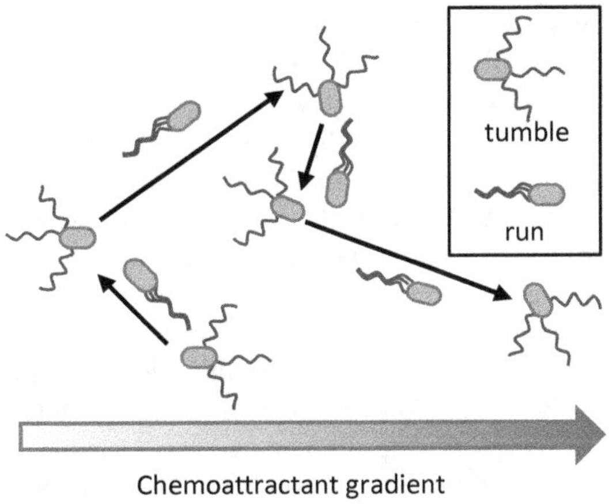

Figure 9.1 The biased random walk. This cartoon illustrates the biased random walk undertaken by chemotactic bacteria in the presence of a chemoeffector gradient, using the swimming mode of *E. coli* as an example. In the absence of a chemoeffector gradient, bacteria perform a random walk, alternating between smooth running and reorientation. When an increased concentration of a chemoattractant is detected the length of time between reorientations is lengthened, thus allowing the bacterium to move up the chemoattractant concentration gradient.

of chemoattractant or a reduced concentration of chemorepellent there will be less frequent reorientations. Conversely, in the presence of an increased concentration of chemorepellent or a reduced concentration of chemoattractant the frequency with which the cell changes direction will increase. Frequent reorientation is additionally important because of the temporal nature of gradient sensing (Brown and Berg, 1974; Segall *et al.*, 1986). Bacteria usually move too rapidly and are too small to sense a gradient across their body length, instead they typically compare the current conditions with those a second or so ago. Thus, to achieve net migration towards a more favourable environment, the bacterium must frequently reset its direction, resulting in adoption of a biased random walk as opposed to direct targeted motion.

The chemoeffectors sensed by bacteria comprise an assortment of chemical moieties, including peptides, amino acids, sugars and organic acids, in addition to smaller molecules and ions such as oxygen, nitrate and nickel. The precise molecules that are detected by a cell and the information they relay (i.e. the presence of a harmful or beneficial environment) vary between different bacterial species and, in some instances, within the same species under different environmental conditions. For instance, acetate is a chemoattractant for *R. sphaeroides* but repels *E. coli* cells (Tso and Adler, 1974; Packer and Armitage, 2000), whilst oxygen is an attractant for both species under aerobic conditions, but is a chemorepellent for photosynthetically grown *R. sphaeroides* (Romagnoli *et al.*, 2002).

Bacterial signal transduction systems can be broadly divided into two categories, one- and two-component systems. In a one-component system a single protein both detects the signal and elicits a response. These proteins are typically composed of two domains, a sensor domain, and a regulatory domain which determines the output of the signalling pathway. In a two-component system the sensor domain and regulatory domains are found in separate proteins. The sensor domain is a histidine protein kinase (HPK) which phosphorylates the regulatory domain controlling the response. Although the number and arrangement of domains differs between the sensory proteins, in a canonical two-component system the HPKs phosphorylate a histidine residue within their dimerization and histidine phosphorylation (Dhp) domain. The chemotaxis system is a modified two-component system; in this case the dimerization domain lacks the histidine that is autophosphorylated in other kinases. Instead autophosphorylation occurs at a histidine residue in the N-terminal histidine-containing phosphotransfer domain.

Conserved elements in chemosensory networks

Chemotaxis is an ancient behaviour that most likely became established in the eubacteria and archaea owing to the considerable selective advantage endowed to chemotactic cells in nutrient poor environments (Kelly *et al.*, 1988). Not only is the behaviour itself widespread throughout the bacterial domain, but the basic features of chemotaxis are mediated largely by a conserved battery of proteins that appear to be well adapted to carry out the fundamental requirements of a chemosensory network: sensing, signalling to the flagellar motor, signal termination and adaptation (Fig. 9.2a).

Fundamental steps in signalling

The first critical element in the chemosensory response is the ability to detect specific effectors. Chemoeffectors are sensed by receptor proteins, the most common of which are known as methyl-accepting chemotaxis proteins (MCPs), and have a conserved overall architecture but diverse ligand-binding capabilities. In the majority of chemoreceptors ligand-binding domains, which may be located in the periplasm, cytoplasm or cell membrane, transmit a conformational change through a transmembrane domain to the cytoplasmic region of the protein (Hazelbauer *et al.*, 2008). Thousands of these chemoreceptors are organized into highly cooperative transmembrane clusters, typically located at the poles of the cell (Maddock and Shapiro, 1993; Briegel *et al.*, 2009). Within the same cluster, bacteria contain multiple types of chemoreceptors with different ligand-binding specificities. These arrays of chemoreceptors are capable of not only transducing a single ligand-binding event but also of the integration of signals incident on different receptors. Allosteric interactions give rise to substantial amplification of these inputs such that even minute changes in chemoeffector concentration can elicit significant cellular responses over a wide range of background concentrations (Bray *et al.*, 1998; Goldman *et al.*, 2009).

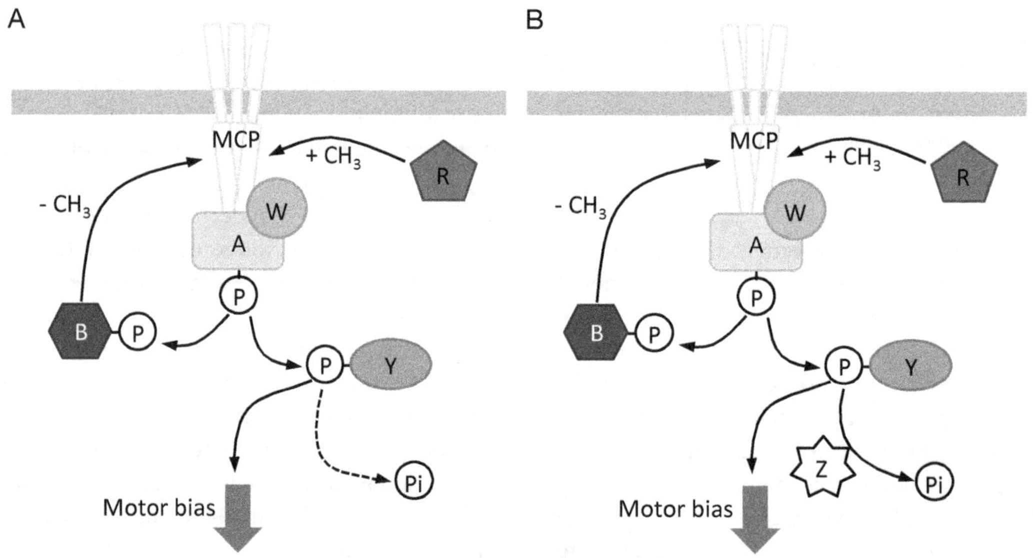

Figure 9.2 Conserved features in chemosensory networks. (a) The proteins CheA, CheW, CheY, CheB and CheR, in addition to transmembrane chemoreceptors, are common to the chemosensory networks of the vast majority of chemotactic bacteria and archaea. Chemoreceptors detect ligand binding events and transduce the signal across the membrane and then, through CheW, to CheA thereby leading to a perturbation in the kinase activity of CheA. CheA-P can phosphotransfer to two response regulators, CheY and CheB. CheY-P interacts with the flagellar motor biasing the time period between reorientations whilst CheB-P, in conjunction with CheR, plays a role in the adaptation to persistent chemoeffector binding. (b) In addition to the conserved core of chemosensory proteins, *E. coli* also encodes a CheY-P specific phosphatase, CheZ. In *E. coli* a reduction in the level of chemoattractant leads to increased CheA kinase activity which in turn raises the concentration of CheY-P in the cell. CheY-P binds to the flagellar motor causing it to switch from counter clockwise to clockwise rotation and, as a result, the cell tumbles. Demethylation of chemoreceptors by CheB-P reduces the ability of the chemoreceptors to activate CheA autophosphorylation whilst CheZ catalysed dephosphorylation of CheY-P ensures rapid signal termination.

Following signal reception, the information transduced by the chemoreceptors must be relayed to the flagellar motor to bring about an appropriate change in motile behaviour. In chemosensory networks this is achieved through a two-component system. The relative simplicity of the chemotaxis two-component system is an important feature; a wide range of potential chemoeffectors converge via this system onto a small set of chemotaxis proteins that act to bring about a response. The histidine protein kinase in the chemotaxis two-component system is CheA. CheA and the adaptor protein CheW localize to the C-terminal region of the chemoreceptors (Sourjik and Berg, 2000). Ligand induced conformational changes in the chemoreceptors are coupled through CheW to bring about changes in CheA kinase activity (Borkovich and Simon, 1990). The next step in the signal transduction process is phosphotransfer from phosphorylated CheA (CheA-P) to a specific aspartate residue on response regulators (RRs), which in almost all chemotaxis systems include one or more homologues of the proteins CheY and CheB. The function of CheY is to transduce signals from the sensory clusters to the flagellar motor(s). CheY-P dissociates from CheA and diffuses to the flagellar motor where it binds to the switch complex components FliM and FliN (Welch *et al.*, 1993; Khan *et al.*, 2000; Sarkar *et al.*, 2010). The effect of CheY-P binding to the switch complex (switching, stopping, slowing etc.) is dependent upon the specific bacterium (Larsen *et al.*, 1974; Armitage and Macnab, 1987; Gotz and Schmitt, 1987; Platzer *et al.*, 1997; Scharf *et al.*, 1998).

An essential requirement of biological signalling systems is their transitory temporal nature; prompt cessation of the intracellular signal that modulates motor bias is necessary for a chemosensory response. In order for a system to exhibit a dynamic response the output (motor bias) must closely parallel the signal input at the receptors, and this in turn necessitates a rapid turnover over of the state of the signalling apparatus. The half-lives of phosphorylated response regulators in two-component systems can vary from seconds to days but are typically of the order of minutes (Stock *et al.*, 2000). CheY-Ps have evolved to undergo rapid auto-dephosphorylation, with the half-life of CheY-P normally being in the region of seconds to tens of seconds. The auto-dephosphorylation half-lives of the CheY-P species are still, however, usually longer than the duration of the chemosensory response. Specific CheY-P phosphatases are therefore utilized by chemosensory apparatus to achieve sufficiently rapid signal termination (Silversmith, 2010). The mechanisms whereby the necessary prompt signal termination is ensured show considerable variation across different bacterial species.

Chemotaxis requires bacteria to have the ability to detect chemoeffector gradients. At rarely more than a few microns in length, bacteria are physically too small to effectively detect spatial gradients across the cell body. Instead, bacteria employ temporal comparisons of the chemoeffector concentrations encountered at different points along the swimming trajectory (Brown and Berg, 1974; Berg and Purcell, 1977). Bacteria therefore sense not the absolute values of chemoeffector concentrations, but rather their rates of change with time, a feat achievable only if the cell is able to remember the levels of chemoeffector encountered in the immediate past (Vladimirov and Sourjik, 2009). Such short-term memory is typically provided by chemical modification of the chemoeffector receptors; the previous occupancy is encoded in the pattern of chemical modification for comparison with the present state of chemoeffector binding. Bacterial memory is by necessity transient. If the cell is to continue detecting changes in chemoeffector concentrations with consistently high sensitivity, over the wide range of chemoeffector concentrations with which it may be faced, it must possess mechanisms for resetting the sensory apparatus to its prestimulus state after each signalling event has passed. In short, the cell must be able to adapt. The importance of adaptation is borne out in the phenotypes of mutant cells deficient in the adaptation machinery which, after an initial response to a chemoeffector gradient, become locked in a smooth swimming or perpetually reorienting behaviour (Goy *et al.*, 1978; Yonekawa *et al.*, 1983).

The second widespread RR of CheA, CheB, plays a vital role in the adaptation process. CheB and another conserved chemotaxis protein CheR function reciprocally to demethylate and methylate, respectively, specific glutamate residues located in the cytoplasmic region of the chemoreceptors (Springer and Koshland, 1977; Stock and Koshland, 1978). Whilst CheR appears to be constitutively active, the rate of demethylation catalysed by CheB is increased when CheB is phosphorylated by CheA (Lupas and Stock, 1989). Chemoreceptor occupancy affects the rate of CheA kinase activity which in turn changes the rate of CheB methylesterase activity. This, combined with the differing affinities of CheR and CheB for active and inactive receptors or in some cases specific methylation sites, results in the generation of chemoreceptor methylation patterns that reflect the composition of chemoeffectors in the environment just encountered. Because the ability of chemoreceptors to bind their cognate ligands and to activate CheA is influenced by methylation/demethylation of the cytoplasmic region of the chemoreceptor this chemical modification cycle serves to reset CheA activity to its prestimulus level even if the elevated or lowered chemoeffector concentration should persist, achieving adaptation. Furthermore, the methylation pattern of chemoreceptors serves to provide the cell with the short-term chemical memory necessary for temporal comparisons, and hence for gradient sensing. This memory results from the typically ~1-second lag between a change in chemoreceptor occupancy and its methylation/demethylation (Brown and Berg, 1974; Berg and Tedesco, 1975); this lag may be attributed to the slow kinetics of CheR and CheB and to their low copy numbers within the cell (Li and Hazelbauer, 2004). Thus, the chemoreceptor modification pattern at any given moment reflects the chemoeffector composition of the environment encountered in the immediate past, facilitating comparison between the past and present environments and hence allows the cell to 'decide' whether or not to bias its motion towards reorientation or continued smooth swimming.

This basic organization of the chemotactic signalling network is evident throughout the eubacteria in all chemotactic species studied to date (Fig. 9.2a). However, over the course of evolution it has become elaborated and embedded in different signalling and metabolic networks of in individual bacterial species (Wuichet and Zhulin, 2010). This differentiation has yielded a plethora of systems in which the same fundamental processes and chemical reactions are implemented in distinct contexts that optimize the survival of each bacterial species within its particular niche. Of those bacteria in which chemotaxis is understood in particular detail, the peritrichously flagellated enteric bacteria *E. coli* and *Salmonella enterica* serovar Typhimurium stand out as the most thoroughly characterized to date (Fig. 9.2b). Although differing in the number and specificity of their chemoreceptors, these species otherwise share the same organization of the chemosensory network and mechanisms for adaptation. For simplicity, we will describe the chemosensory machinery of *E. coli* although insight on the arrangement of this network has been drawn from both species. *E. coli* has five chemoreceptors four of which, Tar, Tsr, Trg and Tap, sense compounds such as amino acids and sugars in addition to responding to changes in temperature and pH. The fifth chemoreceptor, Aer, responds to changes in the redox potential of the cell. *E. coli* has only a single homologue of each of the chemotaxis proteins CheA, CheY, CheW, CheB and CheR and a single CheY-P specific phosphatase, CheZ. The *che* genes, as well as the genes encoding the chemoreceptors Tar and Tap, are encoded in a single operon. In *E. coli*, a reduction in the level of attractant binding to the chemoreceptors leads to a conformational change which results in the stimulation of CheA kinase activity (Hazelbauer *et al.*, 2008; Hazelbauer and Lai, 2010). CheY-P forms rapidly and diffuses to the flagellar motor where it binds to the switch complex causing a change in the direction of rotation of the flagellar motor from counter clockwise to clockwise (the mechanism by which this occurs is reviewed in Sowa and Berry, 2008). When the motor rotates in the counter clockwise direction the flagella filaments form a bundle propelling the cell forward. On switching to clockwise rotation the bundle of filaments is disrupted and the cell tumbles. Rapid signal termination in the absence of continued stimulation is ensured by CheZ. CheB-P forms on a slower timescale than

CheY-P (Bray *et al.*, 2007), and together with CheR is responsible for adaptation through the methylation/demethylation and deamidation of chemoreceptors. Methylation by the constitutively active CheR enhances the ability of the chemoreceptors to activate CheA autophosphorylation. Demethylation is carried out by CheB, the methylesterase activity of which increases by two orders of magnitude when phosphorylated by CheA (Anand and Stock, 2002). In *E. coli*, adaptation is also mediated through the substrate specificity of CheB and CheR; CheB will preferentially demethylate active receptors (those that are stimulating CheA autophosphorylation) whereas CheR preferentially methylates inactive receptors (Stock and Koshland, 1981). In response to a decrease in the concentration of attractant the chemoreceptors stimulate CheA autophosphorylation leading to an increased concentration of CheY-P and consequently to bacterial tumbling. The formation of CheB-P leads to demethylation of active chemoreceptors reducing their ability to activate CheA and thus restoring the system to pre-stimulus levels.

Many other bacterial species have more than one network of chemosensory genes. These networks may include the genes for multiple homologues of the chemosensory proteins found in *E. coli*. In addition chemotaxis–like networks are used by some organisms to control processes other than motility. The more complex chemotaxis apparatus utilized by other bacterial species are now being dissected in some detail revealing commonalities and distinctions relative to the *E. coli* system; such differences may reflect the distinct metabolic capabilities, habitats and lifestyles of these bacterial species. There are many commonly studied model systems for chemotaxis and to discuss all of them is outside the scope of this review. Instead we will describe the common features and variation in sensing, signalling, signal termination and adaptation drawing examples from the chemosensory pathways of *E. coli*, *R. sphaeroides*, *Bacillus subtilis* and *Sinorhizobium meliloti*. We will conclude with a brief discussion of examples from organisms where networks encoding proteins seen in chemotactic pathways are used to control cellular functions other than swimming motility.

The steps dissected

Sensing

Chemoreceptor structure

The fundamental unit of chemoreceptor oligomerization is a homodimer, comprised typically of approximately 60 kDa subunits (Milligan and Koshland, 1988). These homodimers form predominantly helical structures approximately 300 Å in length and can be divided into three main functional regions, for sensing, signal conversion, and kinase control (for detailed reviews on bacterial chemoreceptors see Hazelbauer *et al.*, 2008; Hazelbauer and Lai, 2010). Although the structure of chemoreceptors varies, many of them share the same domain architecture (Fig. 9.3) (Alexander and Zhulin, 2007). The most commonly observed architecture consists of a transmembrane helix with the N-terminus located in the cytoplasm, followed by a periplasmic ligand-binding domain whose structure is dependent upon the type of ligand. This is then followed by a second transmembrane helix which is connected to a HAMP domain; so termed because it is a structural motif found in histidine kinases, adenylyl cyclases, methyl-accepting chemotaxis proteins and some phosphatases (Aravind and Ponting, 1999). The HAMP domain forms a four-helix bundle motif and is implicated in signal transduction between the extracellular and more membrane-distal intracellular regions of the protein. From the HAMP domain extends the long kinase control element of the receptor, which is composed of a dimer of helical hairpins that arrange into a ~250-Å-long antiparallel four-helix bundle (Kim *et al.*, 1999). The kinase control element contains an adaptation domain bearing the methylation sites targeted by CheB and CheR, a conserved glycine hinge and the protein interaction domain that associates with CheA and CheW. To date there is no high-resolution structure available for a complete transmembrane chemoreceptor, and in particular, there is no crystal structure of the transmembrane region. Considerable insights have, however, been made based on the structures of individual chemoreceptor domains and of chemoreceptors which naturally lack a transmembrane region.

The structure of the ligand-binding domain

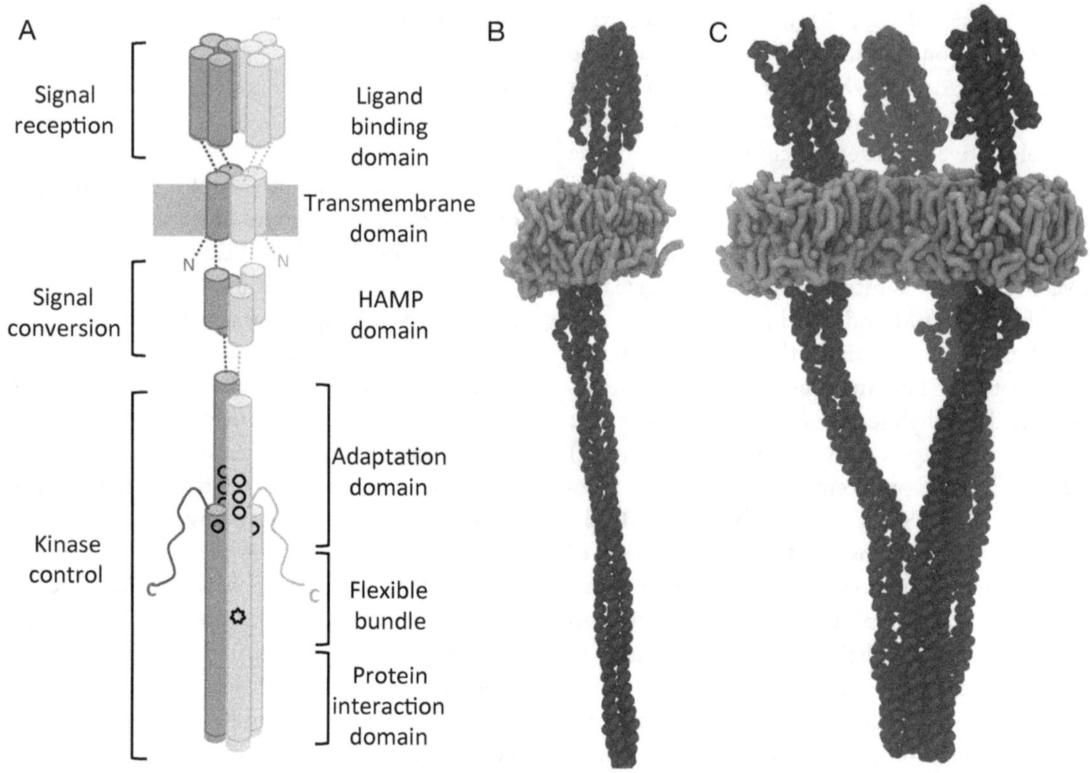

Figure 9.3 The domain architecture of a prototypical chemoreceptor. (a) A schematic diagram illustrating the domain architecture shared by many chemoreceptors and highlighting some of the functionally important features. In this diagram the helices are represented by cylinders, methylation sites by open circles and the location of the glycine hinge by a star (After Hazelbauer *et al.*, 2008)). (b) and (c) Ribbon diagram illustrating models of the *E. coli* receptor Tsr as (b) a dimer and (c) a trimer of dimers. Models were generated from crystal structures of individual domains using the program Modeller v9.9 (Eswar *et al.*, 2001). Positioning of the membrane (shown in light grey) and equilibration of the structures was carried out using coarse-grained molecular dynamics; crystal contacts seen in the structure of the kinase control region of Tsr were enforced during equilibration of the trimer of dimers.

of the *S. typhimurium* aspartate receptor, Tar, has been determined in the presence and absence of ligand giving insights into the conformational changes that occur on aspartate binding (Milburn *et al.*, 1991). The structures show that the aspartate ligand binds at the interface of the two subunits in the homodimer. There are few secondary structural changes, but rather a difference in the orientation of the two subunits between the apo and ligand-bound states. The first structure of a HAMP domain, from an archaeal protein, was determined in 2006 and more recently that of a series of three concatenated HAMP domains from the *Pseudomonas aeruginosa* receptor Aer has also been solved (Hulko *et al.*, 2006; Airola *et al.*, 2010). In each case the four helix bundle motif that had been predicted based on sequence analysis was observed (Singh *et al.*, 1998), but the packing between the helices differs to some extent both between species and between the individual domains in the concatenated structure. As the primary role of the HAMP domain is in signal transduction these differences in packing most likely reflect the population of multiple conformational states (Parkinson, 2010). Finally, the structures of the cytoplasmic region of the *E. coli* serine chemoreceptor, Tsr, and that of a soluble chemoreceptor from *T. maritima* both showed an extended, anti-parallel, four-helix coiled coil (Kim *et al.*, 1999; Pollard *et al.*, 2009). The structure of the *T. maritima* receptor demonstrates asymmetry between the two monomers which is thought to play a role in the propagation of conformational changes from the HAMP domain to the CheA binding region.

There is considerable variation in the number of chemoreceptors found in different species of bacteria, ranging from a single chemoreceptor in organisms such as *Mesorhizobium loti* to over 60 chemoreceptors in *Magnetospirillum magnetotacticum* (Alexander and Zhulin, 2007). Amongst the model organisms referred to in this review *E. coli* has 5 chemoreceptors, *B. subtilis* 10 chemoreceptors, *R. sphaeroides* 13 chemoreceptors and *S. meliloti* 9 chemoreceptors. The greatest structural variation between chemoreceptors is found in the ligand-binding region. Chemoreceptors can be divided into four major classes, the first three of which contain a transmembrane region but differ in the cellular location of the ligand-binding domain (Fig. 9.4) (Wuichet *et al.*, 2007). The ligand-binding domain of class I chemoreceptors is located in the periplasm; this class is the most abundant and includes receptors such as *E. coli* Tar and Tsr. In contrast, class II chemoreceptors, exemplified by *E. coli* Aer, have their ligand-binding domain in the cytoplasm at the N-terminus of the protein. Class III chemoreceptors can be further subdivided into class IIIm and class IIIc; class IIIm chemoreceptors contain transmembrane

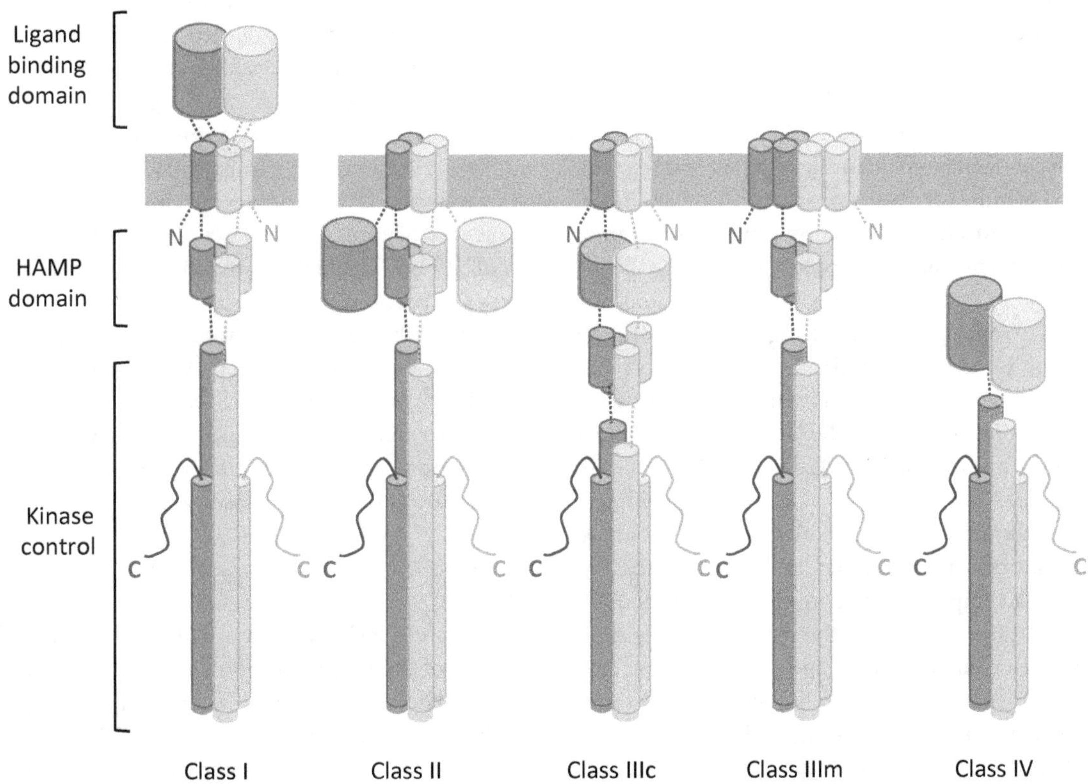

Figure 9.4 Schematic representation of the different classes of chemoreceptors. Chemoreceptors can be divided into four classes based upon their topology. This cartoon illustrates the domain connectivity in each of the four classes, with helices represented by narrow cylinders and ligand binding domains of varying secondary structure by wide cylinders. In class I chemoreceptors the ligand-binding domain is in the periplasm, whilst in class II chemoreceptors is it located in the cytoplasm at the N-terminus of the protein. Class III can be further subdivided into two categories, IIIc and IIIm. In class IIIc there is a ligand-binding domain inserted between the transmembrane and HAMP domains, whilst in class IIIm there is a sensory domain in the membrane. Finally, class IV comprises the soluble chemoreceptors (Figure after Wuichet *et al.* (2007).

ligand sensing regions whilst class IIIc have an N-terminal region embedded in the lipid bilayer followed by a sensor domain in the cytoplasm. Class IV chemoreceptors do not span the lipid bilayer and typically also lack HAMP domains. In most cases these chemoreceptors are thought to localize to the chemoreceptor array at the cell membrane. For example, McpS from *P. aeruginosa* was shown to localize to the cell pole using immunofluorescence microscopy (Bardy and Maddock, 2005). Similarly, fluorescent fusions of both soluble chemoreceptors encoded by *S. meliloti* were shown to localize to the cell pole along with six of the transmembrane chemoreceptors (Meier and Scharf, 2009). In *R. sphaeroides*, however, the soluble chemoreceptors form discrete clusters in the cytoplasm and will be discussed further subsequently (Wadhams *et al.*, 2002, 2003).

The domain architecture of the kinase control region is shared by all chemoreceptors. Identification of the sequences of the cytoplasmic kinase control regions from over 150 different microbial species facilitated identification of features that are essential for structure and function and led to the classification of chemoreceptors based upon the sequence of the cytoplasmic region. Chemoreceptor cytoplasmic domains consist of multiple heptad repeats and can be assigned to one of seven major or five minor classes depending on the number of repeats. The major classes are named 24H, 28H, 34H, 36H, 38H, 40H and 44H indicating the number of heptad repeats (Alexander and Zhulin, 2007).

Although the fundamental unit of the chemoreceptors is a homodimer, higher order structure plays an important role in chemoreceptor function. In the structure of the cytoplasmic region of Tsr determined by Kim and co-workers in 1999 the Tsr homodimers were joined at the C-terminus to form a trimer of dimers (Kim *et al.*, 1999). There is now a wealth of biochemical, structural and computational evidence that chemoreceptors form a trimer of dimers *in vivo*. Individual dimers are capable of performing some chemoreceptor functions, whilst the trimer of dimers is required for others. Experiments using nanodiscs containing a specific number of chemoreceptors revealed that the dimer is competent in ligand-binding, signal transduction and adaptation but that maximal activation of CheA kinase requires a trimer of dimers (Boldog *et al.*, 2006; Amin and Hazelbauer, 2010). Association of the trimers of dimers into larger arrays is important for integration of signals from different chemoeffectors and also for sensitivity. Immunogold labelling combined with microscopy, fluorescence microscopy and cryo-electron tomography have all been used to investigate the location and structures of chemoreceptors within an *E. coli* cell (Maddock and Shapiro, 1993; Sourjik and Berg, 2000; Zhang *et al.*, 2007). The chemoreceptors form large clusters in the membrane near the pole of the cell, in addition to smaller clusters along the cell body. In 2008 cryo-electron tomography of *Caulobacter crescentus* showed that the chemoreceptors were arranged in hexagonal arrays and that trimer of dimer model for chemoreceptor structure fit well with the observed electron density (Briegel *et al.*, 2008; Khursigara *et al.*, 2008). More recently an electron cryotomography study of 13 different species which express chemoreceptors from six different structure classes provided evidence for the hypothesis that hexagonal architecture is a universal feature of chemoreceptor arrays (Briegel *et al.*, 2009). The high membrane curvature at the poles of bacilli such as *E. coli* appears to be incongruent with the lesser curvature of the receptor complexes themselves. This may result in periodic breaking of the hexagonal arrays. Membrane curvature may also play a role in limiting the size of chemoreceptor arrays, a feature that may have implications for cluster allostery.

Chemoreceptor cluster stability and assembly

CheA and CheW associate with the C-terminal region of the chemoreceptors and are visible as a 'base-plate' in cryo-electron tomography images (Miller *et al.*, 2006; Park *et al.*, 2006a; Briegel *et al.*, 2009). It was initially thought that cluster formation was dependent on the presence of CheA and CheW (Sourjik and Berg, 2000). Subsequently, in a study where the chemoreceptor clusters were visualized by fluorescence microscopy of YFP fused to CheR, it was shown that chemoreceptor clusters can form in *E. coli* in the absence of both CheA and CheW. The clusters are, however, more diffuse than in wild-type cells. Thus CheA and

CheW are not essential for chemoreceptor cluster formation but instead enhance cluster formation (Kentner *et al.*, 2006). Core chemosensory components (a chemoreceptor, CheA and CheW) have been shown for *S. typhimurium* and *E. coli* to form ultrastable complexes *in vitro* (Erbse and Falke, 2009). When the components purified in isolation are combined kinase activity is maintained on the timescale of days. This ultrastability is proposed to arise from the multiple protein–protein interactions in the chemoreceptor lattice. The chemoreceptor array has also been shown to be stable on a timescale considerably longer than the chemosensory response *in vivo*. Sourjik and co-workers analysed the exchange dynamics of the *E. coli* chemosensory array using fluorescence recover after photobleaching (Schulmeister *et al.*, 2008). CheZ, CheA, and CheW were shown to exchange slowly, with a characteristic time of 8 (CheZ) and 12 minutes (CheA and CheW). Similarly, only partial recovery of fluorescence was observed for Tar over the 30 minute duration of the experiment indicating that the majority of the protein was immobile on this timescale. Other components of the chemosensory network, CheB, CheR and CheY exchanged much more rapidly. The exchange time for the enzymes responsible for adaptation was approximately 15 seconds, whilst that for CheY was of the order of seconds. The exchange time for the adaptation enzymes is of a similar timescale to the adaptation response which has been hypothesized to be important in ensuring a uniform distribution of these enzymes (Schulmeister *et al.*, 2008).

Components of the *E. coli* chemoreceptor array fused to fluorescent proteins have been used to shed light on the mechanism by which these proteins assemble at the cell pole. Green fluorescent protein labelled Tar was shown to be inserted over the entire length of the cell body before being incorporated into the array at the cell pole (Shiomi *et al.*, 2006). In addition, a study analysing the position of yellow fluorescent protein fused to CheR demonstrated that lateral clusters form with periodic positioning along the cell body. It was proposed that lateral clusters occur at future division sites thereby ensuring that daughter cells will have at least one chemosensory cluster (Thiem *et al.*, 2007). Recent experiments have provided evidence that the assembly of chemoreceptor arrays in *E. coli* is a stochastic process. Sourjik and co-workers investigated how both the number of receptor clusters and the size of these clusters varied with both the expression level of chemotaxis proteins and with cell length (Thiem and Sourjik, 2008). Their data are consistent with a model where receptors inserted throughout the cell membrane will either nucleate a new cluster or join an existing one dependent upon the distance from an existing cluster and the receptor concentration. In another study, photoactivated localization microscopy was used to investigate the cellular localization of three components of the *E. coli* chemotaxis network, including the chemoreceptor Tar, at high resolution (Greenfield *et al.*, 2009). As with conventional fluorescence microscopy both large clusters, with thousands of Tar receptors at the cell pole, and smaller clusters, with hundreds of receptors along the cell body, were observed. Over 38% of labelled Tar was, however, seen to localize in much smaller clusters with the cluster sizes distributed according to a stretched exponential [an equation of the form $f(x) = \exp(-(x)^{\alpha})$] as would be expected for a stochastic assembly process.

Although localization of chemoreceptor clusters in many enterobacteria is thought to be a stochastic process, there is evidence that positioning can, in some bacteria, be an active process mediated by homologues of the DNA-segregation protein ParA. Bacteria utilize a number of different mechanisms to ensure that DNA is accurately segregated upon cell division (reviewed in Gerdes *et al.*, 2010). One of the most common mechanisms comprises the Walker-like ATP-ase ParA, a centromere-binding protein ParB and a *cis*-acting centromere-like region of DNA *parC/parS*. Many bacterial chromosomes encode orphan *parA* genes (lacking a *parB* DNA binding partner), a number of which have been shown to be important in the segregation and subcellular positioning of cellular components other than DNA (Gerdes *et al.*, 2010). *R. sphaeroides* expresses the components of two chemotaxis operons under laboratory conditions, both of which are essential for chemotaxis (Porter *et al.*, 2002 also discussed further subsequently). The proteins encoded by these operons differ in their cellular localization leading to the formation of two clusters of chemosensory proteins, one at the cell pole and the other in the cytoplasm (Wadhams *et al.*, 2003). Proper localization and segregation of the cytoplasmic cluster proteins has been shown to be dependent upon the ParA homologue PpfA, which is encoded in the same operon as the cytoplasmic cluster chemosensory proteins (Thompson *et al.*, 2006). Analysis of cells where the cytoplasmic cluster chemoreceptor TlpT were labelled with YFP showed that the majority of cells contained either a single cytoplasmic cluster located at mid-cell, or two cytoplasmic clusters located at the cell quarters. Deletion of PpfA disrupted the positioning of the cytoplasmic chemosensory clusters and the resulting strain is only weakly chemotactic.

More recently, Waldor and co-workers have shown that the ParA-like ATPase ParC actively promotes the polar localization of chemotaxis cluster II in *Vibrio cholerae* (Ringgaard *et al.*, 2011). *V. cholerae* is monotrichous, with a single flagellum localized to the old cell pole. Fluorescent protein fusions to components of *V. cholerae* chemotaxis cluster II (CheW1, CheY3 and ParC) demonstrated that the localization of these proteins is cell-cycle dependent. In newly divided cells the three proteins are found at the old, flagellated, pole. As the cells grow, ParC appears at the new pole, followed subsequently by CheW1 and CheY3, ensuring that each daughter cell has a chemosensory cluster on cell division. Deletion of ParC leads to cells with mislocalized chemotaxis proteins and reduced chemotactic ability. It is likely that ParA homologues play an important role in the chemotaxis of a variety of bacterial species; in particular a number of γ-proteobacteria encode a ParA homologue within their chemotaxis operon (Ringgaard *et al.*, 2011). Phylogenetic comparison of ParA homologues encoded in a *parAB* locus with those located in chemotaxis operons showed that the proteins could be grouped into three different clades. ParA homologues that are encoded in a *parAB* locus on the chromosome are clustered in clade I, whilst clade III is comprised of orphan ParA homologues encoded in chemotaxis operons. Interestingly, however, *R. sphaeroides* PpfA is a member of clade II which contains both plasmid encoded ParA homologues and ParA homologues encoded in chemotaxis operons. Thus chemotaxis operons encode two differing sets of ParA homologues, potentially differing in their ability to localize membrane bound or cytoplasmic chemosensory clusters (Ringgaard *et al.*, 2011).

Although PpfA and ParC have been shown to play an important role in chemotaxis the molecular mechanism through which localization is achieved is not yet understood. ATPase activity is, however, essential for the ability of each protein to promote correct localization of chemotaxis proteins indicating that localization is an active process. Why is the localization of chemosensory clusters an active process in some bacteria and a stochastic process in others? The clade III ParA homologues are found in monotrichous bacteria whose flagellum is located at the cell pole. Waldor and co-workers hypothesize that the targeted localization of the chemosensory proteins to the cell pole in such organisms is important in coupling the chemotactic response spatially with the flagellum. In *E. coli*, which is peritrichously flagellated, the arrangement of chemosensory proteins resulting from a stochastic assembly process would also facilitate a rapid chemotactic response by minimizing the diffusion time of CheY-P (Ringgaard *et al.*, 2011).

Chemoreceptor function – propagation of structural changes in signalling

Once a chemoreceptor has bound a cognate ligand it must transduce this event to the signal output domain that interacts with CheW and CheA. A two-state model for the mechanism by which chemoreceptors transduce signal has been proposed and is supported by a wide range of mutational, structural and biochemical data (detailed reviews include Hazelbauer *et al.*, 2008; Hazelbauer and Lai, 2010; Parkinson, 2010). Although it is most likely an oversimplification, the model provides a conceptually accessible basis for understanding chemotactic signal transduction. In this model it is held that the chemoreceptor complex can exist in either of two conformational states (each of which would in fact most probably be an ensemble of related conformations) that are in perpetual dynamic equilibrium. One of these, termed the kinase-on state, is competent to activate the kinase activity of associated CheA molecules whereas the kinase-off state inhibits CheA autophosphorylation. At any given time the activity of an associated CheA protein would be modulated by the degree to which each of the two states of the chemoreceptor is populated, which in turn would be influenced by the state of occupancy of the signal reception domain (i.e. whether it is associated with chemoattractant or chemorepellent or is unoccupied). In *E. coli*, where CheY-P promotes tumbling, chemoattractants would be expected to stabilize the kinase-off state, giving rise to smooth swimming.

Switching of the receptor between these two states is thought to begin with conformational changes in the ligand-binding domain being propagated through the membrane. Such propagation is transduced by a piston-like movement of the transmembrane helices by no more than 2 Å towards or away from the cytoplasm (reviewed in detail in Falke and Hazelbauer, 2001). In particular the second transmembrane helix in each monomer (TM2) is thought to play an important role in signalling. There are a wealth of data supporting this model. For example, cysteine cross-linking studies have demonstrated that it is possible to lock the receptor in the kinase-on or kinase-off state by cross-linking TM1 (the first transmembrane helix in each monomer) and TM2, and that alteration of the rate of cross-linking between TM1 and TM2 is consistent with the sliding of TM2 towards the cytoplasm relative to TM1 (Chervitz and Falke, 1995; Beel and Hazelbauer, 2001). In addition, it has been shown that ligand-induced conformational changes in TM2 are reversed by receptor methylation (Lai *et al.*, 2006). The position of transmembrane helices in the lipid bilayer is determined by the distribution of charged, aromatic, and hydrophobic residues along the helix, with charged residues typically found at the membrane–water interface and aromatic residues at the polar–hydrophobic interface (de Planque and Killian, 2003). In particular, two aromatic residues in TM2 have been shown to play an important role in determining its position in the lipid bilayer. Movement of the pair of aromatic residues positioned at the C-terminal end of TM2 towards or away from the cytoplasm simulates attractant and repellent bound states respectively (Draheim *et al.*, 2006). Similarly introduction of charged residues in the interfacial region can stabilize the on- or off-state depending upon the position of the mutation (Miller and Falke, 2004). The transmembrane region of the chemoreceptors is likely to adopt (at least) two major conformations, with some protein–lipid interactions favouring the first state and others the second, such that small perturbations in the ligand-binding domain lead to a change in the conformation of transmembrane helices.

The HAMP signal-conversion module forms a four-helix bundle with two helices contributed by each monomer (AS1 and AS2 from the first monomer, AS1′ and AS2′ from the second). The piston-like movement of the TM2 is expected to induce rearrangements in the HAMP domain via the AS1/AS1′ helices, inducing a repacking of the four-helix bundle and culminating in shifts in the AS2/AS2′ helices that are connected to the kinase control module. Two different models have been developed for the nature of this conformational change. The first was proposed on the basis of the structure of an archaeal HAMP domain in which the interhelical interactions have *x-da* ‘knob-on-knob’ packing rather than the more usual *a-d* ‘knob-in-hole’ packing. Rotation between these two packing modes was hypothesized to lead to signal transduction (Hulko *et al.*, 2006). An alternative model, supported by mutational and structural data, proposes that there are two sets of HAMP conformations, HAMP(A) which is tightly packed and HAMP(B) which is more loosely packed and in which the register of AS2 and AS2′ differ to HAMP(A). Changes in TM2 conformation would shift the population of HAMP(A) and HAMP(B) conformations and the changes in AS2–AS2′ packing would be propagated to the rest of the structure (Zhou *et al.*, 2009; Parkinson, 2010; Zhou *et al.*, 2011).

Signals received at the ligand-binding domain are ultimately transduced to the kinase control module. Insight into the mechanism of signal propagation through this region has been derived from mutational and structural analyses (see, for example, Starrett and Falke, 2005; Coleman *et al.*, 2005; Winston *et al.*, 2005; Swain *et al.*, 2009; Pollard *et al.*, 2009). Falke and co-workers have proposed the ‘yin–yang’ model for signal propagation in which signals are transmitted from the adaptation region to the kinase control region via changes in interhelical packing. In this model the packing of the adaptation and kinase control modules are antisymmetrically coupled such that there can be strong packing interactions in the adaptation region or the kinase control module but not in both regions simultaneously. Strong inter-helical packing interactions in the kinase control region stabilize the kinase-off state; conversely, kinase activity is stimulated when there is tight packing of the adaptation region (Swain *et al.*, 2009). An earlier study analysed the effects of methylation on the adaptation region. In studies of the adaptation region, mutations of glutamate to aspartate mimic the unmethylated state and those to glutamine the methylated state. Negatively charged residues in this region destabilize the local structure, as demonstrated by faster interhelical disulfide formation rates in cysteine cross-linking studies when the four glutamate/glutamine residues in Tar are mutated to aspartate. Methylation of charged residues at the dimer interface reduces interhelical repulsion leading to stronger packing (Starrett and Falke, 2005). The structure of a soluble chemoreceptor from *T. maritima* provides further insights into the mechanisms through which changes in helical packing may be propagated along the length of the kinase control module. The structure shows asymmetry between the two monomers, with one subunit being curved throughout its length and the other being kinked at a bulge in the adaptation region. When compared with the more symmetrical structures of the cytoplasmic domains of Tsr and Tm1143 it can be seen that the net result of this bulge is to shift the kinase control region relative to the flexible bundle region (the position of the kinase control region differs by ~25 Å between the two structures when the adaptation regions are aligned). Thus a change in packing in the adaptation region has been shown to give rise to a significant structural displacement in the kinase control region (Pollard *et al.*, 2009).

Chemoreceptor function – the importance of higher order structures

In vivo, chemoreceptor complexes assemble into higher-order clusters that are stable on the timescale of chemotactic signalling. There may be a number of reasons why the chemoreceptors are organized into large complexes. By localizing the components of a signalling pathways within a defined region of space the efficiency of signal transduction might be increased simply because the components do not have to travel far to interact with each other. Sensory clusters have been demonstrated to facilitate considerable cooperativity between the receptors within the cluster thereby contributing to the high sensitivity and signal amplification that are crucial for effective bacterial chemotactic signalling. The importance of cooperativity in the ability of chemoreceptors to sense very small concentrations of attractant, in addition to responding to concentrations varying by over five orders of magnitude, was first demonstrated *in silico* by Bray and co-workers (Bray *et al.*, 1998; Duke and Bray, 1999). One advantage of clustering is the capacity for integration of signals from different receptors. Cross-linking studies on *E. coli* expressing either Tsr and Tar or Tsr and Trg showed that mixed populations of receptors cluster together in trimers of dimers (Studdert and Parkinson, 2004). Using FRET to measure the concentration of CheY-P in cells where the composition of the receptor population was varied demonstrated that there was a strong effect of receptor homogeneity on both the cooperativity and sensitivity of the chemotactic response (Sourjik and Berg, 2004). In cells where both Tsr and Tar were expressed at native levels kinase activity was completely inhibited by serine, but only partially by an aspartate analogue. In contrast, when Tsr was deleted addition of the aspartate analogue inhibited kinase activity with the response showing both a greater sensitivity and cooperativity. A Monod-Wyman-Changeux-type model has been used to analyse *in vivo* FRET data to determine the number of receptors in a signalling team. In this model tight coupling between members of a team ensures concerted switching between kinase-on and kinase-off states. This approach has shown that 10–20 dimers are coupled together and that the size of the signalling teams is dependent upon the modification state of the receptors (Sourjik and Berg, 2004; Mello and Tu, 2005; Keymer *et al.*, 2006; Endres *et al.*, 2008). In cells expressing only the receptor Tar, the size of the signalling teams was 2- to 3-fold larger in a mutant simulating the kinase-off state than in one simulating the kinase-on state (Endres *et al.*, 2008), suggesting an additional level of sophistication in the adaptation process. The all-or-nothing model for receptor interaction is inconsistent, however, with the observation that methylation within a heterogeneous cluster is ligand specific. Bray and co-workers have proposed a model for the mechanism by which receptor specific methylation can occur (Goldman *et al.* 2009). In their model cooperative interactions occur between CheA molecules rather than between chemoreceptors thus uncoupling signal amplification and adaptation. Each ligand molecule causes a conformational change in a single chemoreceptor leading to the activation of CheA molecules directly associated with that chemoreceptor. Active CheA molecules are then proposed to enhance the activity of neighbouring CheA molecules. As the chemoreceptors associated with neighbouring CheA molecules do not changed conformation their methylation state does not change. An alternative model where receptors adapt locally but respond cooperatively has been developed by Tu and co-workers. This model was able to predict the relative methylation dynamics in a mixed population of receptors and has important implications for sensing in a environment with multiple chemoeffectors. Local adaptation prevents the 'poisoning' of the array by methylation of all receptors on exposure to an extreme level of a particular stimulus (Lan *et al.*, 2011).

Signalling to the flagellar motor

Poised at the output domains of the chemoreceptors, CheA responds to changes in chemoeffector concentrations. The majority of CheAs, referred to as classical CheAs, share a common domain architecture. These CheAs form a homodimer whose monomers are composed of five distinct structural domains designated P1 to P5. The P4 domain displays kinase activity, transferring a phosphoryl group to a conserved histidine residue in the P1 domain (Swanson *et al.*, 1993). The P2 domain is not essential for chemotaxis, but binds CheY and CheB increasing their concentration in the vicinity of the P1 domain thus increasing the rate of phosphor transfer to the RRs (Stewart *et al.*, 2000; Jahreis *et al.*, 2004). The P3 domain is responsible for dimerization of CheA monomers; such interactions are vital for kinase activity (Surette *et al.*, 1996). The P5 domain interacts with both CheW and the chemoreceptors localizing CheA to the chemoreceptor array (Miller *et al.*, 2006; Park *et al.*, 2006a; Zhao and Parkinson, 2006).The P5 domain also couples chemoreceptor signalling to the P4 domain kinase activity, thereby transducing external signals into an intracellular signal (Bourret *et al.*, 1993). The P5 domain is homologous to the adaptor protein CheW (Bilwes *et al.*, 1999), suggesting that CheW may have evolved from CheA enhancing the ability of the kinase to couple simultaneously to multiple different chemoreceptors (Parkinson *et al.*, 2005).

CheA kinase activity is highly sensitive to changes in receptor conformation induced by chemoeffector binding. In some bacteria, including *E. coli*, a decrease in the concentration of attractant leads to stimulation of kinase activity (Borkovich *et al.*, 1989). Conversely, in bacteria such as *B. subtilis* an increase in the concentration of attractant suppresses the kinase activity of CheA; CheY-P has an opposite effect in *B. subtilis* to that seen in *E. coli*, causing the direction of rotation of the flagellar motor to change from clockwise to counterclockwise; in *B. subtilis* an increase in the concentration of CheY-P leads to less frequent tumbling (Bischoff and Ordal, 1992). In either case, activation of the kinase ultimately results in phosphotransfer to one of the CheA response regulators, of which CheY and CheB are the most widespread.

Unlike many of the response regulators seen in canonical two-component systems, CheY is a single domain protein comprising only the receiver component of the archetypal response regulators. A compact, 14 kDa, protein CheY contains a conserved aspartate residue located at the end of a central five-stranded β-sheet flanked by five α-helices (Stock *et al.*, 1989; Volz and Matsumura, 1991). CheA-P is a substrate for the Mg^{2+} dependent phosphotransfer to the conserved aspartate. The structure of CheY-P has yet to be determined because of the rapid autodephosphorylation rate of CheY-P; NMR and crystal structures are, however, available for BeF_3^--activated CheY and phosphono-CheY

(Cho *et al.*, 2000; Halkides *et al.*, 2000; Lee *et al.*, 2001a). Comparison of the active and inactive forms of CheY reveals that, outside loop regions, only modest changes in the backbone conformation occur on activation. One of the key structural changes occurs at the position of Tyr106 which in apo-CheY can adopt one of two rotameric states, internal or external. In the inactive conformation of CheY Tyr106 adopts an external conformation whereas in the activate conformation structural changes involving Lys109, Thr87 and the β4-α4 loop region favour the adoption of an internal conformation by Tyr106 (Fig. 9.5). In the crystal structure of CheY in complex with CheA, the external conformation of Thr106 appears to be stabilized by hydrogen bonding to a glutamate residue (Welch *et al.*, 1998). Phosphorylation of CheY could therefore reduce the affinity for CheA of CheY-P relative to CheY by destabilization of the external conformation of Tyr106. Following dissociation from CheA, CheY-P diffuses through the cytoplasm where it interacts with the switch complex of the flagellar motor. CheY-P has been shown to bind FliM with ~20 fold greater

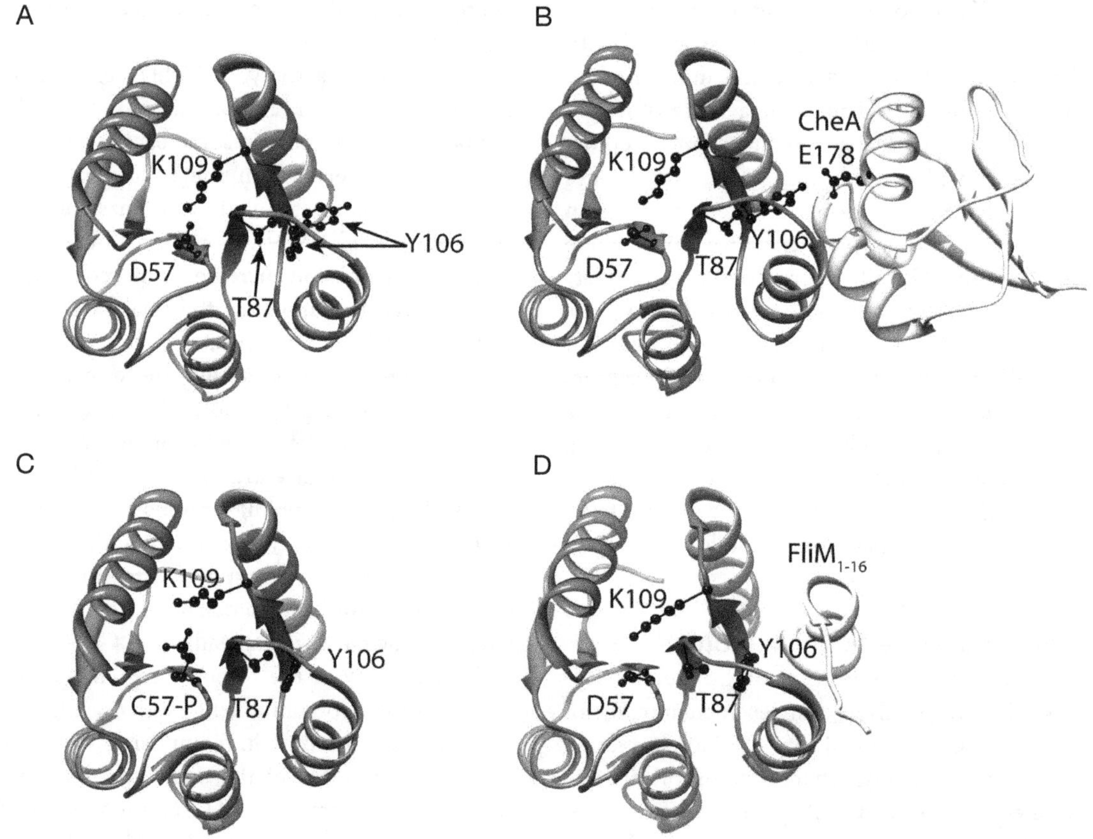

Figure 9.5 Conformational changes occurring on phosphorylation of CheY. The structures of CheY(-P) solved under a variety of conditions are illustrated with the CheY backbone represented as dark grey ribbons, the backbone of binding partners as light grey ribbons and with mechanistically important side-chain residues shown in black. (a) apo-CheY, (b) CheY in complex with the P2 domain of CheA, (c) phosphono-CheY, and d) CheY-BeF_3^- in complex with $FliM_{1-16}$. In the structure of apo-CheY Tyr106 adopted two rotameric states, internal and external (a), whilst in the activated structure the internal conformation is favoured (c). The bias in conformation towards the internal rotamer is thought to play an important role in the increased affinity of CheY-P compared with CheY for $FliM_{1-16}$ (d), and also the reduced of CheY-P for affinity for CheA (b). Images were generated in UCSF chimera using PDB codes 3CHY, 1F4V, 1C4W and 1A0O (Volz and Matsumura, 1991; Welch *et al.*, 1994; Halkides *et al.*, 2000; Lee *et al.*, 2001a; Pettersen *et al.*, 2004).

affinity than CheY (Welch *et al.*, 1994). Mutational analysis mapped these interactions to the N-terminal 16 residues of FliM ($FliM_{1-16}$) (Bren and Eisenbach, 1998). The crystal structure of BeF_3^-–CheY in complex with $FliM_{1-16}$ confirmed that the interaction interface of CheY is primarily composed of residues from the α4-β5-α5 face, as had been suggested by genetic and physiological analysis (Lee *et al.*, 2001b). The structure again highlighted the importance of conformational changes of Tyr106 which, in inactive CheY would be expected to sterically interfere with FliM binding. More recently, it has been shown that *E. coli* CheY-P also binds to FliN. It is proposed that FliM 'catches' CheY-P and that switching of the flagellar motor is triggered by interaction of CheY-P with FliN (Sarkar *et al.*, 2010). The effect of CheY-P binding to the flagellar motor varies between species as described previously and the molecular mechanisms for switching/slowing/stopping of the flagellar motor are not fully understood (Sowa and Berry, 2008).

Phosphorylation is not the only covalent modification undergone by CheYs. It has been shown that *E. coli* CheY can be acetylated at six different lysine residues and that mutations that impair CheY acetylation *in vivo* result in a chemotaxis-defective phenotype (Barak and Eisenbach, 2001; Barak *et al.*, 2004). Acetylation occurs through two separate mechanisms, both of which are important *in vivo*. The lysine residues can either be acetylated through the action of acetyl coenzyme A synthetase, using acetate or acetyl coenzyme A as a substrate, or through autoacetylation using acetyl coenzyme A as the donor of acetyl groups (Barak *et al.*, 2004, 2006). The lysine residues that can be acetylated lie in the region of CheY that is important for binding to CheA, CheZ and FliM. It has recently been shown that when CheY is acetylated through non-enzymatic methods *in vitro* to a level similar to that found *in vivo* (approximately one acetyl group per CheY) binding to CheY, CheZ and FliM is impaired (Liarzi *et al.*, 2010). The authors propose a model where the regulation of CheY occurs on two timescales. Phosphorylation occurs on a rapid timescale modulated by the presence of chemoeffector molecules. In contrast, the proportion of CheY molecules competent for phosphorylation at any one time would depend on the degree of acetylation and hence the metabolic state of the cell. In order for acetylation to be an effective regulatory mechanism it would be expected that there would be a mechanism for deacetylation. It has recently been reported that acetylated CheY is a substrate for the NAD^+-dependent deacetylase CobB *in vitro* and that *cob* deletion strains exhibited reduced chemotaxis *in vivo*; this suggests that CobB may regulate chemotaxis through deacetylation of CheY (Li *et al.*, 2010).

Integration of signals from more than one chemotaxis pathway – *Rhodobacter sphaeroides*

Many bacterial species possess multiple homologues of the chemosensory proteins seen in *E. coli*, in addition to other proteins and different network architectures. Recently, phylogenomic classification has been performed for the components of chemosensory networks from 245 different species. The signalling pathways involving chemotaxis proteins in these species can first be divided into three groups based on their function, those that affect flagellar motility (Fla), those that are involved in control of motility utilizing type-IV pili (TFP), and those whose functions do not involve motility (alternative cellular function or ACF). The Fla group can then be further subdivided into 17 classes with differing network architecture; such classification will facilitate the prediction of chemotactic behaviour in species that have not been studied previously (Wuichet and Zhulin, 2010).

In the study of chemosensory pathways there are a number of model organisms which express multiple homologues of the core conserved proteins, of which one of the most well-characterized species is *R. sphaeroides* (Fig. 9.6) (Porter *et al.*, 2011a). This purple non-sulfur bacterium shows great metabolic versatility and, as such, displays chemotaxis towards a wide range of different stimuli including organic acids such as propionate and succinate, some sugars, oxygen, and light. *R. sphaeroides* has three chemotaxis operons, *che*Op1–3, in addition to genes encoded outside operons, and two sets of genes encoding flagella (Mackenzie, 2001; Porter *et al.*, 2011b). The genes encoded by *che*Op1 control the rotation

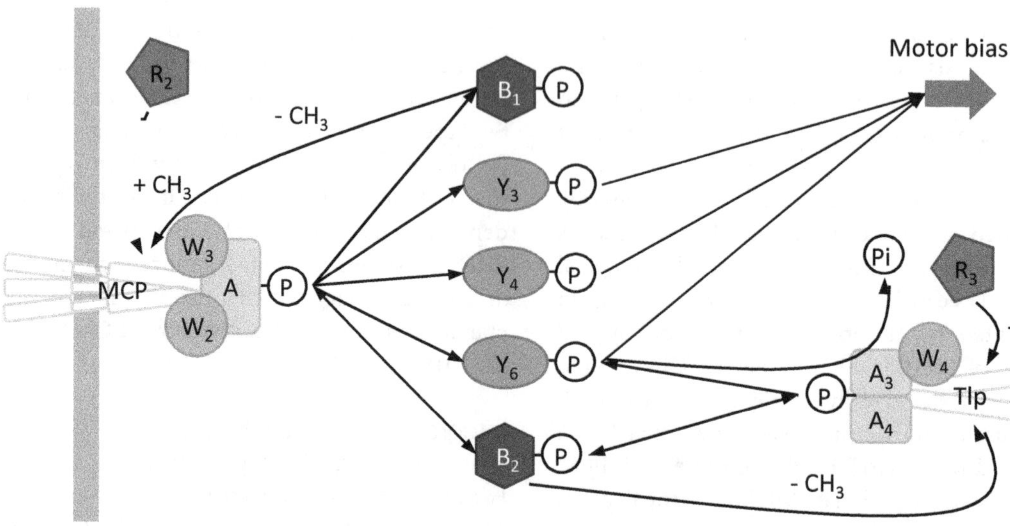

Figure 9.6 The chemosensory network of *R. sphaeroides*. Control of the Fla1 flagellum of *R. sphaeroides* requires two sets of chemosensory genes; one membrane bound at the pole of the cell and the other forming a discrete cluster in the cytoplasm. In response to external signals the polar chemosensory cluster can phosphorylate all of the response regulators. In contrast, the cytoplasmic chemosensory cluster, thought to respond to changes in the metabolic state of the cell, can only phosphorylate a subset of the response regulators. Integration of signals between the two clusters is however possible through the shuttling of phosphoryl groups from the cytoplasmic kinase to the polar kinase by $CheB_2$.

of the multiple unipolar flagella encoded by the *fla*2 genes (del Campo *et al.*, 2007). As neither *che*Op1 nor *fla*2 genes are expressed under laboratory conditions, the *fla*2 system has been less well characterized than that of *fla1*. Recently however, mutants expressing the genes for Fla2 flagella have been isolated (Poggio *et al.*, 2007). The genes encoded by *che*Op2 and *che*Op3 control the rotation of a single, sub polar flagellum encoded by the *fla*1 genes and are essential for chemotaxis under laboratory conditions (Shah *et al.*, 2000; Porter *et al.*, 2002). They comprise one chemoreceptor, three CheAs, three CheWs, two CheRs, two CheBs and three CheYs. In addition there are three chemoreceptors known as Tlps which lack transmembrane regions. Tlps are hypothesized to be important in sensing the metabolic state of the cell and it has been demonstrated that metabolism of some chemoeffectors is required for a chemotactic response to those chemoeffectors (Poole and Armitage, 1989; Poole *et al.*, 1993; Jeziore-Sassoon *et al.*, 1998).

A key feature of chemotaxis in *R. sphaeroides* is the spatial separation of the components necessary for a chemosensory response (Fig. 9.7). The proteins encoded by *che*Op2 and *che*Op3 form two networks whose integrated response controls flagellar rotation (Porter *et al.*, 2002). These networks localize to distinct regions of the cell (Wadhams *et al.*, 2003). Transmembrane chemoreceptors localize to the poles of the cell along with $CheA_2$, $CheW_2$, $CheW_3$ and $CheR_2$ all of which are encoded by *che*Op2. An additional cluster of chemosensory proteins forms in the cytoplasm of the cell. This cluster is composed of the Tlps, and the *che*Op3 encoded $CheA_3$, $CheA_4$, $CheW_4$ and $CheR_3$. As is the case for the polar chemosensory arrays the cytoplasmic cluster contains thousands of Tlps and associated proteins, however the architecture of the cytoplasmic cluster is not known.

The two clusters differ in their specificity for the chemotaxis RRs (Porter and Armitage, 2002). The polar cluster senses signals arising in the periplasm and controls the autophosphorylation rate of $CheA_2$; $CheA_2$ has the domain architecture of a classical CheA and is able to phosphotransfer to the three CheYs ($CheY_3$, $CheY_4$ and $CheY_6$) and both CheBs ($CheB_1$ and $CheB_2$). In contrast, the cytoplasmic cluster

A
B

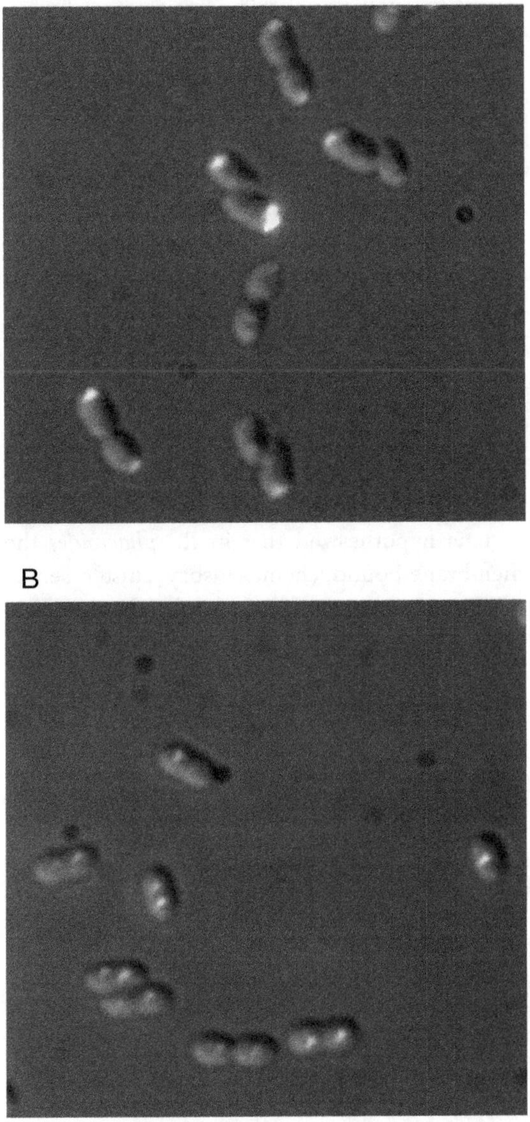

Figure 9.7 The two chemosensory clusters of *R. sphaeroides* are located in different regions of the cell. (a) A fluorescence image showing yellow fluorescent protein tagged $CheA_2$ localizing to the polar chemosensory cluster. (b) A fluorescence image showing yellow fluorescent protein tagged $CheA_4$ localizing to the cytoplasmic chemosensory cluster.

senses intracellular signals and the kinase is able to phosphorylate only a subset of the expressed response regulators. There are two CheAs at the cytoplasmic cluster, $CheA_3$ and $CheA_4$, neither of which contain the full complement of domains necessary for autophosphorylation (Porter and Armitage, 2004). $CheA_3$ contains P1 and P5 domains separated by a 794 amino acid linker which has no homology to any known domain but contains a $CheY_6$ specific phosphatase (Porter *et al.*, 2008). $CheA_4$ consists only of P3, P4 and P5 domains. These proteins work together with the P4 domain of $CheA_4$ receiving a phosphoryl group from ATP that is then transferred to the P1 domain of $CheA_3$. The $CheA_3$/$CheA_4$ kinase specifically transfers phosphoryl groups to $CheY_6$ and $CheB_2$. The reason for the separation of the kinase and phosphotransfer domains between two separate proteins is unknown. One hypothesis is that, since $CheA_3$ contains a phosphatase region, separation of the functions between two proteins allows independent regulation of kinase and phosphatase activity.

What is the role of the multiple CheY homologues in *R. sphaeroides* chemotaxis? It has been shown that $CheY_6$ is indispensable for chemotaxis but that there is functional redundancy between $CheY_3$ and $CheY_4$ (Porter *et al.*, 2006). In the absence of all chemotaxis proteins the *R. sphaeroides* flagellar motor does not stop (Pilizota *et al.*, 2009). $CheY_6$-P is able to induce pauses in the rotation of the motor; but in the absence of one of $CheY_3$ or $CheY_4$ the cells are non-chemotactic. $CheY_6$ has been shown to be diffuse throughout the cytoplasm whilst $CheY_3$ and $CheY_4$ associate partially with both the polar and cytoplasmic clusters with the remaining protein delocalized. Control theory analysis has been employed to probe the possible functions of these proteins. Experimental predictions were made based on four models for the role of $CheY_3$ and $CheY_4$. In these models $CheY_3$ and $CheY_4$ either act solely as a phosphate sink, compete with $CheY_6$ for binding to the flagellar motor or modulate the properties of the cytoplasmic cluster (Roberts *et al.*, 2009). Three of the models failed to adequately describe the experimental data; however, one model, in which $CheY_3$-P and $CheY_4$-P are able to associate with the cytoplasmic cluster modulating kinase activity of $CheA_3$/$CheA_4$, has not yet been invalidated. If this model was correct then signals received and transduced by the polar cluster would be able to influence the response from the cytoplasmic cluster, thereby providing a degree of cross-talk between the two chemotactic systems. Although this model is consistent with the available experimental evidence, it is based on computational studies that trialled only four models; it therefore awaits further experimental support or invalidation.

There is other evidence that the cytoplasmic and polar chemosensory clusters do not partake in a one-way conversation. Ordinary differential equation models of the *R. sphaeroides* chemotactic phosphorylation network formulated on the basis of available experimental data, predicted that phosphotransfer may occur between the two types of sensory cluster (Tindall *et al.*, 2010). In this model, which is supported by experimental data, phosphoryl groups can be transferred from $CheA_3$-P, in the cytoplasmic cluster, to $CheA_2$, localized to the transmembrane cluster, through the intermediary $CheB_2$-P. $CheA_2$ has the capacity to transfer phosphoryl groups to any of the *R. sphaeroides* chemotactic response regulators. This relay would allow $CheA_3$ to indirectly signal to its non-cognate response regulators ensuring integration of signals from the metabolic state of the cell arising from the cytoplasmic chemotaxis cluster and the level of chemoeffectors detected by the polar cluster chemoreceptors.

Within any chemotactic network that possesses multiple homologues of the canonical chemotaxis proteins there is immediately the potential for increased sophistication through interaction specificity (Hamer *et al.*, 2010). The chemosensory pathway of *R. sphaeroides* shows specificity of localization of the components to the appropriate chemosensory cluster and also specificity in phosphotransfer from the CheAs to the RRs. Domain-swapping experiments showed that the P5 domain of each of the CheAs is sufficient to determine to which cluster the protein localizes (Scott *et al.*, 2010). When the P5 domain of $CheA_2$ is replaced with the P5 domain from either $CheA_3$ or $CheA_4$ the resulting protein localizes to the cytoplasmic cluster with the converse also being true. Similarly, the P1 domain contains all of the determinants necessary to specify to which response regulators phosphotransfer can take place (Scott *et al.*, 2010). Insight into how such specificity might be achieved came from a co-crystal structure of $CheA_3$-P1 domain and $CheY_6$, which revealed Met13 of $CheY_6$ to project out of the α1 helix of this protein into a hydrophobic pocket of CheA3-P1 (Bell *et al.*, 2010). This methionine residue is replaced by a serine residue in all the CheY and CheB homologues not cognate to $CheA_3$. Mutagenesis of Met13 and the two adjacent residues in $CheY_6$ abrogates phosphotransfer to the mutant construct from $CheA_3$. Conversely, non-cognate RRs into which Ala12 and Met13 from $CheY_6$ have been engineered can be efficiently phosphorylated by $CheA_3$-P. This demonstrates the molecular principles on which specificity within the chemotactic signalling network might be achieved, revealing subtle modifications of just a few amino acid residues to be pivotal in determining the specificity of intermolecular signalling networks.

It is hypothesized that in *R. sphaeroides* the membrane-bound chemosensory cluster senses periplasmic chemoeffectors whilst the cytoplasmic cluster responds to changes in the metabolic state of the cell. It is not fully understood, however, why the integration of signals from these two separate clusters is necessary for the chemotactic response in *R. sphaeroides*. Other bacteria also encode soluble chemoreceptors, many of which are presumed to detect changes in cytoplasmic chemoeffectors, however these soluble chemoreceptors localize to the membrane-bound chemosensory cluster (see for example Bardy and Maddock, 2005; Meier and Scharf, 2009). Localization of multiple chemoreceptors to a single array facilitates integration of signals from a variety of chemoeffectors. *R. sphaeroides* exhibits great metabolic diversity, it is possible that the presence of two chemosensory clusters is important in ensuring that the chemotactic response is appropriate for the current growth conditions or metabolic need, perhaps by giving greater weight to cytoplasmic chemoeffectors than if all of the chemoreceptors were localized to a single array.

Signal termination

Whilst signal termination is an essential event in bacterial chemotaxis it is one of the less conserved aspects in terms of the molecular details of its implementation. Chemoeffector dissociation can terminate the signal being relayed into the cell, as can covalent modification of the receptor by the adaptation apparatus. The rapidity of phosphotransfer from CheA-P to the RRs ensures prompt termination of the histidine kinase signal (Hess *et al.*, 1988; Stewart, 1997), in marked contrast to the stable phosphorylation of receptors often employed in eukaryotic signalling. Diversity in signal termination arises through the mechanisms utilized to cease signalling by CheY-P and hence ensure close correspondence between CheY-P levels and chemoreceptor occupancy. A significant feature of signal termination in chemotaxis is the localization of CheY-P phosphatases. In most bacterial species it appears that at least one phosphatase is localized to the site of CheY phosphorylation. Simulations in which the cellular localization of *E. coli* CheZ was varied predicted that localization of CheZ to the chemoreceptor array would flatten the intracellular gradient of CheY-P thereby ensuring that all flagellar motors experienced a similar concentration of CheY-P. In contrast, simulations with diffuse CheZ showed an exponential gradient of CheY-P radiating from the cell pole which would lead to flagellar motors further from the receptor cluster experiencing a reduced stopping bias (Lipkow *et al.*, 2005). These are in agreement with studies utilizing FRET imaging of single cells (Vaknin and Berg, 2004).

Signal termination through CheY-P-specific phosphatase proteins

The autodephosphorylation half-life of *E. coli* CheY-P is approximately 20 seconds, longer than the duration of the entire chemosensory response. In the presence of CheZ however, the half-life of CheY-P is significantly reduced (Hess *et al.*, 1988). *E. coli* CheZ is a dimer comprised of two 214 residue monomers the core of which consists of two long helical hairpins interacting to form a four-helix bundle (Fig. 9.8) (Zhao *et al.*, 2002). X-ray crystallography of the CheY–CheZ complex showed there to be two interaction regions between CheZ and CheY, one located at the midpoint of the helical bundle core and containing the catalytic residues for dephosphorylation, and the second at the C-terminus of CheZ (the region connecting the core to the C-terminal helix was not resolved in the structure). The C-terminus of CheZ shows significant

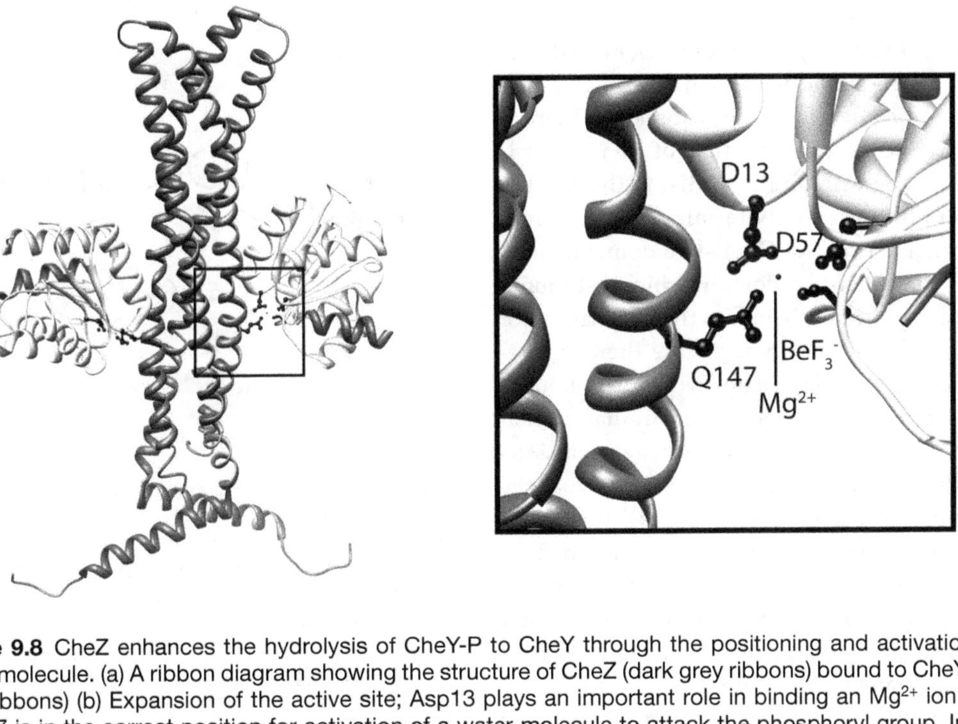

Figure 9.8 CheZ enhances the hydrolysis of CheY-P to CheY through the positioning and activation of a water molecule. (a) A ribbon diagram showing the structure of CheZ (dark grey ribbons) bound to CheY (light grey ribbons) (b) Expansion of the active site; Asp13 plays an important role in binding an Mg^{2+} ion whilst Gln147 is in the correct position for activation of a water molecule to attack the phosphoryl group. Images were generated in UCSF chimera using PDB code 1KMI (Zhao *et al.*, 2002; Pettersen *et al.*, 2004).

sequence identity to $FliM_{1-16}$ and interacts with a similar region of CheY. Together these two regions of CheZ serve to 'clamp' CheY in place in the complex. There had been some debate as to the mechanism by which CheZ enhanced the dephosphorylation of CheY-P. From the crystal structure of the CheY–CheZ complex in the presence of the phosphate analogue BeF_3^- it was proposed that residue Gln147 of CheZ enhances CheY-P hydrolysis by positioning and activating a water molecule for attack on the phosphoryl group (Zhao *et al.*, 2002).

CheZ is found only in β- and γ-proteobacteria, many other bacteria utilize members of the CheC phosphatase family for signal termination. The CheC family can be divided into three subgroups, CheC, CheX and FliY (reviewed in Muff and Ordal, 2008). The three subgroups share a conserved sequence motif defining the phosphatase active site but show a low level of sequence similarity in other regions. They are, however, structurally related with each protein containing at least one CheC-like domain (Park *et al.*, 2004, 2006b). The structures of CheC and CheX from *T. maritima* have been determined by X-ray crystallography (Park *et al.*, 2004). CheC and CheX share a similar fold (with changes in some secondary structural elements) but CheC is monomeric, whilst CheX forms a dimer composed of two CheC-like domains. There is currently no crystal structure of FliY but it is thought to be structurally similar to the flagellar protein FliM, with an N-terminal CheY-P binding domain, a central CheC-like domain, and a C-terminal FliN-like domain which is thought to localize the phosphatase to the flagellar motor (Park *et al.*, 2006b). In each case the enhancement of CheY-P dephosphorylation is thought to occur through a mechanism similar to that of CheZ with a conserved asparagine residue orienting a water molecule for attack on the phosphoryl group (Park *et al.*, 2004). The CheC, CheX and FliY families are thought to have evolved from a common ancestor, but now play different roles in chemotaxis. FliY and CheX are most important as phosphatases for CheY-P, whilst many CheC proteins are important in adaptation or in pathways other than chemotaxis (Muff and Ordal, 2008).

Signal termination through phosphate sinks

An alternative mechanism for signal termination is exemplified by the α-proteobacterium *S. meliloti* (Schmitt, 2002). The chemosensory network of *S. meliloti* is similar to that of *E. coli* but lacks a CheZ phosphatase and has two CheYs: $CheY_1$ and $CheY_2$ (Fig. 9.9) (Greek *et al.*, 1995). Both $CheY_1$ and $CheY_2$ can be phosphorylated by CheA, but only $CheY_2$ is capable of interacting with the flagellar motor and inducing the change in speed of flagellar motor rotation necessary for reorientation in *S. meliloti* (Sourjik and Schmitt, 1996). The role of $CheY_1$ is to act as a phosphate sink; in the absence of stimulation of the kinase activity of CheA, phosphoryl groups from $CheY_2$-P can be transferred back to CheA and then to $CheY_1$ (Sourjik and Schmitt, 1998). This shifts the equilibrium away from the formation of $CheY_2$-P to the formation of $CheY_1$-P and hence

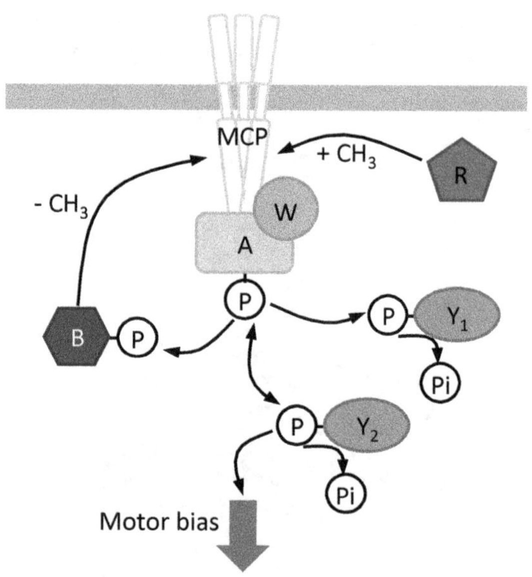

Figure 9.9 The chemosensory network of *S. meliloti*. The chemosensory network of *S. meliloti* does not include a CheY-P specific phosphatase. Instead there are two homologues of CheY, one of which acts as a phosphate sink. Upon stimulation of CheA kinase activity phosphotransfer occurs to both $CheY_1$ and $CheY_2$, however only $CheY_2$-P can bind to the flagellar motor. On termination of CheA kinase activity phosphoryl groups are transferred from $CheY_2$ to $CheY_1$ via CheA thus ensuring rapid reduction in the concentration of $CheY_2$-P.

leads to signal termination. Scharf and co-workers have analysed the molecular mechanism of the interaction of $CheY_1$ and $CheY_2$ with CheA (Riepl *et al.*, 2008). They have demonstrated that both $CheY_2$ and $CheY_2$-P interact with the P2 domain of CheA, but that $CheY_1$ and $CheY_1$-P do not. The binding sites for $CheY_2$ and $CheY_2$-P were distinct, but overlapping. As the sites overlap, this has the potential to favour the binding of one or other species dependent upon the state of the kinase; this will be particularly important given that, unlike in *E. coli*, phosphotransfer can occur in both directions.

Adaptation

The enzymes CheR and CheB, which respectively catalyse the methylation and demethylation of chemoreceptors, are found in all but one of the chemosensory network architectures identified by Zhulin and co-workers, and occur in ~90% of genomes with chemotaxis components (Wuichet and Zhulin, 2010). However, comparison of the adaptation systems of *E. coli* and *B. subtilis* reveals that homologues of the same protein can mediate adaptation in quite different ways.

CheB and CheR in *E. coli*

In *E. coli* receptors are methylated in response to the addition of attractant and demethylated in response to removal of attractant (Toews *et al.*, 1979). CheR catalyses the transfer of methyl groups from S-adenosyl methionine to specific glutamate residues at four principle sites in the kinase control module of chemoreceptors. In some cases these glutamate residues are translated as glutamine and subsequently deamidated by CheB, apparently to ensure that a new chemoreceptor is inserted into the membrane in a neutral signalling conformation (Kehry *et al.*, 1983). Analysis of the methylation patterns of receptors from *E. coli* and *S. typhimurium* showed that the consensus sequence for methylation sites in these organisms is Glx-Glx-X-X-Ala-Ser/Thr, with methylation occurring on the second Glx residue (Kehry and Dahlquist, 1982; Terwilliger and Koshland, 1984). The consensus sequence in other organisms is not necessarily identical to that seen for *E. coli*. For example, in *T. maritima* the consensus is Ala/Ser-sm-X-Glx-Glu-X-sm-Ala/Ser (where sm is any small amino acid) with methylation occurring primarily at the first Glx residue (Perez *et al.*, 2006). In *E. coli*, methylation of a chemoreceptor achieves two results; desensitisation of the receptor to chemoattractant, and enhancement of CheA kinase activity. As described previously, the dynamic nature of the kinase control module of chemoreceptors plays an important role in signalling. Neutralization of charged residues in the adaptation region of the chemoreceptor stabilizes the interface between homodimers and stimulates CheA kinase activity; a mutant where all eight methyl-accepting glutamate residues in the homodimer were neutralized by mutation to glutamine exhibited a locked 'kinase-on' phenotype (Starrett and Falke, 2005).

Adaptation is directly coupled to chemoeffector binding by at least two feedback mechanisms. The first mechanism is the stimulation of demethylation by CheB through phosphotransfer from CheA-P. CheB has two domains, an N-terminal response regulator domain and a C-terminal methylesterase domain, joined by a linker region (Djordevic *et al.*, 1998). Phosphorylation of the response regulator domain leads to a 100-fold increase in methylesterase activity (Lupas and Stock, 1989). In the crystal structure of unphosphorylated CheB the N-terminal domain blocks the active site of the methylesterase domain. Phosphorylation is thought to induce conformational changes that disrupt the interface between the two domains, a process in which the linker region plays an important role (Djordevic *et al.*, 1998; Anand *et al.*, 2000; Anand and Stock, 2002). The second feedback mechanism arises from the different selectivity of CheB and CheR with the former preferentially demethylating kinase-on chemoreceptors and the latter being selective for those in the kinase-off state (Stock and Koshland, 1981). The two feedback mechanisms ensure that CheA activity, and hence CheY-P levels, are reset after each stimulus.

In the *E. coli* system, both CheB and CheR interact with the membrane-associated sensory clusters through a conserved Asn-Trp-Glu-Thr-Phe pentapeptide located in the C-terminal sequences of the dominant chemoreceptors Tsr and Tar (Wu *et al.*, 1996; Barnakov *et al.*, 1999, 2002). CheR has a higher affinity for the pentapeptide than CheB, reflecting the differing roles of pentapeptide binding for these two enzymes. Binding of CheR increases the concentration of this enzyme in the vicinity of the chemoreceptors whilst binding of CheB leads to enhancement of methylesterase activity (Barnakov *et al.*, 2002). The flexible linker that connects the pentapeptide motif to the rest of the chemoreceptor is important for chemoreceptor function; systematic deletion of the Tar linker region showed that methylation and demethylation decreased as the number of residues deleted was increased *in vitro*, and that mutants showed either reduced or no chemotaxis *in vivo* (Li and Hazelbauer, 2006). The role of the linker is to extend the reach of a single adaptation enzyme to several nearby chemoreceptors creating an 'adaptational-assistance neighbourhood' that facilitates modification of glutamate residues in ~6 receptor dimers (Li and Hazelbauer, 2005). This allows a more graded mode of chemoreceptor modification and prevents saturation of the adaptation enzymes, contributing to the high precision of adaptation exhibited by bacterial cells over a wide range of chemoeffector concentrations (Endres and Wingreen, 2006).

CheB and CheR in *B. subtilis*

CheB and CheR from *E. coli* and *B. subtilis* share functional homology; CheRs from either organism can methylate receptors from both organisms *in vitro*, likewise CheBs from either organism can demethylate receptors from both organisms *in vitro* (Burgess-Cassler and Ordal, 1982; Nettleton and Ordal, 1989). Despite this, the roles of receptor methylation differ substantially between the two organisms. In *B. subtilis* rapid methanol release, corresponding to receptor demethylation, is observed on both the addition and removal of attractant (Kirby *et al.*, 1999). Following demethylation the receptors are remethylated on a slower timescale to almost prestimulus levels. The methylation pattern of the asparagine receptor McpB has been analysed in detail (Zimmer *et al.*, 2000). Three methylation sites were identified, positioned at Glu371, Glu630 and Glu637. Mutation of these residues to Aspartate, thereby preventing methylation, revealed that methylation at some sites decreased the activity of the receptors, methylation at others increased receptor activity, and that some residues appeared to be demethylated both upon the addition and removal of attractant. It has been proposed that selective modification of different combinations of glutamate residues might serve to specifically promote or inhibit the activities of chemoreceptors, although the molecular basis for the differential methylation of these sites is not fully understood (Zimmer *et al.*, 2000). Some insight into how methylation may alter kinase activity has, however, been proposed based on the structure of the cytoplasmic region of MCP_{1143} from *T. maritima* (Park *et al.*, 2006a; Rao *et al.*, 2008). In a homology model of McpB, based on the *T. maritima* structure, the three residues that are methylated cluster together at the dimer interface suggesting that charge–charge interactions could perturb the stability of this interface. Modification of these residues could potentially induce rotation of the helices relative to each other leading to changes in kinase activity.

More complex adaptation networks – *B. subtilis*

Whilst many bacterial species apparently utilize mechanisms akin that of *E. coli* for remembering their past environmental situations a multitude of others diverge from this prototypical behaviour. *B. subtilis* provides the most thoroughly interrogated case of such alternative adaptation modes (Fig. 9.10) (for a detailed review see Rao *et al.*, 2008). In addition to chemoreceptors, CheA, CheW, CheY, CheB and CheR *B. subtilis* encodes a CheY-P specific phosphatase, FliY, and the proteins CheC, CheD and CheV. Altogether, there are three integrated adaptation pathways, involving CheB/CheR, CheC/CheD/CheY-P, and CheZ, that in combination are thought to give rise to two distinct adaptation modes.

The roles of *B. subtilis* CheC, CheD and CheY-P in adaptation

The proteins CheC and CheD function together with CheA-P to promote the adaptation of CheA kinase activity via a negative feedback mechanism (Muff and Ordal, 2007). As described earlier, CheC is a CheY specific phosphatase composed of a single CheC-like domain. CheD is a 16 kDa protein with an α/β/β sandwich fold that has been shown to catalyse the deamidation of specific

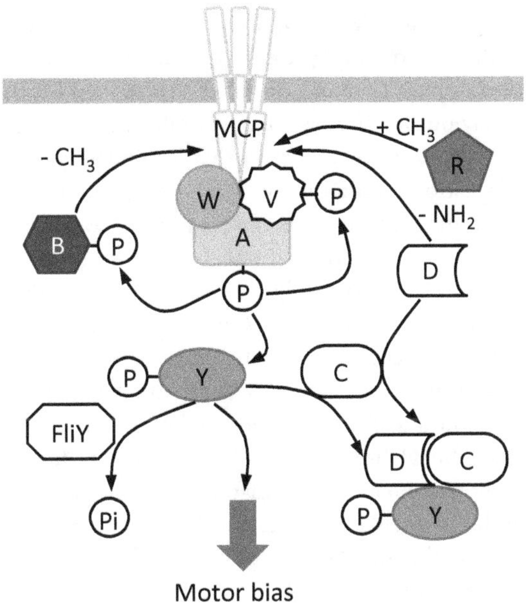

Figure 9.10 The *B. subtilis* chemosensory network has three integrated adaptation pathways. The response of *B. subtilis* to chemoeffectors is inverted with respect to that of *E. coli*; when *B. subtilis* is exposed to an increase in chemoattractant CheA kinase activity is increased. The increase in CheA kinase activity leads to a greater concentration of CheY-P in the cell which promotes counter clockwise rotation of the flagellar motor and smooth swimming. Rapid signal termination is ensured by the phosphatase, FliY, which localizes to the flagellar motor. There are three separate adaptation pathways in the *B. subtilis* chemosensory network; one utilizing CheB and CheR, the second CheV, and the third CheC, CheD and CheY-P. The CheB/CheR pathway mediates adaptation through methylation of chemoreceptors, the CheV pathway is thought to lead to conformational changes in CheA kinase inhibiting its activity whilst the CheC/CheD/FliY pathway controls the level of receptor deamidation.

glutamine residues to glutamate and to stimulate CheA kinase activity (Garrity and Ordal, 1997; Kristich and Ordal, 2002; Chao *et al.*, 2006). Although CheC exhibits CheY-P phosphatase activity *in vitro*, deletion of CheC does not lead to a completely smooth swimming phenotype (Szurmant *et al.*, 2004). In contrast, mutants in which FliY has been deleted are smooth swimming, indicating a high level of CheY-P in the cell and suggesting that FliY plays the dominant role in signal termination (Szurmant *et al.*, 2003). Instead, deletion of either CheC or CheD leads to defects in adaptation (Rosario *et al.*, 1995). When CheC is deleted cells do not adapt to positive stimuli to the same extent as wild-type, displaying a prolonged smooth-swimming bias following addition of attractant, and having increased methylation of chemoreceptors. Conversely, when CheD is deleted the cells respond poorly to chemoattractant; a tumbling phenotype is observed and chemoreceptors show reduced methylation.

The molecular mechanism by which CheC and CheD interact to promote adaptation has been extensively analysed (Rao *et al.*, 2008). CheC and CheD were initially shown to interact using yeast two-hybrid and co-precipitation assays (Rosario and Ordal, 1996), and the subsequent crystal structure of the CheC–CheD complex from *T. maritima* allowed analysis of this interaction at atomic resolution (Chao *et al.*, 2006). The structure reveals that CheC binds to CheD in a manner that mimics the interaction of CheD with the chemoreceptor, with the CheD-binding region of CheC differing only in the mutation of a glutamate to an alanine from the CheD recognition motif common to the chemoreceptors. Binding of CheC to CheD thus blocks the active site for deamidation of glutamine residues in the chemoreceptor. This is not the only role of the CheC–CheD interaction however; CheD is known to increase CheC phosphatase activity 5-fold *in vitro* (Szurmant *et al.*, 2004). Analysis of the CheC–CheD complex indicates that CheD is likely to activate CheC by providing an increased interaction surface for CheY, rather than through allosteric interactions. (Chao *et al.*, 2006). Importantly, it has also been shown that CheY-P increases the affinity of CheC for CheD (Chao *et al.*, 2006; Muff and Ordal, 2007). Mutational analysis was used to further probe the interactions of CheC with CheY and CheD; mutations were made in the regions of CheC important for CheD binding, CheY-P binding, and CheY-P phosphatase activity. In addition to *in vitro* characterization of the mutant CheC proteins, the proteins were tested for their ability to complement a Δ*cheC* mutation (Muff and Ordal, 2007). Strains in which either residues important for CheD or CheY-P binding were mutated demonstrated a similar phenotype to the *cheC* deletion strain. In contrast, partial complementation was observed when residues involved in CheY-P phosphatase activity (but not binding) were mutated suggesting that it is the interaction of CheC with CheD and CheY-P that is important for adaptation.

A model for the role of CheC, CheD and CheY-P in adaptation has been proposed (Muff and Ordal, 2007; Rao *et al.*, 2008). In the absence of chemoattractant CheD interacts with the chemoreceptor, causing an enhanced tendency for the receptors to adopt a conformation which stimulates CheA kinase activity. Upon binding a chemoattractant, the chemoreceptor will stimulate CheA kinase activity, resulting in a rise in CheY-P concentration in the cell and a concomitant increase in the proportion of CheC in complex with CheY-P. CheY-P binding to CheC increases its affinity for CheD thus recruiting CheD away from the chemoreceptor thereby decreasing the ability of CheD to activate CheA and countering the effects of chemoattractant binding. Conversely, when the level of attractant is decreased the concentration of CheY-P will drop and a smaller proportion of CheC will be in complex with CheY-P. This will lead to an increase in the amount of CheD free to bind to, and activate, the chemoreceptors.

The role of CheV in *B. subtilis* adaptation

CheV is an adaptor protein composed of an N-terminal coupling domain, homologous to that of CheW, and a C-terminal response regulator receiver domain that can be phosphorylated on an aspartate residue by CheA-P (Fredrick and Helmann, 1994; Rosario *et al.*, 1994; Karatan *et al.*, 2001). Mutants where CheV is deleted show a reduced ability to respond to attractants in a capillary assay compared with wild-type cells but were shown to respond a large step-change in azetidine-2-carboxylic acid concentration in a tethered cell assay (Rosario *et al.*, 1994). *In vitro* analysis showed that CheA can phosphotransfer to Asp253 and that CheV-P has a significantly longer half-life than *B. subtilis* CheB-P. Mutants expressing $CheV_{D253A}$ were able to respond to chemoattractant, showing that phosphorylation of the response regulator domain is not required for conformational coupling of receptor state to kinase activity by the N-terminal domain. Similarly, the N-terminal domain was not required for phosphorylation by CheA *in vitro*. Although mutants expressing $CheV_{D253A}$ or the isolated N-terminal domain of CheV show wild-type behaviour in the absence of stimuli and in the initial response to attractant, they exhibit impaired adaptation with an increase in the duration free-swimming events. This observation is consistent with an increase in the intracellular concentration of CheY-P, presumably due to increased CheA kinase activity (Karatan *et al.*, 2001). It has been proposed that, on chemoattractant induced CheA activation, CheV presents a potential target for phosphotransfer, alongside CheB and CheY. Phosphorylation of this adaptor could induce a conformational change that alters its interactions with CheA so as to inhibit its kinase activity and oppose the initial effects of chemoeffector stimulation (Rao *et al.*, 2008).

Why does *B. subtilis* need three adaptation systems? When only one of the adaptation systems is deleted cells are still capable of chemotaxis, albeit to a reduced level relative to wild-type cells; deletion of two or more of the adaptation networks is required to severely inhibit chemotaxis (Rao *et al.*, 2008). It has been proposed that different adaptation systems respond to differently sized chemoeffector gradients. When CheB is deleted, the cells respond in a wild-type fashion to small chemoattractant gradients, thus the CheV and CheC/CheD/CheY-P systems are capable of facilitating adaptation under these conditions. The dominance of the CheB/CheR system in response to large chemoeffector gradients is supported by the exponential increase in methanol detected with increased chemoattractant (as release of methanol occurs only on the action of the CheB/CheR adaptation system). The observation that chemoreceptor demethylation occurs within a minute of exposure to chemoattractant, a timescale consistent with the rate of adaptation of *B. subtilis*, whilst remethylation appears not to be completed until 20 minutes following stimulation has led to the proposal of a further distinctive feature of the *B. subtilis* adaptation system. It has been postulated that gradual remethylation may serve to alter the sensitivity of chemoreceptors over time to allow the CheC-CheD-CheY and CheV systems to reset, thereby maintaining their sensitivities over a wide range of chemoeffector concentrations. Unlike the *E. coli* system, this would independently modulate the affinity of chemoreceptors for their cognate ligands and their abilities to activate CheA, potentially providing greater versatility to the adaptation network than would be possible if the two processes were coupled.

Chemotactic networks lacking CheB and CheR.

Over 90% of genomes that contain chemoreceptors, CheA and CheW also encode CheB and CheR, suggesting that CheB and CheR mediated methylation/demethylation is the dominant mechanism for adaptation in chemosensory networks (Wuichet and Zhulin, 2010). CheB and CheR appear, however, to be absent from a number of organisms which have been shown to have a chemotactic response. The mechanisms of adaptation in these organisms are not yet well understood as those of *E. coli* and *B. subtilis*. In some cases, such as the phototactic response of type IV pilus dependent motility in the cyanobacterium *Synechocystis* sp. PCC6803, the speed of the response is sufficiently slow that spatial rather than temporal sensing may possibly be used in adaptation (Bhaya *et al.*, 2001). Others, such as *Helicobacter pylori*, encode homologues of CheV which is known to be important in adaptation in *B. subtilis* as discussed previously (Pittman *et al.*, 2001). In addition, even when CheB and CheR are present they are not necessarily essential for the response to all external stimuli. For example, *E. coli* cells encoding the aerotaxis receptor Aer as the sole chemoreceptor and lacking CheB and CheR display Aer mediated aerotaxis. Aer detects changes in environmental oxygen levels indirectly, using the FAD moiety of its PAS domain as a redox sensor and it is possible that CheB and CheR are not essential for aerotaxis owing to the transient nature of cellular redox changes (Bibikov *et al.*, 2004).

Other chemotaxis-like networks

The networks that control motility in free-swimming bacteria are the most thoroughly characterized chemosensory systems to date; however, free-swimming is not the only behaviour that is controlled by chemotaxis-like proteins. Here we give a brief overview of some of the work on chemosensory systems controlling other cellular behaviour.

Chemotaxis-like networks control the gliding motility of *Myxococcus xanthus*

There are a number of different mechanisms by which bacteria move along surfaces. Some species become hyperflagellate, forming bundles containing multiple flagella which rotate together thereby increasing motive force. Other species are pushed forwards by slime extruded from the cell pole, or pulled forwards by pili. The Gram-negative soil bacterium *Myxococcus xanthus* is a model organism for the study of motility on solid surfaces (reviewed in detail in Mauriello *et al.*, 2010a). *M. xanthus* has a complex, multicellular, life-cycle that includes vegetative growth (with cells undergoing replication, predation and swarming) and sporulation under low-nutrient conditions. *M. xanthus* has multiple pathways utilizing homologues of chemotaxis proteins and also exhibits two distinct forms of motility.

Adventurous motility is undertaken by isolated *M. xanthus* cells and will not be discussed here. Social, or S-, motility is the mechanism by which groups of *M. xanthus* cells travel together and is mediated by type IV pili (TFP). These TFP are typically 5–7 nm in diameter and can extend for several body lengths from the pole at leading edge of the cell. They bind to the surface along which a bacterium is moving; retraction of the TFP through depolymerization at the cell membrane pulls the cell forwards. S-motility is slow, with cells moving at a speed of ~2 μm/min. Approximately every 10 minutes the direction of motility is reversed; this reversal is a highly complex process involving the relocation of a number of proteins controlling the TFP to the opposite cell pole. Reorientation is thought to play a role analogous to changes in flagellar rotation in free-swimming bacteria, facilitating changes of direction in response to the prevailing environmental conditions.

M. xanthus has multiple chemosensory systems that are important in controlling motility, with cross-talk between these systems playing a role in co-ordinating the two different types of motility. Here, we will focus on the Frz system which acts as the master regulator controlling cell reversal (for detail on the other chemosensory networks see Mauriello *et al.*, 2010a). The *frz* operon encodes many genes with homology to those seen in the canonical chemotaxis systems of free-swimming bacteria (Fig. 9.11). FrzCD is a cytoplasmic chemoreceptor, FrzA and FrzB are homologous to CheW, whilst FrzE is a HPK–RR hybrid protein consisting of fused CheA- and CheY-like domains. Adaptation is mediated through FrzF and FrzG which are homologous to CheR and CheB; FrzF, however, also encodes a C-terminal domain containing three tetratricopeptide repeats (TPRs). Finally FrzZ is a response regulator composed of two tandem CheY-like domains.

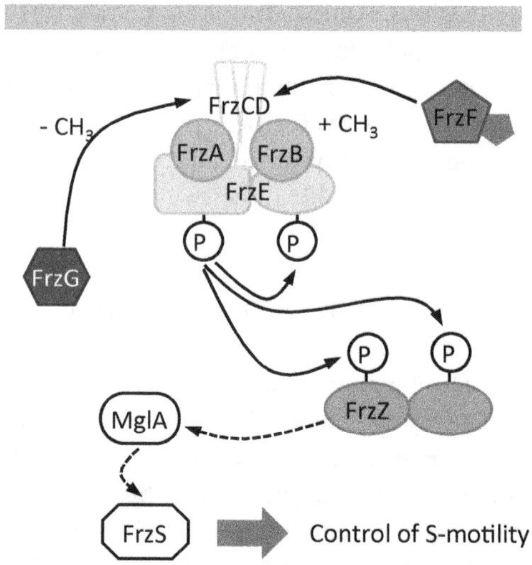

Figure 9.11 The chemotaxis-like pathway controlling Social motility in *M. xanthus*. S-motility in *M. xanthus* utilizes type IV pili extending from the pole at the leading edge of the cell. *M. xanthus* cells reverse the direction of motility approximately every 10 minutes, a process important in the chemotactic response of groups of cells. These cell-reversals require movement of the pili from the leading to the trailing pole of the cell and are regulated by the Frz chemotaxis-like pathway. Activation of FrzCD leads to an increase in the kinase activity of FrzE. Three CheY-like domains can receive phosphoryl groups from FrzE, one is directly fused to the kinase and two lie in the RR FrzZ. Phosphorylation of the CheY-like domain of FrzE inhibits kinase activity, whilst phosphorylation of FrzZ is thought to activate the protein. The proteins MglA and FrzS are known to be downstream effectors, however the molecular mechanism for the interaction between these proteins and for their activation by FrzZ is not known.

The Frz system differs from the chemotaxis networks seen in free-swimming bacteria in a number of ways including the fusion of a CheY domain to the kinase, the nature of the RR, and the requirement of a TPR domain for adaptation. The kinase activity of FrzE is stimulated by FrzCD upon chemoattractant binding resulting in the transfer of phosphoryl groups from FrzE to FrzZ. Phosphorylation of the FrzE CheY-like domain, however, negatively regulates the activity of the kinase domain (Inclán *et al.*, 2008). The mechanism of this inhibition is not well understood, but it has been suggested that the phosphorylated CheY domain may physically occlude the active site of the kinase. FrzZ is phosphorylated on two residues, Asp52 and Asp220 (Inclán *et al.*, 2007). The downstream effectors of the Frz pathway are thought to be the GTPase MglA, and FrzS, a protein comprised of an N-terminal receiver domain lacking the conserved aspartate residue and a C-terminal coiled-coil. Analysis of *mglA* and *frzS* mutant strains shows that both proteins play an important role in the regulation of cell reversal in S-motility (Spormann and Kaiser, 1999; Ward *et al.*, 2000). FrzS localizes to the leading pole of the cell and fluorescent fusions to FrzS have been shown to oscillate from pole to pole during cell reversals (Mignot *et al.*, 2005). It is known that MglA is essential for FrzS localization and that MglA interacts with both FrzZ and FrzS *in vitro*; however, the molecular mechanism by which FrzZ may interact with MglA to regulate FrzS activity is not yet understood (Mauriello *et al.*, 2010b). The role played by methylation in adaptation of the Frz pathway is also an area of active research. FrzCD can be methylated at three glutamate residues (Glu168, Glu175, and Glu182) and the FrzF TPRs are thought to play a key role in regulating site-specific methylation. A mutant of FrzF lacking the TPRs is capable of methylating only one of these glutamate residues *in vitro*, whilst full-length FrzF can methylate all three glutamate residues. *In vivo* analyses of mutant strains in which Glu168, Glu175 and Glu182 were mutated both singly and in combination has shown that methylation of each site plays a specific role in cellular behaviour. Glu168 appears to play a role in the regulation of social swarming, Glu175 in the control of single cell reversal frequency whilst Glu182 appears to be critical for controlling the kinase of FrzE activity (Scott *et al.*, 2008).

Chemotaxis-like networks controlling behaviour other than motility

The primary difference between the core components of chemosensory networks controlling flagellar rotation and those controlling alternate cellular functions lies in the response regulators; these networks typically include a RR composed of a phosphorylatable receiver domain fused to an additional output domain. Such chemotaxis-like pathways control diverse behaviours, including developmental gene expression in *M. xanthus*, production of extracellular polysaccharides in *Azospirillum brasilense*, and biofilm formation in *P. aeruginosa*.

The Che3 system of *M. xanthus* includes most of the core proteins seen in a canonical chemotaxis system (chemoreceptors, a hybrid CheA-RR, CheW, CheB, and CheR) in addition to the RR CrdA and a periplasmic lipoprotein CrdB (Fig. 9.12a). CrdA is a transcription factor comprising an N-terminal, phosphorylatable, receiver domain and a C-terminal helix–turn–helix DNA binding domain. Deletion of *che3* genes did not lead to defects in motility; instead the deletion strains were deficient in developmental regulation. Under starvation conditions and at high cell density, wild-type *M. xanthus* cells aggregate into mounds and then form fruiting bodies. Strains from which

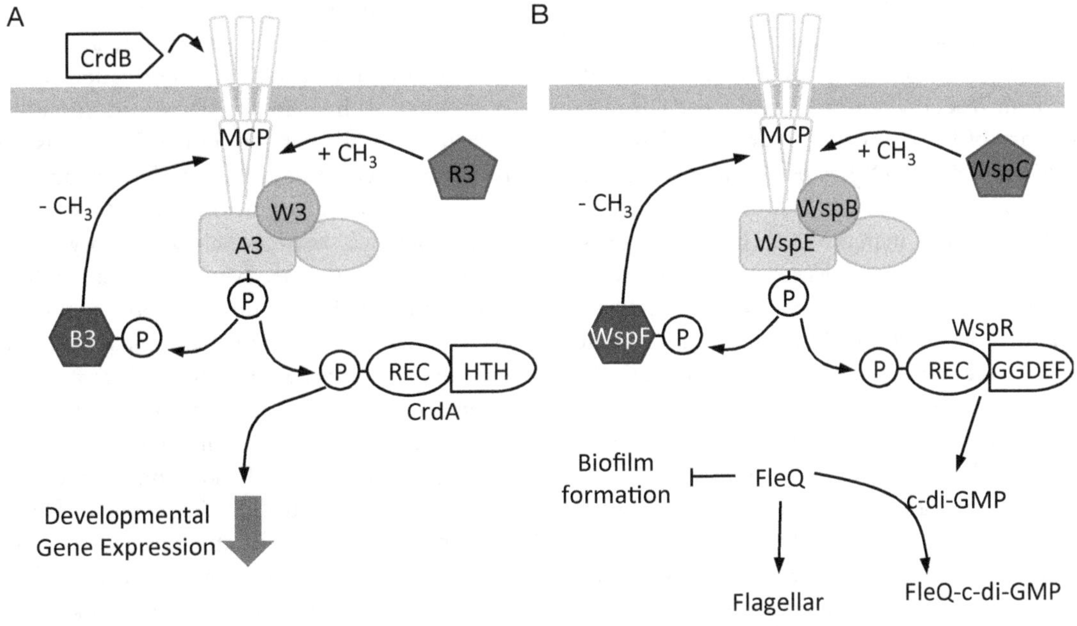

Figure 9.12 Chemotaxis-like pathways controlling gene expression. Many chemotaxis-like pathways that control behaviour other than motility differ from the canonical systems in the architecture of their response regulators. Typically, the RRs in these non-canonical systems are comprised of a receiver (REC) domain fused to an additional output domain which often has enzymatic or DNA binding activity. The Che3 system of *M. xanthus* (a) includes an RR with a helix–turn–helix DNA binding domain. This pathway controls the changes in gene expression necessary for the development of fruiting bodies under starvation conditions. (b) The Wsp system of pseudomonas, in contrast, controls gene expression by altering the levels of the secondary messenger cyclic-di-GMP. In the Wsp system the RR REC domain is fused to a domain with diguanylyl cyclase activity (a GGDEF domain). Stimulation of WspE leads to activation of the diguanylyl cyclase and the derepression of genes necessary for biofilm formation.

the *che3* genes have been deleted make fruiting bodies at lower cell density than wild-type cells and are defective in sporulation. CrdA was shown to interact with CheA3 in a yeast two-hybrid assay and analysis of Δ*crdA* strains led to the conclusion that CrdA was the output for the Che3 pathway. The Che3 pathway thus activates the genes necessary for development of fruiting bodies in the absence of nutrients (Kirby and Zusman, 2003).

The α-proteobacterium *A. brasilense* can colonize the roots of grass and cereal species and, under conditions of limiting nitrogen but high aeration, can also aggregate into flocs. *A. brasilense* has four chemotaxis-like pathways the first of which, Che1, contains homologues of CheA, CheW, CheY, CheB and CheR. Although the *che1* genes were initially identified through their ability to complement non-chemotactic strains, a strain with the entire operon deleted is capable of chemotaxis and exhibits only subtle changes in motility. The phenotypes of strains with a single *che1* gene deleted are more complex; deletion of CheB1 and CheR1 has a more pronounced effect on motility than deletion of CheA1 and CheY1, and the severity of these defects is attractant dependent (Stephens *et al.*, 2006). Δ*cheA1* mutants aggregate and flocculate significantly more than wild-type cells and it was hypothesized that the dominant role of the Che1 pathway is in the production of exopolysaccharide necessary for flocculation (Bible *et al.*, 2008). This hypothesis is supported by the recent characterization of the extracellular matrix of *che1* pathway mutants by atomic force microscopy which demonstrates that the fibrillar extracellular structures associated with flocculation appear earlier in mutant than in wild-type strains (Edwards *et al.*, 2011).

Chemotaxis proteins are known to play a role in regulation of biofilm formation in *P. aeruginosa* through the Wsp pathway, one of numerous pathways modulating the concentration of the second messenger cyclic-di-GMP in the cell (Hickman *et al.*, 2005). It is known that in many bacteria high levels of cyclic-di-GMP are associated with biofilm formation, whilst free-swimming cells have a lower cellular concentration of cyclic-di-GMP. The Wsp pathway includes a chemoreceptor (WspA) two CheW homologues (WspB and WspD), a hybrid CheA-RR (WspE), a CheR (WspC), and a CheB (WspF). There is an additional response regulator with diguanylyl cyclase activity, WspR, in which a receiver domain is fused to a GGDEF domain (Fig. 9.12b). It is proposed that the Wsp pathway responds to external signals such as changes in surface attachment (Guvener and Harwood, 2007). Activation of WspE leads to phosphorylation of WspR and a concomitant increase in cellular concentration of cyclic-di-GMP. High levels of cyclic-di-GMP inhibit the transcriptional activator FleQ, preventing the transcription of the genes necessary for flagellar gene expression and depressing the genes necessary for biofilm formation (Hickman and Harwood, 2008).

Prototypical two-component systems comprised of a membrane-bound histidine protein kinase and a soluble response regulator detect a wide range of external environmental signals including osmotic stress, cell density and the presence of nutrients. In the majority of these systems the response regulator acts as a transcription factor controlling processes ranging from osmoregulation to biofilm formation and sporulation (Stock *et al.*, 2000). Why then do some organisms utilize a chemotaxis-like network rather than a classical two-component system to control developmental gene expression or biofilm formation? One important difference between chemosensory signal transduction and other two-component systems is the organization of the chemoreceptors into arrays. This allows the integration of signals from multiple different chemoeffectors which may play an important role in, for example, determining whether the external environment is favourable for developmental changes. In addition the chemoreceptor arrays facilitate detection of small changes in chemoeffector concentration over a wide range of background signal intensities. The second key feature of many chemotaxis-like networks is the inclusion of enzymes catalysing the methylation and demethylation of chemoreceptors. Changes in the methylation state of methyl-accepting chemoreceptors are important both for temporal sensing and in adaptation to persistent stimuli. Chemotaxis-like networks are therefore likely to play an important role in cellular processes which require the detection of time-varying signals from multiple external stimuli over a wide range of background signal intensities.

Summary

The ability to receive, decipher, and respond to prevailing environmental signals with a behavioural change is a fundamental property of all living organisms, from non-cognitive bacteria to sentient organisms with complex multicellular data-processing apparatus. All of these species, regardless of their level of organizational and behavioural complexity, employ the same fundamental steps to achieve these feats; detecting and interpreting sensory impressions and deciding the most appropriate behaviour with which to respond. Bacteria have evolved a core network of interacting proteins that are optimized to control and regulate a chemotactic response. This elegant behaviour requires accurate perception of the environment, short term memory and adaptation to changes in the concentration of chemoeffectors.

References

Airola, M.V., Watts, K.J., Bilwes, A.M., and Crane, B.R. (2010). Structure of concatenated HAMP domains provides a mechanism for signal transduction. Structure *18*, 436–448.

Alexander, R.P., and Zhulin, I.B. (2007). Evolutionary genomics reveals conserved structural determinants of signaling and adaptation in microbial chemoreceptors. Proc. Natl. Acad. Sci. U.S.A. *104*, 2885–2890.

Amin, D.N., and Hazelbauer, G.L. (2010). The chemoreceptor dimer is the unit of conformational coupling and transmembrane signaling. J. Bacteriol. *192*, 1193–1200.

Anand, G.S., and Stock, A.M. (2002). Kinetic basis for the stimulatory effect of phosphorylation on the methylesterase activity of CheB. Biochemistry *41*, 6752–6760.

Anand, G.S., Goudreau, P.N., Lewis, J.K., and Stock, A.M. (2000). Evidence for phosphorylation-dependent conformational changes in methylesterase CheB. Protein Sci. *9*, 898–906.

Aravind, L., and Ponting, C.P. (1999). The cytoplasmic helical linker domain of receptor histidine kinase and methyl-accepting proteins is common to many prokaryotic signalling proteins. FEMS Microbiol. Lett. *176*, 111–116.

Armitage, J.P., and Macnab, R.M. (1987). Unidirectional, intermittent rotation of the flagellum of *Rhodobacter sphaeroides*. J. Bacteriol. *169*, 514–518.

Barak, R., and Eisenbach, M. (2001). Acetylation of the response regulator, CheY, is involved in bacterial chemotaxis. Mol. Microbiol. *40*, 731–743.

Barak, R., Prasad, K., Shainskaya, A., Wolfe, A.J., and Eisenbach, M. (2004). Acetylation of the chemotaxis response regulator CheY by acetyl-coA synthetase purified from *Escherichia coli*. J. Mol. Biol. *342*, 383–401.

Barak, R., Yan, J., Shainskaya, A., and Eisenbach, M. (2006). The chemotaxis response regulator CheY can catalyze its own acetylation. J. Mol. Biol. *359*, 251–265.

Bardy, S.L., and Maddock, J.R. (2005). Polar localization of a soluble methyl-accepting protein of *Pseudomonas aeruginosa*. J. Bacteriol. *187*, 7840–7844.

Barnakov, A.N., Barnakova, L.A., and Hazelbauer, G.L. (1999). Efficient adaptational demethylation of chemoreceptors requires the same enzyme-docking site as efficient methylation. Proc. Natl. Acad. Sci. U.S.A. *96*, 10667–10672.

Barnakov, A.N., Barnakova, L.A., and Hazelbauer, G.L. (2002). Allosteric enhancement of adaptational demethylation by a carboxyl-terminal sequence on chemoreceptors. J. Biol. Chem. *277*, 42151–42156.

Beel, B.D., and Hazelbauer, G.L. (2001). Signalling substitutions in the periplasmic domain of chemoreceptor Trg induce or reduce helical sliding in the transmembrane domain. Mol. Microbiol. *40*, 824–834.

Bell, C.H., Porter, S.L., Strawson, A., Stuart, D.I., and Armitage, J.P. (2010). Using structural information to change the phosphotransfer specificity of a two-component chemotaxis signalling complex. PLoS Biol. *8*, e1000306.

Berg, H.C., and Brown, D.A. (1972). Chemotaxis in *Escherichia coli* analysed by three-dimensional tracking. Nature *239*, 500–504.

Berg, H.C., and Purcell, E.M. (1977). Physics of chemoreception. Biophys. J. *20*, 193–219.

Berg, H.C., and Tedesco, P.M. (1975). Transient response to chemotactic stimuli in *Escherichia coli*. Proc. Natl. Acad. Sci. U.S.A. 72, 3235–3239.

Bhaya, D., Takahashi, A., and Grossman, A.R. (2001). Light regulation of type IV pilus-dependent motility by chemosensor-like elements in *Synechocystis* PCC6803. Proc. Natl. Acad. Sci. U.S.A. *98*, 7540–7545.

Bibikov, S.I., Miller, A.C., Gosink, K.K., and Parkinson, J.S. (2004) Methylation-independent aerotaxis mediated by the *Escherichia coli* Aer protein. J. Bacteriol. *186*, 3730–3737.

Bible, A.N., Stephens, B.B., Ortega, D.R., Xie, Z., and Alexandre, G. (2008). Function of a chemotaxis-like signal transduction pathway in modulating motility, cell clumping, and cell length in the alphaproteobacterium *Azospirillum brasilense*. J. Bacteriol. *190*, 6365–6375.

Bilwes, A.M., Alex, L.A., Crane, B.R., and Simon, M.I. (1999). Structure of CheA, a signal-transducing histidine kinase. Cell *96*, 131–141.

Bischoff, D.S., and Ordal, G.W. (1992). *Bacillus subtilis* chemotaxis: a deviation from the *Escherichia coli* paradigm. Mol. Microbiol. *6*, 23–28.

Boldog, T., Grimme, S., Li, M., Sligar, S.G., and Hazelbauer, G.L. (2006). Nanodiscs separate chemoreceptor oligomeric states and reveal their signaling properties. Proc. Natl. Acad. Sci. U.S.A. *103*, 11509–11514.

Borkovich, K.A., Kaplan, N., Hess, J.F., and Simon, M.I. (1989). Transmembrane signal transduction in bacterial chemotaxis involves ligand-dependent activation of phosphate group transfer. Proc. Natl. Acad. Sci. U.S.A. *86*, 1208–1212.

Borkovich, K.A., and Simon, M.I. (1990). The dynamics of protein phosphorylation in bacterial chemotaxis. Cell *63*, 1339–1348.

Bourret, R.B., Davagnino, J., and Simon, M.I. (1993). The carboxy-terminal portion of the CheA kinase mediates regulation of autophosphorylation by transducer and CheW. J. Bacteriol. *175*, 2097–2101.

Bray, D., Levin, M.D., and Morton-Firth, C.J. (1998). Receptor clustering as a cellular mechanism to control sensitivity. Nature *393*, 85–88.

Bray, D., Levin, M.D., and Lipkow, K. (2007). The chemotactic behavior of computer-based surrogate bacteria. Curr. Biol. *17*, 12–19.

Bren, A., and Eisenbach, M. (1998). The N terminus of the flagellar switch protein, FliM, is the binding domain for the chemotactic response regulator, CheY. J. Mol. Biol. *278*, 507–514.

Briegel, A., Ding, H.J., Li, Z., Werner, J., Gitai, Z., Dias, D.P., Jensen, R.B., and Jensen, G.J. (2008). Location and architecture of the *Caulobacter crescentus* chemoreceptor array. Mol. Microbiol. *69*, 30–41.

Briegel, A., Ortega, D.R., Tocheva, E.I., Wuichet, K., Li, Z., Chen, S., Muller, A., Iancu, C.V., Murphy, G.E., Dobro, M.J., Zhulin, I.B., and Jensen, G.J. (2009). Universal architecture of bacterial chemoreceptor arrays. Proc. Natl. Acad. Sci. U.S.A. *106*, 17181–17186.

Brown, D.A., and Berg, H.C. (1974). Temporal stimulation of chemotaxis in *Escherichia coli*. Proc. Natl. Acad. Sci. U.S.A. *71*, 1388–1392.

Burgess-Cassler, A., and Ordal, G.W. (1982). Functional homology of *Bacillus subtilis* methyltransferase II and *Escherichia coli* cheR protein. J. Biol. Chem. *257*, 12835–12838.

del Campo, A.M., Ballado, T., de la Mora, J., Poggio, S., Camarena, L., and Dreyfus, G. (2007). Chemotactic control of the two flagellar systems of *Rhodobacter sphaeroides* is mediated by different sets of CheY and FliM proteins. J. Bacteriol. *189*, 8397–8401.

Chao, X., Muff, T.J., Park, S.Y., Zhang, S., Pollard, A.M., Ordal, G.W., Bilwes, A.M., and Crane, B.R. (2006). A receptor-modifying deamidase in complex with a signaling phosphatase reveals reciprocal regulation. Cell *124*, 561–571.

Chen, S., Beeby, M., Murphy, G.E., Leadbetter, J.R., Hendrixson, D.R., Briegel, A., Li, Z., Shi, J., Tocheva, E.I., Muller, A., *et al.* (2011). Structural diversity of bacterial flagellar motors. EMBO J. *30*, 2972–2981.

Chervitz, S.A., and Falke, J.J. (1995). Lock on/off disulfides identify the transmembrane signaling helix of the aspartate receptor. J. Biol. Chem. *270*, 24043–24053.

Cho, H.S., Lee, S.Y., Yan, D., Pan, X., Parkinson, J.S., Kustu, S., Wemmer, D.E., and Pelton, J.G. (2000). NMR structure of activated CheY. J. Mol. Biol. *297*, 543–551.

Coleman, M.D., Bass, R.B., Mehan, R.S., and Falke, J.J. (2005). Conserved glycine residues in the cytoplasmic domain of the aspartate receptor play essential roles in kinase coupling and on–off switching. Biochemistry *44*, 7687–7695.

Djordevic, S., Goudreau, P.N., Xu, Q., Stock, A.M., and West, A.H. (1998). Structural basis for methylesterase CheB regulation by a phosphorylation-activated domain. Proc. Natl. Acad. Sci. U.S.A. *95*, 1381–1386.

Draheim, R.R., Bormans, A.F., Lai, R.Z., and Manson, M.D. (2006). Tuning a bacterial chemoreceptor with protein–membrane interactions. Biochemistry *45*, 14655–14664.

Duke, T.A.J. and Bray, D. (1999). Heightened sensitivity of a lattice of membrane receptors. Proc. Natl. Acad. Sci. U.S.A. *96*, 10104–10108.

Edwards, A.N., Siuti, P., Bible, A.N., Alexandre, G., Retterer, S.T., Doktycz, M.J., and Morrell-Falvey, J.L. (2011). Characterization of cell surface and extracellular matrix remodeling of *Azospirillum brasilense* chemotaxis-like 1 signal transduction pathway mutants by atomic force microscopy. FEMS Microbiol. Lett. *314*, 131–139.

Endres, R.G., Oleksiuk, O., Hansen, C.H., Meir, Y., Sourjik, V., and Wingreen, N.S. (2008). Variable sizes of *Escherichia coli* chemoreceptor signaling teams. Mol. Syst. Biol. *4*, 211.

Endres, R.G., and Wingreen, N.S. (2006). Precise adaptation in bacterial chemotaxis through 'assistance neighborhoods'. Proc. Natl. Acad. Sci. U.S.A. *103*, 13040–13044.

Erbse, A.H., and Falke, J.J. (2009). The core signaling proteins of bacterial chemotaxis assemble to form an ultrastable complex. Biochemistry *48*, 6975–6987.

Eswar, N., Webb, B., Marti-Renom, M.A., Madhusudhan, M.S., Eramian, D., Shen, M., Pieper, U., and Sali, A. (2007). Comparative protein structure modeling using MODELLER. Curr. Protocols Protein Sci. *50*, 2.9.1–2.9.31.

Falke, J.J., and Hazelbauer, G.L. (2001). Transmembrane signaling in bacterial chemoreceptors. Trends Biochem. Sci. *26*, 257–265.

Fredrick, K.L., and Helmann, J.D. (1994). Dual chemotaxis signaling pathways in *Bacillus subtilis*: a sigma D-dependent gene encodes a novel protein with both CheW and CheY homologous domains. J. Bacteriol. *176*, 2727–2735.

Garrity, L.F., and Ordal, G.W. (1997). Activation of the CheA kinase by asparagine in *Bacillus subtilis* chemotaxis. Microbiology *143*, 2945–2951.

Gerdes, K., Howard, M., and Szardenings, F. (2011). Pushing and pulling in prokaryotic DNA segregation. Cell *141*, 927–942.

Gest, H. (2004). The discovery of microorganisms by Robert Hooke and Antoni van Leeuwenhoek, Fellows of the Royal Society. Notes Rec. Roy. Soc. *58*, 187–201.

Goldman, J.P., Levin, M.D., and Bray, D. (2009). Signal amplification in a lattice of coupled protein kinases. Mol. Biosyst. *5*, 1853–1859.

Gotz, R., and Schmitt, R. (1987). *Rhizobium meliloti* swims by unidirectional, intermittent rotation of right-handed flagellar helices. J. Bacteriol. *169*, 3146–3150.

Goy, M.F., Springer, M.S., and Adler, J. (1978). Failure of sensory adaptation in bacterial mutants that are defective in a protein methylation reaction. Cell *15*, 1231–1240.

Greek, M., Platzer, J., Sourjik, V., and Schmitt, R. (1995). Analysis of a chemotaxis operon in *Rhizobium meliloti*. Mol. Microbiol. *15*, 989–1000.

Greenfield, D., McEvoy, A.L., Shroff, H., Crooks, G.E., Wingreen, N.S., Betzig, E., and Liphardt, J. (2009). Self-organization of the *Escherichia coli* chemotaxis network imaged with super-resolution light microscopy. PLoS Biol. *7*, e1000137.

Guvener, Z.T., and Harwood, C.S. (2007). Subcellular location characteristics of the *Pseudomonas aeruginosa* GGDEF protein, WspR, indicate that it produces cyclic-di-GMP in response to growth on surfaces. Mol. Microbiol. *66*, 1459–1473.

Halkides, C.J., McEvoy, M.M., Casper, E., Matsumura, P., Volz, K., and Dahlquist, F.W. (2000). The 1.9 Å resolution crystal structure of phosphono-CheY, an analogue of the active form of the response regulator, CheY. Biochemistry *39*, 5280–5286.

Hamer, R., Chen, P.Y., Armitage, J.P., Reinert, G., and Deane, C.M. (2010). Deciphering chemotaxis pathways using cross species comparisons. BMC Syst. Biol. *4*, 3.

Hazelbauer, G.L., and Lai, W.C. (2010). Bacterial chemoreceptors: providing enhanced features to two-component signaling. Curr. Opin. Microbiol. *13*, 124–132.

Hazelbauer, G.L., Falke, J.J., and Parkinson, J.S. (2008). Bacterial chemoreceptors: high-performance signaling in networked arrays. Trends Biochem. Sci. *33*, 9–19.

Hess, J.F., Oosawa, K., Kaplan, N., and Simon, M.I. (1988). Phosphorylation of three proteins in the signaling pathway of bacterial chemotaxis. Cell *53*, 79–87.

Hickman, J.W., and Harwood, C.S. (2008). Identification of FleQ from *Pseudomonas aeruginosa* as a c-di-GMP-responsive transcription factor. Mol. Microbiol. *69*, 376–389.

Hickman, J.W., Tifrea, D.F., and Harwood, C.S. (2005). A chemosensory system that regulates biofilm formation through modulation of cyclic diguanylate levels. Proc. Natl. Acad. Sci. U.S.A. *102*, 14422–14427.

Hulko, M., Berndt, F., Gruber, M., Linder, J.U., Truffault, V., Schultz, A., Martin, J., Schultz, J.E., Lupas, A.N., and Coles, M. (2006). The HAMP domain structure implies helix rotation in transmembrane signaling. Cell *126*, 929–940.

Inclán, Y.F., Vlamakis, H.C., and Zusman, D.R. (2007). FrzZ, a dual CheY-like response regulator, functions as an output for the Frz chemosensory pathway of *Myxococcus xanthus*. Mol. Microbiol. *65*, 90–102.

Inclán, Y.F., Laurent, S., and Zusman, D.R. (2008). The receiver domain of FrzE, a CheA–CheY fusion protein, regulates the CheA histidine kinase activity and downstream signalling to the A- and S-motility systems of *Myxococcus xanthus*. Mol. Microbiol. *68*, 1328–1339.

Jahreis, K., Morrison, T.B., Garzon, A., and Parkinson, J.S. (2004). Chemotactic signaling by an *Escherichia coli* CheA mutant that lacks the binding domain for phosphoacceptor partners. J. Bacteriol. *186*, 2664–2672.

Jeziore-Sassoon, Y., Hamblin, P.A., Bootle, W.C., Poole, P.S., and Armitage, J.P. (1998). Metabolism is required for chemotaxis to sugars in *Rhodobacter sphaeroides*. Microbiology *144*, 229–239.

Karatan, E., Saulmon, M.M., Bunn, M.W., and Ordal, G.W. (2001). Phosphorylation of the response regulator chev is required for adaptation to attractants during *Bacillus subtilis* chemotaxis. J. Biol. Chem. *276*, 43618–43626.

Kehry, M.R., and Dahlquist, F.W. (1982). The methyl-accepting chemotaxis proteins of *Escherichia coli*. Identification of the multiple methylation sites on methyl-accepting chemotaxis protein I. J. Biol. Chem. *257*, 10378–10386.

Kehry, M.R., Bond, M.W., Hunkapiller, M.W., and Dahlquist, F.W. (1983). Enzymatic deamidation of methyl-accepting chemotaxis proteins in *Escherichia coli* catalyzed by the cheB gene product. Proc. Natl. Acad. Sci. U.S.A. *80*, 3599–3603.

Kelly, F.X., Dapsis, K.J., and Lauffenburger, D.A. (1988). Effect of bacterial chemotaxis on dynamics of microbial competition. Microb. Ecol. *16*, 115–131.

Kentner, D., Thiem, S., Hildenbeutel, M., and Sourjik, V. (2006). Determinants of chemoreceptor cluster formation in *Escherichia coli*. Mol. Microbiol. *61*, 407–417.

Keymer, J.E., Endres, R.G., Skoge, M., Meir, Y., and Wingreen, N.S. (2006). Chemosensing in *Escherichia coli*: two regimes of two-state receptors. Proc. Natl. Acad. Sci. U.S.A. *103*, 1786–1791.

Khan, S., Pierce, D., and Vale, R.D. (2000). Interactions of the chemotaxis signal protein CheY with bacterial flagellar motors visualized by evanescent wave microscopy. Curr. Biol. *10*, 927–930.

Khursigara, C.M., Wu, X., and Subramaniam, S. (2008). Chemoreceptors in *Caulobacter crescentus*: trimers of receptor dimers in a partially ordered hexagonally packed array. J. Bacteriol. *190*, 6805–6810.

Kim, K.K., Yokota, H., and Kim, S.H. (1999). Four-helical-bundle structure of the cytoplasmic domain of a serine chemotaxis receptor. Nature *400*, 787–792.

Kirby, J.R., and Zusman, D.R. (2003). Chemosensory regulation of developmental gene expression in *Myxococcus xanthus*. Proc. Natl. Acad. Sci. U.S.A. *100*, 2008–2013.

Kirby, J.R., Saulmon, M.M., Kristich, C.J., and Ordal, G.W. (1999). CheY-dependent methylation of the asparagine receptor, McpB, during chemotaxis in *Bacillus subtilis*. J. Biol. Chem. *274*, 11092–11100.

Kristich, C.J., and Ordal, G.W. (2002). *Bacillus subtilis* CheD is a chemoreceptor modification enzyme required for chemotaxis. J. Biol. Chem. *277*, 25356–25362.

Lai, W.C., Beel, B.D., and Hazelbauer, G.L. (2006). Adaptational modification and ligand occupancy have opposite effects on positioning of the transmembrane signalling helix of a chemoreceptor. Mol. Microbiol. *61*, 1081–1090.

Lan, G., Schulmeister, S., Sourjik, V., and Tu, Y. (2011). Adapt locally and act globally: strategy to maintain high chemoreceptor sensitivity in complex environments. Mol. Syst. Biol. *7*, 475.

Larsen, S.H., Reader, R.W., Kort, E.N., Tso, W.W., and Adler, J. (1974) Change in direction of flagellar rotation is the basis of the chemotactic response in *Escherichia coli*. Nature *249*, 74–77.

Lee, S.Y., Cho, H., Pelton, J.G., Yan, D., Berry, E.A., and Wemmer, D.E. (2001a). Crystal structure of activated CheY. J. Biol. Chem. *276*, 16425–16431.

Lee, S.Y., Cho, H.S., Pelton, J.G., Yan, D., Henderson, R.K., King, D.S., Huang, L.S., Kustu, S., Berry, E.A., and Wemmer, D.E. (2001b). Crystal structure of an activated response regulator bound to its target. Nat. Struct. Mol. Biol. *8*, 52–56.

Li, M., and Hazelbauer, G.L. (2004). Cellular stoichiometry of the components of the chemotaxis signaling complex. J. Bacteriol. *186*, 3687–3694.

Li, M., and Hazelbauer, G.L. (2005). Adaptational assistance in clusters of bacterial chemoreceptors. Mol. Microbiol. *56*, 1617–1626.

Li, M., and Hazelbauer, G.L. (2006). The carboxyl-terminal linker is important for chemoreceptor function. Mol. Microbiol. *60*, 469–479.

Li, R., Gu, J., Chen, Y.Y., Xiao, C.L., Wang, L.W., Zhang, Z.P., Bi, L.J., Wei, H.P., Wang, X.D., Deng, J.Y., *et al.* (2010). CobB regulates *Escherichia coli* chemotaxis by deacetylating the response regulator CheY. Mol. Microbiol. *76*, 1162–1174.

Liarzi, O., Barak, R., Bronner, V., Dines, M., Sagi, Y., Shainskaya, A., and Eisenbach, M. (2010). Acetylation represses the binding of CheY to its target proteins. Mol. Microbiol. *77*, 1606.

Lipkow, K., Andrews, S.S., and Bray, D. (2005). Simulated diffusion of phosphorylated CheY through the cytoplasm of *E. coli*. J. Bacteriol. *187*, 45–53.

Lupas, A., and Stock, J. (1989). Phosphorylation of an N-terminal regulatory domain activates the CheB methylesterase in bacterial chemotaxis. J. Biol. Chem. *264*, 17337–17342.

Mackenzie, C. (2001). The home stretch, a first analysis of the nearly completed genome of *Rhodobacter sphaeroides* 2.4.1. Photosynth. Res. *70*, 19–41.

Maddock, J.R., and Shapiro, L. (1993). Polar location of the chemoreceptor complex in the *Escherichia coli* cell. Science *259*, 1717–1723.

Mauriello, E.M.F., Mignot, T., Yang, Z., and Zusman, D.R. (2010a). Gliding motility revisited: how do the myxobacteria move without flagella? Microbiol. Mol. Biol. Rev. *74*, 229–249.

Mauriello, E.M.F., Mouhamar, F., Nan, B., Ducret, A., Dai, D., Zusman, D.R., and Mignot, T. (2010b). Bacterial motility complexes require the actin-like protein, MreB and the Ras homologue, MglA. EMBO J. *29*, 315–326.

Meier, V.M., and Scharf, B.E. (2009). Cellular localization of predicted transmembrane and soluble chemoreceptors in *Sinorhizobium meliloti*. J. Bacteriol. *191*, 5724–5733.

Mello, B.A., and Tu, Y. (2005). An allosteric model for heterogeneous receptor complexes: understanding bacterial chemotaxis responses to multiple stimuli. Proc. Natl. Acad. Sci. U.S.A. *102*, 17354–17359.

Mignot, T., Merlie, J.P., and Zusman, D.R. (2005). Regulated pole-to-pole oscillations of a bacterial gliding motility protein. Science *310*, 855–857.

Milburn, M.V., Prive, G.G., Milligan, D.L., Scott, W.G., Yeh, J., Jancarik, J., Koshland, D.E., and Kim, S.H. (1991). Three-dimensional structures of the ligand-binding domain of the bacterial aspartate receptor with and without a ligand. Science *254*, 1342–1347.

Miller, A.S., and Falke, J.J. (2004). Side chains at the membrane–water interface modulate the signaling state of a transmembrane receptor. Biochemistry *43*, 1763–1770.

Miller, A.S., Kohout, S.C., Gilman, K.A., and Falke, J.J. (2006). CheA kinase of bacterial chemotaxis: chemical mapping of four essential docking sites. Biochemistry *45*, 8699–8711.

Milligan, D.L., and Koshland, D.E. (1988). Site-directed cross-linking. Establishing the dimeric structure of the aspartate receptor of bacterial chemotaxis. J. Biol. Chem. *263*, 6268–6275.

Muff, T.J., and Ordal, G.W. (2007). The CheC phosphatase regulates chemotactic adaptation through CheD. J. Biol. Chem. *282*, 34120–34128.

Muff, T.J., and Ordal, G.W. (2008). The diverse CheC-type phosphatases: chemotaxis and beyond. Mol. Microbiol. *70*, 1054–1061.

Nettleton, D.O., and Ordal, G.W. (1989). Functional homology of chemotactic methylesterases from *Bacillus subtilis* and *Escherichia coli*. J. Bacteriol. *171*, 120–123.

Packer, H.L., and Armitage, J.P. (2000). Behavioral responses of *Rhodobacter sphaeroides* to linear gradients of the nutrients succinate and acetate. Appl. Environ. Microbiol. *66*, 5186–5191.

Park, S.Y., Chao, X., Gonzalez-Bonet, G., Beel, B.D., Bilwes, A.M., and Crane, B.R. (2004). Structure and function of an unusual family of protein phosphatases: The bacterial chemotaxis proteins CheC and CheX. Mol. Cell *16*, 563–574.

Park, S.Y., Borbat, P.P., Gonzalez-Bonet, G., Bhatnagar, J., Pollard, A.M., Freed, J.H., Bilwes, A.M., and Crane, B.R. (2006a). Reconstruction of the chemotaxis receptor-kinase assembly. Nat. Struct. Mol. Biol. *13*, 400–407.

Park, S.Y., Lowder, B., Bilwes, A.M., Blair, D.F., and Crane, B.R. (2006b). Structure of FliM provides insight into assembly of the switch complex in the bacterial flagella motor. Proc. Natl. Acad. Sci. U.S.A. *103*, 11886–11891.

Parkinson, J.S. (2010). Signaling mechanisms of HAMP domains in chemoreceptors and sensor kinases. Annu. Rev. Microbiol. *64*, 101–122.

Parkinson, J.S., Ames, P., and Studdert, C.A. (2005). Collaborative signaling by bacterial chemoreceptors. Curr. Opin. Microbiol. *8*, 116–121.

Perez, E., Zheng, H., and Stock, A.M. (2006). Identification of methylation sites in *Thermotoga maritima* chemotaxis receptors. J. Bacteriol. *188*, 4093–4100.

Pettersen, E.F., Goddard, T.D., Huang, C.C., Couch, G.S., Greenblatt, D.M., Meng, E.C., and Ferrin, T.E. (2004). UCSF Chimera visualization system for exploratory research and analysis. J. Comput. Chem. *25*, 1605–1612.

Pfeffer, W. (1888). Ueber chemotaktische bewegungen von bakterien, flagellaten und volvocineen. Unters. Botan. Inst. Tubingen *2*, 582–661.

Pilizota, T., Brown, M.T., Leake, M.C., Branch, R.W., Berry, R.M., and Armitage, J.P. (2009). A molecular brake, not a clutch, stops the *Rhodobacter sphaeroides* flagellar motor. Proc. Natl. Acad. Sci. U.S.A. *106*, 11582–11587.

Pittman, M.S., Goodwin, M., and Kelly, D.J. (2001) Chemotaxis in the human gastric pathogen *Helicobacter pylori*: different roles for CheW and the three CheV paralogues, and evidence for CheV2 phosphorylation. Microbiology *147*, 2493–2504.

de Planque, M.R.R., and Killian, J.A. (2003). Protein_lipid interactions studied with designed transmembrane peptides: role of hydrophobic matching and interfacial anchoring. Mol. Membr. Biol. *20*, 271–284.

Platzer, J., Sterr, W., Hausmann, M., and Schmitt, R. (1997). Three genes of a motility operon and their role in flagellar rotary speed variation in *Rhizobium meliloti*. J. Bacteriol. *179*, 6391–6399.

Poggio, S., Abreu-Goodger, C., Fabela, S., Osorio, A., Dreyfus, G., Vinuesa, P., and Camarena, L. (2007). A complete set of flagellar genes acquired by horizontal transfer coexists with the endogenous flagellar system in *Rhodobacter sphaeroides*. J. Bacteriol. *189*, 3208–3216.

Pollard, A.M., Bilwes, A.M., and Crane, B.R. (2009). The structure of a soluble chemoreceptor suggests a mechanism for propagating conformational signals. Biochemistry *48*, 1936–1944.

Poole, P.S., and Armitage, J.P. (1989). Role of metabolism in the chemotactic response of *Rhodobacter sphaeroides* to ammonia. J. Bacteriol. *171*, 2900–2902.

Poole, P.S., Smith, M.J., and Armitage, J.P. (1993). Chemotactic signalling in *Rhodobacter sphaeroides* requires metabolism of attractants. J. Bacteriol. *175*, 291–294.

Porter, S.L., and Armitage, J.P. (2002). Phosphotransfer in *Rhodobacter sphaeroides* chemotaxis. J. Mol. Biol. *324*, 35–45.

Porter, S.L., and Armitage, J.P. (2004). Chemotaxis in *Rhodobacter sphaeroides* requires an atypical histidine protein kinase. J. Biol. Chem. *279*, 54573–54580.

Porter, S.L., Warren, A.V., Martin, A.C., and Armitage, J.P. (2002). The third chemotaxis locus of *Rhodobacter sphaeroides* is essential for chemotaxis. Mol. Microbiol. *46*, 1081–1094.

Porter, S.L., Wadhams, G.H., Martin, A.C., Byles, E.D., Lancaster, D.E., and Armitage, J.P. (2006). The CheYs of *Rhodobacter sphaeroides*. J. Biol. Chem. *281*, 32694–32704.

Porter, S.L., Roberts, M.A.J., Manning, C.S., and Armitage, J.P. (2008). A bifunctional kinase-phosphatase in bacterial chemotaxis. Proc. Natl. Acad. Sci. U.S.A. *105*, 18531–18536.

Porter, S.L., Wadhams, G.H., and Armitage, J.P. (2011a). Signal processing in complex chemotaxis pathways. Nat. Rev. Micro. *9*, 153–165.

Porter, S.L., Wilkinson, D.A., Byles, E.D., Wadhams, G.H., Taylor, S., Saunders, N.J., and Armitage, J.P. (2011b). Genome sequence of *Rhodobacter sphaeroides* strain WS8N. J. Bacteriol. *193*, 4027–4028.

Rao, C.V., Glekas, G.D., and Ordal, G.W. (2008). The three adaptation systems of *Bacillus subtilis* chemotaxis. Trends Microbiol. *16*, 480–487.

Riepl, H., Maurer, T., Kalbitzer, H.R., Meier, V.M., Haslbeck, M., Schmitt, R., and Scharf, B. (2008). Interaction of CheY2 and CheY2-P with the cognate CheA kinase in the chemosensory-signalling chain of *Sinorhizobium meliloti*. Mol. Microbiol. *69*, 1373–1384.

Ringgaard, S., Schirner, K., Davis, B.M., and Waldor, M.K. (2011). A family of ParA-like ATPases promotes cell pole maturation by facilitating polar localisation of chemotaxis proteins. Gene Dev. *25*, 1544–1555.

Roberts, M., August, E., Hamadeh, A., Maini, P.K., McSharry, P.E., Armitage, J.P., and Papachristodoulou, A. (2009). A model invalidation-based approach for elucidating biological signalling pathways, applied to the chemotaxis pathway in *R. sphaeroides*. BMC Syst. Biol. *3*, 105.

Romagnoli, S., Packer, H.L., and Armitage, J.P. (2002). Tactic responses to oxygen in the phototrophic bacterium *Rhodobacter sphaeroides* WS8N. J. Bacteriol. *184*, 5590–5598.

Rosario, M.M., and Ordal, G.W. (1996). CheC and CheD interact to regulate methylation of *Bacillus subtilis* methyl-accepting chemotaxis proteins. Mol. Microbiol. *21*, 511–518.

Rosario, M.M., Fredrick, K.L., Ordal, G.W., and Helmann, J.D. (1994). Chemotaxis in *Bacillus subtilis* requires either of two functionally redundant CheW homologs. J. Bacteriol. *176*, 2736–2739.

Rosario, M.M., Kirby, J.R., Bochar, D.A., and Ordal, G.W. (1995). Chemotactic methylation and behavior in *Bacillus subtilis*: role of two unique proteins, CheC and CheD. Biochemistry *34*, 3823–3831.

Sarkar, M.K., Paul, K., and Blair, D. (2010). Chemotaxis signaling protein CheY binds to the rotor protein FliN to control the direction of flagellar rotation in *Escherichia coli*. Proc. Natl. Acad. Sci. U.S.A. *107*, 9370–9375.

Scharf, B.E., Fahrner, K.A., Turner, L., and Berg, H.C. (1998). Control of direction of flagellar rotation in bacterial chemotaxis. Proc. Natl. Acad. Sci. U.S.A. *95*, 201–206.

Schmitt, R. (2002). Sinorhizobial chemotaxis: a departure from the enterobacterial paradigm. Microbiology *148*, 627–631.

Schulmeister, S., Ruttorf, M., Thiem, S., Kentner, D., Lebiedz, D., and Sourjik, V. (2008). Protein exchange dynamics at chemoreceptor clusters in *Escherichia coli*. Proc. Natl. Acad. Sci. U.S.A. *105*, 6403–6408.

Scott, A.E., Simon, E., Park, S.K., Andrews, P., and Zusman, D.R. (2008). Site-specific receptor methylation of FrzCD in *Myxococcus xanthus* is controlled by a tetra-trico peptide repeat (TPR) containing regulatory domain of the FrzF methyltransferase. Mol. Microbiol. *69*, 724–735.

Scott, K.A., Porter, S.L., Bagg, E.A., Hamer, R., Hill, J.L., Wilkinson, D.A., and Armitage, J.P. (2010). Specificity of localization and phosphotransfer in the CheA proteins of *Rhodobacter sphaeroides*. Mol. Microbiol. *76*, 318–330.

Segall, J.E., Block, S.M., and Berg, H.C. (1986). Temporal comparisons in bacterial chemotaxis. Proc. Natl. Acad. Sci. U.S.A. *83*, 9486–9493.

Shah, D.S.H., Porter, S.L., Martin, A.C., Hamblin, P.A., and Armitage, J.P. (2000). Fine tuning bacterial chemotaxis: analysis of *Rhodobacter sphaeroides* behaviour under aerobic and anaerobic conditions by mutation of the major chemotaxis operons and cheY genes. EMBO J. *19*, 4601–4613.

Shiomi, D., Yoshimoto, M., Homma, M., and Kawagishi, I. (2006). Helical distribution of the bacterial chemoreceptor via colocalization with the Sec protein translocation machinery. Mol. Microbiol. *60*, 894–906.

Silversmith, R.E. (2010). Auxiliary phosphatases in two-component signal transduction. Curr. Opin. Microbiol. *13*, 177–183.

Singh, M., Berger, B., Kim, P.S., Berger, J.M., and Cochran, A.G. (1998). Computational learning reveals coiled coil-like motifs in histidine kinase linker domains. Proc. Natl. Acad. Sci. U.S.A. *95*, 2738–2743.

Sourjik, V., and Berg, H.C. (2000). Localization of components of the chemotaxis machinery of *Esherichia coli* using fluorescent protein fusions. Mol. Microbiol. *37*, 740–751.

Sourjik, V., and Berg, H.C. (2004). Functional interactions between receptors in bacterial chemotaxis. Nature *428*, 437–441.

Sourjik, V., and Schmitt, R. (1996). Different roles of CheY1 and CheY2 in the chemotaxis of *Rhizobium meliloti*. Mol. Microbiol. 22, 427–436.

Sourjik, V., and Schmitt, R. (1998). Phosphotransfer between CheA, CheY1, and CheY2 in the chemotaxis signal transduction chain of *Rhizobium meliloti*. Biochemistry 37, 2327–2335.

Sowa, Y., and Berry, R.M. (2008). Bacterial flagellar motor. Q. R. Biophys. *41*, 103–132.

Spormann, A.M., and Kaiser, D. (1999). Gliding mutants of *Myxococcus xanthus* with high reversal frequencies and small displacements. J. Bacteriol. *181*, 2593–2601.

Springer, W.R., and Koshland, D.E. (1977). Identification of a protein methyltransferase as the cheR gene product in the bacterial sensing system. Proc. Natl. Acad. Sci. U.S.A. 74, 533–537.

Starrett, D.J., and Falke, J.J. (2005). Adaptation mechanism of the aspartate receptor: electrostatics of the adaptation subdomain play a key role in modulating kinase activity. Biochemistry 44, 1550–1560.

Stephens, B.B., Loar, S.N., and Alexandre, G. (2006). Role of CheB and CheR in the complex chemotactic and aerotactic pathway of *Azospirillum brasilense*. J. Bacteriol. *188*, 4759–4768.

Stewart, R.C. (1997). Kinetic characterization of phosphotransfer between CheA and CheY in the bacterial chemotaxis signal transduction pathway. Biochemistry 36, 2030–2040.

Stewart, R.C., Jahreis, K., and Parkinson, J.S. (2000). Rapid phosphotransfer to CheY from a CheA protein lacking the CheY-binding domain. Biochemistry *39*, 13157–13165.

Stock, A.M., Mottonen, J.M., Stock, J.B., and Schutt, C.E. (1989). Three-dimensional structure of CheY, the response regulator of bacterial chemotaxis. Nature *337*, 745–749.

Stock, A.M., Robinson, V.L., and Goudreau, P.N. (2000). Two-component signal transduction. Annu. Rev. Biochem. *69*, 183–215.

Stock, J.B., and Koshland, D.E. (1978). A protein methylesterase involved in bacterial sensing. Proc. Natl. Acad. Sci. U.S.A. *75*, 3659–3663.

Stock, J.B., and Koshland, D.E. (1981). Changing reactivity of receptor carboxyl groups during bacterial sensing. J. Biol. Chem. *256*, 10826–10833.

Studdert, C.A., and Parkinson, J.S. (2004). Cross-linking snapshots of bacterial chemoreceptor squads. Proc. Natl. Acad. Sci. U.S.A. *101*, 2117–2122.

Surette, M.G., Levit, M., Liu, Y., Lukat, G., Ninfa, E.G., Ninfa, A., and Stock, J.B. (1996). Dimerization is required for the activity of the protein histidine kinase chea that mediates signal transduction in bacterial chemotaxis. J. Biol. Chem. *271*, 939–945.

Swain, K.E., Gonzalez, M.A., and Falke, J.J. (2009). Engineered socket study of signaling through a four-helix bundle: evidence for a yin–yang mechanism in the kinase control module of the aspartate receptor. Biochemistry *48*, 9266–9277.

Swaney, K.F., Huang, C.H., and Devreotes, P.N. (2010). Eukaryotic chemotaxis: A network of signaling pathways controls motility, directional sensing, and polarity. Annu. Rev. Biophys. *39*, 265–289.

Swanson, R.V., Schuster, S.C., and Simon, M.I. (1993). Expression of CheA fragments which define domains encoding kinase, phosphotransfer, and CheY binding activities. Biochemistry *32*, 7623–7629.

Szurmant, H., Bunn, M.W., Cannistraro, V.J., and Ordal, G.W. (2003). *Bacillus subtilis* hydrolyzes CheY-P at the location of its action: the flagellar switch. J. Biol. Chem. *278*, 48611–48616.

Szurmant, H., Muff, T.J., and Ordal, G.W. (2004). *Bacillus subtilis* CheC and FliY are members of a novel class of CheY-P-hydrolyzing proteins in the chemotactic signal transduction cascade. J. Biol. Chem. *279*, 21787–21792.

Terwilliger, T.C., and Koshland, D.E. (1984). Sites of methyl esterification and deamination on the aspartate receptor involved in chemotaxis. J. Biol. Chem. *259*, 7719–7725.

Thiem, S., and Sourjik, V. (2008). Stochastic assembly of chemoreceptor clusters in *Escherichia coli*. Mol. Microbiol. *68*, 1228–1236.

Thiem, S., Kentner, D., and Sourjik, V. (2007). Positioning of chemosensory clusters in *E. coli* and its relation to cell division. EMBO J. *26*, 1615–1623.

Thompson, S.R., Wadhams, G.H., and Armitage, J.P. (2006). The positioning of cytoplasmic protein clusters in bacteria. Proc. Natl. Acad. Sci. U.S.A. *103*, 8209–8214.

Tindall, M.J., Porter, S.L., Maini, P.K., and Armitage, J.P. (2010). Modeling chemotaxis reveals the role of reversed phosphotransfer and a bi-functional kinase-phosphatase. PLoS Comput. Biol. *6*, e1000896.

Toews, M.L., Goy, M.F., Springer, M.S., and Adler, J. (1979). Attractants and repellents control demethylation of methylated chemotaxis proteins in *Escherichia coli*. Proc. Natl. Acad. Sci. U.S.A. *76*, 5544–5548.

Tso, W.W., and Adler, J. (1974). Negative chemotaxis in *Escherichia coli*. J. Bacteriol. *118*, 560–576.

Ueno, T., Oosawa, K., and Aizawa, S.I. (1992). M ring, S ring and proximal rod of the flagellar basal body of *Salmonella* typhimurium are composed of subunits of a single protein, FliF. J. Biol. Chem. *227*, 672–677.

Ulrich, L.E., Koonin, E.V., and Zhulin, I.B. (2005). One-component systems dominate signal transduction in prokaryotes. Trends Microbiol. *13*, 52–56.

Vaknin, A., and Berg, H.C. (2004). Single-cell FRET imaging of phosphatase activity in the *Escherichia coli* chemotaxis system. Proc. Natl. Acad. Sci. U.S.A. *101*, 17072–17077.

Vladimirov, N., and Sourjik, V. (2009). Chemotaxis: how bacteria use memory. Biol. Chem. *390*, 1097–1104.

Volz, K., and Matsumura, P. (1991). Crystal structure of *Escherichia coli* CheY refined at 1.7-Å resolution. J. Biol. Chem. *266*, 15511–15519.

Wadhams, G.H., Martin, A.C., Porter, S.L., Maddock, J.R., Mantotta, J.C., King, H.M., and Armitage, J.P. (2002). TlpC, a novel chemotaxis protein in *Rhodobacter sphaeroides*, localizes to a discrete region in the cytoplasm. Mol. Microbiol. *46*, 1211–1221.

Wadhams, G.H., Warren, A.V., Martin, A.C., and Armitage, J.P. (2003). Targeting of two signal transduction pathways to different regions of the bacterial cell. Mol. Microbiol. *50*, 763–770.

Ward, M.J., Lew, H., and Zusman, D.R. (2000). Social motility in *Myxococcus xanthus* requires FrzS, a protein with an extensive coiled-coil domain. Mol. Microbiol. *37*, 1357–1371.

Welch, M., Oosawa, K., Aizawa, S.I., and Eisenbach, M. (1993). Phosphorylation-dependent binding of a signal molecule to the flagellar switch of bacteria. Proc. Natl. Acad. Sci. U.S.A. *90*, 8787–8791.

Welch, M., Oosawa, K., Aizawa, S.I., and Eisenbach, M. (1994). Effects of phosphorylation, Mg^{2+}, and conformation of the chemotaxis protein CheY on its binding to the flagellar switch protein FliM. Biochemistry *33*, 10470–10476.

Welch, M., Chinardet, N., Mourey, L., Birck, C., and Samara, J.P. (1998). Structure of the CheY-binding domain of histidine kinase CheA in complex with CheY. Nat. Struct. Biol. *5*, 25–29.

Winston, S.E., Mehan, R., and Falke, J.J. (2005). Evidence that the adaptation region of the aspartate receptor is a dynamic four-helix bundle: Cysteine and disulfide scanning studies. Biochemistry *44*, 12655–12666.

Wu, J., Li, J., Li, G., Long, D.G., and Weis, R.M. (1996). The receptor binding site for the methyltransferase of bacterial chemotaxis is distinct from the sites of methylation. Biochemistry *35*, 4984–4993.

Wuichet, K., Alexander, R.P., and Zhulin, I.B. (2007). Comparative genomic and protein sequence analyses of a complex system controlling bacterial chemotaxis. In Methods in Enzymology Two-Component Signaling Systems, Part A, Melvin, I.S., ed. (Academic Press), pp. 1, 3–1,31.

Wuichet, K., and Zhulin, I.B. (2010). Origins and diversification of a complex signal transduction system in prokaryotes. Sci. Signal. 3, ra50.

Yonekawa, H., Hayashi, H., and Parkinson, J.S. (1983). Requirement of the cheB function for sensory adaptation in *Escherichia coli*. J. Bacteriol. *156*, 1228–1235.

Zhang, P., Khursigara, C.M., Hartnell, L.M., and Subramaniam, S. (2007). Direct visualization of *Escherichia coli* chemotaxis receptor arrays using cryo-electron microscopy. Proc. Natl. Acad. Sci. U.S.A. *104*, 3777–3781.

Zhao, J., and Parkinson, J.S. (2006). Cysteine-scanning analysis of the chemoreceptor-coupling domain of the *Escherichia coli* chemotaxis signaling kinase CheA. J. Bacteriol. *188*, 4321–4330.

Zhao, R., Collins, E.J., Bourret, R.B., and Silversmith, R.E. (2002). Structure and catalytic mechanism of the *E. coli* chemotaxis phosphatase CheZ. Nat. Struct. Biol. *9*, 570–575.

Zhou, Q., Ames, P., and Parkinson, J.S. (2009). Mutational analyses of HAMP helices suggest a dynamic bundle model of input_output signalling in chemoreceptors. Mol. Microbiol. 73, 801–814.

Zhou, Q., Ames, P., and Parkinson, J.S. (2011). Biphasic control logic of HAMP domain signalling in the *Escherichia coli* serine chemoreceptor. Mol. Microbiol. *80*, 596–611.

Zimmer, M.A., Tiu, J., Collins, M.A., and Ordal, G.W. (2000). Selective methylation changes on the *Bacillus subtilis* chemotaxis receptor McpB promote adaptation. J. Biol. Chem. 275, 24264–24272.

Regulation of Iron Homeostasis in Bacteria

10

Pierre Cornelis and Simon C. Andrews

Abstract

Iron is an essential nutrient for the growth of bacteria, but excess cellular iron can be hazardous because of its capacity to induce formation of reactive oxygen species (ROS) via the Fenton/Haber-Weiss reaction. For these reasons, bacteria must exert strict control over uptake and storage of iron so as to maintain sufficient levels of iron whilst limiting the accumulation of ROS. In this chapter we review how global regulators belonging to the extensive 'ferric uptake regulator family' (Fur, DtxR, Mur, IdeR, Irr) control the acquisition of iron from a range of sources. We also consider how Fur-like proteins not only control expression of their targets directly, but also mediate indirect control via alternative sigma factors or small regulatory RNAs.

Introduction

For the vast majority of bacteria, iron is an essential element that is not readily available owing to the poor solubility of the oxidized Fe^{3+} form that prevails aerobically (Andrews *et al.*, 2003). Because of this, bacteria inhabiting aerobic niches often suffer deficiencies in iron supply. Pathogenic bacteria experience a particularly acute form of iron-restriction. This arises from the host's 'iron-withdrawal response' to infection, whereby iron availability is constrained by increasing lactoferrin (an iron-chelating, bacteriostatic, extracellular glycoprotein) levels and reducing the degree of iron saturation for the circulating iron-transport protein, transferrin (Finkelstein *et al.*, 1983; Andrews *et al.*, 2003). The importance of iron to bacteria stems from its multiple metabolic roles. Examples of its crucial metabolic involvement include redox-stress resistance (e.g. haem-bearing catalases) and DNA manufacture (di-Fe containing ribonucleotide reductases). Although indispensable, iron is an unfriendly, hazardous metal as the Fe^{2+}-triggered Fenton reaction produces destructive reactive oxygen species (ROS) such as superoxide (O_2^-), hydrogen peroxide (H_2O_2) as well as the highly reactive hydroxyl radical ($OH^\bullet$).

Siderophore-mediated iron uptake in bacteria

Ferrous iron is readily oxidized by molecular oxygen at physiological pH generating highly insoluble ferric iron which precipitates forming iron oxy hydroxide mineral complexes. To compensate for this lack of soluble iron, many bacteria synthesize powerful extracellular Fe^{3+} chelators, known as siderophores, which have the capacity to rescue iron from precipitates through the generation of ferri-siderophore complexes (Neilands, 1987; Andrews *et al.*, 2003; Bagg and Neilands, 1987; Braun and Killmann, 1999). The general mechanism of siderophore-mediated high-affinity iron uptake by Gram-positive and Gram-negative bacteria is well reviewed (Andrews *et al.*, 2003; Wandersman and Delepelaire, 2004; Cornelis *et al.*, 2009). For Gram-positive bacteria, there is only one membrane and so only a single receptor (membrane-tethered), together with an archetypal ATP-Binding Cassette (ABC) permease, are required. In the case of Gram-negative bacteria there are two membranes to negotiate so a receptor located within the outer membrane, a periplasmic ferrisiderophore-binding protein and an ABC-type permease are employed (Andrews *et al.*, 2003). The internalized ferrisiderophore complex is dissociated by degradation of the siderophore (e.g. enterobactin is cleaved by an esterase) or by reduction of the complexed iron (ferrous iron has low affinity for siderophores) which allows reuse of the apo-siderophore (Andrews *et al.*, 2003). For the Gram-negative *Pseudomonas aeruginosa*, iron is released from the siderophore, pyoverdine, by an as yet unidentified periplasmic reductase (Schalk, 2008).

Haem-mediated iron uptake in bacteria

Haem is an important source of iron for bacteria as it is abundant within animal hosts (Wandersman and Delepelaire, 2004). However, haem must first be released from haemoglobin, haemopexin or other haemoproteins found within the host before it can be accessed by bacteria (Wyckoff *et al.*, 2005). There are at least two distinct types of pathway for haem extraction and uptake, one involves an extracellular 'haemophore' protein that scavenges haem for delivery to a OM-receptor-dependent transport system, and the other pathway utilizes OM receptors that bind haemoproteins on the cell surface where haem is then extracted (Wandersman and Stojiljkovic, 2000; Wandersman and Delepelaire, 2004). Once in the cytoplasm, iron is released from haem by haem-oxygenase-mediated degradation of the protoporphyrin ring into biliverdin and CO or, as described recently, iron can be removed from haem, without cleavage of the tetrapyrrole ring haem, by periplasmic or cytosolic haem-iron retrieving enzymes (Wandersman and Delepelaire, 2004; Letoffe *et al.*, 2009). Since free haem is toxic and poorly soluble, it has been proposed that haem-binding chaperones are employed to prevent oxidative damage and precipitation (Wyckoff *et al.*, 2005; Lansky *et al.*, 2006).

Iron storage proteins

Most bacteria, including some anaerobes, have the ability to store excess iron. This ability is bestowed upon them by the presence of iron-storage proteins, of which there are three types: ferritins, bacterioferritins, and the DNA-binding protein, Dps ('DNA-binding proteins from starved cells') (Andrews, 1998; Bereswill *et al.*, 2000; Carrondo, 2003; Chiancone *et al.*, 2004). All three types are evolutionarily related and have similar structures consisting of 12 or 24 subunits arranged to form a large spherical protein shell surrounding a hollow centre capable of sequestering 500–3000 ferric-iron atoms.

Regulation of iron uptake and storage in bacteria: the Fur paradigm

As suggested above, careful regulation of iron acquisition is necessary since the buildup of free iron in the cell can trigger Fenton chemistry whereas lack of iron restricts growth (Imlay *et al.*, 1988; Imlay and Linn, 1988; Winterbourn, 1995; Jang and Imlay, 2007). This control is achieved homeostatically. When faced with iron scarcity, the iron-uptake systems are up-regulated and iron stores are released. However, iron sufficiency triggers the reverse processes: down-regulation of iron uptake and deposition of iron stores. This regulatory response is coordinated by the ferric uptake regulator (Fur) – a conserved protein found in many Gram-negative and Gram-positive bacteria. Fur functions largely as a simplistic repressor, switching off the transcription of iron-transport and siderophore-biosynthesis genes under high iron conditions. Fur acts as a homodimeric protein and it generally contains one structural Zn^{2+} and one regulatory Fe^{2+} per monomer (Bsat *et al.*, 1998; Escolar *et al.*, 1999; Pohl *et al.*, 2003; Lee and Helmann, 2007). Both the precise identity and number of genes targeted by Fur is species dependent, but Fur often acts on a large set of genes with functions stretching well beyond iron homeostasis. In addition, Fur frequently exerts its regulatory control indirectly through secondary regulators such as small regulatory RNAs, extracytoplasmic σ factors (ECFs), AraC regulators, and two-component sensor-regulators. An example of such secondary control is seen in *P. aeruginosa* where the Fe^{2+}–Fur complex represses eleven ECF sigma factors as well as three other transcriptional regulators, PfeR,

PirR (controlling uptake of the siderophore enterobactin) and PchR (regulating import of another siderophore, pyochelin) (Potvin *et al.*, 2008; Cornelis *et al.*, 2009). As an illustration of indirect regulation by Fur, control of *P. aeruginosa* pyoverdine and pyochelin biosynthesis and uptake are described in Fig. 10.1 (Cornelis *et al.*, 2009). The pyoverdine siderophore binds Fe^{3+} outside the cell to form ferripyoverdine which binds to the outer membrane receptor, FpvA. This association triggers a conformational change in FpvA which causes a cleavage of the membrane anti-sigma protein FpvR, liberating two ECF sigma factors, PvdS and FpvI (Beare *et al.*, 2003; Ravel and Cornelis, 2003; Redly and Poole, 2003, 2005; Visca *et al.*, 2007). PvdS is needed for the transcription of pyoverdine biosynthesis genes and of other genes such as those which code for virulence factors (exotoxine A, PrpL protease). FpvI associates with the RNA polymerase to transcribe the *fpvA* gene. In the case of pyochelin, ferripyochelin is transported into the cytoplasm where it binds to the AraC regulator, PchR (Michel *et al.*, 2005). PchR activates the transcription of the genes required for biosynthesis of pyochelin and of the *fptA* gene encoding the receptor. PchR also acts as a negative autoregulator. Fur-Fe^{2+} represses the expression of the two ECF sigma factors and of *pchR* (Michel *et al.*, 2005).

Fe^{2+}-Fur normally represses by binding to the promoter region of cognate genes to block access by RNA polymerase. This binding is dependent on the presence of a Fur-binding site (the Iron or Fur box) that conforms to a 'minimal' consensus sequence of 19 bp which can be subdivided into three identical and adjacent 6 bp sites. It is believed that two Fe^{2+}-Fur dimers bind to each Fur box in a tandem head-to-tail fashion. This architecture allows the Fur box to be extended by addition of further adjacent 6 bp sites. A few such extended versions have been identified and are thought to bind multiple (above two) Fur dimers (Escolar *et al.*, 1999; van Oeffelen *et al.*, 2008; Cornelis *et al.*, 2009). The reason for such extended Fur boxes is unclear.

In addition to its behaviour as a repressor, Fur also acts positively to induce the expression of various genes specifying iron proteins (e.g. bacterioferritin and the iron co-factored superoxide dismutase, SodB) in response to high-iron conditions (Andrews *et al.*, 2003). This positive regulation is generally mediated indirectly by Fur-Fe^{2+}-repressed small-regulatory RNA molecules. In *E. coli*, the Fur-responsive regulatory RNA is called RyhB. This small RNA molecule cooperates with an RNA chaperone (Hfq) to hybridize to targeted messages and thus operates post-transcriptionally by sponsoring degradation of the targeted mRNA by ribonuclease (RNAse E) (Massé and Gottesman, 2002; Massé *et al.*, 2005; Prévost *et al.*, 2007). RyhB can also enhance expression by preventing degradation of its target (Prévost *et al.*, 2007). Other bacteria also employ small regulatory RNA molecules to mediate repression of iron–protein genes. For *P. aeruginosa*, three transcripts specified by the *prrF* locus fulfil this function: the closely identical and co-transcribed small RNA molecules, PrrF1 and PrrF2, together with PrrH which is a regulatory RNA molecule under haem and iron regulation. PrrH is involved in haem-dependent control of target genes (Wilderman *et al.*, 2004; Oglesby-Sherrouse and Vasil, 2010).

Not all Fur-Fe^{2+} induced genes are dependent on regulatory RNA molecules. For instance, Fur-Fe $^{2+}$ induction of the ferritin gene (*ftnA*) of *E. coli* requires the histone-like protein, H-NS (Nandal *et al.*, 2010). The transcription of *ftnA* is 'silenced' by H-NS, which occupies multiple sites upstream and within the *ftnA* promoter. Occupation of the downstream sites (on the promoter) appears to be cooperative and dependent upon occupation of the upstream sites. Fur-Fe^{2+} binds to the upstream sites (under high iron) and thus displaces H-NS from these positions. This, in turn, causes loss of H-NS from the downstream sites (on the promoter) due to the withdrawal of cooperative H-NS binding. The promoter is thus free to engage RNA polymerase when Fur-Fe^{2+} levels are high (Nandal *et al.*, 2010). See Fig. 10.2 for the different types of Fur-mediated regulation in the model bacterium *E. coli*.

Other examples of direct iron-induction of transcription by Fur are found in *Helicobacter pylori* and *Neisseria meningitidis* (Delany *et al.*, 2001a,b, 2003, 2004; Grifantini *et al.*, 2004; Danielli and Scarlato, 2010). Fur is able to act as

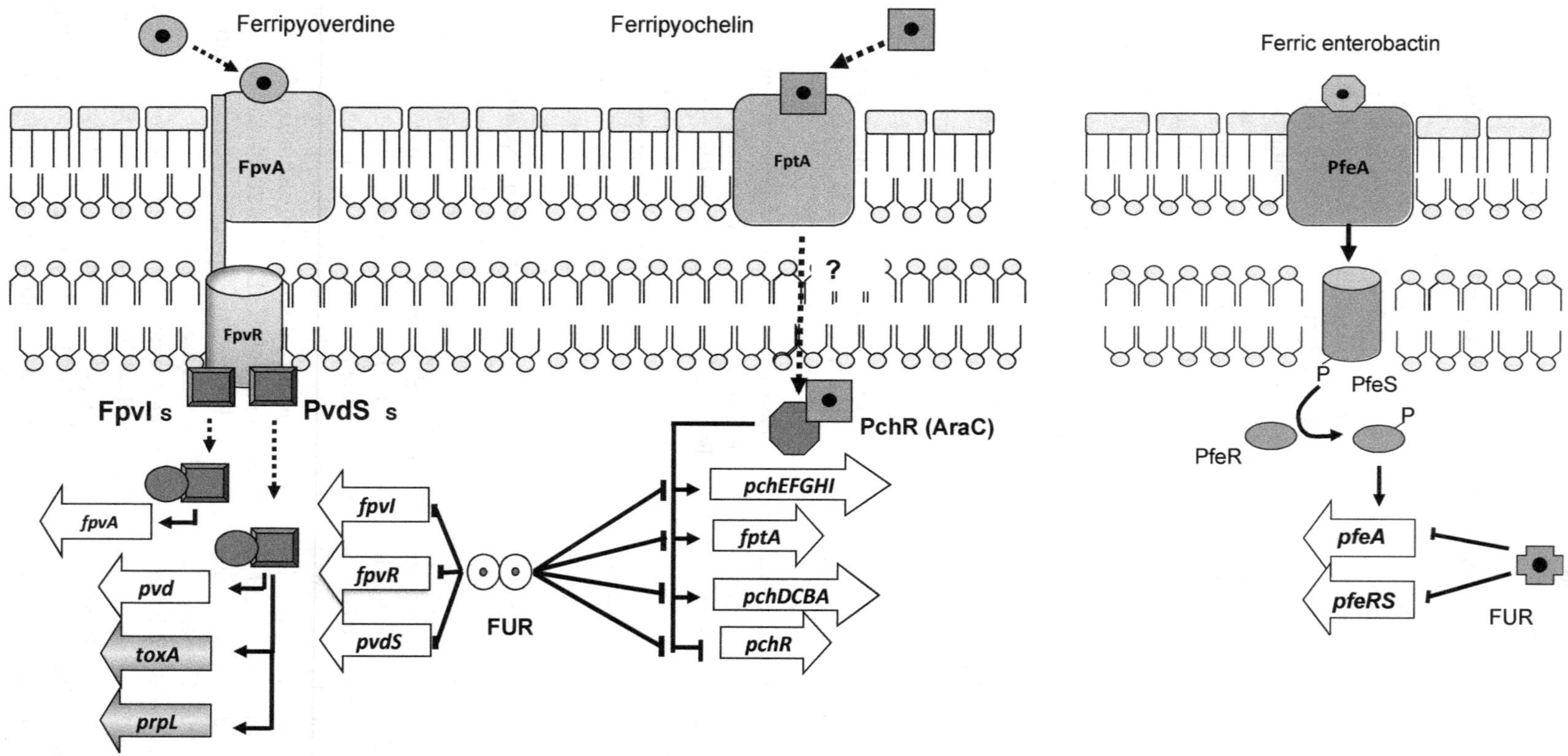

Figure 10.1 Indirect control by Fur of iron uptake mediated by pyoverdine, pyochelin and enterobactin in *P. aeruginosa*. Ferripyoverdine is bound at the level of the outer membrane by the FpvA receptor, which triggers a conformational change and a degradation of the anti-sigma factor FpvR, liberating the two ECF σ PvdS and FpvI, which associate with the core RNA polymerase to activate the transcription of the pyoverdine biosynthesis genes (*pvd*) and the extracellular virulence factors PrpL and exotoxin A (PvdS) and the receptor gene, *fpvA* (FpvI). Fur-Fe^{2+} represses the transcription of both *pvdS* and *fpvI*. Ferripyochelin transport in the cell occurs after binding of ferripyochelin to the FptA receptor. Pyochelin binds to the AraC regulator PchR which then activates the transcription of pyochelin biosynthesis genes and of the receptor gene *fptA*. PchR exerts a negative autoregulation. Fur-Fe^{2+} represses the transcription of *pchR*. Ferric enterobactin binding to the PfeA receptor triggers the phosphorylation of the PfeS sensor, which in turn phosphorylates the response regulator PfeR. Phosphorylated PfeR activates the transcription of pfeA and pfeRS.

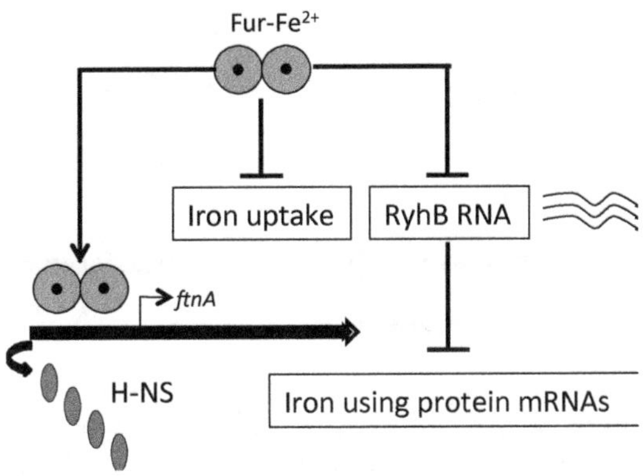

Figure 10.2 Mechanisms of Fur-mediated gene regulation in *E. coli*: In the majority of the cases Fur-Fe^{2+} binds an operator sequence preventing the RNA polymerase from recognizing the promoter sequence. The repressor in its holo form also represses the transcription of the *ryhB* gene encoding a small regulatory RNA. RyhB is an antisense RNA which first associates with the Hfq protein and then binds the mRNAs of iron-containing proteins (18 target genes known in *E. coli*) recruiting RNAse E which destroys the messages, together with RyhB. In the case of the *ftnA* gene encoding the ferritin protein, Fur-Fe^{2+} displaces H-NS from binding sites upstream of the *ftnA* promoter. This prevents cooperative downstream binding of H-NS directly on the *ftnA* promoter which in turn relieves H-NS mediated silencing of *ftnA* transcription.

a repressor in its iron-free apo-form in *H. pylori*. This allows iron-dependent induction (or more accurately, de-repression) of the *sodB* and *pfr* genes, encoding the Fe co-factored SodB and ferritin (Carpenter *et al.*, 2009; Danielli and Scarlato, 2010). Interestingly, Fur proteins from various other bacteria were unable to replicate apo-Fur repression in a *fur* mutant of *H. pylori* although they could engage in Fur-Fe^{2+} repression. This indicates that the apo-Fur repression seen in *H. pylori* is not a general feature of Fur proteins (Miles *et al.*, 2010). In addition to its repression effects in its apo and Fe forms, Fur can also function as a transcriptional activator in *H. pylori* in its ferrated form. Fur-Fe^{2+} induces the *oorDABC* operon, which encodes the essential ferredoxin-containing enzyme 2-oxoglutarate oxidoreductase, by binding just upstream of the promoter (Danielli and Scarlato, 2010). The four modes of Fur-mediated regulation in *H. pylori* are summarized in Fig. 10.3.

Other regulators of iron uptake in Gram-negative bacteria

Although Fur is the most common and best known coordinator of bacterial iron control, many bacteria do not utilize Fur-like proteins to achieve global iron regulation and instead employ regulators quite different from Fur (Lee and Helmann, 2007). Indeed, Fur-like proteins may be present but often have roles beyond iron-dependent transcriptional control. Some respond to manganese (Mur), others sense zinc (Zur), peroxides (PerR) or nickel (Nur) (Patzer and Hantke, 1998; Chao *et al.*, 2004; Ahn *et al.*, 2006). The closest Fur homologue in the Rhizobiales is Mur. It has been demonstrated that Mur is a Mn^{2+}-dependent repressor of the manganese uptake transporter, MntH in *Bradyrhizobium japonicum* (Hohle and O'Brian, 2009) and of the manganese uptake transporter MntABCD/SitABCD in *Sinorhizobium meliloti*, and *Rhizobium leguminosarum* (Chao *et al.*, 2004; Platero *et al.*, 2004). The iron-regulatory requirement in Rhizobiales is met by two regulators, RirA (rhizobial iron regulator) and Irr (iron response regulator) (Hamza *et al.*, 1998; Johnston *et al.*, 2007). RirA is part of a Rrf2 group of transcription regulators (Pfam, PF001022) that include NsrR, the NO-responsive activator, and IscR, the [Fe–S] cluster biosynthesis regulator (Yeo *et al.*, 2006). RirA is a transcriptional repressor controlling siderophore-dependent iron uptake in *Rhizobium leguminosarum* and in *Sinorhizobium meliloti* (Todd *et al.*, 2002; Yeoman *et al.*, 2004; Chao *et al.*, 2005; Viguier *et al.*, 2005). RirA appears to respond to cellular [Fe–S] cluster level and not cytosolic free iron levels.

Irr is the other iron-regulatory protein used by rhizobia. It belongs to the Fur family of regulators and is unlike RirA. Irr is a global regulator of iron homeostasis, haem synthesis and metabolism in the α-Proteobacteria. However, Irr does not directly sense free-iron levels but responds to the relative levels of haem and its precursor, protoporphyrin (Hamza *et al.*, 1998; Martinez *et al.*, 2005, 2006; Rodionov *et al.*, 2006; Todd *et al.*, 2006). Under high iron conditions, Irr combines with ferrochelatase and haem; this stimulates loss of Irr by proteolysis. Inactivation of Irr therefore requires the interaction with both ferrochelatase

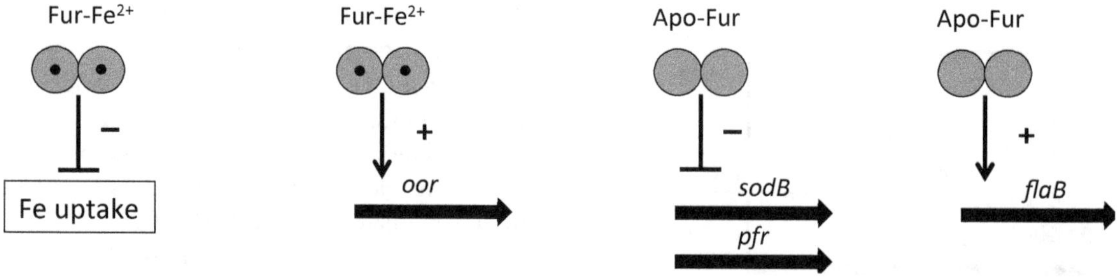

Figure 10.3 Mechanisms of gene regulation by the *H. pylori* Fur protein: Fur-Fe^{2+} can classically act as a repressor of iron uptake genes or it can be an activator of the expression of the *oor* genes. Apo-Fur can either activate the flagellin *flaB* gene or repress the gene for the Fe-cofactored superoxide dismutase SodB and the ferritin *pfr* gene.

and its product, haem. Under low-iron conditions, haem concentrations are reduced, and the protoporphyrin IX–ferrochelatase complex does not bind Irr, preventing its degradation and enabling it to control (it functions as an activator a well as a repressor) a number of genes with roles in iron homeostasis (Fig. 10.4). In this way, the rhizobia use two distinct regulators to control iron through sensing [Fe–S] cluster and haem manufacture, rather than by monitoring free iron levels.

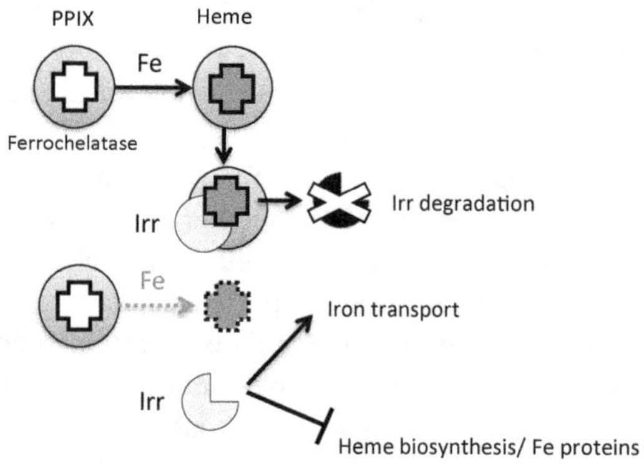

Figure 10.4 Regulation of iron homeostasis by Irr in *B. japonicum*: the rhizobacterial Irr regulator binds both haem after iron incorporation into protoporphyrin IX (PPIX) and the ferrochelatase enzyme – this promotes Irr degradation. When the level of iron is too low to ensure enough metallation of PPIX, the ferrochelatase–PPIX complex cannot bind Irr, Irr is free and can activate the transcription of genes involved in the uptake of iron while down-regulating genes for the biosynthesis of haem and the storage of iron.

The DtxR family of iron uptake regulators in Gram-positive bacteria

In the high-GC Gram-positive bacteria, the role of Fur is taken by DtxR-like proteins (Diphtheria toxin repressor). These proteins include DtxR in *Corynebacterium* species and IdeR in *Mycobacterium tuberculosis* (Schmitt *et al.*, 1995; Dussurget *et al.*, 1996; Gold *et al.*, 2001; Qian *et al.*, 2002; Rodriguez *et al.*, 2002; Kunkle and Schmitt, 2005; Pennella and Giedroc, 2005; Brune *et al.*, 2006). Such proteins control iron-dependent gene expression by acting as transcriptional repressors in a manner that is similar to that of Fur, although these two types of regulator are not related. In *M. tuberculosis*, IdeR can also activate transcription by binding upstream of the promoter, as occurs for the bacterioferritin and ferritin genes (Rodriguez, 2006). The structures of DtxR and IdeR in complex with DNA have been obtained (Pohl *et al.*, 1999a,b; Wisedchaisri *et al.*, 2004).

Regulation of iron–sulfur cluster biogenesis by the IscR regulator

Iron-sulfur proteins have important functions in the bacterial cell since some are enzymes involved in the central metabolism (TCA cycle) while others are involved in the transport of electrons in the respiration process or in oxido-reduction reactions (ferredoxins). Some regulatory proteins also have Fe–S clusters, such as the oxidative-stress regulator SoxR, the redox sensor Fnr, and IscR, the regulator of Fe–S cluster biogenesis (Schwartz *et al.*, 2001; Py and Barras, 2010). *E. coli* has two pathways for the biosynthesis and insertion of Fe–S clusters, Isc and Suf (Py and Barras, 2010). The Suf pathway is only used to repair Fe–S clusters that have been damaged owing to oxidation and is a stress-induced salvage pathway (Lee *et al.*, 2008). IscR regulates both the *iscR-iscSUA hscBA-fdx* and the *sufABCDSE* gene clusters, but in different ways (Fig. 10.5). Holo-IscR (with intact Fe–S cluster) represses its own transcription and the transcription of the *isc hsc* genes, a situation likely to occur when iron is not limiting (Py and Barras, 2010). When the cell faces an iron shortage, the amount of apo-IscR rises, resulting in the production of *iscR-iscSUA hscBA-fdx* transcripts, which are then targeted by the small RNA RyhB, resulting in the degradation of the transcript downstream of the *iscR* region (Fig. 10.5). The resulting increase in apo-IscR positively regulates the transcription of the *sufABCDSE* genes (Fig. 10.5). Interestingly, the regulation of the transcription of the *suf* operon is even more complex since it involves the OxyR regulator, Fur and the nucleoid protein, integration host factor (Ihf) (Fig. 10.5). In this case Fur-Fe^{2+} represses the *suf* operon when iron-sufficiency is attained. In addition it was found that IscR regulates *hya* and *hyb* operons for the synthesis of Fe–S cluster-containing periplasmic hydrogenases 1 and 2, and the gene for the periplasmic nitrate reductase (Giel *et al.*, 2006). IscR also regulates the expression of the *soxS* gene encoding an oxidative stress response regulator, and the *sodA* gene for the manganese co-factored superoxide dismutase (Giel *et al.*, 2006; Wu and Outten, 2009). An intriguing observation is the involvement of IscR in biofilm formation via the control of the *fimE* gene which encodes a recombinase involved in type I fimbriae formation (Wu and Outten, 2009).

Importance of the Fur family of regulators for resistance to oxidative stress

Iron control is an important aspect of redox stress resistance since high levels of free cytosolic iron

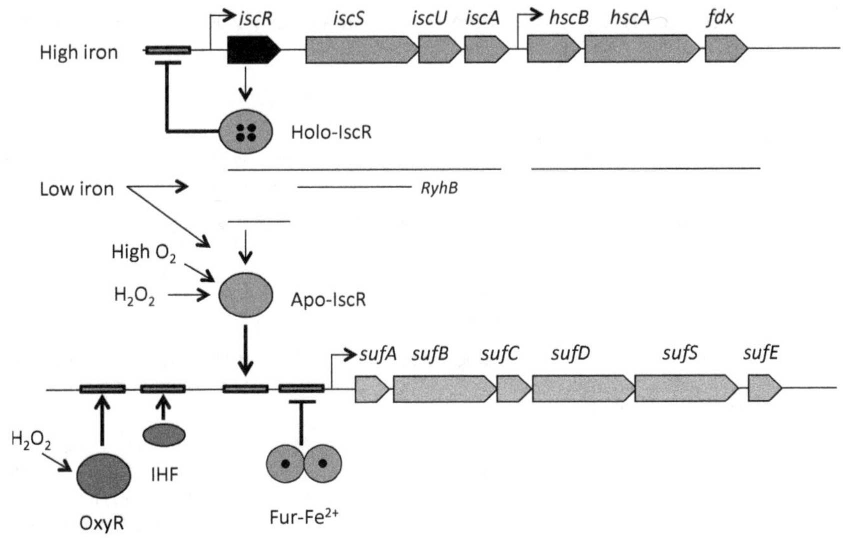

Figure 10.5 IscR-mediated regulation of Fe–S cluster assembly in *E. coli*. When sufficient iron and sulfur are present in the cell IscR is in the [Fe–S]-containing holo-form and is working as a repressor of the *isc-hsc* gene cluster. Under conditions of iron deficiency, Fur-Fe^{2+} cannot repress the transcription of the small RNA *ryhB*, resulting in post-transcriptional degradation of the *iscSUA-hscBAfdx* mRNA, leaving the *iscR* gene transcript intact; this results in the production of apo-IscR. Apo-IscR acts as an activator of the transcription of the *suf* operon together with IHF and OxyR. The presence of high oxygen levels or of H_2O_2 can also result in the destruction of the IscR [Fe–S] cluster, which results in the apo-form of IscR. When iron sufficiency is restored, Fur-Fe^{2+} represses the transcription of the *suf* operon, shifting [Fe–S] cluster assembly to the *isc*-encoded system.

promote iron-catalysed production of ROS (Touati, 2000; Cornelis *et al.*, 2011). Loss of iron control can therefore lead to redox stress sensitivity. In some cases the global-iron regulator appears essential and so cannot be inactivated (e.g. Fur in *P. aeruginosa* and IdeR in *M. tuberculosis*); this might well be because of redox-stress related toxicity problems (Barton *et al.*, 1996; Rodriguez *et al.*, 2002). The ability of Fur-Fe^{2+} to at once activate the transcription of iron-proteins gene as well as repress iron transport indicates that loss of *fur* causes enhanced acquisition of iron together with decreased assembly of iron into bio-matter (Andrews *et al.*, 2003). The impact of these two effects is raised levels of 'free' cellular iron promoting Fenton chemistry leading to ROS generation. Indeed, *E. coli fur* mutants have lower amounts of total cellular iron but have raised levels of reactive free Fe^{2+}. This is a result of lower levels of iron-containing proteins, like ferritin, such that imported iron is not readily assimilated and so remains unincorporated (in a 'free' form) in the cytosol (Abdul-Tehrani *et al.*, 1999). This effect is also seen in mutants of *P. aeruginosa* with low Fur activity (Hassett *et al.*, 1996) and in other bacteria where lack of *fur* results in sensitivity to oxidative stress (Lopez-Gomollon *et al.*, 2009; da Silva Neto *et al.*, 2009; Jittawuttipoka *et al.*, 2010). Also, in *Agrobacterium tumefaciens* and *Sinorhizobium meliloti*, *rirA* mutants display a growth deficiency in iron-replete media due to enhanced accumulation of iron causing increased sensitivity to H_2O_2 (Chao *et al.*, 2005; Ngok-Ngam *et al.*, 2009).

In *E. coli*, transcription of the *fur* gene is under the positive control of the LysR-like regulator OxyR and of SoxR-SoxS, which together trigger the production of enzymes involved in the removal of ROS (catalases, alkylhydroperoxidases, superoxide dismutases), establishing a supplementary link between the response to oxidative stress and iron management (Zheng *et al.*, 1999, 2001; Cornelis *et al.*, 2011).

It should be noted that in some bacteria, a Fur-like protein (PerR, Peroxide regulator) plays a major role in redox-stress management, rather than iron homeostasis. PerR is found in Gram-positive bacteria such as *Bacillus subtilis*, *Clostridium acetobutylicum*, *Streptococcus*, and *Staphylococcus aureus* (Bsat *et al.*, 1998; Horsburgh *et al.*, 2001b; Gryllos *et al.*, 2008; Hillmann *et al.*, 2009), as well as some Gram-negative species such as *Campylobacter jejuni* and *Neisseria gonorrhoeae* (Wu *et al.*, 2006; Palyada *et al.*, 2009). A list of metal-responsive regulators is presented in Table 10.1.

Table 10.1 Examples of bacterial metal-responsive regulators

Regulator	Metals	Function	Organisms	Reference
Fur	Fe^{2+}	Iron uptake repression and iron storage activation	*E. coli, P. aeruginosa, B. subtilis, S. aureus, H. pylori, X. campestris, N. meningitidis*	Hassett *et al.* (1996), Escolar *et al.* (1999), Lee and Helmann (2007)
RirA	Fe^{2+}	Iron uptake repression and iron storage activation	α-Proteobacteria	Todd *et al.* (2002), Yeoman *et al.* (2004), Chao *et al.* (2005)
DtxR/IdeR	Fe^{2+}	Iron uptake	*Corynebacterium, Mycobacterium*	Schmitt *et al.* (1995), Pohl *et al.* (1999a,b), Gold *et al.* (2001), Qian *et al.* (2002), Rodriguez *et al.* (2002), Wisedchaisri *et al.* (2004), Kunkle and Schmitt (2005), Brune *et al.* (2006)
PerR	Zn^{2+}	Oxidative stress response	*Bacillus subtilis, Campylobacter jejuni*	Bsat *et al.* (1998), van Vliet *et al.* (1999), Horsburgh *et al.* (2001a), Lee and Helmann (2006), Wu *et al.* (2006), Hillmann *et al.* (2009), Palyada *et al.* (2009), Duarte and Latour (2010)
IscR	Fe–S cluster	Fe–S cluster assembly	*E. coli, P. aeruginosa, Erwinia chrisanthemi*	Schwartz *et al.* (2001), Giel *et al.* (2006), Yeo *et al.* (2006), Lee *et al.* (2008), Rincon-Enriquez *et al.* (2008), Py and Barras (2010)

Conclusions

As revealed in this short review, the concept of regulation of iron homeostasis in bacteria has advanced considerably since the first discovery of the role played by Fur as repressor of iron uptake (Hantke, 1987). One of the most significant developments has been the finding that Fur can indirectly and post-transcriptionally activate the expression of many Fe-containing proteins, either directly or via small regulatory RNAs, by H-NS displacement or by repressing in the 'apo' form. A further important progression has been the realization that Fur is not the only global Fe-regulation option available to bacteria. Some bacteria use non-Fur proteins (DtxR, Irr, IdeR and RirA) to discharge the role of central iron regulator. Thus, the repertoire of iron homeostasis regulation factors has been very much extended beyond the Fur paradigm, and there can be little doubt that this exciting story will continue to develop as further discoveries materialize.

References

Abdul-Tehrani, H., Hudson, A.J., Chang, Y.S., Timms, A.R., Hawkins, C., Williams, J.M., Harrison, P.M., Guest, J.R., and Andrews, S.C. (1999). Ferritin mutants of *Escherichia coli* are iron deficient and growth impaired, and *fur* mutants are iron deficient. J. Bacteriol. *181*, 1415–1428.

Ahn, B.E., Cha, J., Lee, E.J., Han, A.R., Thompson, C.J., and Roe, J.H. (2006). Nur, a nickel-responsive regulator of the Fur family, regulates superoxide dismutases and nickel transport in *Streptomyces coelicolor*. Mol. Microbiol. *59*, 1848–1858.

Andrews, S.C. (1998). Iron storage in bacteria. Adv. Microb. Physiol. *40*, 281–351.

Andrews, S.C., Robinson, A.K., and Rodriguez-Quinones, F. (2003). Bacterial iron homeostasis. FEMS Microbiol. Rev. *27*, 215–237.

Bagg, A., and Neilands, J.B. (1987). Molecular mechanism of regulation of siderophore-mediated iron assimilation. Microbiol. Rev. *51*, 509–518.

Barton, H.A., Johnson, Z., Cox, C.D., Vasil, A.I., and Vasil, M.L. (1996). Ferric uptake regulator mutants of *Pseudomonas aeruginosa* with distinct alterations in the iron-dependent repression of exotoxin A and siderophores in aerobic and microaerobic environments. Mol. Microbiol. *21*, 1001–1017.

Beare, P.A., For, R.J., Martin, L.W., and Lamont, I.L. (2003). Siderophore-mediated cell signalling in *Pseudomonas aeruginosa*: divergent pathways regulate virulence factor production and siderophore receptor synthesis. Mol. Microbiol. *47*, 195–207.

Bereswill, S., Greiner, S., van Vliet, A.H., Waidner, B., Fassbinder, F., Schiltz, E., Kusters, J.G., and Kist, M. (2000). Regulation of ferritin-mediated cytoplasmic iron storage by the ferric uptake regulator homolog (Fur) of *Helicobacter pylori*. J. Bacteriol. *182*, 5948–5953.

Braun, V., and Killmann, H. (1999). Bacterial solutions to the iron-supply problem. Trends Biochem. Sci. *24*, 104–109.

Brune, I., Werner, H., Huser, A.T., Kalinowski, J., Puhler, A., and Tauch, A. (2006). The DtxR protein acting as dual transcriptional regulator directs a global regulatory network involved in iron metabolism of *Corynebacterium glutamicum*. BMC Genomics 7, 21.

Bsat, N., Herbig, A., Casillas-Martinez, L., Setlow, P., and Helmann, J.D. (1998). *Bacillus subtilis* contains multiple Fur homologues: identification of the iron uptake (Fur) and peroxide regulon (PerR) repressors. Mol. Microbiol. *29*, 189–198.

Carpenter, B.M., Gancz, H., Gonzalez-Nieves, R.P., West, A.L., Whitmire, J.M., Michel, S.L., and Merrell, D.S. (2009). A single nucleotide change affects fur-dependent regulation of *sodB* in *H. pylori*. PLoS One *4*, e5369.

Carrondo, M.A. (2003). Ferritins, iron uptake and storage from the bacterioferritin viewpoint. EMBO J. *22*, 1959–1968.

Chao, T.C., Becker, A., Buhrmester, J., Puhler, A., and Weidner, S. (2004). The *Sinorhizobium meliloti fur* gene regulates, with dependence on Mn(II), transcription of the *sitABCD* operon, encoding a metal-type transporter. J. Bacteriol. *186*, 3609–3620.

Chao, T.C., Buhrmester, J., Hansmeier, N., Puhler, A., and Weidner, S. (2005). Role of the regulatory gene *rirA* in the transcriptional response of *Sinorhizobium meliloti* to iron limitation. Appl. Environ. Microbiol. *71*, 5969–5982.

Chiancone, E., Ceci, P., Ilari, A., Ribacchi, F., and Stefanini, S. (2004). Iron and proteins for iron storage and detoxification. Biometals *17*, 197–202.

Cornelis, P., Matthijs, S., and Van Oeffelen, L. (2009). Iron uptake regulation in *Pseudomonas aeruginosa*. Biometals 22, 15–22.

Cornelis, P., Wei, Q., Andrews, S.C., and Vinckx, T. (2011). Iron homeostasis and management of oxidative stress response in bacteria. Metallomics 3, 540–549.

Danielli, A., and Scarlato, V. (2010). Regulatory circuits in *Helicobacter pylori*: network motifs and regulators involved in metal-dependent responses. FEMS Microbiol. Rev. *34*, 738–752.

Delany, I., Pacheco, A.B., Spohn, G., Rappuoli, R., and Scarlato, V. (2001a). Iron-dependent transcription of the *frpB* gene of *Helicobacter pylori* is controlled by the Fur repressor protein. J. Bacteriol. *183*, 4932–4937.

Delany, I., Rappuoli, R., and Scarlato, V. (2004). Fur functions as an activator and as a repressor of putative virulence genes in *Neisseria meningitidis*. Mol. Microbiol. *52*, 1081–1090.

Delany, I., Spohn, G., Rappuoli, R., and Scarlato, V. (2001b). The Fur repressor controls transcription of iron-activated and -repressed genes in *Helicobacter pylori*. Mol. Microbiol. *42*, 1297–1309.

Delany, I., Spohn, G., Rappuoli, R., and Scarlato, V. (2003). An anti-repression Fur operator upstream of the promoter is required for iron-mediated transcriptional autoregulation in *Helicobacter pylori*. Mol. Microbiol. *50*, 1329–1338.

Duarte, V., and Latour, J.M. (2010). PerR vs OhrR: selective peroxide sensing in *Bacillus subtilis*. Mol. Biosyst. *6*, 316–323.

Dussurget, O., Rodriguez, M., and Smith, I. (1996). An *ideR* mutant of *Mycobacterium smegmatis* has derepressed siderophore production and an altered oxidative-stress response. Mol. Microbiol. *22*, 535–544.

Escolar, L., Perez-Martin, J., and de Lorenzo, V. (1999). Opening the iron box: transcriptional metalloregulation by the Fur protein. J. Bacteriol. *181*, 6223–6229.

Finkelstein, R.A., Sciortino, C.V., and McIntosh, M.A. (1983). Role of iron in microbe–host interactions. Rev. Infect. Dis. 5 *Suppl 4*, S759–777.

Giel, J.L., Rodionov, D., Liu, M., Blattner, F.R., and Kiley, P.J. (2006). IscR-dependent gene expression links iron–sulphur cluster assembly to the control of O_2-regulated genes in *Escherichia coli*. Mol. Microbiol. *60*, 1058–1075.

Gold, B., Rodriguez, G.M., Marras, S.A., Pentecost, M., and Smith, I. (2001). The *Mycobacterium tuberculosis* IdeR is a dual functional regulator that controls transcription of genes involved in iron acquisition, iron storage and survival in macrophages. Mol. Microbiol. *42*, 851–865.

Grifantini, R., Frigimelica, E., Delany, I., Bartolini, E., Giovinazzi, S., Balloni, S., Agarwal, S., Galli, G., Genco, C., and Grandi, G. (2004). Characterization of a novel *Neisseria meningitidis* Fur and iron-regulated operon required for protection from oxidative stress: utility of DNA microarray in the assignment of the biological role of hypothetical genes. Mol. Microbiol. *54*, 962–979.

Gryllos, I., Grifantini, R., Colaprico, A., Cary, M.E., Hakansson, A., Carey, D.W., Suarez-Chavez, M., Kalish, L.A., Mitchell, P.D., White, G.L., *et al.* (2008). PerR confers phagocytic killing resistance and allows pharyngeal colonization by group A *Streptococcus*. PLoS Pathog. *4*, e1000145.

Hamza, I., Chauhan, S., Hassett, R., and O'Brian, M.R. (1998). The bacterial irr protein is required for coordination of heme biosynthesis with iron availability. J. Biol. Chem. *273*, 21669–21674.

Hantke, K. (1987). Selection procedure for deregulated iron transport mutants (*fur*) in *Escherichia coli* K 12: fur not only affects iron metabolism. Mol. Gen. Genet. *210*, 135–139.

Hassett, D.J., Sokol, P.A., Howell, M.L., Ma, J.F., Schweizer, H.T., Ochsner, U., and Vasil, M.L. (1996). Ferric uptake regulator (Fur) mutants of *Pseudomonas aeruginosa* demonstrate defective siderophore-mediated iron uptake, altered aerobic growth, and decreased superoxide dismutase and catalase activities. J. Bacteriol. *178*, 3996–4003.

Hillmann, F., Doring, C., Riebe, O., Ehrenreich, A., Fischer, R.J., and Bahl, H. (2009). The role of PerR in O_2-affected gene expression of *Clostridium acetobutylicum*. J. Bacteriol. *191*, 6082–6093.

Hohle, T.H., and O'Brian, M.R. (2009). The *mntH* gene encodes the major $Mn^{(2+)}$ transporter in *Bradyrhizobium japonicum* and is regulated by manganese via the Fur protein. Mol. Microbiol. *72*, 399–409.

Horsburgh, M.J., Clements, M.O., Crossley, H., Ingham, E., and Foster, S.J. (2001a). PerR controls oxidative stress resistance and iron storage proteins and is required for virulence in *Staphylococcus aureus*. Infect. Immun. *69*, 3744–3754.

Horsburgh, M.J., Ingham, E., and Foster, S.J. (2001b). In *Staphylococcus aureus*, fur is an interactive regulator with PerR, contributes to virulence, and Is necessary for oxidative stress resistance through positive regulation of catalase and iron homeostasis. J. Bacteriol. *183*, 468–475.

Imlay, J.A., and Linn, S. (1988). DNA damage and oxygen radical toxicity. Science *240*, 1302–1309.

Imlay, J.A., Chin, S.M., and Linn, S. (1988). Toxic DNA damage by hydrogen peroxide through the Fenton reaction *in vivo* and *in vitro*. Science *240*, 640–642.

Jang, S., and Imlay, J.A. (2007). Micromolar intracellular hydrogen peroxide disrupts metabolism by damaging iron–sulfur enzymes. J. Biol. Chem. *282*, 929–937.

Jittawuttipoka, T., Sallabhan, R., Vattanaviboon, P., Fuangthong, M., and Mongkolsuk, S. (2010). Mutations of ferric uptake regulator (*fur*) impair iron homeostasis, growth, oxidative stress survival, and virulence of *Xanthomonas campestris* pv. *campestris*. Arch. Microbiol. *192*, 331–339.

Johnston, A.W., Todd, J.D., Curson, A.R., Lei, S., Nikolaidou-Katsaridou, N., Gelfand, M.S., and Rodionov, D.A. (2007). Living without Fur: the subtlety and complexity of iron-responsive gene regulation in the symbiotic bacterium *Rhizobium* and other alpha-proteobacteria. Biometals *20*, 501–511.

Kunkle, C.A., and Schmitt, M.P. (2005). Analysis of a DtxR-regulated iron transport and siderophore biosynthesis gene cluster in *Corynebacterium diphtheriae*. J. Bacteriol. *187*, 422–433.

Lansky, I.B., Lukat-Rodgers, G.S., Block, D., Rodgers, K.R., Ratliff, M., and Wilks, A. (2006). The cytoplasmic heme-binding protein (PhuS) from the heme uptake system of *Pseudomonas aeruginosa* is an intracellular heme-trafficking protein to the delta-regioselective heme oxygenase. J. Biol. Chem. *281*, 13652–13662.

Lee, J.W., and Helmann, J.D. (2006). Biochemical characterization of the structural Zn^{2+} site in the *Bacillus subtilis* peroxide sensor PerR. J. Biol. Chem. *281*, 23567–23578.

Lee, J.W., and Helmann, J.D. (2007). Functional specialization within the Fur family of metalloregulators. Biometals *20*, 485–499.

Lee, K.C., Yeo, W.S., and Roe, J.H. (2008). Oxidant-responsive induction of the *suf* operon, encoding a Fe-S assembly system, through Fur and IscR in *Escherichia coli*. J. Bacteriol. *190*, 8244–8247.

Letoffe, S., Heuck, G., Delepelaire, P., Lange, N., and Wandersman, C. (2009). Bacteria capture iron from heme by keeping tetrapyrrol skeleton intact. Proc. Natl. Acad. Sci. U.S.A. *106*, 11719–11724.

Lopez-Gomollon, S., Sevilla, E., Bes, M.T., Peleato, M.L., and Fillat, M.F. (2009). New insights into the role of Fur proteins: FurB (All2473) from *Anabaena* protects DNA and increases cell survival under oxidative stress. Biochem. J. *418*, 201–207.

Martinez, M., Ugalde, R.A., and Almiron, M. (2005). Dimeric *Brucella abortus* Irr protein controls its own expression and binds haem. Microbiology *151*, 3427–3433.

Martinez, M., Ugalde, R.A., and Almiron, M. (2006). Irr regulates brucebactin and 2,3-dihydroxybenzoic acid biosynthesis, and is implicated in the oxidative stress resistance and intracellular survival of *Brucella abortus*. Microbiology *152*, 2591–2598.

Massé, E., and Gottesman, S. (2002). A small RNA regulates the expression of genes involved in iron metabolism in *Escherichia coli*. Proc. Natl. Acad. Sci. U.S.A. *99*, 4620–4625.

Massé, E., Vanderpool, C.K., and Gottesman, S. (2005). Effect of RyhB small RNA on global iron use in *Escherichia coli*. J. Bacteriol. *187*, 6962–6971.

Michel, L., Gonzalez, N., Jagdeep, S., Nguyen-Ngoc, T., and Reimmann, C. (2005). PchR-box recognition by the AraC-type regulator PchR of *Pseudomonas aeruginosa* requires the siderophore pyochelin as an effector. Mol. Microbiol. *58*, 495–509.

Miles, S., Carpenter, B.M., Gancz, H., and Merrell, D.S. (2010). *Helicobacter pylori* apo-Fur regulation appears unconserved across species. J. Microbiol. *48*, 378–386.

Nandal, A., Huggins, C.C., Woodhall, M.R., McHugh, J., Rodriguez-Quinones, F., Quail, M.A., Guest, J.R., and Andrews, S.C. (2010). Induction of the ferritin gene (*ftnA*) of *Escherichia coli* by Fe($^{2+}$)-Fur is mediated by reversal of H-NS silencing and is RyhB independent. Mol. Microbiol. *75*, 637–657.

Ngok-Ngam, P., Ruangkiattikul, N., Mahavihakanont, A., Virgem, S.S., Sukchawalit, R., and Mongkolsuk, S. (2009). Roles of *Agrobacterium tumefaciens* RirA in iron regulation, oxidative stress response, and virulence. J. Bacteriol. *191*, 2083–2090.

van Oeffelen, L., Cornelis, P., Van Delm, W., De Ridder, F., De Moor, B., and Moreau, Y. (2008). Detecting cis-regulatory binding sites for cooperatively binding proteins. Nucleic Acids Res. *36*, e46.

Oglesby-Sherrouse, A.G., and Vasil, M.L. (2010). Characterization of a heme-regulated non-coding RNA encoded by the *prrF* locus of *Pseudomonas aeruginosa*. PLoS One *5*, e9930.

Palyada, K., Sun, Y.Q., Flint, A., Butcher, J., Naikare, H., and Stintzi, A. (2009). Characterization of the oxidative stress stimulon and PerR regulon of *Campylobacter jejuni*. BMC Genomics *10*, 481.

Patzer, S.I., and Hantke, K. (1998). The ZnuABC high-affinity zinc uptake system and its regulator Zur in *Escherichia coli*. Mol. Microbiol. *28*, 1199–1210.

Pennella, M.A., and Giedroc, D.P. (2005). Structural determinants of metal selectivity in prokaryotic metal-responsive transcriptional regulators. Biometals *18*, 413–428.

Platero, R., Peixoto, L., O'Brian, M.R., and Fabiano, E. (2004). Fur is involved in manganese-dependent regulation of *mntA* (*sitA*) expression in *Sinorhizobium meliloti*. Appl. Environ. Microbiol. *70*, 4349–4355.

Pohl, E., Holmes, R.K., and Hol, W.G. (1999a). Crystal structure of a cobalt-activated diphtheria toxin repressor–DNA complex reveals a metal-binding SH3-like domain. J. Mol. Biol. *292*, 653–667.

Pohl, E., Holmes, R.K., and Hol, W.G. (1999b). Crystal structure of the iron-dependent regulator (IdeR) from *Mycobacterium tuberculosis* shows both metal binding sites fully occupied. J. Mol. Biol. *285*, 1145–1156.

Pohl, E., Haller, J.C., Mijovilovich, A., Meyer-Klaucke, W., Garman, E., and Vasil, M.L. (2003). Architecture of a protein central to iron homeostasis: crystal structure and spectroscopic analysis of the ferric uptake regulator. Mol. Microbiol. *47*, 903–915.

Potvin, E., Sanschagrin, F., and Levesque, R.C. (2008). Sigma factors in *Pseudomonas aeruginosa*. FEMS Microbiol. Rev. *32*, 38–55.

Prévost, K., Salvail, H., Desnovers, G., Jacques, J.F., Phaneuf, E., and Masse, E. (2007). The small RNA RyhB activates the translation of *shiA* mRNA encoding a permease of shikimate, a compound involved in siderophore synthesis. Mol. Microbiol. *64*, 1260–1273.

Py, B., and Barras, F. (2010). Building Fe-S proteins: bacterial strategies. Nat. Rev. Microbiol. *8*, 436–446.

Qian, Y., Lee, J.H., and Holmes, R.K. (2002). Identification of a DtxR-regulated operon that is essential for siderophore-dependent iron uptake in *Corynebacterium diphtheriae*. J. Bacteriol. *184*, 4846–4856.

Ravel, J., and Cornelis, P. (2003). Genomics of pyoverdine-mediated iron uptake in pseudomonads. Trends Microbiol. *11*, 195–200.

Redly, G.A., and Poole, K. (2003). Pyoverdine-mediated regulation of FpvA synthesis in *Pseudomonas aeruginosa*: involvement of a probable extracytoplasmic-function sigma factor, FpvI. J. Bacteriol. *185*, 1261–1265.

Redly, G.A., and Poole, K. (2005). FpvIR control of *fpvA* ferric pyoverdine receptor gene expression in *Pseudomonas aeruginosa*: demonstration of an interaction between FpvI and FpvR and identification of mutations in each compromising this interaction. J. Bacteriol. *187*, 5648–5457.

Rincon-Enriquez, G., Crete, P., Barras, F., and Py, B. (2008). Biogenesis of Fe/S proteins and pathogenicity: IscR plays a key role in allowing *Erwinia chrysanthemi* to adapt to hostile conditions. Mol. Microbiol. *67*, 1257–1273.

Rodionov, D.A., Gelfand, M.S., Todd, J.D., Curson, A.R., and Johnston, A.W. (2006). Computational reconstruction of iron- and manganese-responsive transcriptional networks in alpha-proteobacteria. PLoS Comput. Biol. *2*, e163.

Rodriguez, G.M. (2006). Control of iron metabolism in *Mycobacterium tuberculosis*. Trends Microbiol. *14*, 320–327.

Rodriguez, G.M., Voskuil, M.I., Gold, B., Schoolnik, G.K., and Smith, I. (2002). *ideR*, An essential gene in *Mycobacterium tuberculosis*: role of IdeR in iron-dependent gene expression, iron metabolism, and oxidative stress response. Infect. Immun. *70*, 3371–3381.

Schalk, I.J. (2008). Metal trafficking via siderophores in Gram-negative bacteria: specificities and characteristics of the pyoverdine pathway. J. Inorg. Biochem. *102*, 1159–1169.

Schmitt, M.P., Predich, M., Doukhan, L., Smith, I., and Holmes, R.K. (1995). Characterization of an iron-dependent regulatory protein (IdeR) of *Mycobacterium tuberculosis* as a functional homolog of the diphtheria toxin repressor (DtxR) from *Corynebacterium diphtheriae*. Infect. Immun. *63*, 4284–4289.

Schwartz, C.J., Giel, J.L., Patschkowski, T., Luther, C., Ruzicka, F.J., Beinert, H., and Kiley, P.J. (2001). IscR, an Fe–S cluster-containing transcription factor, represses expression of *Escherichia coli* genes encoding Fe–S cluster assembly proteins. Proc. Natl. Acad. Sci. U.S.A. *98*, 14895–14900.

da Silva Neto, J.F., Braz, V.S., Italiani, V.C., and Marques, M.V. (2009). Fur controls iron homeostasis and oxidative stress defense in the oligotrophic alpha-proteobacterium *Caulobacter crescentus*. Nucleic Acids Res. *37*, 4812–4825.

Todd, J.D., Sawers, G., Rodionov, D.A., and Johnston, A.W. (2006). The *Rhizobium leguminosarum* regulator IrrA affects the transcription of a wide range of genes in response to Fe availability. Mol. Genet. Genomics *275*, 564–577.

Todd, J.D., Wexler, M., Sawers, G., Yeoman, K.H., Poole, P.S., and Johnston, A.W. (2002). RirA, an iron-responsive regulator in the symbiotic bacterium *Rhizobium leguminosarum*. Microbiology *148*, 4059–4071.

Touati, D. (2000). Iron and oxidative stress in bacteria. Arch. Biochem. Biophys. *373*, 1–6.

Viguier, C., P, O.C., Clarke, P., and O'Connell, M. (2005). RirA is the iron response regulator of the rhizobactin 1021 biosynthesis and transport genes in *Sinorhizobium meliloti* 2011. FEMS Microbiol. Lett. *246*, 235–242.

Visca, P., Imperi, F., and Lamont, I.L. (2007). Pyoverdine siderophores: from biogenesis to biosignificance. Trends Microbiol. *15*, 22–30.

van Vliet, A.H., Baillon, M.L., Penn, C.W., and Ketley, J.M. (1999). *Campylobacter jejuni* contains two fur homologs: characterization of iron-responsive regulation of peroxide stress defense genes by the PerR repressor. J. Bacteriol. *181*, 6371–6376.

Wandersman, C., and Delepelaire, P. (2004). Bacterial iron sources: from siderophores to hemophores. Annu. Rev. Microbiol. *58*, 611–647.

Wandersman, C., and Stojiljkovic, I. (2000). Bacterial heme sources: the role of heme, hemoprotein receptors and hemophores. Curr. Opin. Microbiol. *3*, 215–220.

Wilderman, P.J., Sowa, N.A., FitzGerald, D.J., FitzGerald, P.C., Gottesman, S., Ochsner, U.A., and Vasil, M.L. (2004). Identification of tandem duplicate regulatory small RNAs in *Pseudomonas aeruginosa* involved in iron homeostasis. Proc. Natl. Acad. Sci. U.S.A. *101*, 9792–9797.

Winterbourn, C.C. (1995). Toxicity of iron and hydrogen peroxide: the Fenton reaction. Toxicol. Lett. *82–83*, 969–974.

Wisedchaisri, G., Holmes, R.K., and Hol, W.G. (2004). Crystal structure of an IdeR–DNA complex reveals a conformational change in activated IdeR for base-specific interactions. J. Mol. Biol. *342*, 1155–1169.

Wu, H.J., Seib, K.L., Srikhanta, Y.N., Kidd, S.P., Edwards, J.L., Maguire, T.L., Grimmond, S.M., Apicella, M.A., McEwan, A.G., and Jennings, M.P. (2006). PerR controls Mn-dependent resistance to oxidative stress in *Neisseria gonorrhoeae*. Mol. Microbiol. *60*, 401–416.

Wu, Y., and Outten, F.W. (2009). IscR controls iron-dependent biofilm formation in *Escherichia coli* by regulating type I fimbria expression. J. Bacteriol. *191*, 1248–1257.

Wyckoff, E.E., Lopreato, G.F., Tipton, K.A., and Payne, S.M. (2005). *Shigella dysenteriae* ShuS promotes utilization of heme as an iron source and protects against heme toxicity. J. Bacteriol. *187*, 5658–5664.

Yeo, W.S., Lee, J.H., Lee, K.C., and Roe, J.H. (2006). IscR acts as an activator in response to oxidative stress for the *suf* operon encoding Fe-S assembly proteins. Mol. Microbiol. *61*, 206–218.

Yeoman, K.H., Curson, A.R., Todd, J.D., Sawers, G., and Johnston, A.W. (2004). Evidence that the *Rhizobium* regulatory protein RirA binds to cis-acting iron-responsive operators (IROs) at promoters of some Fe-regulated genes. Microbiology *150*, 4065–4074.

Zheng, M., Doan, B., Schneider, T.D., and Storz, G. (1999). OxyR and SoxRS regulation of *fur*. J. Bacteriol. *181*, 4639–4643.

Zheng, M., Wang, X., Doan, B., Lewis, K.A., Schneider, T.D., and Storz, G. (2001). Computation-directed identification of OxyR DNA binding sites in *Escherichia coli*. J. Bacteriol. *183*, 4571–4579.

Anaerobic Regulatory Networks in Bacteria

11

Petra Tielen, Max Schobert, Elisabeth Härtig and Dieter Jahn

Abstract

Survival and growth during periods of low oxygen tension are essential for the successful colonization of natural habitats by bacteria. For the coordination of the necessary biochemical adaption processes upon oxygen deprivation bacteria employ a fine tuned interplay of various regulatory proteins and sRNAs. The iron sulfur cluster containing oxygen sensor Fnr and its multiple variants are often found involved in the corresponding regulatory networks throughout the bacterial kingdom. Similarly, the alternative electron acceptor nitrate is usually detected by the two-component system NarXL and its derivatives. In contrast, other systems including the quinone pool responsive two component system ArcBA, the ResDE system or the regulatory sRNAs FnrS and ArcZ are limited to certain bacterial groups. Here we describe the regulatory networks and their components underlying the adaption processes of the model bacteria *Escherichia coli*, *Pseudomonas aeruginosa* and *Bacillus subtilis* to an anaerobic lifestyle.

Introduction

In nature, oxygen serves as major electron acceptor during respiratory energy conservation. This is the most effective mode of energy generation currently known and consequently wide spread in all forms of life. Moreover, several metabolic enzymatic reactions rely on the participation of oxygen. However, bacteria often encounter periods of oxygen deprivation in their natural habitats. These organisms employ two fundamentally different forms of alternative anaerobic energy generation: anaerobic respiration and fermentation. During anaerobic respiration, oxygen can be replaced by alternative electron acceptors including nitrate, nitrite, trimethylamine *N*-oxide (TMAO), dimethyl sulfoxide (DMSO), fumarate, sulfate, several metals (Mn(IV) Co(III)), CO_2 and many more compounds. Similar to oxygen respiration, membrane localized protein complexes transfer electrons from appropriate donors often via quinones to these acceptors and utilize the differences in redox potential for a proton translocation process via the membrane. The generated proton motive force drives ATP generation via ATP synthases. Anoxygenic anaerobic photosynthesis provides an additional way for proton gradient formation in the absence of oxygen. Fermentation generates ATP at the substrate level. Energy, released from substrate molecules by bond cleavage or oxidation, is directly utilized for ATP synthesis. Well-studied examples are the ATP formation during glycolysis in combination with various forms of mixed acid fermentation. Here, formation of the still energy rich fermentation products like lactate, 2,3-butanediol or ethanol is necessary for NAD^+ regeneration owing to missing or reduced $NADH+H^+$ consuming respiratory processes. Only acetate fermentation yields an additional ATP molecule. Usually fermentations products are excreted and may be re-imported for further utilization.

Clearly, the different forms of aerobic and anaerobic energy generation are characterized by significant differences in their energy yields. The most effective way of ATP production is oxygen respiration, which is preferred over the various mentioned other anaerobic respiratory processes.

The lowest energy recovery is provided by the multiple fermentation processes. Consequently, a bacterium capable of various aerobic and anaerobic lifestyles is tightly coordinating its energy metabolism and the corresponding enzymatic systems. Owing to the various ecological niches inhabited by the bacteria a broad spectrum of combinations of the various strategies is found. Correspondingly, the regulatory networks, their components, and regulatory strategies differ significantly. Here, we describe the regulatory networks of *E. coli*, *P. aeruginosa* and *B. subtilis* for the adaptation to oxygen limitation.

Escherichia coli

E. coli is a Gram-negative rod shaped enterobacterium usually found in the lower intestine of warm-blooded animals. The majority of *E. coli* strains are harmless, however, some cause infections. *E. coli* has the ability to survive and grow outside its host. During these various lifestyles, *E. coli* encounters seven distinct modes of growth: growth with excessive aeration, sufficient aeration, limited aeration, anaerobic growth with ample nitrate, anaerobic growth with nitrite or limited availability of nitrate, anaerobic growth with electron acceptors others than nitrate or nitrite, and fermentation.

Energy metabolism

E. coli has developed a versatile metabolism for energy generation to fully utilize the resources of its environment. A broad spectrum of respiratory and fermentative strategies is available to sustain growth under conditions of changing oxygen tension and available alternative electron acceptors (Fig. 11.1). During respiration, electrons from various donors are shuttled at the inner membrane via primary dehydrogenases

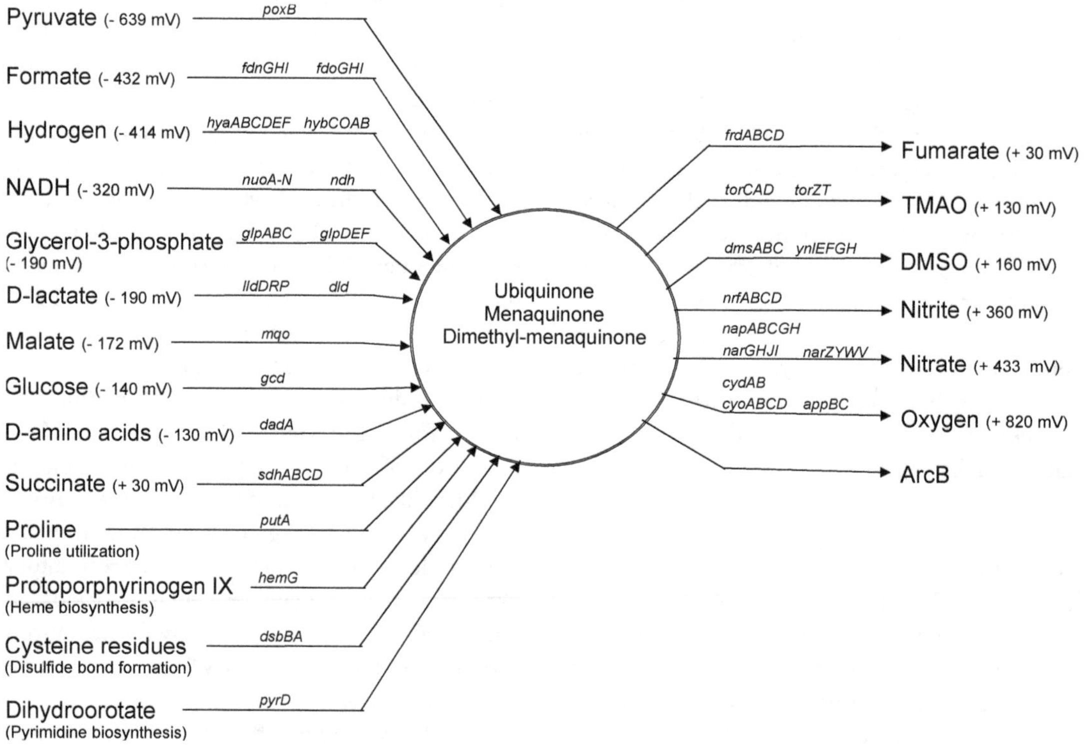

Figure 11.1 Respiratory pathways of energy generation in *Escherichia coli*. On the left side electron donor with their redox potential are given. Moreover, enzymatic reactions donating electrons to the quinone pool are shown. Transfer of electron occurs via three different quinones to the electron acceptor outlined on the right with their corresponding redox potential. Involved enzymes are depicted by their coding genes or operons. The processes are used for proton gradient formation and subsequent ATP synthesis.

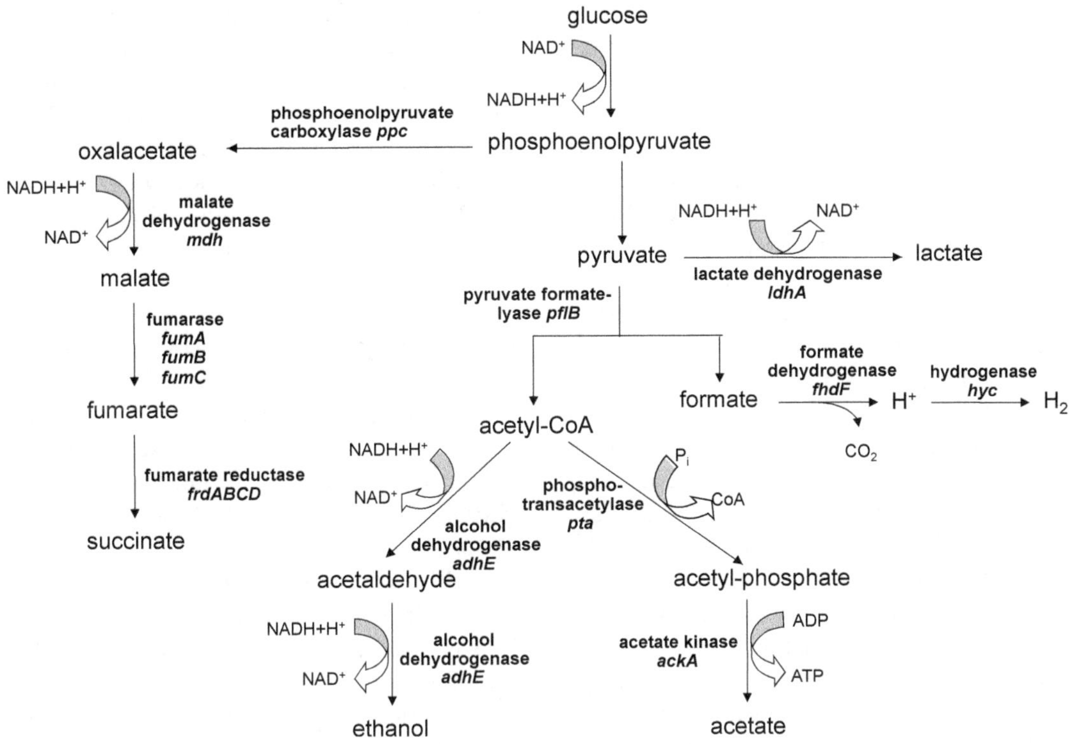

Figure 11.2 Mixed acid fermentation in *Escherichia coli*. Metabolic reactions involved in substrate level ATP generation and NAD^+ regeneration are shown. Involved enzymes and encoding genes are indicated.

into different electron transport chains (Fig. 11.1). Ten different electron donors (NADH, formate, hydrogen, glucose, D-lactate, succinate, glycerol-3-phosphate, D-amino acids, malate, pyruvate) are known for *E. coli* respiration (Forzi and Sawers, 2007; Friedrich and Pohl, 2007; Vignais and Billoud, 2007; Unden and Dünnwald, 2008). Electrons are transferred via three different types of quinones (ubiquinone, menaquinone, demethylmenaquinone) to terminal reductases, which catalyse the reduction of multiple electron acceptors (Fig. 11.1). Six catabolic electron acceptors (oxygen, nitrate, nitrite, fumarate, TMAO, DMSO) are known (Cheng and Weiner, 2007; Tomasiak *et al.*, 2007; Cole and Richardson, 2008; Borisov and Verkhovsky, 2009; Möbius *et al.*, 2010). Additionally, enzymes of proline utilization (PutA), haem biosynthesis (HemG), disulfide bond (DsbBA) and pyrimidine formation (PyrD) donate electrons to the quinone pool (Inaba, 2009; Kow *et al.*, 2009; Möbius *et al.*, 2010; Srivastava *et al.*, 2010). Finally, the redox regulatory protein ArcB receives electrons from the guinone pool. As the result of electron transfer some of the membrane bound primary dehydrogenases and terminal reductases are proton pumping or are involved in redox loops for proton gradient formation. The resulting electrochemical gradient drives ATP formation via ATP synthase (Vik, 2007). The various primary dehydrogenases, their electron donors, the terminal reductases and corresponding electron acceptors are shown in Fig. 11.1. Multiple paths for the electrons can be constructed, limited by the redox potential of the participating components. *E. coli* also performs under anaerobic conditions an advanced mixed acid fermentation with the central enzyme pyruvate-formate lyase, which replaces pyruvate dehydrogenase (Fig. 11.2). Formed products are lactate, ethanol, succinate, acetate, H_2 and CO_2. Most reactions allow for NAD^+ regeneration while acetate formation yields ATP. The fine-tuned regulation in response to oxygen tension and the presence of alternative electron acceptors is mainly mediated by the complex interplay of the regulators Fnr, ArcBA, NarXL, NarQP and the regulatory sRNAs FnrS and ArcZ. Consequently, the review is focusing on the function of these regulators. However, other regulatory networks including those for nitric oxide (NorR) oxygen stress (SoxRS, OxyR), nitrogen metabolism (RpoN) and fermentative growth (FhlA) are closely interconnected (Reitzer, 2003; Tucker *et al.*, 2006; Imlay, 2008; Lee *et al.*, 2009; Salim and Feig, 2010; Dong *et al.*, 2011; Gu and Imlay, 2011).

However, the recovery of pumped protons and generated ATP molecules varies significantly between the multiple possible electron transfer chains. Electron transfer from the NADH ($E_0'=-320$ mV) to oxygen ($E_0'=820$ mV) using both the proton pumping NADH dehydrogenase NuoA-N and the terminal reductase CyoA-D systems yield six to eight protons pumped per two electrons transported which translates into one to two molecules of ATP. In contrast electron transfer from formate $E_0'=-432$ mV) to nitrate $E_0'=433$ mV) via formate dehydrogenase FdnGHI and nitrate reductase NarGHI yields four protons pumped per two electrons transferred and from hydrogen $E_0'=-414$ mV) to fumerate ($E_0'=30$ mV) via dydrogenase HybCOAB and fumerate reductase FrdA-D only two protons per two electrons with the consequence of lower ATP recovery.

Additionally, *E. coli* performs a mixed acid fermentation with the end products succinate, lactate, ethanol, acetate, hydrogen and carbon dioxide. Here, energy recovery occurs with ATP formation at the substrate level. Consequently, the energy recovery is drastically lower compared with respiratory processes.

For an economical and therefore competitive lifestyle *E. coli* should form only the systems with the highest energy yield under conditions where the appropriate electron acceptors are available. This results in a preference for oxygen respiration over nitrate respiration and under anaerobic conditions of nitrate respiration over other respiratory and fermentative processes. The cells are confronted with a complex regulatory task in which during a shift from aerobic to anaerobic conditions aerobic systems should be repressed and anaerobic systems induced. Moreover, a differential anaerobic response in dependence of nitrate availability has to be mediated. In *E. coli* the redox/oxygen sensors Fnr and ArcAB in concert with the nitrate/nitrite sensors NarXL and NarPQ and with several other electron donor/acceptor specific regulators are mediating this complex cellular adaptation process.

Oxygen regulator Fnr and the small RNA FnrS

The transcriptional regulator of fumarate and nitrate reduction (Fnr) directly senses molecular oxygen via lability of an attached $[4Fe–4S]^{2+}$ cluster (Guest *et al.*, 1996; Kiley and Beinert, 1998; Beinert and Kiley, 1999; Green *et al.*, 2009; Fleischhacker and Kiley, 2011). The structure of Fnr and its homologues is depicted in Fig. 11.3. The N-terminus of the protein contains four essential Cys residues (positions 20, 23, 29, 122), that function as ligands for the iron–sulfur cluster (Khoroshilova *et al.*, 1995). The middle part of Fnr consists of a long dimerization helix before the protein forms a DNA binding helix–turn–helix motif at its C-terminus (Moore and Kiley, 2001). The regulator is activated by iron sulfur cluster formation via the Isc system (Schwartz *et al.*, 2000; Mettert *et al.*, 2008). Formation of the $[4Fe–4S]^{2+}$ cluster promotes dimerization of the regulator (Lazazzera *et al.*, 1996; Jervis and Green, 2007) by an induced conformational change which masks negative charge repulsions present between two monomeric Fnr molecules without iron sulfur clusters (Moore *et al.*, 2006). In the presence of oxygen, the $[4Fe–4S]^{2+}$ cluster is most probably oxidized in two steps, first to a $[3Fe–4S]^{1+}$ cluster, Fe(II) plus hydrogen peroxide from the dismutation of superoxide, and finally to a $[2Fe–2S]^{2+}$ cluster with the release of two sulfide (S^{2-}) ions and an Fe(III) ion. Fnr with the $[2Fe–2S]^{2+}$ cluster becomes monomeric and inactive (Khoroshilova *et al.*, 1995; Crack *et al.*, 2006, 2007, 2008). Fnr isolated from aerobic *E. coli* cultures is in its apo-form without an iron–sulfur-cluster (Sutton *et al.*, 2004; Achebach *et al.*, 2005; Reinhart *et al.*, 2008). Apo-Fnr is subject to ClpXP-dependent proteolytic degradation (Mettert and Kiley, 2005). Moreover,

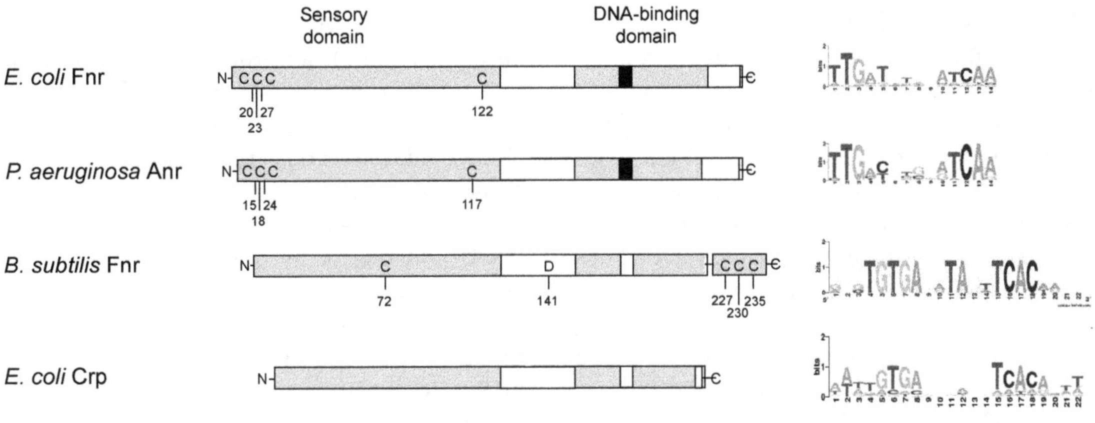

Figure 11.3 Structure of Fnr type regulators. The schematic structure of *E. coli* Fnr, *P. aeruginosa* Anr, *B. subtilis* Fnr and *E. coli* Crp are depicted. Cysteine residues involved in $[4Fe–4S]^{2+}$ cluster formation and the location of the DNA binding domain are shown. Sequence logos for the corresponding DNA binding are provided. A colour version of this figure is available in the plate section at the back of the book.

the *fnr* gene is negatively autoregulated. Consequently, a constant cellular concentration of Fnr is permanently measuring environmental oxygen tensions via cycling between its active and inactive states (Engel *et al.*, 1991; Dibden and Green, 2005). The Fnr cycle has recently been kinetically modelled and the dynamics of the aerobic to anaerobic transition with the contribution of transcriptional to post-translational regulatory levels was predicted (Tolla and Savageau, 2010, 2011).

Fnr recognizes a highly conserved palindromic DNA binding site (5′-TTGATnnnnATCAA-3′, Fig. 11.3). Two different basic types of Fnr-dependent promoters exist. Class I promoters carry Fnr binding sites at positions −61,5, −71,5, −82,5 or −92,5 upstream the transcriptional start site (Wing *et al.*, 1995) permitting only one protein–protein contact between the regulator and the C-terminal domain of RNA polymerase subunit α (Browning *et al.*, 2002). More often found are class II type promoters with a Fnr binding site centred at position −41,5 upstream of the transcriptional start site allowing for multiple Fnr-RNA polymerase contacts (Lee *et al.*, 2000; Wing *et al.*, 2000; Browning *et al.*, 2002). During this process, activating region 1 (AR1) of Fnr contacts the C-terminal domain of the α-subunit of RNA polymerase and prevents inhibition of transcriptional activation by an untethered α subunit. Contact of activation region 3 (AR3) with σ^{70} activates transcription (Bell and Busby, 1994). Fnr is also a repressor of various genes. However, Fnr repressed promoters do not show a clear-cut pattern of Fnr binding sites. Most contain multiple Fnr binding sites upstream of the transcriptional start site (Guest *et al.*, 1996). A simple promoter occlusion mechanism was excluded, rather a displacement of RNA polymerase α subunit via contact with AR1 was proposed to cause transcriptional inhibition (Green *et al.*, 1998; Green and Marshall, 1999).

A model of the Fnr helix–turn–helix motif based on the crystal structure of the Crp–cAMP–DNA complex and deduced mutational studies suggested Ser 212 interaction with the T-A base-pair in position 1 and Glu 209 interaction with G –C in position 3 of one palindrome half site (Guest *et al.*, 1996).

The transcription of the Hfq-binding sRNA FnrS is strongly induced during the shift from aerobic to anaerobic conditions by Fnr, but is also affected by ArcAB and Crp. Over 30 mRNAs are negatively regulated in their levels upon FnrS production (Boysen *et al.*, 2010; Durand and Storz, 2010). Many of the corresponding genes were known before to be controlled by Fnr, but were lacking Fnr binding sites. Base pairing of some mRNAs with FnrS was predicted. Clearly, FnrS mediates Fnr-dependent gene regulation at the post transcriptional level.

The Fnr and FnrS regulons

The oxygen-responsive Fnr regulator is responsible for the induction of multiple anaerobic metabolic, especially catabolic pathways and corresponding stress responses (Figs. 11.4 and 11.5). To a minor part Fnr represses genes related to aerobic cellular functions. Repression of aerobic genes is mainly mediated via the FnrS sRNA at the post transcriptional level (Figs. 11.4 and 11.5). Global transcriptional profiling for Fnr regulon determination was described at least five times (Salmon *et al.*, 2003; Kang *et al.*, 2005; Constantinidou *et al.*, 2006; Grainger *et al.*, 2007; Kumar and Shimizu, 2011). Many genetic mutant and reporter gene based investigation complete the picture (Sawers *et al.*, 1988; Gunsalus and Park, 1994; Ravcheev *et al.*, 2007). The results of the various investigations are summarized in Fig. 11.5. Obviously, Fnr is the major oxygen-responsive regulator activating operons involved in anaerobic respiration including multiple electron donating dehydrogenases (*fdnGHI, fdoGHI, hybOABCDEFG, glpABC*) and electron accepting terminal reductases (*narGHJI, napFDAGHBC, nrfABCDEFG, frdABCD, dmsABC*). Moreover, genes encoding fermentation enzymes (*pfl, ack, adhE*) are induced by Fnr. Corresponding transporters and cofactor biosynthetic pathways are co-induced (*moaABCDE, narK, nikABCDE, focA, glpTQ, feoAB*). Additional regulatory systems involved in oxygen- (*aer*) and chemotaxis (*tsr, tar-tap-cheRBYZ*), anaerobic carnitine utilization (*caiF*), cell density (*fliZY*) redox sensing (*arcA*) and cyclic di-GMP signalling (*yfgF*) are also co-induced. Parts of the anaerobic reductive branch of the citrate cycle (*ansB, aspA, dcuB, fumB*), other related functions (*citCDEFXG, pykA*) and the nitrite reductase (*nirBDC*) are also subject to Fnr induction. The transcription of the genes for the anaerobic class III ribonucleotide reductase (*nrdDG*) is induced by Fnr, while the genes for the class Ia oxygen-dependent reductase (*nrdAB*) are repressed. Other Fnr repressed operons encode electron donating NADH dehydrogenases (*nuoA-N, ndh*) and glucose dehydrogenase (gcd), ubiquinone and haem biosynthetic genes (*ubiCA, hemA*) and the terminal electron-accepting cytochrome d oxidase (*cydAB*). Typical aerobic functions of the central metabolism including pyruvate dehydrogenase (*pdhR, aceEF, lpd*), pyruvate oxidase (*poxB*), aconitate hydratase (*acnA*) and NAD-linked aldehyde dehydrogenase (*aldA*) are also negatively controlled by Fnr. Moreover, some stress responses are Fnr repressed, including genes for catalase and Mn-superoxide dismutases (*katE, sodA*), for pH adaptation (*gadAB, gadXW*), for flavohaemoglobin (*hmp*) and for the cold shock response (*cspA*). The repression of the Fe-superoxide dismutase (*sodB*) is mediated post transcriptionally by the

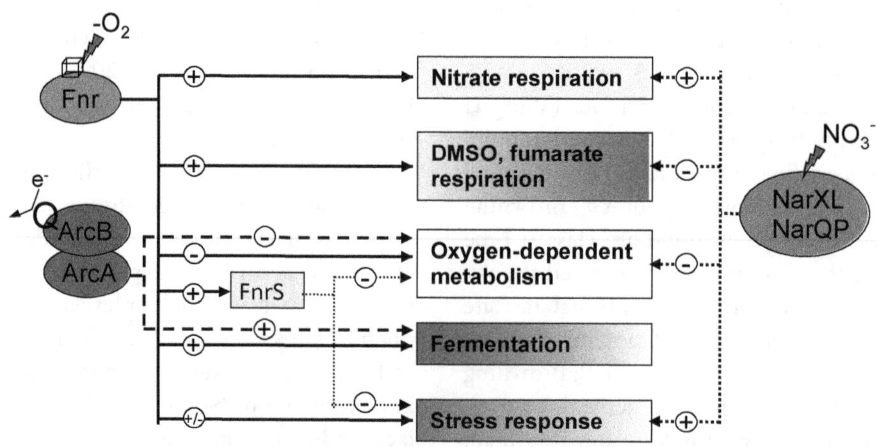

Figure 11.4 Overview over the anaerobic regulatory network of *E. coli* using Fnr, the sRNA FnrS, ArcBA and NarXL/NarQP. Detected environmental stimuli and regulated mayor metabolic processes are given. A colour version of this figure is available in the plate section at the back of the book.

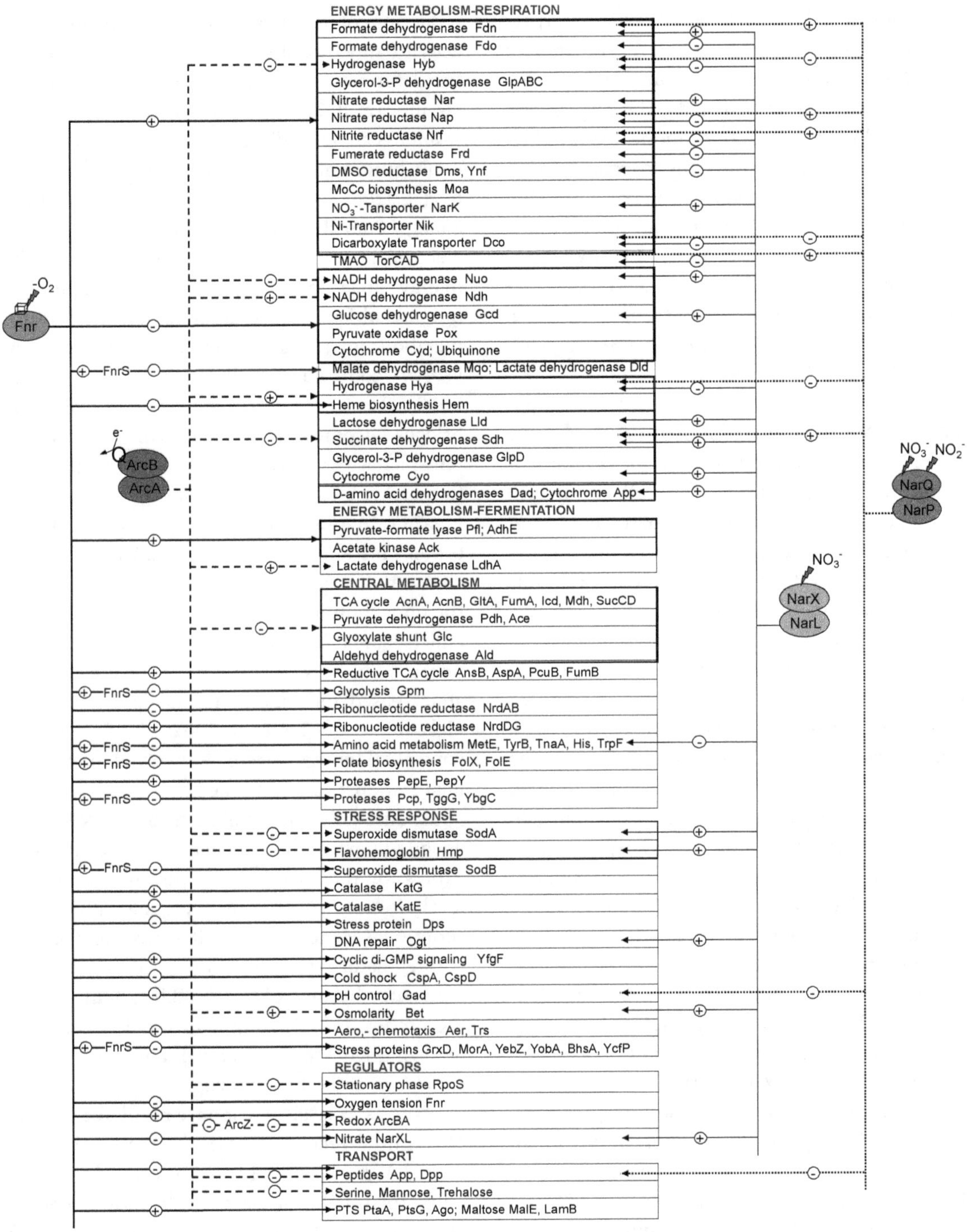

Figure 11.5 Detailed regulatory network of *E. coli* for the adaptation to anaerobic growth conditions. The interplay of oxygen regulator Fnr, the sRNA FnrS, the redox regulator ArcBA, and the nitrate/nitrite regulatory systems NarXL and NarQP is outlined. In the case of arrows touching the outside of a set of genes boxed with bold lines regulation of all genes by the corresponding regulator is indicated. Arrows entering a boxed gene or a set of genes reveal a single regulator-target gene interaction. Regulated metabolic processes are given in blue letters. Regulatory systems and corresponding stimuli are shown in different colours. A colour version of this figure is available in the plate section at the back of the book.

Fnr induced sRNA FnrS. Further FnrS targets include the cytochrome-related transporter (*cydCD*), a cytochrome b_{561} homologue (*yceI*), the malic enzyme (*sfcA-maeA*), the respiratory, electron donating malate-quinone oxidoreductase (*mqo*) and D-lactate dehydrogenase (*dld*), the fermentative propanol-preferring alcohol dehydrogenase (*adhP*) and the glycolytic phosphoglycerol mutase 1 (*gpmA*). Finally, folate biosynthesis (*folX*, *folE*), amino acid formation (*metE*, *tyrB*), various proteases (*dcp*, *yggG*, *ybjC*), stress proteins (*sodB*, *ydh-grxD*, *marA*, *yebZ*, *yobA*, *bhsA-ycfR*) are repressed by FnrS.

The redox regulatory system ArcBA

ArcBA (anoxic redox control) is a two-component regulatory system (Fig. 11.6). ArcB is a hybrid membrane-associated sensor kinase and ArcA the corresponding response regulator. During the shift from aerobic to anaerobic growth conditions major aerobic metabolic functions are repressed and some anaerobically metabolic functions are induced by the system (Malpica *et al.*, 2006; Alvarez *et al.*, 2010).

The hybrid sensor kinase ArcB consists of a typical transmitter domain with the conserved His residue in position 292, a receiver domain with the conserved Asp residue in position 576 and a secondary C-terminal phosphotransfer domain (HPt) with the conserved His residue at position 717 (Alvarez *et al.*, 2010). In the linker region, which connects these domains with the transmembrane part, two redox-active Cys residues (positions 180, 241) are found in a Per-Arnt-Sim (PAS) domain (Fig. 11.4). Moreover, a leucine zipper motif was described (Georgellis *et al.*, 1998; Taylor and Zhulin, 1999).

Under low oxygen tension conditions, ArcB performs an ATP-dependent autophosphorylation of His-292. Subsequently, the phosphoryl group is first transferred intramolecularly to Asp-576 and from there to His-717 of the phosphoryl transfer domain and finally to ArcA (Georgellis *et al.*, 1997; Kwon *et al.*, 2000). The process is stimulated by typical metabolites of the anaerobic metabolism like D-lactate, pyruvate and acetate (Georgellis *et al.*, 1999; Rodriguez *et al.*, 2004). Intermolecular disulfide bridge formation between two Cys-180 residues of two ArcB subunits resulted in 85% reduction of the kinase activity. Residual 15% activity are controlled via inter subunit disulfide bond formation between two Cys-241 residues (Malpica *et al.*, 2004). Furthermore, these cysteine residues can be oxidized by quinones with a major contribution of the menaquinone pool (Georgellis *et al.*, 2001; Malpica *et al.*, 2004; Perrenoud and Sauer, 2005; Bekker *et al.*, 2010). Consequently, electron flux via transport chains containing the mainly anaerobically employed menaquinone, but also ubiquinone in combination with typical metabolic intermediates of the anaerobic metabolism are controlling ArcB activity (Portnoy *et al.*, 2010). Under aerobic condition, the disulfide bridges become oxidized converting ArcB into a phosphatase that catalyses the dephosphorylation of ArcA-P by a reverse phosphate transfer via His-717, Asp-576 and His-292. Moreover, the protein histidine phosphatase SixA exhibits specific activity against phosphorylated His-717 of the phosphotransfer domain of ArcB (Matsubara and Mizuno, 2000). The structure of the phosphatase was solved and revealed a nucleophilic attack-based mechanism (Hamada *et al.*, 2005; Hakoshima and Ichihara, 2007).The *sixA* gene is controlled by sigma factor E involved in extracellular responses and extreme heat shock.

Additionally, the small regulatory RNA ArcZ, encoded next to the *arcB* gene, is repressing *arcB* expression most likely at the post transcriptional level. In turn ArcA-P is repressing *arcZ* transcription (Mandin and Gottesman, 2010). This autoregulatory loop is adding to the complex regulatory balance of the ArcBA system. Finally, the *arcB* mRNA is subject to rapid degradation involving RNAse E with a half-life below 2 minutes (Aiso and Ohki, 2003).

ArcA is a classical response regulator with an N-terminal receiver domain carrying the conserved Asp-54 for phosphorylation and a winged helix–turn–helix domain for promoter binding (Fig. 11.6). ArcA belongs to the OmpR/PhoB subfamily of response regulators, accounting for 15 of the 34 presently identified response regulators in *E. coli*. Biochemical studies and the crystal structure of the regulatory domain of ArcA in the presence and absence of a phosphate analogue BeF_3^- revealed controlled dimerization as regulatory principle (Jeon *et al.*, 2001; Toro-Roman *et*

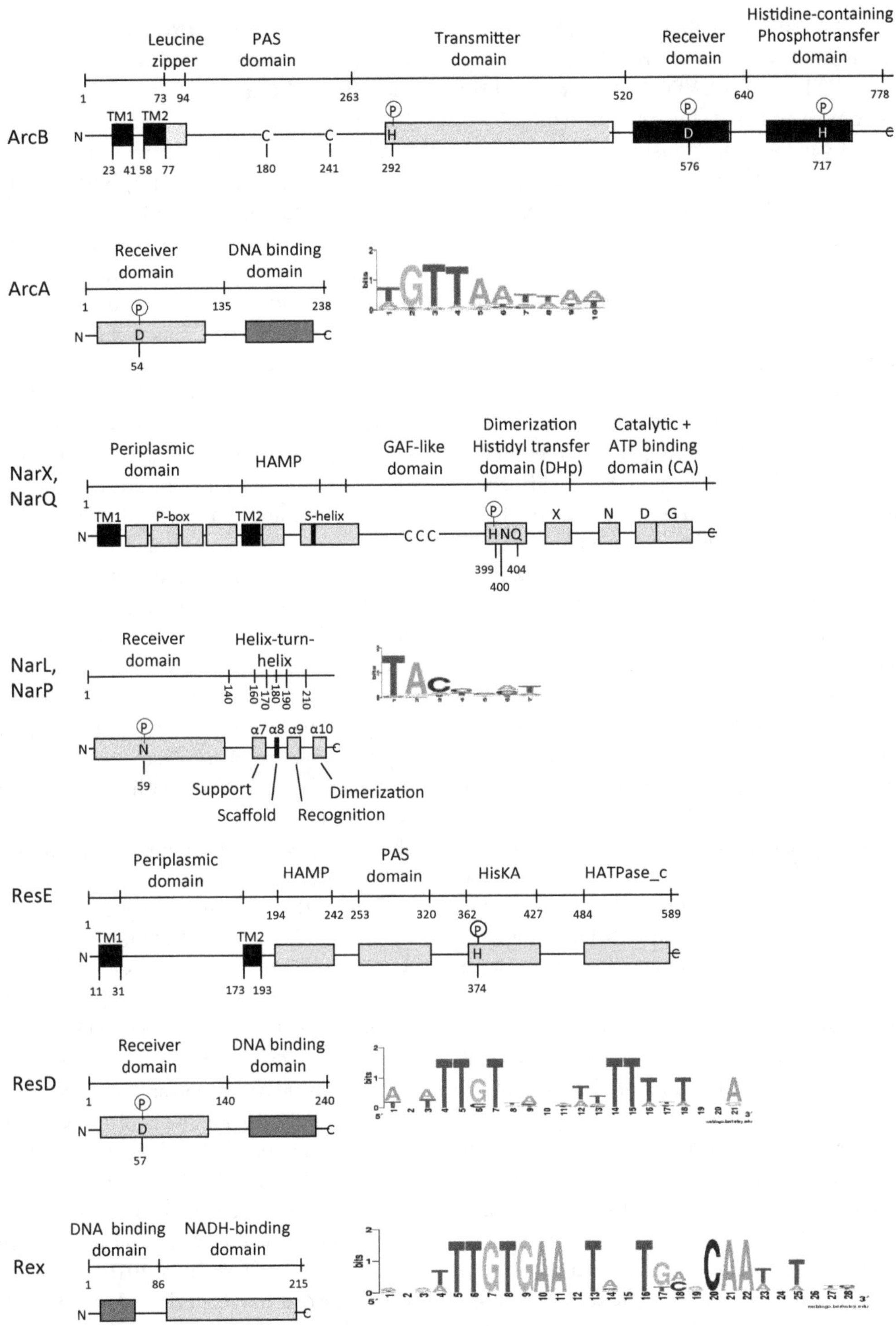

Figure 11.6 Two-component regulatory systems involved in the adaptation to anaerobic conditions in *E. coli*, *P. aeruginosa* and *B. subtilis*. The schematic structure of the proteins for the two component regulatory systems ArcBA (*E.coli*), NarXL (*E.coli*), NarQP (*E.coli*), ResDE (*B. subtilis*) and the NADH sensor Rex (*B. subtilis*) are shown. Functionally important domains and amino acid residues are labelled. Sequence logos provide insights to the corresponding DNA binding sites. A colour version of this figure is available in the plate section at the back of the book.

al., 2005). Position weight matrix analyses of the experimentally confirmed ArcA-P DNA binding sites (5′-GTTAATTAAATGTTA-3′) defined the most conserved bases (Liu and De Wulf, 2004). The *arcA* gene was previously known as the *dye* gene since growth of the corresponding *arcA* mutant is sensitive to the photosensitizer redox dyes toluidine blue and methylene blue. The observed phenotype requires ArcA-P regulated *cydA* expression (Alvarez *et al.*, 2010).

The ArcAB regulon

Many genetic investigations and five different transcriptome analyses were employed for the elucidation of the ArcA-P regulon (see Fig. 11.5; Oshima *et al.*, 2002; Liu and De Wulf, 2004; Salmon *et al.*, 2005; Portnoy *et al.*, 2010; Rolfe *et al.*, 2011). The overlap between the various transcriptome studies was surprisingly low, most probably owing to strain and growth condition differences. However, combination of all available data revealed some clear-cut functions of the ArcAB system (Figs. 11.4 and 11.5). During the shift from aerobic to anaerobic conditions, formation of the enzymes of the TCA cycle (*pdhR, aceEF, acnA, acnB, gltA, fumA, icd, mdh, sucCD*), the glyoxylate path (*glcDEFGB*) and of typical oxygen respiratory chains (*nuoA-N, sdhABCD, lldPRD, hycABCDEF, cyoABCD, glpD*) including cofactor synthesis are repressed. Moreover, the genes of the general stress response including *rpoS* for the alternative stress sigma factor and *sodA* for superoxide dismutase are also down-regulated. Interestingly, ArcB controls via ArcA-P *rpoS* transcription competing with activating Crp-cAMP, while ArcB phosphorylation of the proteolytic targeting factor RssB induces RpoS proteolysis (Mika and Hengge, 2005). Central genes for fermentative processes (*ldhA*), a few proteins of anaerobic/microaerobic respiratory chains (*ndh, cyaAB, hyaABCDEF*) and for carnitine metabolism (*caiTABCDF*) are up-regulated. Interestingly, several PTS systems and a serine transporter are found up-regulated (*treB, manXYZ, sda*), while various other PTS genes (*agoCD, gluC, ptsA, ptsG*) and transporters for maltose (*malE, lamB*), rhamnose (*rhaI*) and oligopeptides (*oppABCDF*) are repressed.

The nitrate and nitrite responsive two component regulatory systems NarXL and NarQP

Nitrate is the preferred anaerobic electron acceptor. The NarXL and the NarQP two-component regulatory systems (Fig. 11.4) both detect nitrate and mediate an appropriate gene regulatory response for optimal metabolic adaptation (Stewart and Rabin, 1995; Darwin and Stewart, 1996). NarQP is additionally responding to nitrite (Noriega *et al.*, 2010). This way *E.coli* is equipped with a sensitive system for the growth conditions in the human body (NarQP) and one for conditions of nitrate excess as found in a sewage farm (NarXL). The sensor kinases NarX and NarQ show a highly homologous structure (Fig. 11.6). At the N-terminus, a periplasmic domain localized between two transmembrane helices contains a P-box (amino acid residues between 40 and 60) responsible for nitrate and nitrite recognition with amino residues involved in nitrate versus nitrite discrimination (Cavicchioli *et al.*, 1995; Williams and Stewart, 1997a,b). Obviously, NarQP is better equipped to nitrite perception than NarXL. The crystal structure of the NarX periplasmic sensor domain of *E. coli* NarX with nitrate and its apo state revealed a four-helix bundle. Conformational changes upon nitrate binding result in a piston-type displacement between the N- and C- terminal helices of the domain inducing transmembrane signalling (Cheung and Hendrickson, 2009). It was suggested that the signal is transmitted via the membrane using a conserved signal transduction element called HAMP (<u>h</u>istidine kinase, <u>a</u>denylyl cyclase, <u>m</u>ethyl-accepting chemotaxis protein, <u>p</u>hosphatase) (Appleman and Stewart, 2003). The approximately 50 amino acids HAMP domain is a four-helix parallel coiled-coil built from two α-helices (Fig. 11.4). The second helix overlaps with a signalling helix (S-helix) with heptad stutter element, which is important for the signalling process (Stewart and Chen, 2010). Little is known about the central GAF-like domain containing three characteristic Cys residues. The C-terminal transmitter module consists of the DHp domain (dimerization, histidyl transfer) consisting of an H- and X-box sequence and a CA domain (catalytic and ATP binding) consisting of N-, D-,

F- and G-box sequences. The H-box contains His-399 (NarX) which gets autophosphorylated upon signalling, utilizing ATP from the catalytic and ATP (CA) binding domain (Schröder *et al.*, 1994; Cavicchioli *et al.*, 1995; Noriega *et al.*, 2008). In the presence of ADP, a strong dephosphorylation of NarX-P and NarQ-P was observed ensuring that only a small proportion of the proteins remain phosphorylated in the absence of nitrate (Noriega *et al.*, 2008).

Next, the phosphate at His-399 of the NarX transmitter module gets transferred to an Asp residue of the cognate response regulator, like Asp-59 of NarL (Fig. 11.6). In the absence of the inducer molecule nitrate, the transmitter module of NarX accelerates the dephosphorylation of NarL-P. This activity is called transmitter phosphatase activity and depends on Asp-400 and Gln-404 of the DHp domain adjacent to phosphorylated His-399 (Huynh *et al.*, 2010).

The crystal structures of native NarL and the NarL DNA binding domain complexed with DNA provide insights into the structure–function relationships of the response regulator (Baikalov *et al.*, 1996, 1998; Maris *et al.*, 2002, 2005). The N-terminal 131 amino acid receiver domain is folded into a five-strand β-sheet flanked by five α-helices. The C-terminal 62 amino acids DNA-binding domain is a bundle of four α-helices. The middle two helices form the helix–turn–helix motif for DNA binding. In the unphosphorylated form of NarL the receiver domain is turned towards the C-terminal DNA binding domain in a way that DNA binding is permitted. Consequently, phosphorylation of Asp 59 must transmit the information via the interdomain interface for movement of the DNA binding C-terminal domain. An important prerequisite for DNA binding is the dimerization of NarL via the C-terminal dimerization domain. Interestingly, DNA recognition is achieved by concerted effects of solvation, van der Waals forces and inherent DNA deformability and only to a minor extent via major groove hydrogen bonding. Consequently, during DNA binding the rather small DNA binding domain of NarL perturbs the DNA helix with the result of major DNA curvature and the transition of B- to A-form DNA. In this context especially Val 189 enforces distortions and interacts with the polar floor of the major groove (Maris *et al.*, 2002, 2005). NarL and NarP are two highly homologous response regulators (Fig. 11.4); however, they differentially regulate gene expression in *E. coli* (Constantinidou *et al.*, 2006). Some operons are controlled by NarL or NarP alone, whereas others are regulated by both (Fig. 11.5). The binding site for NarL was defined as 5′-TAC C/T C/T A/C T-3′ usually arranged as an inverted pair of tail-to-tail heptamers with 2 bp spacing (7–2–7 = 5′-TAC C/T C/T A/C T nn T A/C C/T C/T CAT-3′). While NarP seemed to require this binding site arrangement NarL showed more flexibility and recognized 7–4–7, up to 7–10–7 constructs. Interestingly, genetic investigation suggested that NarL can be activated by both, NarX and NarQ (Stewart and Rabin, 1995). An asymmetric cross-regulation between NarX-NarL and NarQ-NarP was observed (Noriega *et al.*, 2010). NarX preferentially phosphorylates NarL, while NarQ revealed only a slight kinetic preference for NarP over NarL. Finally, the *narX* mRNA is rapidly degraded by RNAse E, ensuring rapid fine-tuned signal perception and response (Aiso and Ohki, 2003).

The NarXL and NarQP regulons

The major task of both two-component regulatory systems is the differentiation of the anaerobic response mainly initiated by Fnr and ArcBA (Fig. 11.5). In the presence of nitrate, less efficient modes of energy generation are suppressed by the two systems. Typical additional cellular functions accompanying nitrate respiration are induced. The NarQP system is additionally responding to nitrite, allowing for a further fined tuned adaption (Fig. 11.5). In contrast to the other regulatory systems, less transcriptome information is available (Constantinidou *et al.*, 2006). Most knowledge comes from classical genetic investigations (Stewart, 1994; Stewart and Rabin, 1995; Darwin and Stewart, 1996; Constantinidou *et al.*, 2006; Lin *et al.*, 2007; Cole and Richardson, 2008; Squire *et al.*, 2009).

The operons for the electron donating primary dehydrogenases (*nuoA-N, fdnGHI, gcd, lldPRD, sdhABCD, dadA*) most likely interacting with the nitrate/nitrite respiratory system are induced by NarXL. Some other important systems (*fdnGHI*, *sdhABCD*) are enhanced additionally by NarQP. In this context, most likely systems not sustaining nitrate/nitrite respiration are repressed by NarXL, NarQP or both, including formate dehydrogenase (*fdoGHI*) and various hydrogenases (*hybO-ABCDEFG*, *hyaABCDEFG*, *hycABDEFGH*). An important contribution of both regulatory systems to the differentiation of the anaerobic response is the control of terminal reductase formation and fermentative gene expression (Fig. 11.5). Here, the regulation of NarXL and NarQP clearly differs. The *narGHJI* operon for the proton pumping nitrate reductase is induced by NarXL. The operons for the second membrane localized nitrate reductase *narZYV*, for the periplasmatic enzyme (*napABCGH*) including required cytochrome c maturation (*ccmABCDEFGH*), respiratory nitrite reductase (*nrfABCD*) and TMAO reductase (*torDCA*) are all induced by NarQP, but repressed by NarXL. Importantly, NarXL represses also the genes for fumarate reductase function (*frdABCD*,*dcuSR*), DMSO reductases (*dmsABC*, *ynfEFGH*), cytochrome *bd* oxidase II (*appBC*), fermentative pyruvate-formate lyase (*pfl*) and ethanol dehydrogenase (*adhP*) focusing the energy metabolism of *E. coli* on nitrate respiration. Several stress responsive systems are induced by NarXL including DNA repairing 0^6-alkylguanine DNA alkyltransferase (*ogt*), Mn-superoxide dismutase (*sodA*), betaine and carnitine transport, synthesis and regulation (*betI*, *betT*, *caiTABCDF*) and the NO-specific flavohaemoglobin (hmp). Others, like cold shock (*cspD*) are repressed by NarXL. Often a co-activation of promoters by NarL with Fnr was observed. However, an Fnr-independent mechanism of NarL activation was established for the *yeaR* and *ogt* promoters (Lin *et al.*, 2007; Squire *et al.*, 2009).

Regulation of DMSO and TMAO reductase formation

E. coli possesses two operons encoding DMSO reductase called *dmsABC* and *ynfEFGHI*. The *dmsA* promoter is activated by Fnr binding to a single site localized between basepairs −41 and −42 upstream the transcriptional start site (Cotter and Gunsalus, 1989; Lamberg and Kiley, 2000; Bearson *et al.*, 2002). NarL represses *dmsABC* transcription in response to nitrate via multiple binding sites. Fnr is a classical class II activator of *ynfEFGHI* (Xu *et al.*, 2009). NarL represses the operon via a single binding site overlapping the −10 region of the promoter (Xu *et al.*, 2009).

In contrast, the TMAO encoding *torCAD* operon is expressed under anaerobic conditions independently of Fnr and ArcBA (Pascal *et al.*, 1984; Simon *et al.*, 1994). DNA arrays indicated the repression by NarXL and induction by NarQP. The operon is also transcribed under aerobic conditions sustaining TMAO reduction during the aerobic exponential growth phase (Ansaldi *et al.*, 2007). Transcription of *torCAD* strongly responds to TMAO availability (Simon *et al.*, 1994). The complex TorTSR phosphorelay mediates this regulation. TorT is a periplasmic TMAO binding protein. TorT interacts tightly with the N-terminal region of the membrane bound TorS sensor (Baraquet *et al.*, 2006). TorS is part of a family of unorthodox sensors with three phorphorylation sites. TorR is the response regulator containing the fourth phosphorylation site of the system. TMAO binding to TorT induces conformational changes to the TorT–TorS complex. This results in a four-step phosphorylation cascade with the result of active phosphorylated TorR, which binds and activates the *tor* operon (Simon *et al.*, 1995a,b; Jourlin *et al.*, 1996a,b, 1997). The structure of the TorS sensor domain was solved (Moore and Hendrickson, 2009). It has a novel fold consisting of a membrane-proximal right-handed four-helical bundle and a membrane-distal left-handed four-helical bundle with significantly different structural dispositions. A mechanism similar to the NarX sensor was postulated.

The oxygen-dependent regulatory network of *E. coli*

With a fine-tuned interplay of various regulatory proteins and sRNAs, *E. coli* achieves a fast and efficient adaption during the shift from aerobic to anaerobic conditions (Fig. 11.4). First, the oxygen-sensing iron–sulfur-cluster containing Fnr regulator broadly induces microaerobic and anaerobic metabolic functions. A set of primary electron donating dehydrogenases (for formate, hydrogen, glycerol-3-phosphate), the corresponding terminal reductases (for nitrite, nitrate, fumarate, DMSO) and necessary cofactor biosynthesis as well as transporters are induced. Aerobic respiratory functions including primary dehydrogenases (for NADH, glucose, pyruvate, malate) and a terminal reductase (for oxygen) are repressed by Fnr or FnrS. Furthermore, fermentation processes (formate, acetate formation) are enhanced by Fnr. Further repression of aerobic respiration (lactate, succinate, glycerol-3-phosphate, oxygen utilization), central metabolism (pyruvate dehydrogenase, TCA cycle, glyoxylate shunt) and parallel induction of fermentation (lactate) is mediated by ArcBA. Similarly, Fnr, FnrS and ArcBA coordinate a fine–tuned complex stress response (oxygen stress, cold shock, acid stress, osmolarity changes, aero- and chemotaxis). Furthermore, typical oxygen-dependent catalytic processes like ribonucleotide reduction are replaced by oxygen-independent systems via Fnr-dependent gene regulation. In the presence of nitrate or nitrite NarXL and NarQP are differentiating this general anaerobic adaptation mediated by Fnr, FnrS and ArcBA towards a preferred nitrate respiring lifestyle.

Consequently, other alternative electron acceptor utilizing systems (for fumarate, DMSO, TMAO) are repressed while the membrane bound nitrate reductase Nar gets induced. The periplasmic nitrate reductase Nap and other nitrite utilizing systems are repressed by NarXL and induced by NarQP.

Additionally, NarXL induces a set of most likely nitrate and nitrite reductase coupling primary dehydrogenases (for glucose, lactate, D-amino acids, NADH). Moreover, NarXL participates in the fine-tuning of the stress response. The presented complex regulatory network (Fig. 11.5) is further closely linked to other important regulons including the general stress response (RpoS), catabolite regulation (Crp-CAMP), nitrogen regulation and many other regulatory circuits.

Phylogeny of Fnr, ArcBA, NarXL and NarQP

E. coli Fnr is part of the large super family of Crp-Fnr transcription regulators with 21 distinct protein subfamilies (Körner *et al.*, 2003; Rinaldo *et al.*, 2006). They are involved in the sensing of multiple intra- and extracellular parameters from the catabolic status of the cell up to the detection of oxygen tension. Current bioinformatics approaches using the 2011 version of NCBI database and the PRODORIC bioinformatics tool identified several thousand Crp-Fnr type proteins throughout the bacterial kingdom. The regulons for various α-proteobacterial Crp-Fnr type regulations were reconstructed (Dufour *et al.*, 2010).

In contrast, the NarXL/NarQP analogues/proteins have been only described for proteobacteria (Stewart, 2003). Active ArcBA was described for *E. coli*, *Haemophilus influenzae*, *Actinobacillus pleuropneumoniae* and *Vibrio fischeri* (Bose *et al.*, 2007; Wong *et al.*, 2007; Buettner *et al.*, 2008).

Pseudomonas aeruginosa

The anaerobic life of *P. aeruginosa*

P. aeruginosa is a versatile opportunistic human pathogen, which is able to colonize a broad spectrum of different aquatic and soil habitats as well as eukaryotic hosts. In the environment and during pathogenesis, *P. aeruginosa* encounters micro-milieus with oxygen-limited to anaerobic conditions (Schobert and Tielen, 2010). The investigation of *P. aeruginosa* anaerobic metabolism received increased attention because of its important role during chronic infection of the cystic fibrosis (CF) lung (Worlitzsch *et al.*, 2002; Yoon *et al.*, 2002). CF is a lethal genetic disorder in humans and results in an increased mucus production in the upper airways. CF mucus is rich in nutrients and colonized by a diverse range of different bacteria among them *P. aeruginosa* (Harrison, 2007). It has been shown that CF mucus provides an oxygen-limited environment and contains potentially anaerobic niches (Worlitzsch *et al.*, 2002; Yoon *et al.*, 2002). The physiological changes of *P. aeruginosa* in response to oxygen limitation contribute to increased antibiotic tolerance and support persistent infection (Schobert and Jahn, 2010; Schobert and Tielen, 2010).

The anaerobic metabolism of *P. aeruginosa*

As a facultative anaerobe *P. aeruginosa* is able to grow and survive anaerobically via anaerobic respiration of nitrate or nitrite, which is termed denitrification (Fig. 11.7). Moreover, *P. aeruginosa* has been shown to generate energy under anaerobic conditions by two unusual fermentation pathways using arginine or pyruvate (Carlson and Ingraham, 1983; Davies *et al.*, 1989; Zumft, 1997; Eschbach *et al.*, 2004; Williams *et al.*, 2006).

During denitrification nitrate is reduced in four sequential steps to dinitrogen (Fig. 11.7). Thereby, protons are translocated to the periplasm generating the proton motive force, which is used for ATP synthesis catalysed by the ATP synthase (Zumft and Körner, 1997; Williams *et al.*, 2006). The first step of denitrification, the reduction of nitrate to nitrite, is mediated by two dissimilatory nitrate reductases, the membrane bound NarGHI encoded by *narGHJI* and the periplasmic NapAB encoded by the *napEFDABC* gene cluster. The *narGHJI operon* contains an additional gene encoding the NarJ protein, which is important for the assembly of a functional nitrate reductase (Blasco *et al.*, 1998). The operon includes the $narK_1K_2$ genes at the 5′-end. Physiological data indicate that $narK_2$ encodes a nitrate/nitrite antiporter, while $narK_1$ might encode a high affinity nitrate uptake system as observed for *Paracoccus denitrificans* (Sharma *et al.*, 2006; Goddard *et al.*, 2008).

The second step of denitrification, the reduction of nitrite to nitric oxide (NO) is catalysed by the periplasmic cytochrome cd_1 type nitrite reductase, NirS (Silvestrini *et al.*, 1994). The NirS protein and two c-type cytochromes, NirM and NirC are encoded in the *nirSMCFDLGHJEN operon,* which includes also genes encoding enzymes for haem d_1 synthesis (Kawasaki *et al.*, 1997). Electrons are transferred to NirS from the cytochrome bc_1 complex of the electron transfer chain via the c-type cytochromes NirM, NirC and the copper protein azurin (Hasegawa *et al.*, 2001, 2003; Vijgenboom *et al.*, 1997). The *nirQOP* gene cluster is divergently transcribed with respect to the the *nirSMCFDLGHJEN operon.* The first gene of this operon, *nirQ*, has been shown to encode an ATP-binding protein, which activates NirS post translationally (Hayashi *et al.*, 1998).

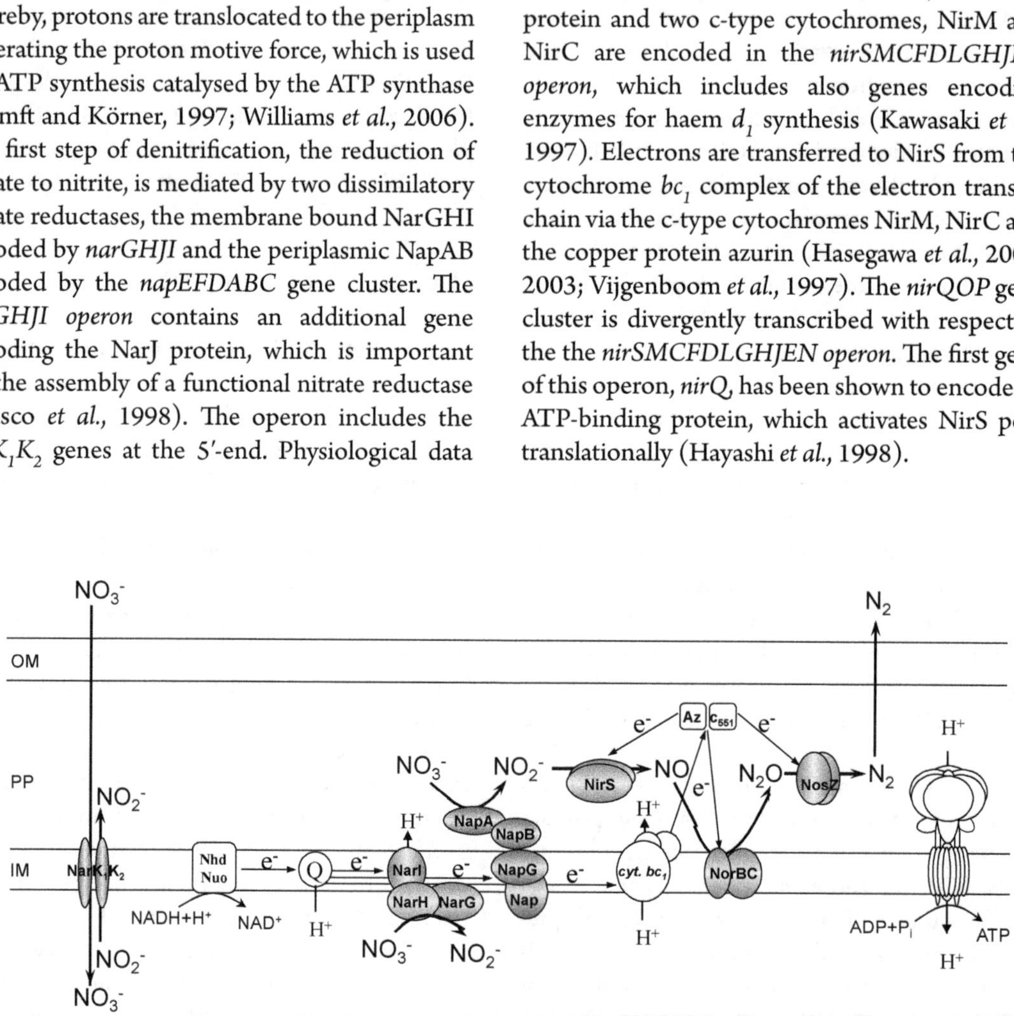

Figure 11.7 The *P. aeruginosa* denitrification pathway. The inner membrane-bound nitrate reductases NarGHI and NapAB, the periplasmic nitrite reductase NirS, the membrane-bound NO reductase NorCB, and the periplasmic nitrous-oxide reductase NosZ catalyse the reduction of nitrate to dinitrogen. The nitrate and nitrite transport is mediated by the putative antiporter $NarK_1K_2$. NADH provides the required electrons via the NADH:quinone oxidoreductase I Nuo and the NADH:quinone oxidoreductase II Ndh. The electrons are transferred to the quinone pool Q, to the cytochrome bc_1 (cyt. bc_1) or directly to the nitrate reductases NarGHI and further to azurin (Az) or to the cytochrome c_{551}. As a consequence, H^+ is translocated to the periplasm. The proton motive force is used for ATP generation by the ATPase. PP, periplasm; IM, inner membrane; OM, outer membrane. A colour version of this figure is available in the plate section at the back of the book.

The periplasmic nitric oxide reductase NorCB catalyses the third step of the denitrification pathway, the reduction of nitric oxide to nitrous oxide (Kumita *et al.*, 2004). The *norCBD* cluster contains the genes encoding this reductase. The operon is located in close proximity to the *nir* operon (Arai *et al.*, 1995b). Again, electrons are transferred to the NorC subunit from the cytochrome bc_1 complex via a c-type cytochrome (Zumft, 2005). Nitric oxide is an important intermediate in denitrification. It is a signal molecule but also toxic to the cell. Therefore, the controlled production and reduction of this intermediate is crucial during denitrification (Yoon *et al.*, 2007; Schobert and Tielen, 2010).

The nitrous oxide reductase NosZ, encoded by the *nosRZDFYL operon* catalyses the reduction of nitrous oxide to dinitrogen, which comprises the last step in the denitrification pathway. NosZ forms a homodimer containing two copper centres Cu_A and Cu_Z (Zumft and Kroneck, 2007). The insertion of copper into NosZ is mediated by the *nosDFY* encoded proteins. NosR is a putative membrane-bound N_2O sensor, which is involved in the regulation of the *nos* operon (Arai *et al.*, 2003; Zumft and Kroneck, 2007).

P. aeruginosa can also grow very moderately under anaerobic conditions by arginine fermentation, which generates ATP via substrate level formation (Vander Wauven *et al.*, 1984). Arginine is converted to ornithine with the simultaneous generation of ATP by three enzymes of the arginine deiminase pathway, the arginine deiminase ArcA, the catabolic ornithine carbamoyltransferase ArcB, and the carbamate kinase ArcC (Luthi *et al.*, 1986; Baur *et al.*, 1989; Luthi *et al.*, 1990). In addition, an arginine-ornithine antiporter ArcD is essential for uptake of arginine and the secretion of the formed ornithine. The genes of the enzymes and the antiporter ArcD are organized in one operon *arcDABC* (Luthi *et al.*, 1986, 1990).

The second fermentation pathway observed in *P. aeruginosa* converts pyruvate to acetate, lactate and minor amounts of succinate (Eschbach *et al.*, 2004). In contrast to arginine fermentation, fermentation of pyruvate does not sustain anaerobic growth but anaerobic long-term survival (Eschbach *et al.*, 2004). ATP is formed when pyruvate is oxidized to acetate in three steps involving a pyruvate dehydrogenase Pdh, the phosphotransacetylase Pta and the acetate kinase AckA (Eschbach *et al.*, 2004). In a second branch of pyruvate fermentation, pyruvate is reduced to lactate by the lactate dehydrogenase LdhA. This reaction recycles NAD from $NADH+H^+$, which is generated by the pyruvate dehydrogenase of the first branch. Similar to *E. coli*, the various modes of energy generation yield different amounts of ATP. Consequently, a tight regulation in dependence of oxygen tension, N-oxides and arginine is required. The involved regulators are described below.

Oxygen-sensing by Anr

Anr belongs to the Fnr-Crp regulator family (Fig. 11.3) and shows considerable similarity to Fnr of *E. coli* (Beinert and Kiley, 1999; Unden *et al.*, 2002). As Fnr, the active Anr of *P. aeruginosa* contains a $[4Fe–4S]^{2+}$ cluster, which is oxygen labile and is partly destroyed in the presence of oxygen (Spiro, 1994; Unden *et al.*, 2002; Yoon *et al.*, 2007). Active Anr binds to a conserved sequence within the target promoters known as Anr box (Fig. 11.3). Determined Anr boxes like in the *arcD*, *uspN* and *hemN* promoter regions indicate an optimal distance of about −41.5 bp to the transcriptional start site, similar to Fnr class II promoters in *E. coli* (Gamper *et al.*, 1991; Rompf *et al.*, 1998; Boes *et al.*, 2006). However, in contrast to *fnr* of *E. coli*, the *anr* gene of *P. aeruginosa* obviously lacks the corresponding binding site for negative autoregulation (Sawers, 1991).

A recent publication determined the Anr-dependent regulon in *P. aeruginosa* using a combination of experimental and bioinformatics approaches (Trunk *et al.*, 2010). This approach predicted 170 Anr binding sites in the *P. aeruginosa* genome. Interestingly, experimental data indicate that Anr acts predominantly as a transcriptional activator, with only a few examples of transcriptional repression via Anr. This is in clear contrast to the role of Fnr in *E. coli*. Anr is on top of a regulatory cascade regulating the induction of the anaerobic energy metabolism, which is described in detail below (Fig. 11.8).

NO-sensing via Dnr

The Dnr regulator (dissimilatory nitrate respiration regulator) also belongs to the Fnr-Crp

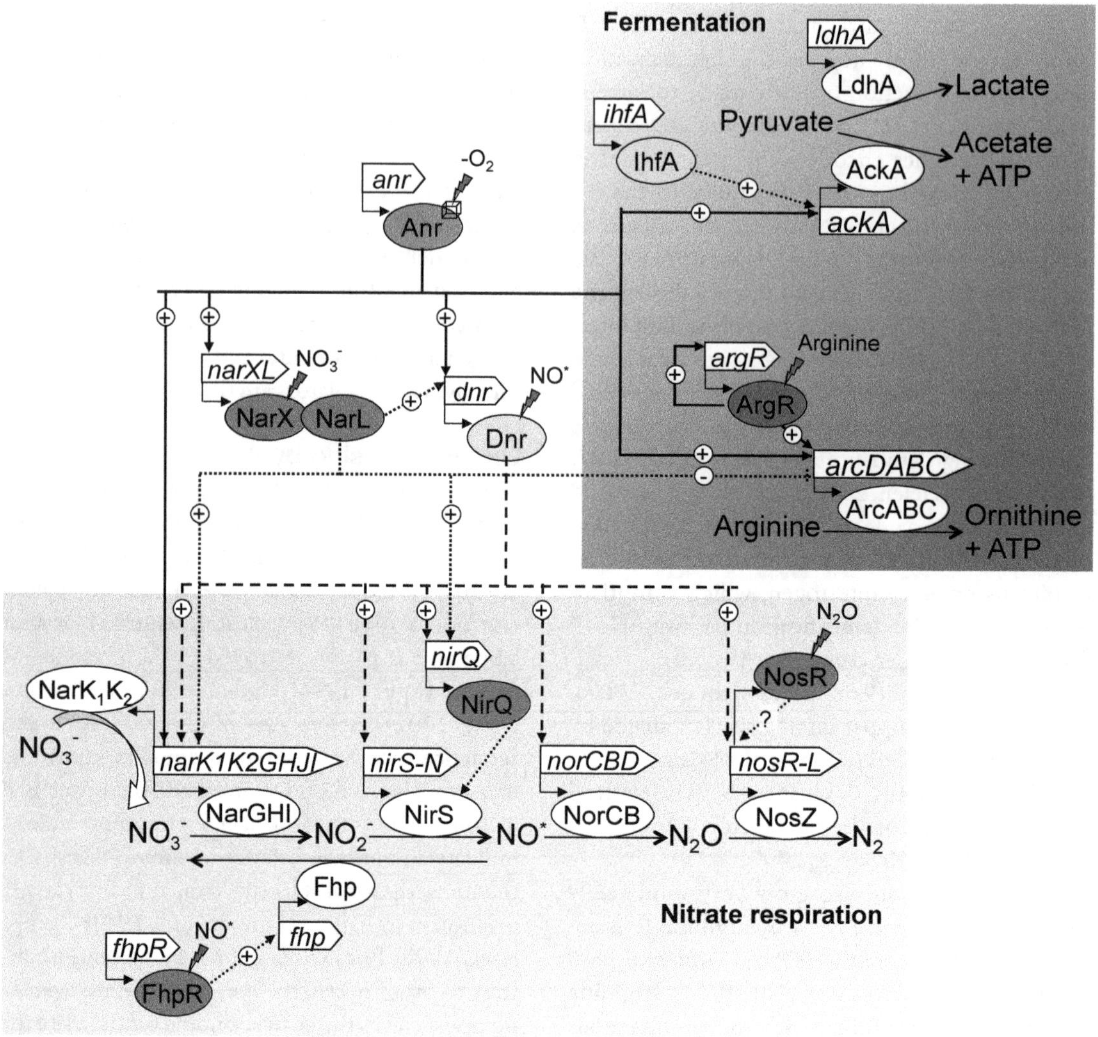

Figure 11.8 Regulatory network controlling the anaerobic metabolism of *P. aeruginosa*. In interplay of the oxygen regulator Anr, the NO-sensors Dnr and FhpR, the N_2O- dependent regulator NosR, the arginine regulator ArgR with InfA and NirQ for the control of genes involved in denitrification (yellow background), arginine and pyruvate fermentation (red background) is shown. A colour version of this figure is available in the plate section at the back of the book.

regulator family (Fig. 11.3). However, in contrast to Anr, Dnr lacks the four cysteine residues essential for formation of the $[4Fe–4S]^{2+}$ cluster. The major signal detected by the Dnr regulator via a bound haem is NO (Giardina *et al.*, 2008; Castiglione *et al.*, 2009). In response to NO, Dnr activates the expression of *nirS*, *nirQ*, *norC* and *nosR*, the gene products of which are involved in the denitrification pathway (Arai *et al.*, 2003). The Dnr recognition site is indistinguishable from the Anr box (Arai *et al.*, 1995a, 1997, 2003). Anr is not able to directly activate gene expression of the *nir*, *nor* and *nos* genes. However, Anr activates transcription of the *dnr* gene and, therefore, forms a regulatory cascade (Arai *et al.*, 1997). Currently, it is still unknown how the Anr and Dnr regulators discriminate between the indistinguishable binding sites. However, several promoters depend on both Anr and Dnr, for example the three genes involved in haem biosynthesis, *hemA*, *hemN* and *hemF* and the nitrate reductase operon *narK*$_1$*K*$_2$*GHJI* (Fig.

11.8) (Rompf *et al.*, 1998; Krieger *et al.*, 2002; Schreiber *et al.*, 2007).

NO-sensing via FhpR

An additional NO-responsive regulatory system of *P. aeruginosa* is the FhpR regulator, which is a NorR type RpoN-dependent activator important for detoxification of NO (Fig. 11.8). The *fhpR* gene is similar to *norR* of *E. coli* (Pohlmann *et al.*, 2000; Hutchings *et al.*, 2002). FhpR activates the expression of *fhp* encoding a flavohaemoglobin (Fig. 11.7). The Fhp/FhpR system form an NO-sensing and detoxifying system (Koskenkorva *et al.*, 2008). Moreover, it is capable of protecting cells against nitrosative stress and is implicated in the oxidative stress response (Koskenkorva-Frank and Kallio, 2009). The *fhp* and *fhpR* genes are divergently oriented and share their promoter regions (Arai *et al.*, 2005; Koskenkorva *et al.*, 2008). A binding motif for RpoN and three binding sites for FhpR were found in the promoter region of *fhp*, but an Anr box was missing, indicating that *fhp* is expressed independently of Anr and Dnr even under oxygen-limiting conditions (Arai *et al.*, 2005; Trunk *et al.*, 2010).

Nitrate-sensing via NarXL

The classical two-component system NarXL is additionally required for the onset of denitrifying growth (Schreiber *et al.*, 2007). NarX shares 38% and 34% amino acid sequence identity to NarX from *E. coli* and *Pseudomonas stutzeri*, respectively (Härtig *et al.*, 1999). This histidine sensor kinase contains a typical C-region and a P-box, which are conserved among nitrate and nitrite responsive kinases. The P-box in NarX of *E. coli* is responsible for discrimination between nitrate and nitrite (Williams and Stewart, 1997a). However, it is still unknown if *P. aeruginosa* NarX also discriminates between nitrate and nitrite. The NarL protein encodes the corresponding response regulator and shares amino acid identities of 63% and 58% to NarL from *P. stutzeri* and *E. coli*, respectively (Härtig *et al.*, 1999). Only a few target promoters have been identified in *P. aeruginosa*, which are activated or repressed by NarL (Fig. 11.7). Transcription of the genes *hemA*, *dnr*, *nirQ* and the *narK*$_1$*K*$_2$*GHJI* operon is activated by NarL, while the genes encoding the arginine fermentation pathway *arcDABC* and the periplasmic nitrate reductase *napEFDABC* are repressed by NarL (Benkert *et al.*, 2008; Van Alst *et al.*, 2009).

Arginine-sensing by ArgR

The transcriptional regulator ArgR (arginine-responsive regulatory protein) *of P. aeruginosa* is involved in the regulation of arginine biosynthesis and catabolism (Fig. 11.7). ArgR is active in the presence of arginine and belongs to the AraC/XylS family of transcriptional regulators (Gallegos *et al.*, 1997; Park *et al.*, 1997b). It does not exhibit any amino acid sequence homology to the arginine regulatory proteins from enteric bacteria (Lim *et al.*, 1987; Lu *et al.*, 1992) or *B. subtilis* (Czaplewski *et al.*, 1992). *P. aeruginosa* ArgR is a homodimer of two subunits, each with a molecular mass of 37 kDa. The ArgR regulon has been determined and consists of more than 30 genes (Lu *et al.*, 2004). The transcription of *carAB*, *argG* and *argF* involved in arginine biosynthesis is repressed by ArgR while the *arcDABC* and the *aruCFGDBE* operons, the latter encoding enzymes of the arginine succinyltransferase pathway are induced. ArgR negatively controls expression of the glutamate biosynthesis genes *gdhA* and *gltBD* and positively controls the *aotJQMOP-argR operon*, which encodes a system for the uptake of arginine and ornithine as well as ArgR itself (Park *et al.*, 1997a,b; Nishijyo *et al.*, 1998; Lu *et al.*, 1999).

N_2O-sensing by NosR

The NosR regulator of *P. stutzeri* is a membrane-bound N_2O-sensing regulator required for the expression of the *nosRZDFYL* operon and, therefore, for the production of the nitrous oxide reductase NosZ (Cuypers *et al.*, 1992; Vollack and Zumft, 2001). A similar gene arrangement of the *nos* operon is present in *P. aeruginosa*. Moreover, *nosR* gene of *P. aeruginosa* showed 75% sequence identity to *nosR* of *P. stutzeri* (Arai *et al.*, 2003). However, the nitrous oxide reductase of *P. aeruginosa* is not expressed in the presence of N_2O. Furthermore, the transcription of *P. aeruginosa nosRZDFYL operon* is dependent on Dnr and does not require NosR. Therefore, the regulatory role of *P. aeruginosa* NosR is uncertain.

The regulator NirQ

The *nirQOP* gene cluster located upstream of the structural gene for nitrite reductase *nirS* encodes the regulator NirQ (Fig. 11.7), which is necessary for the activation of the nitrite and nitric oxide reductase (Arai *et al.*, 1994). It was shown that NirQ activates NirS posttranslationally (Hayashi *et al.*, 1998). The 30.5 kDa ATP-binding protein of *P. aeruginosa* belongs to the NtrC protein family of transcriptional activators and its amino acid sequence shows 76% identity to the *P. stutzeri* NirQ (Jüngst and Zumft, 1992). NirQ is produced in an Anr- and Dnr-dependent manner (Arai *et al.*, 1994). Moreover, NarL positively influences *nirQ* expression (Schreiber *et al.*, 2007).

Anaerobic regulatory network controlling denitrification in *P. aeruginosa*

P. aeruginosa the oxygen-sensing Anr regulator has been shown to be essential for the entire anaerobic energy metabolism, including denitrification, arginine and pyruvate fermentation (Eschbach *et al.*, 2004; Ye *et al.*, 1995). Anr is on top of the anaerobic regulatory network, which is organized in a hierarchical manner (Fig. 11.8).

When *P. aeruginosa* enters an oxygen-limited to anaerobic environment which contains nitrate as alternative electron acceptors a regulatory cascade induces transcription of all four reductases (*nar, nir, nor, nos*) essential for nitrate reduction (Fig. 11.8).

In *P. aeruginosa* denitrification has been also observed under microaerobic conditions simultaneously to oxygen respiration (Davies *et al.*, 1989). Consequently, denitrification requires an active Anr regulator but not strictly anaerobic conditions since Anr increases transcription of the *dnr* gene encoding the NO-sensing Dnr regulator and the *narXL* genes encoding the nitrate sensing two-component system NarXL. (Arai *et al.*, 1997; Schreiber *et al.*, 2007). All three regulators, Anr, Dnr and NarL have been shown to activate transcription of the *narK$_1$K$_2$GHJI operon* encoding the nitrate reductase and nitrate/nitrite antiporter (Sharma *et al.*, 2006; Schreiber *et al.*, 2007). The respective promoter region contains one Anr binding site at −40.5 bp, which is recognized by both Anr and Dnr. Moreover, three putative NarL binding sites have been identified at position −108, −113 and −208 bp of the transcriptional start site (Schreiber *et al.*, 2007). The promoter region of the *narK$_1$K$_2$GHJI* region and *narXL* are divergent and overlapping. The Anr binding site is shared between both promoters and is located at −60.5 upstream the transcriptional start site of *narX* (Schreiber *et al.*, 2007).

Several additional targets of the NarL regulator have been identified in *P. aeruginosa.* NarL directly activates transcription of the *dnr* and *nirQ* genes, which both contain putative NarL binding sites in the respective promoter regions (Schreiber *et al.*, 2007). Moreover, NarL also represses gene expression of the periplasmic nitrate reductase NapAB and the arginine deiminase operon essential for arginine fermentation (Benkert *et al.*, 2008; Van Alst *et al.*, 2009). In both cases repression via NarL does not completely abolish gene expression. But the repression of the *napEFDABC* operon and significant reduction of arginine fermentation pathway reduces the presence of less favourable energy metabolic pathways.

The NO-sensing Dnr regulator induces expression of *nirS, nirQ, nor-* and *nosR* promoters (Arai *et al.*, 2003). Although each promoter contains a conserved Anr box the Anr regulator does not activate expression of the respective genes. Therefore, a strict cascade regulation is observed, where Anr and NarL activate transcription of the *dnr* gene (Arai *et al.*, 1997; Schreiber *et al.*, 2007). The active Dnr regulator in turn activates transcription of *nirS, norC* and *nosR* promoters alone and the *nirQOP operon* together with an active NarL (Arai *et al.*, 2003; Schreiber *et al.*, 2007). The NirQ protein binds ATP and activates the nitrite reductase at the posttranslational level (Arai *et al.*, 1994; Hayashi *et al.*, 1998). This way, Dnr together with NarL and NirQ activates the lower denitrification pathway consisting of the nitrite, nitric oxide and nitrous oxide reductases. In addition, Dnr increases transcription of the *hemA, hemN* and *hemF* genes, which encode enzymes of haem biosynthesis (Rompf *et al.*, 1998; Krieger *et al.*, 2002; Schreiber *et al.*, 2007). Dnr increases transcription of the nitrate reductase operon as outlined above (Schreiber *et al.*, 2007). In contrast to Anr, which has been shown to directly activate 29 transcriptional units in a microarray experiment, the Dnr regulon consists of only four members, which are under direct control of Dnr alone (Trunk *et al.*, 2010).

Regulation of denitrification via quorum sensing

The Anr, Dnr, NarXL regulators form the classic anaerobic regulatory network controlling expression of all four gene clusters (*nar, nir, nor, nos*) encoding the four reductases for denitrification. Recently, additional impact of quorum sensing and the RpoN sigma factor has been published. *P. aeruginosa* possesses the quorum-sensing systems LasI–LasR responding to the signalling molecule *N*-(3-oxo-dodecanoyl)-homoserine lactone and the RhlI–RhlR system using *N*-(butanoyl)-L-homoserine lactone. These systems are organized in a hierarchical manner and have been reported to control expression of more than 300 genes in response to population density. In addition to these homoserine lactone-dependent systems *P. aeruginosa* contains the *Pseudomonas* quinolone signal (PQS), which functions through the action of 2-heptyl-3-hydroxyl-4-quinolone and is linked to the Las and Rhl systems (Calfee *et al.*, 2001; Girard and Bloemberg, 2008; Juhas *et al.*, 2005; Latifi *et al.*, 1996; Pesci *et al.*, 1997; Venturi, 2006).

The RhlR system was found to moderately repress the *narK*1, *nirS*, *norC* and *nosR* promoters (Toyofuku *et al.*, 2007). However, it is still unclear if RhlR interacts directly with the respective promoter regions or with another part of the regulatory network (Toyofuku *et al.*, 2007). It was shown that PQS influences denitrification. PQS seems to directly modulate enzyme activities of the nitrate, nitrite, nitric oxide reductases while the activity on the nitrous oxide reductase is influenced via the proteins PqsR and PqsE (Toyofuku *et al.*, 2008). The effects of PQS may be important under microaerobic conditions since PQS biosynthesis requires molecular oxygen and is down-regulated under strict anaerobic conditions. The exact mechanism of this regulation is still unclear and awaits further investigation.

NO detoxification systems during denitrification

One intermediate of denitrification is NO, which is, on one hand, toxic to the cell and, on the other hand, a signal molecule for biofilm dispersal in *P. aeruginosa* (Barraud *et al.*, 2006, 2009). During denitrification, the accumulation of toxic NO levels have to be avoided, which requires tight control of the NO-producing nitrite reductase and the NO-reducing NO reductase. It has been shown that a mutation in the *rhlR* gene deregulates the nitrate, nitrite and nitric oxide reductase activities, which in turn leads to the accumulation of toxic NO. These observations underscore the importance of quorum sensing for a functional denitrification pathway (Yoon *et al.*, 2002). While the accumulation of NO during denitrification is controlled by the nitric oxide reductase NorCB, there are several other systems, which contribute to NO detoxification. The Fhp/FhpR system forms an additional NO-sensing and detoxifying system in *P. aeruginosa*. The RpoN-dependent FhpR regulator activates the *fhp* genes in response to NO (Koskenkorva *et al.*, 2008). The corresponding *fhp* gene product is a flavohaemoglobin, which carries NO dioxygenase activity (Arai *et al.*, 2005).

Regulation of arginine and pyruvate fermentation pathways.

Arginine and pyruvate fermentation are dominantly regulated via the oxygen sensing Anr (Fig. 11.7). Arginine fermentation further depends on the activation of the *arcDABC* operon by the arginine-sensing regulator ArgR (Lu *et al.*, 1999). In the presence of nitrate and arginine, the nitrate-responsive NarL regulator competes for the activation mediated by ArgR, since the NarL binding site is overlapping with the ArgR recognition sequence in the *arcDABC* promoter region (Benkert *et al.*, 2008). Therefore, expression of arginine fermentation genes is reduced but not completely repressed during denitrification, which was also confirmed by a transcriptome analysis (Platt *et al.*, 2008).

The Anr regulator also controls pyruvate fermentation by activation of the expression of the *ackA-pta* operon. In addition to Anr the DNA-bending integration host factor (IHF) is needed for transcription of the *ackA-pta* operon (Eschbach *et al.*, 2004).

Bacillus subtilis

The anaerobic regulatory network of *B. subtilis*

The Gram-positive soil bacterium *B. subtilis* continuously adapts its physiology to oxygen tension in its natural environment. Consequently, *B. subtilis* is able to grow under anaerobic conditions using nitrate respiration and fermentation processes (Nakano and Hulett, 1997; Nakano and Zuber, 2001). During the process of ammonification nitrate serves as alternative electron acceptor and is reduced to nitrite by the dissimilatory nitrate reductase encoded by the *narGHJI* operon. Nitrite is further converted into ammonia by the assimilatory nitrite reductase NasDE (Fig. 11.9A) (Hoffmann *et al.*, 1995, 1998). In the absence of nitrate, *B. subtilis* is able to perform a mixed acid fermentation where pyruvate is transformed into the end products lactate, acetoin, 2,3-butanediol, ethanol and acetate (Fig. 11.9B) (Nakano *et al.*, 1997; Cruz-Ramos *et al.*, 2000). NAD^+ regeneration is primarily mediated by a cytoplasmic lactate dehydrogenase, encoded by *ldh* that converts pyruvate to lactate (Cruz-Ramos *et al.*, 2000; Nicholson, 2008). Acetoin is synthesized from pyruvate in a two-step reaction catalysed by acetolactate synthase and acetolactate decarboxylase, encoded by the *alsSD* operon (Renna *et al.*, 1993; Cruz-Ramos *et al.*, 2000). Subsequently, acetoin is converted to 2,3-butanediol by the 2,3-butanediol dehydrogenase encoded by *bdhA*, comprising a second step for NAD^+ regeneration (Nicholson, 2008). The third major fermentation product,

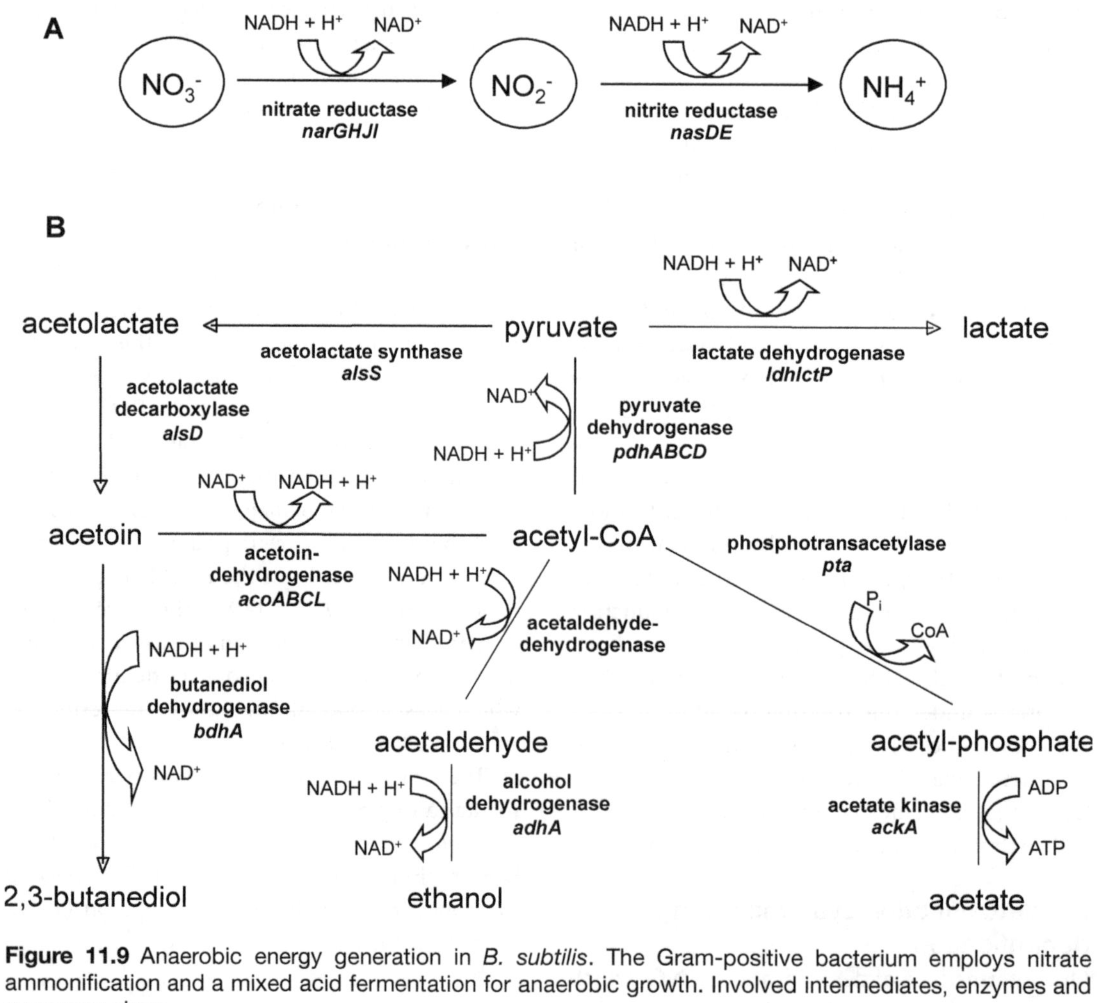

Figure 11.9 Anaerobic energy generation in *B. subtilis*. The Gram-positive bacterium employs nitrate ammonification and a mixed acid fermentation for anaerobic growth. Involved intermediates, enzymes and genes are given.

acetate, is formed from acetyl-coenzyme A in a two-step reaction catalysed by phosphotransacetylase and acetate kinase, encoded by *pta* and *ackA*, respectively. The latter step usually leads to the formation of ATP. Production of acetate is also performed under aerobic growth conditions as part of the carbon overflow metabolism. Anaerobic respiration and fermentation are less efficient with respect to the energy yield than aerobic respiration. Therefore, oxygen tension is continuously monitored to coordinate expression of various pathways for optimal energy recovery. The anaerobic stimulon of *B. subtilis* comprises several hundred genes and operons. On the transcriptional level the anaerobic adaptation of *B. subtilis* is the result of the interplay of three major regulators: the two-component regulatory system ResDE, the redox regulator Fnr and the redox-sensing repressor Rex (Cruz-Ramos *et al.*, 1995; Sun *et al.*, 1996; Larsson *et al.*, 2005).

The two-component regulatory system ResDE

The two-component regulatory system ResDE consists of a sensor histidine kinase and the response regulator ResD (Geng *et al.*, 2004). ResE undergoes autophosphorylation presumably at the conserved histidine residue at position 374 and promotes phosphorylation of ResD (Fig. 11.6). The exact nature of the signal and the mechanism of signal perception by the sensor kinase ResE is still unknown. ResE is predicted to consist of two transmembrane helices flanking an extracytoplasmic loop. The second transmembrane region is followed by a HAMP, a PAS and a kinase domain in the cytoplasm. Deletion of the PAS domain leads to a loss of function indicating its essential role in the activation mechanism of ResE (Baruah *et al.*, 2004).

The ResD response regulator gets activated by phosphate transfer from ResE to aspartate residue 57 (Fig. 11.6). Phosphorylated ResD binds to the promoter regions of its target genes. The ResD regulon consists of 27 genes or operons (Ye *et al.*, 2000). ResD was shown to regulate directly the expression of the *ldhlctP* operon, encoding lactate dehydrogenase and lactate permease. ResDE also regulates the expression of *nasDEF*, encoding the assimilatory nitrite reductase and the *hmp* gene coding for a flavohaemoglobin (Fig. 11.9). Moreover, ResD activates the expression of the oxygen regulator gene *fnr* (Nakano *et al.*, 2000; Geng *et al.*, 2007). ResD binding sites within the promoter of *nasD, hmp, fnr, yclJK* and *ctaA* were defined by DNaseI-footprinting analyses and a consensus binding site (TTGTGAAN$_3$TTTNTN$_2$A) was deduced. The *yclJK* operon encodes a two-component regulatory system found anaerobically induced by ResDE. However, the signal recognized by the YclJK system and its target genes are still unknown (Härtig *et al.*, 2004).

ResDE orthologues have been found in *Bacilli*, *Listeria*, and *Staphylococci* and were shown to function in virulence gene regulation. SrrAB in *Staphylococcus aureus* regulates energy production and production of virulence factors in response to oxygen limitation (Throup *et al.*, 2001; Yarwood *et al.*, 2001; Pragman *et al.*, 2004). ResD from *Listeria monocytogenes* is required for repression of virulence gene expression in the presence of easily fermentable carbon sources (Larsen *et al.*, 2006). For *Bacillus cereus* ResDE-dependent regulation of the enterotoxin gene was reported (Duport *et al.*, 2006).

The oxygen regulator Fnr of *B. subtilis*

The anaerobic expression of the *fnr* gene is dependent on ResDE activation (Nakano and Zuber, 2001). The anaerobic regulator Fnr of *B. subtilis* belongs to the Crp/Fnr family of transcription factors (Körner *et al.*, 2003). Under anaerobic conditions, Fnr gets activated by binding of a [4Fe–4S]$^{2+}$ presumably followed by structural rearrangements that enables DNA binding of the transcription factor via a helix–turn–helix domain (Fig. 11.3). In contrast to *E. coli* Fnr the [4Fe–4S]$^{2+}$ cluster of *B. subtilis* Fnr is coordinated by three C-terminally located cysteine residues at position 227, 230 and 235 and one aspartate residue at position 141 (Fig. 11.3) (Reents *et al.*, 2006; Gruner *et al.*, 2010). Furthermore, *B. subtilis* Fnr also forms a dimer in the presence of oxygen and absence of the iron–sulfur-cluster (Fig. 11.3). These significant differences suggest an activation mechanism different to the *E. coli* protein.

The Fnr regulon was defined using global transcriptional profiling and genetic promoter analyses (Fig. 11.9). It comprises three transcriptional units: the *narGHJI* operon encoding the nitrate reductase, the *narKfnr* operon encoding a nitrate transporter and the *fnr* gene itself, and *arfM* encoding a modulator of anaerobic respiration and fermentation (Fig. 11.9) (Cruz-Ramos *et al.*, 1995; Marino *et al.*, 2001). Mutation of the *arfM* gene leads to a reduced anaerobic expression of the fermentative *ldhlctP* (lactate fermentation) and *alsSD* (acetoin formation) operons and of the haem biosynthesis genes *hemN* and *hemZ* indicating a role of ArfM as modulator of anaerobic respiration and fermentation (Homuth *et al.*, 1999; Marino *et al.*, 2001).

All three Fnr-dependent promoters have a highly conserved *B. subtilis* Fnr-binding site (TGTGA-N_6-TCACA) at position −41.5 and −40.5 with respect to the transcriptional start site, respectively.

During transcriptome analyses, the fermentative genes *ldhlctP*, *alsSD*, *ywcJ* (potential nitrite transporter), and the *cydABCD* operon (high-affinity cytochrome *bd*) were found to be repressed in the presence of nitrate (Fig. 11.9). Mutational analysis of the Fnr-binding sites found in the corresponding promoters excluded their function in Fnr-mediated repression. Instead, these genes were part of the regulon of the redox repressor Rex of *B. subtilis* (Larsson *et al.*, 2005). Fermentative expression of the *alsSD* operon is dependent on AlsR, a regulator of the family of LysR-type transcriptional regulators (Fig. 11.9) (Renna *et al.*, 1993).

The redox regulator Rex

The redox regulator Rex responds indirectly to changes in oxygen availability by reacting to changes in the NADH+H^+/NAD^+ ratio (Fig. 11.4). During aerobic and anaerobic respiration NADH+H^+ gets rapidly re-oxidized by the electron transport chain leading to a low NADH+H^+/NAD^+ ratio. The redox regulator Rex of *B. subtilis* consists of a N-terminal helix–turn–helix DNA binding domain and a C-terminal NADH-binding domain (Fig. 11.4) At low NADH+H^+/NAD^+ ratios Rex dimers bind to specific binding sites and repress expression of the corresponding genes (Wang *et al.*, 2008). During fermentative growth conditions NADH+H^+ levels increase and NADH binds with higher affinity than NAD^+ to Rex inducing a domain re-arrangement followed by the release of the repressor from the DNA (Wang *et al.*, 2011). Subsequently, anaerobic transcription of *ldhlctP*, *cydABCD* and *ywcJ* can be activated by ResD. DNase I footprinting analysis revealed three binding sites of Rex in the *cydA* promoter region and a consensus sequence (5′TTTGTGAANTN_2TN_3CAA′3) was deduced (Schau *et al.*, 2004). Potential Rex binding sites were also found in the *ywcJ*, *ldh* and *alsS* promoters overlapping non-functional potential Fnr binding site (Fig. 11.9). A point mutation in the consensus sequence of Rex in the *alsS* promoter leads to a drastically reduced *in vitro* binding affinity of the Rex repressor (Wang *et al.*, 2008). The Rex repressor was also found in other Gram-positive bacteria like *Streptomyces* and *Staphylococcus* (Brekasis and Paget, 2003; Pagels *et al.*, 2010). Rex of *S. aureus* was found to regulate a multitude of anaerobically induced genes and directly regulates the expression of the ResDE homologue SrrAB (Pagels *et al.*, 2010).

The anaerobic regulatory network of *B. subtilis*

Anaerobic adaptation in *B. subtilis* is a fine-tuned interplay of mainly three different regulatory proteins (Fig. 11.10). First, the two-component regulatory system ResDE induces several genes of the anaerobic metabolism and the oxygen regulator gene *fnr*. Fnr in turn activates genes needed for nitrate respiration. As long as NADH+H^+ is effectively re-oxidized by the electron transport chain, Rex is activated and represses expression of fermentative genes (Fig. 11.10). Each regulator involved in the anaerobic adaptation process responds differently to the signal of reduced oxygen tension. For the ResDE two-component regulatory system the exact signal and mode of signal perception are still unknown, but an indirect oxygen sensing mechanism is assumed. Fnr is directly sensing oxygen tension using an oxygen sensitive $[4Fe–4S]^{2+}$ cluster as a cofactor and the repressor activity of Rex is dependent on the NADH+H^+/NAD^+ ratio that shows characteristic variations in response to oxygen.

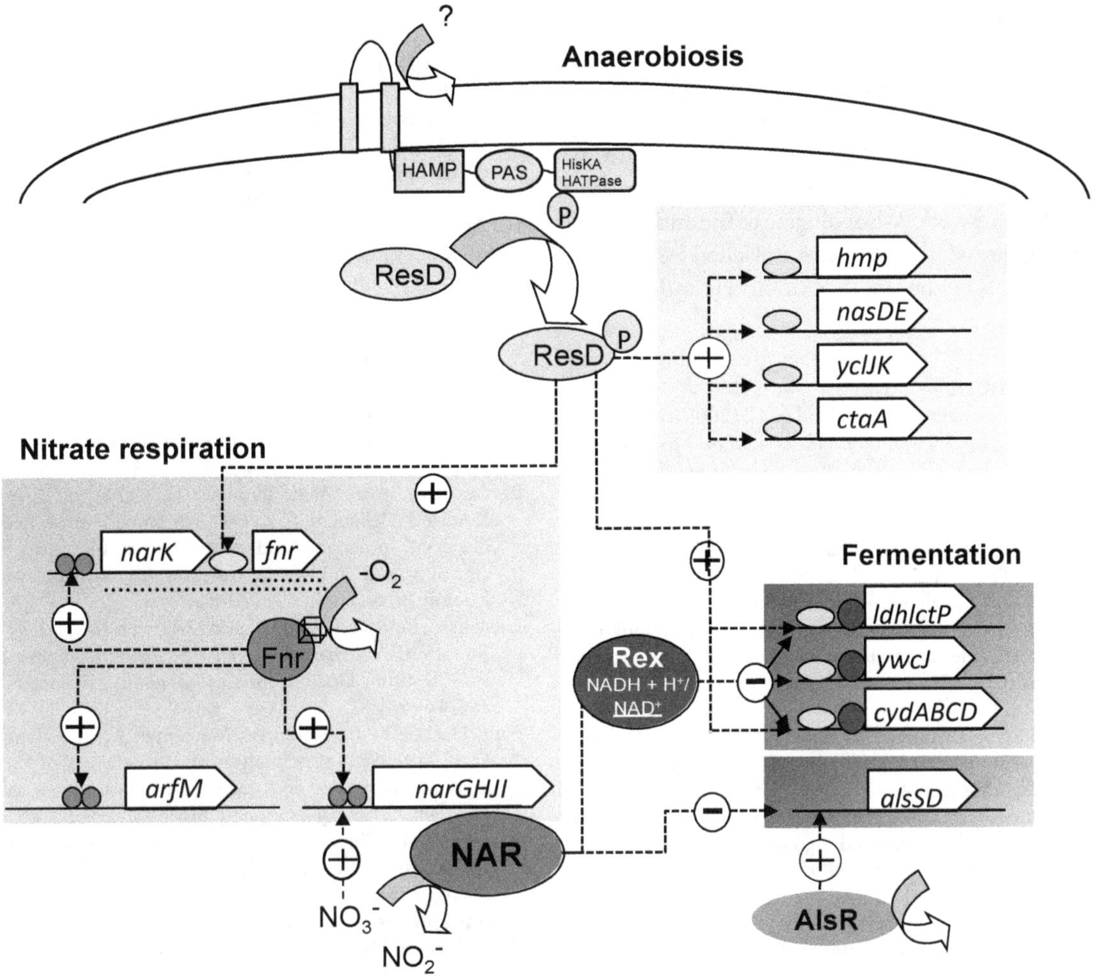

Figure 11.10 Regulatory network for the adaptation of *B. subtilis* to anaerobic growth. The differential anaerobic gene expression for enzymes involved in nitrate ammonification (green background), mixed acid fermentation (red background) and related processes (yellow background) by ResDE, Fnr, Rex and AlsR are shown. Where known, corresponding stimuli of the regulators are given. A colour version of this figure is available in the plate section at the back of the book.

Summary and future prospects

The analysis of the regulatory networks of three model bacteria (*E. coli, P. aeruginosa, B. subtilis*) for the adaptation to anaerobic growth conditions revealed some common regulatory proteins (Fnr/Anr, NarXL), some specialized factors (ArcA, ResDE, AlsR) and species specific regulatory strategies. Employed scenarios reflect the different anaerobic metabolic capacities and environments inhabited by these organisms. For some of the systems, fundamental structural and functional information is missing for the understanding of gene regulation at the molecular and atomic level. Furthermore, high-throughput omics technologies once combined with bioinformatical modelling approaches provide the basis for the future understanding of these various regulatory circuits at the systems biology level. Models used for predictions about the regulation of one gene by one factor, extended to complete regulons and interconnected to regulatory networks will serve in the future for the deeper understanding of the complex bacterial adaptation processes to anaerobic environments.

Acknowledgements

Research in the authors laboratory is funded by the Deutsche Forschungsgemeinschaft (DFG), Bundesministerium für Bildung und Forschung (BMBF) and the Mukoviszidose e.V. The authors would like to thank all coworkers who contributed to this work. We apologize to the authors and colleagues whose work was not cited because of limitations on the length of the article and number of references.

References

Achebach, S., Selmer, T., and Unden, G. (2005). Properties and significance of apoFNR as a second form of air-inactivated [4Fe–4S].FNR of *Escherichia coli*. FEBS J. *272*, 4260–4269.

Aiso, T., and Ohki, R. (2003). Instability of sensory histidine kinase mRNAs in *Escherichia coli*. Genes Cells *8*, 179–187.

Alvarez, A.F., Malpica, R., Contreras, M., Escamilla, E., and Georgellis, D. (2010). Cytochrome *d* but not cytochrome *o* rescues the toluidine blue growth sensitivity of *arc* mutants of *Escherichia coli*. J. Bacteriol. *192*, 391–399.

Ansaldi, M., Theraulaz, L., Baraquet, C., Panis, G., and Mejean, V. (2007). Aerobic TMAO respiration in *Escherichia coli*. Mol. Microbiol. *66*, 484–494.

Appleman, J.A., and Stewart, V. (2003). Mutational analysis of a conserved signal-transducing element: the HAMP linker of the *Escherichia coli* nitrate sensor NarX. J. Bacteriol. *185*, 89–97.

Arai, H., Igarashi, Y., and Kodama, T. (1994). Structure and ANR-dependent transcription of the *nir* genes for denitrification from *Pseudomonas aeruginosa*. Biosci. Biotechnol. Biochem. *58*, 1286–1291.

Arai, H., Igarashi, Y., and Kodama, T. (1995a). Expression of the *nir* and *nor* genes for denitrification of *Pseudomonas aeruginosa* requires a novel CRP/FNR-related transcriptional regulator, DNR, in addition to ANR. FEBS Lett. *371*, 73–76.

Arai, H., Igarashi, Y., and Kodama, T. (1995b). The structural genes for nitric oxide reductase from *Pseudomonas aeruginosa*. Biochim. Biophys. Acta *1261*, 279–284.

Arai, H., Kodama, T., and Igarashi, Y. (1997). Cascade regulation of the two CRP/FNR-related transcriptional regulators (ANR and DNR) and the denitrification enzymes in *Pseudomonas aeruginosa*. Mol. Microbiol. *25*, 1141–1148.

Arai, H., Mizutani, M., and Igarashi, Y. (2003). Transcriptional regulation of the *nos* genes for nitrous oxide reductase in *Pseudomonas aeruginosa*. Microbiology *149*, 29–36.

Arai, H., Hayashi, M., Kuroi, A., Ishii, M., and Igarashi, Y. (2005). Transcriptional regulation of the flavohemoglobin gene for aerobic nitric oxide detoxification by the second nitric oxide-responsive regulator of *Pseudomonas aeruginosa*. J. Bacteriol. *187*, 3960–3968.

Baikalov, I., Schroder, I., Kaczor-Grzeskowiak, M., Grzeskowiak, K., Gunsalus, R.P., and Dickerson, R.E. (1996). Structure of the *Escherichia coli* response regulator NarL. Biochemistry 35, 11053–11061.

Baikalov, I., Schroder, I., Kaczor-Grzeskowiak, M., Cascio, D., Gunsalus, R.P., and Dickerson, R.E. (1998). NarL dimerization? Suggestive evidence from a new crystal form. Biochemistry *37*, 3665–3676.

Baraquet, C., Theraulaz, L., Guiral, M., Lafitte, D., Mejean, V., and Jourlin-Castelli, C. (2006). TorT, a member of a new periplasmic binding protein family, triggers induction of the Tor respiratory system upon trimethylamine N-oxide electron-acceptor binding in *Escherichia coli*. J. Biol. Chem. *281*, 38189–38199.

Barraud, N., Hassett, D.J., Hwang, S.H., Rice, S.A., Kjelleberg, S., and Webb, J.S. (2006). Involvement of nitric oxide in biofilm dispersal of *Pseudomonas aeruginosa*. J. Bacteriol. *188*, 7344–7353.

Barraud, N., Storey, M.V., Moore, Z.P., Webb, J.S., Rice, S.A., and Kjelleberg, S. (2009). Nitric oxide-mediated dispersal in single- and multi-species biofilms of clinically and industrially relevant microorganisms. Microb. Biotechnol. *2*, 370–378.

Baruah, A., Lindsey, B., Zhu, Y., and Nakano, M.M. (2004). Mutational analysis of the signal-sensing domain of ResE histidine kinase from *Bacillus subtilis*. J. Bacteriol. *186*, 1694–1704.

Baur, H., Luethi, E., Stalon, V., Mercenier, A., and Haas, D. (1989). Sequence analysis and expression of the arginine-deiminase and carbamate-kinase genes of *Pseudomonas aeruginosa*. Eur. J. Biochem. *179*, 53–60.

Bearson, S.M., Albrecht, J.A., and Gunsalus, R.P. (2002). Oxygen and nitrate-dependent regulation of *dmsABC* operon expression in *Escherichia coli*: sites for Fnr and NarL protein interactions. BMC Microbiol. *2*, 13.

Beinert, H., and Kiley, P.J. (1999). Fe-S proteins in sensing and regulatory functions. Curr. Opin. Chem. Biol. *3*, 152–157.

Bekker, M., Alexeeva, S., Laan, W., Sawers, G., Teixeira de Mattos, J., and Hellingwerf, K. (2010). The ArcBA two-component system of *Escherichia coli* is regulated by the redox state of both the ubiquinone and the menaquinone pool. J. Bacteriol. *192*, 746–754.

Bell, A., and Busby, S. (1994). Location and orientation of an activating region in the *Escherichia coli* transcription factor, FNR. Mol. Microbiol. *11*, 383–390.

Benkert, B., Quack, N., Schreiber, K., Jaensch, L., Jahn, D., and Schobert, M. (2008). Nitrate-responsive NarX-NarL represses arginine-mediated induction of the *Pseudomonas aeruginosa* arginine fermentation *arcDABC* operon. Microbiology *154*, 3053–3060.

Blasco, F., Dos Santos, J.P., Magalon, A., Frixon, C., Guigliarelli, B., Santini, C.L., and Giordano, G. (1998). NarJ is a specific chaperone required for molybdenum cofactor assembly in nitrate reductase A of *Escherichia coli*. Mol. Microbiol. *28*, 435–447.

Boes, N., Schreiber, K., Hartig, E., Jaensch, L., and Schobert, M. (2006). The *Pseudomonas aeruginosa* universal stress protein PA4352 is essential for surviving anaerobic energy stress. J. Bacteriol. *188*, 6529–6538.

Borisov, V.B., and Verkhovsky, M.I. (2009). Oxygen as Acceptor. In EcoSal-*Escherichia coli* and *Salmonella*: Cellular and Molecular Biology, Stewart, V., ed. (ASM Press, Washington, DC).

Bose, J.L., Kim, U., Bartkowski, W., Gunsalus, R.P., Overley, A.M., Lyell, N.L., Visick, K.L., and Stabb, E.V. (2007). Bioluminescence in *Vibrio fischeri* is controlled by the redox-responsive regulator ArcA. Mol. Microbiol. *65*, 538–553.

Boysen, A., Moller-Jensen, J., Kallipolitis, B., Valentin-Hansen, P., and Overgaard, M. (2010). Translational regulation of gene expression by an anaerobically induced small non-coding RNA in *Escherichia coli*. J. Biol. Chem. *285*, 10690–10702.

Brekasis, D., and Paget, M.S. (2003). A novel sensor of NADH/NAD+ redox poise in *Streptomyces coelicolor* A3(2). EMBO J. *22*, 4856–4865.

Browning, D.F., Lee, D.J., Green, J., and Busby, S. (2002). Secrets of bacterial transcription initiation taught by the *Escherichia coli* FNR protein. In Signals, Switches, Regulons And Cascades: Control Of Bacterial Gene Expression, Hodgson, D.A., and Thomas, C.M., eds (Cambridge University Press, Cambridge), pp. 127–142.

Buettner, F.F., Bendallah, I.M., Bosse, J.T., Dreckmann, K., Nash, J.H., Langford, P.R., and Gerlach, G.F. (2008). Analysis of the *Actinobacillus pleuropneumoniae* ArcA regulon identifies fumarate reductase as a determinant of virulence. Infect. Immun. *76*, 2284–2295.

Calfee, M.W., Coleman, J.P., and Pesci, E.C. (2001). Interference with *Pseudomonas* quinolone signal synthesis inhibits virulence factor expression by *Pseudomonas aeruginosa*. Proc. Natl. Acad. Sci. U.S.A. *98*, 11633–11637.

Carlson, C.A., and Ingraham, J.L. (1983). Comparison of denitrification by *Pseudomonas stutzeri*, *Pseudomonas aeruginosa*, and *Paracoccus denitrificans*. Appl. Environ. Microbiol. *45*, 1247–1253.

Castiglione, N., Rinaldo, S., Giardina, G., and Cutruzzola, F. (2009). The transcription factor DNR from *Pseudomonas aeruginosa* specifically requires nitric oxide and haem for the activation of a target promoter in *Escherichia coli*. Microbiology *155*, 2838–2844.

Cavicchioli, R., Schroder, I., Constanti, M., and Gunsalus, R.P. (1995). The NarX and NarQ sensor-transmitter proteins of *Escherichia coli* each require two conserved histidines for nitrate-dependent signal transduction to NarL. J. Bacteriol. *177*, 2416–2424.

Cheng, V., and Weiner, J. (2007). *S*- and *N*-Oxide Reductases. In EcoSal-*Escherichia coli* and *Salmonella*: Cellular and Molecular Biology, Stewart, V., ed. (ASM Press, Washington, DC).

Cheung, J., and Hendrickson, W.A. (2009). Structural analysis of ligand stimulation of the histidine kinase NarX. Structure *17*, 190–201.

Cole, J., and Richardson, D.J. (2008). Respiration of Nitrate and Nitrite. In EcoSal-*Escherichia coli* and *Salmonella*: Cellular and Molecular Biology, Stewart, V., ed. (ASM Press, Washington DC).

Constantinidou, C., Hobman, J.L., Griffiths, L., Patel, M.D., Penn, C.W., Cole, J.A., and Overton, T.W. (2006). A reassessment of the FNR regulon and transcriptomic analysis of the effects of nitrate, nitrite, NarXL, and NarQP as *Escherichia coli* K12 adapts from aerobic to anaerobic growth. J. Biol. Chem. *281*, 4802–4815.

Cotter, P.A., and Gunsalus, R.P. (1989). Oxygen, nitrate, and molybdenum regulation of *dmsABC* gene expression in *Escherichia coli*. J. Bacteriol. *171*, 3817–3823.

Crack, J.C., Gaskell, A.A., Green, J., Cheesman, M.R., Le Brun, N.E., and Thomson, A.J. (2008). Influence of the environment on the $[4Fe–4S]^{2+}$ to $[2Fe–2S]^{2+}$ cluster switch in the transcriptional regulator FNR. J. Am. Chem. Soc. *130*, 1749–1758.

Crack, J.C., Green, J., Cheesman, M.R., Le Brun, N.E., and Thomson, A.J. (2007). Superoxide-mediated amplification of the oxygen-induced switch from [4Fe–4S] to [2Fe–2S] clusters in the transcriptional regulator FNR. Proc. Natl. Acad. Sci. U.S.A. *104*, 2092–2097.

Crack, J.C., Green, J., Le Brun, N.E., and Thomson, A.J. (2006). Detection of sulfide release from the oxygen-sensing [4Fe–4S] cluster of FNR. J. Biol. Chem. *281*, 18909–18913.

Cruz-Ramos, H., Boursier, L., Moszer, I., Kunst, F., Danchin, A., and Glaser, P. (1995). Anaerobic transcription activation in *Bacillus subtilis*: identification of distinct FNR-dependent and -independent regulatory mechanisms. EMBO J. *14*, 5984–5994.

Cruz-Ramos, H., Hoffmann, T., Marino, M., Nedjari, H., Presecan-Siedel, E., Dreesen, O., Glaser, P., and Jahn, D. (2000). Fermentative metabolism of *Bacillus subtilis*: physiology and regulation of gene expression. J. Bacteriol. *182*, 3072–3080.

Cuypers, H., Viebrock-Sambale, A., and Zumft, W.G. (1992). NosR, a membrane-bound regulatory component necessary for expression of nitrous oxide reductase in denitrifying *Pseudomonas stutzeri*. J. Bacteriol. *174*, 5332–5339.

Czaplewski, L.G., North, A.K., Smith, M.C., Baumberg, S., and Stockley, P.G. (1992). Purification and initial characterization of AhrC: the regulator of arginine metabolism genes in *Bacillus subtilis*. Mol. Microbiol. *6*, 267–275.

Darwin, A.J., and Stewart, V. (1996). The NAR Modulon Systems: Nitrate and Nitrite Regulation of Anaerobic Gene Expression. In Regulation of Gene Expression in *Escherichia coli*, Lin, E.C., and Lynch, A.S., eds (R.G. Landes Company, Austin, TX), pp. 343–359.

Davies, K.J., Lloyd, D., and Boddy, L. (1989). The effect of oxygen on denitrification in *Paracoccus denitrificans* and *Pseudomonas aeruginosa*. J. Gen. Microbiol. *135*, 2445–2451.

Dibden, D.P., and Green, J. (2005). *In vivo* cycling of the *Escherichia coli* transcription factor FNR between active and inactive states. Microbiology *151*, 4063–4070.

Dong, T., Yu, R., and Schellhorn, H. (2011). Antagonistic regulation of motility and transcriptome expression by RpoN and RpoS in *Escherichia coli*. Mol. Microbiol. *79*, 375–386.

Dufour, Y.S., Kiley, P.J., and Donohue, T.J. (2010). Reconstruction of the core and extended regulons of global transcription factors. PLoS Genet. *6*, e1001027.

Duport, C., Zigha, A., Rosenfeld, E., and Schmitt, P. (2006). Control of enterotoxin gene expression in *Bacillus cereus* f4430/73 involves the redox-sensitive resde signal transduction system. J. Bacteriol. *188*, 6640–6651.

Durand, S., and Storz, G. (2010). Reprogramming of anaerobic metabolism by the FnrS small RNA. Mol. Microbiol. *75*, 1215–1231.

Engel, P., Trageser, M., and Unden, G. (1991). Reversible interconversion of the functional state of the gene regulator FNR from *Escherichia coli in vivo* by O2 and iron availability. Arch. Microbiol. *156*, 463–470.

Eschbach, M., Schreiber, K., Trunk, K., Buer, J., Jahn, D., and Schobert, M. (2004). Long-term anaerobic survival of the opportunistic pathogen *Pseudomonas aeruginosa* via pyruvate fermentation. J. Bacteriol. *186*, 4596–4604.

Fleischhacker, A.S., and Kiley, P.J. (2011). Iron-containing transcription factors and their roles as sensors. Curr. Opin. Chem. Biol. *15*, 335–341.

Forzi, L., and Sawers, R.G. (2007). Maturation of [NiFe]-hydrogenases in *Escherichia coli*. Biometals *20*, 565–578.

Friedrich, T., and Pohl, T. (2007). NADH as Donor. In EcoSal-*Escherichia coli* and *Salmonella*: Cellular and Molecular Biology, Stewart, V., ed. (ASM Press , Washington DC).

Gallegos, M.T., Schleif, R., Bairoch, A., Hofmann, K., and Ramos, J.L. (1997). Arac/XylS family of transcriptional regulators. Microbiol. Mol. Biol. Rev. *61*, 393–410.

Gamper, M., Zimmermann, A., and Haas, D. (1991). Anaerobic regulation of transcription initiation in the *arcDABC* operon of *Pseudomonas aeruginosa*. J. Bacteriol. *173*, 4742–4750.

Geng, H., Nakano, S., and Nakano, M.M. (2004). Transcriptional activation by *Bacillus subtilis* ResD: tandem binding to target elements and phosphorylation-dependent and -independent transcriptional activation. J. Bacteriol. *186*, 2028–2037.

Geng, H., Zhu, Y., Mullen, K., Zuber, C.S., and Nakano, M.M. (2007). Characterization of ResDE-dependent fnr transcription in *Bacillus subtilis*. J. Bacteriol. *189*, 1745–1755.

Georgellis, D., Lynch, A.S., and Lin, E.C. (1997). *In vitro* phosphorylation study of the arc two-component signal transduction system of *Escherichia coli*. J. Bacteriol. *179*, 5429–5435.

Georgellis, D., Kwon, O., De Wulf, P., and Lin, E.C. (1998). Signal decay through a reverse phosphorelay in the Arc two-component signal transduction system. J. Biol. Chem. *273*, 32864–32869.

Georgellis, D., Kwon, O., and Lin, E.C. (1999). Amplification of signaling activity of the arc two-component system of *Escherichia coli* by anaerobic metabolites. An *in vitro* study with different protein modules. J. Biol. Chem. *274*, 35950–35954.

Georgellis, D., Kwon, O., and Lin, E.C. (2001). Quinones as the redox signal for the arc two-component system of bacteria. Science *292*, 2314–2316.

Giardina, G., Rinaldo, S., Johnson, K.A., Di Matteo, A., Brunori, M., and Cutruzzola, F. (2008). NO sensing in *Pseudomonas aeruginosa*: structure of the transcriptional regulator DNR. J. Mol. Biol. *378*, 1002–1015.

Girard, G., and Bloemberg, G.V. (2008). Central role of quorum sensing in regulating the production of pathogenicity factors in *Pseudomonas aeruginosa*. Future Microbiol. *3*, 97–106.

Goddard, A.D., Moir, J.W., Richardson, D.J., and Ferguson, S.J. (2008). Interdependence of two NarK domains in a fused nitrate/nitrite transporter. Mol. Microbiol. *70*, 667–681.

Grainger, D.C., Aiba, H., Hurd, D., Browning, D.F., and Busby, S.J. (2007). Transcription factor distribution in *Escherichia coli*: studies with FNR protein. Nucleic Acids Res. *35*, 269–278.

Green, J., and Marshall, F.A. (1999). Identification of a surface of FNR overlapping activating region 1 that is required for repression of gene expression. J. Biol. Chem. *274*, 10244–10248.

Green, J., Baldwin, M.L., and Richardson, J. (1998). Down-regulation of *Escherichia coli yfiD* expression by FNR occupying a site at −93.5 involves the AR1-containing face of FNR. Mol. Microbiol. *29*, 1113–1123.

Green, J., Crack, J.C., Thomson, A.J., and LeBrun, N.E. (2009). Bacterial sensors of oxygen. Curr. Opin. Microbiol. *12*, 145–151.

Gruner, I., Frädrich, C., Böttger, L.H., Trautwein, A.X., Jahn, D., and Härtig, E. (2010). Aspartate 141 is the fourth ligand of the oxygen sensing $[4Fe–4S]^{2+}$ cluster of *Bacillus subtilis* transcriptional regulator Fnr. J. Biol. Chem. *286*, 2017–2021.

Gu, M., and Imlay, J.A. (2011). The SoxRS response of *Escherichia coli* is directly activated by redox-cycling drugs rather than by superoxide. Mol. Microbiol. *79*, 1136–1150.

Guest, J.R., Green, J., Irvine, A.S., and Spiro, S. (1996). The FNR Modulon and FNR-Regulated Gene Expression. In Regulation of Gene Expression in *Escherichia coli*, Lin, E.C., and Lynch, A.S., eds (R.G. Landes Company, Austin, TX), pp. 317–342.

Gunsalus, R.P., and Park, S.J. (1994). Aerobic-anaerobic gene regulation in *Escherichia coli*: control by the ArcAB and Fnr regulons. Res. Microbiol. *145*, 437–450.

Hakoshima, T., and Ichihara, H. (2007). Structure of SixA, a histidine protein phosphatase of the ArcB histidine-containing phosphotransfer domain in *Escherichia coli*. Methods Enzymol. *422*, 288–304.

Hamada, K., Kato, M., Shimizu, T., Ihara, K., Mizuno, T., and Hakoshima, T. (2005). Crystal structure of the protein histidine phosphatase SixA in the multistep His-Asp phosphorelay. Genes Cells *10*, 1–11.

Harrison, F. (2007). Microbial ecology of the cystic fibrosis lung. Microbiology *153*, 917–923.

Härtig, E., Schiek, U., Vollack, K.U., and Zumft, W.G. (1999). Nitrate and nitrite control of respiratory nitrate reduction in denitrifying *Pseudomonas stutzeri* by a two-component regulatory system homologous to NarXL of *Escherichia coli*. J. Bacteriol. *181*, 3658–3665.

Härtig, E., Geng, H., Hartmann, A., Hubacek, A., Münch, R., Ye, R.W., Jahn, D., and Nakano, M.M.

(2004). *Bacillus subtilis* ResD induces expression of the potential regulatory genes *yclJK* upon oxygen limitation. J. Bacteriol. *186*, 6477–6484.

Hasegawa, N., Arai, H., and Igarashi, Y. (2001). Two c-type cytochromes, NirM and NirC, encoded in the *nir* gene cluster of *Pseudomonas aeruginosa* act as electron donors for nitrite reductase. Biochem. Biophys. Res. Commun. *288*, 1223–1230.

Hasegawa, N., Arai, H., and Igarashi, Y. (2003). Need for cytochrome bc1 complex for dissimilatory nitrite reduction of *Pseudomonas aeruginosa*. Biosci. Biotechnol. Biochem. *67*, 121–126.

Hayashi, N.R., Arai, H., Kodama, T., and Igarashi, Y. (1998). The *nirQ* gene, which is required for denitrification of *Pseudomonas aeruginosa*, can activate the RubisCO from *Pseudomonas hydrogenothermophila*. Biochim. Biophys. Acta *1381*, 347–350.

Hoffmann, T., Troup, B., Szabo, A., Hungerer, C., and Jahn, D. (1995). The anaerobic life of *Bacillus subtilis*: cloning of the genes encoding the respiratory nitrate reductase system. FEMS Microbiol. Lett. *131*, 219–225.

Hoffmann, T., Frankenberg, N., Marino, M., and Jahn, D. (1998). Ammonification in *Bacillus subtilis* utilizing dissimilatory nitrite reductase is dependent on resDE. J. Bacteriol. *180*, 186–189.

Homuth, G., Rompf, A., Schumann, W., and Jahn, D. (1999). Transcriptional control of *Bacillus subtilis hemN* and *hemZ*. J. Bacteriol. *181*, 5922–5929.

Hutchings, M.I., Mandhana, N., and Spiro, S. (2002). The NorR protein of *Escherichia coli* activates expression of the flavorubredoxin gene *norV* in response to reactive nitrogen species. J. Bacteriol. *184*, 4640–4643.

Huynh, T.N., Noriega, C.E., and Stewart, V. (2010). Conserved mechanism for sensor phosphatase control of two-component signaling revealed in the nitrate sensor NarX. Proc. Natl. Acad. Sci. U.S.A. *107*, 21140–21145.

Imlay, J.A. (2008). Cellular defenses against superoxide and hydrogen peroxide. Annu. Rev. Biochem. *77*, 755–776.

Inaba, K. (2009). Disulfide bond formation system in *Escherichia coli*. J Biochem *146*, 591–597.

Jeon, Y., Lee, Y.S., Han, J.S., Kim, J.B., and Hwang, D.S. (2001). Multimerization of phosphorylated and non-phosphorylated ArcA is necessary for the response regulator function of the Arc two-component signal transduction system. J. Biol. Chem. *276*, 40873–40879.

Jervis, A.J., and Green, J. (2007). *In vivo* demonstration of FNR dimers in response to lower O(2) availability. J. Bacteriol. *189*, 2930–2932.

Jourlin, C., Bengrine, A., Chippaux, M., and Mejean, V. (1996a). An unorthodox sensor protein (TorS) mediates the induction of the *tor* structural genes in response to trimethylamine N-oxide in *Escherichia coli*. Mol. Microbiol. *20*, 1297–1306.

Jourlin, C., Simon, G., Pommier, J., Chippaux, M., and Mejean, V. (1996b). The periplasmic TorT protein is required for trimethylamine N-oxide reductase gene induction in *Escherichia coli*. J. Bacteriol. *178*, 1219–1223.

Jourlin, C., Ansaldi, M., and Mejean, V. (1997). Transphosphorylation of the TorR response regulator requires the three phosphorylation sites of the TorS unorthodox sensor in *Escherichia coli*. J. Mol. Biol. *267*, 770–777.

Juhas, M., Eberl, L., and Tummler, B. (2005). Quorum sensing: the power of cooperation in the world of *Pseudomonas*. Environ. Microbiol. 7, 459–471.

Jüngst, A., and Zumft, W.G. (1992). Interdependence of respiratory NO reduction and nitrite reduction revealed by mutagenesis of *nirQ*, a novel gene in the denitrification gene cluster of *Pseudomonas stutzeri*. FEBS Lett. *314*, 308–314.

Kang, Y., Weber, K.D., Qiu, Y., Kiley, P.J., and Blattner, F.R. (2005). Genome-wide expression analysis indicates that FNR of *Escherichia coli* K-12 regulates a large number of genes of unknown function. J. Bacteriol. *187*, 1135–1160.

Kawasaki, S., Arai, H., Kodama, T., and Igarashi, Y. (1997). Gene cluster for dissimilatory nitrite reductase (*nir*) from *Pseudomonas aeruginosa*: sequencing and identification of a locus for heme d1 biosynthesis. J. Bacteriol. *179*, 235–242.

Khoroshilova, N., Beinert, H., and Kiley, P.J. (1995). Association of a polynuclear iron–sulfur center with a mutant FNR protein enhances DNA binding. Proc. Natl. Acad. Sci. U.S.A. *92*, 2499–2503.

Kiley, P.J., and Beinert, H. (1998). Oxygen sensing by the global regulator, FNR: the role of the iron–sulfur cluster. FEMS Microbiol. Rev. *22*, 341–352.

Körner, H., Sofia, H.J., and Zumft, W.G. (2003). Phylogeny of the bacterial superfamily of Crp-Fnr transcription regulators: exploiting the metabolic spectrum by controlling alternative gene programs. FEMS Microbiol. Rev. *27*, 559–592.

Koskenkorva, T., Aro-Karkkainen, N., Bachmann, D., Arai, H., Frey, A.D., and Kallio, P.T. (2008). Transcriptional activity of *Pseudomonas aeruginosa fhp* promoter is dependent on two regulators in addition to FhpR. Arch. Microbiol. *189*, 385–396.

Koskenkorva-Frank, T.S., and Kallio, P.T. (2009). Induction of *Pseudomonas aeruginosa fhp* and *fhpR* by reactive oxygen species. Can. J. Microbiol. *55*, 657–663.

Kow, R.L., Whicher, J.R., McDonald, C.A., Palfey, B.A., and Fagan, R.L. (2009). Disruption of the proton relay network in the class 2 dihydroorotate dehydrogenase from *Escherichia coli*. Biochemistry *48*, 9801–9809.

Krieger, R., Rompf, A., Schobert, M., and Jahn, D. (2002). The *Pseudomonas aeruginosa hemA* promoter is regulated by Anr, Dnr, NarL and Integration Host Factor. Mol. Genet. Genomics *267*, 409–417.

Kumar, R., and Shimizu, K. (2011). Transcriptional regulation of main metabolic pathways of *cyoA*, *cydB*, *fnr*, and *fur* gene knockout *Escherichia coli* in C-limited and N-limited aerobic continuous cultures. Microb. Cell Fact. *10*, 3.

Kumita, H., Matsuura, K., Hino, T., Takahashi, S., Hori, H., Fukumori, Y., Morishima, I., and Shiro, Y. (2004). NO reduction by nitric-oxide reductase from denitrifying bacterium *Pseudomonas aeruginosa*: characterization of reaction intermediates that appear in the single turnover cycle. J. Biol. Chem. *279*, 55247–55254.

Kwon, O., Georgellis, D., and Lin, E.C. (2000). Phosphorelay as the sole physiological route of signal transmission by the arc two-component system of *Escherichia coli*. J. Bacteriol. *182*, 3858–3862.

Lamberg, K.E., and Kiley, P.J. (2000). FNR-dependent activation of the class II *dmsA* and *narG* promoters of *Escherichia coli* requires FNR-activating regions 1 and 3. Mol. Microbiol. *38*, 817–827.

Larsen, M.H., Kallipolitis, B.H., Christiansen, J.K., Olsen, J.E., and Ingmer, H. (2006). The response regulator ResD modulates virulence gene expression in response to carbohydrates in *Listeria monocytogenes*. Mol. Microbiol. *61*, 1622–1635.

Larsson, J.T., Rogstam, A., and von Wachenfeldt, C. (2005). Coordinated patterns of cytochrome bd and lactate dehydrogenase expression in *Bacillus subtilis*. Microbiology *151*, 3323–3335.

Latifi, A., Foglino, M., Tanaka, K., Williams, P., and Lazdunski, A. (1996). A hierarchical quorum-sensing cascade in *Pseudomonas aeruginosa* links the transcriptional activators LasR and RhIR (VsmR) to expression of the stationary-phase sigma factor RpoS. Mol. Microbiol. *21*, 1137–1146.

Lazazzera, B.A., Beinert, H., Khoroshilova, N., Kennedy, M.C., and Kiley, P.J. (1996). DNA binding and dimerization of the Fe-S-containing FNR protein from *Escherichia coli* are regulated by oxygen. J. Biol. Chem. *271*, 2762–2768.

Lee, D.J., Wing, H.J., Savery, N.J., and Busby, S.J. (2000). Analysis of interactions between Activating Region 1 of *Escherichia coli* FNR protein and the C-terminal domain of the RNA polymerase alpha subunit: use of alanine scanning and suppression genetics. Mol. Microbiol. *37*, 1032–1040.

Lee, P.E., Demple, B., and Barton, J.K. (2009). DNA-mediated redox signaling for transcriptional activation of SoxR. Proc. Natl. Acad. Sci. U.S.A. *106*, 13164–13168.

Lim, D.B., Oppenheim, J.D., Eckhardt, T., and Maas, W.K. (1987). Nucleotide sequence of the *argR* gene of *Escherichia coli* K-12 and isolation of its product, the arginine repressor. Proc. Natl. Acad. Sci. U.S.A. *84*, 6697–6701.

Lin, H.Y., Bledsoe, P.J., and Stewart, V. (2007). Activation of *yeaR-yoaG* operon transcription by the nitrate-responsive regulator NarL is independent of oxygen- responsive regulator Fnr in *Escherichia coli* K-12. J. Bacteriol. *189*, 7539–7548.

Liu, X., and De Wulf, P. (2004). Probing the ArcA-P modulon of *Escherichia coli* by whole genome transcriptional analysis and sequence recognition profiling. J. Biol. Chem. *279*, 12588–12597.

Lu, C.D., Houghton, J.E., and Abdelal, A.T. (1992). Characterization of the arginine repressor from *Salmonella typhimurium* and its interactions with the *carAB* operator. J. Mol. Biol. *225*, 11–24.

Lu, C.D., Winteler, H., Abdelal, A., and Haas, D. (1999). The ArgR regulatory protein, a helper to the anaerobic regulator ANR during transcriptional activation of the *arcD* promoter in *Pseudomonas aeruginosa*. J. Bacteriol. *181*, 2459–2464.

Lu, C.D., Yang, Z., and Li, W. (2004). Transcriptome analysis of the ArgR regulon in *Pseudomonas aeruginosa*. J. Bacteriol. *186*, 3855–3861.

Luthi, E., Mercenier, A., and Haas, D. (1986). The *arcABC* operon required for fermentative growth of *Pseudomonas aeruginosa* on arginine: Tn5-751-assisted cloning and localization of structural genes. J. Gen. Microbiol. *132*, 2667–2675.

Luthi, E., Baur, H., Gamper, M., Brunner, F., Villeval, D., Mercenier, A., and Haas, D. (1990). The *arc* operon for anaerobic arginine catabolism in *Pseudomonas aeruginosa* contains an additional gene, *arcD*, encoding a membrane protein. Gene *87*, 37–43.

Malpica, R., Franco, B., Rodriguez, C., Kwon, O., and Georgellis, D. (2004). Identification of a quinone-sensitive redox switch in the ArcB sensor kinase. Proc. Natl. Acad. Sci. U.S.A. *101*, 13318–13323.

Malpica, R., Sandoval, G.R., Rodriguez, C., Franco, B., and Georgellis, D. (2006). Signaling by the *arc* two-component system provides a link between the redox state of the quinone pool and gene expression. Antioxid. Redox. Signal *8*, 781–795.

Mandin, P., and Gottesman, S. (2010). Integrating anaerobic/aerobic sensing and the general stress response through the ArcZ small RNA. EMBO J. *29*, 3094–3107.

Marino, M., Ramos, H.C., Hoffmann, T., Glaser, P., and Jahn, D. (2001). Modulation of anaerobic energy metabolism of *Bacillus subtilis* by *arfM* (*ywiD*). J. Bacteriol. *183*, 6815–6821.

Maris, A.E., Sawaya, M.R., Kaczor-Grzeskowiak, M., Jarvis, M.R., Bearson, S.M., Kopka, M.L., Schroder, I., Gunsalus, R.P., and Dickerson, R.E. (2002). Dimerization allows DNA target site recognition by the NarL response regulator. Nat. Struct. Biol. *9*, 771–778.

Maris, A.E., Kaczor-Grzeskowiak, M., Ma, Z., Kopka, M.L., Gunsalus, R.P., and Dickerson, R.E. (2005). Primary and secondary modes of DNA recognition by the NarL two-component response regulator. Biochemistry *44*, 14538–14552.

Matsubara, M., and Mizuno, T. (2000). The SixA phospho-histidine phosphatase modulates the ArcB phosphorelay signal transduction in *Escherichia coli*. FEBS Lett. *470*, 118–124.

Mettert, E.L., and Kiley, P.J. (2005). ClpXP-dependent proteolysis of FNR upon loss of its O2-sensing [4Fe–4S] cluster. J. Mol. Biol. *354*, 220–232.

Mettert, E.L., Outten, F.W., Wanta, B., and Kiley, P.J. (2008). The impact of O(2) on the Fe–S cluster biogenesis requirements of *Escherichia coli* FNR. J. Mol. Biol. *384*, 798–811.

Mika, F., and Hengge, R. (2005). A two-component phosphotransfer network involving ArcB, ArcA, and RssB coordinates synthesis and proteolysis of sigmaS (RpoS) in *E. coli*. Genes Dev. *19*, 2770–2781.

Möbius, K., Arias-Cartin, R., Breckau, D., Hannig, A.L., Riedmann, K., Biedendieck, R., Schroder, S., Becher, D., Magalon, A., Moser, J., *et al.* (2010). Heme biosynthesis is coupled to electron transport chains for energy generation. Proc. Natl. Acad. Sci. U.S.A. *107*, 10436–10441.

Moore, J.O., and Hendrickson, W.A. (2009). Structural analysis of sensor domains from the TMAO-responsive histidine kinase receptor TorS. Structure *17*, 1195–1204.

Moore, L.J., and Kiley, P.J. (2001). Characterization of the dimerization domain in the FNR transcription factor. J. Biol. Chem. *276*, 45744–45750.

Moore, L.J., Mettert, E.L., and Kiley, P.J. (2006). Regulation of FNR dimerization by subunit charge repulsion. J. Biol. Chem. *281*, 33268–33275.

Nakano, M., and Zuber, P. (2001). Anaerobiosis. In *Bacillus subtilis* and its Closest Relatives: from Genes to Cells, Sonenshein, A.L., Hoch, J.A., and Losick, R., eds (ASM Press, Washington DC), pp. 393–404.

Nakano, M.M., and Hulett, F.M. (1997). Adaptation of *Bacillus subtilis* to oxygen limitation. FEMS Microbiol. Lett. *157*, 1–7.

Nakano, M.M., Dailly, Y.P., Zuber, P., and Clark, D.P. (1997). Characterization of anaerobic fermentative growth of *Bacillus subtilis*: identification of fermentation end products and genes required for growth. J. Bacteriol. *179*, 6749–6755.

Nakano, M.M., Zhu, Y., Lacelle, M., Zhang, X., and Hulett, F.M. (2000). Interaction of ResD with regulatory regions of anaerobically induced genes in *Bacillus subtilis*. Mol. Microbiol. *37*, 1198–1207.

Nicholson, W.L. (2008). The *Bacillus subtilis* ydjL (bdhA) gene encodes acetoin reductase/2,3-butanediol dehydrogenase. Appl. Environ. Microbiol. *74*, 6832–6838.

Nishijyo, T., Park, S.M., Lu, C.D., Itoh, Y., and Abdelal, A.T. (1998). Molecular characterization and regulation of an operon encoding a system for transport of arginine and ornithine and the ArgR regulatory protein in *Pseudomonas aeruginosa*. J. Bacteriol. *180*, 5559–5566.

Noriega, C.E., Lin, H.Y., Chen, L.L., Williams, S.B., and Stewart, V. (2010). Asymmetric cross-regulation between the nitrate-responsive NarX-NarL and NarQ-NarP two-component regulatory systems from *Escherichia coli* K-12. Mol. Microbiol. *75*, 394–412.

Noriega, C.E., Schmidt, R., Gray, M.J., Chen, L.L., and Stewart, V. (2008). Autophosphorylation and dephosphorylation by soluble forms of the nitrate-responsive sensors NarX and NarQ from *Escherichia coli* K-12. J. Bacteriol. *190*, 3869–3876.

Oshima, T., Aiba, H., Masuda, Y., Kanaya, S., Sugiura, M., Wanner, B.L., Mori, H., and Mizuno, T. (2002). Transcriptome analysis of all two-component regulatory system mutants of *Escherichia coli* K-12. Mol. Microbiol. *46*, 281–291.

Pagels, M., Fuchs, S., Pané-Farré, J., Kohler, C., Menschner, L., Hecker, M., McNamarra, P.J., Bauer, M.C., Von Wachenfeldt, C., Liebeke, M., *et al.* (2010). Redox sensing by a Rex-family repressor is involved in the regulation of anaerobic gene expression in *Staphylococcus aureus*. Mol. Microbiol. *76*, 1142–1161.

Park, S.M., Lu, C.D., and Abdelal, A.T. (1997a). Cloning and characterization of *argR*, a gene that participates in regulation of arginine biosynthesis and catabolism in *Pseudomonas aeruginosa* PAO1. J. Bacteriol. *179*, 5300–5308.

Park, S.M., Lu, C.D., and Abdelal, A.T. (1997b). Purification and characterization of an arginine regulatory protein, ArgR, from *Pseudomonas aeruginosa* and its interactions with the control regions for the *car*, *argF*, and *aru* operons. J. Bacteriol. *179*, 5309–5317.

Pascal, M.C., Burini, J.F., and Chippaux, M. (1984). Regulation of the trimethylamine N-oxide (TMAO) reductase in *Escherichia coli*: analysis of *tor*::Mud1 operon fusion. Mol. Gen. Genet. *195*, 351–355.

Perrenoud, A., and Sauer, U. (2005). Impact of global transcriptional regulation by ArcA, ArcB, Cra, Crp, Cya, Fnr, and Mlc on glucose catabolism in *Escherichia coli*. J. Bacteriol. *187*, 3171–3179.

Pesci, E.C., Pearson, J.P., Seed, P.C., and Iglewski, B.H. (1997). Regulation of *las* and *rhl* quorum sensing in *Pseudomonas aeruginosa*. J. Bacteriol. *179*, 3127–3132.

Platt, M.D., Schurr, M.J., Sauer, K., Vazquez, G., Kukavica-Ibrulj, I., Potvin, E., Levesque, R.C., Fedynak, A., Brinkman, F.S., Schurr, J., *et al.* (2008). Proteomic, microarray, and signature-tagged mutagenesis analyses of anaerobic *Pseudomonas aeruginosa* at pH 6.5, likely representing chronic, late-stage cystic fibrosis airway conditions. J. Bacteriol. *190*, 2739–2758.

Pohlmann, A., Cramm, R., Schmelz, K., and Friedrich, B. (2000). A novel NO-responding regulator controls the reduction of nitric oxide in *Ralstonia eutropha*. Mol. Microbiol. *38*, 626–638.

Portnoy, V.A., Scott, D.A., Lewis, N.E., Tarasova, Y., Osterman, A.L., and Palsson, B.O. (2010). Deletion of genes encoding cytochrome oxidases and quinol monooxygenase blocks the aerobic-anaerobic shift in *Escherichia coli* K-12 MG1655. Appl. Environ. Microbiol. *76*, 6529–6540.

Pragman, A.A., Yarwood, J.M., Tripp, T.J., and Schlievert, P.M. (2004). Characterization of Virulence Factor Regulation by SrrAB, a Two-Component System in *Staphylococcus aureus*. J. Bacteriol. *186*, 2430–2438.

Ravcheev, D.A., Gerasimova, A.V., Mironov, A.A., and Gelfand, M.S. (2007). Comparative genomic analysis of regulation of anaerobic respiration in ten genomes from three families of gamma-proteobacteria (*Enterobacteriaceae*, *Pasteurellaceae*, *Vibrionaceae*). BMC Genomics *8*, 54.

Reents, H., Gruner, I., Harmening, U., Böttger, L.H., Layer, G., Heathcote, P., Trautwein, A.X., Jahn, D., and Härtig, E. (2006). *Bacillus subtilis* Fnr senses oxygen via a $[4Fe–4S]^{2+}$ cluster coordinated by three cysteine residues without change in the oligomeric state. Mol. Microbiol. *60*, 1432–1445.

Reinhart, F., Achebach, S., Koch, T., and Unden, G. (2008). Reduced apo-fumarate nitrate reductase regulator (apoFNR) as the major form of FNR in aerobically growing *Escherichia coli*. J. Bacteriol. *190*, 879–886.

Reitzer, L. (2003). Nitrogen assimilation and global regulation in *Escherichia coli*. Annu. Rev. Microbiol. *57*, 155–176.

Renna, M.C., Najimudin, N., Winik, L.R., and Zahler, S.A. (1993). Regulation of the *Bacillus subtilis alsS*, *alsD*, and *alsR* genes involved in post-exponential-phase production of acetoin. J. Bacteriol. *175*, 3863–3875.

Rinaldo, S., Giardina, G., Brunori, M., and Cutruzzola, F. (2006). N-oxide sensing and denitrification: the

DNR transcription factors. Biochem. Soc. Trans. *34*, 185–187.

Rodriguez, C., Kwon, O., and Georgellis, D. (2004). Effect of D-lactate on the physiological activity of the ArcB sensor kinase in *Escherichia coli*. J. Bacteriol. *186*, 2085–2090.

Rolfe, M.D., Beek, A.T., Graham, A.I., Trotter, E.W., Asif, H.M., Sanguinetti, G., de Mattos, J.T., Poole, R.K., and Green, J. (2011). Transcript profiling and inference of *Escherichia coli* k-12 arca activity across the range of physiologically relevant oxygen concentrations. J. Biol. Chem. *286*, 10147–10154.

Rompf, A., Hungerer, C., Hoffmann, T., Lindenmeyer, M., Romling, U., Gross, U., Doss, M.O., Arai, H., Igarashi, Y., and Jahn, D. (1998). Regulation of *Pseudomonas aeruginosa hemF* and *hemN* by the dual action of the redox response regulators Anr and Dnr. Mol. Microbiol. *29*, 985–997.

Salim, N.N., and Feig, A.L. (2010). An upstream Hfq binding site in the *fhlA* mRNA leader region facilitates the OxyS–fhlA interaction. PLoS One *5*, pii: e13028.

Salmon, K., Hung, S.P., Mekjian, K., Baldi, P., Hatfield, G.W., and Gunsalus, R.P. (2003). Global gene expression profiling in *Escherichia coli* K12. The effects of oxygen availability and FNR. J. Biol. Chem. *278*, 29837–29855.

Salmon, K.A., Hung, S.P., Steffen, N.R., Krupp, R., Baldi, P., Hatfield, G.W., and Gunsalus, R.P. (2005). Global gene expression profiling in *Escherichia coli* K12: effects of oxygen availability and ArcA. J. Biol. Chem. *280*, 15084–15096.

Sawers, R.G. (1991). Identification and molecular characterization of a transcriptional regulator from *Pseudomonas aeruginosa* PAO1 exhibiting structural and functional similarity to the FNR protein of *Escherichia coli*. Mol. Microbiol. *5*, 1469–1481.

Sawers, R.G., Zehelein, E., and Bock, A. (1988). Two-dimensional gel electrophoretic analysis of *Escherichia coli* proteins: influence of various anaerobic growth conditions and the fnr gene product on cellular protein composition. Arch. Microbiol. *149*, 240–244.

Schau, M., Chen, Y., and Hulett, F.M. (2004). *Bacillus subtilis* YdiH is a direct negative regulator of the *cydABCD* operon. J. Bacteriol. *186*, 4585–4595.

Schobert, M., and Jahn, D. (2010). Anaerobic physiology of *Pseudomonas aeruginosa* in the cystic fibrosis lung. Int. J. Med. Microbiol. *300*, 549–556.

Schobert, M., and Tielen, P. (2010). Contribution of oxygen-limiting conditions to persistent infection of *Pseudomonas aeruginosa*. Future Microbiol. *5*, 603–621.

Schreiber, K., Krieger, R., Benkert, B., Eschbach, M., Arai, H., Schobert, M., and Jahn, D. (2007). The anaerobic regulatory network required for *Pseudomonas aeruginosa* nitrate respiration. J. Bacteriol. *189*, 4310–4314.

Schröder, I., Wolin, C.D., Cavicchioli, R., and Gunsalus, R.P. (1994). Phosphorylation and dephosphorylation of the NarQ, NarX, and NarL proteins of the nitrate-dependent two-component regulatory system of *Escherichia coli*. J. Bacteriol. *176*, 4985–4992.

Schwartz, C.J., Djaman, O., Imlay, J.A., and Kiley, P.J. (2000). The cysteine desulfurase, IscS, has a major role in *in vivo* Fe–S cluster formation in *Escherichia coli*. Proc. Natl. Acad. Sci. U.S.A. *97*, 9009–9014.

Sharma, V., Noriega, C.E., and Rowe, J.J. (2006). Involvement of NarK1 and NarK2 proteins in transport of nitrate and nitrite in the denitrifying bacterium *Pseudomonas aeruginosa* PAO1. Appl. Environ. Microbiol. *72*, 695–701.

Silvestrini, M.C., Falcinelli, S., Ciabatti, I., Cutruzzola, F., and Brunori, M. (1994). *Pseudomonas aeruginosa* nitrite reductase (or cytochrome oxidase): an overview. Biochimie *76*, 641–654.

Simon, G., Mejean, V., Jourlin, C., Chippaux, M., and Pascal, M.C. (1994). The *torR* gene of *Escherichia coli* encodes a response regulator protein involved in the expression of the trimethylamine N-oxide reductase genes. J. Bacteriol. *176*, 5601–5606.

Simon, G., Jourlin, C., Ansaldi, M., Pascal, M.C., Chippaux, M., and Mejean, V. (1995a). Binding of the TorR regulator to cis-acting direct repeats activates tor operon expression. Mol. Microbiol. *17*, 971–980.

Simon, G., Mejean, V., Jourlin, C., Chippaux, M., and Pascal, M.C. (1995b). The *torR* gene of *Escherichia coli* encodes a response regulator protein involved in the expression of the trimethylamine N-oxide reductase genes. J. Bacteriol. *177*, 275.

Spiro, S. (1994). The FNR family of transcriptional regulators. Antonie Van Leeuwenhoek *66*, 23–36.

Squire, D.J., Xu, M., Cole, J.A., Busby, S.J., and Browning, D.F. (2009). Competition between NarL-dependent activation and Fis-dependent repression controls expression from the *Escherichia coli yeaR* and *ogt* promoters. Biochem. J. *420*, 249–257.

Srivastava, D., Zhu, W., Johnson, W.H., Jr., Whitman, C.P., Becker, D.F., and Tanner, J.J. (2010). The structure of the proline utilization a proline dehydrogenase domain inactivated by N-propargylglycine provides insight into conformational changes induced by substrate binding and flavin reduction. Biochemistry *49*, 560–569.

Stewart, V. (1994). Regulation of nitrate and nitrite reductase synthesis in enterobacteria. Antonie Van Leeuwenhoek *66*, 37–45.

Stewart, V. (2003). Biochemical Society Special Lecture. Nitrate- and nitrite-responsive sensors NarX and NarQ of proteobacteria. Biochem. Soc. Trans. *31*, 1–10.

Stewart, V., and Chen, L.L. (2010). The S helix mediates signal transmission as a HAMP domain coiled-coil extension in the NarX nitrate sensor from *Escherichia coli* K-12. J. Bacteriol. *192*, 734–745.

Stewart, V., and Rabin, R.S. (1995). Dual sensors and dual response regulators interact to control nitrate- and nitrite-responsive gene expression in *Escherichia coli*. In Two-Component Signal Transduction, Hoch, J.A., and Silhavy, T.J., eds (American Society for Microbiology, Washington DC), pp. 233–252.

Sun, G., Sharkova, E., Chesnut, R., Birkey, S., Duggan, M.F., Sorokin, A., Pujic, P., Ehrlich, S.D., and Hulett, F.M. (1996). Regulators of aerobic and anaerobic respiration in *Bacillus subtilis*. J. Bacteriol. *178*, 1374–1385.

Sutton, V.R., Stubna, A., Patschkowski, T., Munck, E., Beinert, H., and Kiley, P.J. (2004). Superoxide destroys the [2Fe–2S]2+ cluster of FNR from *Escherichia coli.* Biochemistry *43*, 791–798.

Taylor, B.L., and Zhulin, I.B. (1999). PAS domains: internal sensors of oxygen, redox potential, and light. Microbiol. Mol. Biol. Rev. *63*, 479–506.

Tolla, D.A., and Savageau, M.A. (2010). Regulation of aerobic-to-anaerobic transitions by the FNR cycle in *Escherichia coli.* J. Mol. Biol. *397*, 893–905.

Tolla, D.A., and Savageau, M.A. (2011). Phenotypic repertoire of the FNR regulatory network in *Escherichia coli.* Mol. Microbiol. *79*, 149–165.

Tomasiak, T., Cecchini, G., and Iverson, T.M. (2007). Succinate as donor; fumarate as acceptor. In EcoSal-*Escherichia coli* and *Salmonella*: Cellular and Molecular Biology, Stewart, V., ed. (ASM Press, Washington DC).

Toro-Roman, A., Mack, T.R., and Stock, A.M. (2005). Structural analysis and solution studies of the activated regulatory domain of the response regulator ArcA: a symmetric dimer mediated by the alpha4-beta5-alpha5 face. J. Mol. Biol. *349*, 11–26.

Toyofuku, M., Nomura, N., Fujii, T., Takaya, N., Maseda, H., Sawada, I., Nakajima, T., and Uchiyama, H. (2007). Quorum sensing regulates denitrification in *Pseudomonas aeruginosa* PAO1. J. Bacteriol. *189*, 4969–4972.

Toyofuku, M., Nomura, N., Kuno, E., Tashiro, Y., Nakajima, T., and Uchiyama, H. (2008). Influence of the *Pseudomonas* quinolone signal on denitrification in *Pseudomonas aeruginosa.* J. Bacteriol. *190*, 7947–7956.

Trunk, K., Benkert, B., Quack, N., Munch, R., Scheer, M., Garbe, J., Jansch, L., Trost, M., Wehland, J., Buer, J., *et al.* (2010). Anaerobic adaptation in *Pseudomonas aeruginosa*: definition of the Anr and Dnr regulons. Environ. Microbiol. *12*, 1719–1733.

Throup, J.P., Zappacosta, F., Lunsford, R.D., Annan, R.S., Carr, S.A., Lonsdale, J.T., Bryant, A.P., McDevitt, D., Rosenberg, M., and Burnham, M.K.R. (2001). The srhsr gene pair from *Staphylococcus aureus*:genomic and proteomic approaches to the identification and characterization of gene function. Biochemistry *40*, 10392–10401.

Tucker, N.P., D'Autreaux, B., Spiro, S., and Dixon, R. (2006). Mechanism of transcriptional regulation by the *Escherichia coli* nitric oxide sensor NorR. Biochem. Soc. Trans. *34*, 191–194.

Unden, G., Achebach, S., Holighaus, G., Tran, H.G., Wackwitz, B., and Zeuner, Y. (2002). Control of FNR function of *Escherichia coli* by O_2 and reducing conditions. J. Mol. Microbiol. Biotechnol. *4*, 263–268.

Unden, G., and Dünnwald, P. (2008). The aerobic and anaerobic respiratory chain of *Escherichia coli* and *Salmonella enterica*: enzymes and energetics. In EcoSal-*Escherichia coli* and *Salmonella*: Cellular and Molecular Biology, Steart, V., ed. (ASM Press, Washington DC).

Van Alst, N.E., Sherrill, L.A., Iglewski, B.H., and Haidaris, C.G. (2009). Compensatory periplasmic nitrate reductase activity supports anaerobic growth of *Pseudomonas aeruginosa* PAO1 in the absence of membrane nitrate reductase. Can. J. Microbiol. *55*, 1133–1144.

Vander Wauven, C., Pierard, A., Kley-Raymann, M., and Haas, D. (1984). *Pseudomonas aeruginosa* mutants affected in anaerobic growth on arginine: evidence for a four-gene cluster encoding the arginine deiminase pathway. J. Bacteriol. *160*, 928–934.

Venturi, V. (2006). Regulation of quorum sensing in *Pseudomonas.* FEMS Microbiol. Rev. *30*, 274–291.

Vignais, P.M., and Billoud, B. (2007). Occurrence, classification, and biological function of hydrogenases: an overview. Chem Rev *107*, 4206–4272.

Vijgenboom, E., Busch, J.E., and Canters, G.W. (1997). *In vivo* studies disprove an obligatory role of azurin in denitrification in *Pseudomonas aeruginosa* and show that azu expression is under control of *rpoS* and ANR. Microbiology *143*, 2853–2863.

Vik, S.B. (2007). ATP Synthesis by Oxidative Phosphorylation. In EcoSal-*Escherichia coli* and *Salmonella*: Cellular and Molecular Biology, Stewart, V., ed. (ASM Press, Washington DC).

Vollack, K.U., and Zumft, W.G. (2001). Nitric oxide signaling and transcriptional control of denitrification genes in *Pseudomonas stutzeri.* J. Bacteriol. *183*, 2516–2526.

Wang, E., Bauer, M.C., Rogstam, A., Linse, S., Logan, D.T., and Von Wachenfeldt, C. (2008). Structure and functional properties of the *Bacillus subtilis* transcriptional repressor Rex. Mol. Microbiol. *69*, 466–478.

Wang, E., Ikonen, T.P., Knaapila, M., Svergun, D., Logan, D.T., and von Wachenfeldt, C. (2011). Small-angle X-ray scattering study of a Rex family repressor: Conformational response to NADH and NAD^+ binding in solution. J. Mol. Biol. *408*, 670–683.

Williams, H.D., Zlosnik, J.E., and Ryall, B. (2006). Oxygen, cyanide and energy generation in the cystic fibrosis pathogen *Pseudomonas aeruginosa.* Adv. Microb. Physiol. *52*, 1–71.

Williams, S.B., and Stewart, V. (1997a). Discrimination between structurally related ligands nitrate and nitrite controls autokinase activity of the NarX transmembrane signal transducer of *Escherichia coli* K-12. Mol. Microbiol. *26*, 911–925.

Williams, S.B., and Stewart, V. (1997b). Nitrate- and nitrite-sensing protein NarX of *Escherichia coli* K-12: mutational analysis of the amino-terminal tail and first transmembrane segment. J. Bacteriol. *179*, 721–729.

Wing, H.J., Williams, S.M., and Busby, S.J. (1995). Spacing requirements for transcription activation by *Escherichia coli* FNR protein. J. Bacteriol. *177*, 6704–6710.

Wing, H.J., Green, J., Guest, J.R., and Busby, S.J. (2000). Role of activating region 1 of *Escherichia coli* FNR protein in transcription activation at class II promoters. J. Biol. Chem. *275*, 29061–29065.

Wong, S.M., Alugupalli, K.R., Ram, S., and Akerley, B.J. (2007). The ArcA regulon and oxidative stress resistance in Haemophilus influenzae. Mol. Microbiol. *64*, 1375–1390.

Worlitzsch, D., Tarran, R., Ulrich, M., Schwab, U., Cekici, A., Meyer, K.C., Birrer, P., Bellon, G., Berger,

J., Weiss, T., *et al.* (2002). Effects of reduced mucus oxygen concentration in airway *Pseudomonas* infections of cystic fibrosis patients. J. Clin. Invest. *109*, 317–325.

Xu, M., Busby, S.J., and Browning, D.F. (2009). Activation and repression at the *Escherichia coli ynfEFGHI* operon promoter. J. Bacteriol. *191*, 3172–3176.

Yarwood, J.M., McCormick, J.K., and Schlievert, P.M. (2001). Identification of a novel two-component regulatory system that acts in global regulation of virulence factors of *Staphylococcus aureus*. J. Bacteriol. *183*, 1113–1123.

Ye, R.W., Haas, D., Ka, J.O., Krishnapillai, V., Zimmermann, A., Baird, C., and Tiedje, J.M. (1995). Anaerobic activation of the entire denitrification pathway in *Pseudomonas aeruginosa* requires Anr, an analog of Fnr. J. Bacteriol. *177*, 3606–3609.

Ye, R.W., Tao, W., Bedzyk, L., Young, T., Chen, M., and Li, L. (2000). Global gene expression profiles of *Bacillus subtilis* grown under anaerobic conditions. J. Bacteriol. *182*, 4458–4465.

Yoon, S.S., Hennigan, R.F., Hilliard, G.M., Ochsner, U.A., Parvatiyar, K., Kamani, M.C., Allen, H.L., DeKievit, T.R., Gardner, P.R., Schwab, U., *et al.* (2002). *Pseudomonas aeruginosa* anaerobic respiration in biofilms: relationships to cystic fibrosis pathogenesis. Dev. Cell 3, 593–603.

Yoon, S.S., Karabulut, A.C., Lipscomb, J.D., Hennigan, R.F., Lymar, S.V., Groce, S.L., Herr, A.B., Howell, M.L., Kiley, P.J., Schurr, M.J., *et al.* (2007). Two-pronged survival strategy for the major cystic fibrosis pathogen, *Pseudomonas aeruginosa*, lacking the capacity to degrade nitric oxide during anaerobic respiration. EMBO J. *26*, 3662–3672.

Zumft, W.G. (1997). Cell biology and molecular basis of denitrification. Microbiol. Mol. Biol. Rev. *61*, 533–616.

Zumft, W.G. (2005). Nitric oxide reductases of prokaryotes with emphasis on the respiratory, heme-copper oxidase type. J. Inorg. Biochem. *99*, 194–215.

Zumft, W.G., and Körner, H. (1997). Enzyme diversity and mosaic gene organization in denitrification. Antonie Van Leeuwenhoek *71*, 43–58.

Zumft, W.G., and Kroneck, P.M. (2007). Respiratory transformation of nitrous oxide (N_2O) to dinitrogen by *Bacteria* and *Archaea*. Adv. Microb. Physiol. *52*, 107–227.

Take It or Leave It: Mechanisms Underlying Bacterial Bistable Regulatory Networks

12

Jeroen Siebring, Robin A. Sorg, Martijn Herber, Oscar P. Kuipers

Abstract

Bistable switches occur in regulatory networks that can exist in two distinct stable states. Such networks allow distinct switching of individual cells. In bacteria these switches coexist with regulatory networks that respond gradually to environmental input. Bistable switches play key roles in high investment processes such as cellular differentiation in which only the end-result of the process is functional. Bistable switches are involved in development of phenotypical traits, such as virulence, bacterial persistence, sporulation and production of valuable compounds like antibiotics. The phenomenon of bistable networks is here explained and discussed. Additionally, natural and evolutionary solutions for creating all-or-none regulatory decisions are described as well as the use of bistable networks in synthetic biology.

Glossary

Before commencing the discussion of bistable switches in bacterial signalling networks, a glossary of terms will be given in alphabetical order, to outline terms that are regularly used throughout this manuscript, their definition and how we interpret them.

- **Bet hedging strategies** are strategies in which risks are spread. For example, suboptimal phenotypes may be allowed in order to increase survival in case of unpredictably fluctuating environments with high penalty alterations. Phenotypical diversification can be a bet hedging strategy.
- **Bifurcation** is the process leading to a bimodal distribution.
- In a **bimodally** distributed population two distinct states of its identical members (e.g. cells, systems, molecules) can be identified at the same time in the same space.
- **Bistability** is a specific property of a system which means it can exist in two distinct stable states. Intermediate states are unstable.
- **Degradation** is here used to describe all processes that lower protein concentration. This also includes, for example dilution.
- **Diversification** is the process by which individuals of a population develop into different phenotypes.
- **Gene expression** includes all processes from gene transcription, to translation, to post-translational modification.
- The term **heterogeneous** is used to describe a genetically identical population containing different phenotypes.
- **Noise** is used to denote infidelity in processes that can lead to inter-individual differences in the outcome of that process. For example, equal levels of expression can lead to different absolute protein numbers between two genetically identical cells.
- **Stochasticity** is a term used for chance processes.

Introduction

Until recently and mostly because of technical limitations, isogenic cultures were treated as a homogeneous population in microbial science. This was in spite of early experimental observations of phenotypical bifurcation (Novick and Weiner, 1957). With the emergence of new techniques, in particular the widespread use of fluorescent proteins and single cell techniques, it became possible to study phenomena in individual cells and study inter-individual differences within an isogenic culture. Such research provided, for example new insights in cellular signalling patterns. It was found that phenotypical development is not in all cases a deterministic process, but can also be stochastic, i.e. the result of a chance process (Paulsson, 2004; Kaern *et al.*, 2005). Some questions that arise are how such probabilistic events influence the cellular development and why an organism would depend on chance for its fate.

In this chapter we will address phenotypical heterogeneity driven by underlying bistable regulatory networks. Bistable switches are elements in genetic circuits that have two distinct stable states of either 'Low' or 'High' expression states and can cause bifurcation within a clonal population (Kaern *et al.*, 2005; Smits *et al.*, 2006; Veening *et al.*, 2008; Tiwari *et al.*, 2011).

The phenomenon of phenotypical heterogeneity is of great interest to the biotechnological industry, where cell type diversification of producer strains hamper optimal production efficiency. Understanding the basics of this diversification may help to 'steer' a culture towards more efficient phenotype maintenance. Another field with high demand for knowledge about phenotypic heterogeneity is medicine. Bacterial persistence of pathogens, notably making pathogens escape antibiotic treatment, is linked to bistable signalling patterns. Also, pathogenicity development in bacteria, where only a part of the population becomes virulent, is a process showing complex regulation patterns linked to this phenomenon.

In the following, the functioning of bistable switches will be discussed, as well as examples to illustrate the role of bistable switches in (bacterial) processes. A key element in genetic bistable networks is gene expression noise and this will therefore also be addressed.

Bistable switches: the system level

The mechanism of bistability is best described by using an engineering analogy and terminology. Therefore, a more abstract explanation of the topic from an engineering point of view will be given. The applied biological impact shall be elucidated in more detail when translating this knowledge to bacterial populations, and furthermore with examples found in real biological systems.

Bistable switches: a gene expression control mechanism

In bacterial signalling networks, adapted biochemical response patterns to all kinds of environmental conditions can be observed. This response of signalling networks results in distinct gene expression patterns. The interplay of processes in a given genotype generates a particular phenotype. Gene expression networks are tuned to respond to environmental input factors in distinct manners. The simplest systems rely on single input parameters such as the substrate concentration for an enzyme activity. A set of input values (like substrate concentrations) for such a simplistic system can give rise to a continuous response such as saturation (also referred to as hyperbolic) or sigmoidal activity curves. With more complex regulatory networks, other response patterns may emerge such as discontinuous ones. This situation occurs for positive feedback loops that generate bistable switches. In this case, signalling networks do not respond in a graded manner, but reside either in a 'Low' or in a 'High' expression state.

It is safe to assume that for every single gene a certain level of gene expression regulation has evolved. In the succession of regulation possibilities, the first level of control is at the process of gene transcription. Regulation of transcription is found at any imaginable step in this process, including RNA polymerase binding, open loop formation, promoter clearance and elongation. The first three actions mentioned take place at the promoter; this seems to be the hot spot of gene expression control (Alberts, 2007; Van Hijum *et al.*, 2009).

Representing the first step in gene expression regulation, transcription control is also a good choice from an energy budget point of view, preventing energy expenditure on production of unneeded protein. A disadvantage of regulation at the transcriptional level is slow response behaviour. Both build-up and especially degradation of the signal (for example a transcriptional activator) are time consuming. Control of translation, and more importantly protein modification, offers faster regulation possibilities. Processes like phosphorylation are very fast and thus preferred for regulation patterns where immediate responses are required.

Bistable switches are found at all regulation levels, from transcription to protein modification and may also combine different levels. However, they are most abundant at the transcriptional level, although switches at the protein modification level are also known, such as transitions within the cell cycle (Santos and Ferrell, 2008).

In the following, the main focus shall be on examples of switches regulated by transcription, bearing in mind that the general principles also account for other regulation levels.

Bistable switches: basic examples

The different requirements for gene expression responses are intuitive to understand, for example by comparing scenarios for the tetracycline resistance gene TetA and for bacterial flagellum development.

The first example deals with responses to the antibiotic tetracycline. Expression of the resistance gene *tetA* is optimized to follow a graded response. TetA, encoded on the *Escherichia coli* Tn10 transposon, confers tetracycline resistance by an active efflux of the antibiotic from the cytoplasm. This transposon additionally encodes transcriptional regulators responsible for controlling an efficient expression level of the resistance gene. When increasing the concentration of tetracycline, a corresponding increase in TetA concentration is required to facilitate more antibiotic efflux (Yamaguchi *et al.*, 1991). Permanent high levels of TetA are not desirable because of interference with the cellular metabolism and for energy budget reasons.

This straightforward example of a graded (and thus continuous) response is in contrast to a reasonable reaction pattern for the construction of a flagellum, for example needed for motility development of *Bacillus subtilis*. Growing this organism to mid-exponential phase, the formation of a subpopulation of swimming cells can be observed (Kearns and Losick, 2005). The regulation pattern for this phenotype requires a more complex architecture as it represents a binary – an all-or-nothing – decision. Construction of an incomplete flagellum would be a waste of energy as only a completed flagellum is functional. In this example, input factors are already much more difficult to determine precisely and probably depend on the interplay between a set of factors such as nutrient availability and quorum sensing. The likelihood of switching to a motile phenotype increases for each individual cell when the environment develops towards having the right conditions for triggering motility development. On the molecular level, this is reflected by the emergence of the transcription factor σ^D, the key initiator for flagellum development (Mirel *et al.*, 1992). On the population level, motile and non-motile subpopulations of *B. subtilis* can be observed within the same environment (Vlamakis *et al.*, 2008). The ratio of these two cell types depends on the progression of the triggering conditions. Consequently, it also reflects the increase in chance for individual cells to switch their phenotype to the motile phenotype. The response to such a difficult to estimate environmental input is thus discontinuous; cells either do not express flagellum related genes or they 'jump' to a state where they do.

Bistable switches: the basic model

To understand the operating mode of a bistable switch, it is best to illustrate the individual key factor activities with the help of a diagram. The most reduced functional example of such a network is a positive feedback loop of a transcriptional activator cooperatively regulating its own expression (Fig. 12.1).

When plotting promoter activity against activator concentration, a sigmoidal curve emerges (Fig. 12.2A). Low activator concentrations lead to low transcription rates. Accordingly, high activator concentrations lead to high transcription rates. Even without any transcriptional activator present there is minor promoter activity because of promoter leakiness. On the other side, promoter activity has a theoretical maximum. This means that there exists a maximal transcription rate where increasing activator concentration has no additional enhancing effect. At this point,

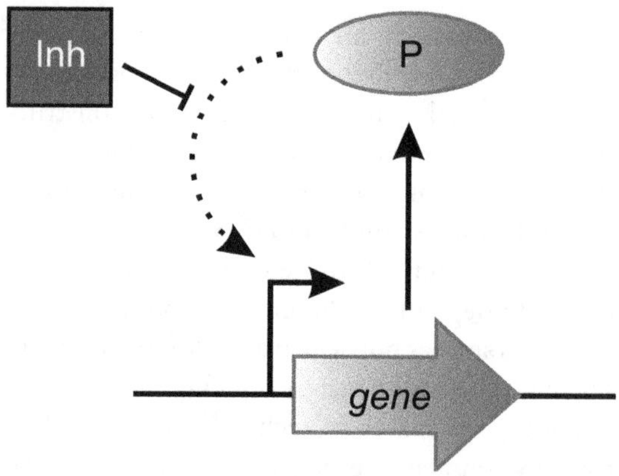

Figure 12.1 A positive feedback system potentially leading to a bistable switch; the dashed line represents a net positive induction of the gene, in spite of the fact that intermediary components may have a repressing effect; the inhibitor (Inh; square) allows for external control of the network. Often, oligomerization of the gene product (P) is key in autoregulation in natural networks.

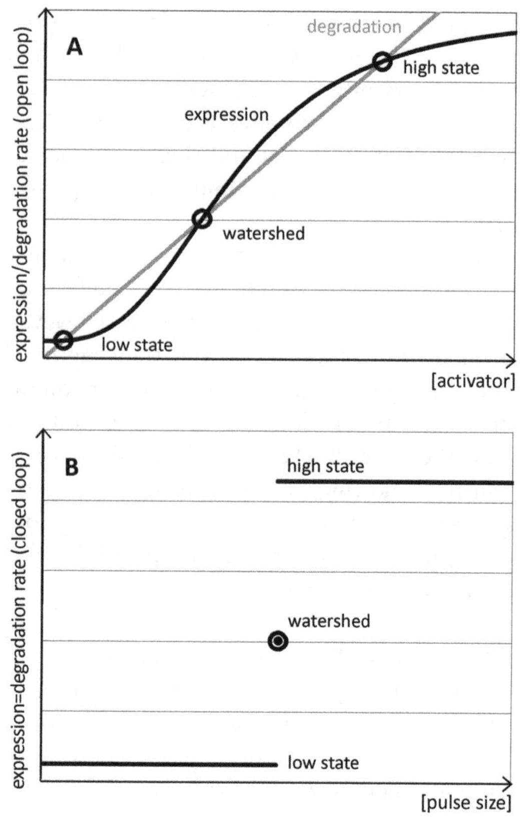

Figure 12.2 Open (A) and closed (B) loop representation of a regulatory network creating a bistable switch; (A) the black curve represents expression, the grey line degradation; intersections form the three steady states; (B) different pulse sizes of additional activator expression allow for switching of the system to the 'High' state.

other limitations become dominant, for example limitation of RNA polymerase availability. The sigmoidal shape, instead of a saturation curve, is caused by an assumed cooperative behaviour of the transcription factor. This is for example the case when there are two binding sites at the promoter and the functional activator units are capable of dimerizing when binding to these operators. The sigmoidal shape is the outcome of an enhanced binding affinity for the second activator unit upon occupation of the first binding site. The expression curve illustrated (Fig. 12.2A) is known as open loop; positive feedback is not yet taken into consideration here.

Antagonistic processes, limiting the gain of activator molecules, consist of active decay by proteolysis and passive decay due to cell growth mediated dilution. The combination of these two processes is in the following referred to as degradation, whereas in the case of stable proteins the dominant effect is dilution. This results in a linear decrease in activator concentration, with a slope starting from the origin, when assuming a constant growth speed. The loss of activator is proportional to the current activator concentration; the gradient represents the dilution rate (Fig. 12.2A).

The described situation changes when the gene product activates expression of its own gene. In this case the so-called closed loop of the positive feedback system emerges. The degradation line intersects the sigmoidal expression curve (Fig. 12.2A), illustrating steady states where the system resides at equilibrium (protein turnover equals protein production). Whenever the gain is higher than the loss (production overcomes degradation), the net effect will be a build-up of activator. In contrast, more loss than gain will consequently result in loss of activator. In the given example, the system is tuned to have three intersections and thus three equilibriums, of which only the higher and the lower ones are stable. When increasing or decreasing activator concentration slightly to simulate naturally occurring fluctuations, the system always falls back to these two stable points. In contrast, the central intersection is unstable; increasing activator concentrations pulls the system towards the higher stable point while decreasing activator concentrations pulls the system towards the lower one. Therefore, this point represents an unstable steady state that is called the 'watershed' or also referred to as the threshold of the system. The position of the watershed in relation to the two stable points is also of importance when taking fluctuations into consideration. In case of close proximity to one of the stable points, or more precisely within the fluctuation amplitude's reach, the system will eventually switch to the other stable state (Alon, 2006).

Bistable switches: environmental input

In an idealized model without fluctuations, a bistable switch changes its output value discontinuously from the 'Low' to the 'High' state, or vice versa, for a critical environmental input. For a system in which an activator regulates its own expression, the input can be mediated using a second copy of the same activator gene under control of an inducible promoter. Starting from a system in the 'Low' state, offering induction pulses of alternating size would temporarily create a correlating accumulation of activator. In case of weak pulses, the amount of activator expressed stays below the threshold (watershed) and the system rebalances back to the 'Low' state. In contrast, relatively strong pulses push the amount of activator above the threshold, after which the system adjusts to the 'High' state. In this scenario, the switching event and residing in the 'High' state would be irreversible in case of no additional control mechanism to antagonize the activator. The plot for the input factor of increasing pulse strengths would thus result in a discontinuous response curve for the post-pulse re-equilibrated systems. In our model, this looks like a flat line corresponding to the 'Low' state jumping at the threshold matching pulse strength to a second flat line corresponding to the 'High' state (Fig. 12.2B).

Another and more relevant example for an environmental input is the case where the input itself is a crucial part of tuning a functional bistable system. The scenario of an inhibitor hampering the functionality of a transcriptional activator matches this requirement. This example furthermore allows for regulation in both directions, i.e. from 'Low' to 'High' and also from a 'High' to a 'Low' state. In an open loop representation without any inhibition, the gain curve and degradation line cross only once representing the 'High' state (Fig. 12.3A). Increasing inhibitor concentration either stretches (for low inhibitor affinity) or shifts (for high inhibitor affinity) the gain curve lower, as less of the abundant activator is actually functional. At high inhibitor concentrations, there is consequently also just one single crossing, this time representing the 'Low' state (Fig. 12.3A). Only at intermediate inhibitor amounts do two stable points emerge (Fig. 12.3A), illustrating the

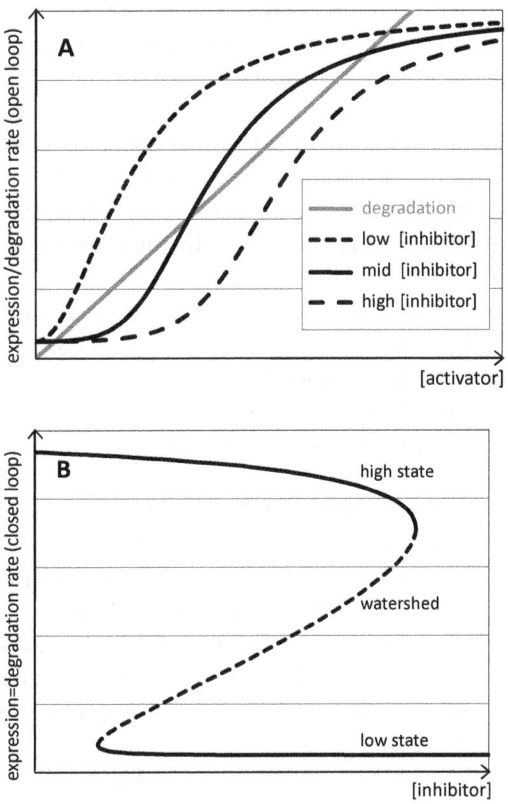

Figure 12.3 Open (A) and closed (B) loop representation of a network showing conditional bistable behaviour at intermediate inhibitor concentration. (A) Expression curves (black) shift depending on the level of inhibition. (B) The closed loop graph reveals the parameter framework of the bistable system.

dependency of this bistable network on a specific parameter framework. Presuming fluctuation-free conditions, switching from the 'High' state to 'Low' when gradually increasing inhibitor concentrations does not occur until the gain curve is stretched and the upper part outruns the degradation line and only one single intersection remains: the 'Low' state (Fig. 12.3B). Decreasing the inhibitor concentration again, however, triggers no instant switching back for the same concentration. The reason lies in the different amount of activator molecules present in the 'Low' state. The effect of activator abundance combined with their percentage of inhibition determines the expression activity of the open loop system. Therefore, one has to again decrease the level of inhibitor until one single intersection is left, this time representing the 'High' state. The two critical inhibitor concentrations for triggering the switching consequently determine the input framework for this bistable system (Alon, 2006).

Both examples of environmental regulation also illustrate the history dependency of these systems. Cells that experienced the critical environmental input for changing their expression pattern switch and remain in the new state; cells without this input still reside in the original state, even though they may be in the precise same environment at the moment of observation. This phenomenon is called hysteresis: without knowing the history of the process, one cannot predict the current state (Ninfa *et al.*, 2004).

Bistable switches: requirements

In terms of network architecture, the most important characteristic of a bistable switch is the presence of a positive feedback loop. To be more accurate, the product of all parts of the feedback needs to be positive and can therefore also be, for example, double negative. In the biological world usually more complex designs are observed. The elements of these loops can be very specific factors serving uniquely for the purpose of this regulatory pattern, but the network may also include participants of general metabolism.

A second requirement relies in an ultrasensitive response, characterized by a sigmoidal activity curve of the open loop system. These response types are based on cooperative oligomerization of two or four regulators, meaning a stronger binding of the second regulator in comparison to the first one. However, it has been shown that the prerequisite of ultrasensitive responses may be circumvented when non-linear degradation of the activator occurs (Tan *et al.*, 2009). Therefore, the second key requirement is better defined as an interaction pattern that creates three steady states, whereas two are stable and the middle one marks the watershed.

It should be stressed that bistable behaviour of a system is conditional; it only emerges over an environmental input gradient for a given expression pattern. Some genes consequently may sometimes be part of a bistable system and sometimes not, depending on the fulfilment of environmental conditions. This requirement distinguishes bistable regulatory networks from the phenomenon of phase variation (high frequency, random phenotype switching). Phase variation employs genetic mechanisms such as strand slippage and small-scale inversions to generate different phenotypes within a population. However, the line separating the two concepts of bistability and phase variation becomes blurred when considering a mechanism such as DNA methylation, which has been shown to be associated with phase variation, yet is considered to be an epigenetic mechanism (Van der Woude, 2011). This illustrates that there might be an intersection between the two phenomena, describing once observations from a systems and once from a population point of view.

The term bistability is often used in the context of bistable systems. According to the physical definition, bistability simply refers to the possibility of having two stable states within a network. This, however, accounts for all theoretically possible bistable systems, independently of actual population characteristics displayed. In a microbiological context, the term bistability is rather used when the coexistence of two distinct phenotypes is observed. However, interpreting the underlying reason for so-called bimodal distributions needs to be treated with caution; it has been shown that also other phenomena can lead to bimodal distribution patterns without the presence of a bistable regulation system (To and Maheshri, 2010). Such a system, for example phosphorylation of regulators involved in sporulation, is discussed later in this chapter.

When a cell population bifurcates within the same environment, stochastic fluctuations of the relevant regulators are held responsible. Whenever the mean environmental conditions reside in proximity to the threshold, these fluctuations push individual cells into switching to the other stable state. If the watershed is also within the range of fluctuations of the other stable state, switching in both directions is possible. In that case, the ratio of 'Low' to 'High' state cells reflects the ratio of the two switching kinetics. Regulatory systems making up bistable switches that lead to population bifurcations might also be described as 'non-sharp' switches; fluctuations create microenvironments of different switching conditions, thus only a subpopulation switches. These microenvironments may be completely randomly distributed or they may emerge as position-dependent gradients as, for example, in the case of biofilms. The second case also points out the conceptual relatedness between microbial bifurcations and differentiation in higher organisms. Multicellular eukaryotes create very tightly controlled environmental conditions as information signals for correct cell type differentiation. For unicellular organisms, environmental information leading to cellular specification is less deterministic. Here, fluctuations represent the critical element for triggering bistable switches, allowing for the coexistence of two different cell types fulfilling different functions within the same environment.

System level: noise

For bifurcation of (bacterial) populations, stochastic fluctuations of the relevant regulators are a driving force. In gene regulatory networks such fluctuations are the result of gene expression noise. So, even though it is not a topic exclusively linked to bistability, the next section is dedicated to noise in gene expression.

Role of stochasticity in gene transcription

An engineering approach to bistable switches allows for clearly defined and clean mathematical simulation. In nature, gene expression is not as deterministic, but a 'noisy' process in itself. Many biochemical processes in a cell involve low molecule numbers and infrequent interactions and are therefore sensitive to stochastic fluctuations. Gene expression is an example of such a biochemical process (Paulsson, 2004; Kaern *et al.*, 2005; Eldar and Elowitz, 2010). As the nature of each step of gene expression shows a probabilistic behaviour, the number of mRNA and protein molecules of every gene fluctuates randomly (Kaern *et al.*, 2005; Tripathi and Chowdhury, 2008).

Gene expression noise is the term used for inter-individual differences in mRNA and/or protein level at a certain time point. Three key concepts add to expression noise: transcription noise, translation noise and propagation.

Transcription noise arises from concentration differences of local transcription factors, sigma factor, RNA polymerase or nucleotides. Additionally, both quality and quantity of mature mRNA can be affected by, for example, mismatching, RNA polymerase slippage and consequential frame shifts (Penno *et al.*, 2006; Tamas *et al.*, 2008; Meyerovich *et al.*, 2010). The mRNA half-life time is thereupon subject to stochasticity as a consequence of degradation. This process is influenced by the formation of secondary structures and both specific and unspecific degradation pathways.

A source of translational noise relies in stochasticity of interactions between ribosome, template mRNA and aminoacyl-tRNA (Alberts, 2007). The sum of all errors leads to inter-individual differences in protein concentration when comparing clonal cells.

Terms often used in the context of gene expression noise are intrinsic and extrinsic noise. Intrinsic noise refers to the differences in expression rate over time between two identical copies of a promoter in the same cell (Fig. 12.4A,B), while extrinsic noise is shown by the similarities in expression rate changes over time with two identical promoters within one cell (Fig. 12.4C, D). The latter effect is caused by fluctuations of components of the transcription and translation machinery, equally influencing expression of both genes (Elowitz, 2002; Swain *et al.*, 2002).

The sources of noise during transcription of a gene have implications for transcription efficiency

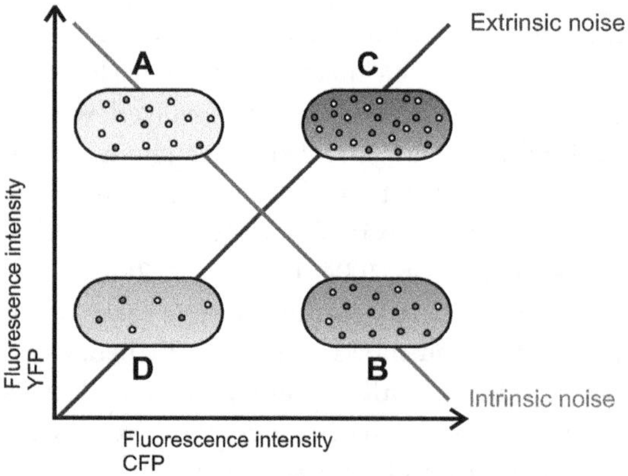

Figure 12.4 Intrinsic and extrinsic noise can be quantified and distinguished using two genes controlled by identical regulatory sequences. In this cartoon a situation is sketched where a cell harbours a yellow fluorescent protein (YFP) and a cyan fluorescent protein (CFP) under control of identical promoters. Cells expressing more of either fluorescent protein appear yellow or blue (A and B), whereas cells with the same amount of each protein appear green (C and D). In situation A and B differences in amount of each product occurs, even though these genes are under the control of an identical promoter. Here the differences are caused by intrinsic noise. In situation C and D a difference in total amount of protein produced arose, caused by extrinsic noise (Elowitz, 2002; Swain *et al.*, 2002). The cartoons are superimposed over a hypothetical flow cytometry graph, where relative fluorescence intensity of each fluorophore is plotted. A colour version of this figure is available in the plate section at the back of the book.

and therefore offer a level of transcription control. Depending on the properties of transcription factors and promoters, genes can be in an active state for varying periods. In the inactive state in contrary, when no transcription factors are bound, varying thresholds of activation may be observed. The properties of all parts involved in transcription allow for a concept called transcriptional bursting (Eldar and Elowitz, 2010).

Transcriptional bursting

Usage of MS2-coat proteins and fluorescence *in situ* hybridization (FISH) allows studying of gene expression. Several studies have revealed the occurrence of distinct 'bursts' of transcription: infrequent periods of transcriptional activity that produce many transcripts within a short time. Such transcription bursts were shown to lead to large variability in mRNA numbers (Golding *et al.*, 2005; Kaern *et al.*, 2005; Raj *et al.*, 2006; Zenklusen *et al.*, 2008; Youk *et al.*, 2010).

Transcriptional bursting can be explained on a mechanistic level. Binding of a transcription factor improves RNA polymerase holoenzyme binding to a promoter region by either altering the conformation of RNA polymerase or the target DNA (Coulombe and Burton, 1999). Binding and release of trans-acting factors such as activator or repressor molecules can give rise to 'On' and 'Off' switching of a bacterial gene. Once the transcription factor is bound, it is likely that the promoter region is accessible for a longer time period allowing more than one (preassembled) RNA polymerase complex to bind. In this way, more than one RNA polymer can be transcribed from an activated gene, creating a burst of mRNA molecules before the gene switches back to an inactive state. Different trans-acting factors and promoter elements including the TATA-box have been shown to be important for stabilizing promoter complexes and allow efficient transcription, for example, by rapid re-initiation on an assembled promoter complex (Kaern *et al.*, 2005). The stability of promoter complexes and their assembly efficiency will therefore influence the amplitude and frequency of a transcription response. High and frequent transcriptional burst can increase the switching probability in case of bistable switches as well as lowering thresholds.

It should be noted that transcriptional bursts in bacteria are much weaker than in mammalian cells and have only been shown in bacteria using inducible genes. In a study visualizing single mRNA molecules in yeast using FISH, transcriptional bursting was not observed. The authors explain that this is probably due to the relatively short protein lifetime (~2× mRNA lifetime) in which case maintaining constant expression is better achieved by non-bursting, low-variation expression that constantly produces new proteins. Constant protein production is then achieved by efficient translation, as most mRNAs (~70%) in a yeast cell are also polysome-associated (Zenklusen *et al.*, 2008; Youk *et al.*, 2010). Constitutive expression without bursts was also observed in bacteria (Yu, 2006). It is tempting to speculate that transcriptional bursting is a gene property rather than a general gene expression feature.

Noise propagation

The sum of transcriptional and translational noise influences the levels of resulting proteins. However, differences between mRNA and protein lifetime can cause protein fluctuations not displaying the same amplitude or frequency of fluctuations, as do transcriptional bursts. Protein lifetime can 'buffer' transcriptional bursts (Fig. 12.5). Accumulation of proteins over time tends to average out the variability generated by bursty expression (Eldar and Elowitz, 2010). When protein lifetime on average is short, such as in the yeast system mentioned above, this buffering capacity is not present and less bursty expression is required (Zenklusen *et al.*, 2008).

Additionally, during cell division, random partitioning of proteins between daughter cells will occur, which can effectively increase or decrease cellular concentrations of protein in the progeny (Fig. 12.6) (Rosenfeld *et al.*, 2005; Tyagi, 2010).

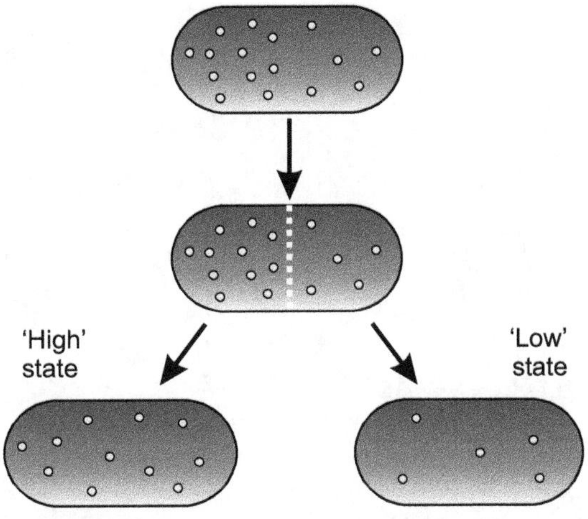

Figure 12.6 Upon cell division molecules (intracellular spheres) can be unequally partitioned, causing noisy inheritance. When differences occur in RNA or transcription factors, this noise can affect the phenotype of the progeny, stimulating phenotypical heterogeneity (Taniguchi *et al.*, 2010).

Population level: inter-individual stochasticity

Visualizing bimodality: different responses in an isogenic population

Bistable switches can cause diversification between individuals of a clonal population. This phenotypic diverging development was described on the system level by a part of the cell population switching to a 'High' state whereas the rest remain in a 'Low' state. Such bifurcations can be visualized using fluorescent protein (FP) expression under control of one of the components of the bistable switch of interest. Using simple wide field epi-fluoresence microscopy, the distribution of the different states the cell may enter can be observed. This bimodality is also clearly measured using flow cytometry. With flow cytometry cells are rapidly flushed through a light source capillary or flow cell one by one. Properties of the individual cells can be recorded, such as their fluorescent properties. These data can be visualized as histograms of cell count versus fluorescence intensity. In case of a bimodally distributed population a clear distribution of low-fluorescence cells and high-fluorescence cells can be distinguished (Fig. 12.7).

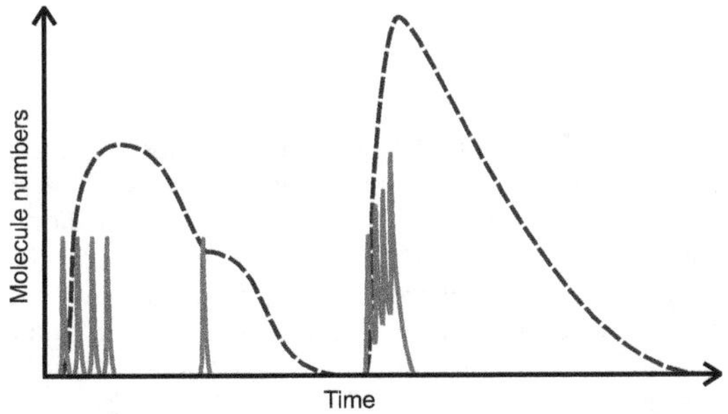

Figure 12.5 A cartoon representation, showing how mRNA bursting (solid line) and short mRNA lifetime shapes protein levels (dashed line) in time. The long protein lifetime can function as a buffer time-averaging out noisy mRNA expression (Eldar and Elowitz, 2010).

Bet hedging and kin selection

Noise in a bistable regulatory network can drive all-or-none switching between phenotypic states. Why would an organism rely on noise and stochasticity for its cell fate decision? And why would evolution allow the emergence of multiple phenotypes instead of only the best phenotype for that environment?

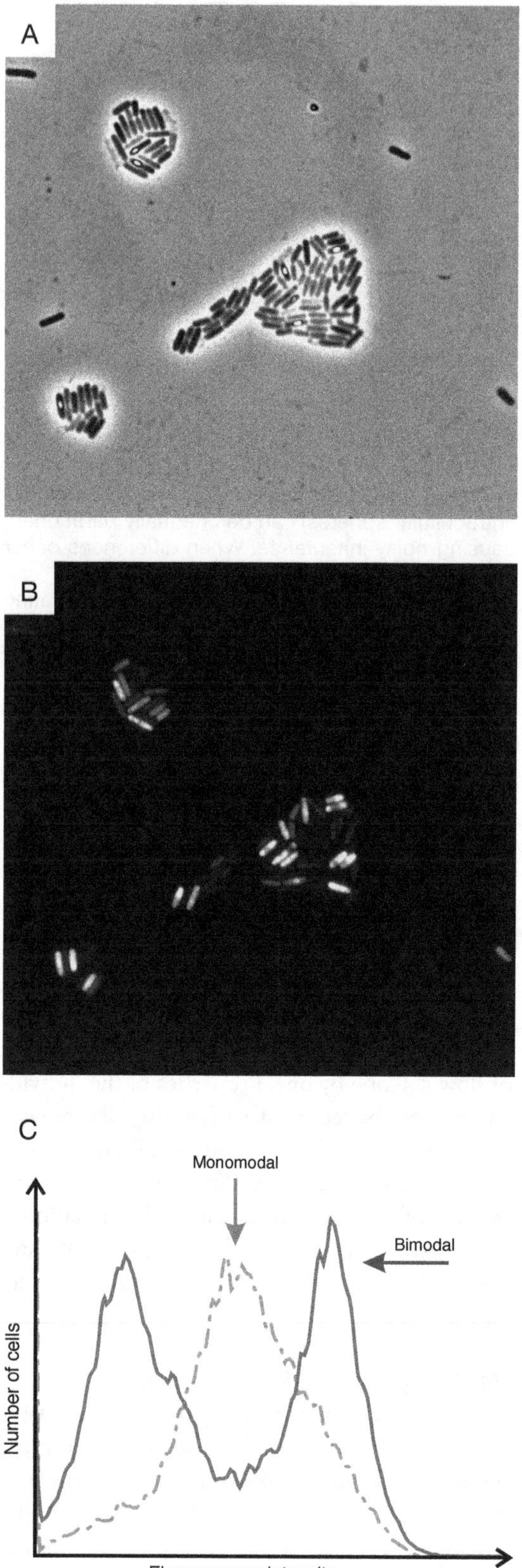

Phenotypical diversification within an isogenic culture can increase fitness of this genotype. Fitness-increasing strategies based on phenotypical diversification can include, amongst others, (i) introduction of maladapted phenotypes that might be favourable upon sudden and rapid changes in the environment or (ii) dividing the energetic burden of certain pathways by dedicating a subpopulation to such a pathway, sharing the outcome amongst genetically identical organisms. These two strategies are referred to as bet hedging and kin selection respectively (De Jong *et al.*, 2011).

Under lab conditions, such as in a fermenter with optimal conditions, it would be most advantageous for an organism to maximize growth rates. In nature the environment is neither homogeneous, nor stable. Under natural conditions, diversification driven by stochastic gene expression, can optimize growth speeds of the total population during random events of environmental change; bet hedging (Thattai and Van Oudenaarden, 2004; de Jong *et al.*, 2011). In case of an isogenic bacterial culture, spreading of investments in costly phenotypes can lower individual survival rates caused by maladaptation, but can increase species fitness in a longer time frame for the environmental selection landscape. For example, a bacterial culture may have evolved to produce a minority population of slow or non-growing cells with a persister phenotype that is resistant to cellular stresses such as exposure to antibiotics, and is present in the population even when antibiotics are absent.

Figure 12.7 With techniques such as fluorescence microscopy (A and B) and flow cytometry (C) it is possible to study gene expression in populations (Smits *et al.*, 2006). (A) A phase-contrast and (B) a fluorecence microscopy image of sporulating and non-sporulating *Bacillus subtilis*. The marker P*spoIIA*-mCherry (B) acts as a signal for entry into the sporulation pathway. This results in a bimodal distribution of red-fluorescent cells, because *B. subtilis* either enters sporulation or not (Errington, 1993). (C) Flow-cytometry results showing typical monomodal (dashed line) and bimodal (solid line) fluorescence distributions. The bimodal distribution shows two subpopulations; one with low and one with high expression of a fluorescent marker.

Persister cells lower the overall growth rate of the population and thus lower colonization success, but these cells have a higher chance of survival upon a rapid and lethal influx of antibiotic, increasing total population fitness compared with strains that do not form persister cells. A bet hedging strategy only increases fitness in a rapidly and unpredictably fluctuating environment where the environmental change selects against maladapted phenotypes (Olofsson *et al.*, 2009; de Jong *et al.*, 2011). Convincing experimental data in yeast on phenotype switching rates and environmental fluctuation frequencies has been reported (Acar *et al.*, 2008), demonstrating that fast-switching populations outgrow slow-switchers when the environment fluctuates rapidly and vice versa.

A bet hedging strategy differs from kin selection, in which individuals sacrifice individual fitness for fitness increase of 'kin' (Veening *et al.*, 2008; De Jong *et al.*, 2011). Biofilm formation, where only a portion of all individuals produce extracellular matrix for local clonal cells to benefit from, is a kin selection strategy (Veening *et al.*, 2008; De Jong *et al.*, 2011). Kin selection is fitness increasing as long as this form of bacterial altruism benefits genetically identical cells.

Example of organized colonies with specialized cell types

In complex colony structures such as a biofilm, the microenvironments inside a biofilm – a result of nutrient and oxygen gradients, and self-generated signals – can push the individuals towards phenotypic heterogeneity, allowing spatiotemporal organization of phenotypes (Vlamakis *et al.*, 2008).

Noisy expression in bistable networks allows this spatiotemporal phenotypical development. If evolved under certain constraints it is possible to fine-tune the underlying genetic regulatory network in such a way that distinct 'tissues' develop within a microbial culture. This is rather intuitive for multicellular organisms; however, multiple tissues with distinct functions are also found within a biofilm of unicellular organisms as demonstrated in *B. subtilis*.

Promoter studies of specific biofilm phenotypes by Vlamakis *et al.* (2008) demonstrated the topological organization within a biofilm. They found, for example, that spore-forming bacilli reside in the centre of the colony, furthest away from the food source (the agar) (Fig. 12.8), which is to be expected given that this phenotype is associated with starvation conditions. It was already shown, using LacZ-fusions, that *B. subtilis* spores form at the tips of complex biofilm structures, creating co-called 'fruiting bodies' (Branda *et al.*, 2001; Veening *et al.*, 2006). Additionally, a chronological sequence in phenotype development was identified. Sporulating cells develop from the transient motile and matrix producing phenotypes (Vlamakis *et al.*, 2008). A more important finding is that mutants defective in extracellular matrix production, resulting in unstructured biofilms, are defective in formation of these fruiting bodies (Vlamakis *et al.*, 2008; Lopez *et al.*, 2009; López and Kolter, 2010).

Indeed, phenotypes in *B. subtilis* biofilms – such as endospores, extracellular matrix-producing cells, motile cells and competent cells – are mutually exclusive and result from bistable regulatory networks (Lopez *et al.*, 2009). In fact, competence development in *B. subtilis* is a classic case study for bistable regulatory networks (Smits *et al.*, 2005, 2006).

Spatiotemporal development of phenotypes allows for distributing the cost of gene expression over the population in order to optimize metabolic pathway utilization. For recent reviews on bacterial biofilm formation see Lopez *et al.* (2009) and Abee *et al.* (2011).

Recently described bistable systems and refinements to classical bistable systems

The architecture of bistable switches in bacteria is diverse and of varying complexity. The selected examples are discussed in order of increasing complexity, starting with a simple positive feedback loop in *Vibrio cholerae* toxin production.

Vibrio cholerae virulence

The gammaproteobacterium *Vibrio cholerae* is the causative agent of human cholera. This diarrhoeal illness spreads rapidly and can have life threatening consequences in situations of poor

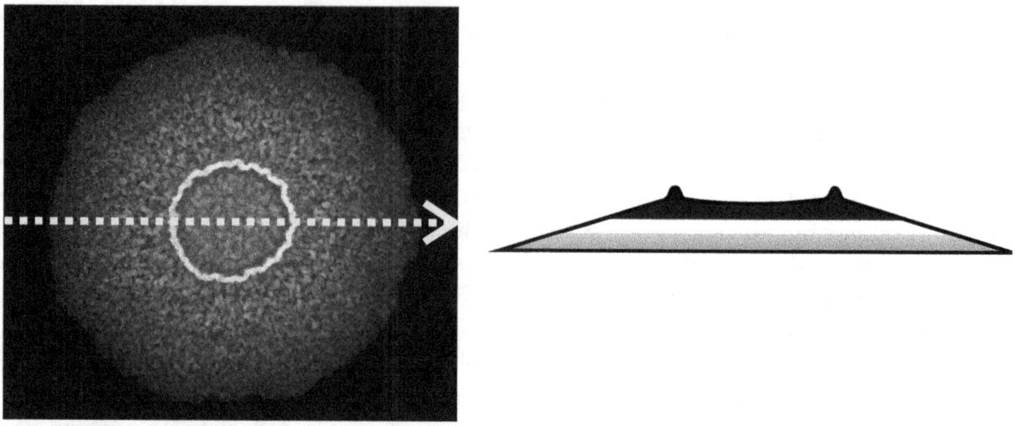

Figure 12.8 The bacterial community encased within a biofilm differentiates into distinct subpopulations of specialized cells. The figure shows a *Bacillus* biofilm with a cartoon of a cross section. Sporulating cells are located at the top of the colony and in the pili (black region), motile cells are found at the base of a colony (grey region) colocalized with biofilm matrix producing cells (grey and white regions) (Branda *et al.*, 2001; Vlamakis *et al.*, 2008; López and Kolter, 2010).

hygiene. The cholera outbreak in Haiti after the January 2010 earthquake is a recent example of how rapidly the illness can spread and how deadly infection can be.

Virulence of *V. cholerae* depends on (amongst others) the combined production of self-aggregating toxin-coregulated pilus (TCP), a colonization factor, and cholera toxin (CT) (Krukonis, 2003; Nelson *et al.*, 2009). Both TCP and CT are highly expressed early in the infection process when *V. cholerae* resides adjacent to epithelial surfaces. However, in accumulated fluid in the rabbit ileal loop model, expression of both these virulence factors was not detected using microarrays, despite CT being required for fluid secretion in this model. This lead to the hypothesis that perhaps only a small subpopulation of *V. cholerae* expresses these genes under these conditions (Nielsen *et al.*, 2010).

Expression of both the principal TCP repeating unit encoding gene (*tcpA*) and the gene encoding the A subunit of CT (*ctxA*) are under control of the proximal master regulator of virulence gene expression, ToxT. *toxT* is expressed by read-through from the upstream *tcpA-F* operon. Dimerized ToxT promotes expression of the *tcpA-F* operon creating a positive feedback loop (Nielsen *et al.*, 2010) (Fig. 12.9).

Using a GFP under control of the ToxT induced promoter of *tcpA*, Nielsen and coworkers could demonstrate a bimodal distribution of low and high *tcpA* expressing cells. Gene expression of this subpopulation got overlooked using microarrays because of dilution in the total mRNA batch. Nielsen *et al.* also show that this switch is reversible, epigenetic and persists after removal of the virulence induction stimulus, bicarbonate. The researchers speculate that the underlying genetic regulator allows continued production of virulence genes that may induce emergence of a hyper infectious subpopulation of bacteria in the stool of cholera patients. Furthermore, they argue that the heterogeneous expression of *tcpA* and *ctxA* in *V. cholerae* is the first example of bistable regulation of a major virulence pathway in a pathogenic organism (Nielsen *et al.*, 2010).

The temperate phage lambda

The temperate bacteriophage lambda has been one of the paradigms of bistable genetic networks for 60 years; it was first characterized in 1951 (Lederberg, 1951). It is one of the best understood decision circuits in biochemistry, yet the precise nature of the genetic network that produces these stable states is subject of on-going research. One intriguing and controversial question is what the minimal components of the network are which produce the observed behaviour.

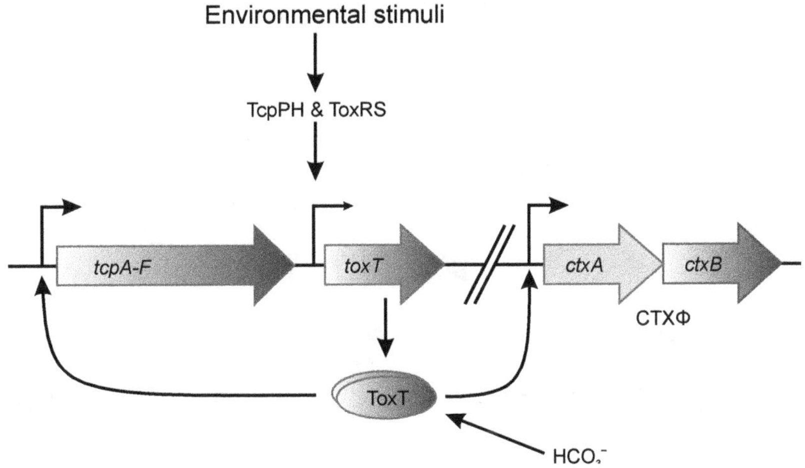

Figure 12.9 A schematic representation of the bistable network regulating expression of *tcpA-F* and *ctxAB*. Transcriptional regulator ToxT promotes expression of the *tcpA-F* operon and its own expression by means of read-through from the *tcpA-F* operon, effectively creating a positive feedback loop and a bistable switch. This bistable switch supports bimodal expression of both virulence factors TCP and the vibriophage CTXΦ encoded cholera toxin (CT) genes (Nielsen *et al.*, 2010).

Phage lambda is almost uniquely suited to the study of evolutionary influences on a single genetic component since the circuit *is* the organism. This provides a tractable framework to study the interplay between the molecular genetics and the life history and evolutionary trade-offs of an organism. Even though, strictly speaking, the phage lambda is not a separate form of life nor a bacterium, this regulatory network still merits a place in this chapter because it is, just like the *lac* operon discussed below, a bistable paradigm.

The genetic mechanism of phage lambda is a gene regulatory network for selection of either a lytic or lysogenic strategy for the virus in its host. The lytic strategy causes wide propagation of the virus through overproduction of phage particles, followed by death of the host and release of the phages. The lysogenic strategy is a long-term bet that the phage's survival is better ensured by remaining within the host. This is achieved by integrating the phage genome into that of the host, where it lies dormant (De Paepe and Taddei, 2006; Avlund *et al.*, 2009; Singh and Weinberger, 2009). The states are mutually exclusive and extremely stable.

The classical view describes two mutually repressing transcription factors, Cro and CI, binding to the same operator sites upstream of the genes associated with either strategy (Fig. 12.10). This double feedback loop drives the network into one of two stable states, partially by random fluctuations and partially by secondary factors influencing the production of the transcription factors (Arkin *et al.*, 1998). The formation of octamers by the two transcription factors at the operator sites is thought to occur in order to stabilize either expression state, resulting in a locked-in decision. The binding affinity of CI and Cro are different for each of the operators, yet the overall effect is to block or provide access to the antagonists' promoters for RNA polymerase. In addition, CI stimulates its own production by binding to its promoter, P_{RM}. The two most important external factors influencing the decision are host stress (through RecA) and the number of phages infecting the host (through phage encoded CII) (Kobiler *et al.*, 2005; Court *et al.*, 2007). Note that only a minimal set of all currently known phage lambda decision circuit elements is described here for clarity.

Which strategy the phage will select after infecting its host can be considered from an evolutionary point of view. The classical view is that this is mainly a function of the number of phages that

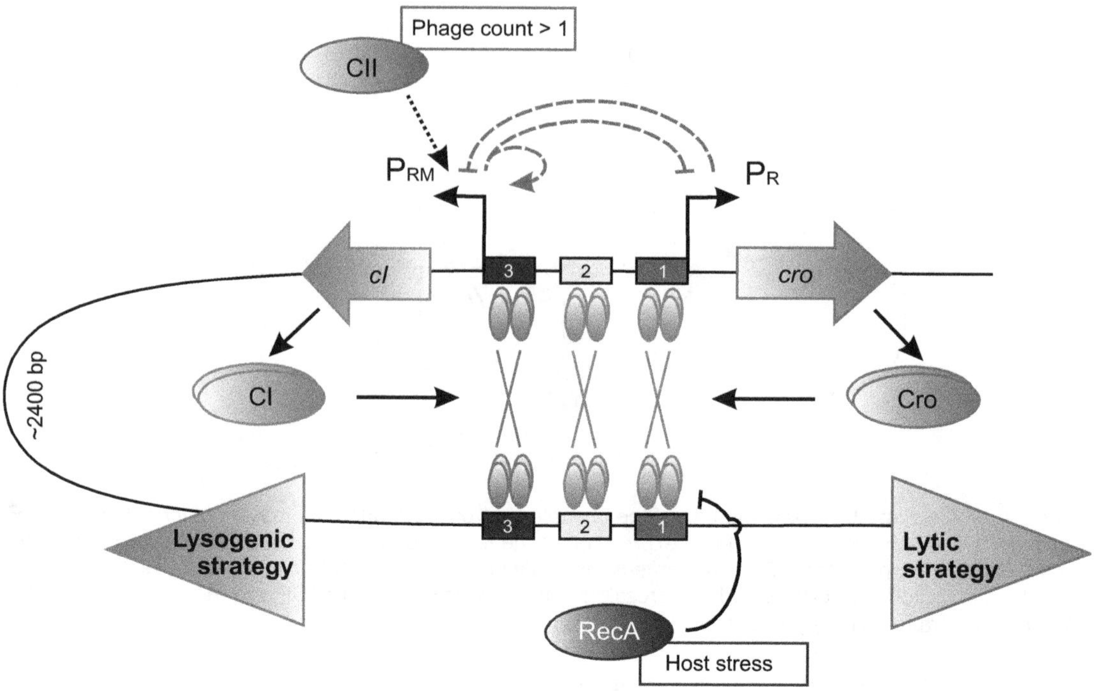

Figure 12.10 The products of the lambda phage encoded genes *cI* and *cro* bind to operator sites (boxes 1–3 on the DNA strand) in their own promoter regions, repressing each other's expression. The net effect of CI and Cro activity is illustrated by the dashed arrows. Activity of CI is inhibited by RecA (of host origin) preventing CI binding/multimer formation. The action of RecA – a sign of stress/lower fitness of the host – pushes the system towards the lytic strategy. CII, of phage origin, stimulates *cI* expression stimulating a lysogenic strategy (Court *et al.*, 2007; Cao *et al.*, 2010).

have already infected the host (the multiplicity of infection, MOI) (Lieb, 1953; Kourilsky, 1973). The rationale is that the number of phage genomes already present in the host is an indication of the 'phages to hosts' ratio in the environment. Pursuing a lytic strategy, therefore, reduces the chance of finding a new, uninfected host at high MOI (Avlund *et al.*, 2009). When two or more lysogenic phages are already present in the host, the levels of CII exceed the threshold, stimulating expression of CI, driving the phage to lysogeny; if there is none or one other phage present, the infecting phage will enter the lytic pathway (Fig. 12.10). Additionally, the CII regulator inhibits transcription of late lytic genes.

The role of CII has recently been the focus of some scrutiny; some authors have reported that some singly infecting phage enter the lysogenic pathway even without the benefit of CII (Atsumi and Little, 2006; Schubert *et al.*, 2007). These findings support the view of the lambda gene regulatory network as a stochastic system, which results in distributions of responses rather than absolutes.

An additional influence on *cI*/*cre* expression is the architecture of the phage DNA. The operator binding sites in between the promoters of *cI* and *cro* (P_{RM} and P_R, respectively) are also copied approximately 2400 bp downstream of *cI* (Fig. 12.10). Anderson and Yang constructed lambda strains with different spacing between the left and right copies of the operator regions. Measurements of fluorescently labelled CI showed that the presence of a DNA loop caused by binding of CI octamers to both operator copies yielded the highest CI expression (Anderson and Yang, 2008). These results were supported by a modelling study. Morelli *et al* showed that cells infected by lambda remain in the lysogenic state even though the number of free CI molecules are very low at about 200 (Dodd *et al.*, 2004). Low regulator molecule numbers raise the expectation that the system would be very susceptible to noise in *cI* expression. The fact that it is not susceptible to such noise, yet still sensitive enough to switch to the lytic state upon DNA damage (through RecA-induced CI autocleavage) is according to the model because of the DNA looping. Removal of the DNA loop in the model requires a 3–6 times higher basal level of CI in the (simulated) cell to yield bistable behaviour of the phage network (Morelli *et al.*, 2009).

It is a characteristic of this type of stochastic biochemical systems that its unpredictability lies in the low number of regulator molecules involved. The phage lambda circuitry contains an intricate set of mutual repression feedback loops that keep the overall number of proteins low. The number of (local) regulatory molecules is further influenced by the cell's total volume, the effects of macromolecular crowding and non–specific interactions of the regulatory proteins with the DNA. Recent experimental work by St-Pierre and Endy on the phage decision process showed that pre-infection host volume strongly correlates with post-infection outcome (St-Pierre and Endy, 2008). Essentially, large cells had up to a 5-fold lower chance of entering the lysogenic state than small cells. These observations suggest that stochasticity in the gene regulatory network of phage lambda may be less important to the outcome than other, so far unmeasured host-intrinsic factors. Cell volume in that case is an ensemble variable encoding a number of host-specific physiological differences; overall health or the expression of host regulators affecting lambda development. A recent report by Zeng *et al* supports this view of stochastic decision making being a result of 'hidden variables' in host state which otherwise act deterministically; specifically, the importance of host spatial organization of proteins is suggested to such an unmeasured but crucial factor in the lambda decision making process (Zeng *et al.*, 2010). Further refinement of the modelling approaches and experimental measurements (especially single-cell and single-molecule studies) will be necessary for resolving these questions, showing the continuing relevance of the phage lambda decision making network to elucidate the nature of bistable switches.

The lac operon in *E. coli*: a carbon source incorporated feedback circuit

The best known and best studied bacterial bistable switch is the lactose operon in *E. coli* (Monod and Jacob, 1961; Thattai and Van Oudenaarden, 2004; Smits *et al.*, 2006; Veening *et al.*, 2008). It was first described in 1957 (Novick and Weiner, 1957) and has become the canonical example of a bistable switch in which an environmental signal (lactose) is incorporated in the circuitry and of epigenetic inheritance in bacteria. The basic workings of the *lac* operon are well documented, yet new insights are still continuously obtained (e.g. Robert *et al.*, 2010).

The *lac* operon is repressed by LacI in absence of lactose in the environment. Presence of high lactose or a (non-metabolisable) lactose analogue (e.g. thiomethyl-β-galactoside (TMG) or isopropyl-D-thio-β-galactopyranoside (IPTG)) can relieve LacI repression. To be more precise, LacI repression is inhibited by an intermediate of lactose metabolism, allolactose. The resulting transcriptional burst allows for rapid production of the lactose permease LacY that supports high internal lactose levels. These high internal lactose levels maintain a derepressed state of the *lac* operon, effectively creating a positive feedback loop (Fig. 12.11). Expression of *lacY* displays a bimodal distribution in *E. coli* under certain inducer concentrations (Novick and Weiner, 1957; Choi *et al.*, 2008; Taniguchi *et al.*, 2010). The high expression state of the cell can be epigenetically inherited because of an increased number of lactose permeases in sibling cells (Casadesús and D'Ari, 2002; Veening *et al.*, 2008). Low concentrations of IPTG cannot induce expression of the *lac* operon, unless the *E. coli* cells have been subjected to high IPTG concentrations before (Novick and Weiner, 1957; Casadesús and D'Ari, 2002).

In a recent paper by Robert and co-workers the *lac* system was revisited, studying the switching probability of growing *E. coli* cells using microfluidics. They found that at high inducer concentrations (1) lower LacI concentrations increase switching probability (lower protein numbers increases susceptibility to noise; entrance of only a few inducer molecules can already result in switching) and (2)

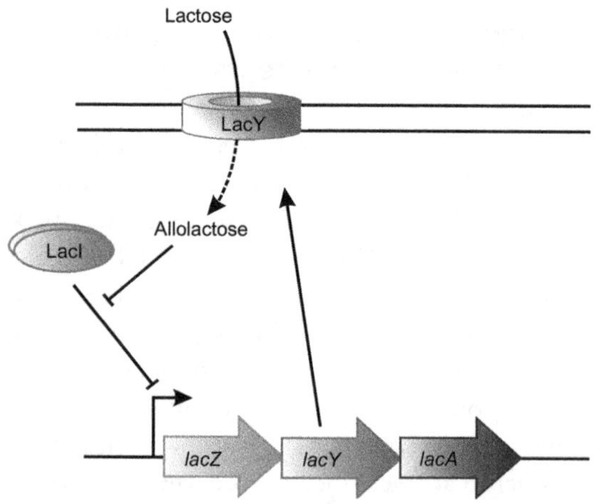

Figure 12.11 The leaky promoter of the *lac* operon offers a basal noisy expression of the *lac* operon. LacZ, β-galactosidase (cleaves disaccharide lactose into glucose and galactose); LacY β-galactoside permease; LacA β-galactoside transacetylase (transfers an acetyl group from acetyl-CoA to β-galactosides). LacI is the consitutivally expressed *lac* repressor that binds to the *lac* operator, downstream of the RNA polymerase docking site. Allolactose binds to LacI dimer in an allosteric manner, altering its conformation and consequently the binding properties of LacI and thus its repression abilities (Swint-Kruse and Matthews, 2009).

slower growth increased switching probability (Robert *et al.*, 2010).

When *E. coli* is grown under conditions that allow a bimodal lactose utilization distribution the switching probability seems to be influenced by growth rate rather than by external factors (Robert *et al.*, 2010). Such a mechanism allows slow-growing, possibly less fit cells to more readily change their gene expression to promote adaptation to the environment. Growth rate would thus be used by the cell as a measure of environmental quality. This finding, however, is counterintuitive for two reasons. First, in contrast to slow growth, faster growth would dilute LacI concentrations, lowering the initiation threshold, increasing the switching possibility to the 'High' state. And secondly, expression noise tends to increase in a smaller volume (Kaern *et al.*, 2005) and for the *lac* operon it was calculated that when noise intensity increases switching lactose metabolism 'Off' becomes easier and switching it 'On' becomes more difficult (Ochab-Marcinek, 2010). Perhaps slower growth allows accumulation of permeases in a smaller membrane surface, increasing inhibitor influx; Robert *et al.* (2010) found no difference in expression variability between slow and fast growing cells. Cell volume thus seems to play a role in switching, an effect also noted for the phage lambda system.

What is interesting about the *lac* system is that bistability is only achieved by induction with non-metabolizable lactose analogues, not with (allo) lactose itself (Van Hoek and Hogeweg, 2006, 2007; Santillán and Mackey, 2008). The positive feedback loop is much stronger for artificial inducers than for lactose, because they are not metabolized (Van Hoek and Hogeweg, 2007). As such, it seems to be a naturally occurring, synthetic switch. Bistability in the *lac* systems seems dependent on many more factors than inducer concentration and expression noise alone. So, bistability of the *lac* system appears conditional rather than the system being bistable by default.

Antibiotic production in *Streptomyces coelicolor*

The genus of *Streptomyces* is highly relevant to (pharmaceutical) industry. Over two-thirds of all naturally derived antibiotics in current use are produced using *Streptomyces* (Bentley *et al.*, 2002). Streptomycetes are high G+C, filamentous, Gram-positive soil bacteria (Bentley *et al.*, 2002). The (feeding) mycelium of a *Streptomyces* colony allows for creation of high local concentrations of secreted digestive enzymes and digestion products (Hodgson, 2000). It has been proposed that the antibiotics produced by a *Streptomyces* colony serve to defend the food source that has become available after investing in production of extracellular digestive enzymes (Challis and Hopwood, 2003). However, antibiotics are often also toxic for its producer. Therefore, it pays off to coordinate antibiotic production so that they are only produced after all cells have invested in appropriate defences (Mehra *et al.*, 2008). One system for coordination of cellular responses is quorum sensing.

Because of transcriptional noise, cells may prematurely alter gene expression without the associated environmental signal or cue being present. Such response noise can be harmful in the described case of antibiotic production. An additional benefit of synchronizing antibiotic production is that functional concentrations of energetically costly (Challis and Hopwood, 2003) antibiotics can be achieved quickly by orchestrated production, efficiently utilizing production costs. Quorum sensing buffers environmental 'misreading'; if large numbers of cells from a colony respond to a cue, then this cue is more likely to be present (Mehra *et al.*, 2008).

The ScbR/ScbA quorum sensing system of *Streptomyces coelicolor* displays bistable characteristics. The distribution of 'Low'/'High' expression is modulated by extracellular γ-butyrolactone (Scb1) concentrations (Mehra *et al.*, 2008). The ScbR dimer directly binds to the promoter region of the gene encoding the pathway-specific regulator CpkO and so represses its expression (Takano, 2006; Mehra *et al.*, 2008). CpkO activates expression of the *cpk* gene cluster that encodes enzymes for polyketide production (Mehra *et al.*, 2008; Gottelt *et al.*, 2010). Scb1 can bind to ScbR in an allosteric manner, alleviating its repressor activity, reminiscent of the lactose-LacI system (Mehra *et al.*, 2008). Additionally, the bistability disappears with very high growth rates also analogous to the *lac* system where low growth rates increase switching frequency (Kaern *et al.*, 2005).

ScbR represses its own transcription as well as the transcription of amplifier protein ScbA. The model proposed in the paper by Mehra *et al.* (2008) is largely consistent with earlier reported observations. In the model, however, the assumption is made that ScbA and ScbR form a complex that activates ScbA expression.

Upon reaching threshold levels of Scb1, repression by ScbR is lowered and expression of ScbA stimulated (Fig. 12.12). Additionally, ScbA is hypothesized to play a direct role in Scb1 production. When enough neighbouring *S. coelicolor* cells produce Scb1 (which can diffuse over the cell's membrane) (Takano *et al.*, 2000; Mehra *et al.*, 2008) the ScbR repression is repressed, allowing expression of the *cpk* cluster.

For more elaborate insights into quorum sensing, the reader is referred to further chapters in this book.

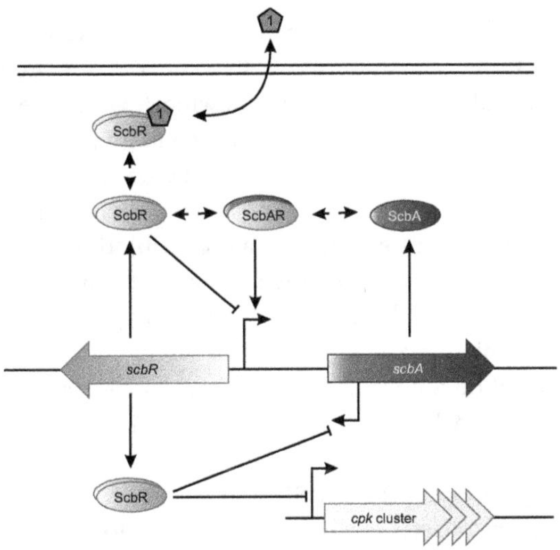

Figure 12.12 The genes encoding for repressor protein ScbR and amplifier protein ScbA are divergently transcribed. ScbR exist mostly in the DNA bound form, repressing its own transcription and that of amplifier protein ScbA. Binding of quorum sensing molecule Scb1 (pentagon; 1) relieves ScbR repression allowing transcription of ScbA. The heterodimer of ScbR and ScbA positively affect transcription of ScbA and, by dilution of ScbR monomer, relief *scbR* repression creating a net positive feedback loop. Additionally ScbA probably plays a direct role in Scb1 production. Relief of ScbR repression allows expression of the *cpk* operon (Mehra *et al.*, 2008).

Bacterial persistence

Persistence in *E. coli*

Bacterial persistence is an elegant example of bet hedging. For example, cells with increased tolerance for antibiotics, at the cost of slower growth, can already emerge before the addition of antibiotics to the culture's environment (Balaban, 2004; Sureka *et al.*, 2008).

In a study by Balaban and co-workers (Balaban, 2004), a very nice example of visualizing a bet hedging *E. coli* culture is shown. Using microfluidics Balaban and co-workers could show the presence of slow growing *E. coli* persister cells, even before conditions of stress. These cells were selected for using exposure of the culture to ampicillin until all vegetative cells were killed. These remaining persister cells would resume growth after removal of ampicillin. These cell's offspring can then again become fast growing ampicillin-sensitive cells, showing that the persister phenotype is transient (Balaban, 2004). Switching to a persister phenotype was predicted to be advantageous in a variable environment with frequent encounters with antibiotics (Kussell, 2005). The underlying regulatory network is not yet (fully) understood, but the phenotype switching bears all characteristics of a decision based on a noise driven bistable switch resulting in bimodality.

Mycobacterial persistence

In mycobacterial persistence, it was found that *rel* expression initiates the stringent response, which leads to a switch to persistence. Rel, which synthesizes the stringent response related 'alarmone' ppGpp, is essential for long-term survival of *Mycobacterium tuberculosis*, the causative agent of tuberculosis. For successful colonization of the host, persistence is one of the primary forces (Sureka *et al.*, 2008; Tiwari *et al.*, 2010).

For *Mycobacterium smegmatis* it was shown that expression of *rel*, *mprA* and sigma factor σ^E was bimodally distributed in 'Low' and 'High' expressing individuals (Sureka *et al.*, 2008; Ghosh *et al.*, 2011). *rel* expression is controlled by sigma factor σ^E, of which the gene expression is activated by phosphorylated MprA, a part of the stress–response network in mycobacteria (Sureka *et al.*, 2008; Tiwari *et al.*, 2010; Ghosh *et al.*, 2011).

Tiwari and co-workers (Tiwari *et al.*, 2010) modelled the mycobacterial stress–response network to study its bistability properties. The mycobacterial stress response network, however, is a rather complex bistable network because the activity of some components is dependent on phosphorylation.

This stress–response network consists of the two-component system MprA/MprB, sigma factor σ^E and anti-sigma factor RseA (Fig. 12.13). σ^E in turn promotes expression of the MprA/MprB operon and of the stringent response regulator, *rel*. Transcriptional regulator MprA was shown to be essential for mycobacterial persistence (Zahrt and Deretic, 2001; Tiwari *et al.*, 2010) and σ^E upregulation is seen under various stress conditions.

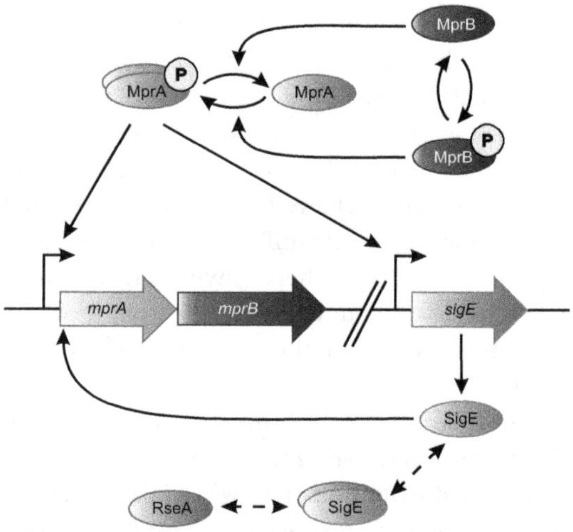

Figure 12.13 Schematic representation of the mycobacterial stress–response network. Expression of *mprA* is induced by its own product and by sigma factor σ^E. MprA also stimulates expression of σ^E. Activity of σ^E (SigE) is modulated by antisigma factor RseA. MprA is only active in a phosphorylated (P) and dimerized state. Phosphorylation of MprA is governed by kinase MprB, which gene is part of the *mprA* transcriptional unit (Tiwari *et al.*, 2010).

MprA and MprB are cotranscribed from one autoregulated operon. The network is triggered by autophosphorylation of sensory kinase MprB in the presence of the phosphate donor polyphosphate. MprB~P can in turn phosphorylate MprA, but MprB also acts as dephosphorylase of MprA~P. MprA~P promotes expression of the MprA/MprB operon. This positive autoregulation, however, was predicted not to be sufficient to sustain bistability in the network. Owing to its bimodal distribution, the average concentration of the sensory kinase MprB is too weak to set up a positive feedback loop. The positive feedback loop via σ^E, however, does offer sufficient impetus for a bistable switch because of the increase in the system's effective cooperativity, arising from the post-translational regulation of σ^E by RseA. The results from the model as presented in the paper match experimental results.

Multilevel regulatory networks: sporulation in *Bacillus subtilis*

The Gram-positive model organism *B. subtilis* produces endospores as a survival strategy under starvation conditions. Only part of the culture will make the decision to produce these highly resistant 'time capsules' that can stay dormant until the environment becomes favourable again.

The decision to sporulate is not taken lightly. The process of sporulation is time and energy consuming and causes fitness loss due to a sporulating subpopulation that cannot divide any more and thus does not create progeny. The regulatory network underlying phenotypical differentiations is a large, complex network with significant cross-talk between separate decision modules. For example, the regulatory module controlling development of the competence phenotype (where cells allow uptake of foreign DNA from the environment) is connected to the sporulation module, blocking entry into the sporulation pathway if competence is developed (Süel *et al.*, 2006; Schultz *et al.*, 2009).

The interesting thing about the decision to sporulate is that the master regulator for sporulation in *B. subtilis*, Spo0A is not bimodally distributed. Spo0A is only active after phosphorylation (like the earlier discussed MprA/MprB operon involved in mycobacterial persistence). A set of kinases and phosphatases make up a phosphate feeding network termed the phosphorelay. Within this network multiple feedback loops have been identified, however so far none of them seems crucial to the development of bimodality. Expression of the individual relay components is also not bimodally distributed (Chastanet *et al.*, 2010; De Jong *et al.*, 2010; Eswaramoorthy *et al.*, 2010).

To add to the complexity, Spo0A~P regulates many more cellular processes than just sporulation and is multistable, rather than bistable. Only when a large portion of Spo0A is phosphorylated is sporulation initiated. At 'medium' levels of phosphorylation, Spo0A~P triggers biofilm formation. This is the result of differing binding affinities of the promoter located Spo0A binding boxes. The promoters of genes triggered at 'intermediate' Spo0A~P levels have a high affinity for Spo0A~P, promoters of genes involved in sporulation (triggered at 'High' Spo0A~P) have low affinities for Spo0A~P (Fujita *et al.*, 2005; Fujita and Losick, 2005).

This is an example of a network architecture with multiple levels of control for which the interactions finally result in distinct stable states without the regulating components (master regulator) showing bimodal distributions themselves (Chastanet *et al.*, 2010; De Jong *et al.*, 2010). This is an interesting phenomenon which demonstrates that, using current molecular detection methods it may not be possible to detect regulator state bimodality. Supporting this conclusion is a recent study that describes a quantitative system-wide analysis of protein and mRNA expression in individual *E. coli* cells with single-molecule sensitivity. A library containing 1018 translational YFP fusions was screened. One very interesting finding was that the researchers did not observe clear bimodal distributions among the 1018 fluorescently labelled strains from their screening library under the growth conditions tested (Taniguchi *et al.*, 2010). This means that bimodality is generally rare, is only induced under specific conditions, or often exists at a different level than at the transcription level such as the post-translational modification level.

Synthetic switches

Synthetic biology is a very promising discipline for which in depth knowledge about regulatory elements and networks is essential. By custom building switches to let bacteria, for example, produce a certain compound many new possibilities for industry and biotechnology are opening up. Below, the current state of engineered bistable switches is summarized.

Harnessing the properties of regulatory networks for production purposes has become a general laboratory approach. For example, for many years the *lac* system has been used in vectors for inducible heterologous gene expression, e.g. in pET vectors. Artificial, non-metabolizable inducers are used to make use of the systems all-or-none switching properties (Van Hoek and Hogeweg, 2006).

A milestone in synthetic biology was the 'repressilator', a regulatory circuit consisting of tetracycline repressor TetR, lambda CI and LacI, each repressing each other's expression. Reporter gene *gfp* was repressed only by TetR. The *E. coli* strain containing this circuit displayed oscillating fluorescence (Elowitz and Leibler, 2000), demonstrating the promise of designed regulatory networks.

By using, amongst others, these *lacI*, lambda *cI* and *tetR* repressors and operators, it is possible to design and build synthetic switches, for example a toggle switch. With a pulse of inducer the network can be switched from either a 'Low' state to a 'High' state of one of the components and vice versa (Fig. 12.14) (Gardner *et al.*, 2000; Kobayashi *et al.*, 2004).

Apart from the construction of regulatory networks as a means of obtaining desirable traits, synthetic networks can help identify lacking knowledge. By observing the differences between *in silico* and *in vivo* behaviour, lacking influences in the network can be observed; in other words, a learning by building approach (Bashor *et al.*, 2010).

Using such synthetically designed networks has already increased our knowledge of network topologies. Different network topologies can yield equivalent network outputs. This has inspired evolutionary studies into whether evolution favours certain topologies. Natural competence in *B. subtilis* is governed by the master regulator ComK. ComK induces its own expression, establishing an ultrasensitive network. ComK levels are reduced by the action of MecA, which binds ComK and so targets ComK for degradation by a multiprotein proteolytic complex. Stress protein ComS competes with ComK for binding with MecA, positively affecting ComK concentrations. ComK (indirectly) inhibits *comS* expression (Süel *et al.*, 2006; Smits *et al.*, 2007). See Fig. 12.15 for a simplified global network topology of *B. subtilis* competence, here termed the native circuit.

This bistable competence regulatory network was used as a case study on possible bias of evolutionary processes to select for a certain network topology (Çağatay *et al.*, 2009). Researchers 'rewired' this network, removing the inhibition of ComK on *comS* and making *mecA* sensitive to ComK induction. This rewired network is referred to as the alternative circuit (Fig. 12.15).

With the natural circuit, there is more cell to cell variability during the time spent in the competent state (referred to as the 'competence window'). With the alternative network, the competence windows were of more uniform length (Bollenbach and Kishony, 2009; Çağatay *et al.*, 2009; Zhang *et al.*, 2010). In the natural system, the concentration ComS is low during the competence window, whereas in the alternative network, MecA correlates positively with ComK, as a result of high ComK concentrations. This consequently means that, owing to protein numbers, the ComS component is noisier than the abundant MecA component. So, the MecA topology allows for more precise regulation, which is advantageous in for example circadian-clock and cell-cycle circuits. In unpredictably fluctuating environments and in the presence of extracellular DNA the noisy characteristics of the native competence network is fitness-increasing. Additionally, the native network is also more robust, meaning that

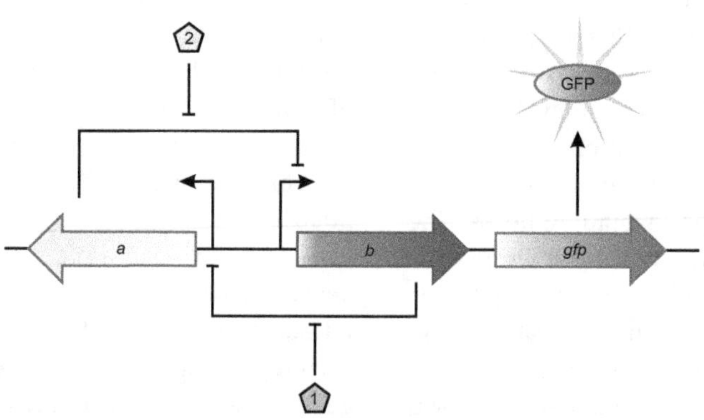

Figure 12.14 Cartoon of a general toggle switch design. The product of the gene *a* inhibits expression of the gene *b* and vice versa. If gene *b* is expressed, *gfp* is co-transcribed. This is regarded the 'High' state of the system. Expression of gene *a* is termed the 'Low' state. By exposing the system to inducer 1 or 2 the system can be switched to each of both stable states.

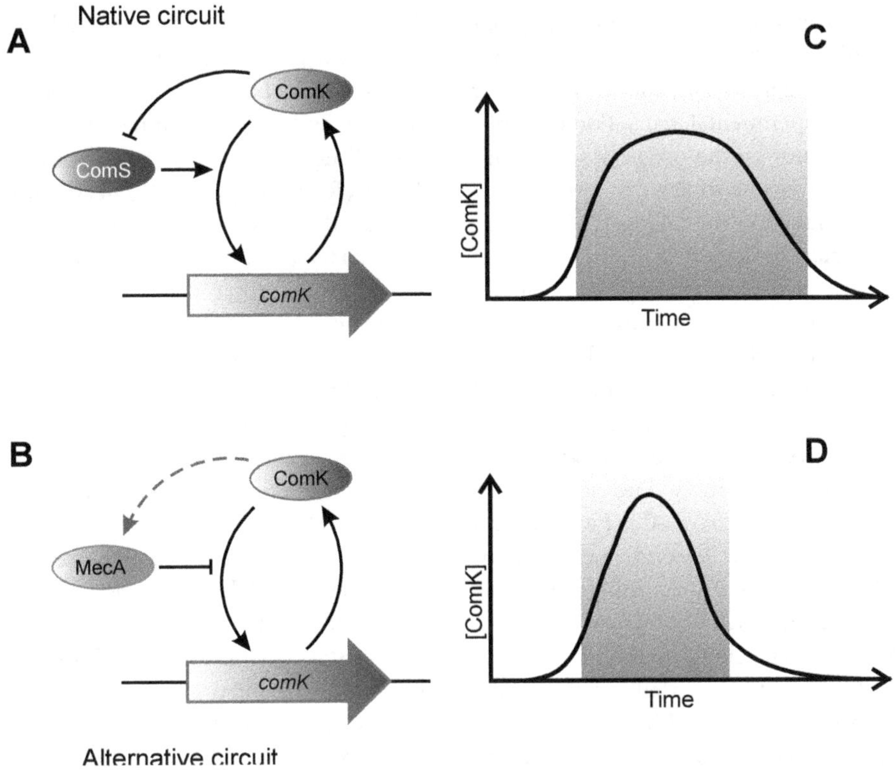

Figure 12.15 Simplified global scheme of two alternative but equivalent regulatory networks. The native circuit (A) resembles the competence regulatory network of *B. subtilis*, the alternative circuit (B) is a 'rewired' version of the native network where the dashed arrow represents an artificially established connection. With the native network topology, the cell to cell variability in the time spent in the competent state, the competence window (shaded area) for the population is wider than for the alternative network because of more cell to cell variability (C and D).

it is harder to switch between stable states (Zhang *et al.*, 2010).

Modelling

Bistable switches have been a attractive subject for trying out mathematical modelling approaches for predicting the behaviour of gene regulatory networks. Reproducing the switching behaviour exhibited in experimental studies is a litmus test of sorts for the validity of the method. Conversely, modelling can help identify factors missing in the experimental understanding of a network as well as suggest the relative importance of the factors to the overall phenomenon. For example, modelling *S. coelicolor* quorum sensing suggests that ScbA-ScbR form an auto–stimulatory complex; a hypothesis that is not yet confirmed by experiment (Mehra *et al.*, 2008; Goryachev, 2011). The advent of single-cell and single-molecule high-resolution experimental techniques now generates data of sufficient resolution to be quantitatively compared with the output of these models or to provide the parameters needed to seed these models.

In general, there is a broad conceptual split between methods based on systems of (1) ordinary differential equations (ODEs) describing classical mass-action reaction kinetics, (2) numerical approximations to stochastic differential equations (SSA, the Gillespie algorithm), (3) Lattice-Gas Automata (LGA) methods and (4) full-scale article-based molecular dynamics models. Whereas 15 years ago only the analytical methods of ODE reaction kinetics were tractable computationally, the increase in computer power over the years has made it practical to carry out simulations of realistically sized gene regulatory networks using all approaches, and compare their predictions to experimental data. For example, the work carried out on the phage lambda lysis/lysogeny choice was one of the first applications of stochastic simulation modelling (Arkin *et al.*, 1998) in molecular biology.

The general theory underlying the ordinary differential equation-based models is based on the assumption that the reactions take place in a well-mixed homogeneous environment and involve a sufficient number of reactants, ensuring that their concentrations are a valid measure. The first assumption generally does not hold for a cell's cytosol (Zhou *et al.*, 2008; Leventis and Silvius, 2010), while the second may hold for some cellular processes but specifically not for the class of processes to which bistable switches generally belong. The switching behaviour of these networks often depends precisely on the scarce number of molecules involved in the binding of a few DNA binding locations. Nonetheless, the wealth and accessibility of mathematical tools to analyse the behaviour of such models ensures that they remain a popular first approximation when modelling biochemical systems. Additionally, their parameters map closely to easily measured, macroscopic enzyme kinetics, a desirable property when trying to link the model to experimental results (Chen *et al.*, 2010).

The stochastic simulation techniques are generally based on the chemical master equation formulation of chemical reaction kinetics (Gillespie, 2007; Kampen, 2007; Ullah and Wolkenhauer, 2007). In short, they are based on a formulation of reactions as probabilities, with the probability of two chemical species reacting at any given time sampled from the Poisson distribution. While still assuming a homogeneous reaction environment, this approach deals well with very low numbers of reacting particles; hence its popularity when describing biochemical systems such as bistable switches which are heavily subject to noise. Although exact mathematical solutions for non-trivial systems of reactions are not possible (Jahnke and Huisinga, 2006), the development and refinement of the stochastic simulation algorithm (Gillespie, 1977; Li *et al.*, 2008) has enabled modelling of biologically relevant phenomena. For end-users, use of one of the software kits implementing the method is highly recommended; for example, modelling with StochKit (Li *et al.*, 2008) and Dizzy (Ramsey *et al.*, 2005) in combination with the SBML (systems biology markup language (Hucka *et al.*, 2003; www.sbml.org) pathway description format are widely used. To overcome some of the inaccuracy imposed by the assumption of a well-mixed cell, *spatial* Gillespie algorithms have been developed. It splits up the cell's reactions in separated volumes, and a chemical species' diffusion from one to the next is treated analogously to a reaction occurring, for example MesoRD (Hattne *et al.*, 2005).

Finally, particle-based modelling offers, in principle, the highest accuracy, but at the price of slower speed. These methods track individual particles as they diffuse through the cell in a continuous representation of space; reactions occur with a probability factor when reactants collide (Andrews *et al.*, 2010). This 'Smoluchowski' biophysical formulation of the intracellular space allows for the incorporation of structural features (membranes, macromolecular crowding) which greatly increases the realism of simulating whole-cell processes. It has been used for modelling wide-ranging phenomena such as *E. coli* chemotaxis (Lipkow and Odde, 2008) and neuron firing. Ready-made packages for molecular-level simulation include Smoldyn (Andrews *et al.*, 2010; www.smoldyn.org), ChemCell (Plimpton and Slepoy, 2003) and MCell (Kerr *et al.*, 2008). These packages are particularly useful in the context of bistable switches when parts of the network are chemical species which enter the cell from the outside, for example lactose must diffuse to LacI and then to the *lac* operon on the DNA or the quorum sensing molecules involved in the *S. coelicolor* network described above. A feature of Smoldyn which is particularly useful for generating realistic models of cell membrane structure is the ability to import meshes of microscope data (Andrews *et al.*, 2010).

A further refinement of the Smoluchowski formulation exists in the form of Green's function reaction dynamics (GFRD). This method exactly solves the Smoluchowski reaction dynamics for pairs of particles (under the assumption that they can be isolated in a small volume of cell space), which allows the simulation to skip from reaction to reaction, greatly speeding it up (Van Zon and Ten Wolde, 2005). One intriguing finding with a model using these methods with regards to the basics of bistable switch operation has been that incorporating diffusion of transcription factors into a model of simple gene expression drastically increases noise, resulting in mRNA bursts (Van Zon *et al.*, 2006). For end-users, these methods too have been integrated in the E-cell package (Takahashi *et al.*, 2004; www.e-cell.org).

Conclusions

The field of study involved in the characterization and analysis of bistable switches and heterogeneity in bacteria is still rapidly expanding, as well as deepening our knowledge, owing to higher-powered experimental and modelling techniques. This chapter's goal was to discuss and summarize the workings and the functionality of bistable switches in bacterial regulatory networks. The basics of the mechanism underlying bimodal decision making in bacteria were discussed and different examples illustrated the myriad forms in which these devices occur in bacteria, and how they increase their fitness.

Bistable switches can occur as the simplest positive feedback loops (*V. cholerae*), to systems in which nutrients are an element of the network's architecture (*lac* system), to systems with regulation on multiple expression levels (*B. subtilis* sporulation). The phenotypical diversification that results from bistable switches can support 'multitissue' colony formation in bacterial biofilms (*B. subtilis*) and can give rise to multicellular organisms such as ourselves, in which cells are programmed to perform specific tasks for the benefit of the entire organism. In all likelihood, many more examples of bistable switches and/or multistable switches will be found in the future, some more complex, some mere variations of the ones described in this chapter. Because expression of genes in a small subpopulation of cells can be overlooked while studying expression using bulk RNA of an entire culture (as has happened in the case of hyperinfectious *V. cholera*) (Nielsen *et al.*, 2010), it is very likely that bistable switches are an even more widespread but not always recognized phenomenon. Additionally, the conditions under which these gene regulatory networks exhibit bistable behaviour may be very specific and hard to find; indeed the canonical *lac* operon is not bistable in natural conditions at all, but only with synthetic analogues of lactose!

The knowledge gained by studying bistable switches will see its way to industry, where construction of desirable switches, such as a toggle switch (Gardner *et al.*, 2000; Kobayashi *et al.*, 2004), can help to programme cells. Knowledge about the function of bistability in development of virulence and disease may yield additional leads for drug development or treatment, as in the case of prevention of persistence development.

Obstacles hindering progression in the construction of programmable cells will be overcome, elucidating elements of regulatory networks that were hitherto undiscovered. This may include even that the role of cell physics is more crucial in generating bistability than the actual network architecture (Morelli *et al.*, 2009). What has already been shown is that bistable switches are not a prerequisite for the generation of a bifurcated population (To and Maheshri, 2010), whereas also bistable switches are no guarantee for bifurcation if conditions only allow existence of one of both stable states. Either way, in spite of the large amount of work done on gene regulatory networks, much more remains to be discovered about them, and computational simulation will probably prove to be a key instrument.

Acknowledgements

The authors would like to thank Dr Maarten Mols for providing us with flow cytometry data and Dr Jan-Willem Veening for critically reviewing the manuscript. JS was financially supported by NWO-CLS (BetNet), projectnr: 600.635.100.08N26.

Note from authors

During the writing of this chapter a review on bistable responses in bacterial genetic networks by Tiwari *et al.* was published (Tiwari *et al.*, 2011). This review is by the authors considered to be very good reading additional to this chapter.

References

Abee, T., Kovács, Á.T., Kuipers, O.P., and Van der Veen, S. (2011). Biofilm formation and dispersal in Gram-positive bacteria. Curr. Opin. Biotechnol. *22*, 172–179.

Acar, M., Mettetal, J.T., and Van Oudenaarden, A. (2008). Stochastic switching as a survival strategy in fluctuating environments. Nat. Genet. *40*, 471–475.

Alberts, B. (2007). Molecular Biology of the Cell, 5th edn. (Garland Science, New York, USA).

Alon, U. (2006). An Introduction To Systems Biology: Design Principles Of Biological Circuits. (Chapman and Hall/CRC Mathematical and Computational Biology, Boca Raton, USA; London, UK).

Anderson, L.M., and Yang, H. (2008). DNA looping can enhance lysogenic CI transcription in phage lambda. Proc. Natl. Acad. Sci. U.S.A. *105*, 5827–5832.

Andrews, S.S., Addy, N.J., Brent, R., and Arkin, A.P. (2010). Detailed simulations of cell biology with Smoldyn 2.1. PLoS Comput. Biol. *6*, e1000705.

Arkin, A., Ross, J., and McAdams, H.H. (1998). Stochastic kinetic analysis of developmental pathway bifurcation in phage λ-Infected *Escherichia coli* cells. Genetics *149*, 1633–1648.

Atsumi, S., and Little, J.W. (2006). A synthetic phage λ regulatory circuit. Proc. Natl. Acad. Sci. U.S.A. *103*, 19045–19050.

Avlund, M., Dodd, I.B., Semsey, S., Sneppen, K., and Krishna, S. (2009). Why do phage play dice? J. Virol. *83*, 11416–11420.

Balaban, N.Q. (2004). Bacterial persistence as a phenotypic switch. Science *305*, 1622–1625.

Bashor, C.J., Horwitz, A.A., Peisajovich, S.G., and Lim, W.A. (2010). Rewiring cells: synthetic biology as a tool to interrogate the organizational principles of living systems. Ann. Rev. Biophys. *39*, 515–537.

Bentley, S.D., Chater, K.F., Cerdeno-Tarraga, A.-M., Challis, G.L., Thomson, N.R., James, K.D., Harris, D.E., Quail, M.A., Kieser, H., Harper, D., *et al.* (2002). Complete genome sequence of the model actinomycete *Streptomyces coelicolor* A3(2). Nature *417*, 141–147.

Bollenbach, T., and Kishony, R. (2009). Quiet gene circuit more fragile than its noisy peer. Cell *139*, 460–461.

Branda, S.S., González-Pastor, J.E., Ben-Yehuda, S., Losick, R., and Kolter, R. (2001). Fruiting body formation by *Bacillus subtilis*. Proc. Natl. Acad. Sci. U.S.A. *98*, 11621–11626.

Çağatay, T., Turcotte, M., Elowitz, M.B., Garcia-Ojalvo, J., and Süel, G.M. (2009). Architecture-dependent noise discriminates functionally analogous differentiation circuits. Cell *139*, 512–522.

Cao, Y., Lu, H.-M., and Liang, J. (2010). Probability landscape of heritable and robust epigenetic state of lysogeny in phage lambda. Proc. Natl. Acad. Sci. U.S.A. *107*, 18445 –18450.

Casadesús, J., and D'Ari, R. (2002). Memory in bacteria and phage. Bioessays *24*, 512–518.

Challis, G.L., and Hopwood, D.A. (2003). Synergy and contingency as driving forces for the evolution of multiple secondary metabolite production by Streptomyces species. Proc. Natl. Acad. Sci. U.S.A. *100*, 14555–14561.

Chastanet, A., Vitkup, D., Yuan, G.-C., Norman, T.M., Liu, J.S., and Losick, R.M. (2010). Broadly heterogeneous activation of the master regulator for sporulation in *Bacillus subtilis*. Proc. Natl. Acad. Sci. U.S.A. *107*, 8486–8491.

Chen, W.W., Niepel, M., and Sorger, P.K. (2010). Classic and contemporary approaches to modeling biochemical reactions. Genes Dev. *24*, 1861–1875.

Choi, P.J., Cai, L., Frieda, K., and Xie, X.S. (2008). A stochastic single-molecule event triggers phenotype switching of a bacterial cell. Science *322*, 442.

Coulombe, B., and Burton, Z.F. (1999). DNA bending and wrapping around RNA Polymerase: A "revolutionary" model describing transcriptional mechanisms. Microbiol. Mol. Biol. Rev. *63*, 457–478.

Court, D.L., Oppenheim, A.B., and Adhya, S.L. (2007). A new look at bacteriophage λ genetic networks. J. Bacteriol. *189*, 298–304.

De Jong, I.G., Veening, J.W., and Kuipers, O.P. (2010). Heterochronic phosphorelay gene expression as a source of heterogeneity in *Bacillus subtilis* spore formation. J. Bacteriol. *192*, 2053–2067.

De Jong, I.G., Haccou, P., and Kuipers, O.P. (2011). Bet hedging or not? A guide to proper classification of microbial survival strategies. Bioessays *33*, 215–223.

De Paepe, M., and Taddei, F. (2006). Viruses' life history: towards a mechanistic basis of a trade-off between survival and reproduction among phages. PLoS Biol. *4*, e193.

Dodd, I.B., Shearwin, K.E., Perkins, A.J., Burr, T., Hochschild, A., and Egan, J.B. (2004). Cooperativity in long-range gene regulation by the λ CI repressor. Genes Dev. *18*, 344 –354.

Eldar, A., and Elowitz, M.B. (2010). Functional roles for noise in genetic circuits. Nature *467*, 167–173.

Elowitz, M.B. (2002). Stochastic gene expression in a single cell. Science *297*, 1183–1186.

Elowitz, M.B., and Leibler, S. (2000). A synthetic oscillatory network of transcriptional regulators. Nature *403*, 335–338.

Errington, J. (1993). *Bacillus subtilis* sporulation: regulation of gene expression and control of morphogenesis. Microbiol. Rev. *57*, 1–33.

Eswaramoorthy, P., Dinh, J., Duan, D., Igoshin, O.A., and Fujita, M. (2010). Single-cell measurement of the levels and distributions of the phosphorelay components in a population of sporulating *Bacillus subtilis* cells. Microbiology *156*, 2294–2304.

Fujita, M., and Losick, R. (2005). Evidence that entry into sporulation in *Bacillus subtilis* is governed by a gradual increase in the level and activity of the master regulator Spo0A. Genes Dev. *19*, 2236–2244.

Fujita, M., González-Pastor, J.E., and Losick, R. (2005). High-and low-threshold genes in the Spo0A regulon of *Bacillus subtilis*. J. Bacteriol. *187*, 1357–1368.

Gardner, T.S., Cantor, C.R., and Collins, J. J. (2000). Construction of a genetic toggle switch in *Escherichia coli*. Nature *403*, 339–342.

Ghosh, S., Sureka, K., Ghosh, B., Bose, I., Basu, J., and Kundu, M. (2011). Phenotypic heterogeneity in mycobacterial stringent response. BMC Syst. Biol. 5, 18.

Gillespie, D.T. (1977). Exact stochastic simulation of coupled chemical reactions. J. Phys. Chem. *81*, 2340–2361.

Gillespie, D.T. (2007). Stochastic simulation of chemical kinetics. Annu. Rev. Phys. Chem. *58*, 35–55.

Golding, I., Paulsson, J., Zawilski, S., and Cox, E. (2005). Real-time kinetics of gene activity in individual bacteria. Cell *123*, 1025–1036.

Goryachev, A.B. (2011). Understanding Bacterial cell–cell communication with computational modeling. Chem. Rev. *111*, 238–250.

Gottelt, M., Kol, S., Gomez-Escribano, J.P., Bibb, M., and Takano, E. (2010). Deletion of a regulatory gene within the cpk gene cluster reveals novel antibacterial activity in *Streptomyces coelicolor* A3 (2). Microbiology *156*, 2343–2353.

Hattne, J., Fange, D., and Elf, J. (2005). Stochastic reaction-diffusion simulation with MesoRD. Bioinformatics *21*, 2923 –2924.

Hodgson, D.A. (2000). Primary metabolism and its control in Streptomycetes: a most unusual group of bacteria. Adv. Microb. Physiol. *42*, 47–238.

Hucka, M., Finney, A., Sauro, H.M., Bolouri, H., Doyle, J.C., Kitano, H., Arkin, A.P., Bornstein, B.J., Bray, D., Cornish-Bowden, A., *et al.* (2003). The systems biology markup language (SBML): a medium for representation and exchange of biochemical network models. Bioinformatics *19*, 524 –531.

Jahnke, T., and Huisinga, W. (2006). Solving the chemical master equation for monomolecular reaction systems analytically. J. Math. Biol. *54*, 1–26.

Kaern, M., Elston, T.C., Blake, W.J., and Collins, J.J. (2005). Stochasticity in gene expression: from theories to phenotypes. Nat. Rev. Genet. *6*, 451–464.

Kampen, N.G.V. (2007). Stochastic Processes in Physics and Chemistry, 3rd edn. (North-Holland Publishing Co., Amsterdam).

Kearns, D.B., and Losick, R. (2005). Cell population heterogeneity during growth of *Bacillus subtilis*. Genes Dev. *19*, 3083 –3094.

Kerr, R.A., Bartol, T.M., Kaminsky, B., Dittrich, M., Chang, J.-C.J., Baden, S.B., Sejnowski, T.J., and Stiles, J.R. (2008). Fast Monte Carlo simulation methods for biological reaction-diffusion systems in solution and on surfaces. SIAM J. Sci. Comput. *30*, 3126.

Kobayashi, H., Kaern, M., Araki, M., Chung, K., Gardner, T.S., Cantor, C.R., and Collins, J.J. (2004). Programmable cells: interfacing natural and engineered gene networks. Proc. Natl. Acad. Sci. U.S.A. *101*, 8414–8419.

Kobiler, O., Rokney, A., Friedman, N., Court, D.L., Stavans, J., and Oppenheim, A.B. (2005). Quantitative kinetic analysis of the bacteriophage λ genetic network. Proc. Natl. Acad. Sci. U.S.A. *102*, 4470 –4475.

Kourilsky, P. (1973). Lysogenization by bacteriophage lambda. I. Multiple infection and the lysogenic response. Mol. Gen. Genet. *122*, 183–195.

Krukonis, E. (2003). From motility to virulence: sensing and responding to environmental signals in *Vibrio cholerae*. Curr. Opin. Microbiol. *6*, 186–190.

Kussell, E. (2005). Bacterial persistence: A model of survival in changing environments. Genetics *169*, 1807–1814.

Lederberg, E.M. (1951). Lysogenicity in *E. coli* K-12. Genetics *36*, 560.

Leventis, R., and Silvius, J.R. (2010). Quantitative experimental assessment of macromolecular crowding effects at membrane surfaces. Biophys. J. *99*, 2125–2133.

Li, H., Cao, Y., Petzold, L.R., and Gillespie, D.T. (2008). Algorithms and software for stochastic simulation of biochemical reacting systems. Biotechnol. Prog. *24*, 56–61.

Lieb, M. (1953). Studies on lysogenization in *Escherichia coli*. Cold Spring Harb. Symp. Quant. Biol *18*, 71–73.

Lipkow, K., and Odde, D.J. (2008). Model for protein concentration gradients in the cytoplasm. Cel. Mol. Bioeng. *1*, 84–92.

López, D., and Kolter, R. (2010). Extracellular signals that define distinct and coexisting cell fates in *Bacillus subtilis*. FEMS Microbiol. Rev. *34*, 134–149.

Lopez, D., Vlamakis, H., and Kolter, R. (2009). Generation of multiple cell types in *Bacillus subtilis*. FEMS Microbiol. Rev. 33, 152–163.

Mehra, S., Charaniya, S., Takano, E., and Hu, W.-S. (2008). A bistable gene switch for antibiotic biosynthesis: The butyrolactone regulon in *Streptomyces coelicolor*. PLoS ONE 3, e2724.

Meyerovich, M., Mamou, G., and Ben-Yehuda, S. (2010). Visualizing high error levels during gene expression in living bacterial cells. Proc. Natl. Acad. Sci. U.S.A. *107*, 11543–11548.

Mirel, D.B., Lustre, V.M., Chamberlin, M.J. (1992). An operon of *Bacillus subtilis* motility genes transcribed by the sigma D form of RNA polymerase. J. Bacteriol. *174*, 4197–4204.

Monod, J., and Jacob, F. (1961). Teleonomic mechanisms in cellular metabolism, growth, and differentiation. Cold Spring Harb. Symp. Quant. Biol. *26*, 389–401.

Morelli, M.J., Ten Wolde, P.R., and Allen, R.J. (2009). DNA looping provides stability and robustness to the bacteriophage λ switch. Proc. Natl. Acad. Sci. U.S.A. *106*, 8101 –8106.

Nelson, E.J., Harris, J.B., Glenn Morris, J., Calderwood, S.B., and Camilli, A. (2009). Cholera transmission: the host, pathogen and bacteriophage dynamic. Nat. Rev. Microbiol. 7, 693–702.

Nielsen, A.T., Dolganov, N.A., Rasmussen, T., Otto, G., Miller, M.C., Felt, S.A., Torreilles, S., and Schoolnik, G.K. (2010). A bistable switch and anatomical site control *Vibrio cholerae* virulence gene expression in the intestine. PLoS Pathog. *6*, e1001102.

Ninfa, A.J., Mayo, A.E. (2004). Hysteresis vs graded responses: the connections make all the difference. Sci. STKE 2004:e20

Novick, A., and Weiner, M. (1957). Enzyme induction as an all-or-none phenomenon. Proc. Natl. Acad. Sci. U.S.A. *43*, 553–566.

Ochab-Marcinek, A. (2010). Extrinsic noise passing through a Michaelis–Menten reaction: A universal response of a genetic switch. J. Theor. Biol. *263*, 510–520.

Olofsson, H., Ripa, J., and Jonzen, N. (2009). Bet-hedging as an evolutionary game: the trade-off between egg size and number. Proc. Biol. Sci. *276*, 2963–2969.

Paulsson, J. (2004). Summing up the noise in gene networks. Nature *427*, 415–418.

Penno, C., Hachani, A., Biskri, L., Sansonetti, P., Allaoui, A., and Parsot, C. (2006). Transcriptional slippage controls production of type III secretion apparatus components in *Shigella flexneri*. Mol. Microbiol. *62*, 1460–1468.

Plimpton, S.J., and Slepoy, A. (2003). ChemCell : a particle-based model of protein chemistry and diffusion in microbial cells. Sandia National Laboratories Technical Report, 2003–4509.

Raj, A., Peskin, C.S., Tranchina, D., Vargas, D.Y., and Tyagi, S. (2006). Stochastic mRNA synthesis in mammalian cells. Plos Biol *4*, e309.

Ramsey, S., Orrell, D., and Bolouri, H. (2005). Dizzy: stochastic simulation of large-scale genetic regulatory networks. J. Bioinform. Comput. Biol. *3*, 415–436.

Robert, L., Paul, G., Chen, Y., Taddei, F., Baigl, D., and Lindner, A.B. (2010). Pre-dispositions and epigenetic inheritance in the *Escherichia coli* lactose operon bistable switch. Mol. Syst. Biol. *6*, 357.

Rosenfeld, N., Young, J.W., Alon, U., Swain, P.S., and Elowitz, M.B. (2005). Gene regulation at the single-cell level. Science's STKE *307*, 1962–1965.

Santillán, M., and Mackey, M.C. (2008). Quantitative approaches to the study of bistability in the lac operon of *Escherichia coli*. J. R. Soc. Interface *5*, S29.

Santos, S.D.M., and Ferrell, J.E. (2008). Systems biology: On the cell cycle and its switches. Nature *454*, 288–289.

Schubert, R.A., Dodd, I.B., Egan, J.B., and Shearwin, K.E. (2007). Cro's role in the CI–Cro bistable switch is critical for λ's transition from lysogeny to lytic development. Genes Dev. *21*, 2461 –2472.

Schultz, D., Wolynes, P.G., Jacob, E.B., and Onuchic, J.N. (2009). Deciding fate in adverse times: Sporulation and competence in *Bacillus subtilis*. Proc. Natl. Acad. Sci. U.S.A. *106*, 21027 –21034.

Singh, A., and Weinberger, L.S. (2009). Stochastic gene expression as a molecular switch for viral latency. Curr. Opin. Microbiol. *12*, 460–466.

Smits, W.K., Eschevins, C.C., Susanna, K.A., Bron, S., Kuipers, O.P., and Hamoen, L.W. (2005). Stripping Bacillus: ComK auto-stimulation is responsible for the bistable response in competence development. Mol. Microbiol. *56*, 604–614.

Smits, W.K., Kuipers, O.P., and Veening, J.-W. (2006). Phenotypic variation in bacteria: the role of feedback regulation. Nat. Rev. Micro. *4*, 259–271.

Smits, W.K., Bongiorni, C., Veening, J.-W., Hamoen, L.W., Kuipers, O.P., and Perego, M. (2007). Temporal separation of distinct differentiation pathways by a dual specificity Rap-Phr system in *Bacillus subtilis*. Mol. Microbiol. *65*, 103–120.

St-Pierre, F., and Endy, D. (2008). Determination of cell fate selection during phage lambda infection. Proc. Natl. Acad. Sci. U.S.A. *105*, 20705 –20710.

Süel, G.M., Garcia-Ojalvo, J., Liberman, L.M., and Elowitz, M.B. (2006). An excitable gene regulatory circuit induces transient cellular differentiation. Nature *440*, 545–550.

Sureka, K., Ghosh, B., Dasgupta, A., Basu, J., Kundu, M., and Bose, I. (2008). Positive feedback and noise activate the stringent response regulator Rel in Mycobacteria. PLoS ONE 3, e1771.

Swain, P.S., Elowitz, M.B., and Siggia, E.D. (2002). Intrinsic and extrinsic contributions to stochasticity in gene expression. Proc. Natl. Acad. Sci. U.S.A. *99*, 12795 –12800.

Swint-Kruse, L., and Matthews, K.S. (2009). Allostery in the LacI/GalR family: variations on a theme. Curr. Opin. Microbiol. *12*, 129–137.

Takahashi, K., Kaizu, K., Hu, B., and Tomita, M. (2004). A multi-algorithm, multi-timescale method for cell simulation. Bioinformatics *20*, 538 –546.

Takano, E. (2006). γ-Butyrolactones: Streptomyces signalling molecules regulating antibiotic production and differentiation. Curr. Opin. Microbiol. *9*, 287–294.

Takano, E., Nihira, T., Hara, Y., Jones, J.J., Gershater, C.J.L., Yamada, Y., and Bibb, M. (2000). Purification and structural determination of SCB1, a γ-Butyrolactone that elicits antibiotic production in *Streptomyces coelicolor* A3 (2). J. Biol. Chem. *275*, 11010–11016.

Tamas, I., Wernegreen, J.J., Nystedt, B., Kauppinen, S.N., Darby, A.C., Gomez-Valero, L., Lundin, D., Poole, A.M., and Andersson, S.G.E. (2008). Endosymbiont gene functions impaired and rescued by polymerase infidelity at poly (A) tracts. Proc. Natl. Acad. Sci. U.S.A. *105*, 14934–14939.

Taniguchi, Y., Choi, P.J., Li, G.-W., Chen, H., Babu, M., Hearn, J., Emili, A., and Xie, X.S. (2010). Quantifying *E. coli* proteome and transcriptome with single-molecule sensitivity in single cells. Science *329*, 533–538.

Thattai, M., and Van Oudenaarden, A. (2004). Stochastic gene expression in fluctuating environments. Genetics *167*, 523–530.

Tiwari, A., Balázsi, G., Gennaro, M.L., and Igoshin, O.A. (2010). The interplay of multiple feedback loops with post-translational kinetics results in bistability of mycobacterial stress response. Phys. Biol. 7, 036005.

Tiwari, A., Ray, J.C.J., Narula, J., and Igoshin, O.A. (2011). Bistable responses in bacterial genetic networks: Designs and dynamical consequences. Math. Biosci. *231*, 76–89.

To, T.-L., and Maheshri, N. (2010). Noise can induce bimodality in positive transcriptional feedback loops without bistability. Science 327, 1142 –1145.

Tripathi, T., and Chowdhury, D. (2008). Transcriptional bursts: A unified model of machines and mechanisms. Europhys. Lett. *84*, 68004.

Tyagi, S. (2010). *E. coli*, what a noisy bug. Science *329*, 518 –519.

Ullah, M., and Wolkenhauer, O. (2007). Family tree of Markov models in systems biology. IET Syst. Biol. *1*, 247–254.

Van Hijum, S.A.F.T., Medema, M.H., and Kuipers, O.P. (2009). Mechanisms and evolution of control logic in prokaryotic transcriptional regulation. Microbiol. Mol. Biol. Rev. 73, 481–509.

Van Hoek, M., and Hogeweg, P. (2007). The effect of stochasticity on the lac operon: an evolutionary perspective. PLoS Comput. Biol. *3*, e111.

Van Hoek, M.J.A., and Hogeweg, P. (2006). In silico evolved lac operons exhibit bistability for artificial inducers, but not for lactose. Biophys. J. *91*, 2833–2843.

Van der Woude, M.W. (2011). Phase variation: how to create and coordinate population diversity. Curr. Opin. Microbiol. *14*, 205–211.

Van Zon, J.S., and Ten Wolde, P.R. (2005). Simulating biochemical networks at the particle level and in time and space: Green's function reaction dynamics. Phys. Rev. Lett. *94*, 128103–128107.

Van Zon, J.S., Morelli, M.J., Tănase-Nicola, S., and Ten Wolde, P.R. (2006). Diffusion of transcription factors can drastically enhance the noise in gene expression. Biophys. J. *91*, 4350–4367.

Veening, J.W., Kuipers, O.P., Brul, S., Hellingwerf, K.J., and Kort, R. (2006). Effects of phosphorelay perturbations on architecture, sporulation, and spore resistance in biofilms of *Bacillus subtilis*. J. Bacteriol. *188*, 3099–3109.

Veening, J.-W., Smits, W.K., and Kuipers, O.P. (2008). Bistability, epigenetics, and bet-bedging in bacteria. Annu. Rev. Microbiol. *62*, 193–210.

Vlamakis, H., Aguilar, C., Losick, R., and Kolter, R. (2008). Control of cell fate by the formation of an architecturally complex bacterial community. Science's STKE 22, 945.

Yamaguchi, A., Adachi, K., Akasaka, T., Ono, N., and Sawai, T. (1991). Metal-tetracycline/H+ antiporter of *Escherichia coli* encoded by a transposon Tn10. Histidine 257 plays an essential role in H+ translocation. J. Biol. Chem. *266*, 6045–6051.

Youk, H., Raj, A., and Van Oudenaarden, A. (2010). Imaging single mRNA molecules in yeast. Methods Enzymol. *470*, 429–446.

Yu, J. (2006). Probing gene expression in live cells, one protein molecule at a time. Science *311*, 1600–1603.

Zahrt, T.C., and Deretic, V. (2001). Mycobacterium tuberculosis signal transduction system required for persistent infections. Proc. Natl. Acad. Sci. U.S.A. *98*, 12706–12711.

Zeng, L., Skinner, S.O., Zong, C., Sippy, J., Feiss, M., and Golding, I. (2010). Decision making at a subcellular level determines the outcome of bacteriophage infection. Cell *141*, 682–691.

Zenklusen, D., Larson, D.R., and Singer, R.H. (2008). Single-RNA counting reveals alternative modes of gene expression in yeast. Nat. Struct. Mol. Biol. *15*, 1263–1271.

Zhang, J., Yuan, Z., Li, H.-X., and Zhou, T. (2010). Architecture-dependent robustness and bistability in a class of genetic circuits. Biophys. J. *99*, 1034–1042.

Zhou, H.-X., Rivas, G., and Minton, A.P. (2008). Macromolecular crowding and confinement: biochemical, biophysical, and potential physiological consequences. Annu. Rev. Biophys. 37, 375–397.

Evolution of DNA-binding Transcription Factors and Regulatory Networks in Prokaryotes

13

Ernesto Perez-Rueda, Nancy Rivera-Gomez, Mario Alberto Martinez-Nuñez and Silvia Tenorio-Salgado

Abstract

The capabilities of organisms to contend with environmental changes depend on their repertoire of genes and their ability to regulate their expression. DNA-binding transcription factors have a fundamental role in this process, because they regulate transcription positively or negatively as a consequence of environmental signals. In this chapter we briefly describe some of the most recent findings on regulatory network evolution from the perspective of DNA-binding transcription factors. We explore diverse elements associated with the evolution of regulatory networks, such as gene duplication, where new interactions can emerge together with their upstream and downstream binding sites. The chapter is divided into sections covering the evolution of transcription factors and their domains, their evolution, and a global analysis. Hypotheses concerning a comprehensive picture of how regulatory networks have evolved in prokaryotes and the role of transcription factors in this organization are discussed.

Introduction

Brian Goodwin has suggested that organisms are more than the sum of their parts. In this regard, we cannot infer the phenotype of one organism by knowing the genes associated with it, because a change in a single gene is not enough to cause a change in the complete phenotype (Goodwin, 1994). Therefore, if we could obtain such information, we could understand how the structure is made. In this direction, organismal development is intimately related to genetic regulation since, for example, organisms or metabolic responses require the concerted action of many regulatory proteins. von Hippel (1998) described in an elegant manuscript that 'regulatory mechanisms developed in all organisms appear to be almost infinite in number, but the basic principles on which they operate are relatively few.' It is plausible that the regulatory elements described in all the organisms change depending on their context; however, they will act in a similar fashion to allow or block gene expression.

In all organisms, it is well known that gene regulation occurs predominantly at the level of transcription initiation, and transcription factors (TFs) play an important role, because they determine when a gene is expressed or repressed, according to the environmental conditions (Martinez-Antonio *et al.*, 2006). Given the importance of this kind of proteins, many authors have evaluated their presence and abundance in diverse organisms. From these studies, it has been observed that the number of TFs increases from a few hundred in archaea and bacteria, such as *Pyrococcus horikoshii*, *Bacillus subtilis* and/or *Escherichia coli* K12, to over 3000 in *Homo sapiens* (Levine and Tjian, 2003; Perez-Rueda *et al.*, 2004; Perez-Rueda and Janga, 2010). This increment correlates with the hypothesis of genome maturation, (Lane and Martin, 2010) where it is proposed that it is necessary for a greater number of regulatory elements to regulate a greater number of genes. Consequently, the number of genetic circuits or regulatory networks that arises also increases (Bhardwaj *et al.*, 2010). Therefore, minor changes in single genes may propagate along such networks and may produce, in the end, quite drastic effects on gene expression in response to external stimuli and change related to development. But how do these changes occur, considering that cellular differentiation, for example, sporulation, requires the concerted interplay between sigma factors, TFs, and their binding sites?

In this chapter we summarize some of the most recent insights from studies on regulatory network evolution from the perspective of DNA-binding TFs, considering that the evolution of regulatory networks requires at least two main mechanisms, gene duplication and gene transfer. Thus, new interactions may emerge together with their upstream and downstream binding sites. We break the subject into sections, covering the evolution of TFs and their domains, the promoters, their evolution, and a global analysis. We finish with some conjectures that attempt to provide a comprehensive picture about how regulatory networks have evolved in prokaryotes and the role of TFs in this organization.

Elements involved in the regulatory process

The regulation of transcription initiation in bacteria is primarily mediated by sigma (σ) factors, which provide most of the specificity for the promoter recognition and DNA melting needed for transcription initiation (Gruber and Gross, 2003; Ishihama, 2000; Wosten, 1998). Indeed, σ factors perform these functions only when bound to the RNA polymerase (RNAP). On the other hand, DNA-binding TFs (Browning and Busby, 2004) affect gene expression, in a wider context, by blocking or allowing the access of the RNAP to the promoter, depending on the operator context and ligand-binding status (Wall *et al.*, 2004; Martinez-Antonio *et al.*, 2006; Miroslavova and Busby, 2006; Janga and Collado-Vides, 2007). Usually, most of the gene transcription in exponentially growing bacteria is initiated by the RNAP carrying a housekeeping σ factor, similar to the *E. coli* σ^{70} or *B. subtilis* σ^{A}. Alternative σ factors typically redirect the transcriptional machinery or RNAP towards a subset of genes required under specific conditions, such as the stress response or growth transitions, among others (Wosten, 1998; Ishihama, 2000; Gruber and Gross, 2003). TFs represent a class of proteins devoted to sensing and binding signals to regulate the response to specific compounds (Martinez-Antonio *et al.*, 2006; Goelzer *et al.*, 2008). In Fig. 13.1 we present a simple regulatory network composed of at least three basic components:

1. the DNA-binding TF, which can be self-regulated;
2. the regulatory region on the DNA, where the TF binds and by which the transcription start is modulated;

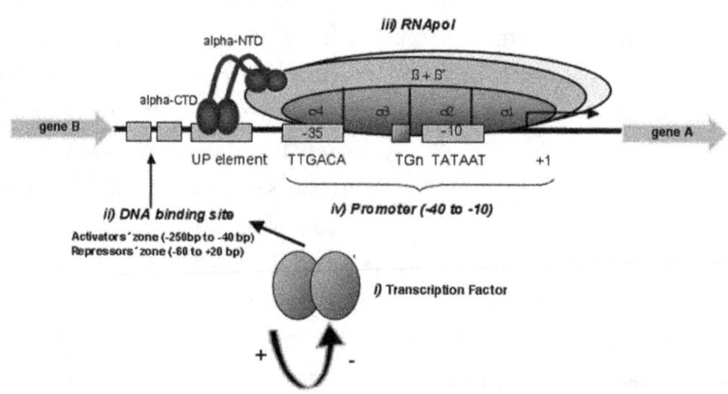

Figure 13.1 Schematic drawing of the basic construction principle of an elementary unit of a hypothetical genetic regulatory network. (i) DNA-binding TF. This protein can activate or repress gene expression. In addition, it can be positive or negatively self-regulated. (ii) A DNA-binding site, which usually is located between the −60 and +20 positions relative to the transcription start. (iii) An RNAP that consists of a protein complex necessary to start the mRNA synthesis. (iv) A sequence promoter the RNAP recognizes specifically over DNA to start the mRNA synthesis.

3 the promoter-binding site, where RNAP binding starts RNA synthesis.

As we will discuss in detail later in this chapter, these links are mostly 'one-to-many,' that is, each TF regulates more than one gene and most genes are controlled by some, albeit generally few, TFs. Every TF is itself regulated by another one. This combinatory perspective gives rise to regulatory networks. Each of these elements can be broken down into regulons. For example, the arabinose regulons in *E. coli* K12, composed of more than 10 different genes involved in the assimilation of arabinose is regulated by two different transcription factors, Crp and AraC (Salgado *et al.*, 2006), where Crp can be associated with a plethora of additional functions. The TF itself typically contains a DNA-binding domain and other regulatory domains, such as multimerization domains that mediate interactions with other proteins or metabolites in order to react to physiological or environmental changes.

Promoter and regulatory region

Transcription starts when the σ factor interacts with the RNAP to recognize its specific sequence promoter (Fig. 13.1). This promoter recognition stage imposes the existence of at least one σ factor per organism, which typically belongs to the σ^{70} family in bacteria (Paget and Helmann, 2003). In this context, bacterial systems could switch between different transcriptional programmes based exclusively on their repertoire of σ factors. Nonetheless, the transcriptional programmes mediated solely via σ factors would be restricted, as a result of their limited repertoire and the small collection of ligands they can recognize, such as guanosine tetraphosphate (ppGpp) (Jores and Wagner, 2003). As a consequence, σ factors exhibit a restricted ability for directly coupling responses to environmental conditions with gene transcription. In addition, σ factors have a constrained DNA-binding region in terms of the lengths and diversity of sequences they recognize, as they need to be structurally coupled to the RNAP on the promoter zone. These DNA restricted zones of action divide the universe of σ factor families into promoters recognized by σ^{70} family and those recognized by σ^{54} family, for instance, the binding zones correspond to about bp −10 to −35 for σ^{70} and bp −12 to −24 for σ^{54}, relative to the transcription start site in the bacterium *E. coli K12* (Gralla, 1996; Lloyd *et al.*, 2001).

On the other hand, TFs define a different regulatory level than do σ factors. These proteins exhibit diverse structural and functional domains, with one of them associated with binding DNA specifically, whereas the second one is devoted to sensing and binding one or more signals from endogenous and/or exogenous sources (Martinez-Antonio *et al.*, 2006). For example, *E. coli* K-12 TyrR binds to three aromatic amino acids and ATP (Pittard *et al.*, 2005); TrmB of the archaeon *Pyrococcus furiosus* binds to three different compounds (Lee *et al.*, 2003, 2005; Perez-Rueda and Janga, 2010). In addition, TFs have the ability to associate combinatorially not only with σ factors but also with a number of other TFs and DNA-binding sites (Adhya, 2003; Barnard *et al.*, 2004), thus allowing the rewiring of a transcriptional network depending on the environmental conditions. For instance, *sodA*, a gene encoding superoxide dismutase in *E. coli*, is regulated by up to eight different TFs responsible for various cellular responses, including Fur (ferric uptake regulation protein), Arc (aerobic respiratory control), and Fnr (fumarate nitrate reduction/regulator of anaerobic respiration) (Compan and Touati, 1993; Salgado *et al.*, 2006). Therefore, the diversity of sequences recognized by TFs is enormous and can occur anywhere from a few bases downstream of the promoter zone to up to hundreds of bases upstream of the transcription start site (Fig. 13.1) (Collado-Vides *et al.*, 1991; Madan Babu and Teichmann, 2003b). For instance, the *E. coli* K-12 global regulator CRP (catabolic repressor protein) can associate with four of the six possible σ factors and coregulate more than 50 different TFs (Salgado *et al.*, 2006). CcpA in *B. subtilis*, a global regulatory protein involved in catabolite repression, may act as a positive regulator of genes involved in excretion of carbon excess and can associate with three different sigma factors (σ^A, σ^L and σ^E) and more than 10 different TFs (Makita *et al.*, 2004; Moreno-Campuzano *et al.*, 2006). In summary, TFs constitute a class of proteins whose space of action is more flexible than σ factors, not only in sensing diverse environmental and endogenous stimuli but also in recognizing a wide range of binding-site sequences over a larger zone on the DNA around the transcription start site.

Evolution of the repertoire of TFs

DNA-binding domains

The structures of more than 30 prokaryotic DNA-binding proteins have now been determined, and hundreds of amino acid sequences are known for many more. In general, the DNA-binding domains associated with TFs are among the most ancient domains, and they have been proposed as derived from a relatively small set of folds (Aravind and Koonin, 1999; Perez-Rueda and Collado-Vides, 2001; Madan Babu and Teichmann, 2003). These domains have been used to classify the TFs in terms of families (Perez-Rueda *et al.*, 2004). From these studies diverse principles have emerged regarding TFs; for example, the most abundant DNA-binding domain in prokaryotes is the helix–turn–helix (HTH), identified in more than 80% of TFs (Perez-Rueda and Collado-Vides, 2001; Perez-Rueda and Janga, 2011) (Fig. 13.2). Most of the archaeal and bacterial HTHs appear to have undergone a common general evolutionary pathway. The HTH might then have been a motif with general nucleic acid-binding functions that appeared early in evolution (Aravind and Koonin, 1999; Roy *et al.*, 2002) and whose subsequent radiation in the archaeal and bacterial lineages might have involved considerable loss and acquisition events. Additionally, lineage-specific duplications resulted in the accumulation of particular families in microbial species, such as the LysR family, whose members have been abundantly identified in almost all organisms. This hypothesis is consistent with the notion that a genome evolves from a set of precursor genes to a mature size by gene duplications and increasing modifications (Yanai *et al.*, 2000). Alternative DNA-binding structures, such as helix–loop–helix motifs, zinc-fingers, and σ-sheet DNA-binding structures, have been also identified, although in lower proportions, and their distributions are constrained to specific organisms. For instance, the β-sheet proteins have been identified almost exclusively in *Gammaproteobacteria*. The distribution of the RNA-binding domains (associated with cold shock proteins) suggests that they might have been acquired after the prokaryotes and eukaryotes split, probably by lateral gene transfer from eukaryotes, based on the high diversity identified in this cellular domain.

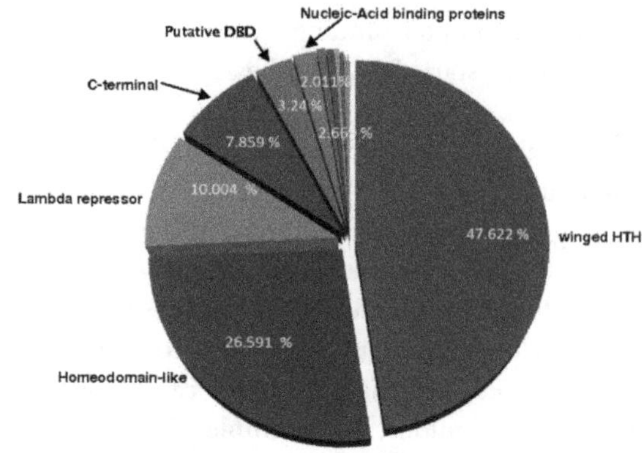

Figure 13.2 Distribution of DNA-binding domains of TFs in bacteria and archaea as defined in the Superfamily database (Madera *et al.*, 2004). The winged HTH represents 47.6% of the total repertoire of DNA-binding domains, being the most abundant structure. The lambda-repressor DNA-binding domain is present at the second highest abundance, as 26% of the repertoire. In minor proportions occur alternative DNA-binding domains, such as the homeodomain-like, with 10.0%, the C-terminal effector domain of the bipartite response regulators, at 7.8%, the putative DNA-binding domain, at 3.24%, and the nucleic acid-binding proteins, at 2.01%. In a low fraction, corresponding to 2.6% of the total DNA-binding domains, are AbrB/MazE/MraZ-like, KorB DNA-binding domain-like, TrpR-like, ACT-like, flagellar transcriptional activator FlhD, haemolysin expression-modulating protein H, a DNA-binding domain in eukaryotic TFs, DNA-binding domain, p53-like TFs, and the DNA-binding domain of the Mlu1 box-binding protein MBP1.

Abundance of TFs correlates with genome size in prokaryotes

It has been documented that organisms respond and adapt to diverse environmental conditions as a consequence of their gene repertoire and regulatory mechanisms, among other elements (Lynch and Conery, 2003; Bengtsson, 2004; Lynch, 2006). Recent studies have shown that the evolutionary events associated with regulatory proteins, such as their expansion and contraction, contribute significantly in shaping the gene repertoire and genome size of the different lineages of prokaryotes (Perez-Rueda *et al.*, 2004; Minezaki *et al.*, 2005; Oguiza *et al.*, 2005; Rodionov, 2007). Based on comparative genomics, it has been shown that transcription factors increase in quadratic proportion with respect to genome size (van Nimwegen, 2003; Cordero and Hogeweg, 2007; Molina and van Nimwegen, 2008). In particular, this proportion is more significant when the repertoire of TFs is compared with the proportion of σ factors, being roughly ten times higher (hundreds of TFs vs. tens of σ factors) when the general profiles in all the genomes analysed are considered, suggesting a proportion on the order of 1 σ factor:10 TFs:100 annotated open reading frames per genome, although some genomes behave exceptionally. This observation suggests that a possible functional relationship between TFs and prokaryote lifestyles influences the observed trend. A plausible hypothesis is that the abundance of TFs increases with an increase in an organisms' complexity (Brown *et al.*, 2002; Changizi, 2001; Levine and Tjian, 2003; van Nimwegen, 2003; West and Brown, 2005) as a consequence of different evolutionary events, such as gene expansion, gene loss, and lateral gene transfer, among others (Levine and Tjian, 2003; Aravind *et al.*, 2005; Madan Babu *et al.*, 2006). In addition, the necessity to regulate responses to variable environments could also contribute to the abundance of TFs.

Lifestyles explain the abundance of σ factors and TFs in larger genomes

In previous sections we suggested that regulatory complexity should increase in larger genomes and might be associated with bacterial lifestyles, as the environment should influence the bacterial genome structure and function. Thus, to understand how the complexity of gene regulation depends on the number of TFs as a function of increasing genome size and how they are associated with the lifestyles, in previous work (Perez-Rueda *et al.*, 2009) we classified in four global lifestyle classes all the bacterial organisms. These included extremophiles, intracellular bacteria, pathogens, and free-living bacteria. From this analysis, it was identified that the increment of regulatory complexity in TFs contributes significantly to the regulatory complexity of prokaryotes belonging to different lifestyle groups. These results agree with previous observations that suggest that a few regulatory elements identified in small genomes would compensate for the regulation of the entire genome with an increase in the number of DNA-binding sites per element, in contrast to the large number of elements identified in large genomes, which control a smaller proportion of DNA-binding sites on average (Molina and van Nimwegen, 2008). In addition, a larger proportion of genes in small genomes are organized in operons, simplifying the transcriptional machinery necessary for gene expression, in contrast to large genomes, which have reduced number of genes in operons, which would influence the proportion of TFs in those organisms (Cherry, 2003), suggesting that complex lifestyles require a higher proportion of TFs and transcription units to better orchestrate a response to changing conditions.

Abundance of TFs does not correlate with diversity of families, and large families are not the most widely distributed

An appealing hypothesis is that the high diversity of TF families contributes significantly to the regulatory plasticity. In line with this hypothesis, the repertoire of TFs identified in bacteria has been classified into families to evaluate their distribution and abundance in all the prokaryotes. This analysis showed a reduced diversity of families in small genomes, with an increasing proportion in larger ones, especially in pathogens and free-living organisms. The diversity of families reaches a maximum in genomes with around 5000 open reading frames. The higher number of TFs in larger genomes does not necessarily imply diversity of families beyond this plateau, but instead an increase in the size of some families of TFs. Congruent with this observation, the average number of TFs per family increases linearly, with a few families of TFs expanding disproportionately (Janga and Perez-Rueda, 2009; Perez-Rueda *et al.*, 2009). These families comprise LysR and TetR, which represent about 25% of the total set of TFs identified. Members of these two families increase abruptly in larger genomes and coincide with the plateauing of the diversity of families in these bacterial genomes. Another feature associated with large families is that they are not widely distributed among bacteria despite their role in controlling important processes, such as cell–cell communication (LuxR), response to external conditions by two-component systems (OmpR), sensing, uptake, and metabolism of external food sources (GntR and LysR), and antibiotic resistance (TetR). Alternatively, families with few copies per genome, such as DnaA, LexA, and IHF, which have been proposed to be essential under standard growth conditions in *E. coli* and in maintaining DNA and nucleoid integrity, (Gerdes *et al.*, 2003; Yamazaki *et al.*, 2008) might be considered universal in bacteria, because they have been identified in at least 80% of the genomes, suggesting gene loss events in bacteria in which they are absent.

In summary, a TF family's abundance and distribution should be associated with the following evolutionary events in bacteria: (i) small families widely distributed among bacteria might be related to ancestral functions beyond transcriptional regulation, such as DNA organization, nucleoid integrity, or DNA salvage; (ii) large families might be associated with the regulation of dispensable or emergent processes in bacterial evolution, such as those involved in quorum sensing, belonging to the members of the LuxR family, which are widely identified in bacteria. Indeed, the evolution of this mechanism in bacteria has been proposed to be one of the early steps in the development of multicellularity (Miller and Bassler, 2001) and may be correlated with bacterial specialization.

Evolution of partner domains

DNA-binding TFs usually make contact with their DNA targets as homodimers or homotetramers, as is the case for the lactose repressor (Lewis, 2005). In this regard, there are diverse questions concerning whether multimerization precedes DNA binding. Therefore, the TFs can act as activators or repressors as a consequence of their multimerization state. Amoutzias *et al.* (2004) concluded that the ancestral TFs were probably the homodimerizing ones and that these proliferated through a series of single-gene duplications. Recent evidence supports this hypothesis, as investigators have described a high abundance of small-sized TFs, or proteins that contain a dimerization domain, but no DNA-binding domain in archaeal genomes (Perez-Rueda and Janga, 2011). The main consequence of gene duplications, mainly involving TFs, is to give rise to a complex interaction network. To the best of our knowledge, there are no studies so far that have linked genetic networks and their regulation with other networks, such as metabolism, although many effector domains are protein interaction domains. These data could help in understanding how the ligand-binding domains have been recruited to regulate gene expression. However, previous studies (Madan Babu and Teichmann, 2003; Aravind *et al.*, 2005) suggested that in the winged HTH superfamily of TFs, the partner domains contribute to the structural differentiation of duplicated genes. Therefore, the partner domains are associated with diverse functions, such as regulating allosterically the function of TFs across binding to a wide variety of functional compounds, in protein–protein interactions, or with enzymatic properties (Madan Babu and Teichmann, 2003), and they are fundamental to linking environmental conditions and the functional conformational changes in the regulators (Taraban *et al.*, 2008). Because there are few exhaustive analyses describing the partner domain repertoire in bacteria, their functional and evolutionary diversity must be evaluated (Madan Babu and Teichmann, 2003; Rivera-Gomez *et al.*, 2011). From this perspective, one might expect that the high diversity of TF families and their associated partner domains contributes significantly to the regulatory plasticity, as we previously mentioned; however, further studies are necessary. Thus, the diversity associated with dimerization domains has allowed the TFs to interact with many different partners and achieve specificity for the target gene and physiological conditions. On the other hand, multimerization domains may have been facilitated by the acquisition of a second protein interaction domain during evolution (Kaufmann *et al.*, 2005).

Evolution of promoters

Unlike coding regions, the evolution of upstream regulatory regions cannot be easily evaluated by means of sequence alignments and comparisons. Indeed, the structure, organization, and function of promoters differ widely from that of the coding sequences, resulting in totally different evolution rates and consequences of sequence variation (Rodriguez-Trelles *et al.*, 2003). In general, promoter organization underlies more variation than the coding sequences, and it is mainly influenced by DNA structure (Olivares-Zavaleta *et al.*, 2006) such as supercoiling (Martinez-Nunez *et al.*, 2010). Bacterial promoter sequences, or upstream regulatory regions, contain TF-binding sites. Several TFs often interact, depending on signals and physiological or environmental requirements. This gives rise to activating or repressing complexes (Koike *et al.*, 2004). Ishihama described that among the set of promoters under the control of the same σ factor, the level of transcription varies depending on the culture conditions or the growth phase. For that reason, it is important to evaluate the diversity of promoters and their regulation associated with all genes in an organism, for instance, in *E. coli* K-12, approximately 50% of the promoters are under the control of one specific regulator, whereas the other 50% of genes are regulated by more than two TFs. Likewise, the promoters for the genes involved in the construction of cell structures are controlled by environmental conditions and specific signals, and each is in turn monitored by a different TF. The binding sites of all these multiple factors are located in a single promoter. The most typical examples of the multifactor promoter system are the promoters for the genes encoding the master regulator for flagellum formation in *E. coli* K12, FlhCD, and the master regulator for biofilm formation, CsgD. The complexities of these promoters reflect the opposite behaviours of bacterial survival, i.e. planktonic growth as single cells (FlhCD) and biofilm formation as a bacterial community under stressful conditions in nature (Soutourina *et al.*, 1999; Ogasawara *et al.*, 2010).

Coevolution of regulatory elements

Since some proteins tend to work together in a functional context, analyses of distributions of different families in the function of the σ factors have been recently performed. Hence, the cooccurrence of the regulatory protein families (TFs and σ factors) in all the prokaryotes was evaluated. From this analysis, it was found that the distribution of the σ^{54} factor and IHF and EBP families are correlated (Fig. 13.3), supporting the functional interdependence discussed above and probable coevolution in which members and mechanisms have been preserved along the course of evolution in bacteria. A second cluster that includes σ^{70}, the ECF family of σ factors, and other highly abundant families (more than 15 members per genome) responsible for regulating diverse mechanisms of stress responses (MarR), antibiotic resistance (TetR), osmotic responses (OmpR), and the quorum-sensing response (LuxR), among other

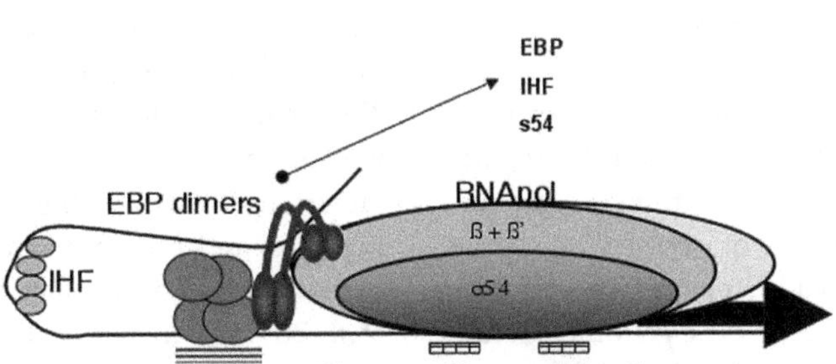

Figure 13.3 Coevolution of σ factors and TF families. A similar occurrence distribution pattern was observed for IHF, EBP, and σ^{54} families, suggesting a functional interdependence between these families. A coregulatory mode of action for these regulatory proteins is also shown. Figure modified from Perez-Rueda *et al.* (2009).

processes, was also found to be clustered, suggesting a strong functional relationship among these σ and TF families. These clusters, in addition, give insights into the functional interdependence between regulatory proteins from different families, which could help in the characterization of regulators in poorly studied organisms.

Evolution of regulatory networks

The regulation of TFs plays a key role in morphological diversity. Simple modifications within the upstream regulation region of a TF can explain both minor and major changes between species, without involving any disruption of gene structure. Therefore, evolution of regulatory regions is thought to be a major source of diversity. (Lozada-Chavez *et al.*, 2008; Perez and Groisman, 2009a,b) Duplication events of TFs are another evolutionary source that can allow diversity, permitting a more versatile adaptation of the functional divergence gained from the duplication of structural genes. Different aspects of the evolution of the regulatory networks have been examined, including the coevolution of the upstream regulatory regions and their corresponding TFs, the likely consequences of gain, loss, and replacement of TFs in the regulatory networks of duplicated genes (Teichmann and Babu, 2004; Gelfand, 2006), and also the topological and dynamic properties of the regulatory networks (Luscombe *et al.*, 2004; Madan Babu *et al.*, 2006; Balaji *et al.*, 2007).

Role of duplication events in regulatory networks

Duplication events of regulatory genes provide new interactions for the transcriptional regulatory network. These new interactions are the raw material for the generation of divergence in gene expression, which should happen in most of the copies that have remained in the genome (Teichmann and Babu, 2004). In this model, the loss and gain of regulatory interactions may occur following the duplication of either a TF or a target gene or following the duplication of both a TF and a target gene (Fig. 13.4). The evolutionary plasticity of the regulatory networks is not only the result of the duplication of TF interactions within a regulatory network, as previously proposed by Teichmann and Babu (2004), but also the result of the divergent effects of the TF interactions in activating or repressing the transcription of duplicated genes, as suggested by Martinez-Nuñez *et al.* (2010). Indeed, examples have been recently identified of regulatory systems where the TF is maintained but a different regulatory role is gained (either activation or repression) in one of the duplicated genes. This evolutionary scenario can be observed in the regulation of the *E. coli* *gntK* and *idnK* gluconate kinase genes, which are involved in 6-phosphogluconate synthesis in the Entner–Doudoroff and pentose phosphate pathways, respectively. Although the same TFs, CRP, GntR, and IdnR, regulate all these genes, IdnR represses the transcription of *gntK*, whereas it activates the transcription of *idnK* (Bausch *et al.*, 2004; Salgado *et al.*, 2006).

This form of diversification in regulation allows plasticity of the transcriptional regulatory network without the need to increase the number of interactions within it, if not only by varying the type of regulation (positive or negative) exerted by the TFs on their targets. It is possible that modulation is one of the first steps towards evolutionary innovation at a biochemical level, perhaps as a step towards the modification of the entire metabolic pathway (Martinez-Nunez *et al.*, 2010).

Concluding remarks

As a consequence of the abundance of data which have become recently available, the idea has emerged to conceptualize genetic networks as directed graphs with nodes corresponding to the TFs, linked by edges to their target genes. The previously mentioned regulatory elements (TFs, σ factors, upstream binding sites, downstream binding sites, and promoters) can be combined at a higher level, into so-called network motifs (Milo *et al.*, 2002; Shen-Orr *et al.*, 2002). The impacts of evolutionary forces on the topological structures of regulatory networks known as motifs have been exhaustively analysed by diverse authors, for example, biological regulatory networks (Milo *et al.*, 2002; Shen-Orr *et al.*, 2002) and duplicated gene networks (Teichmann and Babu, 2004). In

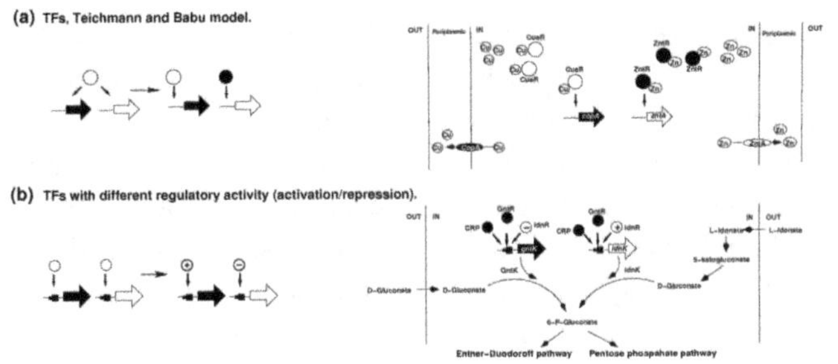

Figure 13.4 Regulatory elements involved in the regulation of the duplicated genes. (a) Simplification of the model of Teichmann and Babu (Teichmann and Babu, 2004), where a new TF is gained to regulate one of the duplicated genes. TFs are shown as circles. (b) Extension of the Teichmann and Babu model. The differential regulation of the duplicated genes depends on the activation or repression mechanism associated with the TF on its target genes. Figure modified from Martínez-Núñez *et al.* (2010).

these studies, the authors reported that duplication of an entirely feed-forward motif (a topological structure in which a TF regulates a second TF and both TFs simultaneously regulate a target gene) has not been observed in the regulatory networks of model organisms, although single genes generated by duplication could be part of a new feed-forward or other kind of motif. For example, in *B. subtilis* the duplicated σ^F and σ^E are part of different feed-forward motifs. In the first case, σ^F forms a feed-forward motif with the anti-anti-σ factor SpoIIAA and the anti-σ factor SpoIIAB, as σ^F regulates SpoIIAA and SpoIIAB expression, whereas SpoIIAA modulates the expression of SpoIIAB. In a similar manner, the second feed-forward loop is formed by σ^E, PhoP, and PhoR, which are involved in phosphate uptake, the post-exponential growth phase, and other stress responses (Pragai *et al.*, 2004). Duplication events, or mutations in the duplicated regulatory genes or in the regulatory target sites, can generate new feed-forward motifs useful for the rewiring of the regulatory networks of the duplicated genes, which would favour the adaptation process of the organism as it responds to changes in its niche (Gelfand, 2006). Probably the most basic motif is the autoregulatory loop: a TF that regulates its own expression (Fig. 13.1).

Generally, these motifs have been considered basic architectures in the regulatory networks, because they often overlap (Browning and Busby, 2004). In this context, diverse authors have analysed the structure of both genetic and protein–protein interaction networks (Yeger-Lotem and Margalit, 2003; Yeger-Lotem *et al.*, 2004). These analyses have shown that certain topologies of small subnets are statistically very much over-represented (Shen-Orr *et al.*, 2002). Conant and Wagner introduced the notion of common ancestry for gene circuits or motifs, where two motifs share a common ancestor if every pair of genes in the two circuits is derived from a common ancestor; all pairs in the circuits must be duplicated genes (Conant and Wagner, 2003). They found that no pairs of motifs with identical topology had common ancestry, and they concluded that their emergence is the result of convergent evolution and not duplication of one or a few ancestral circuits, suggesting that convergent evolution was more likely to be important in module topology than for protein sequences.

A third level of network organization consists of transcriptional modules (Babu *et al.*, 2004). Modules represent collections of TFs that are expressed under distinct experimental or environmental conditions (Ihmels *et al.*, 2002) and are largely controlled by one (or very few) regulators, as was shown by hierarchical clustering of expression profiles (Segal *et al.*, 2003). For instance, the *E. coli* K-12 global regulator CRP can regulate the expression of more than 20 different TFs (Thieffry *et al.*, 1998; Gama-Castro *et al.*, 2008). Exhaustive analyses of global features of genetic networks and protein interaction networks have revealed a scale-free topology (Barabasi and Albert, 1999; Barabasi and Oltvai, 2004); in other words, there are few genes, or so-called hubs, that control many others, and many genes have only a few links. These hubs can be defined as global regulators, such as Crp *in E. coli* K-12. However, regulatory networks are dynamic: it was shown that large-scale topological changes have occurred in the *E. coli, B. subtilis,* and *Saccharomyces cerevisiae* genomes and that although a few TFs serve as permanent hubs, they act transiently only under certain conditions (Luscombe *et al.,* 2004). It is also worth noting that more complex organisms have a higher number of regulatory genes per target gene (van Nimwegen, 2003) suggesting that it is mostly the evolution by duplication and diversification of transcription factors and of their interactions that increases organismic complexity as a whole. Over the last few years, the availability of large numbers of experimental and theoretical data has significantly enhanced our understanding of evolution of complex networks and, at the same time, enabled us to transfer knowledge from better-studied model organisms (such as *E. coli* and *B. subtilis*) to those for which fewer data are available. Basically, the ancestral genetic networks we observe today were probably a small group of DNA-binding domains that, while conserving their structure, diverged into a large variety of TFs. More recently, most proteins, among them TFs, underwent many cycles of domain rearrangements (Amoutzias *et al.,* 2005). Additional dimerization and sensor domains were gained and lost at different times. Further, they evolved across a series of single-gene duplications, thus generating networks of regulatory genes that arrange into these modules. These events may be quite recent and lineage specific, as we have learned from the uneven distribution of some TF families (Perez-Rueda *et al.,* 2004). A growing number of findings suggest that structurally similar or even identical motifs can arise repeatedly and thus represent a simple level of convergent evolution. More complex modules, which may also have preferentially arisen through a series of single-gene duplications, would give rise to similar topologies. The evolution of promoter regions is less well understood, although it is of great importance. Genotypic changes at this level are probably among the main reasons why, despite minor interorganismic differences at the level of proteins, major changes in the topologies of genetic networks during development induce wide morphological differences and diversity in contemporary organisms.

In summary the mechanisms of generating diverse networks can be associated with diverse evolutionary forces, such as gene duplication, gene loss, changes in the regulatory mechanisms (regulatory role modulation), acquisition of new activities, modular rearrangements, and finally, functional divergence. Therefore, we believe that with the availability of more information, we will be able to understand in a more comprehensive fashion the evolutionary dynamics associated with regulatory networks.

Acknowledgements

We gratefully acknowledge to Alejandro Huerta-Saquero, Antonio Ibarra, and the anonymous referee for their helpful comments in the preparation of the manuscript. This work was supported by DGAPA grant IN209511 to EP-R. NR-G, MAM-N, and ST-S were financed by a PhD fellowship from CONACyT.

References

Amoutzias, G.D., Robertson, D.L., and Bornberg-Bauer, E. (2004). The evolution of protein interaction networks in regulatory proteins. Comp. Funct. Genomics *5*, 79–84.

Amoutzias, G.D., Weiner, J., and Bornberg-Bauer, E. (2005). Phylogenetic profiling of protein interaction networks in eukaryotic transcription factors reveals focal proteins being ancestral to hubs. Gene *347*, 247–253.

Aravind, L., and Koonin, E.V. (1999). DNA-binding proteins and evolution of transcription regulation in the archaea. Nucleic Acids Res. *27*, 4658–4670.

Aravind, L., Anantharaman, V., Balaji, S., Babu, M.M., and Iyer, L.M. (2005). The many faces of the helix–turn–helix domain: transcription regulation and beyond. FEMS Microbiol. Rev. *29*, 231–262.

Babu, M.M., Luscombe, N.M., Aravind, L., Gerstein, M., and Teichmann, S.A. (2004). Structure and evolution of transcriptional regulatory networks. Curr. Opin. Struct. Biol. *14*, 283–291.

Balaji, S., Babu, M.M., and Aravind, L. (2007). Interplay between network structures, regulatory modes and sensing mechanisms of transcription factors in the transcriptional regulatory network of *E. coli*. J. Mol. Biol. 372, 1108–1122.

Barabasi, A.L., and Albert, R. (1999). Emergence of scaling in random networks. Science *286*, 509–512.

Barabasi, A.L., and Oltvai, Z.N. (2004). Network biology: understanding the cell's functional organization. Nat. Rev. Genet. *5*, 101–113.

Bausch, C., Ramsey, M., and Conway, T. (2004). Transcriptional organization and regulation of the L-idonic acid pathway (GntII system) in *Escherichia coli*. J. Bacteriol. *186*, 1388–1397.

Bengtsson, B.O. (2004). Modelling the evolution of genomes with integrated external and internal functions. J. Theor. Biol. *231*, 271–278.

Bhardwaj, N., Carson, M.B., Abyzov, A., Yan, K.K., Lu, H., and Gerstein, M.B. (2010). Analysis of combinatorial regulation: scaling of partnerships between regulators with the number of governed targets. PLoS Comput Biol 6, e1000755.

Brown, J.H., Gupta, V.K., Li, B.L., Milne, B.T., Restrepo, C., and West, G.B. (2002). The fractal nature of nature: power laws, ecological complexity and biodiversity. Philos. Trans. R. Soc. Lond. B Biol. Sci. *357*, 619–626.

Browning, D.F., and Busby, S.J. (2004). The regulation of bacterial transcription initiation. Nat. Rev. Microbiol. *2*, 57–65.

Changizi, M.A. (2001). Universal scaling laws for hierarchical complexity in languages, organisms, behaviors and other combinatorial systems. J. Theor. Biol. *211*, 277–295.

Cherry, J.L. (2003). Genome size and operon content. J. Theor. Biol. *221*, 401–410.

Conant, G.C., and Wagner, A. (2003). Asymmetric sequence divergence of duplicate genes. Genome Res. *13*, 2052–2058.

Cordero, O.X., and Hogeweg, P. (2007). Large changes in regulome size herald the main prokaryotic lineages. Trends Genet. *23*, 488–493.

Gama-Castro, S., Jimenez-Jacinto, V., Peralta-Gil, M., Santos-Zavaleta, A., Penaloza-Spinola, M.I., Contreras-Moreira, B., Segura-Salazar, J., Muniz-Rascado, L., Martinez-Flores, I., Salgado, H., *et al.* (2008). RegulonDB (version 6.0): gene regulation model of *Escherichia coli* K-12 beyond transcription, active (experimental) annotated promoters and Textpresso navigation. Nucleic Acids Res. *36*, D120–124.

Gelfand, M.S. (2006). Evolution of transcriptional regulatory networks in microbial genomes. Curr. Opin. Struct. Biol. *16*, 420–429.

Gerdes, S.Y., Scholle, M.D., Campbell, J.W., Balazsi, G., Ravasz, E., Daugherty, M.D., Somera, A.L., Kyrpides, N.C., Anderson, I., Gelfand, M.S., *et al.* (2003). Experimental determination and system level analysis of essential genes in *Escherichia coli* MG1655. J. Bacteriol. *185*, 5673–5684. Goelzer, A., Bekkal Brikci, F., Martin-Verstraete, I., Noirot, P., Bessieres, P., Aymerich, S., and Fromion, V. (2008). Reconstruction and analysis of the genetic and metabolic regulatory networks of the central metabolism of *Bacillus subtilis*. BMC Syst. Biol. *2*, 20.

Goodwin, B. (1994). How the Leopard Changed its Spots: The Evolution of Complexity (Scribner, New York, USA).

Gralla, J.D. (1996). Activation and repression of *E. coli* promoters. Curr. Opin. Genet. Dev. *6*, 526–530.

Gruber, T.M., and Gross, C.A. (2003). Multiple sigma subunits and the partitioning of bacterial transcription space. Annu. Rev. Microbiol. *57*, 441–466.

von Hippel, P.H. (1998). An integrated model of the transcription complex in elongation, termination, and editing. Science *281*, 660–665.

Ihmels, J., Friedlander, G., Bergmann, S., Sarig, O., Ziv, Y., and Barkai, N. (2002). Revealing modular organization in the yeast transcriptional network. Nat. Genet. *31*, 370–377.

Ishihama, A. (2000). Functional modulation of *Escherichia coli* RNA polymerase. Annu. Rev. Microbiol. *54*, 499–518.

Janga, S.C., and Collado-Vides, J. (2007). Structure and evolution of gene regulatory networks in microbial genomes. Res. Microbiol. *158*, 787–794.

Janga, S.C., and Perez-Rueda, E. (2009). Plasticity of transcriptional machinery in bacteria is increased by the repertoire of regulatory families. Comput. Biol. Chem. *33*, 261–268.

Jores, L., and Wagner, R. (2003). Essential steps in the ppGpp-dependent regulation of bacterial ribosomal RNA promoters can be explained by substrate competition. J. Biol. Chem. *278*, 16834–16843.

Kaufmann, K., Melzer, R., and Theissen, G. (2005). MIKC-type MADS-domain proteins: structural modularity, protein interactions and network evolution in land plants. Gene *347*, 183–198.

Koike, H., Ishijima, S.A., Clowney, L., and Suzuki, M. (2004). The archaeal feast/famine regulatory protein: potential roles of its assembly forms for regulating transcription. Proc. Natl. Acad. Sci. U.S.A. *101*, 2840–2845.

Lane, N., and Martin, W. (2010). The energetics of genome complexity. Nature *467*, 929–934.

Levine, M., and Tjian, R. (2003). Transcription regulation and animal diversity. Nature 424, 147–151.

Lewis, M. (2005). The lac repressor. C. R. Biol. *328*, 521–548.

Lloyd, G., Landini, P., and Busby, S. (2001). Activation and repression of transcription initiation in bacteria. Essays Biochem. *37*, 17–31.

Lozada-Chavez, I., Angarica, V.E., Collado-Vides, J., and Contreras-Moreira, B. (2008). The role of DNA-binding specificity in the evolution of bacterial regulatory networks. J. Mol. Biol. *379*, 627–643.

Luscombe, N.M., Babu, M.M., Yu, H., Snyder, M., Teichmann, S.A., and Gerstein, M. (2004). Genomic analysis of regulatory network dynamics reveals large topological changes. Nature *431*, 308–312.

Lynch, M. (2006). Streamlining and simplification of microbial genome architecture. Annu. Rev. Microbiol. *60*, 327–349.

Lynch, M., and Conery, J.S. (2003). The origins of genome complexity. Science *302*, 1401–1404.

Madan Babu, M., and Teichmann, S.A. (2003). Evolution of transcription factors and the gene regulatory network in *Escherichia coli*. Nucleic Acids Res. *31*, 1234–1244.

Madan Babu, M., Teichmann, S.A., and Aravind, L. (2006). Evolutionary dynamics of prokaryotic transcriptional regulatory networks. J. Mol. Biol. *358*, 614–633.

Madera, M., Vogel, C., Kummerfeld, S.K., Chothia, C., and Gough, J. (2004). The SUPERFAMILY database in 2004: additions and improvements. Nucleic Acids Res. 32, D235–239.

Martinez-Antonio, A., Janga, S.C., Salgado, H., and Collado-Vides, J. (2006). Internal-sensing machinery directs the activity of the regulatory network in *Escherichia coli*. Trends Microbiol. *14*, 22–27.

Martinez-Nunez, M.A., Perez-Rueda, E., Gutierrez-Rios, R.M., and Merino, E. (2010). New insights into the regulatory networks of paralogous genes in bacteria. Microbiology 156, 14–22.

Miller, M.B., and Bassler, B.L. (2001). Quorum sensing in bacteria. Annu. Rev. Microbiol. *55*, 165–199.

Milo, R., Shen-Orr, S., Itzkovitz, S., Kashtan, N., Chklovskii, D., and Alon, U. (2002). Network motifs: simple building blocks of complex networks. Science *298*, 824–827.

Minezaki, Y., Homma, K., and Nishikawa, K. (2005). Genome-wide survey of transcription factors in prokaryotes reveals many bacteria-specific families not found in archaea. DNA Res. *12*, 269–280.

Miroslavova, N.S., and Busby, S.J. (2006). Investigations of the modular structure of bacterial promoters. Biochem. Soc. Symp. 1–10.

Molina, N., and van Nimwegen, E. (2008). Universal patterns of purifying selection at noncoding positions in bacteria. Genome Res *18*, 148–160.

van Nimwegen, E. (2003). Scaling laws in the functional content of genomes. Trends Genet. *19*, 479–484.

Ogasawara, H., Yamada, K., Kori, A., Yamamoto, K., and Ishihama, A. (2010). Regulation of the *Escherichia coli* csgD promoter: interplay between five transcription factors. Microbiology *156*, 2470–2483.

Oguiza, J.A., Kiil, K., and Ussery, D.W. (2005). Extracytoplasmic function sigma factors in *Pseudomonas syringae*. Trends Microbiol. *13*, 565–568.

Olivares-Zavaleta, N., Jauregui, R., and Merino, E. (2006). Genome analysis of *Escherichia coli* promoter sequences evidences that DNA static curvature plays a more important role in gene transcription than has previously been anticipated. Genomics *87*, 329–337.

Paget, M.S., and Helmann, J.D. (2003). The sigma70 family of sigma factors. Genome Biol *4*, 203.

Perez, J.C., and Groisman, E.A. (2009a). Evolution of transcriptional regulatory circuits in bacteria. Cell *138*, 233–244.

Perez, J.C., and Groisman, E.A. (2009b). Transcription factor function and promoter architecture govern the evolution of bacterial regulons. Proc. Natl. Acad. Sci. U.S.A. *106*, 4319–4324.

Perez-Rueda, E., and Collado-Vides, J. (2001). Common history at the origin of the position-function correlation in transcriptional regulators in archaea and bacteria. J. Mol. Evol. *53*, 172–179.

Perez-Rueda, E., and Janga, S.C. (2010). Identification and genomic analysis of transcription factors in archaeal genomes exemplifies their functional architecture and evolutionary origin. Mol. Biol. Evol. 27, 1449–1459.

Perez-Rueda, E., and Janga, S.C. (2011). Identification and genomic analysis of transcription factors in archaeal genomes exemplifies their functional architecture and evolutionary origin. Mol. Biol. Evol. *27*, 1449–1459.

Perez-Rueda, E., Collado-Vides, J., and Segovia, L. (2004). Phylogenetic distribution of DNA-binding transcription factors in bacteria and archaea. Comput. Biol. Chem. *28*, 341–350.

Perez-Rueda, E., Janga, S.C., and Martinez-Antonio, A. (2009). Scaling relationship in the gene content of transcriptional machinery in bacteria. Mol. Biosyst. *5*, 1494–1501.

Pragai, Z., Allenby, N.E., O'Connor, N., Dubrac, S., Rapoport, G., Msadek, T., and Harwood, C.R. (2004). Transcriptional regulation of the phoPR operon in *Bacillus subtilis*. J. Bacteriol. *186*, 1182–1190.

Rivera-Gomez, N., Segovia, L., and Perez-Rueda, E. (2011). The diversity and distribution of TFs and their partner domains play an important role in the regulatory plasticity in bacteria. Microbiology. *157*, 2308–2318.

Rodionov, D.A. (2007). Comparative genomic reconstruction of transcriptional regulatory networks in bacteria. Chem. Rev. *107*, 3467–3497.

Rodriguez-Trelles, F., Tarrio, R., and Ayala, F.J. (2003). Evolution of cis-regulatory regions versus codifying regions. Int. J. Dev. Biol. *47*, 665–673.

Roy, S.W., Fedorov, A., and Gilbert, W. (2002). The signal of ancient introns is obscured by intron density and homolog number. Proc. Natl. Acad. Sci. U.S.A. *99*, 15513–15517.

Salgado, H., Gama-Castro, S., Peralta-Gil, M., Diaz-Peredo, E., Sanchez-Solano, F., Santos-Zavaleta, A., Martinez-Flores, I., Jimenez-Jacinto, V., Bonavides-Martinez, C., Segura-Salazar, J., *et al.* (2006). RegulonDB (version 5.0): *Escherichia coli* K-12 transcriptional regulatory network, operon organization, and growth conditions. Nucleic Acids Res. *34*, D394–397.

Segal, E., Shapira, M., Regev, A., Pe'er, D., Botstein, D., Koller, D., and Friedman, N. (2003). Module networks: identifying regulatory modules and their condition-specific regulators from gene expression data. Nat. Genet. 34, 166–176.

Shen-Orr, S.S., Milo, R., Mangan, S., and Alon, U. (2002). Network motifs in the transcriptional regulation network of *Escherichia coli*. Nat. Genet. *31*, 64–68.

Soutourina, O., Kolb, A., Krin, E., Laurent-Winter, C., Rimsky, S., Danchin, A., and Bertin, P. (1999). Multiple control of flagellum biosynthesis in *Escherichia coli*: role of H-NS protein and the cyclic AMP-catabolite activator protein complex in transcription of the flhDC master operon. J. Bacteriol. *181*, 7500–7508.

Taraban, M., Zhan, H., Whitten, A.E., Langley, D.B., Matthews, K.S., Swint-Kruse, L., and Trewhella, J. (2008). Ligand-induced conformational changes and conformational dynamics in the solution structure of the lactose repressor protein. J. Mol. Biol. *376*, 466–481.

Teichmann, S.A., and Babu, M.M. (2004). Gene regulatory network growth by duplication. Nat. Genet. *36*, 492–496.

Thieffry, D., Huerta, A.M., Perez-Rueda, E., and Collado-Vides, J. (1998). From specific gene regulation to genomic networks: a global analysis of transcriptional regulation in *Escherichia coli*. Bioessays *20*, 433–440.

Wall, M.E., Hlavacek, W.S., and Savageau, M.A. (2004). Design of gene circuits: lessons from bacteria. Nat. Rev. Genet. *5*, 34–42.

West, G.B., and Brown, J.H. (2005). The origin of allometric scaling laws in biology from genomes to ecosystems: towards a quantitative unifying theory of biological structure and organization. J. Exp. Biol. *208*, 1575–1592.

Wosten, M.M. (1998). Eubacterial sigma-factors. FEMS Microbiol. Rev. 22, 127–150.

Yamazaki, Y., Niki, H., and Kato, J. (2008). Profiling of *Escherichia coli* Chromosome database. Methods Mol. Biol. *416*, 385–389.

Yanai, I., Camacho, C.J., and DeLisi, C. (2000). Predictions of gene family distributions in microbial genomes: evolution by gene duplication and modification. Phys. Rev. Lett. *85*, 2641–2644.

Yeger-Lotem, E., and Margalit, H. (2003). Detection of regulatory circuits by integrating the cellular networks of protein–protein interactions and transcription regulation. Nucleic Acids Res. *31*, 6053–6061.

Yeger-Lotem, E., Sattath, S., Kashtan, N., Itzkovitz, S., Milo, R., Pinter, R.Y., Alon, U., and Margalit, H. (2004). Network motifs in integrated cellular networks of transcription-regulation and protein–protein interaction. Proc. Natl. Acad. Sci. U.S.A. *101*, 5934–5939.

Index

D

E

S

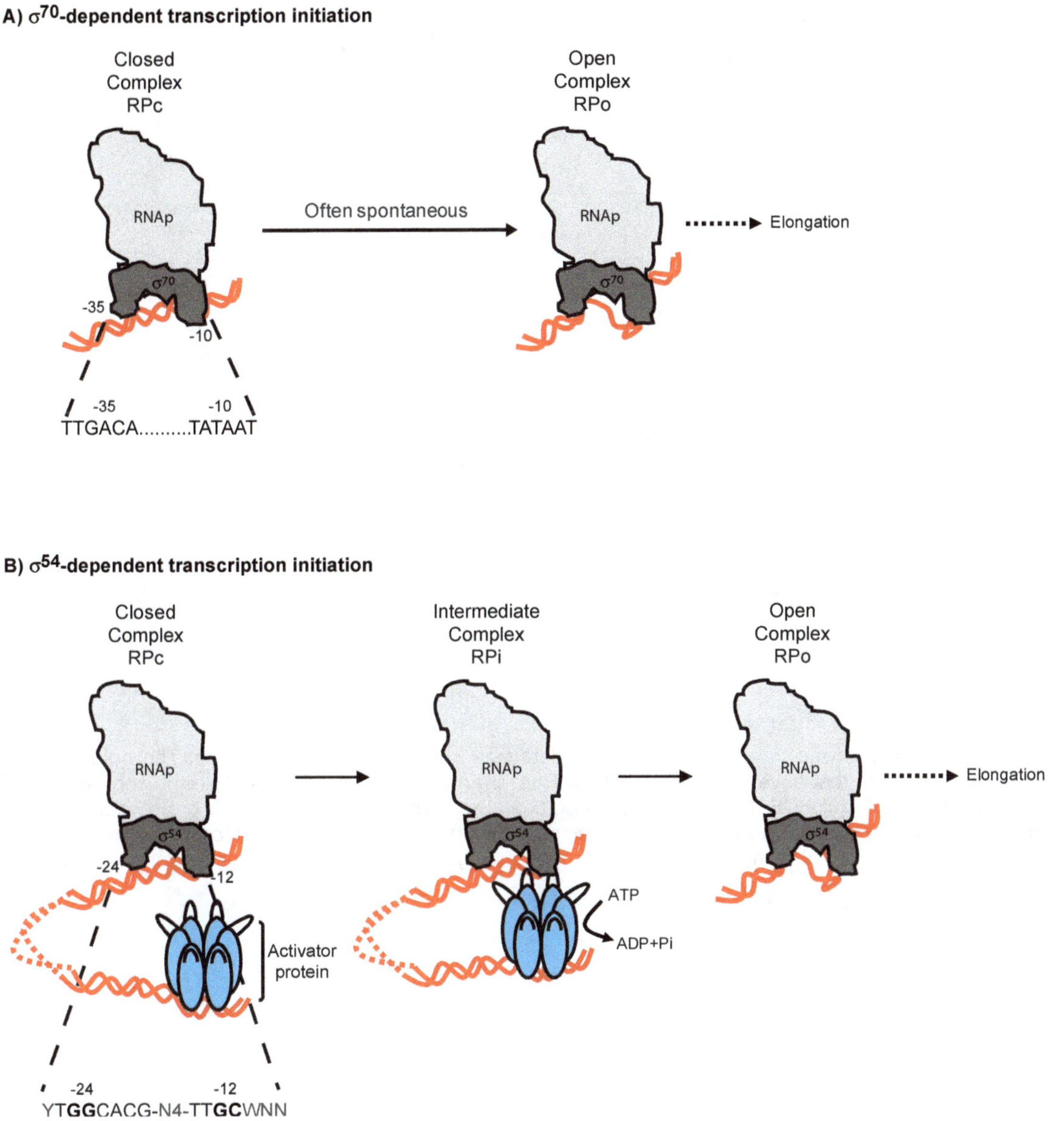

Plate 2.1 Schematic representation of the events leading to open complex formation (RPo) by (A) Eσ^{70} and (B) Eσ^{54}. The typical promoter recognition sequences are given. In the case of the σ^{54} promoter, W represents either A or T; Y represents pyrimidines and N represents non-significant sites (as indicated in Barrios *et al.*, 1999).

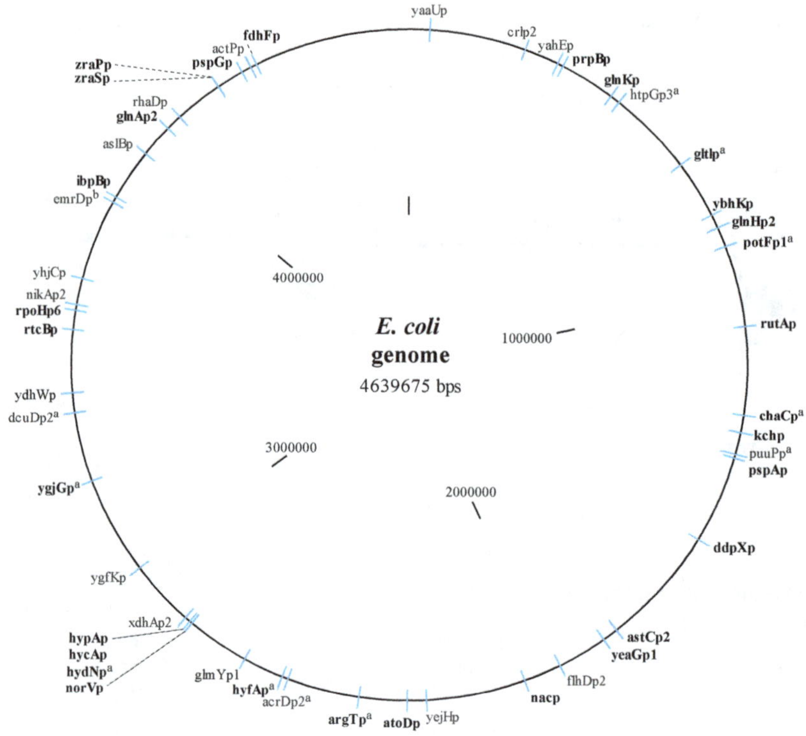

Plate 2.2 Position of the transcription units regulated by $E\sigma^{54}$ in the *E. coli* genome. The bold lettering refers to promoters that have been identified by (Reitzer and Schneider, 2001) and are present as part of the sigmulon on RegulonDB (http://regulondb.ccg.unam.mx/); grey lettering refers to promoters identified in transcriptomic studies (Zhao *et al.*, 2010); black lettering refers to the glmYp1 promoter identified by (Reichenbach *et al.*, 2009); [a] indicates promoters with less than 80% identity with the consensus sequence.

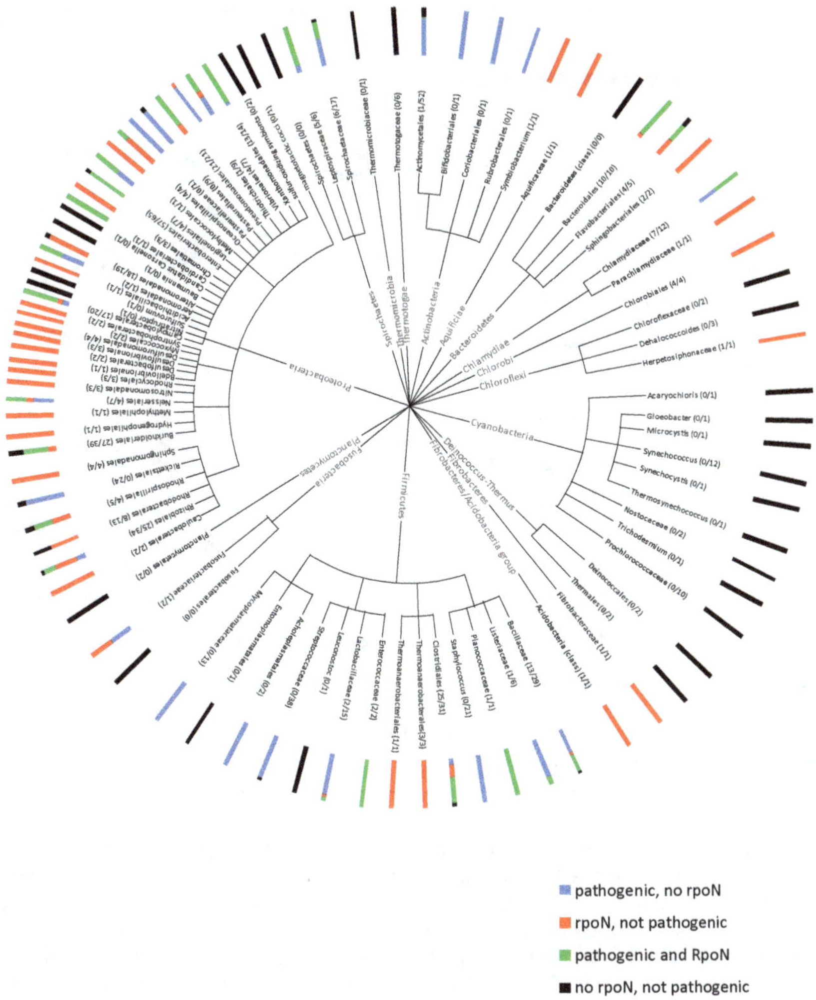

Plate 2.3 A phylogenetic illustration of the class, order and family ranks of the 672 sequenced bacterial genomes (www.jcvi.org). The number of species per family in which the *rpoN* gene was found in relation to the total number of sequenced genomes is given in parentheses. The coloured bars indicate for each family the proportions of species that contain *rpoN* and are pathogens (green) or not (red), and which do not contain *rpoN* and are pathogens (blue) or not (black).

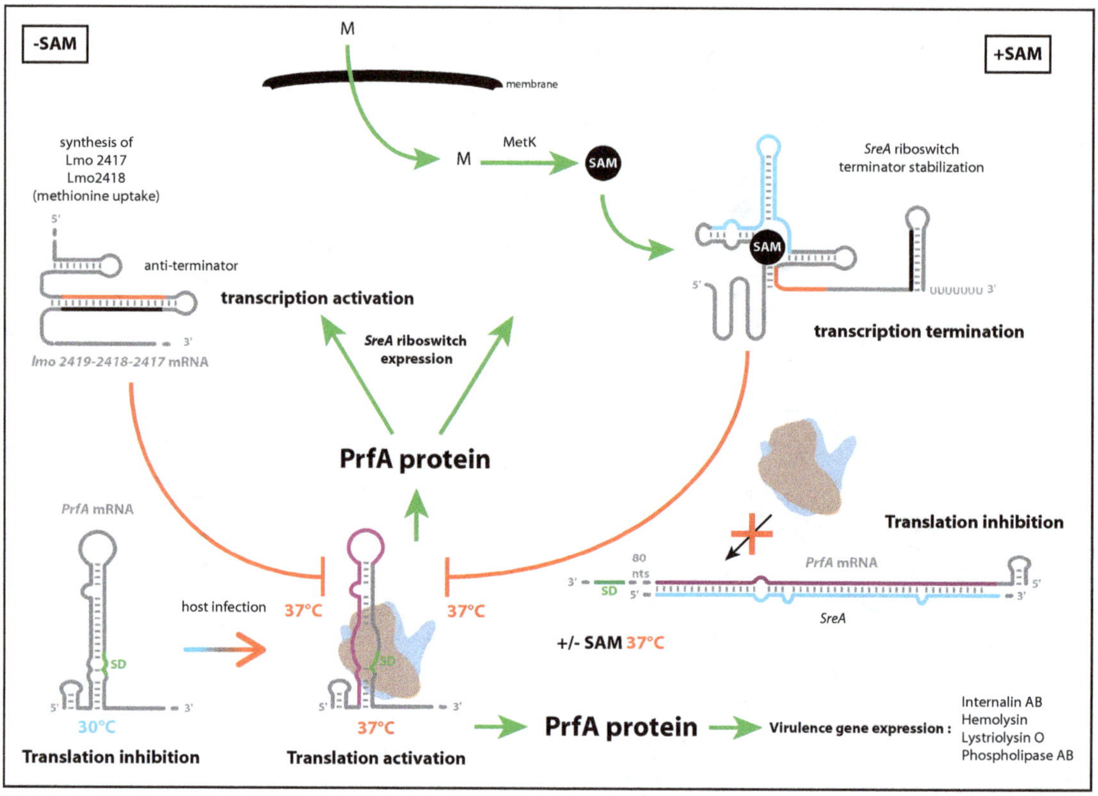

Plate 6.1 Dual regulation of *prfA* mRNA in *L. monocytogenes*. Translation of *prfA* mRNA is inhibited at low temperature (<30°C) because the Shine and Dalgarno sequence (SD) is sequestered into an inhibitory hairpin. When temperature increases to 37°C, the secondary structure 'opens', allowing translation of PrfA. Moreover, PrfA synthesis is regulated *in trans* by a riboswitch responding to the intracellular concentration of S-adenosylmethionine (SAM), SreA, which regulates the expression of the downstream genes (lmo2417–2418) involved in methionine uptake. When nutrients are limiting (low levels of SAM), transcription of the downstream genes takes place owing to the formation of an anti-terminator structure. Conversely, under high levels of SAM, SAM binds to SreA to induce a premature transcription and SreA acts as a sRNA to repress the translation of *prfA* mRNA by binding to the 5′UTR 80 bp upstream of the SD sequence. This interaction is only allowed at high temperature (37°C) where the regulatory sequences of PrfA are exposed for interaction. The complementary regions of SreA and *prfA* mRNA are coloured in blue and purple, respectively.

Plate 6.2 Regulatory networks in virulence gene expression. (A) Quorum sensing regulates virulence gene expression in *S. aureus*. Cell population density is sensed by the *agr* system. When population reaches a critical threshold, AgrB (red), a membrane peptidase matures the auto-inducing peptide AIP encoded by AgrD (grey), which is then secreted in the extracellular environment. Upon threshold concentration of AIP, autophosphorylation of the histidine kinase AgrC (purple) takes place and through a phosphorelay the response regulator, AgrA (light blue) is activated (Geisinger and Novick, 2008). AgrA autoactivates the transcription of its own operon and of RNAIII, which encodes delta-haemolysin (in green). The secondary structure of RNAIII is shown. The 3′UTR (in black) of RNAIII acts as an antisense RNA to repress the synthesis of adhesins (coagulase, protein A, SA1000), of amidases and hydrolases (SA2353, LytM) and of the repressor of toxins Rot, and the 5′UTR (red) activates the translation of *hla* mRNA (see Fig. 6.3). For repression, RNAIII uses two hairpin domains to bind *coa* and *rot* mRNAs. The hairpins H7 and H14 of RNAIII bind to *coa* mRNA to form a duplex stabilized by a loop–loop interaction (upper insert) while they form two loop–loop interactions with the 5′UTR of *rot* mRNA (bottom insert). In both cases, the duplexes prevent

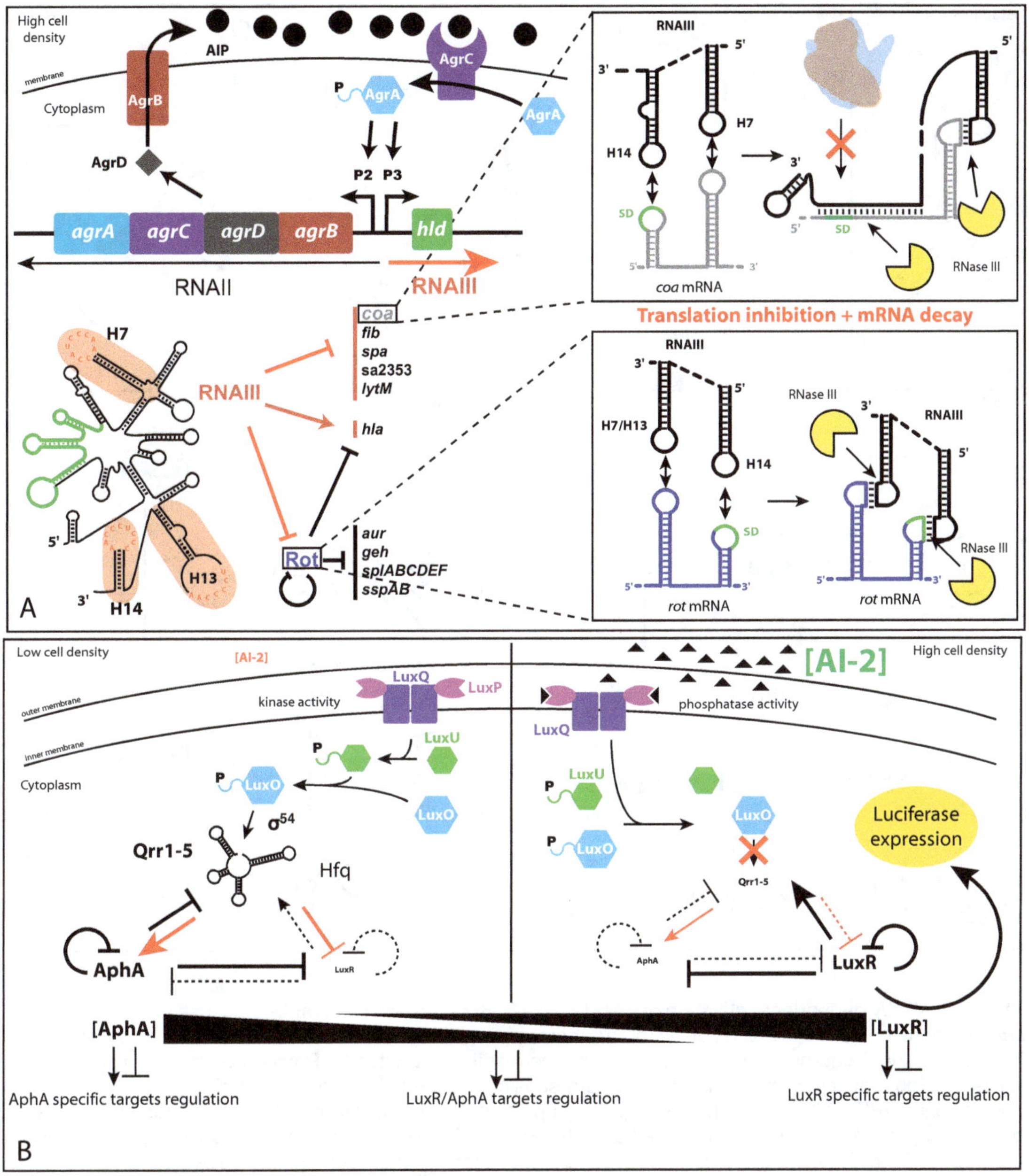

ribosome binding and recruit the RNase III for mRNA degradation. The drawing was adapted from Boisset *et al.* (2007). (B) Several sRNAs regulate QS in *V. harveyi* and *V. cholerae*. At low cell density (LCD), in the absence of the autoinducer (AI-2), the histidine kinase-receptor LuxPQ phosphorylates LuxU (green), which in turn phosphorylates LuxO (light blue). LuxO-P together with the RNAP-σ54 holoenzyme up-regulate expression of 5 sRNAs, Qrr1–5. These sRNAs activate the synthesis of the master regulator AphA through an anti-antisense mechanism and repress the translation of *luxR* mRNA. At high cell density (HCD), and high concentrations of AI-2, the LuxPQ receptor acts as a phosphatase and de-phosphorylates LuxO, thus repressing synthesis of Qrr sRNAs. Under these conditions, LuxR is de-repressed. Maximal levels of AphA or LuxR are attained in LCD or HCD, respectively. A gradient of the two intracellular regulator concentrations is established to provide a large panel of responses. Black and red bars denote transcriptional and post-transcriptional repression, respectively; black and red arrows denote transcriptional and post-transcriptional activation, respectively. The drawing is adapted from Rutherford *et al.* (2011).

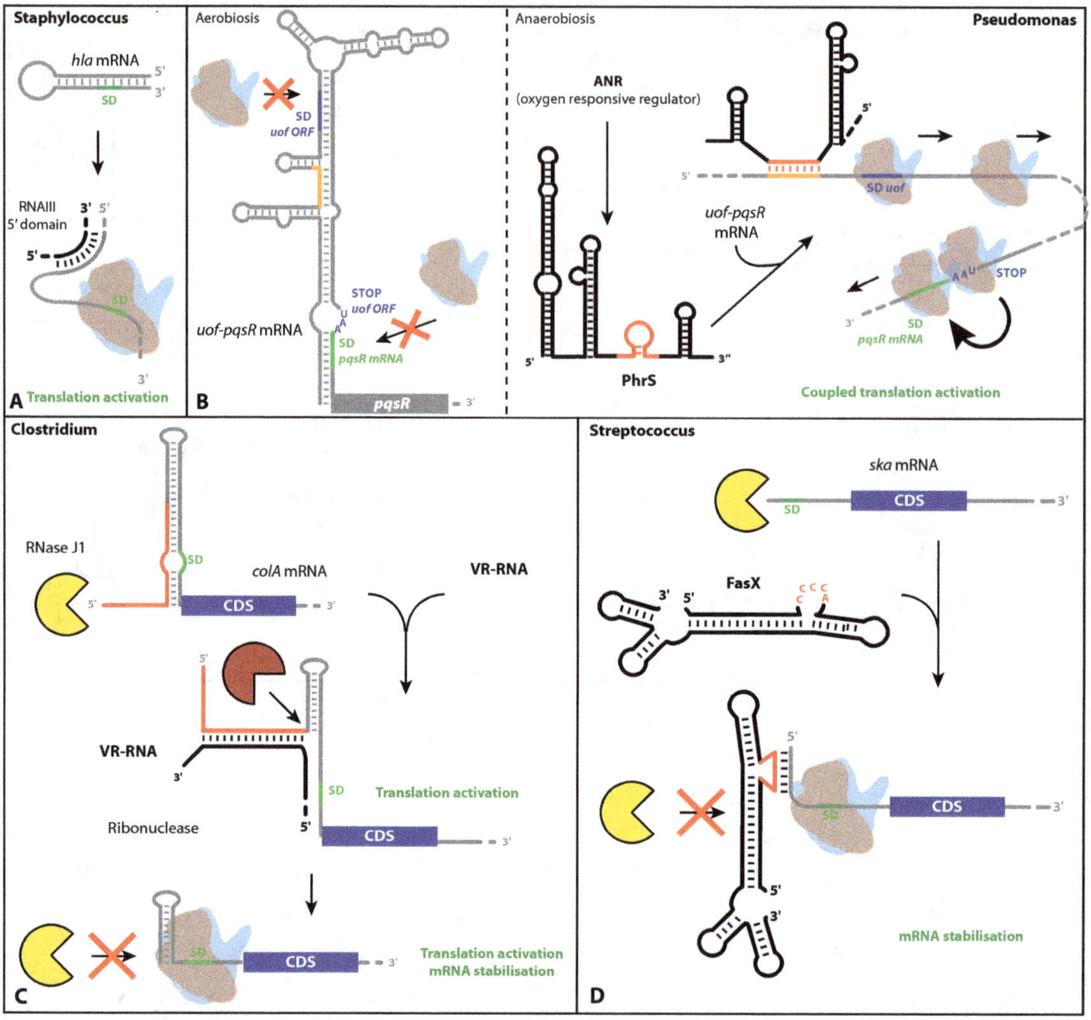

Plate 6.3 sRNA-dependent activation of translation. (A, B) Activation by an anti-antisense mechanism. (A) *S. aureus* RNAIII binds to the 5′UTR of *hla* mRNA thereby preventing the formation of an inhibitory structure. SD is for Shine and Dalgarno sequence. (B) A similar mechanism exploited by PhrS sRNA in *P. aeruginosa*. PhrS sRNA binding induces translation of the quorum-sensing regulator PqsR through activation of translation of an upstream ORF (uof) which is translationally coupled to PqsR. Expression of PhrS is induced by ANR, an oxygen responsive regulator, at high cell density population when oxygen becomes limiting. The drawing is adapted from Sonnleitner *et al.* (2011). (C, D) Activation by stabilization of mRNA structure: (C) In *C. perfringens* VR-RNA binds to the 5′UTR of *colA* mRNA creating a specific site for an unknown RNase (red circle). As a result, a short hairpin structure is generated at the 5′ end, and the SD sequence becomes accessible to the ribosome (Obana *et al.*, 2010). (D) In *S. pyogenes*, FasX binds at the 5′ end of *ska* mRNA encoding streptokinase 30 nts upstream of the SD sequence (Ramirez-Peña *et al.*, 2010). In both cases, the mRNAs are stabilized by preventing access of the RNase J1 (yellow circle).

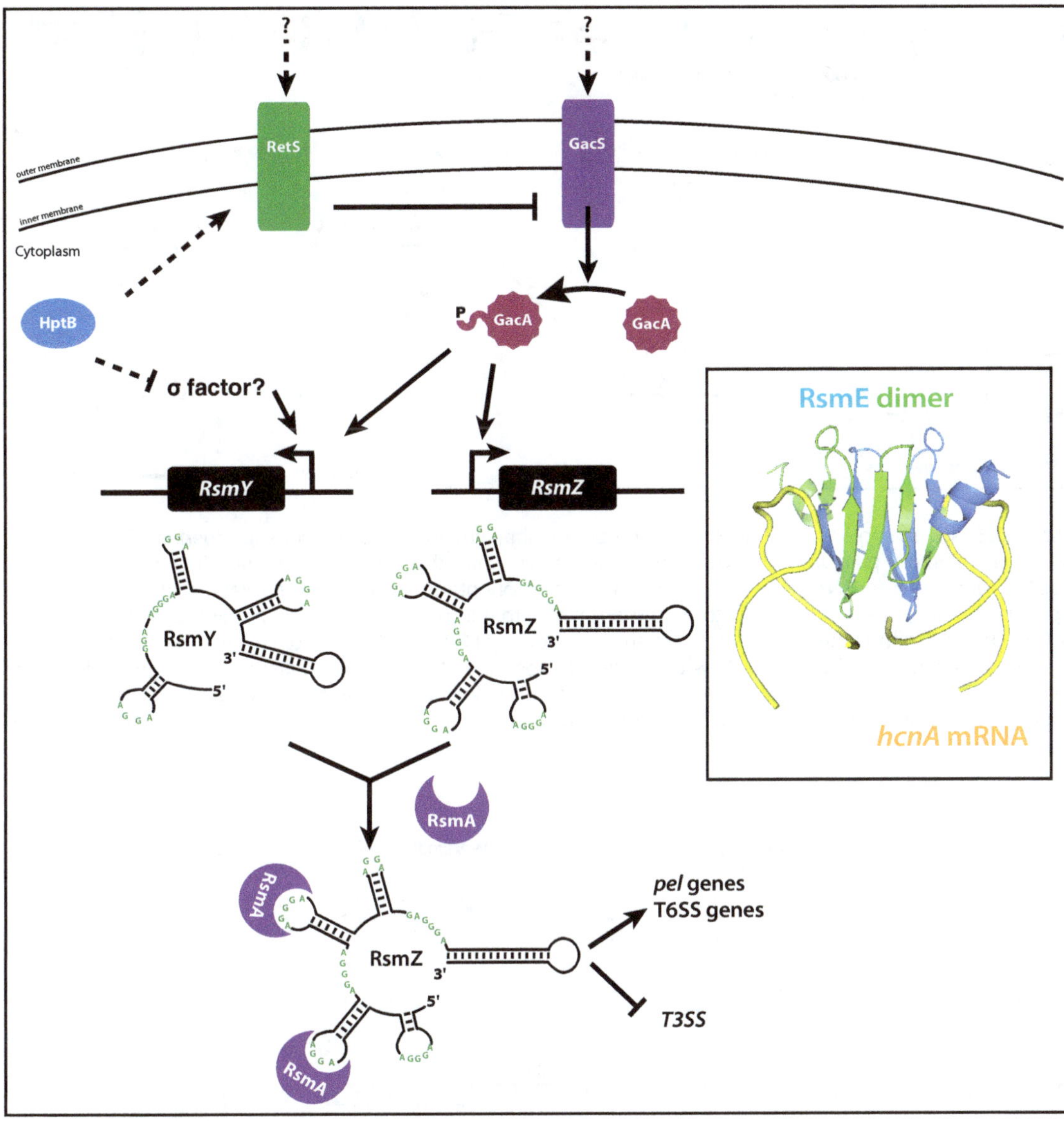

Plate 6.4 Complex regulatory networks controlling virulence gene expression in *P. aeruginosa*. Regulation of expression of RsmY/RsmZ sRNAs is mediated through a two-component system characterized by the histidine kinase GacS (violin) and the response regulator GacA (red). The autophosphorylation of GacS (purple) is inhibited by the sensor protein, RetS (green). The histidine phosphotransfer protein HptB (blue) up-regulates the synthesis of RsmY (Bordi *et al.*, 2010). The synthesis of RsmY/RsmZ out titrates RsmA protein (violin) from its mRNA targets, resulting in activation of *pel* and T6SS genes and repression of T3SS genes. The sRNAs carry a redundant AGGA sequence mimicking the Shine and Dalgarno sequence of the target mRNAs. Inner panel: The solution structure of the complex between RsmE (RsmA homologue) of *P. fluorescens* and a 20-nt fragment of *hcnA* mRNA encompassing the RBS is shown (Schubert *et al.*, 2007).

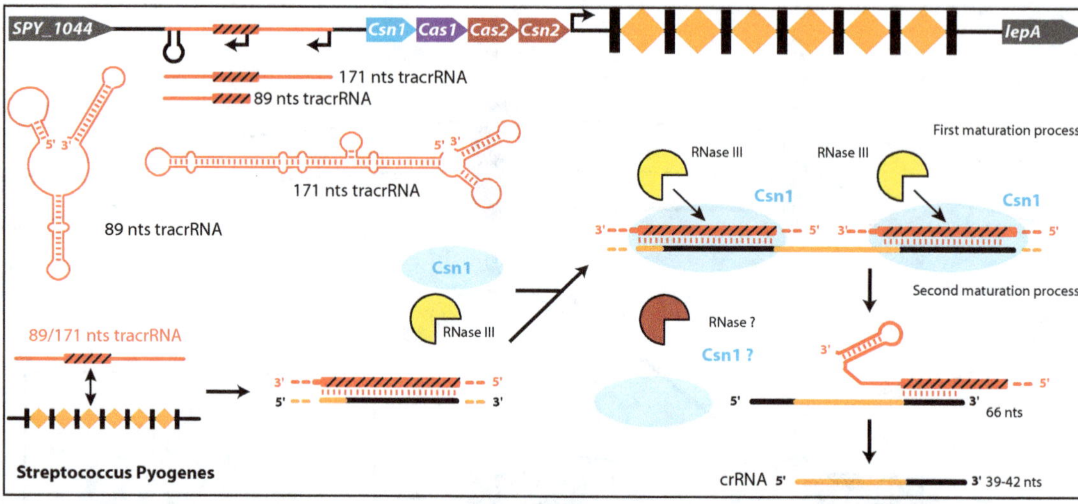

Plate 6.5 An example of CRISPR interference mechanism in *Streptococcus pyogenes*. The regularly interspaced short palindromic repeat locus is shown with the four protein encoding genes (csn1–2, cas1–2) and the tracrRNA gene located in an inverted orientation. The repeats (black rectangles) and the spacers (orange) are transcribed into a long precursor that is processed by the binding of the small sRNA tcacrRNA (red) to the repeat motifs to form short duplexes of 22 nts. These duplexes are then cleaved by the endoribonuclease III (RNase III, yellow circle) in the presence of Csn1 protein (light blue) to generate 66 nts long RNA fragments. A second cleavage induced by an unknown RNase generates a 39–42 nts long RNA fragments. This short RNA fragment is the active form that targets the foreign DNA (Deltcheva *et al.*, 2011).

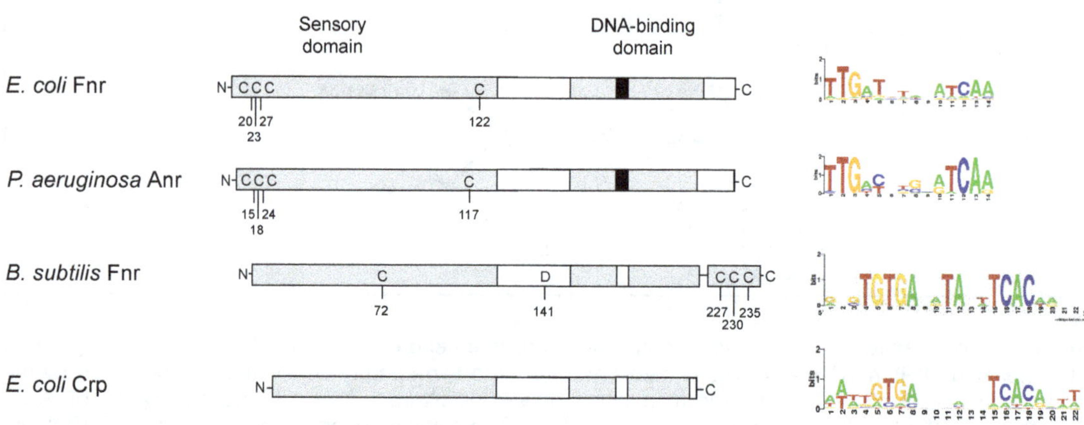

Plate 11.3 Structure of Fnr type regulators. The schematic structure of *E. coli* Fnr, *P. aeruginosa* Anr, *B. subtilis* Fnr and *E. coli* Crp are depicted. Cysteine residues involved in $[4Fe{-}4S]^{2+}$ cluster formation and the location of the DNA binding domain are shown. Sequence logos for the corresponding DNA binding are provided.

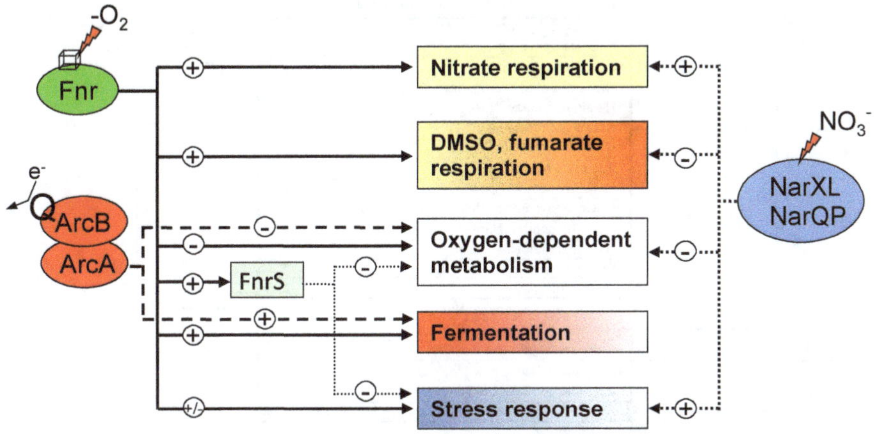

Plate 11.4 Overview over the anaerobic regulatory network of *E. coli* using Fnr, the sRNA FnrS, ArcBA and NarXL/NarQP. Detected environmental stimuli and regulated mayor metabolic processes are given.

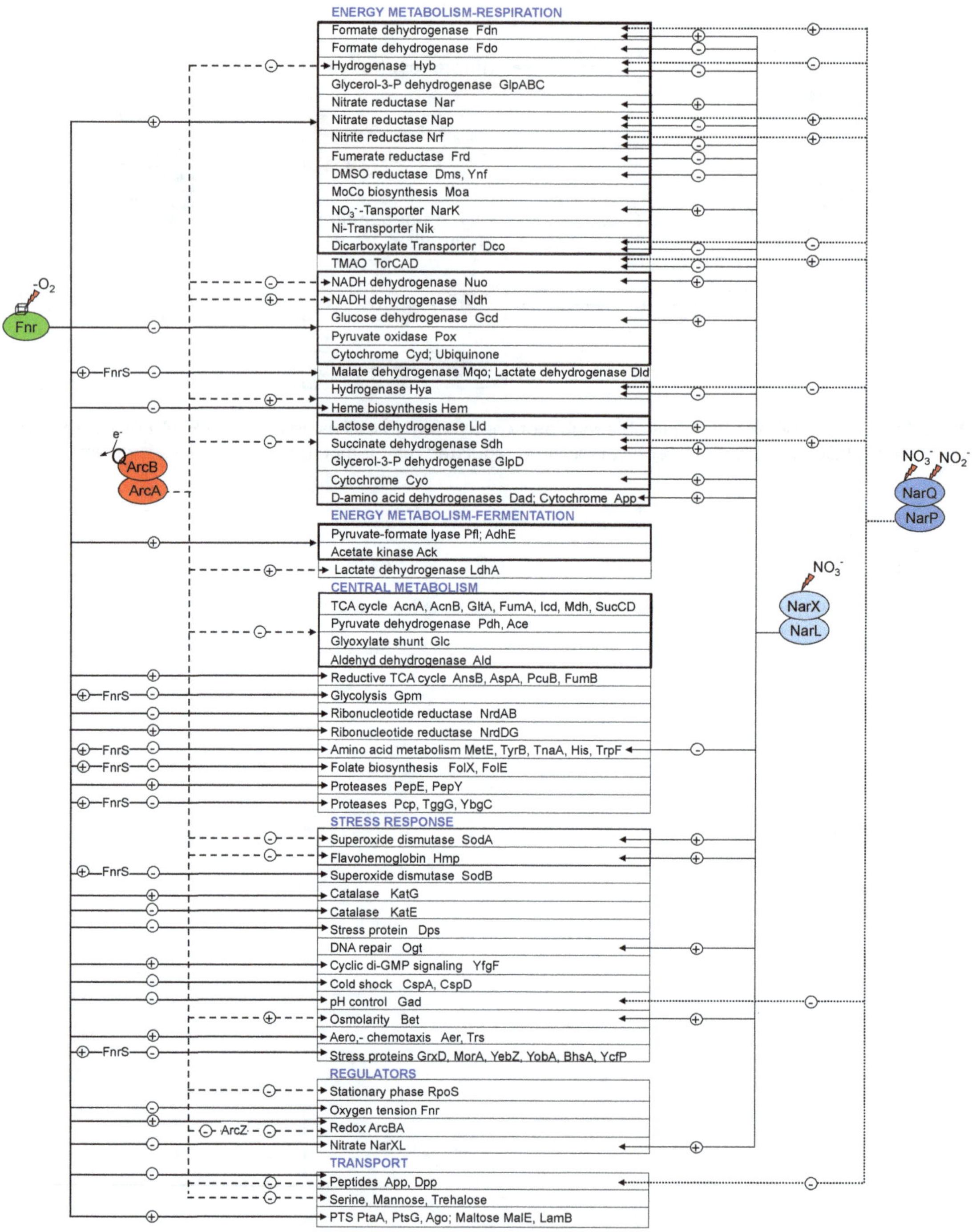

Plate 11.5 Detailed regulatory network of *E. coli* for the adaptation to anaerobic growth conditions. The interplay of oxygen regulator Fnr, the sRNA FnrS, the redox regulator ArcBA, and the nitrate/nitrite regulatory systems NarXL and NarQP is outlined. In the case of arrows touching the outside of a set of genes boxed with bold lines regulation of all genes by the corresponding regulator is indicated. Arrows entering a boxed gene or a set of genes reveal a single regulator-target gene interaction. Regulated metabolic processes are given in blue letters. Regulatory systems and corresponding stimuli are shown in different colours.

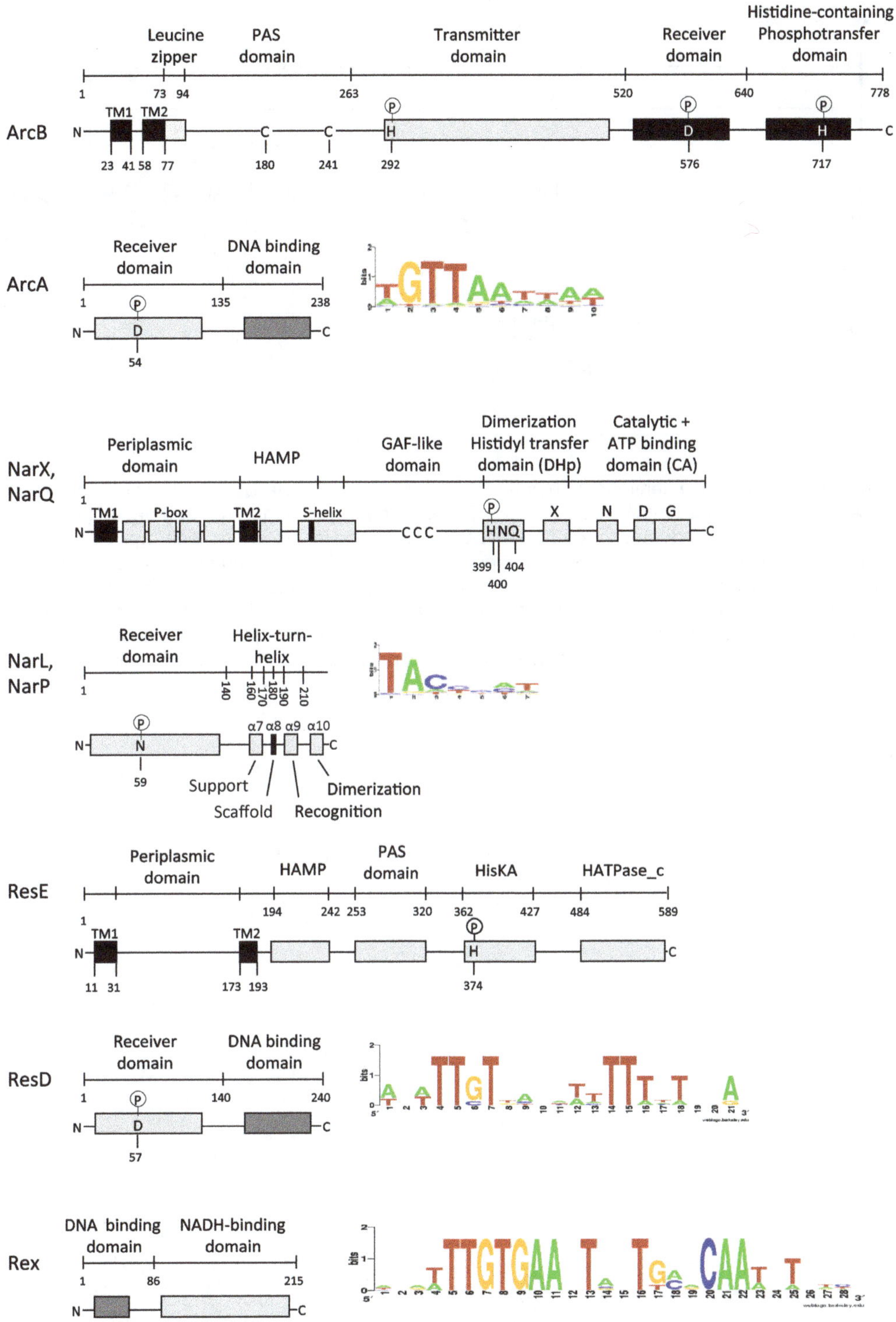

Plate 11.6 Two-component regulatory systems involved in the adaptation to anaerobic conditions in *E. coli*, *P. aeruginosa* and *B. subtilis*. The schematic structure of the proteins for the two component regulatory systems ArcBA (*E.coli*), NarXL (*E.coli*), NarQP (*E.coli*), ResDE (*B. subtilis*) and the NADH sensor Rex (*B. subtilis*) are shown. Functionally important domains and amino acid residues are labelled. Sequence logos provide insights to the corresponding DNA binding sites.

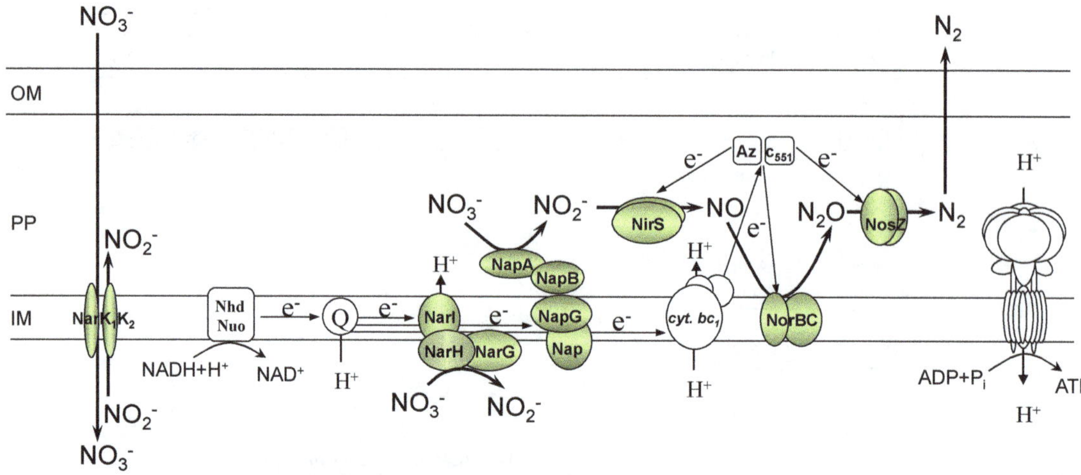

Plate 11.7 The *P. aeruginosa* denitrification pathway. The inner membrane-bound nitrate reductases NarGHI and NapAB, the periplasmic nitrite reductase NirS, the membrane-bound NO reductase NorCB, and the periplasmic nitrous-oxide reductase NosZ catalyse the reduction of nitrate to dinitrogen. The nitrate and nitrite transport is mediated by the putative antiporter NarK$_1$K$_2$. NADH provides the required electrons via the NADH:quinone oxidoreductase I Nuo and the NADH:quinone oxidoreductase II Ndh. The electrons are transferred to the quinone pool Q, to the cytochrome bc_1 (cyt. bc_1) or directly to the nitrate reductases NarGHI and further to azurin (Az) or to the cytochrome c_{551}. As a consequence, H^+ is translocated to the periplasm. The proton motive force is used for ATP generation by the ATPase. PP, periplasm; IM, inner membrane; OM, outer membrane.

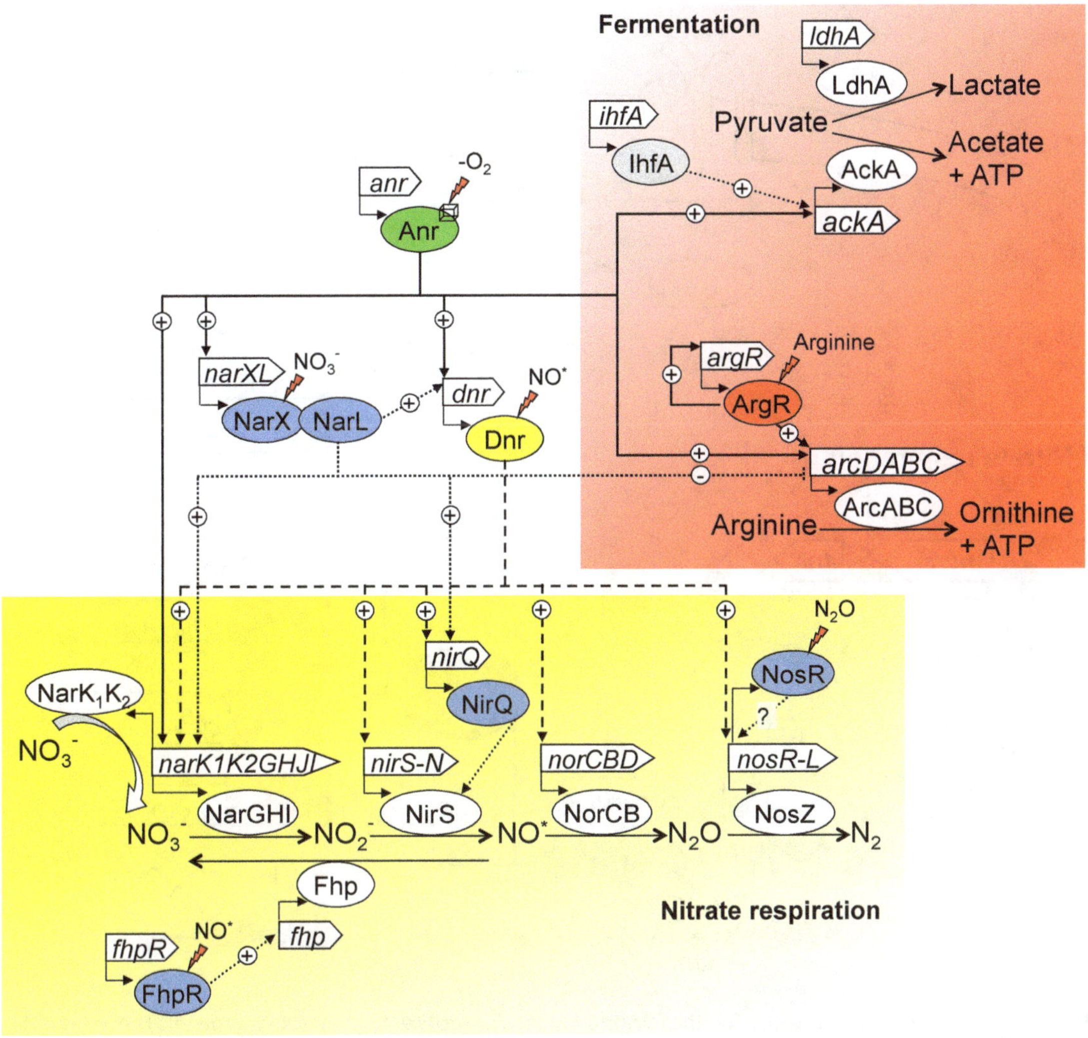

Plate 11.8 Regulatory network controlling the anaerobic metabolism of *P. aeruginosa*. In interplay of the oxygen regulator Anr, the NO-sensors Dnr and FhpR, the N_2O- dependent regulator NosR, the arginine regulator ArgR with InfA and NirQ for the control of genes involved in denitrification (yellow background), arginine and pyruvate fermentation (red background) is shown.

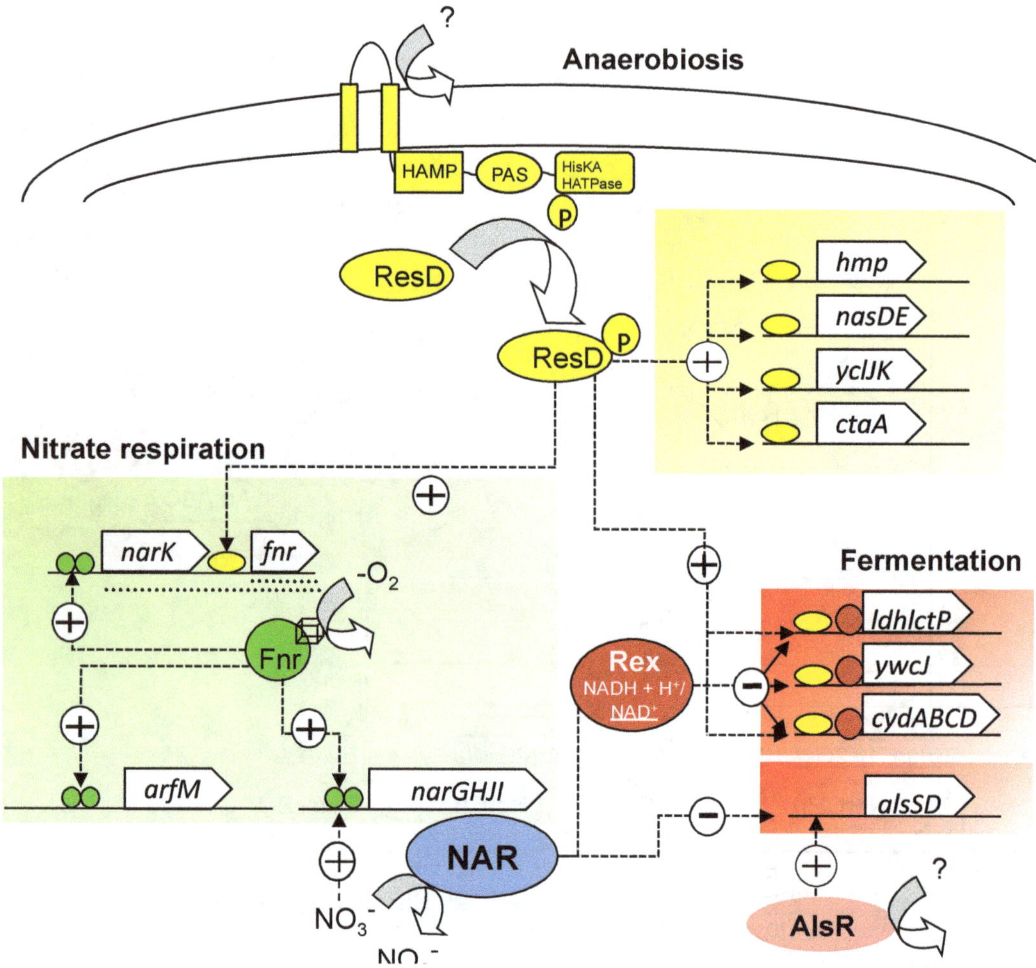

Plate 11.10 Regulatory network for the adaptation of *B. subtilis* to anaerobic growth. The differential anaerobic gene expression for enzymes involved in nitrate ammonification (green background), mixed acid fermentation (red background) and related processes (yellow background) by ResDE, Fnr, Rex and AlsR are shown. Where known, corresponding stimuli of the regulators are given.

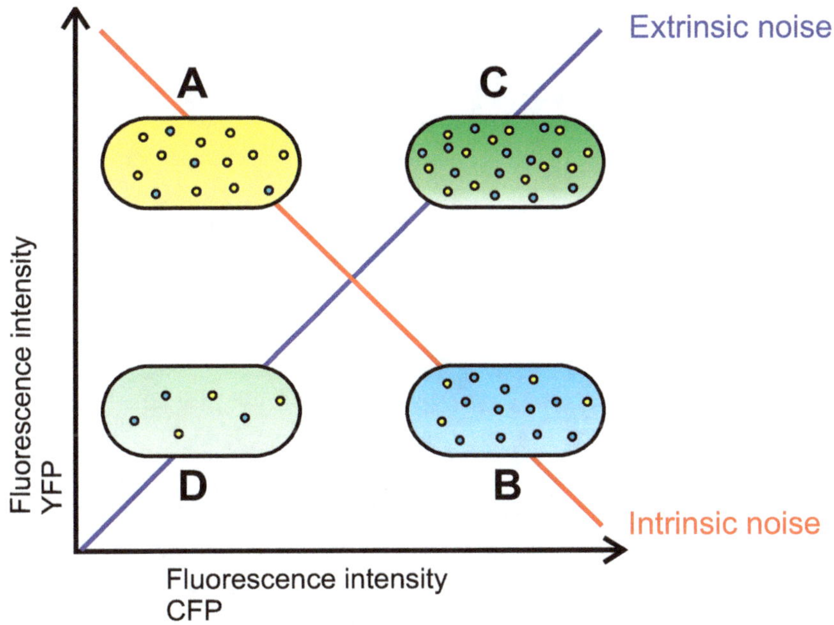

Plate 12.4 Intrinsic and extrinsic noise can be quantified and distinguished using two genes controlled by identical regulatory sequences. In this cartoon a situation is sketched where a cell harbours a yellow fluorescent protein (YFP) and a cyan fluorescent protein (CFP) under control of identical promoters. Cells expressing more of either fluorescent protein appear yellow or blue (A and B), whereas cells with the same amount of each protein appear green (C and D). In situation A and B differences in amount of each product occurs, even though these genes are under the control of an identical promoter. Here the differences are caused by intrinsic noise. In situation C and D a difference in total amount of protein produced arose, caused by extrinsic noise (Elowitz, 2002; Swain *et al.*, 2002). The cartoons are superimposed over a hypothetical flow cytometry graph, where relative fluorescence intensity of each fluorophore is plotted.

www.ingramcontent.com/pod-product-compliance
Lightning Source LLC
LaVergne TN
LVHW062217200926
848484LV00006B/1417